# Renewable Polymers and Polymer-Metal Oxide Composites

# The Metal Oxides Book Series Edited by Ghenadii Korotcenkov

## Forthcoming Titles

- Palladium Oxides Material Properties, Synthesis and Processing Methods, and Applications, Alexander M. Samoylov, Vasily N. Popov, 9780128192238
- Metal Oxides for Non-volatile Memory, Panagiotis Dimitrakis, Ilia Valov, Stefan Tappertzhofen, 9780128146293
- Metal Oxide Nanostructured Phosphors, H. Nagabhushana, Daruka Prasad, S.C. Sharma, 9780128118528
- Nanostructured Zinc Oxide, Kamlendra Awasthi, 9780128189009
- Multifunctional Piezoelectric Oxide Nanostructures, Sang-Jae Kim, Nagamalleswara Rao Alluri, Yuvasree Purusothaman, 9780128193327
- Transparent Conductive Oxides, Mirela Petruta Suchea, Petronela Pascariu, Emmanouel Koudoumas, 9780128206317
- Metal Oxide-Based Nanofibers and Their Applications, Vincenzo Esposito, Debora Marani, 9780128206294
- Metal-Oxides for Biomedical and Biosensor Applications, Kunal Mondal, 9780128230336
- Metal Oxide-Carbon Hybrid Materials, Muhammad Akram, Rafaqat Hussain, Faheem K. Butt, 9780128226940
- Metal Oxide-Based Heterostructures, Naveen Kumar, Bernabe Mari Soucase, 9780323852418
- Metal Oxides and Related Solids for Electrocatalytic Water Splitting, Junlei Qi, 9780323857352
- Advances in Metal Oxides and Their Composites for Emerging Applications, Sagar Delekar, 9780323857055
- Metallic Glasses and Their Oxidation, Xinyun Wang, Mao Zhang, 9780323909976
- Solution Methods for Metal Oxide Nanostructures, Rajaram S. Mane, Vijaykumar Jadhav, Abdullah M. Al-Enizi, 9780128243534
- Metal Oxide Defects, Vijay Kumar, Sudipta Som, Vishal Sharma, Hendrik Swart, 9780323855884
- Renewable Polymers and Polymer-Metal Oxide Composites, Sajjad Haider, Adnan Haider, 9780323851558
- Metal Oxides for Optoelectronics and Optics-Based Medical Applications, Suresh Sagadevan, Jiban Podder, Faruq Mohammad, 9780323858243
- Graphene Oxide-Metal Oxide and Other Graphene Oxide-Based Composites in Photocatalysis and Electrocatalysis, Jiaguo Yu, Liuyang Zhang, Panyong Kuang, 9780128245262

## Published Titles

- Metal Oxides in Nanocomposite-Based Electrochemical Sensors for Toxic Chemicals, Alagarsamy Pandikumar, Perumal Rameshkumar, 9780128207277
- Metal Oxide-Based Nanostructured Electrocatalysts for Fuel Cells, Electrolyzers, and Metal-Air Batteries, Teko Napporn, Yaovi Holade, 9780128184967
- Titanium Dioxide ($TiO_2$) and Its Applications, Leonardo Palmisano, Francesco Parrino, 9780128199602
- Solution Processed Metal Oxide Thin Films for Electronic Applications, Zheng Cui, 9780128149300
- Metal Oxide Powder Technologies, Yarub Al-Douri, 9780128175057
- Colloidal Metal Oxide Nanoparticles, Sabu Thomas, Anu Tresa Sunny, Prajitha V, 9780128133576
- Cerium Oxide, Salvatore Scire, Leonardo Palmisano, 9780128156612
- Tin Oxide Materials, Marcelo Ornaghi Orlandi, 9780128159248
- Metal Oxide Glass Nanocomposites, Sanjib Bhattacharya, 9780128174586
- Gas Sensors Based on Conducting Metal Oxides, Nicolae Barsan, Klaus Schierbaum, 9780128112243
- Metal Oxides in Energy Technologies, Yuping Wu, 9780128111673
- Metal Oxide Nanostructures, Daniela Nunes, Lidia Santos, Ana Pimentel, Pedro Barquinha, Luis Pereira, Elvira Fortunato, Rodrigo Martins, 9780128115121
- Gallium Oxide, Stephen Pearton, Fan Ren, Michael Mastro, 9780128145210
- Metal Oxide-Based Photocatalysis, Adriana Zaleska-Medynska, 9780128116340
- Metal Oxides in Heterogeneous Catalysis, Jacques C. Vedrine, 9780128116319
- Magnetic, Ferroelectric, and Multiferroic Metal Oxides, Biljana Stojanovic, 9780128111802
- Iron Oxide Nanoparticles for Biomedical Applications, Sophie Laurent, Morteza Mahmoudi, 9780081019252
- The Future of Semiconductor Oxides in Next-Generation Solar Cells, Monica Lira-Cantu, 9780128111659
- Metal Oxide-Based Thin Film Structures, Nini Pryds, Vincenzo Esposito, 9780128111666
- Metal Oxides in Supercapacitors, Deepak Dubal, Pedro Gomez-Romero, 9780128111697
- Transition Metal Oxide Thin Film-Based Chromogenics and Devices, Pandurang Ashrit, 9780081018996

Metal Oxides Series

# Renewable Polymers and Polymer-Metal Oxide Composites

## Synthesis, Properties, and Applications

*Series Editor*

*Ghenadii Korotcenkov*

*Edited by*

*Sajjad Haider*
Department of Chemical Engineering,
College of Engineering, King Saud University,
Riyadh, Saud Arabia

*Adnan Haider*
Department of Biological Sciences,
National University of Medical Sciences,
Rawalpindi, Punjab, Pakistan

ELSEVIER

Elsevier
Radarweg 29, PO Box 211, 1000 AE Amsterdam, Netherlands
The Boulevard, Langford Lane, Kidlington, Oxford OX5 1GB, United Kingdom
50 Hampshire Street, 5th Floor, Cambridge, MA 02139, United States



ISBN: 978-0-323-85155-8

For information on all Elsevier publications
visit our website at https://www.elsevier.com/books-and-journals

*Publisher:* Matthew Deans
*Acquisitions Editor:* Kayla Dos Santos
*Editorial Project Manager:* Isabella C. Silva
*Production Project Manager:* Surya Narayanan Jayachandran
*Cover Designer:* Miles Hitchen

Typeset by STRAIVE, India

# Contents

# Contributors

**Ashfaq Ahmad** Department of Chemistry, College of Science, King Saud University Riyadh, Riyadh, Saudi Arabia

**Md. Wasi Ahmad** Department of Chemical Engineering, College of Engineering, Dhofar University, Salalah, Sultanate of Oman

**Muhammad Arshed** Nanoscience and Technology Department (NS & TD), National Center for Physics, Islamabad, Pakistan

**Nausheen Bukhari** Mohammad College of Medicine, Peshawar, Pakistan

**Adib H. Chisty** Department of Applied Chemistry and Chemical Engineering, Faculty of Engineering and Technology, University of Dhaka, Dhaka, Bangladesh

**Huseyin Deveci** Department of Chemical Engineering, Konya Technical University, Konya, Turkey

**Arturo Espinosa-Duque** Research Group in Science with Technological Applications (GI-CAT), Department of Chemistry, Universidad del Valle; Mindtech Research Group (Mindtech-RG), Mindtech s.a.s., Cali, Colombia

**Atiya Fatima** Department of Chemical Engineering, College of Engineering, Dhofar University, Salalah, Sultanate of Oman

**Gulcihan Guzel Kaya** Department of Chemical Engineering, Konya Technical University, Konya, Turkey

**Adnan Haider** Department of Biological Sciences, National University of Medical Sciences, Rawalpindi, Punjab, Pakistan

**Sajjad Haider** Department of Chemical Engineering, College of Engineering, King Saud University, Riyadh, Saudi Arabia

**Papia Haque** Department of Applied Chemistry and Chemical Engineering, Faculty of Engineering and Technology, University of Dhaka, Dhaka, Bangladesh

**Md. Arif Hossain** Department of Chemistry, Faculty of Science, Dhaka University of Engineering & Technology, Gazipur, Bangladesh

**Md. Saddam Hossain** Department of Chemistry, Khulna University of Engineering and Technology, Khulna, Bangladesh

**Mazhar Ul Islam** Department of Chemical Engineering, College of Engineering, Dhofar University, Salalah, Sultanate of Oman

**Md. Minhajul Islam** Department of Applied Chemistry and Chemical Engineering, Faculty of Engineering and Technology, University of Dhaka, Dhaka, Bangladesh

**Md. Mominul Islam** Department of Chemistry, University of Dhaka, Dhaka, Bangladesh

**Md. Sazedul Islam** Department of Applied Chemistry and Chemical Engineering, Faculty of Engineering and Technology, University of Dhaka, Dhaka, Bangladesh

**Sumaya Farhana Kabir** Department of Applied Chemistry and Chemical Engineering, Faculty of Engineering and Technology, University of Dhaka, Dhaka, Bangladesh

**M. Nuruzzaman Khan** Department of Applied Chemistry and Chemical Engineering, Faculty of Engineering and Technology, University of Dhaka, Dhaka, Bangladesh

**Mohammad Ahmed Khan** School of Chemical Engineering, Vellore Institute of Technology, Vellore, Tamilnadu, India

**Mohd Shariq Khan** Department of Chemical Engineering, College of Engineering, Dhofar University, Salalah, Sultanate of Oman

**Muhammad Umar Aslam Khan** Department of Polymer Engineering and Technology, University of the Punjab, Lahore, Pakistan; BioInspired Device and Tissue Engineering Research Group, School of Biomedical Engineering and Health Sciences, Faculty of Engineering, Universiti Teknologi Malaysia, Skudai, Johor, Malaysia; Institute for Personalized Medicine, School of Biomedical Engineering, Shanghai Jiao Tong University, Shanghai, China; Nanoscience and Technology Department (NS & TD), National Center for Physics, Islamabad, Pakistan; Centre for Advanced Composite Materials, Universiti Teknologi Malaysia, Skudai, Johor, Malaysia

**Rawaiz Khan** Department of Polymer Engineering, Faculty of Engineering, School of Chemical and Energy, Universiti Teknologi Malaysia, Johor Bahru; BioInspired Device and Tissue Engineering Research Group, School of Biomedical Engineering and Health Sciences, Faculty of Engineering, Universiti Teknologi Malaysia, Skudai, Johor, Malaysia

**Shaukat Khan** Department of Chemical Engineering, College of Engineering, Dhofar University, Salalah, Sultanate of Oman; School of Chemical Engineering, Yeungnam University, Gyeongsan, South Korea

**Wajid Ali Khan** Department of Chemistry, Abdul Wali Khan University Mardan, Mardan, KP, Pakistan

**Dong Lin** Department of Industrial and Manufacturing Systems Engineering, Kansas State University, Manhattan, KS, United States

**Ram Naresh Mahaling** Laboratory of Polymeric and Materials Chemistry, School of Chemistry, Sambalpur University, Odisha, India

**Abul K. Mallik** Department of Applied Chemistry and Chemical Engineering, Faculty of Engineering and Technology, University of Dhaka, Dhaka, Bangladesh

**Sehrish Manan** Department of Biomedical Engineering, Huazhong University of Science and Technology, Wuhan, PR China

**Muhammad Faheem Maqsood** Institute of Metallurgy and Materials Engineering, University of the Punjab, Lahore, Pakistan

**Rifat Ara Masud** Department of Applied Chemistry and Chemical Engineering, Faculty of Engineering and Technology, Bangabandhu Sheikh Mujibur Rahman Science & Technology University, Gopalganj, Bangladesh

**Srikanta Moharana** School of Applied Sciences, Centurion University of Technology and Management, Odisha, India

**Rozalin Nayak** Laboratory of Polymeric and Materials Chemistry, School of Chemistry, Sambalpur University, Odisha, India

**Andrés Otálora** Research Group in Science with Technological Applications (GI-CAT), Department of Chemistry, Universidad del Valle; Mindtech Research Group (Mindtech-RG), Mindtech s.a.s., Cali, Colombia

**Manuel Palencia** Research Group in Science with Technological Applications (GI-CAT), Department of Chemistry, Universidad del Valle; Mindtech Research Group (Mindtech-RG), Mindtech s.a.s., Cali, Colombia

**Mohammed Mizanur Rahman** Department of Applied Chemistry and Chemical Engineering, Faculty of Engineering and Technology, University of Dhaka, Dhaka, Bangladesh

**Jorge A. Ramírez-Rincón** Research Group in Science with Technological Applications (GI-CAT), Department of Chemistry, Universidad del Valle, Cali, Colombia

**Mohsin Ali Raza** Institute of Metallurgy and Materials Engineering, University of the Punjab, Lahore, Pakistan

**Saiful Izwan Abd Razak** BioInspired Device and Tissue Engineering Research Group, School of Biomedical Engineering and Health Sciences, Faculty of Engineering; Centre for Advanced Composite Materials, Universiti Teknologi Malaysia, Skudai, Johor, Malaysia

**Touseef Rehan** Department of Biochemistry, Shaheed Benazir Bhutto Women University, Peshawar, KP, Pakistan

**Zaeem Ur Rehman** Institute of Metallurgy and Materials Engineering, University of the Punjab, Lahore, Pakistan

**Diego F. Restrepo-Holguín** Research Group in Science with Technological Applications (GI-CAT), Department of Chemistry, Universidad del Valle, Cali, Colombia

**Selvaraj Mohana Roopan** Chemistry of Heterocycles & Natural Research Laboratory, Department of Chemistry, School of Advances Sciences, Vellore Institute of Technology, Vellore, Tamilnadu, India

**Bibhuti B. Sahu** Department of Physics, Veer Surendra Sai University of Technology, Odisha, India

**Nasrullah Shah** Department of Industrial and Manufacturing Systems Engineering, Kansas State University, Manhattan, KS, United States; Department of Chemistry, Abdul Wali Khan University Mardan, Mardan, KP, Pakistan

**Saqlain A. Shah** Materials Science Lab, Department of Physics, Forman Christian College (University), Lahore, Pakistan

**Md. Shahruzzaman** Department of Applied Chemistry and Chemical Engineering, Faculty of Engineering and Technology, University of Dhaka, Dhaka, Bangladesh

**Muhammad Gulraiz Tanvir** Institute of Metallurgy and Materials Engineering, University of the Punjab, Lahore, Pakistan

**Halil Tetik** Department of Industrial and Manufacturing Systems Engineering, Kansas State University, Manhattan, KS, United States

**Salman Ul Islam** School of Life Sciences, College of Natural Sciences, Kyungpook National University, Daegu, Republic of Korea

**Mazhar Ul-Islam** Department of Chemical Engineering, College of Engineering, Dhofar University, Salalah, Sultanate of Oman

**Muhammad Wajid Ullah** Department of Biomedical Engineering, Huazhong University of Science and Technology, Wuhan, PR China

**Mat Uzir Wahit** Department of Polymer Engineering, Faculty of Engineering, School of Chemical and Energy, Universiti Teknologi Malaysia, Johor Bahru; Centre for Advanced Composite Materials, Universiti Teknologi Malaysia, Skudai, Johor, Malaysia

**Farzana Yeasmin** Department of Applied Chemistry and Chemical Engineering, Faculty of Engineering and Technology, Bangabandhu Sheikh Mujibur Rahman Science & Technology University, Gopalganj, Bangladesh

# Editor's biographies

**Dr. Sajjad Haider** is an associate professor at the Department of Chemical Engineering at King Saud University in Riyadh, Saudi Arabia. He received his MSc degree in 1999 and MPhil degree in 2004 from the Institute of Chemical Sciences, University of Peshawar, KPK, Pakistan, and his PhD degree in 2009 from the Department of Polymer Science and Engineering, Kyungpook National University, Taegu, South Korea. His research focuses on electrospun nanofibers, biopolymer composites, and polymer hydrogels to develop scaffolds for tissue regeneration and drug delivery, and metal oxides particles for water treatment application.

**Dr. Adnan Haider** is an assistant professor at the Department of Biological Sciences, National University of Medical Sciences (NUMS), Pakistan. He holds an MSc degree from Kohat University of Science and Technology, Pakistan, and MS leading to PhD degree from Kyungpook National University, Taegu, South Korea. Dr. Haider completed postdoctorate from Yeungnam University in South Korea. His research focuses on electrospun nanofibers, biopolymer composites, and polymer hydrogels to develop scaffolds for tissue regeneration, drug delivery, and water treatment applications.

# Series editor biography

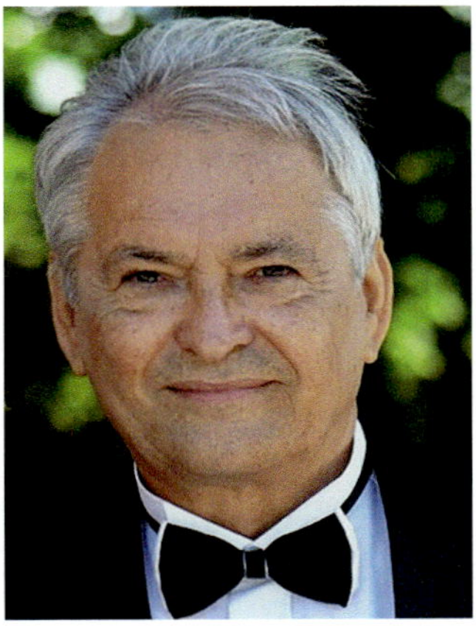

**Ghenadii Korotcenkov** received his PhD in physics and technology of semiconductor materials and devices in 1976 and his Doctor of Science degree (Doc. Hab.) in physics of semiconductors and dielectrics in 1990. He has more than 45 years of experience as a teacher and scientific researcher. For a long time, he was a leader of the gas sensor group and the manager of various national and international scientific and engineering projects carried out in the Laboratory of Micro- and Optoelectronics, Technical University of Moldova, Chisinau, Moldova. During 2007–08, he was an invited scientist at Korea Institute of Energy Research (Daejeon). After this, until 2017, Dr. G. Korotcenkov had been a research professor in the School of Materials Science and Engineering at Gwangju Institute of Science and Technology (GIST), Korea. Currently, he is a chief scientific researcher at Moldova State University, Chisinau, Moldova. Starting from 1995, he has been working on material sciences, focusing on metal oxide film deposition and characterization, surface science, thermoelectric conversion, and design of physical and chemical sensors, including thin-film gas sensors.

Korotcenkov is the author or editor of 39 books and special issues, including the 11-volume "Chemical Sensors" series published by Momentum Press, 15-volume "Chemical Sensors" series published by the Harbin Institute of Technology Press, China; 3-volume "Porous Silicon: From Formation to Application" issue published by CRC Press; 2-volume "Handbook of Gas Sensor Materials" published by Springer; and 3-volume "Handbook of Humidity Measurements" published by CRC Press. Currently, he is the book series editor of "Metal Oxides" published by Elsevier.

Korotcenkov is the author or coauthor of more than 650 scientific publications, including 31 review papers, 38 book chapters, and more than 200 peer-reviewed articles published in scientific journals (h-factor = 41 (Web of Science), h = 42 (Scopus), and h = 56 (Google scholar citation)). He holds 18 patents to his credit. He has presented more than 250 reports in various national and international conferences, including 17 invited talks. As a cochair or a member of program, scientific and steering committees, Korotcenkov participated in more than 30 international scientific conferences. Dr. G. Korotcenkov is a member of editorial boards of five international scientific journals. His name and activities have been listed by many biographical publications, including *Who's Who*. His research activities are honored by the Honorary Diploma of the Government of the Republic of Moldova (2020); Award of the Academy of Sciences of Moldova (2019); Award of the Supreme Council of Science and Advanced Technology of the Republic of Moldova (2004); the Prize of the Presidents of the Ukrainian, Belarus, and Moldovan Academies of Sciences (2003); Senior Research Excellence Award of Technical University of Moldova (2001, 2003, and 2005); and the National Youth Prize of the Republic of Moldova in the field of science and technology (1980); among others. Korotcenkov also received a fellowship from the International Research Exchange Board (IREX, United States, 1998), Brain Korea 21 Program (2008–12), and BrainPool Program (Korea, 2015–17).

# Preface to the series

Synthesis, study, and application of metal oxides are the most rapidly progressing areas of science and technology. Metal oxides are one of the most ubiquitous compound groups on Earth and are available in a large variety of chemical compositions, atomic structures, and crystalline shapes. In addition, they are known to possess unique functionalities that are absent or inferior in other solid materials. In particular, metal oxides represent an assorted and appealing class of materials, exhibiting a full spectrum of electronic properties—from insulating to semiconducting, metallic, and superconducting. Moreover, almost all the known effects, including superconductivity, thermoelectric effects, photoelectrical effects, luminescence, and magnetism, can be observed in metal oxides. Therefore, metal oxides have emerged as an important class of multifunctional materials with a rich collection of properties, which have great potential for numerous device applications. Availability of a wide variety of metal oxides with different electrophysical, optical, and chemical characteristics, their high thermal and temporal stabilities, and their ability to function in harsh environments make them highly suitable materials for designing transparent electrodes, high-mobility transistors, gas sensors, actuators, acoustical transducers, photovoltaic and photonic devices, photo- and heterogeneous catalysts, solid-state coolers, high-frequency and micromechanical devices, energy harvesting and storage devices, and nonvolatile memories, among others in the electronics, energy, and health sectors. In these devices, metal oxides can be successfully used as sensing or active layers, substrates, electrodes, promoters, structure modifiers, membranes, or fibers, i.e., they can be used as active and passive components.

Among other advantages of metal oxides are their low fabrication cost and robustness in practical applications. Furthermore, metal oxides can be prepared in various forms such as ceramics, thick films, and thin films. At that for thin film deposition can be used deposition techniques that are compatible with standard microelectronic technology. This factor is very important for large-scale production, because the microelectronic approach ensures low cost for mass production, offers the possibility of manufacturing devices on a chip, and guarantees good reproducibility. Various metal oxide nanostructures including nanowires, nanotubes, nanofibers, core–shell structures, and hollow nanostructures can also be synthesized. As it is known, the field of metal-oxide nanostructured morphologies (e.g., nanowires, nanorods, nanotubes, etc.) has become one of the most active research areas within the nanoscience community.

The ability to both create a variety of metal oxide-based composites and synthesize various multicomponent compounds significantly expands the range of properties that metal oxide-based materials can offer, making them a truly versatile multifunctional material for widespread use. As it is known, small changes in their chemical composition and atomic structure can result in a spectacular variation in properties and behavior of metal oxides. Current advances in synthesizing and characterizing techniques reveal numerous new functions of metal oxides.

Taking into account the importance of metal oxides for progress in microelectronics, optoelectronics, photonics, energy conversion, sensor, and catalysis, a large number of books devoted to this class of materials have been published. However, one should note that some books from this list are too general, some are collections of various original works without any generalizations, and yet others were published many years ago. However, during the past decade, great progress has been made on the synthesis as well as on the structural, physical, and chemical characterization and application of metal oxides in various devices, and a large number of papers have been published on metal oxides. In addition, till now, many important topics related to the study and application of metal oxides have not been discussed. To remedy the situation in this area, we decided to generalize and systematize the results of research in this direction and to publish a series of books devoted to metal oxides.

The proposed book series "Metal Oxides" is the first of its kind devoted exclusively to metal oxides. We believe that combining books on metal oxides in a series could help readers in searching the required information on the subject. In particular, we hope that the books from our series, with its clear specialization by content, will provide interdisciplinary discussion for various oxide materials with a wide range of topics, from material synthesis and deposition to characterizations, processing, and then to device fabrications and applications. This book series was prepared by a team of highly qualified experts, which guarantees its high quality.

I hope that our books will be useful and easy to navigate. I also hope that readers will consider this "Metal Oxides" book series an encyclopedia of metal oxides that enables to understand the present status of metal oxides, to estimate the role of multifunctional metal oxides in design of advanced devices, and then based on observed knowledge to formulate new goals for further research.

This book series is intended for scientists and researchers, working or planning to work in the field of materials related to metal oxides, i.e., scientists and researchers whose activities are related to electronics, optoelectronics, energy, catalysis, sensors, electrical engineering, ceramics, biomedical designs, etc. I believe that the book series will also be interesting for practicing engineers or project managers in industries and national laboratories involved in designing metal oxide-based devices, helping them with the process and in selecting optimal metal oxide for specific applications. With many references to the vast resource of recently published literature on the subject, this book series will be serving as a significant and insightful source of valuable information, providing scientists and engineers with new insights for understanding and improving existing metal oxide-based devices and for designing new metal oxide-based materials with new and unexpected properties.

I believe that the book series would be very helpful for university students, post docs, and professors. The structure of these books offers a basis for courses in the field of material sciences, chemical engineering, electronics, electrical engineering, optoelectronics, energy technologies, environmental control and many others. Graduate students would also find the book series very useful in their research and understanding the synthesis of metal oxides and the study and applications of these multifunctional materials in various devices. We are sure that each of these audiences will find information useful for their activity.

Finally, I thank all contributing authors and book editors involved in the creation of these books. I am thankful that they agreed to participate in this project and for their efforts in the preparation of these books. Without their participation, this project would have not been possible. I also express my gratitude to Elsevier for giving us the opportunity to publish this series. I especially thank the team at the editorial office at Elsevier for their patience during the development of this project and for encouraging us during the various stages of preparation.

**Ghenadii Korotcenkov**

# Composite materials: Concept, recent advancements, and applications

1

*Md. Sazedul Islam, Md. Shahruzzaman, M. Nuruzzaman Khan, Md. Minhajul Islam, Sumaya Farhana Kabir, Abul K. Mallik, Mohammed Mizanur Rahman, and Papia Haque*
Department of Applied Chemistry and Chemical Engineering, Faculty of Engineering and Technology, University of Dhaka, Dhaka, Bangladesh

## 1 Introduction

### *1.1 Composites*

Composite naturally exists in the form of wood where long cellulose fibers are held together by lignin. Manmade composites of mud and straw for making house by Egyptian or Mesopotamian had reported back on 1500 BCE.

In modern technology, composite materials had emerged in the mid of the twentieth century as one of the promising classes of engineering materials. Composite materials are sought from the necessity of lightweight structures with sufficient high-temperature strength, stiffness, and toughness. In modern age, applications of composites are found in almost every sector from aerospace, marine, and automotive areas to biomedical implants [1].

In general, composite is a multiphase material that has unique and different properties in bulk compared to the properties of any of its component materials [1,2]. The component phases can be minimum of two to a few in numbers in any composites. The one component must be the matrix, which is a primary or continuous phase, and it holds the other one or more components within it to give the final shape of the composite. One or more components remain dispersed in the matrix and separated from the matrix by distinct interface. The matrix phase is usually ductile, and the dispersed phase is the main reinforcing agent in composites. One component by volume is usually retained higher than that of the other one to tailor a desirable property in a composite. However, when the matrix material is same, the properties of composites are largely influenced by relative volume, size, and shape, and by dispersion pattern of the reinforcing agents [3].

Composites are classified either based on the type of matrix materials or based on the form of reinforcing materials. Considering the matrix material, composites can be like metal matrix composites (MMCs), ceramic matrix composites (CMCs), and polymer matrix composites (PMCs). On the basis of reinforcing materials, composites are classified as particulate composites and fibrous composites.

**Renewable Polymers and Polymer-Metal Oxide Composites. https://doi.org/10.1016/B978-0-323-85155-8.00001-7**

### 1.1.1 Metal matrix composites (MMCs)

MMCs possess stiff and hard reinforcement (ceramic or metallic, e.g., lead, tungsten, molybdenum, carbon, silicon carbide, boron) phase within a metallic matrix. Light metals like aluminum (Al), titanium (Ti), iron (Fe), cobalt (Co), copper (Cu), and magnesium (Mg), and alloys of them, are commonly applied matrix materials.

MMCs of high stiffness and strength, good electrical and thermal conductivity, high corrosion and wear resistance, improved fracture and fatigue properties, and enhanced thermal stability can be obtained by the appropriate selection of matrix materials, reinforcements, and production processes [4]. Inclusion of reinforcements within a metal actually creates a crystal defect in it. This is why, in MMCs, the reinforcement usually improves the physical properties rather than enhancing its mechanical properties [5,6]. MMCs can be manufactured in solid or semi-solid state (powder metallurgy), in liquid state (casting, electroplating, spray deposition, eutectic alloy etc.), and also in vapor state (vapor deposition). These composites are expensive due to cost of materials and production. MMCs are frequently used in automobiles parts, cutting tools, and military tank armors.

### 1.1.2 Polymer matrix composites (PMCs)

PMCs must have a polymer matrix either thermoplastic (polycarbonate, polyvinylchloride, nylon, polystyrene) or thermoset (unsaturated polyester, epoxy) and a dispersed/reinforced phase of glass, carbon, ceramic, and other polymers in the form of fibers or particulates. In PMCs, the reinforcement materials add the strength to the composites. The uniqueness of this type of composites is their high strength-to-weight ratio. In fiber-embedded composites, the alignment of fiber is very crucial to achieve a desired strength and stiffness, whereas in particulate composites, distribution and type of particulate to create some interfacial linkages to the matrix is mostly important. They can be manufactured by solvent casting, blending, and in situ techniques. Thermoset PMCs have a great use in electrical equipment and thermoplastic PMCs are of high in demand in biomedical applications. However, one of the problems associated with the composites is their slow degradation effect to environment.

### 1.1.3 Ceramic matrix composites (CMCs)

CMCs are made of ceramic reinforcements (silicon carbide, alumina, carbon, etc.) in the form of fibers within a ceramic matrix. The main task of CMCs is to increase the crack resistance than that of the matrix ceramic under a mechanical or thermomechanical load. Nonetheless, they have low impact resistance and limitations in variations of geometry of the reinforcements. They can be manufactured by pyrolysis, electrophoretic deposition, sintering, etc. They have good applications in space shuttles and in high-temperature and high load-bearing equipment parts.

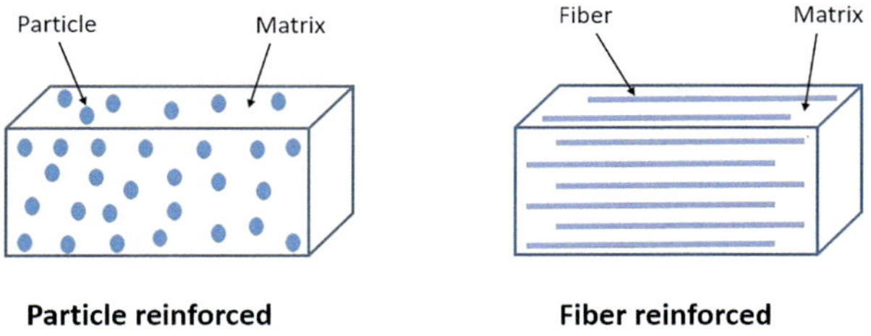


**Fig. 1** Different types of reinforcements (schematic).

### *1.1.4 Reinforcement of composites*

Two types of reinforcements are used in composites, particles and fibers (Fig. 1). For particulate composites, the dispersed phases are in form of particles and can be randomly oriented or of preferred orientation and consist of two-dimensional flat platelets oriented parallel to each other.

Fibrous composites may be divided into short fiber dispersed composites and long fiber dispersed composites. The short fiber reinforced composites may be divided into the following two types.

**(a)** Random orientation of fibers
**(b)** Preferred orientation of fibers.

The long fiber reinforcements can be divided into two categories. They are as follows:

**(a)** Fibers of unidirectional orientation.
**(b)** Fibers of bidirectional orientation (woven).

## *1.2 The rule of mixture*

The Rule of Mixture (ROM) is a classical and simple method to assume some of the mechanical properties of fiber-reinforced polymer-based composites (FRP) from the weighted average of the properties of the different phases along with their volume fraction [3].

To estimate the tensile performance of composite materials using the rule of mixture (ROM), the following assumptions are made [7].

**(1)** Any ply is homogeneous and orthotropic in microscale level when no stress is applied on it and it shows linear elasticity.
**(2)** The fiber is homogenous, unidirectional, and linear elastic.
**(3)** The matrix is homogenous and isotropic and shows linear elasticity.
**(4)** Void content is zero.
**(5)** Applied stress is uniformly distributed over the composites.

Based on these theories, the tensile performance of FRP can be attained by merging linearly the tensile properties and the volume fraction of the fiber and the matrix as follows [7]:

$$\sigma = \sigma_f V_f + \sigma_m V_m \tag{1}$$

$$E = E_f V_f + E_m V_m \tag{2}$$

where

$\sigma=$ the tensile strength of composites,
$\sigma_f=$ the tensile strength of reinforcing fiber,
$\sigma_m=$ the tensile strength of the matrix,
$E=$ the elastic modulus of composites,
$E_f=$ the elastic modulus of reinforcing fiber,
$E_m=$ the elastic modulus of matrix.

The ROM may also be applied for the determination of tensile properties in transverse direction, which can be represented by Eq. (3).

$$E_T = \frac{E_f E_m}{E_f V_m + V_f E_m} \tag{3}$$

However, there are a few nonrealities in the assumptions made for the ROM: The fibers may not be always (not in most of the cases) homogeneously aligned, stress is not always uniformly distributed, and composites are hardly free from any void. These mismatches of assumptions with the practical situations show some deviations in the expected results. Therefore, the elastic modulus as calculated from Eq. (1) is known as the upper-bound modulus and related to a parallel loading to the fibers and the elastic modulus estimated from Eq. (3) is called the lower-bound modulus and is related to a transverse loading to the direction of the reinforced fiber.

## 1.3 Renewable polymers for metal oxide-reinforced composites

Biopolymers are biocompatible, biodegradable, renewable source-based, environmental-friendly, and naturally abundant polymeric materials [8,9]. Biopolymers undergo degradation by the naturally occurring microorganisms and produce $CO_2$ and $H_2O$ byproducts [9]. This renewable source-based biopolymer includes polysaccharides, such as starch, cellulose, chitin/chitosan, and alginate, and also animal protein-based biopolymers, such as gelatin wool, collagen, and silk [3]. The main advantage of natural biopolymers is probably that they have strong influence in cell function and cell adhesion; however, the mechanical properties reduce fast over time due to moisture absorption. Chitin, for example, is limited in application due to its poor thermal stability, mechanical properties, and rigidity [9].

There are various methods that have so far been used to improve the mechanical properties of natural biopolymers [10]. Inclusion of nanofillers has been applied by several researchers to enhance mechanical, thermal, electrical, and biomedical properties of composites. Different sizes ($<$ 100 nm) and types of nanofillers (metals, metal oxides, ceramic, clay, organic polymers, etc) have been investigated to improve environmental decontamination, to make more efficient biomedical applications, to prepare edible packaging, etc. [11]. Other than some common applications of PMCs, some nanofillers have made them to use in some less-known field for PMCs. For instance, clay or silicate-based polymer composites have a good use in effluent treatment, semiconductor metal-based polymers are used in solar cells, and magnetic particle-based PMCs have vast applications in targeted drug delivery systems. [9].

Renewable biopolymer-based composites help in the nucleation and growth of metal oxides within it [12]. The polymers cause a dispersion of metal cations at a comparatively high temperature and thus control the nucleation and growth of the oxides during solid-state synthesis [13]. Chemical and physical bonding of the functional groups of the polymers with metal cations can help them in good dispersion. For example, alginate that contains blocks of guluronate monomers undergoes cross-linking by the electrostatic bonds to metal cations with variable valency and thus dispersed metal cations throughout the structure [14].

Metals have electrical conductivity, antimicrobial properties, optical polarizability, thermal stability, and good chemical reactivity [11]. Metal- or their oxide-based composites have already been used in various fields of applications such as drug delivery, probes for electron microscopy, diagnosis and therapy of diseases, etc. For instance, gold nanoparticle-based nanocomposites are reported to find their applications in optical and photonic fields [15]. Similarly, silver (Ag) is known to have an excellent antibacterial property and have suitably been applied in traditional medicines [16,17].

Metal- or metal oxide nanoparticle-dispersed PMCs are gaining attention to be applied in various fields of electrical, electronics, optics, and photonics. The end uses of these composites, however, greatly depend on the geometry, homogeneity, and distribution uniformity of the particles within the matrix. Ramesan et al. [10] produced silver-doped zinc oxide NP (Ag–ZnO)-based chitin (CT) and cashew gum (CG) composites. They reported that the addition of the particles in biopolymer matrices enhanced the thermal, mechanical, flame resistance, and electrical properties of polymer nanocomposites.

Various metal-based renewable polymer composites are good entities for food packaging applications. The antioxidant and antimicrobial agents in the packaging can interact with the oxygen, water vapor, and microbes present in foods to give it a longer shelf life [9]. Kanmani et al. [18] prepared composite films with gelatin and Ag NPs and studied the antimicrobial properties of the prepared films. Rafieian et al. [19] also produced composite films from wheat gluten matrix and cellulose nanofibrils and thus examined the morphological and thermomechanical properties of the films for food packaging applications.

# 2 Experimental characterization of composites

## 2.1 Chemical properties

The chemical characterization of composite materials possesses a big challenge as composite materials are not soluble in suitable solvents. Chemical properties of composite materials are determined by measuring chemical resistance of the material.

Both MMCs and PMCs can be adversely affected by chemicals, leading to a compromise in structure, strength, flexibility, impact resistance, and appearance. For PMCs, the functional groups in the polymer chain can be attacked by chemicals through oxidation reaction or depolymerization of long molecules, causing unwanted damages. Moreover, some polymers may swell or soften by coming into contact of certain chemicals, and some of the chemicals even cause stress cracking. Factors such as temperature, concentration of chemicals, exposure time, and pressure determine the chemical resistance of a composite [20].

Many common chemicals applied for testing include sulfuric acid, formic acid, tree resin, gasoline, caustic soda, and water. Even common household chemicals such as red wine, coffee, mustard, or ball pen ink are used to carry out the test at room temperature [21]. Mostly used chemical resistance test is against acid and caustic soda. To carry out chemical resistance against these chemicals, samples of proper size are dried and placed in 100mL of 1N NaOH and 1N HCl for different time intervals. On the completion of this process, the samples are filtered out, dried, and weighed. By determining the weight loss, the percent chemical resistance (Pcr) of the sample is calculated [22],

$$\text{Percent chemical resistance, } P_{cr} = \frac{T_i - W_{aci}}{T_i} 100$$

where $T_i$ = weight taken initially and $W_{aci}$ = weight taken after certain interval.

In case of MMCs, immersion in salt solution of 3.5 wt% NaCl is also carried out to determine the chemical resistance as well as corrosion susceptibility of the composite [23].

## 2.2 Thermal properties

Thermal properties of composite materials have become an important consideration in designing electronic materials. Modern electronic devices like light-emitting diodes (LEDs), transistors, semiconductors, and optoelectronics can operate at higher efficiency if unwarranted heat can be instantly transported to heat sinks [24]. This is particularly true for both PMC- and MMC-based field effect transistors due to miniaturization and high-power density, resulting in shortened lifespan. This predicament makes the determination of thermal properties of composite materials a necessary step in designing efficient devices. Thermal conductivity, a desirable property to reduce heat buildup, is a measure of heat transferred across a particular medium. For PMCs, thermal conductivity is determined experimentally by using Debye equation [25],

$$\kappa = \frac{C_p v l}{3}$$

where $C_p$ in the equation stands for the specific heat capacity per unit volume, $l$ is the mean free path of phonon, and $v$ is the velocity of the phonon. In case of PMCs, composites exhibit very low thermal conductivity due to impedance of phonon propagation through the continuous path of chemically bonded atoms because of scattering by polymers. Some defects in polymer's structure like chain ends, entanglements, random orientation voids, and impurities also contribute to lower thermal conductivity of PMCs.

But thermal conductivity of composites, for both PMCs and MMCs, is well predicted by Nielsen's theory over a wide range of conditions. The significance of this theory is that it considers the composite geometry, matrix, and particle properties along with the volume loading of the filler. According to Nielsen [26],

$$\frac{K_c}{K_p} = \frac{1 + AB\varphi}{1 - B\psi\varphi}$$

where $K_c$ and $K_p$ represent the thermal conductivity of the composite and that of the polymer respectively, and $\varphi$ represents the volume fraction of the filler.

$$B = \frac{\frac{K_f}{K_p} - 1}{\frac{K_f}{K_p} + A}$$

where $K_f$ is the filler's thermal conductivity. $A$ is a geometrical value based on three factors: the particle's shape, aspect ratio, and degree of aggregation.

$$\psi = 1 + \frac{1 - \varphi_m}{\varphi_m^2}\varphi$$

where $\varphi_m$ is the maximum packing fraction of the filler determined experimentally.

Using the equations, many methods have been developed to measure thermal conductivity. The steady-state methods and transient or non-steady-state methods are two most popular techniques used to calculate thermal conductivity of a substance [27]. Both methods have their limitations, but they follow the same fundamental laws of conduction of heat and electrical analogy.

For PMCs, a simple experimental method can be used to determine thermal conductivity using thermocouples [28]. This method needs two thermocouples, a heater, a thin foil of indium, specimen, another thin foil of indium, and a heat sink at the bottom.

The thermocouples are inserted in the indium foils to measure the change in temperature of the specimen. The total thermal resistivity is determined using the equation,

$$R = {}^{\Delta T}/_{Q}$$

where $R$ is the summation of the resistivity of the sample ($R_s$) and interface thermal resistance ($R_{\text{int}}$), $Q$ represents the total heat flux generated by the heater, and $\Delta T$ is the temperature difference across the sample. A multiplied by $R$ is plotted against specimen thickness, where $A$ is the specimen area. Thermal resistance ($R_{\text{int}}$) is the intercept of the graph, and thermal conductivity ($K$) is the reciprocal of the slope of the graph. These values can be used in Nielsen's equation to calculate the thermal conductivity of the composites.

In case of MMCs, thermal conductivity can be determined from thermal diffusivity, which describes the rate of temperature spread through a material [29]. Thermal diffusivity is equivalent to the ratio of thermal conductivity to the density and specific heat capacity at constant pressure [30]. The density of specimens is calculated by the Archimedes method where distilled water is used as the medium of immersion.

Thermogravimetric analysis (TGA) in combination with differential scanning calorimetry (DSC) shows thermal decomposition and phase change of the composite materials with increasing temperature [31]. The sample is heated at a constant rate along with an equal mass of reference under inert nitrogen medium from room temperature up to >1000°C. A heating curve shows the change in temperature as heat is supplied to the sample. If the temperature of the sample suddenly stops increasing but the reference's temperature continues to rise, a phase change is indicated by the electrical signal [32]. Specific heat and glass transition temperature ($T_g$) can be measured by DSC using the slope of the heating curve.

## 2.3 Optical properties

Polymer and metal matrix-based composites have been developed for optical wave guiding, sensing, and other applications. Moreover, polymers with high refractive index have found diverse applications in solar cells as antireflection coatings and high refractive index lenses. With the expansion of optical applications, the need for innovative optically functional and transparent materials increases alongside the requirement of experimental characterizations.

As an optical parameter, refractive index of a material is very important as it discloses optical properties of the materials. Refractive index can be measured by several methods, like Abelés method, reflectance and transmittance method, interference fringes method, prism spectrometer method, Kramers-Kronig relation method, and ellipsometric method. Among them, Abelés method is the most simple and accurate for measuring refractive index of composites films [33]. This method is based on the idea that when a polarized light with its electric vector is incident on the plane of incidence, the reflectance of the film ($n_f$) deposited on a substrate ($n_s$) at an angle $\tan\alpha = n_f/n_0$ is the same as the reflectance of the bare substrate. The medium of incidence is air ($n_o$).

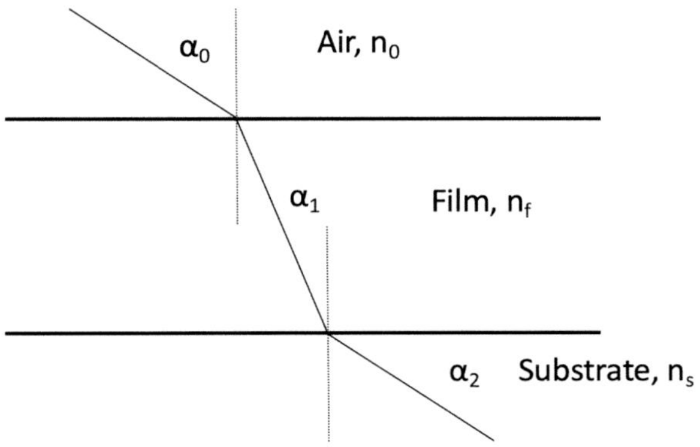


**Fig. 2** Determination of the refractive index of a transparent film on a transparent substrate by Abele's method.

This requires a spectrometer capable of achieving a telescope rotation at a twice the rate of the specimen and focusing on this specimen instead of at infinity along with a monochromatic source of radiation. When polarized monochromatic light having the same intensity is reflected from the film bare substrate, the angle of incident is recorded, and the refractive index of the film ($n_f$) is obtained from tangent tables. For incident light polarized with (Fig. 2), the reflectance from a dielectric layer is

$$R_p = \frac{\tan^2(\alpha_2 - \alpha_0)}{\tan^2(\alpha_1 - \alpha_0)}$$

For both PMCs and MMCs, UV/visible spectrophotometer can be used to determine the optical transmission of samples. The experimental results can be used to predict the parameters such as band gap $E_g$, dielectric constants $\varepsilon_1$, $\varepsilon_2$, volume and surface energy loss function, and optical conductivity using the pointwise unconstrained minimization approach (PUMA) [34–36].

## 2.4 Biochemical properties

Biochemical analysis of composite requires the determination of protein, double-stranded deoxyribonucleic acid (DNA), proteoglycan, and collagen content present in the composite material. For this purpose, several spectroscopy-based methods have been developed. Spectroscopy is an analytical technique based on the interaction of molecules with electromagnetic (EM) radiations. EM radiations can be classified according to its wavelength or energy. Near-ultraviolet (UV)- and visible (vis)-range EM radiations contain sufficient energy to excite electrons for making the jump from the ground state to an excited state. A spectrum can be obtained by plotting the absorption of light as a function of its frequency or wavelength. As proteins have molecules with electrons in delocalized aromatic systems, they can absorb EM radiations in the near-UV (150–400 nm) or the visible (400–800 nm) region with absorption maxima between 275 and 280 nm. The presence of the two aromatic amino acids tryptophan and tyrosine and the disulfide bonds of cystine are responsible for this absorption behavior. The absorbance of tryptophan and tryptophan depends on the microenvironment of their chromophores. Their energy decreases when they move from a polar to a nonpolar environment. As a result, there is a red-shift and the residues exposed to solvent contribute differently to the absorption coefficient [37]. A protein's absorption coefficient ε can be calculated from the numbers of tryptophan (Trp) and tyrosine (Tyr) and cystine (Cys) disulfide bonds denoted as nTrp, nTyr, and nSS, respectively [38]. Similarly, nucleic acids show a strong absorbance peak at 260 nm, which can be utilized to measure their concentrations in solution. The amounts of nucleic acid are often denoted as A260 units. When DNA is double-stranded, one A260 unit is considered as 50 micrograms of DNA, while for single-stranded DNA, it is equivalent to 33 micrograms of DNA. In case of single-stranded ribonucleic acid (RNA), it is equivalent to 40 micrograms of RNA.

## 2.5 Electrical properties

Electrically conductive polymer composites have found applications in sensors, circuit devices, fuel cell, battery, electrodes, and fuel cell bipolar plates [39]. The electrical conductivity of any composite containing metal particles depends on interfacial shells between the polymer matrix and the metal particles as PMCs are considered an assembly of sporadically oriented ellipsoidal metal particles ingrained in a homogeneous polymer matrix. The development of interfacial shells is because of diffusion, chemical reaction, quantum mechanical tunneling of electrons through the insulating medium between metal particles. These interfacial shells interact like an electrical channel without any physical contact among the metal particles. It is thought that the interfacial shell and the metal particle form a type of complex particles. As a result, the metal–polymer composite could be considered as the complex particles embedded in the polymer matrix [40]. Using Maxwell theory, it is possible to derive the electric conductivity component along $k$-axis of the $j$th complex particle,

$$\sigma_{\mathrm{cj,k}} = \sigma_1 \frac{(1 - B_{2j,k})\sigma_1 + B_{2j,k}\sigma_2 + (1 - B_{2j,k})\lambda(\sigma_2 - \sigma_1)}{(1 - B_{2j,k})\sigma_1 + B_{2j,k}\sigma_2 - (1 - B_{2j,k})\lambda(\sigma_2 - \sigma_1)}$$

where $B_{2j,k}$ represents the depolarization factors of the $j$th metal particle along $k$-axis ($k = x, y, z$). $\sigma_1$ and $\sigma_2$ symbols stand for the electric conductivity of the interfacial shell and the metal, respectively. Finally, $a_j$, $b_j$, and $c_j$ are the half-radii of the $j$th elliptical metal particle, respectively, and $t$ is the thickness of the interfacial shell. The conductivity of PMCs depends on the size and the axial ratio of the metal particles. It is possible to avoid the requirement of excess metal particles by adding smaller-sized metal particles with larger axial ratio.

In MMCs, conductivity depends on the microstructural state of the matrix alloy and is inversely proportional to the volume of fibers. The electrical conductance of MMCs surface can be measured by the four-point probe method [41]. This method uses direct current (DC) to circumvent skin effects caused by AC current in the matrix and transient polarization phenomena. Repeatedly changing the current direction and using the absolute value average of both voltage drops, it is possible to avoid capacity and thermoelectric effects of the setup [42].

For PMCs, the measurement of electrical impedance is more common to characterize electrical properties of composites [43]. It is usually measured by an electrochemical potentiostat. Samples are put in between gold electrodes to determine the impedance and dielectric constant with respect to a selected frequency with an applied AC wave amplitude of 500 mV. The impedance is measured using the Eq. [44],

$$|Z| = \sqrt{R^2 + X_c^2}$$

where $R$ is the electrical resistance and $X_c$ is the capacitive reactance. The impedance analyzer records the variation between the maximum current and the minimum current and calculates the electrical resistance of the composite specimen. One advantage of impedance measurement is that both resistance and capacitance of the composite can be determined simultaneously [45].

## 2.6 Thermomechanical properties

Thermomechanical properties can be analyzed using thermal and mechanical analytical techniques that deliver important information on thermal expansion, glass transitions temperature ($T_g$), softening points, composition, and phase changes of phase in materials by using a constant force as a function of temperature [46]. In these techniques, both PMCs and MMCs composites are subjected to non-oscillating stress undergoing distortion against either temperature or time in a fixed atmosphere. Thermomechanical analysis is also useful to determine creep and recovery of PMCs by loading and unloading the samples over periods of time. Thermomechanical analysis instrument consists of a furnace where sample is placed and connected with a thermocouple to gauge the temperature. There is connection with a force generator and linear variable differential transformer (LVDT) for length detector. Any change in temperature in the sample after applying the force from generator is recorded by the probe. LVDT detects any deformation like thermal expansion and softening in sample with changing temperature. Thermomechanical analysis can apply stresses at changeable rates and can also be employed for the determination of flexural strength and Young's modulus of composites.

Thermomechanical analysis is a very effective method for the determination of thermal expansion and contraction of metal matrix and PMCs [47]. Thermal expansion coefficient reveals the dimensional change of polymers with a change in temperature. Thermal expansion can be linear or nonlinear. When the polymer does not exhibit phase transition with temperature, the expansion is called coefficient of thermal expansion and denoted by $\alpha_L$. The equation for coefficient of thermal expansion (CTE) is,

$$\alpha_L = \frac{\Delta L}{L \Delta T}$$

where $L$ is the initial length of sample, $\Delta L$ is the change in length, and $\Delta T$ is the change in temperature.

Glass transition temperature ($T_g$) can be determined by thermomechanical analysis. $T_g$ of a polymer composite can be determined from the coefficient of thermal expansion graph by extrapolating the intercept of a plot between temperature against displacement [48]. For MMCs, it is possible to measure the coefficient of thermal expansion by a thermomechanical analyzer—vertical dilatometer. Samples having required size and shape are cut from the composite and are subjected to heating up and cooling with the rate of 20°C/min in an inert atmosphere of high-purity nitrogen having a flow rate of 30 $cm^3$/min [42,49]. In case of PMCs, a similar method can be used to determine the coefficient of thermal expansion of the composite [28].

# 3 Structural analysis of composites

Structural analysis is very important during the manufacture of advanced composite materials when the materials are designed for any specific application. These analyses are performed to identify the constituent elements, structural shape, size, surface area, porosity, impurities, defects, and crystallinity of numerous composite materials. The structural analyses are accomplished with special importance to maintain the desirable structural properties of the nanocomposites. The analyses are carried out by different analytical techniques such as scanning electron microscopy (SEM), transmission electron microscopy (TEM), scanning tunneling microscopy (STM), optical coherence tomography (OCT) atomic force microscopy (AFM), X-ray studies such as X-ray diffractometer (XRD), small-angle X-ray scattering (SAXS), and wide-angle X-ray scattering (WAXS) etc.

## 3.1 Scanning electron microscopy (SEM)

The SEM analysis is performed to identify the surface morphology of the composite materials. It is a suitable technique to investigate the surface roughness, smoothness, micro-defects, cracking, fracture, and worn. The micro-cracks of the PMCs are generated by the interface decohesions, which are visible in SEM observations. Moreover, the void spaces, responsible for the fragmentation of the wear surface of the PMCs, are also clearly visualized in SEM photographs. The observations help the manufacturer to correlate the density of micro-defects with the applied stress/strain level. Furthermore, the analysis is also able to detect the orientation and bonding between the reinforcing agent and the matrix in the composites. The suitable proportion of the reinforcing fibers to the matrix material can be fixed by successive SEM analysis for several volume fractions. These give information about the de-bonding, delamination, and micro-cracks of the composite materials, which helps to prepare the proper mixing for the desirable material. The performance of the MMCs depends on the uniform distribution of the reinforced particles in the matrix material. The SEM analysis assists to determine the distribution of reinforced particles so that the agglomeration and sedimentation of the particles can be ceased. The homogeneity of the composite can also be confirmed by the successive SEM analysis. The characteristics of the reinforcing materials greatly influence the mechanical properties of the composites. For example, the reasons of fracture in MMCs are particle rupture and interface cracking, which can be clearly identified by SEM images. Furthermore, the presence of intermetallics can be detected that are brittle in nature and deteriorate the service life of the composites. In a research, Pazhouhanfar et al. prepared a MMC from titanium diboride ($TiB_2$) reinforced with Al 6061 matrix and evaluated its microstructural and mechanical properties [50]. The SEM images (Fig. 3) of the fracture surfaces of composites are captured after the tensile tenting of materials. The images show that the fracture surfaces contain dimples, which indicates the ductile rupture as the main reason of fracture. The analysis reveals that the depths of dimples are in a decreasing rate with increasing percentage of reinforced particles. These findings indicate the gradual decrease in plastic deformation before fracture, which defines the decrease in ductile property of the composites.

## 3.2 Transmission electron microscopy (TEM)

The TEM analysis is carried out to measure the size of the nano- and micro-components used in the composite materials. Furthermore, it can identify the continuous or discontinuous dispersion of the reinforcing agent in the matrix material of the PMC and MMCs. Moreover, the bonding at the interface of the reinforcing material and matrix can also be detected from the micrographs. The analysis strongly assists to maintain the desirable mechanical properties of the composite materials. It supplies valuable details about the dislocation density, grain boundary, and grain size of the material. In a research, Liu et al. studied the unique defect evolution during the plastic deformation of Al-5%$TiB_2$/TiC MMC [51]. The TEM images of this composite shown in Fig. 4 indicate the evolution of dislocation density and the reduction of grain size with increasing the shear strain.

## 3.3 Scanning tunneling microscopy (STM)

The STM analysis is performed to produce three-dimensional and high-resolution topographic images of MMC composite surfaces at the atomic level. It can provide electronic structural information about the conductor, semiconductor, and metal surfaces, which has made the research easier in the field of nanoscience and nanotechnology. The STM analysis can determine the electric and magnetic properties of the composite materials. Furthermore, it can measure the electromagnetic interaction between two metallic constituents of MMC. Moreover, STM analysis has been utilized to reveal the mechanism of pitting corrosion and passive film formation as well

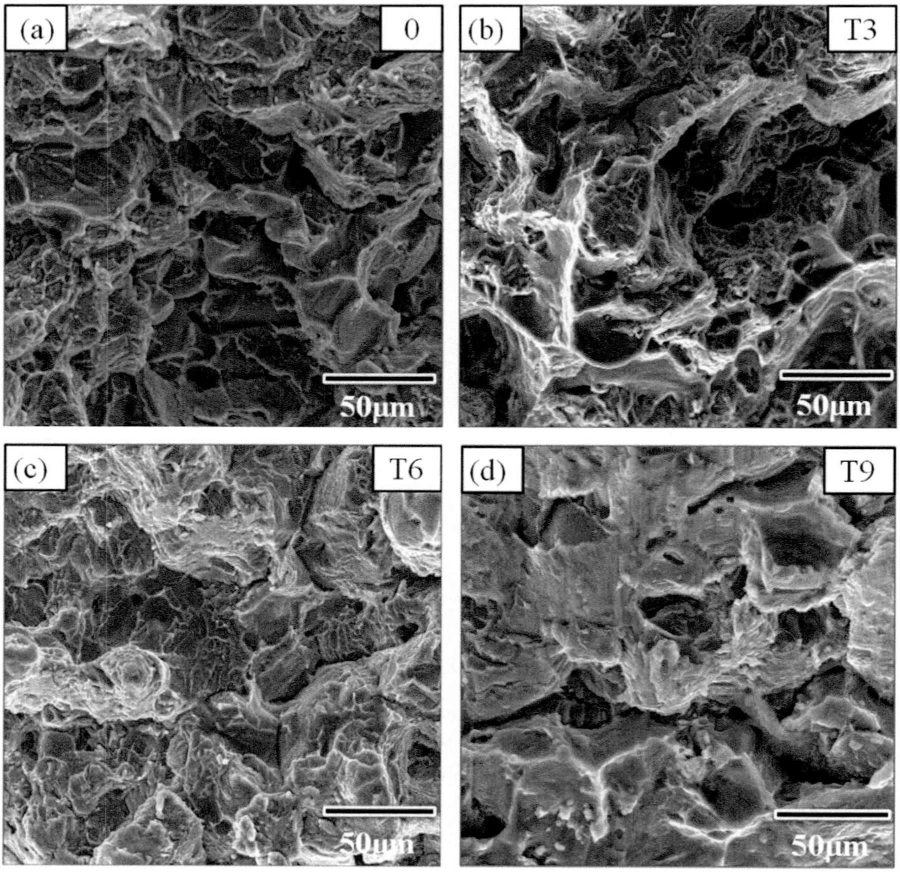


**Fig. 3** SEM images of the fracture surfaces of samples after tensile testing: (A) Al 6061 matrix, (B) Al 6061–3 wt% $TiB_2$, (C) Al 6061–6 wt% $TiB_2$, and (D) Al 6061–9 wt% $TiB_2$.
*Source*: Reprinted with permission from Y. Pazhouhanfar, B. Eghbali, Microstructural characterization and mechanical properties of TiB2 reinforced Al6061 matrix composites produced using stir casting process, Mater. Sci. Eng. A 710 (2018) 172–180.

as the process of inhibition in numerous MMC materials containing different metals such as iron, copper, and nickel.

## 3.4 Atomic force microscopy (AFM)

The AFM analysis is very important to determine the size and locations of the nanoparticle incorporated in the matrix material of composite. The technique is advantageous to capture the three-dimensional images of composite materials containing almost all types of constituents like glass, ceramics, and natural and synthetic polymers etc. It can reveal the surface morphology of the materials especially the

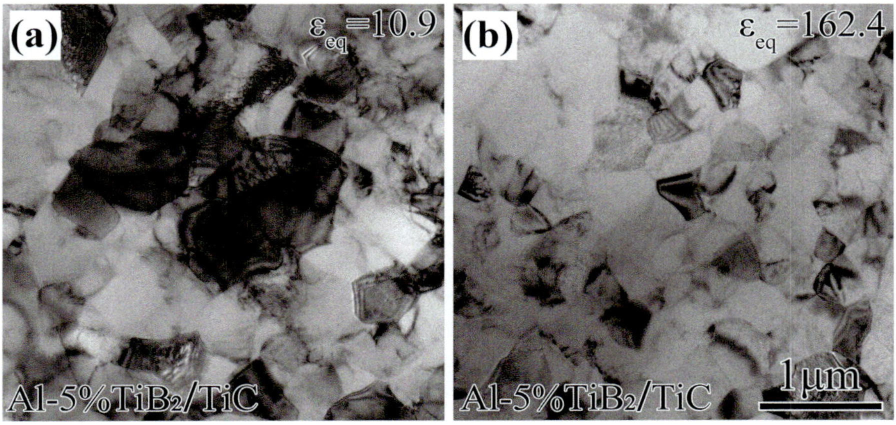


**Fig. 4** TEM images (A) an Al-5%$TiB_2$/TiC MMC at the strain level □$_{eq}$≈10.9 and (B) an Al-5%$TiB_2$/TiC MMC at the strain level □$_{eq}$≈162.4.
*Source*: Reprinted with permission from Y. Liu, F. Wang, Y. Cao, J. Nie, H. Zhou, H. Yang, X. Liu, X. An, X. Liao, Y. Zhao, Unique defect evolution during the plastic deformation of a metal matrix composite, Scr. Mater. 162 (2019) 316–320.

smoothness or roughness. It assists to interpret the adhesion properties as well as the interactions between the constituents of the composite materials. The AFM technique helps to demonstrate the technological problems associated with corrosion, abrasion, friction, and etching etc. In a research, AFM analysis was used to evaluate the microstructural, anticorrosion, and mechanical properties of a MMC coating material (Zn-Al-$SnO_2$) on mild steel [52]. The AFM micrographs (Fig. 5) were presented

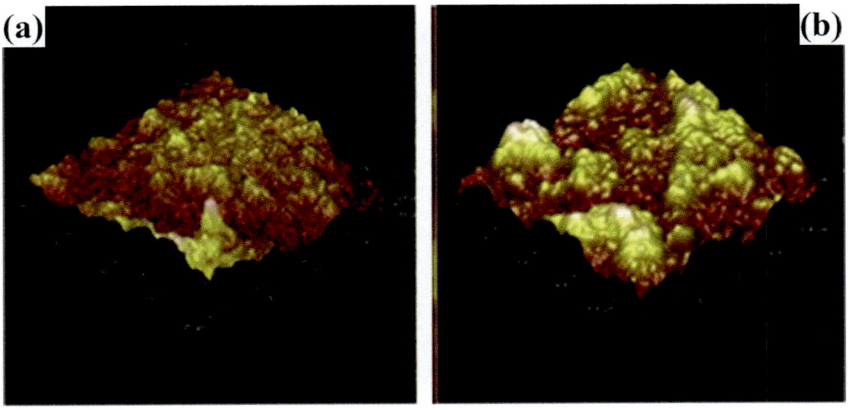


**Fig. 5** AFM images of the (A) Zn-Al-7Sn-S-0.3 V and (B) Zn-Al-13Sn-S-0.5 V.
*Source*: Reprinted with permission from O. Fayomi, A. Popoola, V.S. Aigbodion, Investigation on microstructural, anti-corrosion and mechanical properties of doped Zn–Al–SnO2 metal matrix composite coating on mild steel, J. Alloys Compd. 623 (2015) 328–334.

for two different compositions of the composite among which the Zn-Al-13 $SnO_2$ at 0.5 V sulfate coating performed a uniform distribution of intermetallic particles having identical grains and crystal size. The analysis also explains that the surface roughness of the material increases considerably with increasing the thickness of the film.

### 3.5 Optical coherence tomography (OCT)

The OCT analysis is a noninvasive, noncontact, and nondestructive imaging technique for a detailed characterization of PMCs. It can reveal the microstructure of the materials that assists to predict the permeability and mechanical properties of the composites. The analysis can detect the subtle damages held in composite surfaces, including micro-defect, fracture, void, abrasion, and spot etc. Furthermore, it can find out the cracking, delamination, and empty spaces in the composite materials. The size and location of the constituents in the PMC composites can also be determined by this technique. The phase information obtained from the OCT analysis helps to determine the spatial distribution of the mechanical stresses induced within the composites.

### 3.6 X-ray studies

The X-ray studies allow to reveal the atomic and domain structures and the chemical composition of the composites. The X-ray investigation methods can be several types, such as X-ray diffraction analysis (XRD), small-angle X-ray scattering (SAXS), and wide-angle X-ray scattering (WAXS). The XRD can determine the level of crystallinity for semi-crystalline or amorphous composite materials by measuring the crystal lattice parameters. The crystallinity provides information about the structural compactness of the material, which indicates its hardness. The SAXS can measure the size of micro- and nanomaterials as well as the crystal grain, clusters, and micro-pores. Furthermore, it can find out the geometry of the internal structures of composite materials. Moreover, it can determine the thickness of the crystal layer and inter-planar distances. This method helps to maintain the uniform distribution of the constituents throughout the composite. The WAXS technique provides information not only about the crystallographic structure but also about the chemical stoichiometry. Overall, the X-ray studies provide valuable information that is very important for the development of a desirable composite.

## 4 Mechanical properties of composites

Composite materials are frequently chosen for structural applications, and for this purpose, composites with superior mechanical properties, e.g., high strength, hardness, resistance to wear, are desired. While the role of matrix is to provide ductility and toughness, the reinforcement phase imparts strength and rigidity, and improves wear properties [53]. The mechanical properties of a composite largely depend on several factors: first, the composition and microstructure of the matrix in which reinforcement materials are dispersed; second, the type, size, shape, volume fraction, alignment, and distribution of reinforcing agents, which usually carry the major amount of load; and finally, the interactions, i.e., bonding at the interface of the matrix and the reinforcement materials [54].

## 4.1 Strength

Strength of a material measures the amount of load it can withstand before failure or plastic deformation, and parameters like tensile strength, shear strength, compressive strength, yield strength, and flexural strength determine the mechanical strength of a material. Tensile strength of a material represents its capacity to withstand the static load in tension, compressive strength determines a material's ability to withstand the load in compression, and shear strength is the ability of a material to resist the force that causes its internal structure to slide against itself (Fig. 6). Yield strength of a material can be defined as the stress at which it changes from elastic deformation to plastic deformation permanently. Beyond this point, the materials cease to be elastic and becomes permanently plastic. Flexural strength is the resistance to bending deflection of a material when force is applied to it. It gives idea about the behavior of a material when simple bending load is applied on it [55].

High-strength composites are usually obtained by reinforcing ceramic, carbon, or other high-stiffness particles, whiskers, nanotubes, short or long fibers in metal or polymer matrix [56]. Graphene is a very good reinforcing material to enhance the strength of Al-, Mg-, Ni-, Ti-, and Cu-based MMCs because its breaking strength is about 200 times higher than that of steel [57]. MMCs strengthened with grapheme are lightweight and also possess good toughness. In graphene-reinforced Cu MMC, graphene was found to disperse uniformly within the Cu matrix without any change in structure. The yield strength and the tensile strength of this composite were found to be 144 MPa and 274 MPa, respectively, which were 177% and 27.4% higher than those of pure Cu [58]. Load transfer and dislocation strengthening mechanisms of

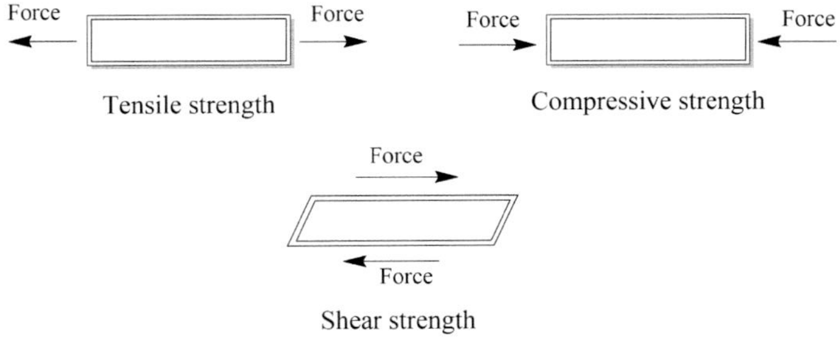


**Fig. 6** Depiction of tensile, compressive, and shear strength of a material.

in situ grown graphene were mainly responsible for this enhancement in strength of grapheme/Cu MMC. Again, 6% graphene-reinforced AA 7075 aluminum alloy showed an increase in tensile strength by 19% as compared to pure AA 7075 aluminum material [59].

Carbon fiber is another reinforcement material that can significantly enhance mechanical properties of MMCs, especially tensile strength, compressive strength, and high-temperature strength [60]. Carbon fiber-reinforced composites are lightweight and suitable for structural applications. It is also compatible with various metal matrices like Al, Cu, Mg, Ni, Pb, Co, and Ti and forms interfacial bonding with them, which leads to superior mechanical and thermal properties. Carbon fiber-reinforced Al matrix composite showed good fiber-matrix interfacial bonding, which led to excellent mechanical properties [61]. Li et al. found that the ultimate tensile strength of AE44 Mg alloy was increased by 127% (up to 412 MPa) when carbon fiber is used as reinforcing material [62]. In case of PMCs, carbon fibers are distributed uniformly in a number of polymer matrices, like polypropylene, polyethylene, polycarbonate, polyamide, polyester, polyimide, polyphenylene sulfide, polyethersulfone, polyetherimide, polyetheretherketone, epoxy resins etc., and improve their mechanical properties [63]. Gupta et al. investigated the mechanical properties of 3D-printed short carbon fiber-reinforced polycarbonate PMC and found good interlocking between the fibers and polymer, which improved the tensile, flexural, and compressive strength [64]. Similar results were obtained with polypropylene composites reinforced by short carbon fiber [65].

Another remarkable reinforcing agent is carbon nanotube that possesses excellent mechanical properties like strength (up to 100 GPa), ultra-high aspect ratios ($\sim$100–100,000), and can be used in both MMCs and PMCs to develop high-strength and lightweight composites for aerospace, automotive, and structural applications [53]. In case of MMCs, CNTs significantly improve the strength of metal matrices having low stiffness like Al, Mg, and Ti. In CNT-reinforced Cu—Cr composite, CNT showed a homogenous distribution in the matrix and formed good interfacial bonding by creating a transition layer of $Cr_3C_2$ at the interface between CNTs and Cu—Cr matrix. As a result, the yield strength of CNT-reinforced Cu—Cr composite increased by 135% as compared to CNT/Cu composite [66]. In another study, it is showed that the ultimate tensile strength of pure Al, which is 133 MPa, can be increased up to 368 MPa by reinforcing it with 1% CNT [67].

SiC is a very common reinforcing material used to enhance the mechanical properties of different metal matrices like Al, Ni, Cu, Mg, and stainless steel. Sivananthan et al. studied the effect of SiC particles on mechanical properties of aluminum 6061alloy matrix and reported an increase of 25.6% tensile strength and 12% compressive strength, respectively, in comparison with aluminum 6061 alloy [68]. SiC had significantly increased the compressive and tensile strength of Al matrix when used as reinforcement [69]. Apart from these, reinforcements like $Al_2O_3$, TiC, $ZrB_2$, $SiO_2$, WC, and $B_4C$ have also been used for years to improve mechanical properties of MMCs and PMCs [70–72].

## 4.2 Modulus

Modulus represents how much a specified property is possessed by a material. For the characterization of mechanical properties of a composite, moduli like Young's modulus, shear modulus, and Poisson's ratios give information about hardness, elasticity, and rigidity.

Young's modulus or elastic modulus indicates the stiffness of a substance and is defined as the ratio of stress to strain. It measures the amount of force needed to deform a material. A higher value of Young's modulus indicates high stiffness of the material. The ratio of shear stress to shear strain is defined as shear modulus or modulus of rigidity of a material. It characterizes the ability of a material to resist any change in its shape while maintaining its volume. Poisson's ratio is expressed by the ratio of radial strain to corresponding axial strain on a material when the stress is applied along one axis. It is required to determine Poisson's ratio of a substance to characterize its stress and deflection properties. It gives information about how the cross-sectional area of a material changes when it is subjected to longitudinal stretching or compression. The higher the value of Poisson's ratio, the harder the material to fracture [73,74].

$$\text{Young's modulus} = \frac{\text{Tensile stress}}{\text{Tensile strain}}$$

$$\text{Shear modulus} = \frac{\text{Shear stress}}{\text{Shear strain}}$$

$$\text{Poison's ratio} = \frac{\text{Radial strain}}{\text{Axial strain}}$$

The spatial orientation of reinforcing agents in polymer matrices remarkably affects the mechanical properties of resulting composites. Li et al. carried out a research to observe the effect of orientation of graphene and graphene oxide nanoplatelets in polymer matrices like PVA, PMMA, and epoxy by determining the Young's modulus of the resulting PMCs [75]. They concluded that random orientation of reinforcing agents reduced the Young's modulus of the composites by two times than the fully aligned fillers. However, the alignment of nanoplatelets in PMC is less significant than that of nanotubes, and better levels of modulus enhancement are possible to achieve with nanoplatelets compared with nanotubes from a geometrical point of view.

The Young's modulus of CNT is in the tera Pascal (TPa) range [53], which indicates that if used as a reinforcing material, it can enhance mechanical properties of both MMCs and PMCs. Yu et al. studied the consequence of matrix-reinforcement interfacial reaction on Young's modulus in CNT-reinforced Al nanocomposite. They found a notable enhancement in Young's modulus from 70.6 GPa in Al matrix to 81.4 GPa in CNT/Al composite material, which was resulted from a strong reinforcing effect of reaction-formed Al4C3 and the effect of expansion of volume during CNTs-Al4C3 transformation [76]. In another research, the effective Young's modulus of single- and multiwalled CNT-reinforced epoxy composites was investigated. The effect of length, layer number, volume percentage, and interspace shear modulus between adjacent CNT layers on Young's modulus was also studied. It was observed that the Young's modulus of multiwalled CNT-reinforced epoxy PMC increased by increasing the abovementioned parameters. High amount of long, multiwalled CNT in polymer matrix with good interaction between adjacent layers was found to enhance the Young's modulus of the PMC [77]. The Young's modulus was found as high as 152 GPa with ten-walled and 8200-nm-long CNT-reinforced epoxy PMC.

## *4.3 Hardness and wear resistance*

Hardness indicates a material's capacity to resist permanent deformation like cutting, bending, indentation, or scratching. Wear resistance property of a material is its ability to resist the progressive loss of material from its surface due to mechanical interaction (abrasion, erosion, adhesion, impact, etc.) with other substances which are in relative motion [78]. Usually, composites with high volume fraction of hard reinforcing phases show high wear resistance.

To improve the hardness and wear resistance of metal and polymer matrices, reinforcements like CNT, carbon fibers, $Al_2O_3$, SiC, B4C, TiC, $TiO_2$, graphene, $ZrB_2$, $ZrSiO_4$, $TiB_2$ etc. and their combinations are widely used [78–80]. As a reinforcing material, CNT was found to enhance hardness of Cu—Cr matrix by 128% in relation to CNT/Cu composite. This enhancement was due to the fact that CNTs were stiffer and stronger than the copper matrix [66]. The hardness of polycarbonate polymer can be increased from 76 MPa to 160 MPa by reinforcing it with 10% short carbon fiber [64]. This enhancement of hardness in PMC is resulted from uniform distribution of fibers and their high stiffness, which resist the deformation of polymer.

Aluminum matrix is known to have poor wear resistance and inadequate hardness, which limit its uses in structural applications. However, both wear resistance and hardness of aluminum can be improved by reinforcing it with ceramic particles [81]. In a study, the wear behavior of Al 7075 aluminum alloy, reinforced with both $Al_2O_3$ and $B_4C$, was investigated, and their results suggested that the wear resistance of the prepared MMC increased with increasing amount of $Al_2O_3$ in the composite up to 15% [82]. Similar results were obtained with SiC-reinforced AA5083 MMC as the wear resistance was found proportional to SiC content in the composite up to 2% [83]. In graphene-reinforced epoxy resin PMC, the addition of 5% graphene increased the wear resistance of pure epoxy resin by 628 times [84]. This unprecedented improvement of wear resistance was resulted from the fact that exfoliated graphene flakes acted as a lubricant by adhering on the surface of a stainless-steel ball during sliding tests and thus enhanced the wear resistance of the composite.

## *4.4 Fatigue*

Fatigue of a material is defined as a progressive structural damage when the material is subjected to cyclic or fluctuating loads. In fatigue damage, crack initiation, crack growth, and propagation phenomena occur, which ultimately lead to complete failure of that material. Fatigue life of a composite material is the number of loading cycles (stress) that it sustains before failure and must be calculated prior to its use in structural applications. The microstructure, interaction between matrix and reinforcing material, size, shape, volume fraction, and distribution of reinforcement and processing route affect the fatigue behavior of a composite [85].

Micro- and nano-sized $Al_2O_3$ particles are investigated widely as reinforcement in MMCs and PMCs to observe their effect in fatigue life and strength. $Al_2O_3$ particles with a particle size of 50nm were used to reinforce 6061 Al matrix in a research [86], and it was found that both fatigue life and strength of constant and cumulative fatigue were improved in the composite. Divagar et al. studied the influence of particle size of SiC and $Al_2O_3$ reinforcement on the fatigue life and strength of AA7075-T651-grade aluminum alloy-based MMC, which was produced by stir casting process [85]. From their results, it was found that the fatigue strength of the prepared MMC was directly proportional to the loading percentage of nanoparticle reinforcement. The MMC with 10% SiC and 5% $Al_2O_3$ exhibited an increase in fatigue strength of 12.13% than the base metal, and this enhancement of fatigue strength was attributed to the fine nano-sized particles used as reinforcement.

# 5 Metal matrix composites

## 5.1 Materials for MMCs

MMCs, like polymer composites, can be classified on the basis of shape of reinforcements such as particles, laminates, fibers, and penetration composite materials with different orientation of them, such as aligned, random, unidirectional, multidirectional. Matrix materials for MMCs are either metals or their alloys. The common matrix materials are aluminum, magnesium, titanium, copper, or their alloys. A list of different alloys along with their composition is given in Table 1.

Advantages of different metal-based matrix materials are given in Table 2.

Al is the most widely used metal matrix for MMCs. However, there are some other good criteria to few other metals to be the matrix of MMCs. Mg-based MMCs require high production cost and complicated manufacturing processes. Some super alloy matrix-based composites (Nb-silicide- or Mo-silicide-based alloys, etc.) are recently used for some special applications such as jet engines and turbine blades. They have a

**Table 1** Common alloys for the matrix of MMCs [87].

| Conventional cast alloys | Conventional wrought alloys | Special alloys |
|---|---|---|
| $AlSi_2CuMgNi$ | AlMgSiCu | AlCuMgNiFe |
| $AlSi_9Mg$ | AlCuSiMn | AlCuMgLi |
| $AlSi_7$ | AlZnMgCu | |
| | $TiAl_6V_4$ | |

**Table 2** Advantages of different metal-based matrix materials.

| Types of metal matrix | Advantages |
|---|---|
| Al-based matrix | • Greater strength, stiffness 70–240 GPa, and wear resistance<br>• Reduced density<br>• Controlled thermal expansion<br>• Cost-effective and environment-friendly |
| Mg-based matrix | • Lightweight and high performance at elevated temperature<br>• Greater bending strength (1000 MPa) |
| Ti-based matrix | • High performance at elevated temperatures<br>• Solid-state processing methods such as foil-fiber-foil method, matrix-coated mono-tape, or fiber method are available<br>• Maximum fiber volume fraction can be up to 80% |
| Cu-based matrix | • High thermal and electrical conductivities<br>• High resistance to annealing<br>• Low coefficient of thermal expansion<br>• Good machinability |

high melting temperature up to 1700°C. Their densities are in the range of 6.6–7.2 g/cm$^3$.

In general, inorganic component of nonmetallic origin (i.e., ceramic particles, glass and carbon fibers, SiC, $Al_2O_3$, C, B, W, Nb) are used as reinforcement in MMCs. They can be of different sizes and shapes such as particles from 1 to 25 mm, whiskers from 10 to 1000 mm, continuous fibers from 15 to 41,000 mm, NPs from 1 to 100 nm, nanotubes from 41,000 to 100 nm, [88].

## 5.2 Consolidation and shaping of MMCs

The main types of consolidation techniques of MMCs are powder blending and consolidation, diffusion bonding, and physical vapor deposition [89].

**(a)** *Powder blending and consolidation*:

Metal powders are consolidated and given to the final shape of MMCs by using various methods like powder metallurgy (PM), direct powder rolling, cold or hot isostatic pressing, injection molding, forging, and sintering. Among all these processes, PM is the easiest and suitable technique to fabricate MMCs.

In the PM method, metal powder of matrix is uniformly blended with the reinforcing metal particulates either by dry mixing or by liquid suspension methods. The blend is then cold-pressed for compaction, which is called the green body, and it is then subjected to remove entrapped gases and moisture. The green body is then consolidated at high temperature to form the desired shape of the condensed composites. Finally, the compacted composites are sintered [90,91]. In this process, the shape of the targeted composites can be achieved precisely from powders; however, for other cast techniques, metals are first melted and casted, and then the finished shape is finally given by cutting or deforming.

**(b)** *Diffusion bonding*:

In diffusion bonding of MMCs, the metal matrix and the reinforcing fibers form bonds at high temperature through the interdiffusion bonding of atoms of them [89]. A wide variety of metal matrices can be processed through this technique; however, it is mainly used for producing aluminum- or magnesium-based MMCs where continuous/discontinuous fibers are used as a reinforcing agent [90].

**(c)** *Physical vapor deposition*:

In this process, first the matrix metal is made to vapor and then condensed on to the reinforcing fibers. The fibers are thus coated by the matrix metal at a rate of about 5–10 μm per minute [90,92]. The coated fibers are then arranged to give the final shape and fused either by hot pressing alone or by hot isostatic pressing.

## 5.3 Advantages and disadvantages of MMCs over PMCs

### 5.3.1 Advantages of MMCs over PMCs

1. MMCs possess very high modulus and high strength, whereas PMCs possess medium modulus and very high strength.
2. MMCs are of higher density than PMCS.
3. MMCs have higher electric and thermal conductivities in comparison with PMCs
4. MMCs can endure higher processing temperature than PMCs.
5. MMCs are insensitive to moisture, whereas PMCs are usually sensitive to moisture.

### 5.3.2 Disadvantages of MMCs over PMCs

1. For PMCs, fracture toughness is usually high, whereas for MMCs, fracture toughness is medium.
2. PMCs are easier to fabricate than MMCs.
3. MMCs are comparatively expensive than PMCs.
4. MMCs are of moderate use compared to PMCs.
5. MMCs have lower chemical resistance than PMCs
6. MMCs are expensive than PMCS
7. MMCs possess inferior insulating properties than PMCs

## 5.4 Application of MMCs

MMCs with high specific stiffness and strength have found applications in different sectors, for instance, robots, high-speed machinery, high-speed machinery, robots, high-speed rotating shafts for ships and/or vehicles, for precision machinery, lasers and electronic packaging etc. [93].

**(a)** *Automotive engineering*:

Relatively lightweight and high thermal stability of MMCs have made them to use in a variety of applications in automobile engineering. In the automotive industry, MMCs are used for the braking and transmission components, engine, pistons, piston rod, and piston pin; cylinder head, cover, crankshaft main bearing, partially strengthened cylinder blocks, engine block, chassis, batteries etc. [94]. For example, Toyota Company made their car piston from alumina-silica short fiber-based aluminum composites. This piston offers high-temperature strength and better wear resistance in comparison with the cast iron piston it replaced [93].

**(b)** *Aircraft and aerospace industry*:

In aircraft industry, the main challenge is to use materials of higher strength and stiffness with low weight to improve aircraft's performance and reduce the costs. MMCs, therefore, obviously show a great potential in this area. For aircraft, MMCs have been used for reinforcement components, rotors, axle tubes, housing covers etc. [95].

Similarly, for aerospace industry, the space shuttles require lightweight but of high dimensional stability and high pointing accuracy to withstand the dynamic and thermal stresses to overcome the gravitational force [89,90]. In this regard, particle-reinforced MMCs, being isotropic, show good specific strength and stiffness and also possess improved thermal and electrical shock resistance [96].

**(c)** *Electronics*:

Better thermal conductivity and tunable thermal expansion coefficient of MMCs have made them in use in the field of electronics. The lightweight and thermal stability of MMCs are added qualities in these applications. For instance, particle-reinforced aluminum or chopped carbon fiber-incorporated aluminum composites are found to be used for better thermal conductivity [93].

**(d)** *Sporting goods*:

Many sports equipment use MMCs for their high strength-to-weight ratio. For example, bicycle parts such as frames, wheel rims, etc., horseshoes, golf clubs, tennis racquets are made from MMCs [93]. Aluminum MMCs with particles of boron or silicon carbide have been reported to be very attractive for sporting goods applications.

**(e)** *Defense*:

In defense, MMCs have received great attention to be used in high-temperature-resistant fighter plane engines and structures, missile or other ammunition structures, tank armors, etc.

Beryllium is used to be applied in Trident missiles; however, nowadays, a high volume fraction (40%) of particle-reinforced aluminum composites are replacing it. In comparison with beryllium, the MMC is cheaper and nontoxic. To increase the swiftness of tank tracks, MMCs are now widely used [97].

In short, it can be said that, in comparison with conventional materials and PMCs, MMCs are resistant to fire and radiation damage, can operate in a broader range of temperatures and have better electrical and thermal conductivities. These advantageous properties and performance have made MMCs suitable to get application in various sectors.

# 6 Isotropic vs. anisotropic material properties

The anisotropic materials having physical properties are directional, which implies that the materials having physical response directly depend on the path in which it works [98]. The mechanical properties of anisotropic materials are different in each direction and asymmetrical at all their planes or axes. However, a subclass of anisotropic materials is orthotropic materials, which display symmetry between the two planes. In broad spectrum, these subclass materials having plane parallel to the reinforced fibers have significantly greater properties contrast to the orthotropic perpendicular plane. For example, for the orthotropic material like wood, the properties perpendicular to the fiber axis (tangential and radial) are lower than its parallel counterparts. In this circumstance, the properties of materials are not equal at radial and tangential directions. These properties are comparatively similar and equally lower to the longitudinal direction.

For the potential applications of composite materials, their anisotropic linearly elastic behavior is essentially realized. Soriano [99] and Daniel and Ishai [100] found that the matrix of linear anisotropic material having unrestricted elastic properties causes the material characterization process hard. However, when there is a symmetry between $\sigma_{ij}$ and $\sigma_{ji}$, and between $\varepsilon_{kl}$ and $\varepsilon_{lk}$, the elastic constants decrease from 81 to 36 and the independent coefficients of the stiffness matrix is shown in Eq. (4) [101–103].

$$\begin{bmatrix} \varepsilon_{11} \\ \varepsilon_{22} \\ \varepsilon_{33} \\ \varepsilon_{23} \\ \varepsilon_{13} \\ \varepsilon_{12} \end{bmatrix} = \begin{vmatrix} Q11 & Q12 & Q13 & Q14 & Q15 & Q16 \\ Q21 & Q22 & Q23 & Q24 & Q25 & Q26 \\ Q31 & Q32 & Q33 & Q34 & Q35 & Q36 \\ Q41 & Q42 & Q43 & Q44 & Q45 & Q46 \\ Q51 & Q52 & Q53 & Q54 & Q55 & Q56 \\ Q61 & Q62 & Q63 & Q64 & Q65 & Q66 \end{vmatrix} \begin{bmatrix} \varepsilon_{11} \\ \varepsilon_{22} \\ \varepsilon_{33} \\ 2\varepsilon_{23} \\ 2\varepsilon_{13} \\ 2\varepsilon_{12} \end{bmatrix} \tag{4}$$

On the other hand, the symmetry requisite for anisotropic materials is decreasing the elastic elements to 21, by the equation $Q_{ij}=Q_{ji}$ [33,98,101,103]. The elastic properties of anisotropic material can be characterized by stress–strain ratios, where the longitudinal and transverse moduli of elasticity and the Poisson's coefficient are the main coefficients [104]. According to Vanalli [105], the structural failure analysis of anisotropic materials is complex. In addition, it must be assumed that normal and shear stresses are responsible for the failure. Subsequently, the occurrence of failure is caused by different sets of stress acting on the element.

In contrast, the transverse isotropic material can be described as an orthotropic material that has isotropy in one of the planes of symmetry. In addition, the transverse materials have the identical properties in this plane in all directions [98,100,104,106]. In comparison of an isotropic transverse material with a common orthotropic material, it is detected that in isotropic transverse materials, there is symmetry between the planes $x_1\,x_3$ and $x_1\,x_2$, decreasing 9 to 5 independent coefficients. The following properties were detected for transverse isotropic material other than orthotropic materials [104]. The normal stress $\sigma_{11}$ causing the linear deformations in the planes $x_2\,x_3$ is equal. Further, the normal stress $\sigma_{22}$ responsible for the linear deformations $\varepsilon_{22}$ and $\varepsilon_{33}$ is equal to the deformations $\varepsilon_{33}$ and $\varepsilon_{22}$, respectively, which is originated by a tension $\sigma_{22}=\sigma_{33}$. Each angular deformation is originated by tangential tension only in the plane in which it works. The angular strain $\gamma_{23}$ caused by a stress $\sigma_{23}$ is equal to an angular strain $\gamma_{13}$ caused by stress $\sigma_{13}=\sigma_{23}$. The comprehensive representation of Hooke's law for isotropic transverse material is shown in Eq. (5), where Young's modulus is represented as $E$, $G$ is the shear modulus, and $\sigma$ is the Poisson coefficient [104].

$$\begin{bmatrix} \varepsilon_{11} \\ \varepsilon_{22} \\ \varepsilon_{33} \\ 2\varepsilon_{23} \\ 2\varepsilon_{13} \\ 2\varepsilon_{12} \end{bmatrix} = \begin{vmatrix} \frac{1}{E_1} & -\frac{\mathrm{v}_{21}}{E_2} & -\frac{\mathrm{v}_{13}}{E_1} & 0 & 0 & 0 \\ -\frac{\mathrm{v}_{12}}{E_1} & \frac{1}{E_1} & -\frac{\mathrm{v}_{12}}{E_1} & 0 & 0 & 0 \\ -\frac{\mathrm{v}_{13}}{E_1} & -\frac{\mathrm{v}_{21}}{E_2} & \frac{1}{E_1} & 0 & 0 & 0 \\ 0 & 0 & 0 & \frac{1}{G_{12}} & 0 & 0 \\ 0 & 0 & 0 & 0 & \frac{1}{G_{13}} & 0 \\ 0 & 0 & 0 & 0 & 0 & \frac{1}{G_{12}} \end{vmatrix} \begin{bmatrix} \sigma_{11} \\ \sigma_{22} \\ \sigma_{33} \\ \sigma_{23} \\ \sigma_{13} \\ \sigma_{12} \end{bmatrix} \tag{5}$$

# 7 Composite's modeling

Mechanical properties of composites mostly depend on the properties of composite materials, geometrical shape and arrangement of reinforcements, interfacial bonding of matrix and reinforcements. To simulate the mechanical properties of composites, different models have been applied since 1940. Based on empirical data, some analytical approaches and regression models have been used. Some special models for some particular cases like crack propagation or interface region are also applied. However, composite modeling can be grossly divided into two types.

1. Analytical models
2. Numerical models

## 7.1 *Analytical models*

### 7.1.1 *ROM and Voigt-Reuss bounds*

Einstein's viscosity law is a relation between viscosity of a suspension of solid spheres with viscosity of the dispersion medium and volume (in $cm^3$) occupied by the dispersed solid. In case of reinforced composites, this law can be applied by modifying the equation as per filler particle size and shape, particle surface properties, and its percentage. The mechanical properties of long fiber composites are somewhat between the properties of the matrix and the fibers. Taking the ROM into consideration, Voigt and Reuss derived the equation for aligned long fiber composites in isostrain situation (load applied on the longitudinal direction fibers) (Eq. (4)) and in isostress situation (load applied normal to fiber direction) (Eq. (5)), respectively. These equations are applicable in any composites regardless of the shape and arrangement of fillers. However, the critical fiber length is very important for composite properties. For continuous aligned long fiber composites, it is measured by the following equation.

$$l_c = \frac{\sigma_f^* d}{2\tau_c} \tag{6}$$

where

$l_c$ = critical length of fibers,
$d$ = fiber diameter,
$\sigma_f^*$ = ultimate tensile strength,
$\tau_c$ = the fiber–matrix interfacial bond strength or shear yield strength of the matrix, which one is less in value.

For glass or carbon fiber, the $l_c$ is almost equal to 1 mm. If the length of fiber is less than $l_c$, then fibers behave like particulates in composites.

For aligned and discontinuous fiber composites, the Voigt model (Eq. (4)) will be modified, including $l_c$ as mentioned in Eq. (7).

If $l > l_c$,

$$\sigma_{cd}^* = \sigma_f^* V_f \left(1 - \frac{l_c}{2l}\right) + \sigma_m'(1 - V_f) \tag{7}$$

where $l$ is the length of the fiber, $\sigma_{cd}^*$ is the longitudinal strength of composites, and $\sigma_m'$ is the fracture stress of the matrix.

If $l < l_c$, then Eq. (7) will be changed to Eq. (8) as below:

$$\sigma_{cd}^* = \frac{l\tau_c}{d} V_f + \sigma_m'(1 - V_f) \tag{8}$$

For discontinuous and random fiber composites, the modulus of it will be the same as Eq. (4) with an inclusion of a constant $k$. Eq. (9) shows this composite modulus-modifying Eq. (4).

$$E_{cd} = KE_f V_f + E_m V_m \tag{9}$$

The constant $K$ is called the fiber efficiency parameter, which depends on $V_f$ and the ratio of modulus of fiber to that of matrix.

### 7.1.2 Hashin-Shtrikman model

Hashin and Shtrikman model calculated the bulk and shear modulus of random composites for upper-bound and lower-bound limits, like Voigt-Reuss model [107]. However, they also derived the equation between bulk ($K_b$), shear ($G$), and elastic modulus ($E$) of the composites (Eq. (10)).

$$E = \frac{9K_b}{1 + 3\left(\frac{K_b}{G}\right)} \tag{10}$$

### 7.1.3 Halpin-Tsai model

This is a partial data-based model and developed equation for aligned discontinuous fibers or aligned nanoplatelet-reinforced composites [108]. Their equations are proven sufficient but only when the fiber volume fraction is low (~5%).

$$\frac{E}{E_m} = \frac{1 + \xi\eta Vf}{1 - \eta Vf} \tag{11}$$

$$\eta = \frac{\left(\frac{E_f}{E_m}\right) - 1}{\left(\frac{E_f}{E_m}\right) + \xi} \tag{12}$$

where $\xi$ is the reinforcement constant, which depends on the shape and size of fiber, packing geometry, and loading directions applied to the composites.

When, $V_f = 0$, $E = E_m$ and $\xi \rightarrow 0$ and Eq. (11) becomes similar to Eq. (5) of Reuss model. Again, if $V_f = 1$, then $E = E_f$ and $\xi \rightarrow 1$ and Eq. (11) becomes similar to Eq. (4) of Voigt model.

### 7.1.4 Hui-Shia model

Mori-Tanaka developed equations for determining elastic constants. After them, Hui-Shia modified the equations to calculate the overall moduli of composites with aligned fiber or flake like fillers both in longitudinal and transverse directions ($E_L$ and $E_T$) [109]. The equations are given below, where $\alpha$ is the aspect ratio of reinforcement material.

$$E_L = E_m\left[1 - \left(V_f/\xi\right)\right]^{-1} \tag{13}$$

$$E_T = E_m\left[1 - \left(V_f/4\right)\left((1/\xi) + (3/\xi + \Lambda)\right)\right]^{-1} \tag{14}$$

$$\Lambda = \left(1 - V_f\right)\left[\frac{3(\alpha^2 + 0.25)g - 2\alpha^2}{\alpha^2 - 1}\right]$$

$$g = \left\{\begin{array}{l} \frac{\alpha}{(\alpha^2 - 1)^{\frac{3}{2}}}\left[\alpha\sqrt{\alpha^2 - 1} - \cos^{-1}\alpha\right], \text{when}\, \alpha \geq 1 \\ \frac{\alpha}{(\alpha^2 - 1)^{\frac{3}{2}}}\left[-\alpha\sqrt{1 - \alpha^2} + \cos^{-1}\alpha\right], \text{when}\, \alpha \leq 1 \end{array}\right\}$$

## 7.2 Numerical models

### 7.2.1 Molecular dynamic model

In this model from atomic information, macroscopic properties can be derived. The model requires initial position and velocities of atoms, interatomic forces, periodic boundary conditions, and controlled pressure and temperature in surroundings. It mimics the thermodynamic situations for atoms in case of filler or matrix of composites [110].

The interaction potential among atoms is actually a summation of different force fields such as interatomic bond strength, force due to angle formation among three neighbor atoms, and force field between two planes of neighboring atoms. It also includes van der Waals and electrostatic attractions in the cumulative interaction potential of atoms.

Several software packages like DL-POLY, LAMMPS, and TINKER are usually applied for molecular dynamic simulation purposes. Periodic boundary conditions are used as a unit cell conditions. The model first analyzes the mechanical properties at nonstrain or equilibrium condition and later different strain fields are applied. The model appreciably assumes the mechanical properties of nanocomposites.

### 7.2.2 Finite element model (FEM)

FEM is probably the most popular numerical model for composite analysis. There are some requirements for FEM analysis, such as

**(a)** Consider deformation
**(b)** Discretization of domain into discrete elements using equilibrium equations
**(c)** Find elements and their equations
**(d)** Fix boundary conditions, traction boundary and free surface boundary
**(e)** Fix continuity condition, traction continuity, displacement continuity
**(f)** Fix constitutive equations, relation with stress, strain and point to plane transformation equations

FEM produces 2D or 3D mesh of object and analyzes the mechanical properties of composites as a continuum micromechanics method. Three types of element representation may be used in the FEM method. They are multiscale representative volume element (RVE), unit cell model, and object-oriented model.

## RVE model

The model can be applied for regular fiber-reinforced composites to determine mechanical properties from elasticity theory by applying 3D mesh arrangement. RVE measures the effects of surrounding materials in single/multiple filler-reinforced composites. The representative unit cell in this model is divided into three distinct regions, and the boundary conditions are analyzed using atomistic potential value. RVE considers the unit cell as a rectangular 3D solid of matrix and the filler as a 3D elastic beam. In atomistic level, atoms and bond between atoms are considered as nodes and load-carrying members, respectively. In this case, filler properties are simulated in the progressive fracture model.

## Unit cell model

This is a special form of RVE where a large number of fillers are considered in each unit. The model applies periodic boundary parameters for both aligned and random composites. The periodic boundary conditions for a 2D model are accepted as follows.

$$u(\text{REd}) = u(\text{LEd}) + \delta_1$$

$$v(\text{REd}) = v(\text{LEd})$$

$$u(\text{TEd}) = u(\text{BEd})$$

$$v(\text{TEd}) = v(\text{BEd}) + \delta_2$$

where REd, LEd, TEd, and BEd denote the right, left, top, and bottom edges of the unit cell of the composites, and $\delta_1$ and $\delta_2$ express the axial and transverse displacements, respectively. In case of the symmetrical boundary conditions for 2D models, all edges are approximated to be free of shear strain and the top edge is free of normal traction. Only the displacement at normal to the x direction that means $u(\text{REd}) = \delta$ is considered for symmetrical boundary parameters.

However, 3D models apply only symmetrical boundary conditions for the faces of the unit cell, and they are given below.

$$u(\text{LF}) = 0$$

$$v(\text{BF}) = 0$$

$$w(\text{BKF}) = 0$$

$$u(\text{RF}) = \delta$$

where LF, BF, BKF, and RF stand for left face, bottom face, back face, and right face. All other faces are free of any strain.

### Object-oriented model

RVE and unit cell consider a boundary condition and element deductions for a small unit and then ensemble multiple of such building blocks to simulate the composite properties in macroscale. These two models also assume the geometry of fillers with very few options like ellipsoids, spheres, cylinders, or cubes. The two models do not produce satisfactory result when the filler has an irregular shape. Object-oriented modeling considers the filler's microstructure, and this is why the model can predict the properties well. Actual image of the fillers and their distribution is first taken by the image scanning method, and this image is then fit into the finite element mesh.

## 8 Application of metal oxide-reinforced renewable polymer composites

In recent years, the necessities of renewable biobased products for potential applications have become enormously vibrant with growing attention on environmental problems such as disposal of waste and reducing hazardous resources. The renewable bioresource-based advanced materials could be competitive candidate to substitute fossil-based man-made products, owing to their superior advantages such as moderately small cost, biocompatibility, biodegradability, easy achievable, sustainable, and harmless. The fabrication of new renewable polymer-based composite is a difficult work and is getting growing consideration of scientist due to its extensive applications from healthcare to energy and aerospace [111–113].

Generally, renewable biobased polymers are insulating. So, however, to get conductive polymer, renewable biopolymers must have a group chain of conductive filling materials, which freely clamp electrons and allow comparatively easier delocalization of electrons. There are many renewable biopolymer matrices containing composite materials, which meet desirable performance requirements. However, because of their low mechanical and thermal properties, the practical uses of biodegradable materials have been restricted. Additionally, the filling materials have been designated to support some of the deficiencies of these composite materials. A variety of biopolymer matrices (chitosan, polylactide, starch, cellulose, pectin, etc.) is used with a wide choice of filler materials for the fabrication of these materials. Metal particles such as copper, Ag, aluminum and their oxides (titanium dioxide ($TiO_2$), zinc oxide (ZnO), aluminum oxide ($Al_2O_3$) etc.) are used as filler materials. The incorporation of nanofiller enhances the anticorrosion, thermal properties, mechanical strength, and electrical conductivity in renewable polymer-based composite [114,115].

According to the needful uses, these biocomposite materials can be made in the form of aerogels, thin films, hydrogels, etc. [116,117]. For example, metal and metal oxide-doped composite materials prepared from natural renewable polymers have acknowledged enormous consideration due to their nonviolent property, biocompatibility, hydrophilic nature, and eco-friendly characteristic. These materials are worthy candidates for some prospective uses (Table 3), including as electro-stimulated drug

**Table 3** Some recently developed composites from renewable nanocellulose polymer as matrix and metal oxides as reinforcing agent.

| Polymer matrix | Metal oxide NPs | Preparation method | Applications | Ref. |
|---|---|---|---|---|
| BC | ZnO | NPs synthesis by ex situ process, immersion of BC film and mixing | Wound dressing systems in burns complication | [118] |
| BC | ZnO | NPs synthesis by ex situ process and mixing with BC and dissolved in NMMO | Biomedical applications and bioelectroanalysis | [119] |
| BC | ZnO | BC altered with maleic anhydride template for NPs synthesis by in situ process | Antibacterial wound dressing and tissue regeneration | [120] |
| BC | $TiO_2$ | Ex situ sol–gel method | Antibacterial and photocatalytic applications | [121] |
| BC | $TiO_2$ | NPs synthesis by ex situ process and mixing with BC dissolved in NM-MO | Wound management and tissue regeneration | [122] |
| BC | CuO | Nanohybrids GO-CuO mixed with homogenized BC | Biomedical applications | [123] |
| BC | MgO | Nanohybrids achieved by in situ co-precipitation technique and externally combined MgO-NPs in the BC | Clinical wound healing | [124] |
| CNF | Cu/CuO | Synthesis of Cu/CuO NPs by in situ green reductive method and covering CNF | Surgical bandage material | [125] |
| BC | $Fe_3O_4$ | Incorporation of $Fe_3O_4$ NPs by in situ process inside the BC network in the presence of oleic acid or PEG | Tissue reconstruction at the cerebral aneurysmal neck defect | [126,127] |
| CNC | $Fe_3O_4$ | Synthesis of $Fe_3O_4$ NPs by ex situ process and mixing with CNC-poly (citric acid) by ultrasonication methods | Dual-contrast agent for MRI in biomedical applications | [128] |

**Table 3** Continued

| Polymer matrix | Metal oxide NPs | Preparation method | Applications | Ref. |
|---|---|---|---|---|
| CNC | $CoFe_2O_4$ | Synthesis of $CoFe_2O_4$ NPs by in situ process | Magnetic fluid hyperthermia, magnetically assisted drug delivery | [129] |
| Wood | ZnO | The incorporation of ZnO in PP/wood biocomposites | ability to absorb UV radiation | [130] |

delivery systems, biosensors, bioactuators, bioconductors, and skin-tissue engineering, neuron and muscle [131–135].

The deposition of metal and metal oxide on nanocellulose-based paper materials can be realized for their superior optical clarity than the regular paper substrate. The incorporation of titanium oxide, carbon nanotubes, Ag nanorods, tin-doped indium oxide, boron nitride, silica NPs, quantum dot, and molybdenum disulfide on nanocellulosic paper substrate gives transparent and conductive paper [136–138]. Another example of biopolymer-derived composite is aerogels with the high surface areas and surface charges, which can be used in conductive polymer composites with the incorporation of metal oxide. For the layer-by-layer association of conductive polymer composites, CNT, $TiO_2$, ZnO, and $Al_2O_3$ are used to enhance the mechanical properties, charge capacity, and flexibility for the use in energy storage and other electronic applications [139,140]. The doping of kaolin in nanocellulose composite was found favorable for the cost-effective, flexible, low surface roughness, and porosity substrate for printed electronic uses [141,142]. Another application of metal oxide–based photocatalyst composites is the reduction of pollutant chemicals. These materials must be visible light active, high photostable, and photoactive. Such properties can be achieved by immobilizing the metal oxides with polymers. The combination of metal oxide with a polymeric compound with suitable energy levels is important consideration to design such device. These will enhance the charge relocation between the inorganic metal-oxide and organic polymer materials to decrease the recombination of the charge carriers [143].

There are several reports of fabrication of nanocomposites from conductive renewable polymers and metal oxide NPs with enhanced photocatalytic reactions in the visible region of light. Another important feature of metallic NPs is effective inhibition of the photooxidation procedure of the biobased renewable polymeric materials. The use of metal oxide-based nanofillers, i.e., ZnO and $TiO_2$, also showed progressive results on improving the stability performance of biocomposite materials [130,144]. For example, the combination of ZnO in PP/wood biocomposites declines the mechanical strength and causes the surface to degrade after weathering, which could be attributed to its improved ability to absorb UV radiation [130]. The incorporation of aluminum trihydrate, zinc borate, and $TiO_2$ into PP-wood biocomposites was studied by Turku and Ǩarki [145]. Due to UV light absorption ability of $TiO_2$, it can avoid surface discoloration. However, $TiO_2$ can ease chemical oxidation of the biocomposite materials [145]. The incorporation of small amounts of Ag NPs is capable to increase the UV stability of HDPE [146]. The HDPE damages essentially when open to UV irradiation after 500 h. The damage is reduced remarkably with an increase in Ag NPs content.

Renewable polymer-based composite materials are utilized in several applications with growing attention because of their biodegradable, sustainable nature, flexibility in their handling circumstances, and reasonable cost of their final product.

# 9 Future aspects and conclusion

Composite materials are creating new avenues in the field of research and industries as they have special properties usually better than their individual components. Researchers have been taking great efforts for developing new composite materials based on changing matrix phase, reinforcing materials, nanosize materials, and biomaterials. At the same time, to enhance the performance of the composite materials, various parameters are needed to be optimized and develop advanced technology to manufacture better composites. The developed composite materials have potential uses mainly in sports, aerospace, chemical automobile, and architecture. Nevertheless, nowadays, composites prepared with biomaterials are gaining attention due to their nontoxicity, biocompatibility, biodegradability, and environmentally friendly.

Conventional composite materials are usually based on fossil fuel and can endure in the environment for a long time because of their less possibility recycle. Composites with filler of various kinds of fibers as well as minerals are very common in composite manufacturing industries. Combining bio- and petro-based material biocomposites with hybrid materials are also producing commercially. For example, biobased reinforcing materials from the products of agriculture and wood, such as jute and flax fibers, are using with synthetic polymer-based matrix phases such as epoxy, polypropylene, ethylene. These composites are more environment-friendly compared to only synthetic composites and are using in architecture, decking, consumer products, and automobile parts. Moreover, fibers from natural and recycle sources have uses in consumer products industries. As mentioned earlier, completely green composites with only biomaterials are the most desirable one to save our environment and make it eco-friendly. Unfortunately, till now, we have many drawbacks of composite with only biomaterials to apply in the fields of automobile and architecture due to the lack of strength and durability. However, in packing, 100% biocomposites and plastics with biodegradable properties are applying. In the future, for the sustainable development, fillers originated from industrial waste, biomass, and waste from food have potential as composites with lightweight and sustainability for automobile parts and other raising manufacturing industries. In this way, we can utilize the wastes and undesired industrial byproducts as well as save our environment.

Till now, huge efforts have been done on the manufacture of composite materials. However, still there are lots of room to advance in this field by applying automation and cutting-edge research. Overcoming the limitations of all green composites by incorporating the minimal number of synthetic materials can produce sustainable and environmentally friendly composites for the future applications.

## References


[1] T.-D. Ngo, Composite and Nanocomposite Materials: From Knowledge to Industrial Applications, BoD–Books on Demand, 2020.

[2] W.D. Callister, D.G. Rethwisch, Materials Science and Engineering, John wiley & sons, NY, 2011.

[3] W.D. Callister, D.G. Rethwisch, Materials Science and Engineering: An Introduction, Wiley, New York, 2018.

[4] T. Clyne, P. Withers, An Introduction to Metal Matrix Composites, Cambridge University Press, 1995.

[5] Y.-Y. Zhang, R.-L. Shen, M.-Z. Li, J.-C. Pang, L. Zhang, S.-X. Li, Z.-F. Zhang, Mechanical damage behavior of metal matrix composites with the arbitrary morphology of particles, J. Mater. Res. Technol. 9 (4) (2020) 7002–7012.

[6] A. Slipenyuk, V. Kuprin, Y. Milman, V. Goncharuk, J. Eckert, Properties of P/M processed particle reinforced metal matrix composites specified by reinforcement concentration and matrix-to-reinforcement particle size ratio, Acta Mater. 54 (1) (2006) 157–166.

[7] Y.-J. You, J.-H.J. Kim, K.-T. Park, D.-W. Seo, T.-H. Lee, Modification of rule of mixtures for tensile strength estimation of circular GFRP rebars, Polymers 9 (12) (2017) 682.

[8] M. Okamoto, B. John, Synthetic biopolymer nanocomposites for tissue engineering scaffolds, Prog. Polym. Sci. 38 (10–11) (2013) 1487–1503.

[9] S.H. Othman, Bio-nanocomposite materials for food packaging applications: types of biopolymer and nano-sized filler, Agric. Agric. Sci. Procedia 2 (2014) 296–303.

[10] M. Ramesan, C. Siji, G. Kalaprasad, B. Bahuleyan, M. Al-Maghrabi, Effect of silver doped zinc oxide as nanofiller for the development of biopolymer nanocomposites from chitin and cashew gum, J. Polym. Environ. 26 (7) (2018) 2983–2991.

[11] B. Sharma, P. Malik, P. Jain, Biopolymer reinforced nanocomposites: a comprehensive review, Mater. Today Commun. 16 (2018) 353–363.

[12] Z. Schnepp, S.C. Wimbush, S. Mann, S.R. Hall, Alginate-mediated routes to the selective synthesis of complex metal oxide nanostructures, CrstEngComm 12 (5) (2010) 1410–1415.

[13] J. Nieto, C. Peniche-Covas, G. Padro, Characterization of chitosan by pyrolysis-mass spectrometry, thermal analysis and differential scanning calorimetry, Thermochim. Acta 176 (1991) 63–68.

[14] Z. Schnepp, S.R. Hall, M.J. Hollamby, S. Mann, A flexible one-pot route to metal/metal oxide nanocomposites, Green Chem. 13 (2) (2011) 272–275.

[15] J.-Y. Lee, Y. Liao, R. Nagahata, S. Horiuchi, Effect of metal nanoparticles on thermal stabilization of polymer/metal nanocomposites prepared by a one-step dry process, Polymer 47 (23) (2006) 7970–7979.

[16] R. Salomoni, P. Léo, A. Montemor, B. Rinaldi, M. Rodrigues, Antibacterial effect of silver nanoparticles in Pseudomonas aeruginosa, Nanotechnol. Sci. Appl. 10 (2017) 115.

[17] B. Le Ouay, F. Stellacci, Antibacterial activity of silver nanoparticles: a surface science insight, Nano Today 10 (3) (2015) 339–354.
[18] P. Kanmani, J.-W. Rhim, Physicochemical properties of gelatin/silver nanoparticle antimicrobial composite films, Food Chem. 148 (2014) 162–169.
[19] F. Rafieian, M. Shahedi, J. Keramat, J. Simonsen, Thermomechanical and morphological properties of nanocomposite films from wheat gluten matrix and cellulose nanofibrils, J. Food Sci. 79 (1) (2014) N100–N107.
[20] B.A. Morris, The Science and Technology of Flexible Packaging: Multilayer Films From Resin and Process to End Use, William Andrew, 2016.
[21] R. Schwalm, UV Coatings: Basics, Recent Developments and New Applications, Elsevier, 2006.
[22] A. Singha, V.K. Thakur, Physical, chemical and mechanical properties of Hibiscus sabdariffa fiber/polymer composite, Int. J. Polymer. Mater. 58 (4) (2009) 217–228.
[23] B. Jiang, A. Chen, J. Gu, J. Fan, Y. Liu, P. Wang, H. Li, H. Sun, J. Yang, X. Wang, Corrosion resistance enhancement of magnesium alloy by N-doped graphene quantum dots and polymethyltrimethoxysilane composite coating, Carbon 157 (2020) 537–548.
[24] J. Lu, K. Yuan, F. Sun, K. Zheng, Z. Zhang, J. Zhu, X. Wang, X. Zhang, Y. Zhuang, Y. Ma, Self-assembled monolayers for the polymer/semiconductor interface with improved interfacial thermal management, ACS Appl. Mater. Interfaces 11 (45) (2019) 42708–42714.
[25] H. Chen, V.V. Ginzburg, J. Yang, Y. Yang, W. Liu, Y. Huang, L. Du, B. Chen, Thermal conductivity of polymer-based composites: fundamentals and applications, Prog. Polym. Sci. 59 (2016) 41–85.
[26] L.E. Nielsen, The thermal and electrical conductivity of two-phase systems, Ind. Eng. Chem. Fundam. 13 (1) (1974) 17–20.
[27] N. Yüksel, A. Avcı, M. Kılıç, The effective thermal conductivity of insulation materials reinforced with aluminium foil at low temperatures, Heat Mass Transf. 48 (9) (2012) 1569–1574.
[28] L. Ren, K. Pashayi, H.R. Fard, S.P. Kotha, T. Borca-Tasciuc, R. Ozisik, Engineering the coefficient of thermal expansion and thermal conductivity of polymers filled with high aspect ratio silica nanofibers, Compos. Part B Eng. 58 (2014) 228–234.
[29] S. Ma, N. Zhao, C. Shi, E. Liu, C. He, F. He, L. Ma, Mo2C coating on diamond: different effects on thermal conductivity of diamond/Al and diamond/Cu composites, Appl. Surf. Sci. 402 (2017) 372–383.
[30] Z.-X. Zhang, Effect of temperature on rock fracture, in: Z.-X. Zhang (Ed.), Rock Fracture and Blasting, Butterworth-Heinemann, 2016, pp. 111–133.
[31] C. Huang, H. Yang, Y. Li, Y. Cheng, Characterization of aluminum/poly (vinylidene fluoride) by thermogravimetric analysis, differential scanning calorimetry, and mass spectrometry, Anal. Lett. 48 (13) (2015) 2011–2021.
[32] P.W. Atkins, L. Jones, Chemical Principles: The Quest for Insight/Peter Atkins, Loretta Jones, 2010.
[33] F. Abelès, VI methods for determining optical parameters of thin films, in: Progress in Optics, Elsevier, 1963, pp. 249–288.
[34] A.B. Khatibani, S. Rozati, Optical and morphological investigation of aluminium and nickel oxide composite films deposited by spray pyrolysis method as a basis of solar thermal absorber, Bull. Mater. Sci. 38 (2) (2015) 319–326.
[35] B.-T. Liu, S.-J. Tang, Y.-Y. Yu, S.-H. Lin, High-refractive-index polymer/inorganic hybrid films containing high TiO2 contents, Colloids Surf. A Physicochem. Eng. Asp. 377 (1–3) (2011) 138–143.

[36] M. Zubair, M. Chowdhury, A dynamic optical constant extraction method for thin films with structural and optical-parametric justifications, J. Appl. Phys. 128 (19) (2020) 195301.
[37] S.B. Brown, Introduction to Spectroscopy for Biochemists, Academic Press, 1980.
[38] J. Smith, Encyclopedia of Life Sciences, Macmillan Publishers Ltd, Nature Publishing Group, 2001.
[39] N.A.M. Radzuan, A.B. Sulong, J. Sahari, A review of electrical conductivity models for conductive polymer composite, Int. J. Hydrogen Energy 42 (14) (2017) 9262–9273.
[40] Q.-Z. Xue, Model for effective thermal conductivity of nanofluids, Phys. Lett. A 307 (5–6) (2003) 313–317.
[41] S. Hasegawa, I. Shiraki, T. Tanikawa, C.L. Petersen, T.M. Hansen, P. Boggild, F. Grey, Direct measurement of surface-state conductance by microscopic four-point probe method, J. Phys. Condens. Matter 14 (35) (2002) 8379.
[42] A. Bahrami, N. Soltani, S. Soltani, M. Pech-Canul, L. Gonzalez, C. Gutierrez, A. Möller, J. Tapp, A. Gurlo, Mechanical, thermal and electrical properties of monolayer and bilayer graded Al/SiC/rice husk ash (RHA) composite, J. Alloys Compd. 699 (2017) 308–322.
[43] R.L. Poveda, N. Gupta, Electrical properties of carbon nanofiber reinforced multiscale polymer composites, Mater. Des. 56 (2014) 416–422.
[44] T. Arfin, N. Yadav, Impedance characteristics and electrical double-layer capacitance of composite polystyrene–cobalt–arsenate membrane, J. Ind. Eng. Chem. 19 (1) (2013) 256–262.
[45] P. Margueres, P. Olivier, M. Mounkaila, S. Sassi, T. Camps, Carbon fibres reinforced composites. Electrical impedance analysis: a gateway to smartness, Int. J. Smart Nano Mater. 11 (4) (2020) 417–430.
[46] S. Gaisford, V. Kett, P. Haines, Principles of Thermal Analysis and Calorimetry, Royal Society of Chemistry, 2019.
[47] C.E. Corcione, M. Frigione, Characterization of nanocomposites by thermal analysis, Materials 5 (12) (2012) 2960–2980.
[48] N. Saba, M. Jawaid, O.Y. Alothman, M. Paridah, A review on dynamic mechanical properties of natural fibre reinforced polymer composites, Construct. Build Mater. 106 (2016) 149–159.
[49] V. Oddone, B. Boerner, S. Reich, Composites of aluminum alloy and magnesium alloy with graphite showing low thermal expansion and high specific thermal conductivity, Sci. Technol. Adv. Mater. 18 (1) (2017) 180–186.
[50] Y. Pazhouhanfar, B. Eghbali, Microstructural characterization and mechanical properties of TiB2 reinforced Al6061 matrix composites produced using stir casting process, Mater. Sci. Eng. A 710 (2018) 172–180.
[51] Y. Liu, F. Wang, Y. Cao, J. Nie, H. Zhou, H. Yang, X. Liu, X. An, X. Liao, Y. Zhao, Unique defect evolution during the plastic deformation of a metal matrix composite, Scr. Mater. 162 (2019) 316–320.
[52] O. Fayomi, A. Popoola, V.S. Aigbodion, Investigation on microstructural, anti-corrosion and mechanical properties of doped Zn–Al–SnO2 metal matrix composite coating on mild steel, J. Alloys Compd. 623 (2015) 328–334.
[53] K.S. Munir, Y. Zheng, D. Zhang, J. Lin, Y. Li, C. Wen, Improving the strengthening efficiency of carbon nanotubes in titanium metal matrix composites, Mater. Sci. Eng. A 696 (2017) 10–25.
[54] L. Shufeng, K. Kondoh, H. Imai, B. Chen, L. Jia, J. Umeda, Microstructure and mechanical properties of P/M titanium matrix composites reinforced by in-situ synthesized TiC–TiB, Mater. Sci. Eng. A 628 (2015) 75–83.

[55] P.S. Reddy, R. Kesavan, B.V. Ramnath, Investigation of mechanical properties of aluminium 6061-silicon carbide, boron carbide metal matrix composite, Silicon 10 (2) (2018) 495–502.

[56] Z. Hu, G. Tong, D. Lin, C. Chen, H. Guo, J. Xu, L. Zhou, Graphene-reinforced metal matrix nanocomposites–a review, Mater. Sci. Technol. 32 (9) (2016) 930–953.

[57] Z. Zhao, P. Bai, W. Du, B. Liu, D. Pan, R. Das, C. Liu, Z. Guo, An overview of graphene and its derivatives reinforced metal matrix composites: preparation, properties and applications, Carbon 170 (2020) 302–326.

[58] Y. Chen, X. Zhang, E. Liu, C. He, C. Shi, J. Li, P. Nash, N. Zhao, Fabrication of in-situ grown graphene reinforced Cu matrix composites, Sci. Rep. 6 (1) (2016) 1–9.

[59] C. Saravanan, S. Dinesh, P. Sakthivel, V. Vijayan, B.S. Kumar, Assessment of mechanical properties of Silicon Carbide and Graphene reinforced aluminium composite, Mater. Today: Proc. 21 (2020) 744–747.

[60] K. Shirvanimoghaddam, S.U. Hamim, M.K. Akbari, S.M. Fakhrhoseini, H. Khayyam, A. H. Pakseresht, E. Ghasali, M. Zabet, K.S. Munir, S. Jia, Carbon fiber reinforced metal matrix composites: fabrication processes and properties, Compos. Part A Appl. Sci. Manuf. 92 (2017) 70–96.

[61] H.A. Alhashmy, M. Nganbe, Laminate squeeze casting of carbon fiber reinforced aluminum matrix composites, Mater. Des. 67 (2015) 154–158.

[62] S. Li, L. Qi, T. Zhang, J. Zhou, H. Li, Interfacial microstructure and tensile properties of carbon fiber reinforced Mg–Al-RE matrix composites, J. Alloys Compd. 663 (2016) 686–692.

[63] F.-L. Jin, S.-Y. Lee, S.-J. Park, Polymer matrices for carbon fiber-reinforced polymer composites, Carbon Lett. 14 (2) (2013) 76–88.

[64] A. Gupta, I. Fidan, S. Hasanov, A. Nasirov, Processing, mechanical characterization, and micrography of 3D-printed short carbon fiber reinforced polycarbonate polymer matrix composite material, Int. J. Adv. Manuf. Technol. 107 (7) (2020) 3185–3205.

[65] R. Várdai, T. Lummerstorfer, C. Pretschuh, M. Jerabek, M. Gahleitner, G. Faludi, J. Móczó, B. Pukánszky, Comparative study of fiber reinforced PP composites: effect of fiber type, coupling and failure mechanisms, Compos. Part A Appl. Sci. Manuf. 133 (2020) 105895.

[66] K. Chu, C.-c. Jia, L.-k. Jiang, W.-s. Li, Improvement of interface and mechanical properties in carbon nanotube reinforced Cu–Cr matrix composites, Mater. Des. 45 (2013) 407–411.

[67] B. Chen, J. Shen, X. Ye, L. Jia, S. Li, J. Umeda, M. Takahashi, K. Kondoh, Length effect of carbon nanotubes on the strengthening mechanisms in metal matrix composites, Acta Mater. 140 (2017) 317–325.

[68] S. Sivananthan, K. Ravi, C.S.J. Samuel, Effect of SiC particles reinforcement on mechanical properties of aluminium 6061 alloy processed using stir casting route, Mater. Today: Proc. 21 (2020) 968–970.

[69] M.P. Reddy, R. Shakoor, G. Parande, V. Manakari, F. Ubaid, A. Mohamed, M. Gupta, Enhanced performance of nano-sized SiC reinforced Al metal matrix nanocomposites synthesized through microwave sintering and hot extrusion techniques, Prog. Nat. Sci.: Mater. Int. 27 (5) (2017) 606–614.

[70] S. Hossain, M.M. Rahman, D. Chawla, A. Kumar, P.P. Seth, P. Gupta, D. Kumar, R. Agrawal, A. Jamwal, Fabrication, microstructural and mechanical behavior of Al-Al2O3-SiC hybrid metal matrix composites, Mater. Today: Proc. 21 (2020) 1458–1461.

[71] A.A.K. Suban, M. Perumal, A. Ayyanar, A.V. Subbiah, Microstructural analysis of B 4 C and SiC reinforced Al alloy metal matrix composite joints, Int. J. Adv. Manuf. Technol. 93 (1) (2017) 515–525.

[72] Z. Baig, O. Mamat, M. Mustapha, Recent progress on the dispersion and the strengthening effect of carbon nanotubes and graphene-reinforced metal nanocomposites: a review, Crit. Rev. Solid State Mater. Sci. 43 (1) (2018) 1–46.
[73] H. Belyadi, E. Fathi, F. Belyadi, Rock mechanical properties and in situ stresses, in: Hydraulic Fracturing in Unconventional Reservoirs, Gulf Professional Publishing, 2017, pp. 207–224.
[74] Y.M. Poplavko, Mechanical properties of solids, in: Y.M. Poplavko (Ed.), Electronic Materials, Elsevier, 2019, pp. 71–93.
[75] Z. Li, R.J. Young, N.R. Wilson, I.A. Kinloch, C. Vallés, Z. Li, Effect of the orientation of graphene-based nanoplatelets upon the Young's modulus of nanocomposites, Compos. Sci. Technol. 123 (2016) 125–133.
[76] Z. Yu, Z. Tan, G. Fan, D.-B. Xiong, Q. Guo, R. Lin, L. Hu, Z. Li, D. Zhang, Effect of interfacial reaction on Young's modulus in CNT/Al nanocomposite: a quantitative analysis, Mater Charact 137 (2018) 84–90.
[77] N. Viet, Q. Wang, W. Kuo, Effective Young's modulus of carbon nanotube/epoxy composites, Compos. Part B Eng. 94 (2016) 160–166.
[78] N. Bhadauria, S. Pandey, P. Pandey, Wear and enhancement of wear resistance–a review, Mater. Today: Proc. 26 (2020) 2986–2991.
[79] S.D. Kumar, M. Ravichandran, M. Meignanamoorthy, Aluminium metal matrix composite with zirconium diboride reinforcement: a review, Mater. Today: Proc. 5 (9) (2018) 19844–19847.
[80] V. Sharma, S. Kumar, R.S. Panwar, O. Pandey, Microstructural and wear behavior of dual reinforced particle (DRP) aluminum alloy composite, J. Mater. Sci. 47 (18) (2012) 6633–6646.
[81] R. Singh, M. Shadab, A. Dash, R. Rai, Characterization of dry sliding wear mechanisms of AA5083/B 4 C metal matrix composite, J. Braz. Soc. Mech. Sci. Eng. 41 (2) (2019) 98.
[82] S. Dhanalakshmi, N. Mohanasundararaju, P. Venkatakrishnan, V. Karthik, Optimization of friction and wear behaviour of Al7075-Al2O3-B4C metal matrix composites using Taguchi method, in: IOP Conference Series: Materials Science and Engineering, IOP Publishing, 2018, p. 012025.
[83] A.H. Idrisi, A.-H.I. Mourad, D.T. Thekkuden, J.V. Christy, Wear behavior of AA 5083/SiC nano-particle metal matrix composite: statistical analysis, in: IOP Conference Series: Materials Science and Engineering, IOP Publishing, 2018, p. 012087.
[84] Z. Zhang, Y. Du, C. Zhu, L. Guo, Y. Lu, J. Yu, I.P. Parkin, J. Zhao, D. Guo, Unprecedented enhancement of wear resistance for epoxy-resin graphene composites, Nanoscale 13 (2021) 2855–2867.
[85] S. Divagar, M. Vigneshwar, S. Selvamani, Impacts of nano particles on fatigue strength of aluminum based metal matrix composites for aerospace, Mater. Today: Proc. 3 (10) (2016) 3734–3739.
[86] H.J. Alalkawi, A.A. Hamdany, A.A. Alasadi, Influence of nanoreinforced particles (Al2O3) on fatigue life and strength of aluminium based metal matrix composite, Al-Khwarizmi Eng. J. 13 (3) (2017) 91–99.
[87] K.U. Kainer, Basics of metal matrix composites, in: K.U. Kainer (Ed.), Metal Matrix Composites: Custom-Made Materials for Automotive and Aerospace Engineering, Wiley-VCH, 2006, pp. 1–54.
[88] K.K. Chawla, Metal Matrix Composites, Composite Materials, Springer, 2012, pp. 197–248.
[89] B.C. Kandpal, J. Kumar, H. Singh, Production technologies of metal matrix composite: a review, IJRMET 4 (2) (2014) 27–32.

[90] S.N. Trinh, S. Sastry, Processing and Properties of Metal Matrix Composites, Mechanical Engineering and Materials Science Independent Study, 2016, p. 10.
[91] P.S. Bains, S.S. Sidhu, H. Payal, Fabrication and machining of metal matrix composites: a review, Mater. Manuf. Processes 31 (5) (2016) 553–573.
[92] M. Surappa, Aluminium matrix composites: challenges and opportunities, Sadhana 28 (1–2) (2003) 319–334.
[93] F. Nturanabo, L. Masu, J.B. Kirabira, Novel applications of aluminium metal matrix composites, in: K.O. Cooke (Ed.), Aluminium Alloys and Composites, ASM International, IntechOpen, 2019.
[94] A. Macke, B. Schultz, P. Rohatgi, Metal matrix composites, Adv. Mater. Processes 170 (3) (2012) 19–23.
[95] B.M. Karthik, M.C. Gowrishankar, S. Sharma, P. Hiremath, M. Shettar, N. Shetty, Coated and uncoated reinforcements metal matrix composites characteristics and applications–a critical review, Cogent Eng. 7 (1) (2020) 1856758.
[96] M. Badiey, A. Abedian, Application of metal Matrix composites (Mmcs) in a satellite boom to reduce weight and vibrations as a multidisciplinary optimization, in: 27th International Congress of the Aeronautical Sciences Iran, 2010.
[97] A.R. Begg, Applications for metal matrix composites, Met. Powder Rep. 46 (10) (1991) 42–45.
[98] O.A. Bauchau, J.I. Craig, Structural Analysis: With Applications to Aerospace Structures, Springer Science & Business Media, 2009.
[99] D.K. Rajak, D.D. Pagar, R. Kumar, C.I. Pruncu, Recent progress of reinforcement materials: a comprehensive overview of composite materials, J. Mater. Res. Technol. 8 (6) (2019) 6354–6374.
[100] I.M. Daniel, O. Ishai, I.M. Daniel, I. Daniel, Engineering Mechanics of Composite Materials, Oxford University Press, New York, 2006.
[101] A.K. Kaw, Mechanics of Composite Materials, CRC Press, 2005.
[102] W. Tan, F. Naya, L. Yang, T. Chang, B. Falzon, L. Zhan, J. Molina-Aldareguía, C. González, J. Llorca, The role of interfacial properties on the intralaminar and interlaminar damage behaviour of unidirectional composite laminates: experimental characterization and multiscale modelling, Compos. Part B Eng. 138 (2018) 206–221.
[103] V.B. Tillmann, Análise Estrutural de Elementos Compósitos Com a Utilização do Método de Elementos Finitos, Universidade Federal de Santa Catarina, 2015.
[104] C.A.C.D. Azevedo, Formulação alternativa para análise de domínios não-homogêneos e inclusões anisotrópicas via MEC, Universidade de São Paulo, 2007.
[105] L. Vanalli, O MEC e o MEF aplicados à análise de problemas viscoplásticos em meios anisotrópicos e compostos, Universidade de São Paulo, São Paulo, Brazil, 2004.
[106] R.M. Jones, Mechanics of Composite Materials, CRC Press, 1998.
[107] Z. Hashin, S. Shtrikman, A variational approach to the theory of the elastic behaviour of multiphase materials, J. Mech. Phys. Solids 11 (2) (1963) 127–140.
[108] J.C. Halpin, Effects of Environmental Factors on Composite Materials, Air Force Materials Lab Wright-Patterson AFB OH, 1969.
[109] C. Hui, D. Shia, Simple formulae for the effective moduli of unidirectional aligned composites, Polym. Eng. Sci. 38 (5) (1998) 774–782.
[110] H. Hu, L. Onyebueke, A. Abatan, Characterizing and modeling mechanical properties of nanocomposites-review and evaluation, J. Miner. Mater. Charact. Eng. 9 (04) (2010) 275.
[111] J.D. Schiffman, C.L. Schauer, A review: electrospinning of biopolymer nanofibers and their applications, Polym. Rev. 48 (2) (2008) 317–352.

[112] S.M. Fonseca, T. Moreira, A.J. Parola, C. Pinheiro, C.A. Laia, PEDOT electrodeposition on oriented mesoporous silica templates for electrochromic devices, Sol. Energy Mater. Sol. Cells 159 (2017) 94–101.
[113] S. Pan, H. Xing, X. Fu, H. Yu, Z. Yang, Y. Yang, W. Sun, The effect of photothermal therapy on osteosarcoma with polyacrylic acid–coated gold nanorods, Dose-Response 16 (3) (2018), 1559325818789841.
[114] A. Gupta, W. Simmons, G.T. Schueneman, D. Hylton, E.A. Mintz, Rheological and thermo-mechanical properties of poly (lactic acid)/lignin-coated cellulose nanocrystal composites, ACS Sustain. Chem. Eng. 5 (2) (2017) 1711–1720.
[115] M. Muccini, A bright future for organic field-effect transistors, Nat. Mater. 5 (8) (2006) 605–613.
[116] T. Anirudhan, S.S. Gopal, S. Rejeena, Synthesis and characterization of poly (ethyleneimine)-modified poly (acrylic acid)-grafted nanocellulose/nanobentonite superabsorbent hydrogel for the selective recovery of β-casein from aqueous solutions, Int. J. Polym. Mater. Polym. Biomater. 64 (15) (2015) 772–784.
[117] S. Mondal, P. Memmott, D. Martin, Preparation and characterization of green biocomposites based on modified spinifex resin and spinifex grass fibres, J. Thermoplast. Compos. Mater. 48 (11) (2014) 1375–1382.
[118] A. Khalid, R. Khan, M. Ul-Islam, T. Khan, F. Wahid, Bacterial cellulose-zinc oxide nanocomposites as a novel dressing system for burn wounds, Carbohydr. Polym. 164 (2017) 214–221.
[119] M. Ul-Islam, W.A. Khattak, M.W. Ullah, S. Khan, J.K. Park, Synthesis of regenerated bacterial cellulose-zinc oxide nanocomposite films for biomedical applications, Cellul. 21 (1) (2014) 433–447.
[120] Z. Luo, J. Liu, H. Lin, X. Ren, H. Tian, Y. Liang, W. Wang, Y. Wang, M. Yin, Y. Huang, In situ fabrication of nano zno/bcm biocomposite based on ma modified bacterial cellulose membrane for antibacterial and wound healing, Int. J. Nanomedicine 15 (2020) 1.
[121] R. Brandes, L. de Souza, V. Vargas, E. Oliveira, A. Mikowski, C. Carminatti, H. Al-Qureshi, D. Recouvreux, Preparation and characterization of bacterial cellulose/TiO2 hydrogel nanocomposite, J. Nanopart. Res. (2016) 73–80. Trans Tech Publ.
[122] S. Khan, M. Ul-Islam, W.A. Khattak, M.W. Ullah, J.K. Park, Bacterial cellulose-titanium dioxide nanocomposites: nanostructural characteristics, antibacterial mechanism, and biocompatibility, Cellul. 22 (1) (2015) 565–579.
[123] Y.-Y. Xie, X.-H. Hu, Y.-W. Zhang, F. Wahid, L.-Q. Chu, S.-R. Jia, C. Zhong, Development and antibacterial activities of bacterial cellulose/graphene oxide-CuO nanocomposite films, Carbohydr. Polym. 229 (2020) 115456.
[124] S.S. Mirtalebi, H. Almasi, M.A. Khaledabad, Physical, morphological, antimicrobial and release properties of novel MgO-bacterial cellulose nanohybrids prepared by in-situ and ex-situ methods, Int. J. Biol. Macromol. 128 (2019) 848–857.
[125] S. Barua, G. Das, L. Aidew, A.K. Buragohain, N. Karak, Copper–copper oxide coated nanofibrillar cellulose: a promising biomaterial, RSC Adv. 3 (35) (2013) 14997–15004.
[126] M. Echeverry-Rendon, L.M. Reece, F. Pastrana, S.L. Arias, A.R. Shetty, J.J. Pavón, J.P. Allain, Bacterial nanocellulose magnetically functionalized for neuro-endovascular treatment, Macromol. Biosci. 17 (6) (2017) 1600382.
[127] J.J. Pavón, J.P. Allain, D. Verma, M. Echeverry-Rendón, C.L. Cooper, L.M. Reece, A.R. Shetty, V. Tomar, In situ study unravels bio-nanomechanical behavior in a magnetic

bacterial nano-cellulose (MBNC) hydrogel for neuro-endovascular reconstruction, Macromol. Biosci. 19 (2) (2019) 1800225.

[128] N. Torkashvand, N. Sarlak, Fabrication of a dual T1 and T2 contrast agent for magnetic resonance imaging using cellulose nanocrystals/Fe3O4 nanocomposite, Eur. Polym. J. 118 (2019) 128–136.

[129] T. Nypelö, C. Rodriguez-Abreu, J. Rivas, M.D. Dickey, O.J. Rojas, Magneto-responsive hybrid materials based on cellulose nanocrystals, Cellul. 21 (4) (2014) 2557–2566.

[130] D. Rasouli, N.T. Dintcheva, M. Faezipour, F.P. La Mantia, M.R.M. Farahani, M. Tajvidi, Effect of nano zinc oxide as UV stabilizer on the weathering performance of wood-polyethylene composite, Polym. Degrad. Stab. 133 (2016) 85–91.

[131] E. Kim, T. Gordonov, W.E. Bentley, G.F. Payne, Amplified and in situ detection of redox-active metabolite using a biobased redox capacitor, Anal. Chem. 85 (4) (2013) 2102–2108.

[132] J. Hur, K. Im, S.W. Kim, J. Kim, D.-Y. Chung, T.-H. Kim, K.H. Jo, J.H. Hahn, Z. Bao, S. Hwang, Polypyrrole/agarose-based electronically conductive and reversibly restorable hydrogel, ACS Nano 8 (10) (2014) 10066–10076.

[133] X. Shi, Y. Zheng, G. Wang, Q. Lin, J. Fan, pH-and electro-response characteristics of bacterial cellulose nanofiber/sodium alginate hybrid hydrogels for dual controlled drug delivery, RSC Adv. 4 (87) (2014) 47056–47065.

[134] L. Zhang, Y. Li, L. Li, B. Guo, P.X. Ma, Non-cytotoxic conductive carboxymethyl-chitosan/aniline pentamer hydrogels, React. Funct. Polym. 82 (2014) 81–88.

[135] S. Sayyar, E. Murray, B. Thompson, J. Chung, D.L. Officer, S. Gambhir, G.M. Spinks, G. G. Wallace, Processable conducting graphene/chitosan hydrogels for tissue engineering, J. Mater. Chem. B 3 (3) (2015) 481–490.

[136] L. Hu, G. Zheng, J. Yao, N. Liu, B. Weil, M. Eskilsson, E. Karabulut, Z. Ruan, S. Fan, J. T. Bloking, Transparent and conductive paper from nanocellulose fibers, Energ. Environ. Sci. 6 (2) (2013) 513–518.

[137] Y. Li, H. Zhu, F. Shen, J. Wan, S. Lacey, Z. Fang, H. Dai, L. Hu, Nanocellulose as green dispersant for two-dimensional energy materials, Nano Energy 13 (2015) 346–354.

[138] D.Y. Liu, G. Sui, D. Bhattacharyya, Synthesis and characterisation of nanocellulose-based polyaniline conducting films, Compos. Sci. Technol. 99 (2014) 31–36.

[139] M. Hamedi, E. Karabulut, A. Marais, A. Herland, G. Nyström, L. Wågberg, Nanocellulose aerogels functionalized by rapid layer-by-layer assembly for high charge storage and beyond, Angew. Chem. 125 (46) (2013) 12260–12264.

[140] J.T. Korhonen, P. Hiekkataipale, J. Malm, M. Karppinen, O. Ikkala, R.H. Ras, Inorganic hollow nanotube aerogels by atomic layer deposition onto native nanocellulose templates, ACS Nano 5 (3) (2011) 1967–1974.

[141] A. Penttilä, J. Sievänen, K. Torvinen, K. Ojanperä, J.A. Ketoja, Filler-nanocellulose substrate for printed electronics: experiments and model approach to structure and conductivity, Cellul. 20 (3) (2013) 1413–1424.

[142] K. Torvinen, J. Sievänen, T. Hjelt, E. Hellen, Smooth and flexible filler-nanocellulose composite structure for printed electronics applications, Cellul. 19 (3) (2012) 821–829.

[143] J.-H. Huang, M.A. Ibrahem, C.-W. Chu, Interfacial engineering affects the photocatalytic activity of poly (3-hexylthiophene)-modified TiO 2, RSC Adv. 3 (48) (2013) 26438–26442.

[144] M. Mochane, T.C. Mokhena, T. Mokhothu, A. Mtibe, E. Sadiku, S.S. Ray, I. Ibrahim, O. Daramola, Recent progress on natural fiber hybrid composites for advanced applications: a review, Express Polym. Lett. 13 (2019) 159–198.
[145] I. Turku, T. Kärki, Accelerated weathering of fire-retarded wood–polypropylene composites, Compos. Part A Appl. Sci. Manuf. 81 (2016) 305–312.
[146] I. Grigoriadou, E. Pavlidou, K.M. Paraskevopoulos, Z. Terzopoulou, D.N. Bikiaris, Comparative study of the photochemical stability of HDPE/Ag composites, Polym. Degrad. Stab. 153 (2018) 23–36.

# Manganese oxides/polyaniline composites as electrocatalysts for oxygen reduction

2

*Md. Saddam Hossain*[a] *and Md. Mominul Islam*[b]
[a]Department of Chemistry, Khulna University of Engineering and Technology, Khulna, Bangladesh, [b]Department of Chemistry, University of Dhaka, Dhaka, Bangladesh

## 1 Introduction

Oxygen is the most abundant element in the Earth's crust. Molecular oxygen ($O_2$) has a superior affinity to accept electron because of its high oxidation potential to generate a number of reactive oxygen species (ROS), including $O_2^{-}$, $HO_2^{-}$, OH·, and $HO^{-}$ and harmless product (water) [1,2]. $O_2$ reduction reaction (ORR) plays a fundamental role in some important disciplines such as catalysis, biology, material dissolution, and energy conversion. Hence, it is one of the most widely investigated reactions due to its practical applications in biological, industrial, and environmental areas, and in clean energy systems such as fuel cells and metal-air batteries. The specific applications of ROS include the study of the co-stability of protein, DNA, biological membranes with ROS, mechanism of aging process, treatment of wastewater, organic synthesis etc. [3–13].

The electrode process of ORR is very complicated, and it follows different pathways to form different products depending on electrode material materials, types of media, and solution pH [3]. By controlling the conditions, one may produce desired products of the ORR. On the other hand, an important application of ORR is in the cathodic chamber of fuel cells, wherein a four-electron ORR occurs to produce a desired neutral product, i.e., water. Unfortunately, the sluggish kinetics of ORR seriously limits to develop an efficient fuel cell cathode with the majority of metals except for the noble metals such as platinum (Pt) and Pt-like materials [14]. Nevertheless, these metals have a high capital cost and renewability problem, and they are sensitive to poisoning that adversely affects the efficiency of fuel cells and limits its practical application. Thus, this fact urges to develop alternative catalysts that are easy to construct, low-priced, and have the similar or better catalytic efficiency for the ORR than for the Pt electrode.

In this regard, transition metal oxides are the most promising alternatives [15–19]. Manganese oxides ($Mn_xO_y$), among the various transition metal oxides, have been extensively studied because of their abundance, low cost, nontoxicity, different allotropic forms, and considerable catalytic activity toward ORR [20–23]. Manganese dioxide ($MnO_2$) with different crystal structures ($\alpha$-, $\beta$-, and $\gamma$-$MnO_2$) has been widely investigated as the cathode catalyst for ORR. Although $Mn_xO_y$ shows insufficient

**Renewable Polymers and Polymer-Metal Oxide Composites. https://doi.org/10.1016/B978-0-323-85155-8.00014-5**

stability in acidic media, these materials have been employed as catalyst of ORR in both alkaline fuel cells and metal-air batteries [24,25]. $\gamma$-MnOOH has been reported to exhibit the best catalysis for ORR among the employed manganese oxides, i.e., $Mn_2O_3$, $Mn_3O_4$, and $Mn_5O_8$ [26]. In fact, the ORR catalysis of $Mn_xO_y$ follows the order of $MnOOH > Mn_2O_3 > Mn_3O_4 > Mn_5O_8$ [21,22].

In practice, due to their low conductivity, transition metal oxides show lower catalytic performances as compared to noble metal-based catalysts, leading to the limitation in practical applications [27]. In addition, the poor adhesion of catalysts to the electrode surface limits the long-term applications of the electrode for the ORR. To address these drawbacks, $MnO_2$-based catalysts are usually supported on highly conductive materials such as polypyrrole (Ppy) [28,29], polythiophene [30], polyaniline (PAni) [31–34], or poly(vinyl alcohol) [35] matrix to form composite. These polymers show promise in various applications since they possess sufficient conductivity, low density, and easy processability.

PAni has been considered as the most attractive polymer among the various conductive polymers studied due to its low-cost, redox property, excellent conductivity, and environmental stability [36–39]. PAni has also an electrocatalytic activity toward ORR [40]. PAni undergoes protonation and deprotonation, and its surface offers adsorption via a nitrogen, having a lone pair of electron. Recently, carbonized PAni ($C_{PAni}$) has been revealed to exhibit an efficient ORR catalysis [41,42]. Although PAni itself suffers from the poor electrochemical cyclability [43], its composites with $Mn_xO_y$ have synergistic properties through the combination of a good conductivity of PAni and excellent cyclability of $Mn_xO_y$ [44,45]. Such a composite material showing the properties of its entities has been frequently proposed as the key solution to develop efficient materials to meet practical challenges such as energy supply, storage, and production. Moreover, PAni can be directly deposited on the electrode surface electrochemically. This actually offers an in situ deposition of $Mn_xO_y$/PAni composite on the conducting electrode substrate that can be directly employed for various applications, including ORR. So far, there are only few papers reporting the catalysis of $Mn_xO_y$ and PAni composites in ORR.

In this chapter, the catalytic activity of $Mn_xO_y$/PAni composite toward ORR is discussed. The recent research works on preparation, processing, characterization, and modification of electrocatalysts and their application in ORR are comprehensively reviewed. The catalytic properties are compared by highlighting relevant thermodynamics and kinetics involved in the electrochemical ORR. The future prospects of $Mn_xO_y$/PAni composites in catalytic ORR are also highlighted.

# 2 Fundamentals of electrochemical ORR

## 2.1 Mechanism

The redox chemistry of ORR mainly depends on the electrode material, type of solutions, and pH of the medium. Generally, in aqueous solution, ORR occurs via either a four-electron or two-electron reduction pathway to form $H_2O$ and $H_2O_2$, respectively. In aprotic solution, ORR follows a two-step reduction by accepting one electron in each step (Table 1). These pathways of ORR have their own importance: Four-electron transfer pathway is important in fuel cells, while two-electron pathway is significant for $H_2O_2$ production. On the contrary, one-electron transfer is important to generate ROS that is important particularly to know the mechanism of some biological processes and organic reactions, including waste treatment [1–13,46–48].

The mechanism of ORR is not actually straightforward. Several intermediate steps are associated with the electron transfer process during ORR (Scheme 1). At first, Damjanovic et al. have proposed the most accepted mechanism of ORR [49,50], and later on, Wroblowa et al. have modified this mechanism [51]. According to this mechanism, ORR proceeds through two parallel reaction pathways with comparable rates. During ORR, different intermediates such as oxygenated (O*), hydroxyl (OH*), and superhydroxyl (OOH*) species could be generated. Several possible transformations take place among these intermediates as shown in Scheme 1 [52]. Although tremendous efforts have been devoted to find out the rate-determining step of ORR, there is still no definitive conclusion because the reaction pathway depends on the catalysts, applied electrode potential, nature of the solvent, and temperature of the system. In most of the cases, the overall rate can be determined by one of these three steps: (i) adsorption of $O_2$ molecule through the first electron transfer, (ii) hydration of $O_2$, and (iii) desorption of $H_2O$. In addition, several studies have supported that oxygen coverage plays a critical role in the mechanism of ORR. In the so-called associative mechanism, a high oxygen coverage causes O—O cleavage posterior to OOH* formation, whereas in dissociation mechanism, a low oxygen coverage makes O—O cleavage anterior to OH* formation [53].

**Table 1** Electrode reaction and thermodynamic potential for ORR in different solutions [46–48].

| Type of solution | Reactions involved | Electron involved | Thermodynamic potential (V vs. SHE) |
|---|---|---|---|
| Acidic aqueous | $O_2 + 4H^+ + 4e^- \rightarrow 2H_2O$ | 4 | 1.229 |
| | $O_2 + 2H^+ + 2e^- \rightarrow H_2O_2$ | 2 | 0.70 |
| | $H_2O_2 + 2H^+ + 2e^- \rightarrow 2H_2O$ | 2 | 1.76 |
| Alkaline aqueous | $O_2 + 2H_2O + 4e^- \rightarrow 4OH^-$ | 4 | 0.401 |
| | $O_2 + H_2O + 2e^- \rightarrow HO_2^- + OH^-$ | 2 | −0.065 |
| | $HO_2^- + H_2O + 2e^- \rightarrow 3OH^-$ | 2 | 0.867 |
| Aprotic | $O_2 + e^- \rightarrow O_2^-$ | 1 | [a]2.65 vs. $Li/Li^+$ |
| | $O_2^- + e^- \rightarrow O_2^{2-}$ | 1 | [a]3.27 vs. $Li/Li^+$ |

[a] Value in dimethyl sulfoxide (DMSO). It varies depending on the polarity of solvent.

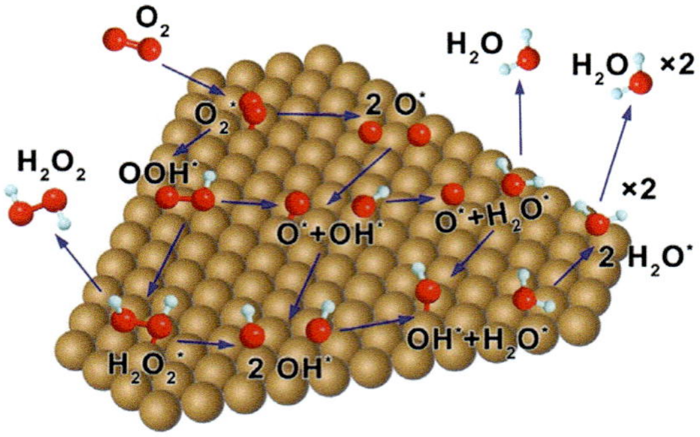


**Scheme 1** Schematic representation of different possible routes of ORR and formation of various intermediate products. Reprinted with permission from Wiley-VCH J.A. Keith, G. Jerkiewicz, T. Jacob, Theoretical investigations of the oxygen reduction reaction on Pt (111), ChemPhysChem, 11 (2010) 2779.

In alkaline solution, ORR can occur through either associative or dissociative mechanism [54–56]. For the associative mechanism, ORR begins with the associative adsorption of $O_2$ and the overall reaction mechanism can be expressed as follows:

$$O_2 + * \rightarrow O_{2(ads)} \tag{1}$$

$$O_{2(ads)} + H_2O + e^- \rightarrow OOH_{(ads)} + OH^- \tag{2}$$

$$OOH_{(ads)} + e^- \rightarrow O_{(ads)} + OH^- \tag{3}$$

$$O_{(ads)} + H_2O + e^- \rightarrow OH_{(ads)} + OH^- \tag{4}$$

$$OH_{(ads)} + e^- \rightarrow OH^- + * \tag{5}$$

where symbol (*) represents a surface free site on the catalyst and subscript (ads) signifies the adsorption state of the species. In this case, four electrons, in total, are accepted by $O_2$, resulting in $4OH^-$ ions being produced to complete a four-electron pathway of ORR. Otherwise, if $OOH_{(ads)}$ accepts an electron, desorption may occur in which peroxide ions ($HO_2^-$) are formed and leave the catalytic site, resulting in the termination of the reaction chain (Eq. 6). This termination actually leads to a two-electron ORR.

$$OOH_{(ads)} + e^- \rightarrow HO_2^{\ -} + * \tag{6}$$

For the dissociative mechanism in alkaline solution, this reaction is simpler and $O_2$ adsorbed on free sites directly dissociates to form two $O_{(ads)}$ species and, in total, consumes four electrons to complete the four-electron transfer pathway of ORR as represented in Eq. (7) followed by the reactions shown in Eqs. (4) and (5).

$$O_2 + 2* \rightarrow 2O*_{(ads)} \tag{7}$$

A similar reaction mechanism occurs in acidic solution, in which, in the presence of protons, the reaction route, especially the reactions expressed in Eqs. (2)–(5), changes to:

$$O_{2(ads)} + H^{+} + e^{-} \rightarrow OOH_{(ads)} \tag{2'}$$

$$OOH_{(ads)} + H^{+} + e^{-} \rightarrow O_{(ads)} + H_2O \tag{3'}$$

$$O_{(ads)} + H^{+} + e^{-} \rightarrow OH_{(ads)} \tag{4'}$$

$$OH_{(ads)} + H^{+} + e^{-} \rightarrow H_2O + * \tag{5'}$$

In these ways, four electrons and $4H^+$ are consumed and $O_2$ is completely reduced to generate two $H_2O$ molecules. In addition, the reaction expressed in Eq. (6) also takes the following form:

$$OOH_{(ads)} + H^{+} + e^{-} \rightarrow H_2O_2 + * \tag{6'}$$

It is worthy to mention that the adsorption of site (*) involving the elementary steps of ORR is crucial and regulates the kinetics of the electrode reaction as well as controls the termination step, leading to the product formation. Thus, it is possible to design electrode substrate at which ORR occurs to form desired product(s) by experimental trials [41–46] and computer modeling [52,55] with various materials, including metals, nonmetals, metal oxides, composites, metal complex etc.

## 2.2 Thermodynamics and kinetics

The thermodynamic potential of ORR and reaction pathway varies depending on the media as represented in Table 1. Generally, to perform ORR electrochemically at a practical rate, one must apply extra potential at the electrode. This extra potential is known as overpotential ($\eta$) that is required to surmount the intrinsic activation barrier of the ORR at the cathode. Fundamentally, the sluggish kinetics of ORR can be improved by performing the ORR at $\eta$ that is larger than the thermodynamic potential.

Electrochemical reactions are reversible redox reactions, and a general form of it may be expressed as:

$$O + ne^{-}\ (\text{at electrode}) \rightleftharpoons R \tag{8}$$

where O and R are the species reduced and oxidized at the electrode surface, respectively. The current flow due to this reversible reaction at $\eta$ that is larger than equilibrium potential per area of electrode surface is current density ($j$). Basically, both forward and backward reactions occur at the applied potential of $\eta$.

The $j$-$\eta$ relationship is, as Eq. (9), known as Butler-Volmer equation [47].

$$j = j^0 \left( e^{\frac{n\alpha F\eta}{RT}} - e^{\frac{n(1-\alpha)F\eta}{RT}} \right) \quad (9)$$

where $j^0$ is the exchange current density, $n$ is the electron number involved at the rate-determining step, $\alpha$ is the transfer coefficient, $F$ is Faraday's constant, $R$ is the gas constant, and $T$ is the temperature (in K). It is noted that $\alpha$ and $n$ are the intrinsic properties and depend on the nature of electrode material. It is evident from the above equation that to achieve low $\eta$ leading to high $j$ and better performance, $j^0$ should be large, while $(RT/\alpha nF)$ should be small.

$j^0$ is defined as the net current density at equilibrium at which it is always zero, because of the equal but opposite amount of current flow in the backward and forward reactions at this condition [47]. In fact, $j^0$ governs the speed of reaction and electrode surface and the nature of reaction affects the magnitude of $j^0$, e.g., $j^0$ of ORR is many folds lower than that of hydrogen oxidation reaction at a Pt electrode. On the contrary, ORR exhibits a low $j^0$ at a gold electrode than that observed at a Pt electrode. Thus, one may get the idea that ORR kinetics is strongly dependent on catalyst or electrode material (see later).

Another thermodynamic term in Eq. (9) is $\alpha$, which is a measure of the symmetry of the energy barrier and is generally a potential-dependent quantity. The value of $\alpha$ determines the fraction of the interfacial potential at an electrode-electrolyte interface. In a narrow potential range in the great majority of measurements, $\alpha$ appears to be constant at a constant $T$ [47].

The most difficult task of electrocatalytic studies is to figure out experimental kinetic data in a rational fashion. The catalytic performance of a particular electrode material can also be revealed by analyzing the voltammetric response using Tafel equation [47], where the dependence of steady state $j$ on $\eta$ is analyzed. The intrinsic activation barriers can be minimized by using highly active catalysts. If the $\eta$ is large, the backward reaction is negligible and Eq. (9) can be simplified to:

$$j = j^0 e^{\frac{n\alpha F\eta}{RT}} \quad (10)$$

$\eta$ logarithmically relates to $j$ and its linear portion is given as Tafel equation ($\eta > 0.05$ V).

$$\eta = a + b \log j \quad (11)$$

where $b$ is the Tafel slope and is related to the mechanism of the electrode reaction. From this equation, it is clear that when $\eta = 0$, the value of $j^0$ can be determined experimentally. $j^0$ on various electrode materials for ORR is depicted in Fig. 1.

The plot of $\eta$ vs. $\log j$ gives a straight line with a slope of $2.303RT/\alpha nF$. This slope is named as Tafel slope. The parameters determining the Tafel slope are actually $\alpha$ and $n$ since the values of all other factors are known. Therefore, for achieving a high current at low $\eta$, the electrochemical reaction should show a low Tafel slope or a large $\alpha n$.

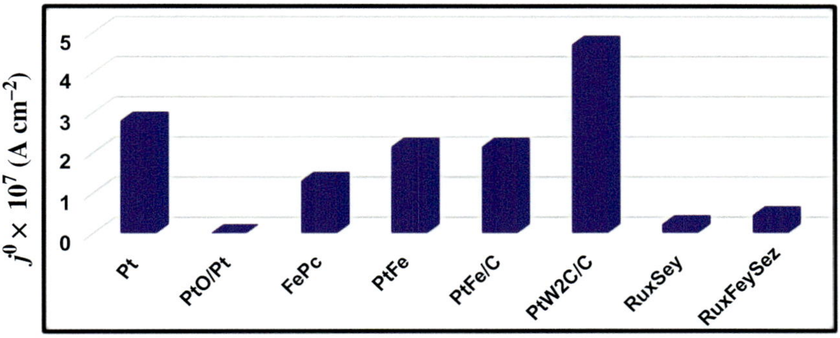


**Fig. 1** ORR exchange current densities on various electrode materials. Data were taken from the different references as cited in [57]. Pt: Platinum, PtO/Pt: Platinum oxide film on Pt surface, FePc: Iron phthalocyanide, PtFe: Pt iron, PtFe/C: PtFe film on carbon surface, $PtW_2C/C$: Pt tungsten carbide film on C surface, $Ru_xSe_y$: Ruthenium selenides, $Ru_xFe_ySe_z$: Ruthenium iron selenides.

Depending on the electrode materials used and potential range, usually two Tafel slopes, namely, $60\,mV\,dec^{-1}$ and $120\,mV\,dec^{-1}$, are obtained for ORR [58,59]. The other factor influencing Tafel slope is mainly $\alpha$ that depends on $T$ as stated above. For ORR at a Pt/C electrode, the value of $\alpha$ increases linearly from 20°C to 250°C. The relationship between $\alpha$ and $T$ is given in Eq. (12) [59,60]:

$$\alpha = \alpha_o T \quad (12)$$

where $\alpha_o$ is a constant with a value of 0.001678. The relative humidity in case of fuel studies of ORR has been reported as another major factor affecting $\alpha$ [59,61].

# 3 Electrocatalysts for ORR

The ORR is the cathodic reaction of fuel cells, wherein a four-electron reduction of $O_2$ to form water is desired [62,63], as stated above. On the contrary, the sluggish kinetics of ORR due to complicated mechanism seriously limits the development of efficient fuel cell cathode. Researchers have been studying various electrode materials at aiming to develop inexpensive and efficient electrode materials for fuel cells technology. The kinetics of ORR are the factors for the performance of proton exchange membrane fuel cells (PEMFCs) [62,63]. Slow reaction kinetics and mixed ORR occurring at the cathode reduce the performance of PEMFCs [64–67]. Currently, Pt having sensible activity and high stability is commercially available electrocatalyst used in PEMFCs [68,69]. However, Pt exhibits $\eta$ of ca. 400 mV over from the thermodynamic potential of 1.19 V at 80°C [70]. This observation has been reported to indicate that the ORR efficiency of Pt is lost through the formation of adsorbed species on the Pt surface [57,70]. Therefore, the researchers have paid much attention to (i) minimize the fuel cells costs using non-Pt catalysts, (ii) enhance the electrocatalytic performance using bimetallic alloy materials such as macrocyanides and chalcogenides of transition metals and metal oxides, and (iii) modify Pt with nanotubes, graphene, and carbon nanofibers. The support of catalysts would significantly tune the activity of the cathode catalyst.

To reveal the mechanism and the energy associated with the ORR occurring on the surface of catalyst, theoretical calculation by density functional theory (DFT) has been carried out [71,72]. It has been observed that the binding energy between oxygen intermediates and the catalyst surface is related to catalytic efficiency of ORR [71,72]. Based on both theoretical calculations and experimental attempts, the so-called volcano plot (Fig. 2) has been constructed comparing catalytic activity to oxygen binding energy ($\Delta E_o$) [73]. Among the catalysts under consideration, although Pt is not at the peak of the volcano, it occupies the most optimal position. It is clear from Fig. 2 that there is still scope for the improvement in designing an ORR catalyst.

It has been reported that Pt(111) single crystal exhibits about ten times higher activity toward ORR than Pt/C catalysts with a mixture of facets on the catalyst surface [74]. Other noble metals such as Pd, Ag, Rh, Ir, Ru, and Au also show similar catalysis [60,75]. However, as a pristine metal, Pt shows the highest catalysis compared with other noble metals (Fig. 2). Moreover, these metals are electrochemically unstable and easily oxidizable compared to Pt.

Pt alloys with transition metals such as Fe, Ni, Co, and Cr offer an improved ORR electrocatalysis [65,76]. PtFe/C catalyst exhibits $j^0$ value of 78.6 mA cm$^{-2}$ for ORR in methanol, which is higher than 65.0 mA cm$^{-2}$ obtained at Pt/C catalyst. In addition, $j^0$ of PtFe/C catalyst for ORR became smaller in methanol-free solution, indicating that PtFe/C is a better methanol-tolerant catalyst than Pt/C. PtFe/C also shows

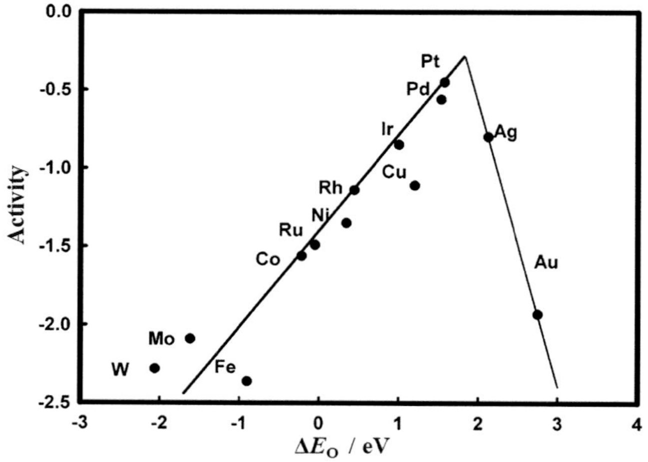


**Fig. 2** ORR activities as a function of oxygen-binding energy.
Reprinted with permission from American Chemical Society J.K. Nørskov, J. Rossmeisl, A. Logadottir, L.R.K.J. Lindqvist, J.R. Kitchin, T. Bligaard, H. Jonsson, Origin of the overpotential for oxygen reduction at a fuel-cell cathode, J. Phys. Chem. B 108 (2004) 17886.

20%–30% more power density than Pt/C [65]. Similarly, excellent performance of PtFe/C in catalyzing ORR has been reported by other researchers [77,78]. PtNi/C offers improved mass and specific activity in ORR compared to Pt/C [79]. The replacement of Ni with Bi in catalyst for ORR, e.g., PtBi/C, exhibits a higher methanol tolerance than Pt/C.

The effects of other noble metals such as Pd, Au, Ir, Rh, and Ru as Pt alloy catalysts for ORR have also been revealed [80–86]. The rate of the ORR on $Pt_3Ni$ catalyst is 90 times faster than that observed on pristine Pt [74]. PtAu/CNTs (carbon nanotubes) show an improved electrocatalytic performance in basic media compared to Pt/CNT [84–86]. However, among these Pt alloys with other noble metals, PdPt alloy exhibits the highest catalytic activity. The improvement of Pt alloy with other noble metals, e.g., Pd, Ag, and Au, possibly occurred due to the fully occupied *d*-orbitals of these metals. The improved ORR kinetics has been reported to result from the effect of the *d*-orbital coupling between metals that in fact decreases the Gibbs free energy for the electron-free steps in ORR [87].

Pd alloys with Cu, Ni, Fe, Co, W, and Mo have also been investigated as promising electrocatalysts for ORR [68,88,89]. In the presence of alcohol, better performance and stability of these alloys than Pd alone have been observed. These catalytic effects of Pd alloys are considered to be due to the change in Pd—Pd bond length, modification of the electron configuration, and change in the surface species and compositions [68,88,89]. Furthermore, applying a Pt monolayer on a Pd or $Pd_3Co$ nanoparticle core increases the rate of ORR [90]. PdSn/C catalyst shows an improved electrocatalytic performance in media containing high methanol concentrations in a fuel cell compared to commercial Pt/C [91]. The effects of Co content in PdCo alloy on the ORR have also been investigated. It has been revealed that the electrocatalysts containing less than 20% Co possess improved activities toward ORR [66].

Although the bimetallic catalysts, e.g., Pt or Pd with Ag, Fe, Sn, Ni, Co, Mo, etc., exhibit a better electrocatalytic activity toward ORR than the pristine Pt or Pd [73], their applications in commercial fuel cell devices are limited due to the lack of suitable methods for a large-scale synthesis [80]. Another major limitation of the bimetallic catalysts is the dissolution of transition metals alloyed in Pt-M or Pd-M catalyst at a potential range between 0.3 and 1.0 V versus NHE in acidic solution [92]. However, the dissolution problem of these transition metals can be overcome by using multicomponent catalysts of Pt or Pd such as PtTiCo, PtTiCr, PtTiCu, PtTiFe, PtTiMn, PtTiMo, PtTiNi, PtTiPd, PtTiTa, PtTiV, PtTiW, PtTiZa, PtCuCo, $PtCoSe_2$, PtIrCo, PdFePt, PtCuCoNi, PdSnPt, and PdCoPt [87]. The components in this type of catalyst gain improved surface activity through the shifting of the *d*-band center that practically minimizes the adsorption energy of surface oxygenated intermediates (Eqs. 2–4).

Macrocyclic complexes and chalcogenides of transition metals have been widely used as ORR electrocatalyst in the last century because of their inactivity for the oxidation of methanol present in the system, PMFCs [93]. The most investigated transition metals are the cobalt and iron phthalocyanides as ORR electrocatalysts [94,95]. These metal chelates favor the chemisorption of $O_2$ during ORR [96]. Fe chelates and Co phthalocyanide lead to a four-electron pathway of ORR to form water and exhibit similar ORR kinetics as obtained with commercial Pt/C [94,97]. Ruthenium chalcogenides ($Ru_xSe_y$) or $Ru_xSe_y$ embedded in a polymetric matrix such as PAni show poor performance toward ORR than Pt [98,99].

$IrO_2$, $MnO_2$, NiO, $CeO_2$, $ZrO_2$, $TiO_2$, and $SnO_2$ metal oxides have been investigated as electrocatalyst for the ORR in different solutions [100–104]. Although metal oxides containing nanoparticles on their surface do not exhibit an improved catalytic performance because of their lower electrolytic conductivity, they play an excellent corrosion resistance property in different electrolytic media [104,105]. However, the carbon-supported metal oxide catalysts possess favorable properties such as a large surface area, high electric conductivity, high stability and excellent catalytic performance as cathode materials for the ORR. The support of carbon in $PtTiO_2$/CNTs shows superior activity than in $PtTiO_2$ [106] due to the high conductivity of CNTs compared to $TiO_2$. $TiO_2$ in Pt/$TiO_2$/C improves the catalytic performance and thermal stability in fuel cells compared to Pt/C [107]. However, most of the metal oxides are unstable in acidic solution. To overcome this problem, conducting polymers such as Ppy and PAni have been used to protect the dissolution of metal oxides. Metal oxide sandwiched between the Ppy layers shows an improved electrochemical stability [66]. $CoFe_2O_4$ oxides sandwiched between Ppy layers offer high electrocatalytic activity toward ORR in acidic media [108].

# 4 Synthesis of $Mn_xO_y$/PAni composites

Various methods, including chemical, electrochemical, and physical methods, have been adopted to synthesize pristine $Mn_xO_y$ [15,20–27,100,103,109–111] and PAni [32–41,112–125] and their composites [33,36,126–133] with tunable properties. Depending on the necessities, different sophisticated techniques have been employed to characterize the material synthesized. These techniques include ultraviolet–visible spectroscopy, Fourier transform infrared spectroscopy, X-ray diffraction, thermogravimetric analysis, Brunauer–Emmett–Teller method, field-emission scanning electron microscopy, energy-dispersive spectroscopy, transmission electron microscopy, high-resolution transmission electron microscopy, electron paramagnetic resonance spectroscopy, X-ray photoelectron spectroscopy etc.

## 4.1 $Mn_xO_y$

The syntheses of nanocrystalline $Mn_xO_y$ have been carried out more widely in the last century. Several techniques, namely, thermolysis from organometallic precursors, direct mixing of potassium permanganate ($KMnO_4$) and polyelectrolyte aqueous solutions, co-precipitation, room temperature synthesis, hydrothermal, solvothermal, biological, wet chemical method, electrospinning, sol–gel, sonochemical, microwave-assisted, complex decomposition, chemical reduction, electrochemical method, direct electrodeposition and sulfur-based reduction followed by acid leaching, are the most common and simple methods to fabricate $Mn_xO_y$ (e.g., MnO, $Mn_3O_4$, $Mn_2O_3$, $MnO_2$,

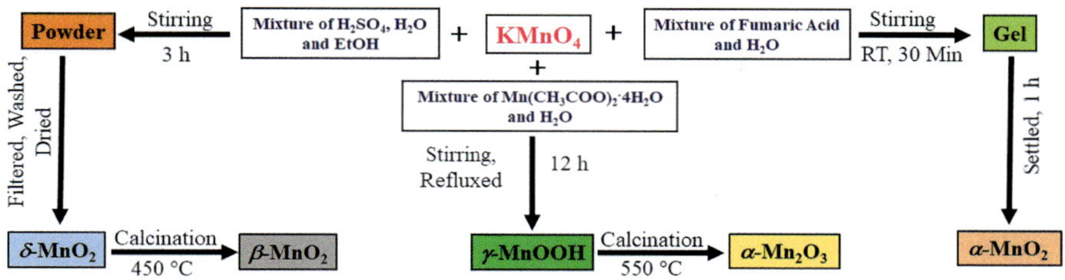


**Scheme 2** Synthetic routes for the preparation of various $Mn_xO_y$ forms from $KMnO_4$ [110]. *RT*, room temperature.

and $Mn_5O_8$) nanocatalyst [15,20–27,100,103,109–111,134]. Different $Mn_xO_y$ can be produced from $KMnO_4$ as can be seen in Scheme 2 [110].

## 4.2 PAni

Chemical, electrochemical, and physical methods have been widely used to synthesize PAni [32–41,112–125]. Chemical synthesis includes in situ oxidative polymerization, counterion-induced processability, emulsion polymerization, inversion emulsion process, dispersion polymerization, blending methods, e.g., solution blending and dry blending, interfacial polymerization, melt processing, and the flux methods. The rapidly mixed reaction is of great significance for the chemical preparation of nanofibers and provides a clearly synthetic mechanism of PAni nanofibers [112–122]. The chemical oxidative polymerization of aniline is of prime choice to synthesize PAni, since it can readily and rapidly prepare bulk quantities of nanostructured PAni [113]. Consequently, in recent years, it has received growing interest. Conventional chemical synthesis of PAni has generally been carried out by polymerizing the monomer with the aid of templates [115,116,135]. In the absence of any template, both interfacial polymerization and rapidly mixed reactions have been used for making pure nanofibers by suppressing the secondary growth of PAni [121,122].

The electrochemical route for the formation of PAni is most advantageous technique, because it offers an in situ fabrication of a film of catalyst on a conducting substrate, i.e., electrode surface. The electropolymerization of aniline is generally carried out in aqueous solutions by applying an external potential at which oxidation of aniline takes place. In the case of the electrochemical synthesis of PAni, potentiostatic, galvanostatic, and potentiodynamic methods have been employed [123–125].

## 4.3 $Mn_xO_y$/PAni composites

Composites of conjugated polymers with metal oxides are promising due to their potential combining properties, which cannot be achieved separately with individual components. Various methods have been followed for the synthesis of $MnO_2$/PAni composites, including traditional template in situ chemical oxidative polymerization [126], stepped chemical [127] and electrochemical routes [34,112,128]. PAni/

$MnO_2$ composites have been synthesized by co-electrodeposition of $MnO_2$ and PAni onto a graphite electrode from an aqueous 0.5 M $H_2SO_4$ solution containing 0.5 M $MnSO_4$ and 0.2 M aniline applying cyclic voltammetric technique [128].

Electrodeposited $MnO_2$ is the product formed by the disproportion of Mn(III) in neutral solution and Mn(III) could not be disproportionated in a strong acid solution [136]. As a result, $MnO_2$ could not be formed on the graphite electrode under the same electrodeposition condition. However, in the presence of aniline, $MnO_2$ could be formed on the graphite electrode surface very well, which reveals that the presence of aniline could accelerate the formation of $MnO_2$ that is produced by the disproportion of Mn(III) under the same condition.

PAni on the surface of $\beta$-$MnO_2$ nanorods has been synthesized by in situ chemical oxidative polymerization to form PAni/$\beta$-$MnO_2$ nanocomposites (see Fig. 3) to investigate the catalytic activity toward ORR as illustrated in Scheme 3 [126]. The observed extraordinary catalysis of PAni/$\beta$-$MnO_2$ has been considered to be resulted from the specific interaction between $\beta$-$MnO_2$ and PAni, increased specific surface area, and enhanced electric conductivity, which altogether favor the electron transfer process during ORR.

PAni/$MnO_2$-MWNTs (multiwalled CNTs) hybrid material has been synthesized by the stepped chemical method. In this method, first, $MnO_2$-MWNTs are prepared by in situ direct coating method, and then PAni is coated onto the $MnO_2$-MWNTs to synthesize PAni/$MnO_2$-MWNTs [127]. Mesoporous $C_{PAni}/Mn_2O_3$ composites with well-controlled diameter and high surface area have been prepared by calcination process [129] to investigate the electrocatalytic activities comparable to those of the benchmark Pt/C toward ORR. Among the composite catalysts with PAni synthesized with

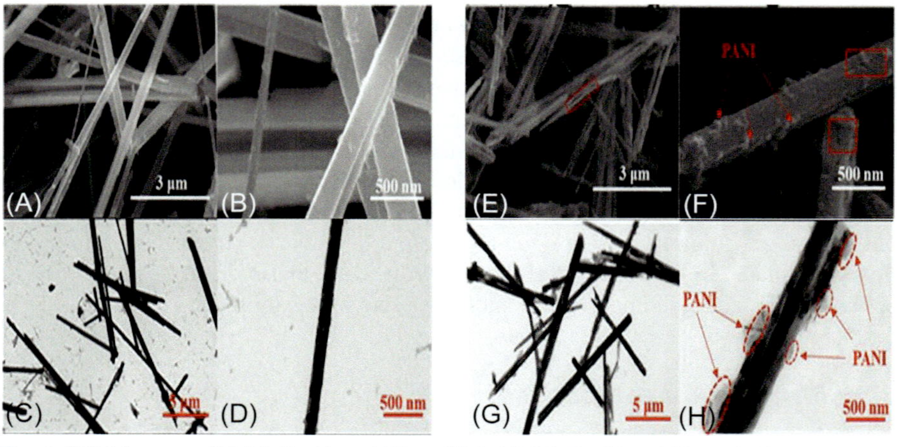


**Fig. 3** SEM images of $\beta$-$MnO_2$ (A and B) and PAni/$\beta$-$MnO_2$ (E and F), and TEM images of $\beta$-$MnO_2$ (C and D) and PAni/$\beta$-$MnO_2$ (G and H).
Reprinted with permissions from Elsevier X. Zhou, Y. Xu, X. Mei, N. Du, R. Jv, Z. Hu, S. Chen, Polyaniline/β-$MnO_2$ nanocomposites as cathode electrocatalyst for oxygen reduction reaction in microbial fuel cells, Chemosphere, 198 (2018) 482.

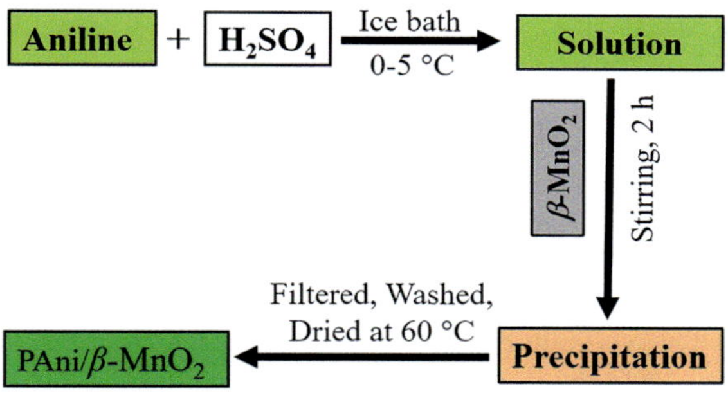


**Scheme 3** In situ chemical oxidative polymerization method to prepare PAni/β-$MnO_2$ nanocomposites [126].

different forms of $Mn_xO_y$, $C_{PAni}/Mn_2O_3$ exhibits higher crystallinity as represented in Fig. 4.

Oxidative polymerization of aniline in acidic medium containing $MnO_2$ as an oxidant has been followed to prepare PAni/$MnO_2$ composites [130] to understand the effect of acids on the degree of crystallinity of the prepared composites. It has been found that adsorbed anions of acid used on the oxide surface work as the charge compensator for positively charged PAni in the formation of PAni/$MnO_2$ composites. The content of PAni in the composite depends on the type of acid used.

PAni and $MnO_2$ have been reported to be simultaneously synthesized using simultaneous oxidation routes. In this method, aniline monomer and $MnCl_2$ are oxidized by $KMnO_4$ to form PAni and $MnO_2$, and meanwhile, $KMnO_4$ is reduced to produce $MnO_2$ [131] by the following reactions:

$$MnO_4^- (aq) + aniline (aq) \rightarrow MnO_2 \cdot nH_2O (s) + PAni (s) \quad (13)$$

$$2MnO_4^- (aq) + 3Mn^{2+} (aq) + 2H_2O (l) \rightarrow 5Mn^{4+} (aq) \quad (14)$$

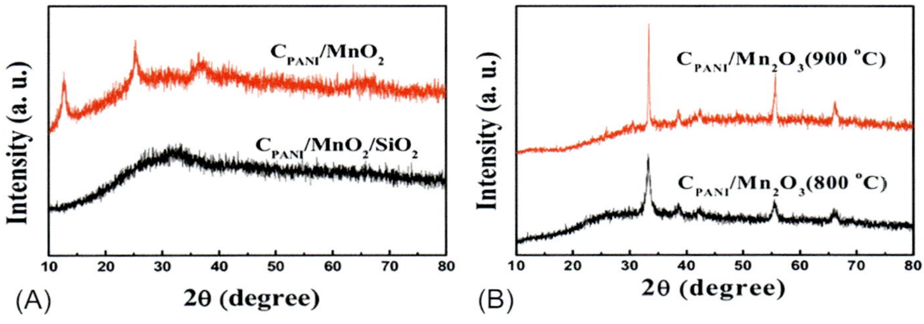


**Fig. 4** (A) XRD patterns of $C_{PAni}/MnO_2/SiO_2$ and $C_{PAni}/MnO_2$ hybrid shells calcined at 560°C. (B) XRD patterns of $C_{PAni}/Mn_2O_3$ hybrid shells derived from PAni/$MnO_2$/$SiO_2$ calcined at different temperature under inert atmospheres.
Reprinted with permission from American Chemical Society S. Cao, N. Han, J. Han, Y. Hu, L. Fan, C. Zhou, R. Guo, Mesoporous hybrid shells of carbonized polyaniline/$Mn_2O_3$ as non-precious efficient oxygen reduction reaction catalyst, ACS Appl. Mater. Interfaces 8 (2016) 6040.

Ni et al. have synthesized surfactant-assisted $MnO_2$/PAni nanocomposite by following an in situ polymerization method to control the size and distribution of $MnO_2$ in the PAni matrix [132]. This method is comprised of three steps: (i) preparation of $MnO_2$ nanoparticles solution in water containing surfactant, (ii) preparation of an acidic aniline solution, and (iii) fabrication of $MnO_2$/PAni nanocomposites by adding $MnO_2$ dispersed solution to aniline solution quickly and thoroughly. In this case, $MnO_2$ nanoparticles are basically prepared with the support of Triton X-100 [132].

Yuan et al. have investigated that in the presence of PAni, PAni/$MnO_2$/MWCNTs composite exhibits a good cyclic voltammetric performance in a mixture of $Na_2SO_4$ and $H_2SO_4$ acid, although $MnO_2$ is not stable in the acidic medium. The stability of $MnO_2$ in acidic solution has been reported to associate with an extraordinary protection by PAni coating formation [133].

# 5 Electrocatalytic activity of $Mn_xO_y$/PAni composites toward ORR

Pristine $Mn_xO_y$ and PAni and their composites have been reported to exhibit an ability to catalyze ORR. The electrocatalytic activity of $Mn_xO_y$/PAni toward ORR depends on their composition, morphology, surface area, porosity, electrochemical stability, the specific interaction between $Mn_xO_y$ and PAni, so on. The different electrochemical parameters such as half-wave potential, charge transfer resistance, ohmic resistance, number of electron involved, Tafel slope, and powder density are the measure of the degree of catalytic performance of a catalyst. A comprehensive discussion on the relation between these parameters and catalytic activity of a material toward ORR is given below.

## 5.1 $MnO_2$/PAni composite

ORR on $MnO_2$-modified electrodes has been investigated extensively in the last century [20–26]. The catalysis of ex situ prepared $MnO_2$/PAni composite toward ORR has been studied for the first time by only cyclic voltammetric technique [128]. Cyclic voltammograms (CVs) measured at a $MnO_2$/PAni-modified graphite electrode in $N_2$- and $O_2$-saturated aqueous solutions are shown in Fig. 5A. A well-defined reduction peak at ca. -0.16 V is observed in $O_2$-saturated solution, while no peak is seen in the reverse scan. Moreover, no reduction peak is actually noticed in $N_2$-saturated solution. Thus, it may be concluded that an irreversible reduction of $O_2$ suitably occurs on $MnO_2$/PAni-modified graphite electrode.

The characteristics of this observed ORR have been further investigated by measuring the CVs at different potential scan rates (see inset of Fig. 5A). As the potential scan rate increases, the current of the peak increases and the peak shifts to negative potentials. The plot of current vs. square root of scan rate fits well with straight line passing through the origin that is the characteristic of a diffusion-controlled process [7,8,47]. Therefore, the ORR at $MnO_2$/PAni-modified graphite electrode is an

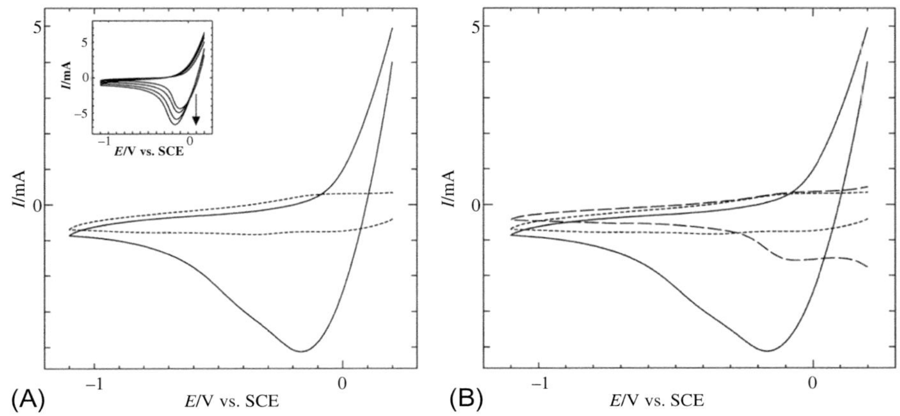


**Fig. 5** (A) CVs obtained on $MnO_2$/PAni-modified graphite electrode in a $N_2$-saturated (dotted line) and $O_2$-saturated (solid line) in 0.1 M $Na_2SO_4$ solutions, in which $MnO_2$ is produced by electrodeposition in 0.5 M $H_2SO_4$ containing 0.5 M $MnSO_4$ and 0.2 M aniline. Potential scan rate: $50\,mV\,s^{-1}$. (B) CVs measured on $MnO_2$-modified (dotted line), PAni-modified (dashed line), and $MnO_2$/PAni-modified (solid line) graphite electrodes in $O_2$-saturated 0.1 M $Na_2SO_4$ solution. Potential scan rate: $50\,mV\,s^{-1}$. Inset: CVs recorded on $MnO_2$/PAni-modified graphite electrode in $O_2$-saturated 0.1 M $Na_2SO_4$ at different potential scan rates of 50, 100, 150, and $200\,mV\,s^{-1}$.
Reprinted with permission from Wiley-VCH K.Q. Ding, Cyclic voltammetrically prepared $MnO_2$-polyaniline composite and its electrocatalysis for oxygen reduction reaction (ORR), J. Chin. Chem. Soc. 56 (2009) 891.

irreversible, diffusion-controlled process as observed at bare metallic and carbon electrodes [7,8,47].

The role of the electrodeposited $MnO_2$ on ORR has been revealed by comparing the CVs obtained at PAni-modified and $MnO_2$/PAni-modified graphite electrodes (Fig. 5B). No reduction peak at the PAni-modified electrode has been noticed as observed at $MnO_2$/PAni-modified graphite electrode. This observation infers that $MnO_2$ is the main substrate in the composite on which ORR could take place. On the contrary, no reduction peak for ORR is interestingly found at graphite electrode modified only with $MnO_2$. Thus, it is clear that both components of the composite are not active for the ORR separately, whereas $MnO_2$ in $MnO_2$/PAni composite can play the role in catalyzing electrochemical ORR.

Although ORR at $MnO_2$-modified electrode has been carried out by many research groups [20–27], its mechanism has not been well studied yet. Ohsaka et al. have reported that ORR occurs at the MnOOH-modified GC electrode through the disproportionation reactions of $O_2^-$ and $HO_2^-$ intermediates to form $O_2$ that undergoes repeated electrode reaction [111]. Yang and Xu have proposed that the catalytic activity of $MnO_2$ toward ORR might be due to its high composition freedom and higher corrosion resistance [137]. However, it has been evident that $MnO_2$ is electrochemically stable during the ORR process as can be seen in Fig. 5A (dotted line).

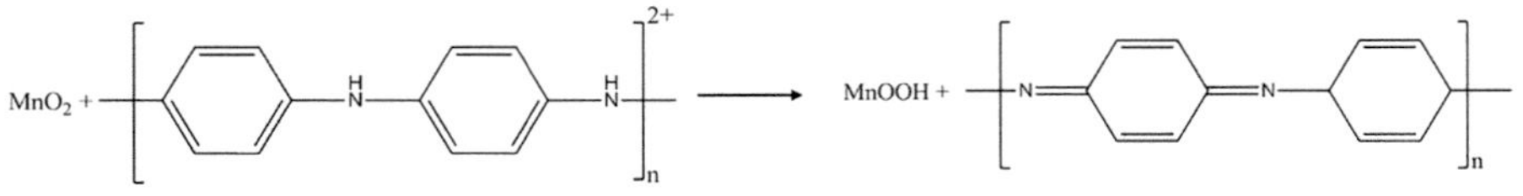


(15)

$$O_2 \rightarrow O_{2,ads}\ (\text{fast}) \tag{16}$$

$$MnOOH + O_{2,ads} \rightarrow MnO_2 + O_{2,ads}^{-}\ (\text{slow}) \tag{17}$$

$$O_{2,ads}^{-} + H_2O + e^{-} \rightarrow HO_2^{-} + OH^{-}\ (\text{fast}) \tag{18}$$

In the catalytic chains of $MnO_2$/PAni toward ORR, a mediation process involving the reduction of Mn(IV) to Mn(III) followed by the electron transfer of Mn(III) to $O_2$ molecule takes place. First, MnOOH is produced by the reaction between PAni and $MnO_2$, and then the electron is transferred from the resultant MnOOH to $O_2$ to generate $O_2^{-}$ via the following reactions as expressed in Eqs. (15)–(17). These processes ultimately enhance the catalytic activity dramatically since $O_2$ molecule might accept electrons from two ways, i.e., from the electrode and from the resultant MnOOH [128].

## 5.2 $Mn_xO_y$/PAni hybrid shells

Catalytic performances of several PAni-based $Mn_xO_y$ composites on the ORR have been evaluated by studying kinetics and mechanism with modern electrochemical methods. Hydrodynamic voltammetry at a rotating ring-disk electrode (RRDE) is the most powerful and informative tool to reveal kinetics and mechanism of the electrode reaction, especially ORR. The comprehensive study of ORR at PAni/$MnO_2$, $C_{PAni}$, $Mn_2O_3$, $C_{PAni}$/$MnO_2$, and $C_{PAni}$/$Mn_2O_3$ catalyst-modified RRDE, as shown in Figs. 6 and 7, is discussed here for example. Different speeds of rotation of RRDE have been necessarily employed during the measurement of linear sweep voltammograms (LSVs) for the ORR (Fig. 6). The half-wave potentials are summarized in Table 2. Moreover, LSVs measured at different catalysts and benchmark Pt/C are compared in Fig. 7. At a glance, the onset potential of ORR at $C_{PAni}$/$Mn_2O_3$ hybrid shells is higher than that of other catalysts considered. The half-wave potential is also much higher than the other electrocatalysts, although it is lower than that of benchmark Pt/C. Moreover, $j$ of ORR at $C_{PAni}$/$Mn_2O_3$ catalyst is also larger than those of other catalysts and is close to $-5.62\,mA\,cm^{-2}$ that is found for the benchmark Pt/C catalyst (see Fig. 7A).

The underlying reason associated with the enhanced catalytic performance of $C_{PAni}$/$Mn_2O_3$ has been explored by further analyzing the voltammograms measured at RRDE. The slopes of characteristic Koutecky-Levich (K-L) plots [47], shown in Fig. 7B especially for $C_{PAni}$/$Mn_2O_3$ and $C_{PAni}$/$MnO_2$ catalysts, are close to the benchmark Pt/C catalyst. This result indicates that ORR at these catalysts is a quasi-four-electron transfer process [3,4,6–8]. In addition, $n$ involves with the ORR, which

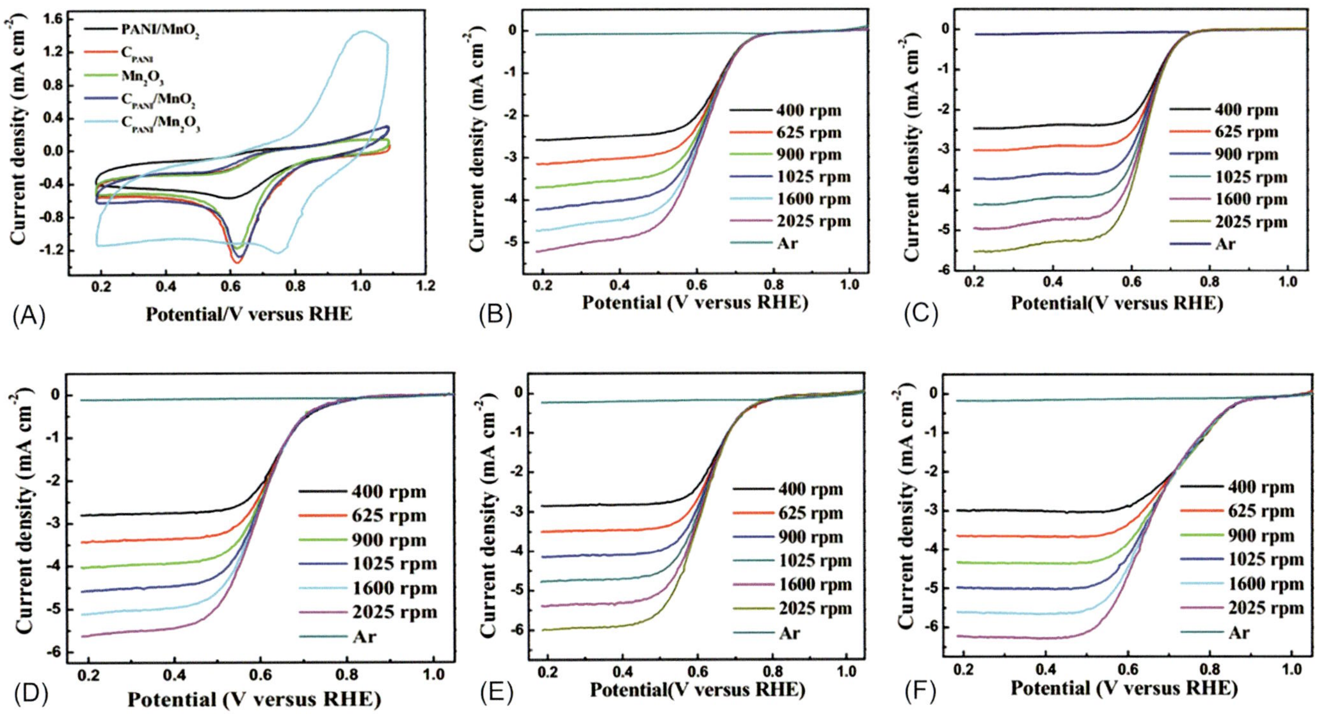


**Fig. 6** (A) CVs obtained on PAni/$MnO_2$-, $C_{PAni}$-, $Mn_2O_3$-, $C_{PAni}/MnO_2$-, and $C_{PAni}/Mn_2O_3$-supported GC electrodes in $O_2$-saturated 0.1 M KOH solution. (B–F) LSV curves recorded on GC RDE modified with (B) PAni/$MnO_2$, (C) $C_{PAni}$, (D) $Mn_2O_3$, (E) $C_{PAni}/MnO_2$, and (F) $C_{PAni}/Mn_2O_3$ in Ar- and $O_2$-saturated 0.1 M KOH solution at different rotation speeds.

Reprinted with permission from American Chemical Society S. Cao, N. Han, J. Han, Y. Hu, L. Fan, C. Zhou, R. Guo, Mesoporous hybrid shells of carbonized polyaniline/$Mn_2O_3$ as non-precious efficient oxygen reduction reaction catalyst, ACS Appl. Mater. Interfaces 8 (2016) 6040.

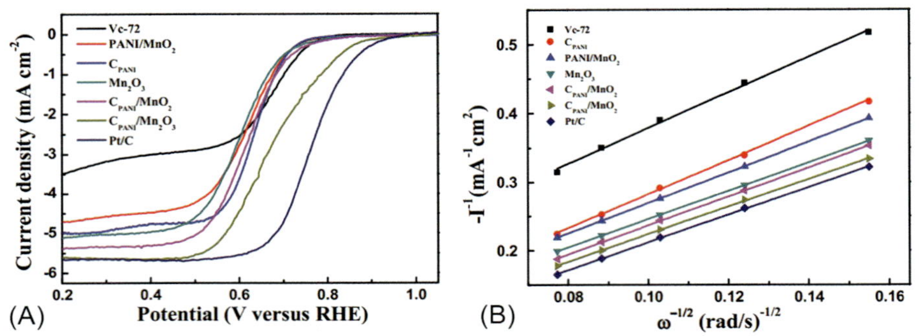


**Fig. 7** (A) LSV curves recorded on PAni/$MnO_2$, $C_{PAni}$, $Mn_2O_3$, $C_{PAni}/MnO_2$, $C_{PAni}/Mn_2O_3$, Vc-72, and Pt/C at a constant rotation of 1600 rpm in $O_2$-saturated 0.1 M KOH solution. (B) K-L plots constructed at 0.3 V for different catalysts used.
Reprinted with permission from American Chemical Society S. Cao, N. Han, J. Han, Y. Hu, L. Fan, C. Zhou, R. Guo, Mesoporous hybrid shells of carbonized polyaniline/$Mn_2O_3$ as non-precious efficient oxygen reduction reaction catalyst, ACS Appl. Mater. Interfaces 8 (2016) 6040.

**Table 2** Half-wave potential of ORR determined at electrocatalysts [129].

| **Catalysts** | **Half-wave potential (V)** |
|---|---|
| Benchmark Pt/C | 0.82 |
| $C_{PAni}/Mn_2O_3$ | 0.78 |
| PAni/$MnO_2$ | 0.71 |
| $C_{PAni}$ | 0.71 |
| $C_{PAni}/MnO_2$ | 0.72 |
| $Mn_2O_3$ | 0.71 |

can be calculated from the slope of K-L plot. It is well accepted that four-electron and two-electron transfer mechanisms of ORR dominate at Pt-based and carbonaceous catalysts, respectively [1–14]. Obviously, the higher value of $n$ dictates superior electrocatalytical performance.

Typical measurements of voltammograms for ORR at RRDE (see Fig. 8A) assist to deeply investigate the mechanism of ORR studied on $Mn_2O_3$, $C_{PAni}/MnO_2$, $C_{PAni}/Mn_2O_3$, and Pt/C catalysts. The extent of mixed ORR, i.e., two- and four-electron ORR, can be evaluated by tracing the current generated at the ring of RRDE. The yield of peroxide ($y_{peroxide}$) generated by the reaction shown in Eq. (6) can be defined as the percentage of peroxide produced with respect to the total reduction products of $O_2$. $y_{peroxide}$ and $n$ can be determined using Eqs. (19) and (20).

$$y_{peroxide} = \frac{200 i_R}{N i_D + i_R} \tag{19}$$

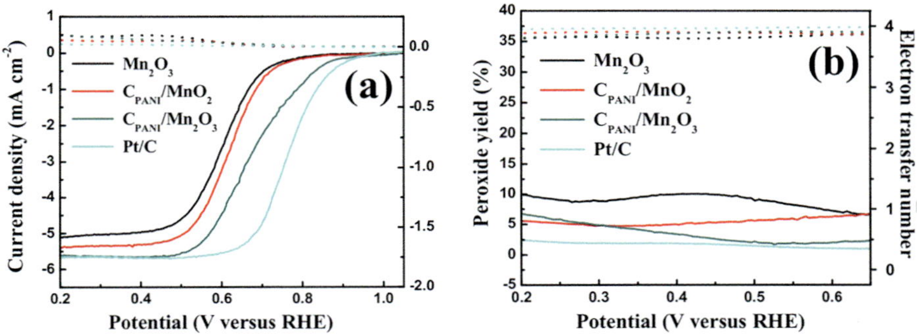


**Fig. 8** (A) Voltammograms measured for ORR at RRDE modified with $Mn_2O_3$, $C_{PAni}/MnO_2$, $C_{PAni}/Mn_2O_3$, and Pt/C: Solid line for disk and dotted line for ring of the RRDE. (B) Peroxide yield (solid line) and the electron transfer number (dotted line) at different potentials. Reprinted with permission from American Chemical Society S. Cao, N. Han, J. Han, Y. Hu, L. Fan, C. Zhou, R. Guo, Mesoporous hybrid shells of carbonized polyaniline/$Mn_2O_3$ as non-precious efficient oxygen reduction reaction catalyst, ACS Appl. Mater. Interfaces 8 (2016) 6040.

$$n = \frac{4Ni_D}{Ni_D + i_R} \tag{20}$$

where $i_D$ and $i_R$ represent the disk current and ring current, respectively, and $N$ is the current collection efficiency of RRDE [47,138,139].

It is noted that the values of $y_{peroxide}$ and $n$ should depend on the potential of the disk electrode [47,140], as can be seen in Fig. 8B. However, the average values of $y_{peroxide}$ of ORR for $Mn_2O_3$, $C_{PAni}/MnO_2$, and $C_{PAni}/Mn_2O_3$ catalysts have been found to be ca. 9.4%, 5.0%, and 5.1%, respectively, determined by considering the disk potential at range from 0.2 to 0.4 V. The corresponding values of $n$ have been evaluated to be 3.81, 3.89, and 3.84 (see Fig. 9). These results are close to those obtained with benchmark Pt/C catalyst. Thus, this supports the results obtained in K-L treatment of voltammograms (discussed above).

The mechanistic of an electrochemical reaction can be revealed by constructing a Tafel plot from the voltammograms measured using the relationship expressed by Eq. (11). One of the important parameters required to construct a Tafel plot is the kinetic current ($i_k$) that can be calculated from the mass transport correction [47,141].

$$i_k = \frac{i \times i_D}{i_D - i} \tag{21}$$

The features of Tafel plot constructed for the ORR at various catalysts are shown in Fig. 10. Basically, all plots possess two slopes. Two slopes, namely, 73 and 114 mV dec$^{-1}$, determined for $C_{PAni}/Mn_2O_3$ hybrid shells are close to 66 and 118 mV dec$^{-1}$, respectively, obtained for Pt/C catalyst. This feature indicates that the first electron reduction of $O_2$ is the rate-determining step at the catalyst surface

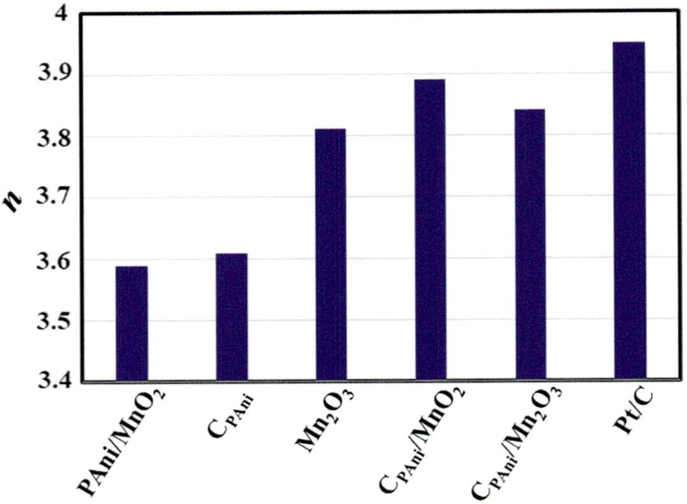


**Fig. 9** Number of electrons involved with ORR at different catalysts [129].

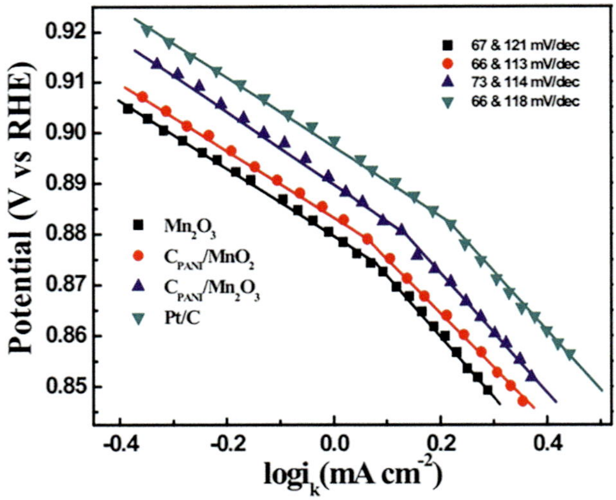


**Fig. 10** Tafel plots of $Mn_2O_3$, $C_{PAni}/MnO_2$, $C_{PAni}/Mn_2O_3$, and Pt/C derived by the mass-transport correction of corresponding LSV data under 1600 rpm.
Reprinted with permission from American Chemical Society S. Cao, N. Han, J. Han, Y. Hu, L. Fan, C. Zhou, R. Guo, Mesoporous hybrid shells of carbonized polyaniline/$Mn_2O_3$ as non-precious efficient oxygen reduction reaction catalyst, ACS Appl. Mater. Interfaces 8 (2016) 6040.

[138]. Thus, the electrocatalytic mechanism of ORR at $C_{PAni}/Mn_2O_3$ hybrid shells might be comprised of the following steps: (i) first, $HO_2^-$ intermediate forms on catalyst surfaces (as shown in Eq. (6)), and (ii) then the redox reaction between Mn species and the introduction of carbon are convinced to favor the charge transfer involved in ORR [23,26,142].

The morphological studies of the catalysts showed that $C_{PAni}/Mn_2O_3$ hybrid shells are mesoporous [129]. The stability of $C_{PAni}/Mn_2O_3$ against the ORR has been investigated to be 91.1%, significantly higher than that of the counterpart Pt/C (79.4%). Moreover, after a long time study of ORR, $C_{PAni}/Mn_2O_3$ electrocatalyst can retain its original mesoporous structure. In fact, the mesoporous $C_{PAni}/Mn_2O_3$ hybrid shells exhibit superior performances in catalyzing ORR than $Mn_xO_y$ and/or carbon-based materials. The specific surface area and average pore diameter of $C_{PAni}/Mn_2O_3$ hybrid have been found to be $185.3\,m^2\,g^{-1}$ and 4.0 nm, respectively. Thus, the higher catalytic activity of $C_{PAni}/Mn_2O_3$ hybrid shells and long-term stability might be due to the following: (i) mesoporous $C_{PAni}/Mn_2O_3$ hybrids enhance the adsorption of $O_2$ and make easier path for transferring electron to adsorbed oxygen during the redox reaction and (ii) mesoporous $Mn_2O_3$ exhibits unique crystal structure in alkaline media. Mn in $Mn_2O_3$ exists as $Mn^{3+}$ state, and the presence of 3*d* metals facilitates the electron conduction (by hopping) and charge transfer (through redox reactions), hence favoring the electrocatalysis. In a Nyquist plot, the smallest semicircle has been found in the middle frequency region by $C_{PAni}/Mn_2O_3$ hybrid shells, followed by $Mn_2O_3$ and $C_{PAni}$, which indicates the reduced electron transfer resistance after the introduction of $C_{PAni}$ [41]. Therefore, due to its robust structure and composition-co-dependent behaviors, the mesoporous $C_{PAni}/Mn_2O_3$ hybrid shells exhibit high electrocatalytic activity toward ORR [129].

## 5.3 Effect of microstates of $MnO_2$ on ORR

It is well known that $Mn_xO_y$, especially $MnO_2$, has different allotropes ($\alpha$-, $\beta$-, $\gamma$-, $\delta$-forms). The dependence of crystallographic microstates on the catalytic activity of a material has been well accepted. Studies of ORR in microbial fuel cells (MFCs) have shown that the activity of $MnO_2$ on the ORR is microstate sensitive and $\beta$-$MnO_2$ exhibits the best effective ORR catalytic activity among $\alpha$, $\beta$, and $\gamma$ forms of $MnO_2$ [143].

The electrocatalytic performance of graphite felt (GF), GF-$\beta$-$MnO_2$–6.0, GF-PAni + $\beta$-$MnO_2$–6.0, and GF-PAni/$\beta$-$MnO_2$–6.0 air cathodes for the ORR has been investigated [126]. The performance of PAni/$\beta$-$MnO_2$ nanocomposites has been investigated by using different loading amounts of 0, 2.2, 4.1, and $6.0\,mg\,cm^{-2}$. Typically, CVs obtained at GF, GF-$\beta$-$MnO_2$–6.0, GF-PAni + $\beta$-$MnO_2$–6.0, and GF-PAni/$\beta$-$MnO_2$–6.0 for ORR are represented in Fig. 11A. Reduction peaks of $O_2$ are found for all the electrodes in the range between −0.1 and − 0.3 V except for GF electrode. The ORR peak is found at −0.17 V on the GF-PAni/$\beta$-$MnO_2$–6.0 electrode, which is relatively more positive than those at GF-PAni + $\beta$-$MnO_2$–6.0 and GF-$\beta$-$MnO_2$–6.0 electrodes. Moreover, a higher *j* is obtained at GF-PAni/$\beta$-$MnO_2$–6.0 electrode than the other materials considered. For instance, GF-PAni/$\beta$-

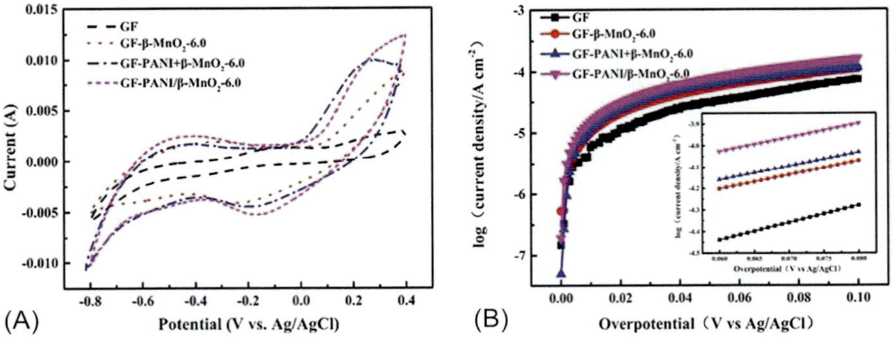


**Fig. 11** (A) CV curves obtained on different catalyst cathodes in air-saturated PBS (pH=7). Scan rate: $50\,mV\,s^{-1}$. (B) Tafel plots. Inset: Linear fit for the Tafel plots.
Reprinted with permissions from Elsevier X. Zhou, Y. Xu, X. Mei, N. Du, R. Jv, Z. Hu, S. Chen, Polyaniline/β-$MnO_2$ nanocomposites as cathode electrocatalyst for oxygen reduction reaction in microbial fuel cells, Chemosphere, 198 (2018) 482.

$MnO_2$–6.0 electrode show 2.7 times higher *j* than that observed at GF electrode. Moreover, among various forms of $MnO_2$ ($\alpha$, $\beta$, $\gamma$, etc.), $\beta$-$MnO_2$ exhibits the best effective ORR catalytic activity as cathode catalyst in MFCs [143].

It has been considered that through the introduction of granular PAni particles onto the external surface of $\beta$-$MnO_2$ nanorods, the catalytic activity of GF-PAni/$\beta$-$MnO_2$–6.0 electrode toward ORR is enhanced dramatically [144]. The higher catalytic performance of PAni/$\beta$-$MnO_2$ nanocomposites toward ORR [145] compared to $\beta$-$MnO_2$ and PAni + $\beta$-$MnO_2$ mixture catalysts may be due to the following causes: (i) PAni itself has the ability to catalyze ORR [40]; (ii) the introduction of PAni granules onto the $\beta$-$MnO_2$ nanorods surface increases the surface area and total pore volume of catalyst, which might facilitate the diffusion, adsorption, and transport of $O_2$ [25]; and (iii) the specific interaction between $\beta$-$MnO_2$ and PAni could enhance the electron delocalization and increase the electrical conductivity that may ultimately facilitate the adsorption and electron transfer processes associated with ORR [125,146].

The other performance tests of a catalyst are the extent of open circuit voltages (OCVs) and power density of the ORR. OCVs of the catalysts studied for different air electrodes have been found in the order of GF-PAni/$\beta$-$MnO_2$–6.0 > GF-PAni + $\beta$-$MnO_2$–6.0 > GF-$\beta$-$MnO_2$–6.0 > GF having the values of 770, 739, 723, and 680 mV, respectively. In addition, GF-PAni/$\beta$-$MnO_2$–6.0 composite electrode shows the highest OCV among the other electrodes used for the ORR with a maximum power density of $248\,mW\,m^{-2}$ (Table 3). It has been observed that both OCV and power density are increased with the increase in loading amount of PAni/$\beta$-$MnO_2$ catalyst. The differences in power output result from cathodes rather than anodes of MFCs [147]. Thus, the differences observed in MFCs performances are mainly due to increasing active sites and accelerating kinetics toward ORR that vary with the variation of the amount of catalyst loaded.

**Table 3** Summary of power density, $R_s$ and $R_{ct}$ values of different cathode materials used in MFCs [126].

| Cathode materials | Powder density ($mW\,m^{-2}$) | $R_s$ ($\Omega$) | $R_{ct}$ ($\Omega$) | Reference |
|---|---|---|---|---|
| GF | 134 | 50.4 | 93.3 | [123] |
| GF-PAni/$\beta$-$MnO_2$–6.0 | 248 | 38.4 | 9.2 | [123] |
| GF-PAni + $\beta$-$MnO_2$–6.0 | 204 | 42.3 | 14.4 | [123] |
| GF-$\beta$-$MnO_2$–6.0 | 183 | 45.5 | 19.8 | [123] |
| GF-PAni/$\beta$-$MnO_2$–4.1 | 191 | 44.2 | 15.6 | [123] |
| GF-PAni/$\beta$-$MnO_2$–2.2 | 152 | 47.2 | 23.9 | [123] |
| CNT-$\beta$-$MnO_2$ | 98.7 | – | – | [143] |
| $MnO_x$/C | 161 | – | – | [144] |
| 0.4 wt% Pt/$MnO_2$ | 165 | – | – | [145] |
| $MnO_2$/CNTs | 210 | – | – | [145] |
| Graphite/$\beta$-$MnO_2$ | 172 | – | – | [146] |

The cathode electrocatalyst should have a high electrical conductivity to ensure a relatively smooth electron(s) transfer [148]. Surprisingly, the resistance of MFC system is found to decrease significantly due to the addition of catalyst for ORR. Generally, both ohmic resistance ($R_s$) and charge transfer resistance ($R_{ct}$) of the GF electrode have been found to be higher than those of the catalysts used (Table 3).

$R_s$ of GF is about 1.1, 1.2, and 1.3 times higher than those of GF-$\beta$-$MnO_2$–6.0, GF-PAni + $\beta$-$MnO_2$–6.0, and GF-PAni/$\beta$-$MnO_2$–6.0 electrodes, respectively. On the other hand, $R_{ct}$ for the ORR at GF-PAni/$\beta$-$MnO_2$–6.0 electrode is much lower than those obtained at GF, GF-$\beta$-$MnO_2$–6.0, and GF-PAni + $\beta$-$MnO_2$–6.0 electrodes (see Table 3). Hence, the lowest $R_s$ and $R_{ct}$ of GF-PAni/$\beta$-$MnO_2$–6.0 electrode could be due to a pronounced synergistic effect between PAni and $\beta$-$MnO_2$ in the composites, resulting in increased conductivity of the electrode and accelerating the electron transfer throughout the electrode process [149]. The favorable adsorption of $O_2$ on the available large number of active sites of PAni/$\beta$-$MnO_2$ catalyst might result in lower values of $R_s$ and $R_{ct}$ and hence enhances the electrocatalytic performance toward ORR.

## 6 Concluding remarks and future prospects

In heterogeneous catalysis, adsorption of the reacting species is a must. The mechanism of ORR involves several intermediate steps, of which some of them are electron transfer associated with adsorption (Eqs. 1–6). Termination of intermediate steps alters four-electron ORR that occurs at Pt and similar metals to a two-electron

ORR producing peroxide species. It has been reported that adsorption orientation of $O_2$ influences the pathway of ORR [150–152]. Two-site (parallel) adsorption results in four-electron reduction to form water, while one-site (end-on) adsorption of $O_2$ leads to a two-electron ORR to form peroxide species (Scheme 4). Depending on the type of metal, crystal structure, and bimetallic site (e.g., alloy), the orientation of adsorption of $O_2$ varies due to different surface energies and affinity of adsorption and hence free energy change associated with the adsorption process [151].

$Mn_xO_y$/PAni composites considered are multifunctional, promising catalyst for ORR, since their surface can be engineered to create sites for desired adsorption orientation of $O_2$ by tuning both dispersed phase and matrix phase. Usually, the compositions of dispersed phase and matrix phase are varied to tune the properties of composites. In case of $Mn_xO_y$/PAni composites, PAni has been mainly studied as matrix phase for many purposes, including ORR catalysis. The study of ORR with $Mn_xO_y$/PAni composite, in which PAni may be used as dispersed phase (small amount of PAni with large amount of $Mn_xO_y$), is yet to be carried out.

The redox forms of PAni are well known (Scheme 5). The surface of composite can be tuned to create suitable sites only by utilizing the redox features of PAni. In this case, knowledge of conductivity of PAni forms would be important to ensure necessary electrical conductivity of the composites. It is worthy to note that only EM form of PAni is conductive. Therefore, one would create mixed PAni surface containing conducting EM and nonconducing LE or PE forms to create PAni surface with desired adsorption sites for $O_2$ adsorption as well as necessary conductivity of the composites. This can be done by the so-called proton doping or dedoping with acid or base treatment of PAni or its composites [112,124].

In addition to various $Mn_xO_y$, the composites formed with different allotropes of $MnO_2$ ($\alpha$-, $\beta$-, $\gamma$-, and $\delta$-forms) that are known to exhibit specific surface properties and catalysis can be easily prepared. Finally, the $Mn_xO_y$/PAni composites can be synthesized chemically by precipitation, thermal, hydrothermal, sol–gel methods, and electrochemical deposition. All of these ideas would open a plenty of opportunity to attempt ORR with various $Mn_xO_y$/PAni composites in designing suitable catalysts for ORR. Most of these are yet to be carried out.

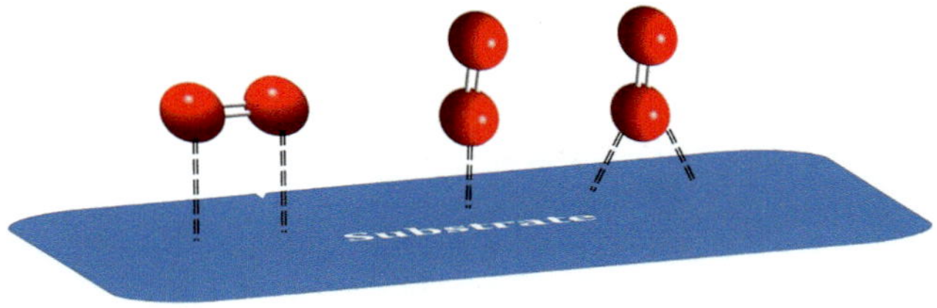


**Scheme 4** Adsorption orientation of $O_2$ on a metal surface. Parallel (two-site) mode, and end-on (one-site) mode and end-one-bimetallic (one-site-bimetallic) mode [150–152].

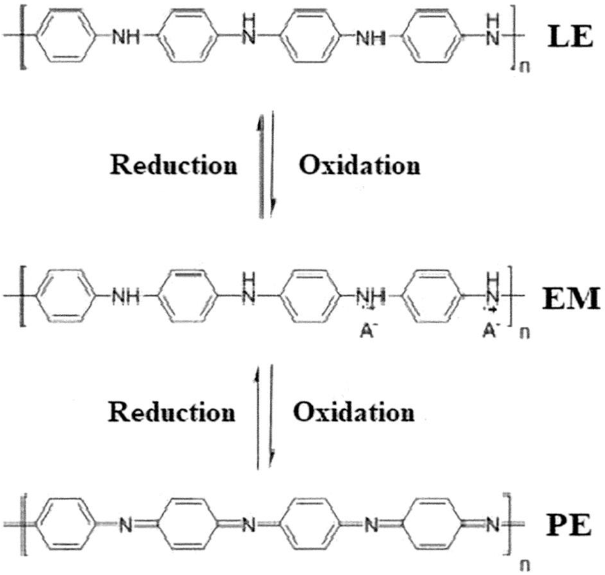


**Scheme 5** Various redox forms of PAni. LE: leucoemeraldine, EM: emeraldine and PE: pernigraniline [112,124].

# References

[1] H. Dong, H. Yu, X. Wang, Catalysis kinetics and porous analysis of rolling activated carbon-PTFE air-cathode in microbial fuel cells, Environ. Sci. Technol. 46 (2012) 13009.

[2] H. Dong, H. Yu, X. Wang, Q. Zhou, J. Feng, A novel structure of scalable air-cathode without Nafion and Pt by rolling activated carbon and PTFE as catalyst layer in microbial fuel cells, Water Res. 46 (2012) 5777.

[3] D.T. Sawyer, G. Chiericato, C.T. Angelis, E.J. Nanni, T. Tsuchiya, Effects of media and electrode materials on the electrochemical reduction of dioxygen, Anal. Chem. 54 (1982) 1720.

[4] D. Zhang, T. Okajima, F. Matsumoto, T. Ohsaka, Electroreduction of dioxygen in 1-n-alkyl-3-methylimidazolium tetrafluoroborate room-temperature ionic liquids, J. Electrochem. Soc. 151 (2004) D31.

[5] D.T. Sawyer, J.S. Valentine, How super is superoxide? Acc. Chem. Res. 14 (1981) 393.

[6] K.A. Striebel, F.R. McLarnon, E.J. Cairns, Oxygen reduction on Pt in aqueous $K_2CO_3$ and KOH, J. Electrochem. Soc. 137 (1990) 3351.

[7] T.I. Farhana, M.Y.A. Mollah, M.A.B.H. Susan, M.M. Islam, Catalytic degradation of an organic dye through electroreduction of dioxygen in aqueous solution, Electochim. Acta 139 (2014) 244.

[8] M.S. Hossain, M.Y.A. Mollah, M.A.B.H. Susan, M.M. Islam, Role of in situ electrogenerated reactive oxygen species towards degradation of organic dye in aqueous solution, Electochim. Acta 344 (2020) 136146.

[9] A.T.Y. Lau, Y. Wang, J.F. Chiu, Reactive oxygen species: current knowledge and applications in cancer research and therapeutic, J. Cell. Biochem. 104 (2008) 657.
[10] Q. Xu, C. He, C. Xiao, X. Chen, Reactive oxygen species (ROS) responsive polymers for biomedical applications, Macromol. Biosci. 16 (2016) 635.
[11] J. Lee, N. Koo, D.B. Min, Reactive oxygen species, aging, and antioxidative nutraceuticals, Compr. Rev. Food Sci. Food Saf. 3 (2004) 21.
[12] M.S. Hossain, A. Sahed, N. Jahan, M.Y.A. Mollah, M.A.B.H. Susan, M.M. Islam, Micelle core as a nest for residence of molecular oxygen–an electrochemical study, J. Electroanal. Chem. 894 (2021) 115361.
[13] M.S. Hossain, A.S.M.H.R. Nixon, M.A.B.H. Susan, M.Y.A. Mollah, M.M. Islam, Electrochemical approach for treatment of textile effluents, in: TRC Book of Papers, 2015, p. 19.
[14] Q. Wen, S. Wang, J. Yan, L. Cong, Z. Pan, Y. Ren, Z. Fan, $MnO_2$–graphene hybrid as an alternative cathodic catalyst to platinum in microbial fuel cells, J. Power Sources 216 (2012) 187.
[15] K.B. Liew, W.R.W. Daud, M. Ghasemi, J.X. Leong, S.S. Lim, M. Ismail, Non-Pt catalyst as oxygen reduction reaction in microbial fuel cells: A review, Int. J. Hydrogen Energy 39 (2014) 4870.
[16] F.C. Anson, C. Shi, B. Steiger, Novel multinuclear catalysts for the electroreduction of dioxygen directly to water, Acc. Chem. Res. 30 (1997) 437.
[17] Y. Feng, T. He, N. Alonso-Vante, In situ free-surfactant synthesis and ORR-electrochemistry of carbon-supported $Co_3S_4$ and $CoSe_2$ nanoparticles, Chem. Mater. 20 (2008) 26.
[18] M. Galib, M.M. Hosen, J.K. Saha, M.M. Islam, S.H. Firoz, M.A. Rahman, Electrode surface modification of graphene-$MnO_2$ supercapacitors using molecular dynamics simulations, J. Mol. Model. 26 (2020) 1.
[19] M.M. Islam, M.Y.A. Mollah, M.A.B.H. Susan, M.M. Islam, Frontier performance of in situ formed $\alpha$-$MnO_2$ dispersed over functionalized multi-walled carbon nanotubes covalently anchored to a graphene oxide nanosheet framework as supercapacitor materials, RSC Adv. 10 (2020) 44884.
[20] W. Xiao, D. Wang, X.W. Lou, Shape-controlled synthesis of $MnO_2$ nanostructures with enhanced electrocatalytic activity for oxygen reduction, J. Phys. Chem. C 114 (2010) 1694.
[21] L. Li, X. Feng, Y. Nie, S. Chen, F. Shi, K. Xiong, W. Ding, X. Qi, J. Hu, Z. Wei, L.J. Wan, Insight into the effect of oxygen vacancy concentration on the catalytic performance of $MnO_2$, ACS Catal. 5 (2015) 4825.
[22] Q. Tang, L. Jiang, J. Liu, S. Wang, G. Sun, Effect of surface manganese valence of manganese oxides on the activity of the oxygen reduction reaction in alkaline media, ACS Catal. 4 (2014) 457.
[23] K. Gong, P. Yu, L. Su, S. Xiong, L. Mao, Polymer-assisted synthesis of manganese dioxide/carbon nanotube nanocomposite with excellent electrocatalytic activity toward reduction of oxygen, J. Phys. Chem. C 111 (2007) 1882.
[24] F. Cheng, J. Shen, W. Ji, Z. Tao, J. Chen, Selective synthesis of manganese oxide nanostructures for electrocatalytic oxygen reduction, ACS Appl. Mater. Interfaces 1 (2009) 460.
[25] F. Cheng, Y. Su, J. Liang, Z. Tao, J. Chen, $MnO_2$-based nanostructures as catalysts for electrochemical oxygen reduction in alkaline media, Chem. Mater. 22 (2010) 898.
[26] (a) L. Mao, D. Zhang, T. Sotomura, K. Nakatsu, N. Koshiba, T. Ohsaka, Mechanistic study of the reduction of oxygen in air electrode with manganese oxides as

electrocatalysts, Electrochim. Acta 48 (2003) 1015. (b) L. Mao, T. Sotomura, K. Nakatsu, N. Koshiba, D. Zhang, T. Ohsaka, Electrochemical characterization of catalytic activities of manganese oxides to oxygen reduction in alkaline aqueous solution, J. Electrochem. Soc. 149 (2002) A504.

[27] Y. Tan, C. Xu, G. Chen, X. Fang, N. Zheng, Q. Xie, Facile synthesis of manganese-oxide-containing mesoporous nitrogen-doped carbon for efficient oxygen reduction, Adv. Funct. Mater. 22 (2012) 4584.

[28] X.M. He, C. Li, F.E. Chen, G.Q. Shi, Polypyrrole microtubule actuators for seizing and transferring microparticles, Adv. Funct. Mater. 17 (2007) 2911.

[29] H.D. Tran, K. Shin, W.G. Hong, J.M. D'Arcy, R.W. Kojima, B.H. Weiller, R.B. Kaner, A template-free route to polypyrrole nanofibers, Macromol. Rapid Commun. 28 (2007) 2289.

[30] X.G. Li, J. Li, Q.K. Meng, M.R. Huang, Interfacial synthesis and widely controllable conductivity of polythiophene microparticles, J. Phys. Chem. B 113 (2009) 9718.

[31] X.G. Li, Q.F. Lü, M.R. Huang, Self-stabilized nanoparticles of intrinsically conducting copolymers from 5-sulfonic-2-anisidine, Small 4 (2008) 1201.

[32] X.G. Li, A. Li, M.R. Huang, Facile high-yield synthesis of polyaniline nanosticks with intrinsic stability and electrical conductivity, Chem. A Eur. J. 14 (2008) 10309.

[33] H.S. Roy, M.M. Islam, M.Y.A. Mollah, M.A.B.H. Susan, Polyaniline-$MnO_2$ composites prepared in-situ during oxidative polymerization of aniline for supercapacitor applications, Mater. Today: Proc. 29 (2020) 1013.

[34] M.G. Rabbani, M.A.B.H. Susan, M.Y.A. Mollah, M.M. Islam, In situ electrodeposition of conducting polymer/metal oxide composites on iron electrode for energy storage applications, Mater. Today: Proc. 29 (2020) 1192.

[35] H.S. Roy, M.Y.A. Mollah, M.M. Islam, M.A.B.H. Susan, Poly (vinyl alcohol)–$MnO_2$ nanocomposite films as UV-shielding materials, Polym. Bull. 75 (2018) 5629.

[36] Y. Wang, Q. Wen, Y. Chen, L. Qi, A novel polyaniline interlayer manganese dioxide composite anode for high-performance microbial fuel cell, J. Taiwan Inst. Chem. Eng. 75 (2017) 112.

[37] N.N. Nova, M.M. Islam, S. Ahmed, M.M. Rahman, M.Y.A. Mollah, M.A.B.H. Susan, Polyaniline-based composite of non-covalently dispersed multiwalled carbon nanotubes for supercapacitor electrode, Int. J. Sci. Technol. Res. 7 (2017) 11.

[38] H.S. Roy, M.M. Islam, M.Y.A. Mollah, M.A.B.H. Susan, Polyaniline-NiO nanocomposites as dielectric materials, Mater. Today: Proc. 5 (2018) 15267.

[39] H.S. Roy, M.M. Islam, M.Y.A. Mollah, M.A.B.H. Susan, Polyaniline-NiO nanocomposites as tunable conducting materials, Mater. Today: Proc. 15 (2019) 380.

[40] V.G. Khomenko, V.Z. Barsukov, A.S. Katashinskii, The catalytic activity of conducting polymers toward oxygen reduction, Electrochim. Acta 50 (2005) 1675.

[41] (a) R. Silva, D. Voiry, M. Chhowalla, T. Asefa, Efficient metal-free electrocatalysts for oxygen reduction: polyaniline-derived N-and O-doped mesoporous carbons, J. Am. Chem. Soc. 135 (2013) 7823. (b) W. Ding, L. Li, K. Xiong, Y. Wang, W. Li, Y. Nie, S. Chen, X. Qi, Z. Wei, Shape fixing via salt recrystallization: a morphology-controlled approach to convert nanostructured polymer to carbon nanomaterial as a highly active catalyst for oxygen reduction reaction, J. Am. Chem. Soc. 137 (2015) 5414.

[42] R. Wu, S. Chen, Y. Zhang, Y. Wang, W. Ding, L. Li, X. Qi, X. Shen, Z. Wei, Template-free synthesis of hollow nitrogen-doped carbon as efficient electrocatalysts for oxygen reduction reaction, J. Power Sources 274 (2015) 645.

[43] M. Higuchi, I. Ikeda, T. Hirao, A novel synthetic metal catalytic system, J. Org. Chem. 62 (1997) 1072.

[44] T. Kurbus, A.M. Le Marechal, D.B. Voncina, Comparison of $H_2O_2$/UV, $H_2O_2/O_3$ and $H_2O_2/Fe^{2+}$ processes for the decolorisation of vinylsulphone reactive dyes, Dyes Pigments 58 (2003) 245.
[45] X.R. Xu, H.B. Li, W.H. Wang, J.D. Gu, Degradation of dyes in aqueous solutions by the Fenton process, Chemosphere 57 (2004) 595.
[46] E. Yeager, Dioxygen electrocatalysis: mechanism in relation to catalyst structure, J. Mol. Catal. 38 (1986) 5.
[47] A.J. Bard, L.R. Faulkner, Electrochemical methods: fundamentals and applications, second ed., Wiley, New York, 2001.
[48] L. Johnson, C. Li, Z. Liu, Y. Chen, S.A. Freunberger, P.C. Ashok, B.B. Praveen, K. Dholakia, J.M. Tarascon, P.G. Bruce, The role of $LiO_2$ solubility in $O_2$ reduction in aprotic solvents and its consequences for Li-$O_2$ batteries, Nat. Chem. 6 (2014) 1091.
[49] A. Damjanovic, M.A. Genshaw, J.M. Bockris, The mechanism of oxygen reduction at platinum in alkaline solutions with special reference to $H_2O_2$, J. Electrochem. Soc. 114 (1967) 1107.
[50] A. Damjanovic, D.B. Sepa, M.V. Vojnovic, New evidence supports the proposed mechanism for $O_2$ reduction at oxide free platinum electrodes, Electrochim. Acta 24 (1979) 887.
[51] H.S. Wroblowa, G. Razumney, Electroreduction of oxygen: a new mechanistic criterion, J. Electroanal. Chem. 69 (1976) 195.
[52] J.A. Keith, G. Jerkiewicz, T. Jacob, Theoretical investigations of the oxygen reduction reaction on Pt (111), ChemPhysChem 11 (2010) 2779.
[53] S. Guo, S. Zhang, S. Sun, Tuning nanoparticle catalysis for the oxygen reduction reaction, Angew. Chem. Int. Ed. 52 (2013) 8526.
[54] J. Zhang, Z. Zhao, Z. Xia, L. Dai, A metal-free bifunctional electrocatalyst for oxygen reduction and oxygen evolution reactions, Nat. Nanotechnol. 10 (2015) 444.
[55] L. Yu, X. Pan, X. Cao, P. Hu, X. Bao, Oxygen reduction reaction mechanism on nitrogen-doped graphene: a density functional theory study, J. Catal. 282 (2011) 183.
[56] L. Zhang, H. Li, J. Zhang, Kinetics of oxygen reduction reaction on three different Pt surfaces of Pt/C catalyst analyzed by rotating ring-disk electrode in acidic solution, J. Power Sources 255 (2014) 242.
[57] C. Song, J. Zhang, Electrocatalytic oxygen reduction reaction, in: PEM Fuel Cell Electrocatalysts and Catalyst Layers, Springer, London, 2008, p. 89.
[58] N. Wakabayashi, M. Takeichi, M. Itagaki, H. Uchida, M. Watanabe, Temperature-dependence of oxygen reduction activity at a platinum electrode in an acidic electrolyte solution investigated with a channel flow double electrode, J. Electroanal. Chem. 574 (2005) 339.
[59] A. Damjanovic, Temperature dependence of symmetry factors and the significance of experimental activation energies, J. Electroanal. Chem. 355 (1993) 57.
[60] C. Song, Y. Tang, J.L. Zhang, J. Zhang, H. Wang, J. Shen, S. McDermid, J. Li, P. Kozak, PEM fuel cell reaction kinetics in the temperature range of 23–120°C, Electrochim. Acta 52 (2007) 2552.
[61] J. Zhang, Y. Tang, C. Song, Z. Xia, H. Li, H. Wang, J. Zhang, PEM fuel cell relative humidity (RH) and its effect on performance at high temperatures, Electrochim. Acta 53 (2008) 5315.
[62] B. Lim, J.W. Kim, S.J. Hwang, S.J. Yoo, E. Cho, T.H. Lim, S.K. Kim, Fabrication and characterization of high-activity Pt/C electrocatalysts for oxygen reduction, Bull. Kor. Chem. Soc. 31 (2010) 1577.

[63] A. Kongkanand, S. Kuwabata, G. Girishkumar, P. Kamat, Single-wall carbon nanotubes supported platinum nanoparticles with improved electrocatalytic activity for oxygen reduction reaction, Langmuir 22 (2006) 2392.
[64] E.H. Yu, U. Krewer, K. Scott, Principles and materials aspects of direct alkaline alcohol fuel cells, Energies 3 (2010) 1499.
[65] W. Yuan, K. Scott, H. Cheng, Fabrication and evaluation of Pt-Fe alloys as methanol tolerant cathode materials for direct methanol fuel cells, J. Power Sources 163 (2006) 323.
[66] B. Wang, Recent development of non-platinum catalysts for oxygen reduction reaction, J. Power Sources 152 (2005) 1.
[67] R.M. Modibedi, M.K. Mathe, R.G. Motsoeneng, L.E. Khotseng, K.I. Ozoemena, E.K. Louw, Electro-deposition of Pd on Carbon paper and Ni foam via surface limited redox-replacement reaction for oxygen reduction reaction, Electrochim. Acta 128 (2014) 406.
[68] H. Meng, D. Zeng, F. Xie, Recent development of Pd-based electrocatalysts for proton exchange membrane fuel cells, Catalysts 5 (2015) 1221.
[69] X.M. Wang, M.E. Wang, D.D. Zhou, Y.Y. Xia, Structural design and facile synthesis of a highly efficient catalyst for formic acid electrooxidation, Phys. Chem. Chem. Phys. 13 (2011) 13594.
[70] D. Thompsett, Catalysts for the Proton Exchange Membrane Fuel cell, Fuel Cell Technology Handbook, Taylor & Francis Group: CRC Press LLC, 2003.
[71] H.A. Hansen, V. Viswanathan, J.K. Nørskov, Unifying kinetic and thermodynamic analysis of $2e^-$ and $4e^-$ reduction of oxygen on metal surfaces, J. Phys. Chem. C 118 (2014) 6706.
[72] I.S. Flyagina, K.J. Hughes, M. Pourkashanian, D.B. Ingham, DFT study of the oxygen reduction reaction on iron, cobalt and manganese macrocycle active sites, Int. J. Hydrogen Energy 39 (2014) 21538.
[73] J.K. Nørskov, J. Rossmeisl, A. Logadottir, L.R.K.J. Lindqvist, J.R. Kitchin, T. Bligaard, H. Jonsson, Origin of the overpotential for oxygen reduction at a fuel-cell cathode, J. Phys. Chem. B 108 (2004) 17886.
[74] V.R. Stamenkovic, B. Fowler, B.S. Mun, G. Wang, P.N. Ross, C.A. Lucas, N.M. Marković, Improved oxygen reduction activity on $Pt_3Ni$ (111) via increased surface site availability, Science 315 (2007) 493.
[75] I. Morcos, E. Yeager, Kinetic studies of the oxygen-peroxide couple on pyrolytic graphite, Electrochim. Acta 15 (1970) 953.
[76] E. Antolini, Palladium in fuel cell catalysis, Energ. Environ. Sci. 2 (2009) 915.
[77] W. Li, Q. Xin, Y. Yan, Nanostructured Pt-Fe/C cathode catalysts for direct methanol fuel cell: The effect of catalyst composition, Int. J. Hydrogen Energy 35 (2010) 2530.
[78] Z. Zhang, M. Li, Z. Wu, W. Li, Ultra-thin PtFe-nanowires as durable electrocatalysts for fuel cells, Nanotechnology 22 (2010), 015602.
[79] H. Yang, W. Vogel, C. Lamy, N. Alonso-Vante, Structure and electrocatalytic activity of carbon-supported Pt-Ni alloy nanoparticles toward the oxygen reduction reaction, J. Phys. Chem. B 108 (2004) 11024.
[80] B. Lim, M. Jiang, P.H. Camargo, E.C. Cho, J. Tao, X. Lu, Y. Zhu, Y. Xia, Pd-Pt bimetallic nanodendrites with high activity for oxygen reduction, Science 324 (2009) 1302.
[81] U.A. Paulus, A. Wokaun, G.G. Scherer, T.J. Schmidt, V. Stamenkovic, N.M. Markovic, P.N. Ross, Oxygen reduction on high surface area Pt-based alloy catalysts in comparison to well defined smooth bulk alloy electrodes, Electrochim. Acta 47 (2002) 3787.

[82] J. Zhang, Y. Mo, M.B. Vukmirovic, R. Klie, K. Sasaki, R.R. Adzic, Platinum monolayer electrocatalysts for $O_2$ reduction: Pt monolayer on Pd (111) and on carbon-supported Pd nanoparticles, J. Phys. Chem. B 108 (2004) 10955.

[83] S.H. Chang, W.N. Su, M.H. Yeh, C.J. Pan, K.L. Yu, D.G. Liu, J.F. Lee, B.J. Hwang, Structural and electronic effects of carbon-supported $Pt_xPd_{1-x}$ nanoparticles on the electrocatalytic activity of the oxygen reduction reaction and on methanol tolerance, Chem. A Eur. J. 16 (2010) 11064.

[84] J.B. Xu, T.S. Zhao, Y.S. Li, W.W. Yang, Synthesis and characterization of the Au-modified Pd cathode catalyst for alkaline direct ethanol fuel cells, Int. J. Hydrogen Energy 35 (2010) 9693.

[85] J.B. Xu, T.S. Zhao, W.W. Yang, S.Y. Shen, Effect of surface composition of Pt-Au alloy cathode catalyst on the performance of direct methanol fuel cells, Int. J. Hydrogen Energy 35 (2010) 8699.

[86] Y. Kim, J.W. Hong, Y.W. Lee, M. Kim, D. Kim, W.S. Yun, S.W. Han, Synthesis of AuPt heteronanostructures with enhanced electrocatalytic activity toward oxygen reduction, Angew. Chem. Int. Ed. 49 (2010) 10197.

[87] J.N. Tiwari, R.N. Tiwari, G. Singh, K.S. Kim, Recent progress in the development of anode and cathode catalysts for direct methanol fuel cells, Nano Energy 2 (2013) 553.

[88] G. Fu, Z. Liu, Y. Chen, J. Lin, Y. Tang, T. Lu, Synthesis and electrocatalytic activity of Au@Pd core-shell nanothorns for the oxygen reduction reaction, Nano Res. 7 (2014) 1205.

[89] W. Li, X. Zhao, A. Manthiram, Room-temperature synthesis of Pd/C cathode catalysts with superior performance for direct methanol fuel cells, J. Mater. Chem. A 2 (2014) 3468.

[90] J.X. Wang, H. Inada, L. Wu, Y. Zhu, Y. Choi, P. Liu, W.P. Zhou, R.R. Adzic, Oxygen reduction on well-defined core-shell nanocatalysts: particle size, facet, and Pt shell thickness effects, J. Am. Chem. Soc. 131 (2009) 17298.

[91] J. Kim, T. Momma, T. Osaka, Cell performance of Pd–Sn catalyst in passive direct methanol alkaline fuel cell using anion exchange membrane, J. Power Sources 189 (2009) 999.

[92] A. Stassi, C. D'urso, V. Baglio, A. Di Blasi, V. Antonucci, A.S. Arico, A.C. Luna, A. Bonesi, W.E. Triaca, Electrocatalytic behaviour for oxygen reduction reaction of small nanostructured crystalline bimetallic Pt–M supported catalysts, J. Appl. Electrochem. 36 (2006) 1143.

[93] (a) Y. Liu, X. Yue, K. Li, J. Qiao, D.P. Wilkinson, J. Zhang, PEM fuel cell electrocatalysts based on transition metal macrocyclic compounds, Coord. Chem. Rev. 315 (2016) 153. (b) N. Alonso-Vante, 5 structure and reactivity of transition metal chalcogenides toward the molecular oxygen reduction reaction, in: Interfacial Phenomena in Electrocatalysis, Springer, New York, 2011, p. 255.

[94] C. Lamy, A. Lima, V. LeRhun, F. Delime, C. Coutanceau, J.M. Léger, Recent advances in the development of direct alcohol fuel cells (DAFC), J. Power Sources 105 (2002) 283.

[95] L. Zhang, J. Zhang, D.P. Wilkinson, H. Wang, Progress in preparation of non-noble electrocatalysts for PEM fuel cell reactions, J. Power Sources 156 (2006) 171.

[96] H. Jahnke, M. Schönborn, G. Zimmermann, Organic dyestuffs as catalysts for fuel cells, physical and chemical applications of dyestuffs, Top. Curr. Chem. 61 (1976) 133.

[97] A. Elzing, A.M.T.P. van der Putten, W. Visscher, E. Barendrecht, The cathodic reduction of oxygen at cobalt phthalocyanine: influence of electrode preparation on electrocatalysis, J. Electroanal. Chem. 200 (1986) 313.

[98] Y. Gochi-Ponce, G. Alonso-Nunez, N. Alonso-Vante, Synthesis and electrochemical characterization of a novel platinum chalcogenide electrocatalyst with an enhanced

tolerance to methanol in the oxygen reduction reaction, Electrochem. Commun. 8 (2006) 1487.
[99] R.G. González-Huerta, J.A. Chávez-Carvayar, O. Solorza-Feria, Electrocatalysis of oxygen reduction on carbon supported Ru-based catalysts in a polymer electrolyte fuel cell, J. Power Sources 153 (2006) 11.
[100] K. Matsuki, H. Kamada, Oxygen reduction electrocatalysis on some manganese oxides, Electrochim. Acta 31 (1986) 13–18.
[101] C.C. Chang, T.C. Wen, H.J. Tien, Kinetics of oxygen reduction at oxide-derived Pd electrodes in alkaline solution, Electrochim. Acta 42 (1997) 557.
[102] E.R. Vago, E.J. Calvo, Oxygen electro-reduction on iron oxide electrodes: III. Heterogeneous catalytic $H_2O_2$ decomposition, J. Electroanal. Chem. 388 (1995) 161.
[103] J.H. Kim, A. Ishihara, S. Mitsushima, N. Kamiya, K.I. Ota, Catalytic activity of titanium oxide for oxygen reduction reaction as a non-platinum catalyst for PEFC, Electrochim. Acta 52 (2007) 2492.
[104] S. Sharma, B.G. Pollet, Support materials for PEMFC and DMFC electrocatalysts-a review, J. Power Sources 208 (2012) 96.
[105] Y.J. Wang, D.P. Wilkinson, J. Zhang, Noncarbon support materials for polymer electrolyte membrane fuel cell electrocatalysts, Chem. Rev. 111 (2011) 7625.
[106] C. Montero-Ocampo, J.V. Garcia, E.A. Estrada, Comparison of $TiO_2$ and $TiO_2$-CNT as cathode catalyst supports for ORR, Int. J. Electrochem. Sci 8 (2013) 12780.
[107] B. Ruiz-Camacho, M.A. Valenzuela, N. Alonso-Vante, The effect of metal oxide-carbon support on electrocatalysts for fuel cell reactions, World J. Eng. 973 (2002).
[108] R.N. Singh, B. Lal, M. Malviya, Electrocatalytic activity of electrodeposited composite films of polypyrrole and $CoFe_2O_4$ nanoparticles towards oxygen reduction reaction, Electrochim. Acta 49 (2004) 4605.
[109] Y. Dessie, S. Tadesse, R. Eswaramoorthy, B. Abebe, Recent developments in manganese oxide based nanomaterials with oxygen reduction reaction functionalities for energy conversion and storage applications: a review, J. Sci.: Adv. Mater. Devices 4 (2019) 353.
[110] F.D. Speck, P.G. Santori, F. Jaouen, S. Cherevko, Mechanisms of manganese oxide electrocatalysts degradation during oxygen reduction and oxygen evolution reactions, J. Phys. Chem. C 123 (2019) 25267.
[111] T. Ohsaka, L. Mao, K. Arihara, T. Sotomura, Bifunctional catalytic activity of manganese oxide toward $O_2$ reduction: novel insight into the mechanism of alkaline air electrode, Electrochem. Commun. 6 (2004) 273.
[112] T. Ohsaka, A.N. Chowdhury, M.A. Rahman, M.M. Islam, Trends in Polyaniline Research, Nova Science Publishers, Inc., USA, 2013.
[113] J. Huang, S. Virji, B.H. Weiller, R.B. Kaner, Polyaniline nanofibers: facile synthesis and chemical sensors, J. Am. Chem. Soc. 125 (2003) 314.
[114] X. Zhang, W.J. Goux, S.K. Manohar, Synthesis of polyaniline nanofibers by "nanofiber seeding", J. Am. Chem. Soc. 126 (2004) 4502.
[115] C.G. Wu, T. Bein, Conducting polyaniline filaments in a mesoporous channel host, Science 264 (1994) 1757.
[116] C. Wang, Z. Wang, M. Li, H. Li, Well-aligned polyaniline nano-fibril array membrane and its field emission property, Chem. Phys. Lett. 341 (2001) 431.
[117] J.C. Michaelson, A.J. McEvoy, Interfacial polymerization of aniline, J. Chem. Soc. Chem. Commun. 1 (1994) 79.
[118] Y. Yang, M. Wan, Chiral nanotubes of polyaniline synthesized by a template-free method, J. Mater. Chem. 12 (2002) 897.

[119] Z. Niu, J. Liu, L.A. Lee, M.A. Bruckman, D. Zhao, G. Koley, Q. Wang, Biological templated synthesis of water-soluble conductive polymeric nanowires, Nano Lett. 7 (2007) 3729.
[120] D. Li, J. Huang, R.B. Kaner, Polyaniline nanofibers: a unique polymer nanostructure for versatile applications, Acc. Chem. Res. 42 (2009) 135.
[121] J. Huang, R.B. Kaner, A general chemical route to polyaniline nanofibers, J. Am. Chem. Soc. 126 (2004) 851.
[122] J. Huang, R.B. Kaner, Nanofiber formation in the chemical polymerization of aniline: a mechanistic study, Angew. Chem. 116 (2004) 5941.
[123] S.K. Mondal, K.R. Prasad, N. Munichandraiah, Analysis of electrochemical impedance of polyaniline films prepared by galvanostatic, potentiostatic and potentiodynamic methods, Synth. Met. 148 (2005) 275.
[124] S. Sultana, M.S. Hossain, M.A.B.H. Susan, M.M. Islam, Electrosorption of heavy metal from aqueous solution on polyaniline modified graphite electrode, Bangladesh J. Sci. Res. 31–33 (2020) 1.
[125] S. Mu, Y. Yang, Spectral characteristics of polyaniline nanostructures synthesized by using cyclic voltammetry at different scan rates, J. Phys. Chem. B 112 (2008) 11558.
[126] X. Zhou, Y. Xu, X. Mei, N. Du, R. Jv, Z. Hu, S. Chen, Polyaniline/$\beta$-$MnO_2$ nanocomposites as cathode electrocatalyst for oxygen reduction reaction in microbial fuel cells, Chemosphere 198 (2018) 482.
[127] K.S. Kim, S.J. Park, Synthesis and high electrochemical performance of polyaniline/$MnO_2$-coated multi-walled carbon nanotube-based hybrid electrodes, J. Solid State Chem. 16 (2012) 2751.
[128] K.Q. Ding, Cyclic voltammetrically prepared $MnO_2$-polyaniline composite and its electrocatalysis for oxygen reduction reaction (ORR), J. Chin. Chem. Soc. 56 (2009) 891.
[129] S. Cao, N. Han, J. Han, Y. Hu, L. Fan, C. Zhou, R. Guo, Mesoporous hybrid shells of carbonized polyaniline/$Mn_2O_3$ as non-precious efficient oxygen reduction reaction catalyst, ACS Appl. Mater. Interfaces 8 (2016) 6040.
[130] A.H. Gemeay, I.A. Mansour, R.G. El-Sharkawy, A.B. Zaki, Preparation and characterization of polyaniline/manganese dioxide composites via oxidative polymerization: Effect of acids, Eur. Polym. J. 41 (2005) 2575.
[131] J. Zhang, D. Shu, T. Zhang, H. Chen, H. Zhao, Y. Wang, Z. Sun, S. Tang, X. Fang, X. Cao, Capacitive properties of PAni/$MnO_2$ synthesized via simultaneous-oxidation route, J. Alloys Compd. 532 (2012) 1.
[132] W. Ni, D. Wang, Z. Huang, J. Zhao, G. Cui, Fabrication of nanocomposite electrode with $MnO_2$ nanoparticles distributed in polyaniline for electrochemical capacitors, Mater. Chem. Phys. 124 (2010) 1151.
[133] C. Yuan, L. Su, B. Gao, X. Zhang, Enhanced electrochemical stability and charge storage of $MnO_2$/carbon nanotubes composite modified by polyaniline coating layer in acidic electrolytes, Electrochim. Acta 53 (2008) 7039.
[134] A.S. Prasad, Green synthesis of nanocrystalline manganese (II, III) oxide, Mater. Sci. Semicond. Process. 71 (2017) 342.
[135] C.R. Martin, Membrane-based synthesis of nanomaterials, Chem. Mater. 8 (1996) 1739.
[136] A. Katafias, J. Fenska, Oxidation of phenothiazine dyes by manganese (III) in sulfuric acid solution, Transit. Met. Chem. 36 (2011) 801.
[137] J. Yang, J.J. Xu, Nanoporous amorphous manganese oxide as electrocatalyst for oxygen reduction in alkaline solutions, Electrochem. Commun. 5 (2003) 306.
[138] I. Roche, E. Chaînet, M. Chatenet, J. Vondrák, Carbon-supported manganese oxide nanoparticles as electrocatalysts for the oxygen reduction reaction (ORR) in alkaline

medium: physical characterizations and ORR mechanism, J. Phys. Chem. C 111 (2007) 1434.
[139] Y. Liang, H. Wang, J. Zhou, Y. Li, J. Wang, T. Regier, H. Dai, Covalent hybrid of spinel manganese–cobalt oxide and graphene as advanced oxygen reduction electrocatalysts, J. Am. Chem. Soc. 134 (2012) 3517.
[140] S. Anantharaj, S. Pitchaimuthu, S. Noda, A review on recent developments in electrochemical hydrogen peroxide synthesis with a critical assessment of perspectives and strategies, Adv. Colloid Interface Sci. 287 (2020) 102331.
[141] X. Han, T. Zhang, J. Du, F. Cheng, J. Chen, Porous calcium–manganese oxide microspheres for electrocatalytic oxygen reduction with high activity, Chem. Sci. 4 (2013) 368.
[142] (a) F.H. Lima, M.L. Calegaro, E.A. Ticianelli, Investigations of the catalytic properties of manganese oxides for the oxygen reduction reaction in alkaline media, J. Electroanal. Chem. 590 (2006) 152. (b) F.H. Lima, M.L. Calegaro, E.A. Ticianelli, Electrocatalytic activity of manganese oxides prepared by thermal decomposition for oxygen reduction, Electrochim. Acta 52 (2007) 3732.
[143] L. Zhang, C. Liu, L. Zhuang, W. Li, S. Zhou, J. Zhang, Manganese dioxide as an alternative cathodic catalyst to platinum in microbial fuel cells, Biosens. Bioelectron. 24 (2009) 2825.
[144] F. Zhao, R.C. Slade, J.R. Varcoe, Techniques for the study and development of microbial fuel cells: an electrochemical perspective, Chem. Soc. Rev. 38 (2009) 1926.
[145] Y. Zhang, G. Mo, X. Li, J. Ye, Iron tetrasulfophthalocyanine functionalized graphene as a platinum-free cathodic catalyst for efficient oxygen reduction in microbial fuel cells, J. Power Sources 197 (2012) 93.
[146] D. Zhou, B. Che, X. Lu, Rapid one-pot electrodeposition of polyaniline/manganese dioxide hybrids: a facile approach to stable high-performance anodic electrochromic materials, J. Mater. Chem. C 5 (2017) 1758.
[147] Q. Huang, P. Zhou, H. Yang, L. Zhu, H. Wu, CoO nanosheets *in situ* grown on nitrogen-doped activated carbon as an effective cathodic electrocatalyst for oxygen reduction reaction in microbial fuel cells, Electrochim. Acta 232 (2017) 339.
[148] L. Hao, J. Yu, X. Xu, L. Yang, Z. Xing, Y. Dai, Y. Sun, J. Zou, Nitrogen-doped $MoS_2$/carbon as highly oxygen-permeable and stable catalysts for oxygen reduction reaction in microbial fuel cells, J. Power Sources 339 (2017) 68.
[149] J.G. Wang, Y. Yang, Z.H. Huang, F. Kang, Rational synthesis of $MnO_2$/conducting polypyrrole@carbon nanofiber triaxial nano-cables for high-performance supercapacitors, J. Mater. Chem. 22 (2012) 16943.
[150] M.S. El-Deab, T. Ohsaka, Manganese oxide nanoparticles electrodeposited on platinum are superior to platinum for oxygen reduction, Angew. Chem. Int. Ed. 45 (2006) 5963.
[151] W.E. Mustain, J. Prakash, Kinetics and mechanism for the oxygen reduction reaction on polycrystalline cobalt–palladium electrocatalysts in acid media, J. Power Sources 170 (2007) 28.
[152] R.A. Sidik, A.B. Anderson, Density functional theory study of $O_2$ electroreduction when bonded to a Pt dual site, J. Electroanal. Chem. 528 (2002) 69.

# Traditional and recently advanced synthetic routes of the metal oxide materials

3

*Atiya Fatima, Mazhar Ul Islam, and Md. Wasi Ahmad*
Department of Chemical Engineering, College of Engineering, Dhofar University, Salalah, Sultanate of Oman

## 1 Introduction

Nanostructures have been part of nature ever since created inside the earth in stunning environmental conditions. Nanomaterials of clays, oxides/hydroxides of *Al*, Fe, Si etc., came into existence long before they were synthesized in the laboratories [1].

Nanomineral materials are the nanoparticles (NPs) of minerals/natural materials with particle size ranging between 100 nm and 0.2 nm. The particle size and shape of the nanomaterial play a crucial role in controlling their properties. Along with this, nanoparticle agglomeration and surface reactivity are also a major factor to be considered while synthesis of new materials [2,3].

In the last few decades, advanced nanomaterials have drawn considerable attention from researchers due to their excellent characteristics and their usage in modern technology. In this context, nanotechnology emerged as an efficient technique to develop nanostructures with special applications in the field of electronics, cosmetics, defense, and biomedical applications [4–6]. Metal-based nanoparticles have emerged as the most widely utilized nanoparticles due to their physiochemical, photothermal, and plasmonic properties, as well as high reactivity and high stability, make metallic nanoparticles an attractive candidate for biomedical applications [7]. These nanostructures display unique properties which are in contrast to their bulk counterparts because of the quantization effect [8]. Nanocrystals of metals such as hematite, magnetite, zincite etc., have found their applications in biotechnology, microelectronics, photovoltaic, and biomedical fields. Nanocrystals of iron oxide or magnetite are used as hypothermic material [9]. Iron pyrite ($FeS_2$) nanosphere was synthesized as a promising solar absorber material for their potential use in photovoltaic applications [10]. Gadolinium oxide nanoparticles have been extensively explored as contrast agents and in theranostic applications [11–13]. ZnO nanoparticles are used in photocatalysis, piezoelectrics, nanocosmetics, etc. [14]. Cobalt nanoparticles are utilized in drug delivery applications and magnetic resonance imaging [15]. Europium hydroxide nanoparticles have been explored for their use in fluorescent imaging [16]. Gold nanoparticles are used in electrochemical, catalysis, and biomedical applications [17–19]. Shah et al. reviewed the role of magnetic core-shell nanocomposites in

**Renewable Polymers and Polymer-Metal Oxide Composites. https://doi.org/10.1016/B978-0-323-85155-8.00004-2**

the removal of organic and inorganic wastes from water [20]. These metal oxide nanoparticles possess a unique feature in terms of their tunability of chemical, physical, and mechanical features as a function of size, which provides a platform to synthesize specific materials with desired properties. These can be made as one-dimensional nanomaterials (thin films), two-dimensional (nanotubes, nanofibers, nanowires etc.), and three-dimensional nanomaterials (nanoparticles, molecular electronics, nanostructured materials, quantum dots, nanoporous materials, etc. [21].

## 2 Metal oxide materials

Metal oxides being crystalline solids are composed of a metal cation and an oxide anion. They have found their applications in several fields of chemical, physical, and material science areas [22,23]. They can form variable geometrical structures possessing an electronic structure, owing to which they exhibit metallic, insulating, and semiconducting characteristics. Technological applications include their use as sensors, fuel cells, piezoelectric devices, etc. Oxide nanoparticles have limited size and high density, which imparts their unique chemical and physical characteristics. A decrease in the size of NPs generates strain and adjoining structural perturbations due to an increase in the number of interface atoms [24]. Alterations in particle size influence the structural characteristics as well as chemical, magnetic, electronic, and conducting properties of the NPs [25]. Structural parameters include the lattice symmetry and cell parameters which are well defined in bulk oxides. However, with the decrease in particle size, stress and surface free energy increase. An increase in stress causes changes in thermodynamic stability which thereby induces structural transformations and modification of cell parameters [26–28] as observed in the nanoparticles of $Al_2O_3$ [29], $Fe_2O_3$, [30], $ZrO_2$, [31], $MoO_3$, $Y_2O_3$ [32], $CeO_2$ [33]. Similarly, with the decrease in size, an increase in surface energy causes the disappearance of nanoparticle, which occurs in the extreme case due to the interaction of NPs with their surrounding environment [29]. Low surface energy in nanoparticles leads to mechanical or structural stability, bulk materials having low stability form stable nanostructures as seen in $TiO_2$, $Al_2O_3$, or $MoO_x$ oxides displaying such phenomenon [29,34]. Nanoparticle interaction with supporting substrate also induces structural perturbations [35].

Electronic properties or conductance is strongly related to the size of the nanostructures, which produce confinement effects or quantum size arising due to the presence of atom-like electronic states as seen in cases of $SnO_2$, $WO_3$, and $In_2O_3$ in gas sensing applications [36]. Another important factor is that the long-range effects of the Madelung field are either absent or limited in nanostructures, which are otherwise present in the electronic properties of a bulk oxide surface [37].

Structural and electronic properties affect the physical and chemical parameters of the solid. Bulk states of many oxides usually have low reactivity and wideband gaps [38], which changes with a decrease in the size of the oxides [39]. A redistribution of charge occurs when going from large structures to smaller aggregates which are significantly larger in case of covalent solids as compared to ionic ones [40]. The degree of ionicity or covalency is strongly dependent on the size, where the increase in iconicity is parallel to a decrease in the size [26]. In addition to this, uncoordinated atoms like corners/edges/O vacancies enhance the chemical reactivity of oxide nanomaterials [41].

It can be deduced that all the listed characteristics and properties of metal oxide nanostructures are strongly dependent on the size, structure, and shape of the nanoparticles. Fabrication of metal oxide nanoparticles (MONPs) plays a crucial role in the structural characteristics of the synthesized NPs. The MONPs properties are controlled by the nucleation, growth, and aging mechanisms. Therefore, controlled synthesis of MONPs is essential for the successful application of MONPs and their performance in functional devices.

# 3 Historical background

Nanoparticles have been used by humans since prehistory, although without the knowledge of their nature. Artisans from Roman designed Lycurgus cup of dichroic glass in the 4th century CE, which was later analyzed to be made of nanoparticles of silver-gold (Ag:Au) alloy, containing 7:3 ratio of Ag:Au in addition with approximately 10% of copper (Cu) dispersed in the glass [42,43]. In the 9th century, AD, lusterware pottery of Mesopotamia contained silver (Ag) and copper (Cu) nanoparticles dispersed in the glassy glaze [44]. During the period of 9th–17th centuries, the Islamic world and later in Europe used glowing ceramic glazes containing Ag, Cu, or other nanoparticles [45]. Ottomans techniques employed cementite nanowires and carbon nanotubes (CNTs) to provide strength and resilience to produce Damascus saber during the 13th–18th century CE [46]. In the 16th century, Italians utilized nanoparticles for making renaissance pottery [47]. The colors and materials were put in use for centuries without knowing the actual cause of these characteristics. Later in the 19th century, Michael Faraday provided a scientific insight into the optical properties of nanometer-scale metals. He demonstrated that the nanostructured gold emitted different colored solutions under certain lighting conditions [48]. The concept of nanotechnology was first introduced by an American physicist Richard Feynman nanotechnology in 1959 and is considered the father of modern nanotechnology [49]. In 1974, a Japanese scientist, Norio Taniguchi, defined the term nanotechnology for the first time [50]. In the early 1970s and 1980s, the term ultrafine particles was first used by scientists during thorough fundamental studies of nanoparticles [51]. However, later in the 1990s, the term nanoparticle came into existence and became quite common. In 1981, physicists Heinrich Rohrer and Gerd Binnig invented Scanning Tunneling Microscope which enabled the visualization of atom for the first time pioneering discoveries in the field of nanotechnology [52,53]. In 1991, Sumio Iijima discovered CNT which exhibited extraordinary strength, electrical, and thermal conductivity [54].

# 4 Prospective advancement in the synthesis

There are several synthetic strategies and techniques which are involved in nanotechnology. Generally, there are two types of approach: (i) bottom-up approach and (ii) top-down approach (Fig. 1) [55].

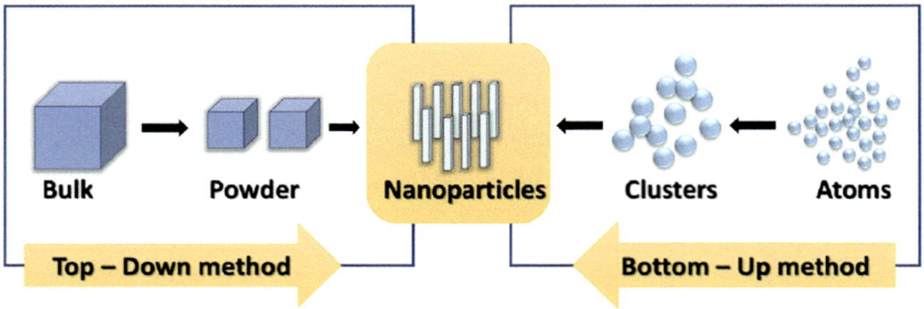


**Fig. 1** Bottom-up and top-down approaches in nanomaterials synthesis processing.
*Source*: Nunes, Daniela, Ana Pimentel, Lidia Santos, P. Barquinha, L. Pereira, E. Fortunato, and R. Martins, Synthesis, design, and morphology of metal oxide nanostructures, Metal Oxide Nanostruct. (2019) 21–57.

The bottom-up approach or self-assembly approach includes nanofabrication utilizing chemical or physical forces to assemble material components at the nanoscale to atomic level further into stable structures. Physical methods involve chemical vapor deposition (CVD), pulse and physical vapor deposition (PVD), pulsed laser deposition (PLD), atomic layer deposition (ALD), ion implantation (II), spray pyrolysis (SP), and molecular beam epitaxy (MBE).

The chemical process for NP synthesis from an aqueous solution includes precipitation and microemulsion routes, hydrothermal, sol–gel, sonochemical, photochemical processes, electrochemical, and microwaved processes [56]. Recently, the green synthesis of NPs has gained considerable attention to avoid the labor-intensive and environmentally detrimental effects of physical and chemical processes [57]. These methods utilize microorganisms such as bacteria, fungi, algae, and plants to synthesize NPs. These methods are eco-friendly, cost-effective, and could be combined with another process such as microwave irradiation methods [58]. A top-down approach rather involves slicing of bulk material or macroscopic structures into nano-sized particles, simultaneously controlling the external factors involved in the NPs formation. These processes involve bulk nano-machining, fragmentation such as nano-milling, spark erosion, etc., and lithography such as optical, UV, X-ray, electron beam, laser, etc. These synthetic strategies are summarized in Fig. 2.

The synthesis of nanoparticulated oxides employs various methods which can modify the characteristics and properties of the obtained MONPs. The functionality of MONPs will depend on their composition, morphology, crystallographic structure, and surface stoichiometry and geometry. The development of systematic studies for nano-oxide synthesis is currently a challenge and can be essentially classified into two streams based on the liquid–solid [59] and gas–solid type of transformations [60].

The liquid–solid transformations follow the bottom-up approach and are most widely used to control the morphological characteristics. Numerous techniques are utilized in this process. The co-precipitation technique involves dissolving a salt precursor (chloride, perchlorate, nitrate, etc.) and base in water or any other solvent to precipitate out the oxo-hydroxide form. This method lacks in achieving control of size

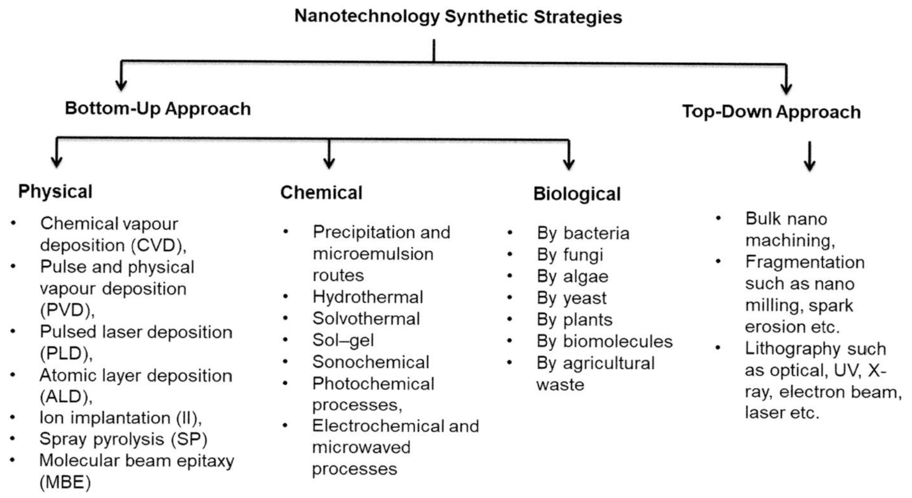


**Fig. 2** Synthetic strategies used in nanotechnology.
Modified from Chavali, Murthy S., and Maria P. Nikolova, Metal oxide nanoparticles and their applications in nanotechnology, SN Appl. Sci. 1 (6) (2019) 1–30.

and chemical homogeneity in the case of mixed-metal oxides. However, solid morphological characteristics could be optimized by the use of surfactants, high-gravity reactive precipitation, and sonochemical methods [61]. Microemulsion involves a ternary mixture comprising water, oil, and a surfactant where metal precursors on water precipitate out as oxo-hydroxides within the aqueous droplets forming monodispersed materials [62]. Solvothermal methods are based on the thermal decomposition of metal complexes added with a suitable surfactant agent to achieve control over the particle size and agglomeration of synthesized MONPs. Hydrothermal synthesis is basically a solution-based reaction where the formation of MONPs can take place in a wide temperature range. Sol–gel processing operates by preparing metal oxides through hydrolysis of precursors in an alcoholic solution forming corresponding oxo-hydroxide. Evaporation of water causes condensation of molecules, forming a network of metal hydroxide where hydroxyl-species undergoes polymerization by condensation to form a dense porous gel. Drying followed by calcination leads to the formation of ultrafine porous oxides [63]. Gas–solid transformation methods are limited to CVD and PLD. The CVD processes utilized for the formation of MONPs include the classical (thermally activated/pyrolytic), plasma-assisted, metal–organic, and photo CVD methodologies [64]. Multiple-pulsed laser deposition involves heating of the target sample to 4000 K leading to instantaneous evaporation, decomposition, and ionization followed by the mixing of desired atoms. Generated gaseous entities absorb radiation from the pulses and acquire kinetic energy to be deposited in a substrate which on heating allows crystalline growth [65].

Among all the techniques, hydrothermal contributes to around 6%, but it has been found to be advantageous in synthesizing the toughest and complex materials with desired physiochemical properties. Although nanomaterials can be prepared by various methods, recently, the hydrothermal method has gained popularity due to advantages related to phase purity and control over the diffusion kinetics, environmental friendly as it involves a closed system.

### 4.1 A brief history of the hydrothermal/solvothermal technique

The term hydrothermal was initially used by a British Geologist (1792–1871), describing the action of water at high temperature and pressure conditions causing changes in the earth crust forming several rocks and minerals [66]. The emergence of hydrothermal technique was first utilized to synthesize submicrometer and nanometer-sized quartz particles dating back to the mid of nineteenth century [67]. However, the lack of sufficient knowledge in hydrothermal solution chemistry and the absence of characterization techniques of nanoscale products restricted the research and application of hydrothermal technique from the 1840s to the early 1990s [8]. Later in the 1980s, the emergence of high-resolution microscopes, followed by insurgency in nanoscale materials in the 1990s, revived the hydrothermal technology [68]. Progress in the knowledge of chemical and physical parameters of hydrothermal systems led to the development of the solvothermal process [69]. These processes gained considerable attention in the twenty-first century in synthesizing nanoparticles with size control, crystallinity, morphology, and crystal phase [70].

## 5 Novel solution routes

Novel hydrothermal solution routes encompass conventional hydrothermal techniques using aqueous solvent, solvothermal techniques using nonaqueous solvent, and supercritical hydrothermal processes using either aqueous or nonaqueous solvents in supercritical conditions.

This book chapter deals with the novel solution routes used in hydrothermal, solvothermal, and supercritical methods to synthesize metal oxides nanomaterials (Table 1). Even though there are various synthetic routes available, hydrothermal and solvothermal methods are preferred solution-based methods for carrying out MONPs synthesis. The novel solution routes used in hydrothermal, solvothermal, and supercritical methods have proved advantageous over conventional technologies in terms of high purity, crystal symmetry, limited particle size distributions, and fast reaction. Advancement in the knowledge of physical chemistry greatly improved the hydrothermal process. Similarly, the concept of green chemistry was introduced in the solvothermal and supercritical process by usage of solvents like organic and organometallic complexes in nanomaterials processing. Utilization of capping agents significantly affected the surface properties from being hydrophilic to hydrophobic and vice versa. The supplementation of external energies like microwave irradiation, electrical, magnetic, and mechanochemical into novel solution routes, particularly in the hydrothermal process, paved a new perspective in nanomaterials processing [8].

**Table 1** Different terminologies under novel solution routes [71].

| Conventional hydrothermal | Solvothermal | Supercritical hydrothermal |
|---|---|---|
| Aqueous solvent<br>Reaction conditions are greater than atmospheric temperature and pressure<br>Suitable for high-quality bulk, fine nanocrystals | Nonaqueous solvents<br>Reaction conditions include low to high temperature<br>Suitable for good quality bulk, fine nanocrystals | Aqueous and nonaqueous Solvents<br>Reaction conditions include critical to supercritical conditions<br>Suitable for fine and nanocrystals |

## 5.1 Soft solution processing

Soft solution processing is a very promising alternative and is a blanket term used for hydrothermal processing and, to some extent, solvothermal and supercritical methods. The soft solution process is a nature-inspired process covering a wide variety of synthetic methods used to synthesize advanced solid materials at a low temperature in the presence of a solution. It encompasses all sets of material processes operable under ambient/near ambient/greater than the ambient conditions to produce the materials [72]. The term soft refers to simple instrumentation techniques and process, low energy consumption, and less environmental charge. Different activation methods such as hydrothermal reactions at low temperature can be combined with the solution treatment of the starting material to synthesize desired materials. Mimicking the earth's natural process, water seems to be the appropriate solvent for soft solution processing. Chemical reactions such as corrosion or precipitation are inevitable in soft solution processing. In addition, an advantage of combining the ability of the solution treatment with various activation methods can extend the applied systems for fabricating materials. Soft solution processing lies in between bioprocessing and artificial processing. Although bioprocessing is helpful in producing materials with low energy requirements, it is only limited to certain species, shape, and size. On the other hand, although artificial processing can have an advantage in producing almost all materials consuming high energy.

### 5.1.1 Hydrothermal

Hydrothermal processes can be described as heterogeneous reactions carried out in closed systems at temperatures exceeding the boiling point of mineralizers at 1 atm pressure involving water as the main mineralizer. Mostly these reactions, along with solvothermal methods, proceed in a sealed reactor called an autoclave (Fig. 3). Autoclaves are pressure vessel, manufactured from metal or steel, a strong alloy which can withstand the high temperature and pressure conditions. Generally, Teflon (PTFE, polytetrafluoroethylene) is used as a liner to provide a chemically inert environment and also to protect the autoclaves from corrosion. Both hydrothermal and

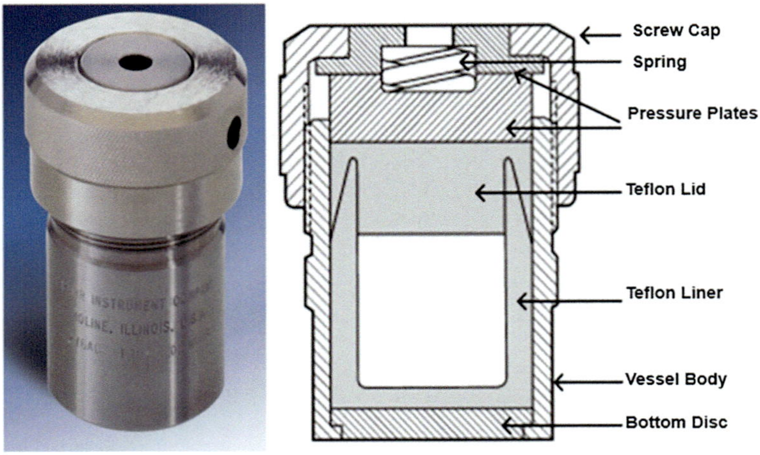


**Fig. 3** Autoclave instrument.
Parr Instrument Company.

solvothermal processes operate in similar reaction conditions, with the only difference of nonaqueous mineralizers in the solvothermal process. These processes operate at relatively low temperature and advantageous in terms of short processing times, fast reaction kinetics, high yield, high crystallinity, phase purity, narrow particle-size distributions, along with being cost-effective, and environmentally benign. Moreover, these processes facilitate multicomponent reactions, phase transformation reactions, heat treatment reactions, crystal growth, dehydration reactions, and decomposition reactions. There are few disadvantages associated with hydrothermal and solvothermal processes, which include expensive Teflon liners and stainless-steel autoclaves, safety issues during the reaction, and the inability to study in-situ reactions due to the closed systems.

Soft solution processing hydrothermal chemistry needs a certain amount of understanding in order to carry out the process under soft and environmentally friendly conditions. It is quite important to study the solvent behavior with respect to temperature and pressure in dealing with aspects like pH variation, density, viscosity, coefficient of expansion, structural variations at critical, subcritical, and supercritical conditions. It is possible to achieve desired particle size and morphologies of MONPs by controlling these variables. Several thermodynamic modeling studies have been established to predict the solubility of species existing in hydrothermal/solvothermal systems [73–75]. Intelligent modeling of the reactions undergoing in hydrothermal process prior to the actual experiments has greatly benefitted in controlling the shape and size of MONPs [76]. These models are based on thermodynamic principles usually performed with commercial software, precisely predicting the required experimental conditions. Hydrothermal methods along with solvothermal techniques synthesize materials directly from solution in two steps, i.e., crystal nucleation followed by subsequent growth. It is already known that the solubility of species in the solution strongly correlates with the solvent density, viscosity, pH, etc. Supersaturation of solution (solubility of the solute becomes greater than its limit) causes nucleation precipitating clusters of crystals [77]. The solubility of different ionic species and their distribution in solution are also crucial factor in the hydrothermal process. These parameters can be predicted by calculating the equilibrium constant ($K$) for each reaction, charge balance, and mass balance. Several thermodynamic models can estimate the species solubility based on the dissociation reaction in the system. Solvent and its role during the hydrothermal crystallization ultimately affect dissolution behavior and solubility (Table 2) [66].

Recent years witnessed the advent of hydrothermal technology reviving this technique by incorporation of various other techniques. Hydrothermal with electrochemical reactions under reduced pressures lesser than 2 MPa and temperatures at 100–200°C produced pure crystalline thin films of $BaTiO_3$, $SrTiO_3$, $LiNbO_3$, $LiNiO_2$, and $LiCoO_2$ [78–80]. Another study reported the synthesis of heteroepitaxial films of lead zirconate titanate (PZT) and $BaTiO_3$ [81]. Supermagnetic ferrite/metal composite thin films were synthesized by directly fabricated thin films of $MFe_2O_4$/Fe (M = Zn, Mg) having controlled grain size and shape on iron (Fe) substrate [82]. Hydrothermal in combination with electrodeposition was used to grow ZnO nanostructures on stainless steel substrates [83]. Other studies reported the synthesis of alumina nanoparticles synthesized from glucose/water-based precursor materials [84], and synthesis of nanostructured NiS nanowhiskers, $Ni_3S_2$ thin films, and $NiS_2$ single crystals in ethylenediamine [85]. Inactivation of endotoxins by following soft hydrothermal process (130°C for 60 min/140°C for 30 min) in the presence of a high-steam saturation ratio or with a flow system was reported by Miyamoto and coworkers. It was also found that endotoxins could be easily removed from water by using soft hydrothermal processing in similar reaction conditions without ultrafiltration, an anion exchanger, or nonselective adsorption with a hydrophobic adsorbent [86]. In another study, the hydrothermal-galvanic couple method was utilized to synthesize barium strontium titanate thin films (BST thin films), having approximately 92% of dielectric tunability [87]. Estevenon and coworkers obtained pure $HfSiO_4$ under soft hydrothermal conditions demonstrating that the increase in temperature and lengthy duration of the hydrothermal reaction elevated the crystallization

**Table 2** Solvent action in hydrothermal crystallization process [66].

| Classification | Action | Application |
|---|---|---|
| Transfer medium | Heat transfer pressure kinetic energy | Erosion, machining, abrasion, hip forming, etc. |
| Adsorbate | Adsorbed or desorbed at the surface | Surface diffusion, catalyst, dispersion crystallization, sintering, ionic exchange, etc. |
| Solvent | Dissolution | Synthesis, extraction, growth, modification, degradation, etching, corrosion, etc. |
| Reactant | Formation of materials quality changing | Formation/decomposition of hydrates, hydroxides, and oxides, corrosion |

state of the $HfSiO_4$ samples [88]. In a recent study, $MoS_2$/PEDOT:PSS hydrogel was synthesized hydrothermally with excellent electrochemical performance [89].

Ramesh and Shivanna [90] synthesized $MoO_3$/ZnO heterostructures through a superficial hydrothermal route. The synthesized $MoO_3$/ZnO was used for the photocatalytic destruction of Eosin Blue (EB) and for the removal of heavy metals Lead (Pb), Chromium (Cr), Zinc (Zn), Iron (Fe), and Copper (Cu) under sunlight. These composites also displayed excellent antimicrobial activity against *Bacillus subtilis* and *with Staphylococcus aureus* and were found to be biocompatible Human embryonic kidney (HEK-293) [90].

### 5.1.2 Solvothermal

The solvothermal technique is a synthetic methodology for nanoparticles synthesis in the closed system requiring high temperature and pressure conditions. It induces a chemical reaction or causes decomposition of the precursors forming the desired material directly from the solution. The choice of solvents in the solvothermal process varies from ethanol, ammonia, hydrochloric acid, hydrofluoric acid, etc. In the case of water as a solvent, the terminology changes to the hydrothermal process. In the solvothermal process, precursors are mixed in a solvent and sealed into an autoclave which is heated at temperatures exceeding the solvent boiling point. Solvothermal reactions can operate at maximum temperatures of 250°C. Pressure can vary depending upon the autoclave filling reaching up to several hundred bars even at low temperatures. Solvothermal reactions in the autoclave can be carried out at smaller volume, from the synthesis of few milliliters to thousands of liters for industrial-scale production. Autoclave being a closed system increases the temperature and pressure of the solution, which leads to the crystallization of the dissolved material. Thus, it is feasible to acquire nanoparticles with high purity and homogeneity by properly selecting the precursor composition and reaction conditions. This method does not require high-temperature conditions. The thermodynamic parameters required in the solvothermal technique are similar to as in the hydrothermal process and have already been discussed in the hydrothermal process.

The temperature and pressure conditions in these methods are suitable for the dissolution of the chemical reagents leading to the production of the MONPs by crystallization, offering a one-step reaction route to complex products. Utilization of solvents provides high diffusivity and increasing mobility of the dissolved ions, which allows efficient mixing of the reagents. The reaction mechanisms of the MONP synthesis are naturally highly system-dependent. Important characteristics, including the density, viscosity, thermal conductivity, ionic product, dielectric constant, and heat capacity, can be exploited in the solvothermal method by increasing the temperature and pressure of the solvent used.

Synthesis of various MONPs has been reported using solvothermal techniques. $Fe_3O_4$ nanoparticles with controlled size (15 to 190 nm) were synthesized by adjusting experimental conditions in the solvothermal method in the presence of sodium dodecyl sulfate and polyethylene glycol acting as protective reagents [91]. Similarly, $MFe_2O_4$ (M = Mn, Co and Ni) nanoparticles (diameters ranging from 5 to 10 nm) were solvothermally synthesized using alcohol (aromatic alcohol or hexanol) as solvent and ligand. It was found that the solvent played an important role in the particle size, with the smallest diameters obtained by the use of hexanol. High crystallinity of $MFe_2O_4$ samples resulting from the solvothermal method resulted in good magnetization values [92]. In a modified approach, $Fe_3O_4$ magnetic NPs were synthesized using a microwave-solvothermal technique in shorter time duration than the traditional solvothermal or hydrothermal process. Ethylene glycol was used as a solvent acting as microwave mediate and reductant; trisodium citrate performed as assistant reductant and electrostatic stabilizer; ammonium acetate ($NH_4Ac$) acted as the nucleating agent. In this process, microwave irradiation created a condition for consistent seeding inside the precursor solution, which accelerated the formation of $Fe_3O_4$ nanocrystals. The synthesized MONPs showed a greater saturation magnetization as compared to the conventionally synthesized sample via the solvothermal method establishing the fact that the former possessed a smaller particle size than the latter [93]. Magnetic nanoparticles containing monodisperse $CuFe_2O_4$ nanoparticles with enhanced heating efficiency as compared to $Fe_3O_4$ NPs were synthesized in triethylene glycol using the solvothermal method [94]. Graphitic carbon nitride (g-$C_3N_4$) and photoanodes made of ZnO modified $TiO_2$ nanotube (TNT) arrays were used to synthesize dye-sensitized solar cells (DSSCs) through solvothermal method using ethylene glycol as solvent. The ZnO + g-$C_3N_4$ modified TNTs photoanode showed a considerable increase in short circuit current (ISC) and open-circuit voltage (VOC) of DSSCs resulting in a 135% increase in the efficiency as compared to the pure TNT arrays photoanode [95]. Cobalt oxide-based NPs were synthesized via the solvothermal method by using Cobalt II acetylacetonate and cobalt III acetylacetonates in $H_2O_2$, acting as a reducing agent. Synthesized $Co_3O_4$ NPs exhibited high electrochemical performance with high specific capacitance [96]. To developed rechargeable Mg-ion batteries, a solvothermal synthesis of $MgMn_2O_4$ spinel NPs was conducted by Yokozaki and coworkers using alcohol solvothermal conditions. The composite exhibited a discharge capacity of 60 mAh $g^{-1}$ while maintaining 80% of capacity retention even after the 10th cycle ([97].

ZnO, *Al*-doped ZnO, *Al*-doped ZnO/CNTs, and *Al*-doped ZnO/graphene composites were synthesize using ethylene glycol as a solvent and reducing agent via a single-step solvothermal route. These composites showed that *Al*-doped ZnO/graphene composite possessed the highest photocatalytic activity (100% within 60 min of simulated sunlight irradiation) as compared with the pure ZnO, *Al*-doped ZnO and *Al*-doped ZnO/CNTs [98].

A facile one-pot solvothermal approach was utilized to synthesize $Cu_2O$, $Cu_2O/TiO_2$ (CT), and $Cu_2O$/ZnO (CZ) heterojunctions photocatalysts for the photodegradation of Congo red (CR) dye. CZ (1.5–1.5) were found to completely degrade 30 ppm CR within 10 min [99].

Newman et al. synthesized nitrogen-doped carbon dots with an average particle size of 2 nm palm kernel shells for sensing and biosensing applications. Ethylenediamine and L-phenylalanine doped CDs were synthesized using hydrothermal and solvothermal approaches via one-pot synthesis techniques. The synthesized N-CDs exhibited excellent photoluminescence, quantum yield (QY) of 13.7% and

8.6% for ethylenediamine (EDA) doped N-CDs (CDs-EDA) and L-phenylalanine (L-Ph) doped N-CDs (CDs-LPh), respectively, with an excitation/emission wavelength of 360 nm/450 nm [100].

### 5.1.3 Supercritical hydrothermal

The supercritical hydrothermal technique is the follow-up step of the conventional hydrothermal process. The hydrothermal method deals with the milder routes (soft solution processing), while the supercritical hydrothermal process operates at a temperature near or above the critical point. The supercritical hydrothermal technique has gained popularity since it can be used to process a variety of materials. Water is the most favored solvent in this method for the synthesis of high melting inorganic compounds and hybrid organic–inorganic NPs. There are various techniques branching out from the supercritical hydrothermal methodology such as supercritical fluid process, supercritical fluids drying, the rapid expansion of supercritical solutions, etc. [1].

Hydrothermal synthesis is normally performed in a batch autoclave where the final product consists of compounds formed during the reaction. A flow-type rapid heating reactor was developed by Adschiri et al. to eliminate these effects (Fig. 3). In this process, the metal salt solution pumped by a high-pressure liquid pump is mixed with supercritical water coming through a high-temperature furnace. In this set-up, the feed material is rapidly heated to the supercritical temperature leading to hydrothermal synthesis in the supercritical water [101] (Fig. 4).

The properties of water change drastically, going from ambient to supercritical conditions. It changes its electrochemical properties, such as decreases in dipole

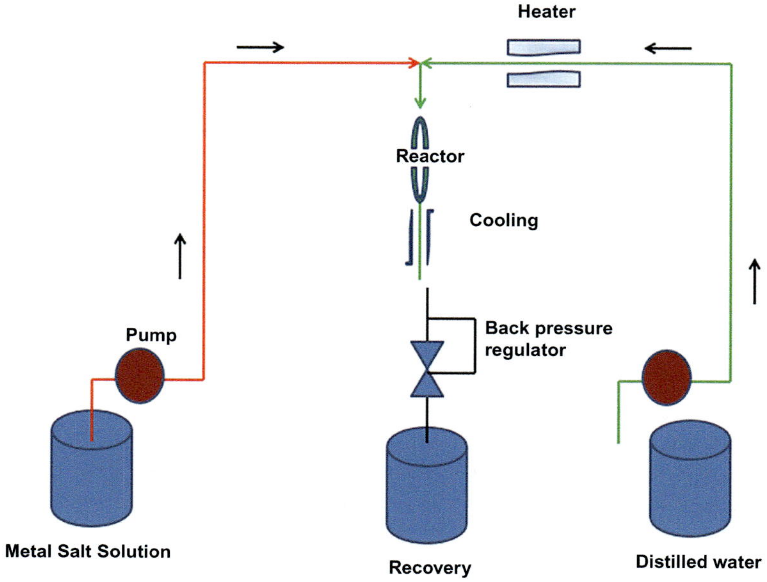


**Fig. 4** Flow chart for supercritical hydrothermal synthesis procedure [101].

moment, decrease in pH value by three units, and increase in water reactivity around critical point [102]. Therefore, the kinetics and equilibrium in water change significantly, displaying unusual behavior, i.e., formation of homogeneous phase between gas and water/organic matter and water [103,104]. Supercritical hydrothermal techniques utilize these unique features for the hydrothermal synthesis of NPs. The conventional hydrothermal synthesis of MONPs is based on the equilibrium kinetics between hydroxides and oxides. At a critical point of water, the reaction kinetics and equilibrium and solubility of metal oxides undergo immense change. These conditions pave way for the supercritical hydrothermal process. The reaction rate is increased two folds above the critical point with a simultaneous decrease in metal oxide water solubility. These changes lead to supersaturation at the mixing point giving rise to a high nucleation rate forming NPs. Particle morphology also is also affected by the changes in water properties around critical point [101]. Rational designing of the supercritical process is based on the information about the kinetics of nanoparticles formation and mixing behavior of the feed material and supercritical water. Mass transfer effects also control the formation and growth of NPs. The resulting particle size and distribution are also dependent on the effects of mixing on the reaction rate [105]. Aoki et al., in their studies, employed *Da* (Damkohler Number, the ratio of reaction rate against mixing rate) to the relationship between particle size and *Re* (Reynolds numberis the ratio of inertial forces to viscous forces inside a flowing fluid and particle size), where high *Da* values gave kinetics control resulting in smaller particle formation [106].

Synthesis of $CoFe_2O_4$ NPs following batch supercritical hydrothermal synthesis method yieldednanoparticles having a diameter of approximately 5 nm. The product characteristic was found to be dependent on $Co^{2+}$ to $Fe^{3+}$ mole ratio (r) and coexisting cations, temperature, and pH [107]. $CeO_2$ nanoparticlesmodified with hydrophilic polymer were synthesized using the supercritical hydrothermal method. $Ce(OH)_4$ was used as a precursor and was treated with supercriticalwater in a batch-type reactor using polyvinyl alcohol or polyacrylic acid as surface modifiers. The reaction yielded nanoparticles with cuboctahedral morphology having a particle size of about 20 nm. Surface modification of NPs by the polymers appeared to control the bandgap of the nanoparticles, which could, in turn, could possibly modify the optical and electronic properties of the MONPs [108]. Another study reported the synthesis of magnesium ferrite ($MgFe_2O_4$) NPs by hydrothermal synthesis in supercritical water. Varying ratios of $Mg(OH)_2$ and $Fe(OH)_3$ were taken to synthesize $MgFe_2O_4$ nanoparticles with different size, among which the particle size of about 20 nm exhibited superparamagnetic behavior [109]. In a similar approach, ZnO particles having high crystallinity were synthesized using zinc nitrate aqueous solution [110]. Later, highly crystalline ZnO nanorods were also reported by following the supercritical hydrothermal synthesis [111]. Kawasaki and coworkers reported the synthesis of highly crystalline, size-controlled anatase titanium dioxide ($TiO_2$) NPs via supercritical hydrothermal reaction. Titanium sulfate ($Ti(SO_4)_2$) solution was used as a precursor, and KOH was utilized to change the solubility of $TiO_2$ and pH and control the particle size. Variation in KOH concentration controlled the average particle size in the range of 13–30 nm while maintaining a coefficient of variation of 0.5 [112].

In a recent study, the reaction kinetics of hydrothermal synthesis for the reaction of nickel nitrate to nickel oxide was observed in a wide range of temperatures and pressures conditions around the critical point of water. It was found that changing the concentration of hydroxyl ion around the critical point strongly affected the reaction kinetics. Also, water properties change greatly around critical point, even with a slight change in temperature and pressure. Intrinsic kinetics was observed to have strong solvent effects owing to a decrease in dielectric constant, causing an increase in intrinsic kinetics above critical temperature or reduced pressure [113].

Highly Cr-substituted $CeO_2$ (Cr-$CeO_2$) NPs were synthesized using a supercritical hydrothermal method involving a subcritical region. It was observed that with the increase in Cr-substitution, the Cr-$CeO_2$ morphology shifted from octahedral geometry to a cluster of small NPs having high lattice strain causing an increase in the oxygen storage capacity (OSC) of Cr-$CeO_2$ NPs. Results indicated that the OSC of Cr-$CeO^2$ affects the catalytic potential of Cr-$CeO_2$ [114].

In a recent study, supercritical $CO_2$ (sc$CO_2$) was used to synthesize a cobalt monoxide/graphene (CoO/G) nanocomposite via the sc$CO_2$-assisted solvothermal technique followed by thermal treatment. It was analyzed that the CoO/G composites possess higher specific capacity (961 mAh $g^{-1}$ at 100 mA $g^{-1}$) and had excellent cyclic stability and rate capability (617 mAh $g^{-1}$ after 500 cycles at 1000 mA $g^{-1}$) when applied as an anode of lithium-ion batteries [115].

François et al. synthesized Y-doped barium zirconate, an electrolyte material having 50 nm diameter with a narrow size distribution for use in protonic Ceramic Fuel cell through the continuous hydrothermal process in supercritical conditions (410°C/ 30.0 MPa) with NaOH reactants and nitrate precursors. Major perovskite phase was obtained with few $BaCO_3$ and YO(OH) impurities, which on thermal treatment (1000°C, 1 h) homogenized the composition forming the compound Ba1.01Zr0.85Y0.15 O3-δ [116].

# 6 Conclusions and future recommendations

Novel solution routes of MONPs synthesis such as hydrothermal, solvothermal, and supercritical hydrothermal techniques have proved to be advantages in terms of their efficacies, including their adaptability for a broad range of materials such as metal oxides, nitrides, etc. These methods have been widely accepted as a classical approach toward the synthesis of various inorganic and metal–organic nanomaterials, including oxides. A rational design of thermodynamic modeling studies facilitates smart engineering of nanocrystals with desired shape, size, distribution, and properties. The selection of an appropriate surface modifier and pH condition of the solution is crucial in the synthesis of nanocrystalline materials. However, it is quite challenging to fully investigate the mechanisms involved in the synthesis of nanocrystals due to the difficulty of sampling in monitoring the whole process. Clearly, new technologies and devices for monitoring the process are required to fully understand hydrothermal crystallization. Hydrothermal and solvothermal processes take place in the subcritical regime while supercritical hydrothermal syntheses are performed in the supercritical regime, i.e., at higher temperatures and pressures, enabling the use of a wide range of compounds as reductants facilitating the synthesis of MONPs with various structures. The shapes of MONPs obtained by these processes are rather limited, thus offering a promising spot for developing new methodologies for the designing of novel nanostructures. The combination of these processes with other techniques such as microwave irradiation offers an appealing research avenue that could deliver important advantages such as reduced reaction times, fast, and controlled heating, etc. Direct synthesis of MONPs with other materials forming hybrid NPs could pave the way toward new developments in the synthesis of advanced MONPs with various applications. The applications of MONPs synthesized by these methods require more applications in the fields of photothermal therapy, photoacoustic tomography, dark-field microscopy measurements, etc. The emerging research in the field of hydrothermal synthesis suggests that these techniques will receive greater attention in the future.

# Acknowledgments

This research was supported through the Block Funding Research Program funded by The Research Council (TRC) of Oman (BFP/RGP/HSS/20/274).

# References

[1] K. Byrappa, S. Ohara, T. Adschiri, Nanoparticles synthesis using supercritical fluid technology–towards biomedical applications, Adv. Drug Deliv. Rev. 60 (3) (2008) 299–327.

[2] R. Asmatulu, Nanotechnology Safety, Newnes, 2013.

[3] R. Asmatulu, P. Nguyen, E. Asmatulu, Nanotechnology safety in the automotive industry, in: Nanotechnology Safety, Elsevier, 2013, pp. 57–72.

[4] M.W. Ahmad, B. Dey, G. Sarkhel, D.S. Bag, A. Choudhury, Exfoliated graphene reinforced polybenzimidazole nanocomposites with high dielectric permittivity at low percolation threshold, J. Mol. Struct. 1177 (2019) 491–498, https://doi.org/10.1016/j.molstruc.2018.10.009.

[5] B. Dey, M.W. Ahmad, A. ALMezeni, G. Sarkhel, D.S. Bag, A. Choudhury, Enhancing electrical, mechanical, and thermal properties of polybenzimidazole by 3D carbon nanotube@graphene oxide hybrid, Compos. Commun. 17 (2020) 87–96, https://doi.org/10.1016/j.coco.2019.11.012.

[6] M.L. Trudeau, J.Y. Ying, Nanocrystalline materials in catalysis and electrocatalysis: structure tailoring and surface reactivity, Nanostruct. Mater. 7 (1–2) (1996) 245–258.

[7] M.N. Sardoiwala, B. Kaundal, S.R. Choudhury, Development of engineered nanoparticles expediting diagnostic and therapeutic applications across blood–brain barrier, in: Handbook of Nanomaterials for Industrial Applications, Elsevier, Handb. Nanomater. Ind. Appl, 2018, pp. 696–709.

[8] K. Byrappa, T. Adschiri, Hydrothermal technology for nanotechnology, Prog. Cryst. Growth Charact. Mater. 53 (2) (2007) 117–166.

[9] A.K. Gupta, M. Gupta, Synthesis and surface engineering of Iron oxide nanoparticles for biomedical applications, Biomaterials 26 (18) (2005) 3995–4021.

[10] M.W. Ahmad, U. Farva, M.A. Khan, Low temperature synthesis of iron pyrite (FeS2) nanospheres as a strong solar absorber material, Mater. Lett. 228 (2018) 129–132.

[11] M.Y. Ahmad, M.W. Ahmad, H. Yue, S.L. Ho, J.A. Park, K.H. Jung, H. Cha, S. Marasini, A. Ghazanfari, S. Liu, Y. Chang, G.H. Lee, In vivo positive magnetic resonance imaging applications of poly(methyl vinyl ether-alt-maleic acid)-coated ultra-small paramagnetic gadolinium oxide nanoparticles, Molecules 25 (5) (2020), https://doi.org/10.3390/molecules25051159.
[12] M.W. Ahmad, W. Xu, S.J. Kim, J.S. Baeck, Y. Chang, J.E. Bae, K.S. Chae, J.A. Park, T.J. Kim, G.H. Lee, Potential dual imaging nanoparticle: Gd2O3 nanoparticle, Sci. Rep. 5 (2015), https://doi.org/10.1038/srep08549.
[13] M.Y. Ahmad, M.W. Ahmad, H. Cha, O. In-Taek, T. Tegafaw, M. Xu, S.L. Ho, S. Marasini, A. Ghazanfari, H. Yue, Cyclic RGD-coated ultrasmall Gd2O3 nanoparticles as tumor-targeting positive magnetic resonance imaging contrast agents, Eur. J. Inorg. Chem. 2018 (26) (2018) 3070–3079.
[14] H. Mirzaei, M. Darroudi, Zinc oxide nanoparticles: biological synthesis and biomedical applications, Ceram. Int. 43 (1) (2017) 907–914.
[15] S.M. Ansari, R.D. Bhor, K.R. Pai, D. Sen, S. Mazumder, G. Kartik, Y.D. Kolekar, C.V. Ramana, Cobalt nanoparticles for biomedical applications: facile synthesis, physiochemical characterization, cytotoxicity behavior and biocompatibility, Appl. Surf. Sci. 414 (2017) 171–187.
[16] K. Kattel, J.Y. Park, W. Xu, B.A. Bony, W.C. Heo, T. Tegafaw, C.R. Kim, M.W. Ahmad, S. Jin, J.S. Baeck, J.Y. Jeong, G.H. Lee, Surface coated Eu(OH)3 nanorods: a facile synthesis, characterization, MR relaxivities and in vitro cytotoxicity, J. Nanosci. Nanotechnol. 13 (11) (2013) 7214–7219, https://doi.org/10.1166/jnn.2013.8081.
[17] D. Kim, S. Jon, Gold nanoparticles in image-guided cancer therapy, Inorg. Chim. Acta 393 (2012) 154–164.
[18] G.Y. Simenyuk, Y.A. Zakharov, N.V. Pavelko, V.G. Dodonov, V.M. Pugachev, A.V. Puzynin, T.S. Manina, C.N. Barnakov, Z.R. Ismagilov, Highly porous carbon materials filled with gold and manganese oxide nanoparticles for electrochemical use, Catal. Today 249 (2015) 220–227.
[19] D.T. Thompson, Using gold nanoparticles for catalysis, Nano Today 2 (4) (2007) 40–43.
[20] N. Shah, F. Claessyns, S. Rimmer, M.B. Arain, T. Rehan, A. Wazwaz, M.W. Ahmad, M. Ul-Islam, Effective role of magnetic core-shell nanocomposites in removing organic and inorganic wastes from water, Recent Pat. Nanotechnol. 10 (3) (2016) 202–212,- https://doi.org/10.2174/1872210510666160429145524.
[21] D.L. Schodek, P. Ferreira, M.F. Ashby, Nanomaterials, nanotechnologies and design, in: An Introduction for Engineers and Architects, Butterworth-Heinemann, 2009.
[22] M. Fernandez-Garcia, A. Martinez-Arias, J.C. Hanson, J.A. Rodriguez, Nanostructured oxides in chemistry: characterization and properties, Chem. Rev. 104 (9) (2004) 4063–4104.
[23] J.A. Rodrígues, M. Fernández-García, Synthesis, Properties and Applications of Oxide Nanoparticles, John Wiley & Sons, 2007.
[24] V. Bansal, P. Poddar, A. Ahmad, M. Sastry, Room-temperature biosynthesis of ferroelectric barium titanate nanoparticles, J. Am. Chem. Soc. 128 (36) (2006) 11958–11963.
[25] W.-T. Liu, Nanoparticles and their biological and environmental applications, J. Biosci. Bioeng. 102 (1) (2006) 1–7.
[26] J.M. McHale, A. Auroux, A.J. Perrotta, A. Navrotsky, Surface energies and thermodynamic phase stability in nanocrystalline aluminas, Science 277 (5327) (1997) 788–791.
[27] N. Millot, D. Aymes, F. Bernard, J.C. Niepce, A. Traverse, F. Bouree, B.L. Cheng, P. Perriat, Particle size dependency of ternary diagrams at the nanometer scale: evidence of TiO2 clusters in Fe-based spinels, J. Phys. Chem. B 107 (24) (2003) 5740–5750.

[28] J. Schoiswohl, G. Kresse, S. Surnev, M. Sock, M.G. Ramsey, F.P. Netzer, Planar vanadium oxide clusters: two-dimensional evaporation and diffusion on Rh (111), Phys. Rev. Lett. 92 (20) (2004) 206103.
[29] V.M. Samsonov, N.Y. Sdobnyakov, A.N. Bazulev, On thermodynamic stability conditions for nanosized particles, Surf. Sci. 532 (2003) 526–530.
[30] P. Ayyub, M. Multani, B. Mustansir, V.R. Palkar, R. Vijayaraghavan, Size-induced structural phase transitions and hyperfine properties of microcrystalline Fe2O3, J. Phys. C Solid State Phys. 21 (11) (1988) 2229.
[31] R.C. Garvie, M.F. Goss, Intrinsic size dependence of the phase transformation temperature in zirconia microcrystals, J. Mater. Sci. 21 (4) (1986) 1253–1257.
[32] G. Skandan, C.M. Foster, H. Frase, M.N. Ali, J.C. Parker, H. Hahn, Phase characterization and stabilization due to grain size effects of nanostructured Y2O3, Nanostruct. Mater. 1 (4) (1992) 313–322.
[33] M.D. Hernández-Alonso, A.B. Hungría, A. Martínez-Arias, J.M. Coronado, J.C. Conesa, J. Soria, M. Fernández-García, Confinement effects in quasi-stoichiometric CeO 2 nanoparticles, Phys. Chem. Chem. Phys. 6 (13) (2004) 3524–3529.
[34] Z. Song, T. Cai, Z. Chang, G. Liu, J.A. Rodriguez, J. Hrbek, Molecular level study of the formation and the spread of MoO3 on au (111) by scanning tunneling microscopy and X-ray photoelectron spectroscopy, J. Am. Chem. Soc. 125 (26) (2003) 8059–8066.
[35] S. Surnev, G. Kresse, M.G. Ramsey, F.P. Netzer, Novel Interface-mediated metastable oxide phases: vanadium oxides on Pd (111), Phys. Rev. Lett. 87 (8) (2001) 86102.
[36] M.E. Franke, T.J. Koplin, U. Simon, Metal and metal oxide nanoparticles in chemiresistors: does the nanoscale matter? Small 2 (1) (2006) 36–50.
[37] J.A. Mejias, A.M. Marquez, J. Fernández Sanz, M. Fernandez-Garcia, J.M. Ricart, C. Sousa, F. Illas, On modelling the interaction of CO on the MgO (100) surface, Surf. Sci. 327 (1–2) (1995) 59–73.
[38] J.A. Rodriguez, G. Liu, T. Jirsak, J. Hrbek, Z. Chang, J. Dvorak, A. Maiti, Activation of gold on titania: adsorption and reaction of SO2 on Au/TiO2 (110), J. Am. Chem. Soc. 124 (18) (2002) 5242–5250.
[39] J.A. Rodriguez, S. Chaturvedi, M. Kuhn, J. Hrbek, Reaction of H2S and S2 with metal/oxide surfaces: band-gap size and chemical reactivity, J. Phys. Chem. B 102 (28) (1998) 5511–5519.
[40] C.A. Scamehorn, N.M. Harrison, M.I. McCarthy, Water chemistry on surface defect sites: chemidissociation versus physisorption on MgO (001), J. Chem. Phys. 101 (2) (1994) 1547–1554.
[41] R. Richards, W. Li, S. Decker, C. Davidson, O. Koper, V. Zaikovski, A. Volodin, T. Rieker, K.J. Klabunde, Consolidation of metal oxide nanocrystals. Reactive pellets with controllable pore structure that represent a new family of porous, inorganic materials, J. Am. Chem. Soc. 122 (20) (2000) 4921–4925.
[42] I. Freestone, N. Meeks, M. Sax, C. Higgitt, The Lycurgus cup—a Roman nanotechnology, Gold Bull. 40 (4) (2007) 270–277.
[43] F.E. Wagner, S. Haslbeck, L. Stievano, S. Calogero, Q.A. Pankhurst, K.P. Martinek, Before striking gold in gold-ruby glass, Nature 407 (6805) (2000) 691–692.
[44] S. Padovani, D. Puzzovio, C. Sada, P. Mazzoldi, I. Borgia, A. Sgamellotti, B.G. Brunetti, L. Cartechini, F. D'acapito, C. Maurizio, XAFS study of copper and silver nanoparticles in glazes of medieval middle-east Lustreware (10th–13th century), Appl. Phys. A 83 (4) (2006) 521–528.
[45] T. Pradell, A. Climent-Font, J. Molera, A. Zucchiatti, M.D. Ynsa, P. Roura, D. Crespo, Metallic and nonmetallic shine in luster: an elastic ion backscattering study, J. Appl. Phys. 101 (10) (2007) 103518.

[46] M. Reibold, P. Paufler, A.A. Levin, W. Kochmann, N. Pätzke, D.C. Meyer, Carbon nanotubes in an ancient Damascus sabre, Nature 444 (7117) (2006) 286.
[47] C.P. Poole Jr, F.J. Owens, Introduction to Nanotechnology, John Wiley & Sons, 2003.
[48] M. Faraday, X. The Bakerian lecture—experimental relations of gold (and other metals) to light, Philos. Trans. R. Soc. Lond. A 147 (1857) 145–181.
[49] R.P. Feynman, There's Plenty of Room at the Bottom, California Institute of Technology, Engineering and Science Magazine, 1960.
[50] J.E. Hulla, S.C. Sahu, A.W. Hayes, Nanotechnology: history and future, Hum. Exp. Toxicol. 34 (12) (2015) 1318–1321.
[51] C.G. Granqvist, R.A. Buhrman, J. Wyns, A.J. Sievers, Far-infrared absorption in ultrafine Al particles, Phys. Rev. Lett. 37 (10) (1976) 625.
[52] G. Binnig, H. Rohrer, C. Gerber, E. Weibel, Tunneling through a controllable vacuum gap, Appl. Phys. Lett. 40 (2) (1982) 178–180.
[53] G. Binnig, H. Rohrer, C. Gerber, E. Weibel, Surface studies by scanning tunneling microscopy, Phys. Rev. Lett. 49 (1) (1982) 57.
[54] S. Iijima, Helical microtubules of graphitic carbon, Nature 354 (6348) (1991) 56–58.
[55] D. Nunes, A. Pimentel, S. Lidia, P. Barquinha, L. Pereira, E. Fortunato, R. Martins, Synthesis, design, and morphology of metal oxide nanostructures, Metal Oxide Nanostruct. (2019) 21–57.
[56] M.S. Chavali, M.P. Nikolova, Metal oxide nanoparticles and their applications in nanotechnology, SN Appl. Sci. 1 (6) (2019) 1–30.
[57] N. Wilson, Nanoparticles: environmental problems or problem solvers? Bioscience 68 (4) (2018) 241–246.
[58] G. Marquis, B. Ramasamy, S. Banwarilal, A.P. Munusamy, Evaluation of antibacterial activity of plant mediated CaO nanoparticles using Cissus Quadrangularis extract, J. Photochem. Photobiol. B Biol. 155 (2016) 28–33.
[59] L. D'Souza, R. Richards, Synthesis of metal-oxide nanoparticles: liquid–solid transformations, in: Synthesis, Properties, and Applications of Oxide Nanomaterials, John Wiley & Sons, Inc, Hoboken, New Jersey, 2007, pp. 81–117.
[60] S. Buzby, R. Franklin, S.I. Shah, Synthesis of metal-oxide nanoparticles: liquid-solid transformations, in: J.A. Rodrıguez, M. Fernández-Garcıa (Eds.), 'Synthesis, Properties and Applications of Oxide Nanoparticles, Whiley, 2007.
[61] K.S. Suslick, S.-B. Choe, A.A. Cichowlas, M.W. Grinstaff, Sonochemical synthesis of amorphous Iron, Nature 353 (6343) (1991) 414–416.
[62] V. Uskoković, M. Drofenik, Synthesis of materials within reverse micelles, Surf. Rev. Lett. 12 (02) (2005) 239–277.
[63] Interrante, V. Leonard, M.J. Hampden-Smith, Chemistry of Advanced Materials: An Overview, John Wiley & Sons, 1997.
[64] M. Ohring, Materials Science of Thin Films, Elsevier, 2001.
[65] G.K. Hubler, Pulsed laser deposition, MRS Bull. 17 (2) (1992) 26–29.
[66] K. Byrappa, M. Yoshimura, Handbook of Hydrothermal Technology, William Andrew, 2012.
[67] K.F.E. Schafthaul, Schafthaul. Pdf, Gelehrt. Anz. Bayer. Akad 20 (1845) 557.
[68] Z.L. Wang, New developments in transmission electron microscopy for nanotechnology, Adv. Mater. 15 (18) (2003) 1497–1514.
[69] E.L. Shock, E.H. Oelkers, J.W. Johnson, D.A. Sverjensky, H.C. Helgeson, Calculation of the thermodynamic properties of aqueous species at high pressures and temperatures. Effective electrostatic radii, dissociation constants and standard partial Molal properties to 1000 C and 5 Kbar, J. Chem. Soc. Faraday Trans. 88 (6) (1992) 803–826.

[70] B.E. Etschmann, W. Liu, D. Testemale, M. Harald, N.A. Rae, O. Proux, J.-L. Hazemann, J. Brugger, An in situ XAS study of copper (I) transport as hydrosulfide complexes in hydrothermal solutions (25–592 C, 180–600 Bar): speciation and solubility in vapor and liquid phases, Geochim. Cosmochim. Acta 74 (16) (2010) 4723–4739.
[71] K. Byrappa, Novel hydrothermal solution routes of advanced high melting nanomaterials processing, J. Cerma. Soc. Jpn. 117 (1363) (2009) 236–244.
[72] M. Yoshimura, Importance of soft solution processing for advanced inorganic materials, J. Mater. Res. 13 (4) (1998) 796–802.
[73] M.M. Łencka, R.E. Riman, Synthesis of lead titanate: thermodynamic modeling and experimental verification, J. Am. Ceram. Soc. 76 (10) (1993) 2649–2659.
[74] K.S. Pitzer, Thermodynamics of electrolytes. I. Theoretical basis and general equations, J. Phys. Chem. 77 (2) (1973) 268–277.
[75] J.F. Zemaitis Jr, D.M. Clark, M. Rafal, N.C. Scrivner, Handbook of Aqueous Electrolyte Thermodynamics: Theory & Application, John Wiley & Sons, 2010.
[76] M.M. Lencka, A. Anderko, R.E. Riman, Hydrothermal precipitation of lead zirconate titanate solid solutions: thermodynamic modeling and experimental synthesis, J. Am. Ceram. Soc. 78 (10) (1995) 2609–2618.
[77] D. Kashchiev, On the relation between nucleation work, nucleus size, and nucleation rate, J. Chem. Phys. 76 (10) (1982) 5098–5102.
[78] M. Yoshimura, K.-S. Han, W. Suchanek, 'Soft solution processing'in situ fabrication of morphology-controlled advanced ceramic materials in low temperature solutions without firing, Bull. Mater. Sci. 22 (3) (1999) 193–199.
[79] M. Yoshimura, W. Suchanek, In situ fabrication of morphology-controlled advanced ceramic materials by soft solution processing, Solid State Ion. 98 (3–4) (1997) 197–208.
[80] M. Yoshimura, W. Suchanek, K.-S. Han, Recent developments in soft, solution processing: one step fabrication of functional double oxide films by hydrothermal-electrochemical methods, J. Mater. Chem. 9 (1) (1999) 77–82.
[81] F.F. Lange, Chemical solution routes to single-crystal thin films, Science 273 (5277) (1996) 903–909.
[82] S.-.H. Yu, M. Yoshimura, Ferrite/metal composites fabricated by soft solution processing, Adv. Funct. Mater. 12 (1) (2002) 9–15.
[83] M. Verde, M. Peiteado, M. Villegas, B. Ferrari, A.C. Caballero, Soft solution processing of ZnO nanoarrays by combining electrophoretic deposition and hydrothermal growth, Mater. Chem. Phys. 140 (1) (2013) 75–80.
[84] M.K. Naskar, Soft solution processing for the synthesis of alumina nanoparticles in the presence of glucose, J. Am. Ceram. Soc. 93 (5) (2010) 1260–1263.
[85] S.-.H. Yu, M. Yoshimura, Fabrication of powders and thin films of various nickel sulfides by soft solution-processing routes, Adv. Funct. Mater. 12 (4) (2002) 277–285.
[86] T. Miyamoto, S. Okano, N. Kasai, Inactivation of Escherichia coli endotoxin by soft hydrothermal processing, Appl. Environ. Microbiol. 75 (15) (2009) 5058–5063.
[87] P.-H. Chan, H.-P. Teng, H.-S. Chan, L. Fu-Hsing, A Facile Control of Epitaxial-Like Barium Strontium Titanate Thin Films with Various Ba/Sr Ratios Synthesized by a Hydrothermal-Galvanic Couple Method, Ceramics International, 2021.
[88] P. Estevenon, T. Kaczmarek, M.R. Rafiuddin, É. Welcomme, S. Szenknect, A. Mesbah, P. Moisy, C. Poinssot, N. Dacheux, Soft hydrothermal synthesis of Hafnon, HfSiO4, Cryst. Growth Des. 20 (3) (2020) 1820–1828.
[89] Y. Chao, G. Yu, Z. Chen, X. Cui, C. Zhao, C. Wang, G.G. Wallace, One-pot hydrothermal synthesis of solution-processable MoS2/PEDOT: PSS composites for high-performance supercapacitors, ACS Appl. Mater. Interfaces 13 (6) (2021) 7285–7296.

[90] A.M. Ramesh, S. Shivanna, Hydrothermal synthesis of MoO3/ZnO heterostructure with highly enhanced photocatalysis and their environmental interest, J. Environ. Chem. Eng. 9 (2) (2021) 105040.
[91] A. Yan, X. Liu, G. Qiu, H. Wu, R. Yi, N. Zhang, X. Jing, Solvothermal synthesis and characterization of size-controlled Fe3O4 nanoparticles, J. Alloys Compd. 458 (1–2) (2008) 487–491.
[92] S. Yáñez-Vilar, S.-A. Manuel, C. Gómez-Aguirre, J. Mira, M.A. Señarís-Rodríguez, S. Castro-García, A simple solvothermal synthesis of MFe2O4 (M= Mn, co and Ni) nanoparticles, J. Solid State Chem. 182 (10) (2009) 2685–2690.
[93] C. Li, Y. Wei, A. Liivat, Y. Zhu, J. Zhu, Microwave-solvothermal synthesis of Fe3O4 magnetic nanoparticles, Mater. Lett. 107 (2013) 23–26.
[94] S.M. Fotukian, A. Barati, M. Soleymani, A.M. Alizadeh, Solvothermal synthesis of CuFe2O4 and Fe3O4 nanoparticles with high heating efficiency for magnetic hyperthermia application, J. Alloys Compd. 816 (2020) 152548.
[95] I. Mohammadi, F. Zeraatpisheh, E. Ashiri, K. Abdi, Solvothermal synthesis of G-C3N4 and ZnO nanoparticles on TiO2 nanotube as photoanode in DSSC, Int. J. Hydrogen Energy 45 (38) (2020) 18831–18839.
[96] A. UmaSudharshini, M. Bououdina, M. Venkateshwarlu, C. Manoharan, P. Dhamodharan, Low temperature solvothermal synthesis of pristine Co3O4 nanoparticles as potential supercapacitor, Surf. Interfaces 19 (2020) 100535.
[97] R. Yokozaki, H. Kobayashi, I. Honma, Reductive solvothermal synthesis of MgMn2O4 spinel nanoparticles for Mg-ion battery cathodes, Ceramics International, 2020.
[98] I. Ahmad, S. Shukrullah, M. Ahmad, E. Ahmed, M.Y. Naz, M.S. Akhtar, N.R. Khalid, A. Hussain, I. Hussain, Effect of Al doping on the photocatalytic activity of ZnO nanoparticles decorated on CNTs and graphene: solvothermal synthesis and study of experimental parameters, Mater. Sci. Semicond. Process. 123 (2021) 105584.
[99] A.M. Mohammed, S.S. Mohtar, F. Aziz, M. Aziz, A. Ul-Hamid, W.N.W. Salleh, N. Yusof, J. Jaafar, A.F. Ismail, Ultrafast degradation of Congo red dye using a facile one-pot solvothermal synthesis of cuprous oxide/titanium dioxide and cuprous oxide/zinc oxide Pn heterojunction photocatalyst, Mater. Sci. Semicond. Process. 122 (2021) 105481.
[100] N. Monday, J.A. Yakubu, N.A. Yusof, S.A. Rashid, R.H. Shueb, Facile hydrothermal and solvothermal synthesis and characterization of nitrogen-doped carbon dots from palm kernel shell precursor, Appl. Sci. 11 (4) (2021) 1630.
[101] T. Adschiri, K. Kanazawa, K. Arai, Rapid and continuous hydrothermal synthesis of Boehmite particles in subcritical and supercritical water, J. Am. Ceram. Soc. 75 (9) (1992) 2615–2618.
[102] G. Brunner, Near critical and supercritical water. Part I. Hydrolytic and hydrothermal processes, J. Supercrit. Fluids 47 (3) (2009) 373–381.
[103] E.U. Franck, Special aspects of fluid solutions at high pressures and sub-and supercritical temperatures, Pure Appl. Chem. 53 (7) (1981) 1401–1416.
[104] T. Yiling, T. Michelberger, E.U. Franck, High-pressure phase equilibria and critical curves of (water+ n-butane) and (water+ n-hexane) at temperatures to 700 K and pressures to 300 MPa, J. Chem. Thermodyn. 23 (1) (1991) 105–112.
[105] K. Sue, M. Suzuki, K. Arai, T. Ohashi, H. Ura, K. Matsui, Y. Hakuta, H. Hayashi, M. Watanabe, T. Hiaki, Size-controlled synthesis of metal oxide nanoparticles with a flow-through supercritical water method, Green Chem. 8 (7) (2006) 634–638.
[106] N. Aoki, A. Sato, H. Sasaki, A.-A. Litwinowicz, G. Seong, T. Aida, D. Hojo, S. Takami, T. Adschiri, Kinetics study to identify reaction-controlled conditions for supercritical hydrothermal nanoparticle synthesis with flow-type reactors, J. Supercrit. Fluids 110 (2016) 161–166.

[107] D. Zhao, X. Wu, H. Guan, E. Han, Study on supercritical hydrothermal synthesis of CoFe2O4 nanoparticles, J. Supercrit. Fluids 42 (2) (2007) 226–233.
[108] M. Taguchi, S. Takami, T. Adschiri, T. Nakane, K. Sato, T. Naka, Supercritical hydrothermal synthesis of hydrophilic polymer-modified water-dispersible CeO 2 nanoparticles, CrstEngComm 13 (8) (2011) 2841–2848.
[109] T. Sasaki, S. Ohara, T. Naka, J. Vejpravova, V. Sechovsky, M. Umetsu, S. Takami, B. Jeyadevan, T. Adschiri, Continuous synthesis of fine MgFe2O4 nanoparticles by supercritical hydrothermal reaction, J. Supercrit. Fluids 53 (1–3) (2010) 92–94.
[110] S. Ohara, T. Mousavand, M. Umetsu, S. Takami, T. Adschiri, Y. Kuroki, M. Takata, Hydrothermal synthesis of fine zinc oxide particles under supercritical conditions, Solid State Ion. 172 (1–4) (2004) 261–264.
[111] S. Ohara, T. Mousavand, T. Sasaki, M. Umetsu, T. Naka, T. Adschiri, Continuous production of fine zinc oxide nanorods by hydrothermal synthesis in supercritical water, J. Mater. Sci. 43 (7) (2008) 2393–2396.
[112] S.-i. Kawasaki, Y. Xiuyi, K. Sue, Y. Hakuta, A. Suzuki, K. Arai, Continuous supercritical hydrothermal synthesis of controlled size and highly crystalline anatase TiO2 nanoparticles, J. Supercrit. Fluids 50 (3) (2009) 276–282.
[113] A. Yoko, Y. Tanaka, G. Seong, D. Hojo, T. Tomai, T. Adschiri, Mixing and solvent effects on kinetics of supercritical hydrothermal synthesis: reaction of nickel nitrate to nickel oxide, J. Phys. Chem. C 124 (8) (2020) 4772–4780.
[114] Y. Zhu, G. Seong, T. Noguchi, A. Yoko, T. Tomai, S. Takami, T. Adschiri, Highly Cr-substituted CeO2 nanoparticles synthesized using a non-equilibrium supercritical hydrothermal process: high oxygen storage capacity materials designed for a low-temperature bitumen upgrading process, ACS Appl. Energy Mater. 3 (5) (2020) 4305–4319.
[115] R. Yuan, H. Wen, L. Zeng, X. Li, X. Liu, C. Zhang, Supercritical CO2 assisted solvothermal preparation of CoO/graphene nanocomposites for high performance lithium-ion batteries, Nanomaterials 11 (3) (2021) 694.
[116] M. François, F. Demoisson, M. Sennour, G. Caboche, Continuous hydrothermal synthesis in supercritical conditions as a novel process for the elaboration of Y-doped BaZrO3, Ceram. Int. 47 (12) (2021) 17799–17803.

# Design and synthesis of metal oxide–polymer composites

4

*Gulcihan Guzel Kaya and Huseyin Deveci*
Department of Chemical Engineering, Konya Technical University, Konya, Turkey

## 1 Polymer composites

Composites are innovative engineering materials comprising at least two components with different properties. By the combination of individual properties of the components, final composite materials are produced with enhanced quality or new features [1]. With the advantages of unique properties, composite materials are commonly used in automobiles, aircraft, construction, marine, electronics, and sporting goods [2]. Composite materials are separated into three groups: (1) metal matrix composites, (2) ceramic matrix composites, and (3) polymer matrix composites. Among them, polymer matrix composites have received great attention owing to their low density, high specific strength, good wear and corrosion resistance, improved gas barrier effects, and flexibility in production in recent years [3,4].

Polymer composites include polymer matrix as a continuous phase that can be separated into three groups: thermoplastics, thermosets, and elastomers. In thermoplastic polymers, molecules are interconnected by intermolecular interactions or van der Waals forces. Intermolecular interactions weaken in the case of heating which makes thermoplastics soft and flexible. Under the cooling process, thermoplastics can be converted into a solid phase. So, thermoplastic polymers can be easily reused or recycled many times. However, repetitive heating–cooling processes can cause adverse effects on the properties of thermoplastics [5]. Widely used thermoplastic polymers are polyethylene (PE), polystyrene (PS), polypropylene (PP), polycarbonate (PC), polyamide 6 (Nylon 6), polyvinyl chloride (PVC), polyethylene terephthalate (PET), and polymethyl methacrylate (PMMA) [6]. Thermosets are three-dimensional cross-linked polymers in which the motion of the polymer chains is restricted. Instead of phase change, thermoset polymers such as unsaturated polyester, epoxy resin, phenol-formaldehyde, melamin-formaldehyde, and urea-formaldehyde decompose with heating [7]. Elastomers are significant group of polymers that can stretch up to 3–10 times of their original dimension under a loading. The polymers also resume their original shape when the loading is removed that provides reversible deformability without permanent flow in the network [8]. There are many types of elastomers including natural rubber, ethylene propylene diene monomer, styrene butadiene, nitrile butadiene rubber, styrene block copolymer, and poly(isobutylene-*co*-isoprene) for general and special purposes [9].

**Renewable Polymers and Polymer-Metal Oxide Composites. https://doi.org/10.1016/B978-0-323-85155-8.00005-4**

Many polymers are synthesized from petroleum-based chemicals. When considered environmental and economic concerns and limited source of petro-chemicals, many researches have focused on the synthesis of renewable polymers [10]. In 2019, the total production of renewable polymers was approximately 3.8 Mt, corresponding to 1% of the total production of polymers derived from petro-chemicals [11]. According to the precursor, renewable polymers are categorized as polymers from biomass (starch, cellulose, sucrose, and fructose), polymers from animal stock (chitosan and chitin), polymers from microbial action (polyhydroxyalkanoates and bacterial cellulose), and polymers from monomers (derived from biomass) through chemical and conventional synthesis (polylactic acid (PLA)) [12,13].

To improve properties of the polymers, different types of filler such as carbon-based materials, clay minerals, $CaCO_3$, metals, metal oxides, boron compounds, silica-based materials, industrial by-products, and organic/agricultural waste have been used [14–20]. Especially for a change in magnetic, optical, electrical, sensing, thermal, and mechanical characteristics of the polymers, metal oxides are preferred in various applications. The metal oxide–polymer composites are easily prepared with the incorporation of metal oxides like $TiO_2$, $Al_2O_3$, CuO, $Fe_2O_3$, NiO, ZnO, $WO_3$, and $SnO_2$ into the polymers for different purposes [21–25].

Studies related to the metal oxide–polymer composites are increasing day by day; so, in this chapter, the metal oxide–polymer composites were overviewed with reflecting the design of the metal oxide–polymer composites, design parameters, synthesis, and properties of the metal oxide–polymer composites.

## 2 Design of metal oxide–polymer composites

Design of the metal oxide–polymer composites is crucial to produce materials with the required properties for tailored applications. Many design parameters considerably affect interactions between a metal oxide and polymer matrix (surface wetting, absorption, electrostatic interaction, diffusion, chemical, and intermolecular bonding) that are effective on the properties of the composite materials [26]. Properties of the metal oxides (shape, size, surface chemistry, and surface area), metal oxide alignment, polymer matrix structure, amount of metal oxide/polymer matrix, surface modifications, and production conditions (process type, pressure, temperature, and time) easily govern metal oxide dispersion [27–30]. The homogenous dispersion of metal oxides in the polymer matrix increases the metal oxide–polymer matrix adhesion that provides synergistic effects on the properties of the metal oxide–polymer composites. Whereas, metal oxide agglomeration due to their high surface energy decreases metal oxide adherence to the polymer matrix [31,32].

Metal oxides are generally synthesized with different sizes and shapes (nanotubes, nanorods, spherical, star, triangular, nanoplate, nanowire, and fiber) in one, two, and three dimensions (1D, 2D, and 3D) [33–35]. As shown in Fig. 1, there are three regimes as a function of metal oxide size depending on their aspect ratio and dimension. In Regime I, quantum chemistry laws are applied to represent the properties of

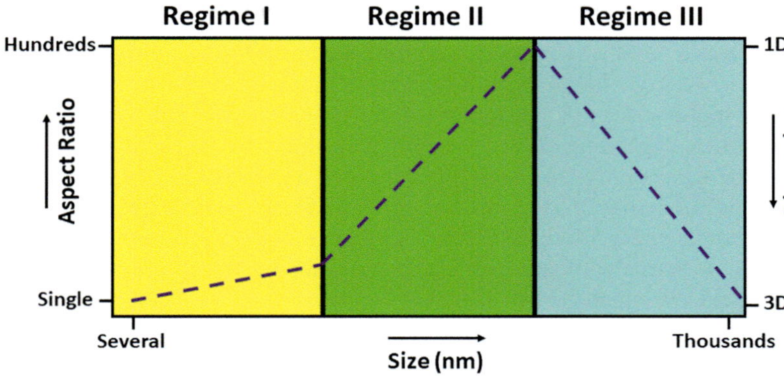


**Fig. 1** Effect of aspect ratio and dimension of metal oxide on the electrochemical properties of the materials based on metal oxide size.

the metal oxides with smaller size. For larger metal oxide particles, solid-state physics laws are dominant in Regime III. One of the laws can be used for metal oxides with the size of 10–100 nm in Regime II. Regarding as electrochemical properties of the metal oxides, minimum response time is achieved in Regime I. In Regime II, the metal oxides show maximum reactivity through their optimized characteristics. However, properties of the metal oxides especially electrochemical features get worse due to agglomerations of the particles in Regime III [36].

Aspect ratio, which states the ratio of longest to the shortest dimension, is a significant parameter. 1D metal oxides have the highest aspect ratio, while the lowest aspect ratio values belong to 3D metal oxide particles [37,38]. According to the application field, the aspect ratio of the metal oxides should be considered. For example, 1D metal oxides are generally chemically and thermally unstable which restricts their utilization in electronics [39,40]. With increasing aspect ratio, 2D metal oxide particles represent better electrochemical properties owing to the anisotropic diffusion of electrons in addition to more path for electrons [41].

The ratio of surface area to volume (SA:V) is another design parameter for metal oxide particles. It is known that 2D metal oxides have the highest SA:V values mostly. Higher SA:V of the metal oxides provides better electrical, optical, and sensing properties due to the more active cites [42,43]. Especially in polymer composites, the metal oxide contact surface to the polymer matrix is directly related to the specific surface area of the metal oxides. The higher surface area of metal oxide increases interfacial attractions between metal oxide and polymer [44].

Controlling metal oxide alignment in the polymer matrix is one of the most substantial design parameters in the preparation of the metal oxide–polymer composites. Metal oxides are dispersed randomly in the polymer matrix by ultrasonication or shear/flow mixing [45]. However, metal oxides with high aspect ratio are aligned using an electric or magnetic field in addition to mechanical stretching to increase the performance of the final composite material [46]. Shear and flow-induced alignments are generally utilized for thin-film production and coatings. Alignment under the electric field is commonly preferred to align 2D fillers in processes in which no solvent is used due to the structure destruction in the case of solvent evaporation. With using higher electric field strength, desirable orientation is achieved in a shorter time. Moreover, depending on electrode position, alignment direction can be designed vertically or horizontally. Magnetic field-induced alignment is an effective strategy for magnetic fillers like $Fe_3O_4$ particles orientation in a polymer matrix. By means of magnetostatic energy derived from anisotropic magnetic susceptibility, especially electrical and thermal properties of the composites including oriented filler under magnetic field are enhanced [47,48].

The dispersion behavior of the metal oxides (size $<100\,nm$) in the polymer matrix is an important design parameter. Due to various interactions between metal oxide particles, agglomerations are inevitable which are generally prevented by surface functionalization of the metal oxides [49]. Surface functionalization processes are separated into two groups: (1) physical and (2) chemical treatments. Physical treatments are carried out by coating particle surfaces with anionic, cationic or polymer surfactants, which minimize particle agglomerations with decreasing surface tension. Chemical treatments provide good dispersion of the metal oxides by covalent bonding between metal oxide surface and coupling agents (silanes, amines, carboxylic acids, thiols, polymers, and organophosphorus materials) [50,51].

The amount of metal oxides considerably affects compatibility between metal oxides and polymer matrix which is directly related to load transfer through composite material. The addition of over the amount of metal oxides generally deteriorates properties of the metal oxide–polymer composites because of nonuniform metal oxide dispersion. Metal oxide particles with high amounts tend to agglomerate because of high surface energy of the particles, which causes poor metal oxide–polymer matrix adhesion [52,53].

Due to the significance of time and energy consumption in large-scale polymer composite manufacturing, process parameters including production method, curing temperature and rate, type of curing, exposure time under heating, and pressure play a crucial role in the design of the metal oxide–polymer composites. There are different curing methods (microwave, ultraviolet, infrared radiation, and radiofrequency) that substantially influence production time, required energy, and quality of the final material. Insufficient curing process generally resulted from heat loss in conventional curing methods leads to lower mechanical strength in polymer composites [54]. Applied pressure during polycondensation reactions affects the molecular weight of the polymer. Additionally, vacuum pressure decreases shrinkage and void problems in polymer composite production compared to ambient atmospheric pressure [55,56]. To design process parameters, the trial and error approach is performed; however, it takes a long time, requires high cost, and causes unreliable results [57]. Therefore, various experimental design methods like Taguchi optimization, artificial neural networks, response surface methodology, and factorial design have been conducted to obtain high-performance polymer composites in recent times [58–61].

# 3 Synthesis of metal oxide–polymer composites

Metal oxide–polymer composites are generally synthesized by three methods: (1) blending or direct mixing, (2) sol–gel process, and (3) in situ polymerization, as shown in Fig. 2 [62].

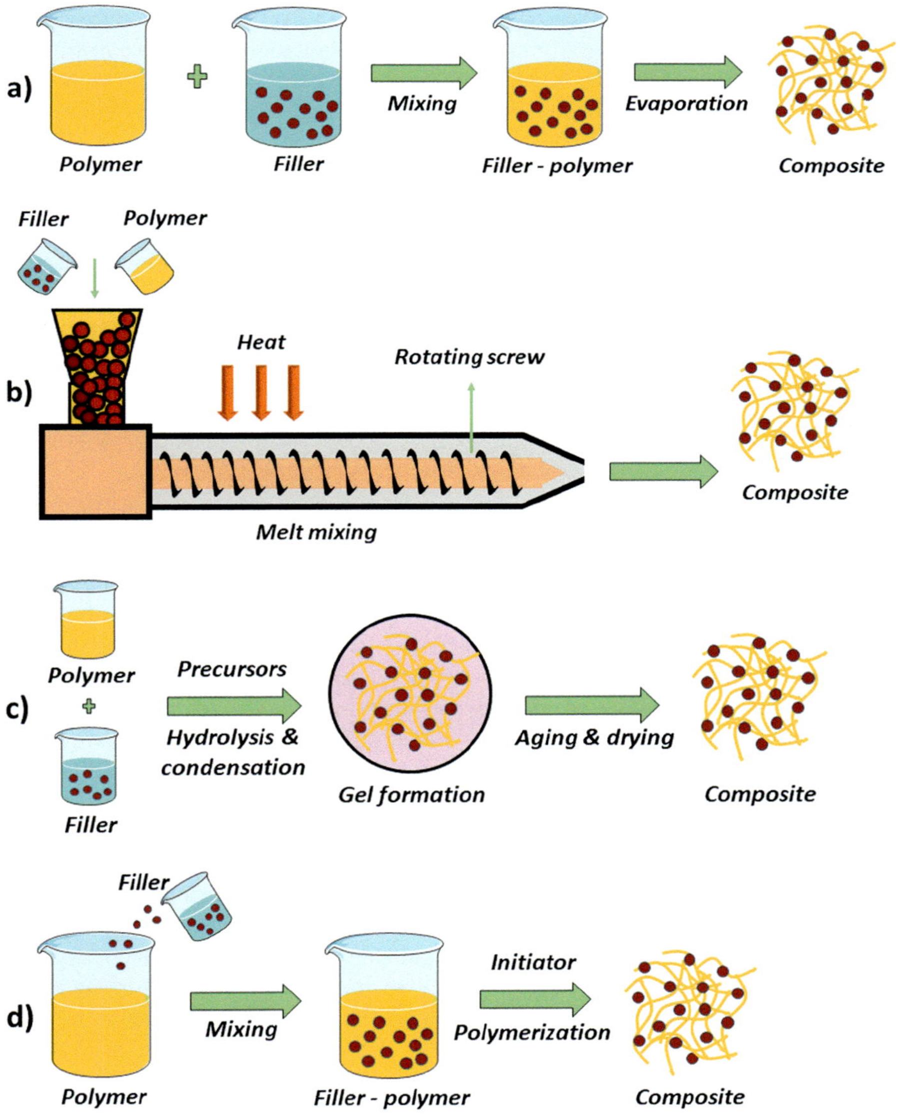


**Fig. 2** Synthesis methods of the metal oxide–polymer composites: (A) solution blending, (B) melt blending, (C) sol–gel method, and (D) in situ polymerization.

## 3.1 Blending

Blending as an ex situ technique is the most popular and simple method for large-scale productions in spite of filler agglomeration problems. The blending process is separated into two groups: (1) solution blending and (2) melt blending [63]. In solution blending, which can be carried out in low temperatures and viscosity, metal oxide particles are first dispersed in a solvent followed by polymer mixing with shear mixing, reflux, magnetic stirring, or ultrasonication [64]. After solvent evaporation, composite materials are prepared through film formation or precipitation. One of the restrictions of this method is that filler loading does not generally exceed 30 wt% due to the filler agglomerations [65]. Regarding environmental concerns and cost issues, using a large amount of solvent and its recovery restricts applications of the solution blending method [66]. If dissolving the polymer is difficult in a solvent, an emulsion or suspension solution can be used, known as emulsion or suspension blending [67]. No solvent requirement makes the melt blending method more convenient in the composite production processes. Metal oxides are mixed with a molten state polymer matrix (usually thermoplastics) by shear mixing or through an extruder or injection in this method. Although melt blending is a low-cost and environmental-friendly synthesis method, high temperature and long process time can cause degradation in the polymer network [68]. Moreover, the difficulty of the filler alignment is another problem in melt blending method [65].

## 3.2 Sol–gel process

Synthesis of metal oxides in the presence of polymer matrix can be facilely provided by sol–gel method. Furthermore, metal oxide synthesis in parallel with polymerization can be conducted by the sol–gel process. The process comprises gel formation, aging of the gel, and drying of the gel, respectively [69]. Hydrolysis and condensation reactions occur using acid, base, or two-step acid–base catalysts resulting in covalent bonding between a metal oxide and polymer matrix during the gel formation. In this step, the solvent is a significant factor to hinder phase separation. Considering functional groups belonging to the polymer matrix and polarity of the polymer matrix, commonly used solvents are acetone, alcohols, tetrahydrofuran, acetic acid, dimethoxymethane, etc. [70]. After strengthening the gel structure in the aging step, solvent molecules are removed from the gel by a drying process. By the help of the sol–gel process, morphology and surface properties of the metal oxides in the polymer matrix can be controlled [71]. Final composite materials with different properties can be produced depending on the temperature, pH, precursor structure, catalyst type and concentration, type of aging solvent, aging time and temperature, and drying conditions. This method enables to carry out the synthesis process at low temperatures and obtain homogeneously dispersed materials. However, undesirable by-products resulted from the reactions, ineffective solvent exchange, and drying can be observed in the sol–gel process [72]. Additionally, brittle composite materials are produced due to the decrease in internal stress through the removal of water and small molecules and solvents from the material structure [73].

## 3.3 In situ polymerization

Better compatibility between a metal oxide and polymer matrix can be achieved by in situ polymerization without polymer degradation under heating. The metal oxide–polymer composites can be synthesized using two types of in situ polymerization which can be initiated by heat, radiation, or different initiators. In the first, metal oxide particles are uniformly dispersed in monomer solution before the polymerization process, and then polymer chains grow around the metal oxide particles [74,75]. The second approach is the synthesis of metal oxides in the polymer matrix in the case of metal salt reduction or oxidation in which interactions between a metal oxide and polymer matrix at the molecular level considerably minimize agglomerations [76]. In addition to these approaches, physical in situ polymerization, which is gas phase technique, can be utilized mainly to synthesize the metal oxide–polymer core-shell composites. After the formation of metal oxides from gas-phase precursors with plasma treatment, metal oxide particles are coated with polymer shells through condensation reactions [77]. In situ polymerization, many factors such as polymer matrix structure, metal oxide precursor type, reactions during the metal oxide formation, composition of the metal oxide and surface modification of the metal oxide significantly influence the properties of the metal oxide–polymer composites [78]. Surface modification of the metal oxides makes easy dispersion of the particles in the polymer matrix. Modification agents can attach to the metal oxide surface with different attractions such as hydrogen bonding, ionic attractions, and coordinative bonding. The interface formation on the surface of the metal oxide prevents the agglomerations. The modification agents are first dissolved in the monomers, and then metal oxides are added to growing polymer in this polymerization method [79].

In the case of using polymer matrix with the disadvantages of thermal instability and insolubility, in situ controlled radical polymerization is one of the best synthesis methods compared to conventional in situ polymerization methods. Dispersion of metal oxides in a monomer solution is followed by grafting of polymer chains onto metal oxide particle surfaces in this method [80]. There are two types of polymer grafting strategies for the metal oxide particles: (1) grafting *from* and (2) grafting *to*. Grafting *from* is explained with polymer chain propagation bonded to initiating sites on the metal oxide surfaces [81]. Grafting *to* refers adherence of preformed polymer molecules onto the active sites of metal oxides covalently or physically. Grafting efficiency in grafting *from* method is higher than that of the grafting *to* method, which is originated from restricted diffusion of small monomers in the grafting *to* method [82]. The other advantage of the grafting *from* method is grafting higher molecular weight polymers; whereas, polymerization conditions, initiator, and monomer content should be controlled during the polymerization in this method [83]. The grafting *to* method is convenient to specify the average number of grafted polymer chains on the metal oxide surfaces in addition to grafted polymer number distribution [84].

Ultrasonication is an effective force to disperse metal oxide particles uniformly in the polymer matrix and promote polymer chains mobility in situ polymerization. Due to higher sonication energy than interaction energy between metal oxide particles, agglomeration of the particles is restricted [85,86]. Ultrasonication provides many advantages such as a higher polymerization rate resulting in a decrease in polymerization time, increased monomer conversion, narrow particle size distribution, and formation of radicals that can initiate the polymerization process [87]. In spite of these advantages, ultrasonication can cause deterioration in the surface properties of the metal oxides.

# 4 Properties of metal oxide–polymer composites

The incorporation of metal oxides with unique properties into the polymers has provided to produce high-performance metal oxide–polymer composites. Mechanical, thermal, sensing, electrical, optical, magnetic, barrier, and antibacterial properties of the metal oxide–polymer composites have been enhanced depending on design parameters. In Table 1, many literature studies related to metal oxide–polymer composites are summarized.

## 4.1 Mechanical properties

Good interfacial interactions between metal oxide particles and polymer matrix provide synergistic effects on the mechanical properties of the metal oxide–polymer composites such as tensile strength and modulus, flexural strength and modulus, elongation at break, impact strength, and wear resistance. Effective load transfer through the composite material is governed by strong metal oxide–polymer matrix adhesion strength, which is directly associated with design parameters.

Mechanical properties of the metal oxide–polymer composites are enhanced with increasing metal oxide volume/weight fraction up to optimum value. Because of the agglomeration tendency of the metal oxide particles in high content, low mechanical strength causes a failure in the composite materials. Evora and Shukla investigated the mechanical properties of $TiO_2$/unsaturated polyester composites as a function of $TiO_2$ content (1–4 vol%) [88]. In the case of 1 vol% $TiO_2$ loading, tensile strength and modulus reached to the maximum value. The compressive strength of the neat polyester was enhanced from ~141 MPa to ~147 MPa with the incorporation of 1 vol % $TiO_2$. And also, fracture toughness increased to ~0.85 MPa $m^{1/2}$ owing to the uniform $TiO_2$ dispersion in the polyester. However, more $TiO_2$ loading deteriorated the mechanical strength of the composite materials that was contributed to agglomerations. Rajesh et al. prepared PVA/PVP nanocomposites including $TiO_2$ (2–16 wt%) by solvent casting method [97]. It was reported that hydrogen bonding between $TiO_2$ and PVA enhanced the mechanical properties of the nanocomposites. PVA/PVP nanocomposite including 16 wt% $TiO_2$ exhibited maximum tensile strength (13.63 MPa) and Young's modulus (1099 MPa). While maximum elongation at break was observed with the addition of 4 wt% $TiO_2$, the amount of $TiO_2$ more than 4 wt% caused a decrease in the elongation at break of the $TiO_2$/PVA/PVP nanocomposites. Ram et al. examined the mechanical properties of $BaTiO_3$/PVDF composites synthesized by the solution casting method [134]. It was observed that incorporation of 25 wt% $BaTiO_3$ improved stress of neat PVDF

**Table 1** Metal oxide–polymer composite studies in the literature

| Metal oxide | Functionalization | Polymer | Synthesis method | Properties | Ref. |
|---|---|---|---|---|---|
| $TiO_2$ (1 vol%) | | Unsaturated polyester | Direct mixing after ultrasonication | An increase in dynamic fracture toughness, tensile and compressive strength | [88] |
| $TiO_2$ (30 wt%) | | PLGA | Solution blending | An enhancement in tensile modulus and strength and compressive modulus | [89] |
| $TiO_2$ (1 vol%) | | Nylon 6,6 | Injection molding | An improvement in creep resistance | [90] |
| $TiO_2$ (0.4 wt%) | Heat treatment | Epoxy | Direct mixing after ultrasonication | An increase in flexural and tensile strength | [91] |
| $TiO_2$ (10 wt%) | | Epoxy | Direct mixing followed by spin coating | An increase in optical transmittance | [92] |
| $TiO_2$ (2.5 wt%) | | PVA | Solvent casting | An increase in tensile strength and elongation at break, high thermal stability | [93] |
| $TiO_2$ (20 wt%) | Silanization | PVA | Solution casting | An increase in UV light absorption | [94] |
| $TiO_2$ (10 wt%) | | PVA | Solvent casting | An increase in refractive index | [95] |
| $TiO_2$ (50 wt%) | | PVA PMMA | Dip coating | Higher transmittance using PMMA matrix | [96] |
| $TiO_2$ (12 wt%) | | PVA/PVP | Solvent casting | An increase in tensile strength and Young modulus, electrical conductivity | [97] |
| $TiO_2$ (15 wt%) | | PMMA | Melt blending | An increase in $T_g$, total heat release (THR), thermal conductivity and thermal diffusivity | [98] |

*Continued*

**Table 1** Continued

| Metal oxide | Functionalization | Polymer | Synthesis method | Properties | Ref. |
|---|---|---|---|---|---|
| $TiO_2$ (20 vol%) | | PMMA | Radiation-induced polymerization | A stronger photoluminescence | [99] |
| $TiO_2$ (10 wt%) | Samarium doping | PBMA | In situ free radical polymerization | An increase in $T_g$, flame retardancy and thermal stability | [100] |
| $TiO_2$ (7 wt%) | | PBMA | In situ free radical polymerization | An improvement in electrical conductivity | [101] |
| $TiO_2$ (5 wt%) (25 nm and 50 nm) | | Polyaniline | In situ polymerization | Higher electrical conductivity with 50 nm $TiO_2$ | [102] |
| $TiO_2$ (4 wt%) | | LDPE | Melt blending | A decrease in $O_2$ permeability | [103] |
| $TiO_2$ (3 wt%) | | PET | Melt blending | A decrease in water vapor permeability | [104] |
| $Al_2O_3$ (30 wt%) | Silanization $TiO_2$ coupling | PP | Compression molding | An increase in tensile strength and impact resistance | [105] |
| $\alpha$-$Al_2O_3$ (4 wt%) | SDBS treatment | PP | Melt blending | An increase in tensile strength and modulus | [106] |
| $Al_2O_3$ (4.5 vol%) | Silanization | Unsaturated polyester | Direct mixing followed by ultrasonication | An increase in fracture toughness | [107] |
| $Al_2O_3$ (18.6 vol%) | | PP | Compression molding | An increase in residual weight at 600°C, high thermal conductivity | [108] |

| | | | | | |
|---|---|---|---|---|---|
| $Al_2O_3$ (3 vol%) | Silanization | Polyvinyl ester | Free radical polymerization | An increase in Young modulus and tensile strength and high stability in acidic and basic solutions | [109] |
| $Al_2O_3$ (1 vol%) (15 nm–0.5 mm) | | Polyvinyl ester | Direct mixing after ultrasonication | An increase in tensile strength and Young modulus with nanoscale $Al_2O_3$ | [110] |
| $Al_2O_3$ (70 vol%) | | Epoxy | Vacuum infiltration | An increase in flexural strength, thermal conductivity and dielectric constant | [111] |
| $Al_2O_3$ (43 vol%) | Silanization | Epoxy | Vacuum infiltration | An improvement in thermal conductivity | [112] |
| $Al_2O_3$ (3 wt%) | | PS | Solution casting | An increase in photoluminescence intensity | [113] |
| $Al_2O_3$ (1 wt%) | Silanization | PMMA | Melt blending | A decrease in refractive index | [114] |
| $Al_2O_3$ (film) | | PLA | Dip coating | A decrease in water vapor transmission rate | [115] |
| Iron oxide (10 vol%) | | Ethylene vinyl acetate | Open mill-mixing | An increase in storage modulus, $T_g$, saturation of magnetization, and electrical conductivity | [116] |
| $Fe_2O_3$ (20 wt%) | | PMMA | Melt blending | An increase in time to ignition and residual weight at 600°C, a decrease in heat release rate | [117] |
| $\gamma$-$Fe_2O_3$ (5 wt%) | | PVA | Solution casting | An increase in refractive index | [118] |
| $Fe_3O_4$ (30 wt%) | | PEDOT | In situ chemical oxidative polymerization | An increase in magnetic saturation and remnant magnetization | [119] |
| $Fe_3O_4$ (11.4 wt%) | CSA<br>TSA<br>HCl | Polyaniline | In situ chemical oxidative polymerization | Higher magnetic saturation in CSA medium | [120] |
| $Fe_3O_4$ (5 wt%) | | PVDF | Spin coating | An increase in magnetic saturation | [121] |
| $Fe_3O_4$ (75.5 wt%) | | Epoxy | Direct mixing | An increase in magnetic saturation | [122] |

*Continued*

**Table 1** Continued

| Metal oxide | Functionalization | Polymer | Synthesis method | Properties | Ref. |
|---|---|---|---|---|---|
| ZnO (30 wt%) (spherical and whisker) | Silanization | PS | Melt blending | Higher electrical conductivity with whisker ZnO | [123] |
| ZnO (1 wt%) | | Resin | Direct mixing | An increase in compressive strength and modulus, a decrease in bacteria growth | [124] |
| ZnO (10 wt%) | | Linear LDPE | Melt blending | An increase in tensile strength, a decrease in elongation at break, an improvement in thermal stability, a decrease in $O_2$ transmission rate | [125] |
| ZnO (1 wt%) | | HDPE | Melt blending | High antibacterial activity | [126] |
| ZnO (10 wt%) | Silanization | PLA | Solvent casting | An increase in tensile strength, a slight decrease in elongation at break, an improvement in $T_g$, good antibacterial activity | [127] |
| ZnO (5 wt%) | | PLA | Melt blending | A decrease in $O_2$ and $CO_2$ permeability, high antibacterial activity | [128] |
| ZnO (4 wt%) | | PHBV | Solution casting | An increase in tensile strength, Young modulus and thermal stability, a decrease in $O_2$ and water vapor permeability, high antibacterial activity | [129] |
| ZnO (5 wt%) | | PHB | Solution casting | An increase in tensile strength, high thermal stability, a decrease in oxygen permeability, good antibacterial activity | [130] |
| ZnO (20 wt%) | | PVC | Solvent casting | An increase in $T_g$, specific heat and thermal stability | [131] |
| ZnO (6 wt%) | Chitosan modification | PVC | Solution casting | An increase in $T_g$, decomposition temperatures and char residue at 800°C | [132] |
| ZnO (20 wt%) | | Polypyrrole | In situ chemical oxidative polymerization | An improvement in electrical conductivity | [133] |
| $BaTiO_3$ (25 wt%) | Silanization | PVDF | Solution casting | An increase in tensile strength and Young modulus | [134] |

| | | | | | |
|---|---|---|---|---|---|
| $BaTiO_3$ (10 wt%) (10 nm and 500 nm) | | PVDF | Solution blending | An increase in thermal stability | [53] |
| CuO nanogranule, nanorods, nanosheet (1.5 wt%) | | Carbon fiber reinforced PTFE | Compression molding | Higher wear resistance with CuO nanosheet | [23] |
| CuO (1 wt%) | | LDPE | Melt blending | High antibacterial activity | [135] |
| $SnO_2$ (4 wt%) | | Epoxy | Direct mixing | A decrease in heat release rate, a slight increase in thermal conductivity | [25] |
| MgO (0.2 wt%) | Hydroxylation | PLA | In situ polymerization | A slight decrease in $T_g$, an increase in thermal stability | [136] |
| NiO (0.5 wt%) | | PVC | Solution casting | An increase in refractive index | [137] |
| $Li_4Ti_5O_{12}$ (7 wt%) | | PVDF | Solution casting | A slight increase in melting temperature | [138] |
| $Co_3O_4$ (0.5 wt%) | | PVA/PEG | Solution casting | An increase in optical transmittance | [139] |
| $NiCoFe_2O_4$ | | PVA (0.2 wt%) | Ultrasonication | An increase in saturation magnetization | [140] |
| $NiFe_2O_4$ (33 wt%) | | Polyaniline | In situ chemical oxidative polymerization | An increase in saturation magnetization | [141] |

CSA: camphorsulfonic acid; HDPE: high-density polyethylene; LDPE: low-density polyethylene; PBMA: poly(n-butyl methacrylate); PEDOT: poly(3,4-ethylenedioxythiophene); PHB: poly(3-hydroxybutyrate); PHBV: poly (3-hydroxybutyrate-*co*-3-hydroxyvalerate); PLGA: poly-lactic-*co*-glycolic acid; PTFE: polytetrafluoroethylene; PVA: polyvinyl alcohol; PVDF: polyvinylidene fluoride; PVP: polyvinylpyrrolidone; SDBS: sodium dodecylbenzene sulfonate; $T_g$: glass transition temperature; TSA: p-toluenesulfonic acid.

by 17% at 5% strain resulting an increase in load-bearing capacity of the composite. Furthermore, the Young's modulus of the $BaTiO_3$/PVDF composite was determined as ~1100 MPa.

Metal oxide properties such as particle size, shape, aspect ratio, and surface area are significant design parameters that considerably influence the mechanical properties of the metal oxide–polymer composites. Small particle size which means higher surface area generally increases interactions of metal oxide particles with the polymer matrix that affect mechanical properties of the composite materials positively. Zhang and Singh reported the effects of $Al_2O_3$ particle size (15 nm, 1 μm, and 35 μm) on the fracture toughness of the unsaturated polyester [107]. It was shown that lower fracture toughness was obtained in the presence of 35 μm $Al_2O_3$ due to the poor $Al_2O_3$–polymer matrix adhesion. In other words, incompatibility between $Al_2O_3$ and unsaturated polyester caused to decrease crack growth resistance resulting in a decrease in fracture toughness. Cho et al. produced polyvinyl ester composites including $Al_2O_3$ with different particle sizes (15 nm, 50 nm, 3 μm, 20 μm, and 70 μm) [110]. The results stated that the tensile strength of the composite materials highly depended on the particle size of the $Al_2O_3$. The tensile strength of the $Al_2O_3$/polyvinyl ester composites increased with decreasing $Al_2O_3$ particle size. Similarly, a slight increase in Young's modulus was specified in the case of smaller particle size; however, it was clear that the amount of $Al_2O_3$ was more dominant than its size regarding as Young's modulus of the $Al_2O_3$/polyvinyl ester composites. Wu et al. used CuO nanosheet, nanogranule, and nanorods as an antiwear additive for carbon fiber-reinforced PTFE composites [23]. The wear resistance of the composites was improved by CuO nanosheet that provided to decrease the wear rate by 51%. The CuO nanosheet played an important role as a frictional interface between carbon fiber and PTFE, leading improved transfer film and bonding strength that increased wear resistance of the composite materials. CuO nanorods and granule were only able to decrease the wear rate by 15% and 11%, respectively.

Surface functionalization of the metal oxides is a promising strategy for better metal oxide adherence to the polymer matrix that positively influences mechanical properties. Arfat et al. showed mechanical properties of the ZnO/PLA nanocomposites synthesized by the solvent casting method [127]. With untreated ZnO incorporation into the PLA matrix, tensile strength and elongation at break of the nanocomposites reduced, which was resulted from PLA degradation owing to transesterification reactions. With the help of surface treatment of ZnO by 3-methacryloxypropyltrimethoxysilane, tensile strength was increased. At 10 wt% ZnO loading, the tensile strength of the untreated ZnO/PLA and treated ZnO/PLA was determined as ~22.6 MPa and ~34.6 MPa, respectively. An improvement in tensile properties was explained with the restriction of catalytic effect of ZnO particles by the surface treatment which prevented transesterification reactions. Mirjalili et al. specified the effect of surface functionalization of $\alpha$-$Al_2O_3$ by SDBS on the tensile and flexural properties of the PP composites [106]. An enhancement in tensile/flexural strength was achieved in the case of using functionalized $\alpha$-$Al_2O_3$. Surfactant behavior of SDBS hindered the agglomeration and increased wettability of $\alpha$-$Al_2O_3$ particles with PP matrix, leading to enhanced mechanical properties of the $\alpha$-$Al_2O_3$/PP composites.

## 4.2 Thermal properties

Depending on design parameters, the combination of metal oxides with a polymer matrix enhances flame retardancy, thermal stability, and thermal conductivity of the metal oxide–polymer composites.

Melting temperature, decomposition temperatures, and char residue of the metal oxide–polymer composites are generally improved using the optimum amount of metal oxide. Elashmawi et al. synthesized ZnO/PVC nanocomposites by the solvent casting method and investigated their thermal properties [131]. As per differential scanning calorimetry analysis, $T_g$ of the nanocomposites increased with 20 wt% ZnO, which was attributed to increased amorphous region of the polymer matrix. Thermogravimetric analysis results showed that mass loss occurred at three steps. Initial mass loss originated from the evaporation of water molecules and solvents. The polymer matrix began to decompose, and then the mass of the material remained constant at high temperatures in which char residue formation was observed in the second step. Char residue of the neat PVC at 600°C increased from ~13% to ~30% in the presence of 20 wt% ZnO. Li and Sun prepared PLA nanocomposites including surface hydroxylated MgO (0.005–0.2 wt%) by in situ polymerization. Thermal stability of the nanocomposites was improved with the increasing amount of MgO. While the onset decomposition temperature ($T_{onset}$) of neat PLA was 278.5°C, it was increased to 296.2°C with 0.2 wt% MgO incorporation. Moreover, the maximum decomposition temperature ($T_{max}$) of the PLA nanocomposite including 0.2 wt% MgO was determined as 324°C.

The flammability of polymeric materials is considerably high that restricts their utilization in many fields. In spite of using flame retardants to overcome this problem, they have some disadvantages such as the requirement of high loading and the release of hazardous volatiles during the combustion. So, metal oxides may be promising flame retardants for polymer matrix composites. Metal oxides can exhibit activity in the gas and condensed phase. Metal oxides minimize radical reactions leading to decelerate combustion in the case of the gas phase. It can be explained with the formation of HOO· radicals which are less reactive than HO· radicals by collision energy. In the condensed phase, metal oxides act as a barrier on the material surface that hinders the release of decomposition products and oxygen diffusion to the material [98]. Hajibeygi et al. examined the effect of chitosan-modified ZnO addition on the flame retardancy of the PVC nanocomposites [132]. It was shown that the first peak heat release rate (pHRR) of neat PVC decreased with the addition of 6 wt% chitosan modified ZnO through the formation of olefinic structure, avoiding HCl release. A decrease in total heat release and heat release capacity indicated improved flame retardancy of the PVC nanocomposites. Laachachi et al. used nano- and micro-$TiO_2$ as a flame retardant for PMMA composites synthesized by melt blending [117]. Although favorable influence on the flame retardancy properties of the neat PMMA were observed with both nano- and micro-size filler, PMMA composites including nano $TiO_2$ showed lower pHRR, time to ignition, and total smoke released owing to the advantage of homogeneous dispersion of nano-sized $TiO_2$ in the PMMA matrix.

It is known that polymeric materials generally exhibit lower thermal conductivity. The utilization of thermally conductive metal oxides such as $Al_2O_3$ improves thethermal conductivity of the polymer composites. For instance, epoxy resin has low intrinsic thermal conductivity (0.15–0.25 $W m^{-1} K^{-1}$). Hao et al. preferred 3-glycidoxypropyltrimethoxy silane functionalized three-dimensional $Al_2O_3$ as filler to increase the thermal conductivity of the epoxy resin [112]. Thermal conductivity of the epoxy composite including 43 vol% $Al_2O_3$ (3.796 $W m^{-1} K^{-1}$) reached to 4.356 $W m^{-1} K^{-1}$ after filler surface functionalization at room temperature. By the help of functionalization through covalent bonding that increases interfacial adhesion between $Al_2O_3$ and epoxy matrix, decreasing interfacial heat resistance and phonon scattering was effective to enhance the thermal conductivity of the epoxy composites. Hu et al. revealed the effect of $Al_2O_3$ content on the thermal conductivity of the epoxy composites [111]. It was clear that the thermal conductivity of the epoxy composites showed an increase with increasing $Al_2O_3$ loading. It was attributed to the close contact of $Al_2O_3$ particle to another one resulting in easily thermal conduction.

## 4.3 Electrical properties

For electronic devices, the electrical properties of the metal oxide–polymer composites are one of the most important parameters to consider. Electrical conductivity, which is conducted by the formation of a continuous conductive network, is achieved by using electrically conductive metal oxides or polymer matrix such as polyacetylene, polypyrrole, polyaniline, polydiaminonapthalene, polythiophene, and polyterthiophene [142,143].

An increase in the amount of metal oxide generally provides effective conductive networks resulting in enhanced electrical conductivity. Rajesh et al. showed the effect of $TiO_2$ loading (2–16 wt%) on the electrical conductivity of PVA/PVP nanocomposites [97]. Until 12 wt% $TiO_2$ loading, the electrical conductivity of PVA/PVP nanocomposites increased depending on the formation of charge transfer complexes in the PVA/PVP matrix that increased electron mobility. However, a slight decrease in electrical conductivity was determined with the addition of 16 wt% $TiO_2$ that was explained with a possible reduction in conduction path length due to agglomerations. Suhailath et al. investigated the electrical conductivity of $TiO_2$/PBMA nanocomposites under the effect of $TiO_2$ loading (3–10 wt%) [101]. The low electrical conductivity of PBMA, which is originated from randomly oriented polymer chains, increased with 7 wt % $TiO_2$ incorporation due to the improved compactness of the PBMA nanocomposites. More $TiO_2$ loading caused to reduce the electrical conductivity of PBMA nanocomposites. Batool et al. synthesized ZnO/polypyrrole composites by chemical oxidation polymerization [133]. The electrical conductivity of neat polypyrrole was increased from 6 $S cm^{-1}$ to 18.52 $S cm^{-1}$ in the presence of 20 wt% ZnO. Effective charge mobility through the conductive network considerably enhanced the electrical conductivity of the ZnO/polypyrrole composites at higher ZnO loading.

Ma et al. synthesized ZnO/PS nanocomposites by the melt blending method. The influence of ZnO particle shape (spherical (s) or whisker (w)) and surface modification of ZnO with phenyltriethoxysilane on the electrical conductivity of the ZnO/PS nanocomposites was revealed. The electrical conductivity of w-ZnO/PS nanocomposites was higher than that of the s-ZnO/PS nanocomposites. It was contributed to a higher aspect ratio and surface electron intensity of the w-ZnO particles that provided an effective conduction network. Furthermore, a decrease in surface resistivity of the ZnO/PS nanocomposites by the surface modification affected electrical conductivity positively.

## *4.4 Optical properties*

When considered the optical properties of the metal oxide–polymer composites, transparency is a key parameter. Depending on applications for different purposes, the degree of transparency can be changed in the metal oxide–polymer composites, which is related to particle size, refractive index, dispersion, and orientation of the metal oxides [144]. Nano-size metal oxides provide higher transparency compared to micro size metal oxides. In other words, larger particles can easily scatter the light that decreases the transparency of the metal oxide–polymer composites. Metal oxide dispersion type is an effective parameter on the specific optical properties of the metal oxide–polymer composites such as UV/visible/infrared light absorption, refractive index, dichroism, and photoluminescence [51]. For instance, UV/infrared absorption, optical transparency, and photoluminescence are promoted through random dispersion of metal oxide particles in the polymer matrix. Ordered dispersion can result in iridescence, while dichroism formation can be observed by uniaxial dispersion in the metal oxide–polymer composites [145].

Chau et al. stated the optical properties of the epoxy nanocomposites including different fractions of $TiO_2$ (10–90 wt%) [92]. Higher transparency ($>90\%$) was achieved for the epoxy nanocomposites in the presence of 10–40 wt% $TiO_2$. Taha et al. synthesized NiO/PVC nanocomposites by the solution casting method and investigated their optical properties as a function of NiO loading [137]. The refractive index of neat PVC increased from 1.42 to 1.79 with the addition of 0.5 wt% NiO. Kruenate et al. revealed the UV blocking ability of the ZnO/PP nanocomposites even at low ZnO loading $\sim$2 wt% [146]. Mallakpour and Barati prepared PVA nanocomposites with the incorporation of $TiO_2$ modified by $\gamma$-aminopropyltriethoxy silane [94]. It was reported that an increase in the amount of $TiO_2$ had synergistic effects on the UV absorption of the nanocomposites. The highest UV absorption percentage was obtained for PVA nanocomposite including 20 wt% $TiO_2$.

Bhavsar et al. exhibited the effect of $Al_2O_3$ loading (1–5 wt%) on the photoluminescence intensity of PS nanocomposites [113]. It was specified that PS nanocomposite including 3 wt% $Al_2O_3$ had higher peak intensity than that of the other nanocomposites. It was related to a decreasing bandgap that enhanced the photoluminescence properties of the $Al_2O_3$/PS nanocomposites. Donya et al. studied on optical properties of the iron oxide/PVA nanocomposites synthesized by the solution casting method [118]. It was revealed that the fluorescence intensity of the nanocomposites decreased with increasing iron oxide loading.

## 4.5 Magnetic properties

Polymer composites can exhibit magnetic properties with the incorporation of metal oxides such as $Fe_3O_4$. Bhatt et al. contributed a ferromagnetic behavior to non-magnetic PVDF with the utilization of magnetic $Fe_3O_4$ particles [121]. With the increasing amount of $Fe_3O_4$, an increase was observed in the magnetic saturation of the $Fe_3O_4$/PVDF composites. Reddy et al. synthesized $Fe_3O_4$/PEDOT composites by in situ chemical oxidative polymerization [119]. Maximum magnetic saturation and remnant magnetization were determined as 13.2 emu $g^{-1}$ and 2.42 emu $g^{-1}$, respectively. There was no hysteresis loop formation for the composites, which was an indication of superparamagnetic $Fe_3O_4$/PVDF composite preparation. In their other study, $Fe_3O_4$/polyaniline nanocomposites were synthesized using inorganic acid (HCl) and organic acid (CSA or TSA) as dopant [120]. The lowest magnetic saturation was 1.21 emu $g^{-1}$ for $Fe_3O_4$/polyaniline/HCl nanocomposite because of HCl leaching that caused to separate $Fe_3O_4$ particles from the nanocomposite structure. Moreover, all types of $Fe_3O_4$/polyaniline nanocomposites showed ferromagnetic behavior under the magnetic field.

## 4.6 Barrier and antibacterial properties

Barrier and antibacterial properties of the metal oxide–polymer composites are significant parameters for specific applications that require protection of the materials against $O_2$, $CO_2$, methanol, water vapor, and various microorganisms. In the case of using metal oxides, barrier properties of the polymer composites can be considerably improved owing to the low permeability of the metal oxide particles which promote a long and tortuous diffusion path. Good compatibility between metal oxide particles and polymer matrix minimizes the interactions of polymer chains with gas or water molecules. Furthermore, in the presence of metal oxides that can exhibit antibacterial activity naturally or by stimuli, the metal oxide–polymer composites can hinder microorganism growth.

Farhoodi et al. examined the effect of $TiO_2$ addition on the barrier properties of the neat PET [104]. Water vapor permeability of the PET was considerably decreased with the incorporation of 3 wt% $TiO_2$ through a longer tortuous path of water molecules. Hirvikorpi et al. represented that the water vapor transmission rate of neat PLA decreased by 47% with $Al_2O_3$ film [115]. Diez-Pascual et al. examined the antibacterial activity of the ZnO/PHBV nanocomposites in addition to their $O_2$ permeability behavior [129]. $O_2$ permeability of the PHBV nanocomposites decreased to 35% in the presence of 4 wt% ZnO. However, more ZnO content caused to increase $O_2$ permeability due to ZnO agglomerations that easily promoted path formation for $O_2$ diffusion. According to bacteria tests, the best antibacterial activity against *Escherichia coli* (*E. coli*) and *Staphylococcus aureus* (*S. aureus*) was observed for PHBV nanocomposite including 5 wt% ZnO which provides bacteria cell disruption. *E. coli* growth was inhibited in a short time in contrast to *S. aureus* that was assigned to the different cell structure of two bacteria. Marra et al. revealed $O_2$ and $CO_2$ permeability and antibacterial activity performance of the ZnO/PLA composites as a function of ZnO loading (1–5 wt%) [128]. $O_2$ and $CO_2$ permeabilities of the neat PLA were decreased with the 5 wt% ZnO addition by 17% and 14%, respectively. And also, *E. coli* inhibition was enhanced with the increasing amount of ZnO. While *E. coli* reduction was ~42% for PLA composite including 1 wt% ZnO in 24 h, ~100% reduction was achieved with the PLA composite including 5 wt% ZnO through the attachment of ZnO particles to the cell membrane.

# 5 Conclusion

Synthesis of the metal oxide–polymer composites is generally carried out by the solution or melt blending, sol–gel method, and in situ polymerization. The metal oxide–polymer composites are designed depending on the intended purpose for different applications, environmental and economic concerns. Many design parameters such as metal oxide and polymer matrix properties, surface modifications, and production conditions significantly influence the properties of the composites. Uniform metal oxide dispersion in the polymer matrix, which is highly related to metal oxide shape, size, aspect ratio, and surface chemistry, is desirable to produce high-performance metal oxide–polymer composites. So, better compatibility between a metal oxide and polymer matrix enhances many properties of the composite materials including mechanical, thermal, electrical, optical, magnetic, barrier, antibacterial, etc.

# References


[1] L. Bruno, Mechanical characterization of composite materials by optical techniques: a review, Opt. Lasers Eng. 104 (2018) 192–203.

[2] N. Shetty, S.M. Shahabaz, S.S. Sharma, S. Divakara Shetty, A review on finite element method for machining of composite materials, Compos. Struct. 176 (2017) 790–802.

[3] G. Guzel, H. Deveci, Properties of polymer composites based on bisphenol A epoxy resins with original/modified steel slag, Polym. Compos. 39 (2) (2018) 513–521.

[4] G. Guzel, H. Deveci, Physico-mechanical, thermal, and coating properties of composite materials prepared with epoxy resin/steel slag, Polym. Compos. 38 (9) (2017) 1974–1981.

[5] G.C. Papanicolaou, S.P. Zaoutsos, Viscoelastic constitutive modeling of creep and stress relaxation in polymers and polymer matrix composites, in: R.M. Guedes (Ed.), Creep and Fatigue in Polymer Matrix Composites, Woodhead Publishing, 2019, pp. 3–59.

[6] A. Adeniyi, O. Agboola, E.R. Sadiku, M.O. Durowoju, P.A. Olubambi, A. Babul Reddy, I.D. Ibrahim, W.K. Kupolati, Chapter 2—thermoplastic-thermoset nanostructured polymer blends, in: S. Thomas, R. Shanks, S. Chandrasekharakurup (Eds.), Design and Applications of Nanostructured Polymer Blends and Nanocomposite Systems, William Andrew Publishing, Boston, 2016, pp. 15–38.

[7] M.R. Vengatesan, A.M. Varghese, V. Mittal, Chapter 3—thermal properties of thermoset polymers, in: Q. Guo (Ed.), Thermosets, second ed., Elsevier, 2018, pp. 69–114.

[8] J. Robert, Young, P.A. Lovell, in: R.J. Young, P.A. Lovell (Eds.), Introduction to polymers, third ed., CRS Press, 2011, pp. 511–529.

[9] R.R. Babu, G.S. Shibulal, A.K. Chandra, K. Naskar, Compounding and vulcanization, in: P.M. Visakh, S. Thomas, A.K. Chandra, A.P. Mathew (Eds.), Advances in Elastomers I, Springer, 2013, pp. 83–133.
[10] Y. Zhu, C. Romain, C.K. Williams, Sustainable polymers from renewable resources, Nature 540 (7633) (2016) 354–362.
[11] A. Pellis, M. Malinconico, A. Guarneri, L. Gardossi, Renewable polymers and plastics: performance beyond the green, N. Biotechnol. 60 (2021) 146–158.
[12] L. Yu, K. Dean, L. Li, Polymer blends and composites from renewable resources, Prog. Polym. Sci. 31 (6) (2006) 576–602.
[13] A. Sharif, M.E. Hoque, Renewable resource-based polymers, in: M.L. Sanyang, M. Jawaid (Eds.), Bio-based Polymers and Nanocomposites: Preparation, Processing, Properties & Performance, Springer International Publishing, Cham, 2019, pp. 1–28.
[14] X. Wang, E.N. Kalali, J.-T. Wan, D.-Y. Wang, Carbon-family materials for flame retardant polymeric materials, Prog. Polym. Sci. 69 (2017) 22–46.
[15] V. Eskizeybek, H. Ulus, H.B. Kaybal, Ö.S. Şahin, A. Avcı, Static and dynamic mechanical responses of $CaCO_3$ nanoparticle modified epoxy/carbon fiber nanocomposites, Compos. Part B Eng. 140 (2018) 223–231.
[16] Y.P. Mamunya, H. Zois, L. Apekis, E.V. Lebedev, Influence of pressure on the electrical conductivity of metal powders used as fillers in polymer composites, Powder Technol. 140 (1-2) (2004) 49–55.
[17] G. Guzel, O. Sivrikaya, H. Deveci, The use of colemanite and ulexite as novel fillers in epoxy composites: Influences on thermal and physico-mechanical properties, Compos. Part B Eng. 100 (2016) 1–9.
[18] G.-W. Lee, M. Park, J. Kim, J.I. Lee, H.G. Yoon, Enhanced thermal conductivity of polymer composites filled with hybrid filler, Compos. A: Appl. Sci. Manuf. 37 (5) (2006) 727–734.
[19] G.G. Kaya, E. Yilmaz, H. Deveci, Sustainable bean pod/calcined kaolin reinforced epoxy hybrid composites with enhanced mechanical, water sorption and corrosion resistance properties, Construct. Build Mater. 162 (2018) 272–279.
[20] M. Bulota, T. Budtova, Valorisation of macroalgae industrial by-product as filler in thermoplastic polymer composites, Compos. A: Appl. Sci. Manuf. 90 (2016) 271–277.
[21] H. Moustafa, A.M. Youssef, N.A. Darwish, A.I. Abou-Kandil, Eco-friendly polymer composites for green packaging: future vision and challenges, Compos. Part B Eng. 172 (2019) 16–25.
[22] D. Gao, X. Wen, Y. Guan, W. Czerwonko, Y. Li, Y. Gao, E. Mijowska, T. Tang, Flame retardant effect and mechanism of nanosized NiO as synergist in PLA/APP/CSi-MCA composites, Compos. Commun. 17 (2020) 170–176.
[23] J. Wu, X. Huang, K. Berglund, X. Lu, X. Feng, R. Larsson, Y. Shi, CuO nanosheets produced in graphene oxide solution: an excellent anti-wear additive for self-lubricating polymer composites, Compos. Sci. Technol. 162 (2018) 86–92.
[24] S.J. Choi, S. Savagatrup, Y. Kim, J.H. Lang, T.M. Swager, Precision pH sensor based on $WO_3$ nanofiber-polymer composites and differential amplification, ACS Sensors 4 (10) (2019) 2593–2598.
[25] W. Cai, W. Guo, Y. Pan, J. Wang, X. Mu, X. Feng, B. Yuan, B. Wang, Y. Hu, Polydopamine-bridged synthesis of ternary h-BN@PDA@$SnO_2$ as nanoenhancers for flame retardant and smoke suppression of epoxy composites, Compos. A: Appl. Sci. Manuf. 111 (2018) 94–105.

[26] E.I. Akpan, X. Shen, B. Wetzel, K. Friedrich, 2—Design and synthesis of polymer nanocomposites, in: K. Pielichowski, T.M. Majka (Eds.), Polymer Composites With Functionalized Nanoparticles, Elsevier, 2019, pp. 47–83.
[27] A.A. Maharramov, M.A. Ramazanov, L. Di Palma, H.A. Shirinova, F.V. Hajiyeva, Influence of magnetite nanoparticles on the dielectric properties of metal oxide/polymer nanocomposites based on polypropylene, Russ. Phys. J. 60 (9) (2018) 1572–1576.
[28] C.V. Garcia, G.H. Shin, J.T. Kim, Metal oxide-based nanocomposites in food packaging: applications, migration, and regulations, Trends Food Sci. Technol. 82 (2018) 21–31.
[29] A.M. Pourrahimi, R.T. Olsson, M.S. Hedenqvist, The role of interfaces in polyethylene/metal-oxide nanocomposites for ultrahigh-voltage insulating materials, Adv. Mater. 30 (4) (2018).
[30] N. Hoogesteijn von Reitzenstein, X. Bi, Y. Yang, K. Hristovski, P. Westerhoff, Morphology, structure, and properties of metal oxide/polymer nanocomposite electrospun mats, J. Appl. Polym. Sci. 133 (33) (2016).
[31] S.J. Jung, T. Lutz, M. Boese, J.D. Holmes, J.J. Boland, Surface energy driven agglomeration and growth of single crystal metal wires, Nano Lett. 11 (3) (2011) 1294–1299.
[32] N. Mandzy, E. Grulke, T. Druffel, Breakage of $TiO_2$ agglomerates in electrostatically stabilized aqueous dispersions, Powder Technol. 160 (2) (2005) 121–126.
[33] V. Polshettiwar, B. Baruwati, R.S. Varma, Self-assembly of metal oxides into three-dimensional nanostructures: synthesis and application in catalysis, ACS Nano 3 (3) (2009) 728–736.
[34] Y.W. Jun, J.S. Choi, J. Cheon, Shape control of semiconductor and metal oxide nanocrystals through nonhydrolytic colloidal routes, Angew. Chem. Int. Ed. 45 (21) (2006) 3414–3439.
[35] C. Pan, D. Zhang, L. Shi, CTAB assisted hydrothermal synthesis, controlled conversion and CO oxidation properties of $CeO_2$ nanoplates, nanotubes, and nanorods, J. Solid State Chem. 181 (6) (2008) 1298–1306.
[36] P.J. Wojcik, L. Pereira, R. Martins, E. Fortunato, Metal oxide nanoparticle engineering for printed electrochemical applications, in: M. Aliofkhazraei, A.S.H. Makhlouf (Eds.), Handbook of Nanoelectrochemistry: Electrochemical Synthesis Methods, Properties and Characterization Techniques, Springer International Publishing, Cham, 2016, pp. 1–29.
[37] L.A. Bauer, N.S. Birenbaum, G.J. Meyer, Biological applications of high aspect ratio nanoparticles, J. Mater. Chem. 14 (4) (2004) 517.
[38] R.Z. Waldman, D.J. Mandia, A. Yanguas-Gil, A.B.F. Martinson, J.W. Elam, S.B. Darling, The chemical physics of sequential infiltration synthesis-A thermodynamic and kinetic perspective, J. Chem. Phys. 151 (19) (2019) 190901.
[39] G. Korotcenkov, B.K. Cho, Instability of metal oxide-based conductometric gas sensors and approaches to stability improvement, Sens. Actuators B 156 (2) (2011) 527–538.
[40] L. Yang, H. Wu, W. Zhang, Z. Chen, J. Li, X. Lou, Z. Xie, R. Zhu, H. Chang, Anomalous oxidation and its effect on electrical transport originating from surface chemical instability in large-area, few-layer 1T'-$MoTe_2$ films, Nanoscale 10 (42) (2018) 19906–19915.
[41] J. Liu, Z. Guo, K. Zhu, W. Wang, C. Zhang, X. Chen, Highly porous metal oxide polycrystalline nanowire films with superior performance in gas sensors, J. Mater. Chem. 21 (30) (2011) 11412–11417.
[42] J.C. Wang, S.P. Hill, T. Dilbeck, O.O. Ogunsolu, T. Banerjee, K. Hanson, Multimolecular assemblies on high surface area metal oxides and their role in interfacial energy and electron transfer, Chem. Soc. Rev. 47 (1) (2018) 104–148.

[43] Z. Sun, T. Liao, Y. Dou, S.M. Hwang, M.S. Park, L. Jiang, J.H. Kim, S.X. Dou, Generalized self-assembly of scalable two-dimensional transition metal oxide nanosheets, Nat. Commun. 5 (2014) 3813.
[44] L.Y. Ng, A.W. Mohammad, C.P. Leo, N. Hilal, Polymeric membranes incorporated with metal/metal oxide nanoparticles: a comprehensive review, Desalination 308 (2013) 15–33.
[45] S. Chen, Y. Cheng, Q. Xie, B. Xiao, Z. Wang, J. Liu, G. Wu, Enhanced breakdown strength of aligned-sodium-titanate- nanowire/epoxy nanocomposites and their anisotropic dielectric properties, Compos. A: Appl. Sci. Manuf. 120 (2019) 84–94.
[46] R. Gholami, H. Khoramishad, L.F.M. da Silva, Glass fiber-reinforced polymer nanocomposite adhesive joints reinforced with aligned carbon nanofillers, Compos. Struct. 253 (2020) 112814.
[47] X. Shen, Q. Zheng, J.-K. Kim, Rational design of two-dimensional nanofillers for polymer nanocomposites toward multifunctional applications, Prog. Mater. Sci. 115 (2021) 100708.
[48] O. Harnack, C. Pacholski, H. Weller, A. Yasuda, J.M. Wessels, Rectifying behavior of electrically aligned ZnO nanorods, Nano Lett. 3 (8) (2003) 1097–1101.
[49] J.-P. Jolivet, S. Cassaignon, C. Chanéac, D. Chiche, O. Durupthy, D. Portehault, Design of metal oxide nanoparticles: control of size, shape, crystalline structure and functionalization by aqueous chemistry, C. R. Chim. 13 (1-2) (2010) 40–51.
[50] F. Ahangaran, A.H. Navarchian, Recent advances in chemical surface modification of metal oxide nanoparticles with silane coupling agents: a review, Adv. Colloid Interface Sci. 286 (2020) 102298.
[51] S. Kango, S. Kalia, A. Celli, J. Njuguna, Y. Habibi, R. Kumar, Surface modification of inorganic nanoparticles for development of organic–inorganic nanocomposites-a review, Prog. Polym. Sci. 38 (8) (2013) 1232–1261.
[52] H. Essabir, M.O. Bensalah, D. Rodrigue, R. Bouhfid, K. Qaiss Ael, Biocomposites based on Argan nut shell and a polymer matrix: effect of filler content and coupling agent, Carbohydr. Polym. 143 (2016) 70–83.
[53] S.F. Mendes, C.M. Costa, C. Caparros, V. Sencadas, S. Lanceros-Méndez, Effect of filler size and concentration on the structure and properties of poly(vinylidene fluoride)/ $BaTiO_3$ nanocomposites, J. Mater. Sci. 47 (3) (2011) 1378–1388.
[54] P.K. Kumar, N.V. Raghavendra, B.K. Sridhara, Optimization of infrared radiation cure process parameters for glass fiber reinforced polymer composites, Mater. Des. 32 (3) (2011) 1129–1137.
[55] M. Komorowska-Durka, G. Dimitrakis, D. Bogdał, A.I. Stankiewicz, G.D. Stefanidis, A concise review on microwave-assisted polycondensation reactions and curing of polycondensation polymers with focus on the effect of process conditions, Chem. Eng. J. 264 (2015) 633–644.
[56] D. Chen, K. Arakawa, C. Xu, Reduction of void content of vacuum-assisted resin transfer molded composites by infusion pressure control, Polym. Compos. 36 (9) (2015) 1629–1637.
[57] A.K. Gopal, S. Adali, V.E. Verijenko, Optimal temperature profiles for minimum residual stress in the cure process of polymer composites, Compos. Struct. 48 (2000) 99–106.
[58] N. Geier, T. Szalay, Optimisation of process parameters for the orbital and conventional drilling of uni-directional carbon fibre-reinforced polymers (UD-CFRP), Measurement 110 (2017) 319–334.
[59] V.K. Vankanti, V. Ganta, Optimization of process parameters in drilling of GFRP composite using Taguchi method, J. Mater. Res. Technol. 3 (1) (2014) 35–41.

[60] Z. Zhang, K. Friedrich, Artificial neural networks applied to polymer composites: a review, Compos. Sci. Technol. 63 (14) (2003) 2029–2044.
[61] T.G. Wakjira, M.L. Nehdi, U. Ebead, Fractional factorial design model for seismic performance of RC bridge piers retrofitted with steel-reinforced polymer composites, Eng. Struct. 221 (2020) 111100.
[62] C. Huang, Q. Cheng, Learning from nacre: constructing polymer nanocomposites, Compos. Sci. Technol. 150 (2017) 141–166.
[63] P. Zarrintaj, R. Khalili, H. Vahabi, M.R. Saeb, M.R. Ganjali, M. Mozafari, Chapter 8—polyaniline/metal oxides nanocomposites, in: M. Mozafari, N.P.S. Chauhan (Eds.), Fundamentals and Emerging Applications of Polyaniline, Elsevier, 2019, pp. 131–141.
[64] A. Sajedi-Moghaddam, E. Saievar-Iranizad, M. Pumera, Two-dimensional transition metal dichalcogenide/conducting polymer composites: synthesis and applications, Nanoscale 9 (24) (2017) 8052–8065.
[65] R. Verdejo, M.M. Bernal, L.J. Romasanta, M.A. Lopez-Manchado, Graphene filled polymer nanocomposites, J. Mater. Chem. 21 (10) (2011) 3301–3310.
[66] S. Mallakpour, N. Nouruzi, Polycaprolactone/metal oxide nanocomposites: an overview of recent progress and applications, in: N.G. Shimpi (Ed.), Biodegradable and Biocompatible Polymer Composites Processing, Properties and Applications, Elsevier, Woodhead Publishing, 2018, pp. 223–228.
[67] R.Y. Hong, B. Feng, G. Liu, S. Wang, H.Z. Li, J.M. Ding, Y. Zheng, D.G. Wei, Preparation and characterization of Fe3O4/polystyrene composite particles via inverse emulsion polymerization, J. Alloys Compd. 476 (1) (2009) 612–618.
[68] S.H. Soytaş, O. Oğuz, Y.Z. Menceloğlu, 9—polymer nanocomposites with decorated metal oxides, in: K. Pielichowski, T.M. Majka (Eds.), Polymer Composites With Functionalized Nanoparticles, Elsevier, 2019, pp. 287–323.
[69] A.D. Pomogailo, Polymer sol-gel synthesis of hybrid nanocomposites, Colloid J. 67 (6) (2005) 658–677.
[70] M. Guglielmi, Synthesis strategies for the preparation of sol-gel nanocomposites, in: M. Guglielmi, G. Kickelbick, A. Martucci (Eds.), Sol-Gel Nanocomposites, New York, NY, Springer New York, 2014, pp. 51–82.
[71] S. Pandey, S.B. Mishra, Sol–gel derived organic–inorganic hybrid materials: synthesis, characterizations and applications, J. Sol-Gel Sci. Technol. 59 (1) (2011) 73–94.
[72] S.R.V.S. Prasanna, K. Balaji, S. Pandey, S. Rana, Chapter 4— metal oxide based nanomaterials and their polymer nanocomposites, in: N. Karak (Ed.), Nanomaterials and Polymer Nanocomposites, Elsevier, 2019, pp. 123–144.
[73] M. Yoshida, M. Lal, N.D. Kumar, P.N. Prasad, $TiO_2$ nano-particle-dispersed polyimide composite optical waveguide materials through reverse micelles, J. Mater. Sci. 32 (15) (1997) 4047–4051.
[74] M.T. Ramesan, V. Nidhisha, P. Jayakrishnan, Synthesis, characterization and conducting properties of novel poly (vinyl cinnamate)/zinc oxide nanocomposites via in situ polymerization, Mater. Sci. Semicond. Process. 63 (2017) 253–260.
[75] G. Kickelbick, The search of a homogeneously dispersed material—the art of handling the organic polymer/metal oxide interface, J. Sol-Gel Sci. Technol. 46 (3) (2008) 281–290.
[76] S.A. Bhat, F. Zafar, A.U. Mirza, A. Mohammad, P. Singh, N. Nishat, Fabrication and biomedical applications of polyvinyl-alcohol-based nanocomposites with special emphasis on the anti-bacterial applications of metal/metal oxide polymer nanocomposites, in: S. Islam, B.S. Butola (Eds.), Advanced Functional Textiles and Polymers, Wiley, Scrivener Publishing, 2019, pp. 309–320.

[77] T. Hanemann, D.V. Szabó, Polymer-nanoparticle composites: from synthesis to modern applications, Materials 3 (6) (2010) 3468–3517.
[78] S. Sarkar, E. Guibal, F. Quignard, A.K. SenGupta, Polymer-supported metals and metal oxide nanoparticles: synthesis, characterization, and applications, J. Nanopart. Res. 14 (2) (2012).
[79] E. Bourgeat-Lami, Organic-inorganic nanostructured colloids, J. Nanosci. Nanotechnol. 2 (1) (2002) 1–24.
[80] H. Roghani-Mamaqani, V. Haddadi-Asl, M. Salami-Kalajahi, In situ controlled radical polymerization: a review on synthesis of well-defined nanocomposites, Polym. Rev. 52 (2) (2012) 142–188.
[81] J.N. Coleman, U. Khan, W.J. Blau, Y.K. Gun'ko, Small but strong: a review of the mechanical properties of carbon nanotube–polymer composites, Carbon 44 (9) (2006) 1624–1652.
[82] S. Hansson, V. Trouillet, T. Tischer, A.S. Goldmann, A. Carlmark, C. Barner-Kowollik, E. Malmstrom, Grafting efficiency of synthetic polymers onto biomaterials: a comparative study of grafting-from versus grafting-to, Biomacromolecules 14 (1) (2013) 64–74.
[83] H. Mahmoodian, O. Moradi, B. Shariatzadeh, Grafting chitosan and polyHEMA on carbon nanotubes surfaces: "grafting to" and "grafting from" methods, Int. J. Biol. Macromol. 63 (2014) 92–97.
[84] H. Liu, H.-Y. Zhao, F. Müller-Plathe, H.-J. Qian, Z.-Y. Sun, Z.-Y. Lu, Distribution of the number of polymer chains grafted on nanoparticles fabricated by grafting-to and grafting-from procedures, Macromolecules 51 (10) (2018) 3758–3766.
[85] R. Zhang, Y. Wang, D. Ma, S. Ahmed, W. Qin, Y. Liu, Effects of ultrasonication duration and graphene oxide and nano-zinc oxide contents on the properties of polyvinyl alcohol nanocomposites, Ultrason. Sonochem. 59 (2019) 104731.
[86] R. Rotaru, M. Savin, N. Tudorachi, C. Peptu, P. Samoila, L. Sacarescu, V. Harabagiu, Ferromagnetic iron oxide–cellulose nanocomposites prepared by ultrasonication, Polym. Chem. 9 (7) (2018) 860–868.
[87] B.A. Bhanvase, S.H. Sonawane, Ultrasound assisted in situ emulsion polymerization for polymer nanocomposite: a review, Chem. Eng. Process. Process Intensif. 85 (2014) 86–107.
[88] V. Evora, A. Shukla, Fabrication, characterization, and dynamic behavior of polyester/$TiO_2$ nanocomposites, Mater. Sci. Eng. A 361 (1-2) (2003) 358–366.
[89] H. Liu, T.J. Webster, Mechanical properties of dispersed ceramic nanoparticles in polymer composites for orthopedic applications, Int. J. Nanomedicine 5 (2010) 299–313.
[90] Z. Zhang, J.-L. Yang, K. Friedrich, Creep resistant polymeric nanocomposites, Polymer 45 (10) (2004) 3481–3485.
[91] M. Khan, A.A. Khurram, T. Li, T. Zhao, T. Subhani, I.H. Gul, Z. Ali, V. Patel, Synergistic effect of organic and inorganic nano fillers on the dielectric and mechanical properties of epoxy composites, J. Mater. Sci. Technol. 34 (12) (2018) 2424–2430.
[92] J.L.H. Chau, C.-T. Tung, Y.-M. Lin, A.-K. Li, Preparation and optical properties of titania/epoxy nanocomposite coatings, Mater. Lett. 62 (19) (2008) 3416–3418.
[93] R. Singh, S.G. Kulkarni, S.S. Channe, Thermal and mechanical properties of nano-titanium dioxide-doped polyvinyl alcohol, Polym. Bull. 70 (4) (2012) 1251–1264.
[94] S. Mallakpour, A. Barati, Efficient preparation of hybrid nanocomposite coatings based on poly(vinyl alcohol) and silane coupling agent modified $TiO_2$ nanoparticles, Prog. Org. Coat. 71 (4) (2011) 391–398.
[95] A. Shehap, D.S. Akil, Structural and optical properties of $TiO_2$ nanoparticles/PVA for different composites thin films, Int. J. Nanoelectron. Mater. 9 (2016) 17–36.

[96] S. Sugumaran, C.S. Bellan, Transparent nano composite PVA–$TiO_2$ and PMMA–$TiO_2$ thin films: optical and dielectric properties, Optik 125 (18) (2014) 5128–5133.
[97] K. Rajesh, V. Crasta, N.B. Rithin Kumar, G. Shetty, P.D. Rekha, Structural, optical, mechanical and dielectric properties of titanium dioxide doped PVA/PVP nanocomposite, J. Polym. Res. 26 (4) (2019).
[98] B. Friederich, A. Laachachi, M. Ferriol, D. Ruch, M. Cochez, V. Toniazzo, Tentative links between thermal diffusivity and fire-retardant properties in poly(methyl methacrylate)–metal oxide nanocomposites, Polym. Degrad. Stab. 95 (7) (2010) 1183–1193.
[99] Z.G. Wang, X.T. Zu, X. Xiang, H.J. Yu, Photoluminescence from $TiO_2$/PMMA nanocomposite prepared by $\gamma$ radiation, J. Nanopart. Res. 8 (1) (2006) 137–139.
[100] K. Suhailath, P. Jayakrishnan, B. Naufal, P. Periyat, V.C. Jasna, M.T. Ramesan, Synthesis by in situ-free radical polymerization, characterization, and properties of poly (n-butyl methacrylate)/samarium-doped titanium dioxide nanoparticles composites, Adv. Polym. Technol. 37 (4) (2018) 1114–1123.
[101] K. Suhailath, M.T. Ramesan, B. Naufal, P. Periyat, V.C. Jasna, P. Jayakrishnan, Synthesis, characterisation and flame, thermal and electrical properties of poly (n-butyl methacrylate)/titanium dioxide nanocomposites, Polym. Bull. 74 (3) (2016) 671–688.
[102] T.-C. Mo, H.-W. Wang, S.-Y. Chen, Y.-C. Yeh, Synthesis and dielectric properties of polyaniline/titanium dioxide nanocomposites, Ceram. Int. 34 (7) (2008) 1767–1771.
[103] A. Nasiri, M.S.-N. Shariaty-Niasar, Z. Akbari, Synthesis of LDPE/Nano $TiO_2$ nanocomposite for packaging applications, Int. J. Nanosci. Nanotechnol. 8 (3) (2012) 165–170.
[104] M. Farhoodi, M.A. Mohammadifar, M. Mousavi, R. Sotudeh-Gharebagh, Z. Emam-Djomeh, Migration kinetics of ethylene glycol monomer from pet bottles into acidic food simulant: effects of nanoparticle presence and matrix morphology, J. Food Process Eng. 40 (2) (2017) e12383.
[105] H.M. Akil, N. Lily, J.A. Razak, H. Ong, Z.A. Ahmad, Effect of various coupling agents on properties of alumina-filled PP composites, J. Reinf. Plast. Compos. 25 (7) (2016) 745–759.
[106] F. Mirjalili, L. Chuah, E. Salahi, Mechanical and morphological properties of polypropylene/nano alpha-$Al_2O_3$ composites, Scientific World Journal 2014 (2014) 718765.
[107] M. Zhang, R.P. Singh, Mechanical reinforcement of unsaturated polyester by $Al_2O_3$ nanoparticles, Mater. Lett. 58 (3–4) (2004) 408–412.
[108] X. Zhang, X. Xia, H. You, T. Wada, P. Chammingkwan, A. Thakur, T. Taniike, Design of continuous segregated polypropylene/$Al_2O_3$ nanocomposites and impact of controlled $Al_2O_3$ distribution on thermal conductivity, Compos. A: Appl. Sci. Manuf. 131 (2020) 105825.
[109] Z. Guo, T. Pereira, O. Choi, Y. Wang, H.T. Hahn, Surface functionalized alumina nanoparticle filled polymeric nanocomposites with enhanced mechanical properties, J. Mater. Chem. 16 (27) (2006) 2800.
[110] J. Cho, M.S. Joshi, C.T. Sun, Effect of inclusion size on mechanical properties of polymeric composites with micro and nano particles, Compos. Sci. Technol. 66 (13) (2006) 1941–1952.
[111] Y. Hu, G. Du, N. Chen, A novel approach for $Al_2O_3$ /epoxy composites with high strength and thermal conductivity, Compos. Sci. Technol. 124 (2016) 36–43.
[112] L.-C. Hao, Z.-X. Li, F. Sun, K. Ding, X.-N. Zhou, Z.-X. Song, Z.-Q. Shi, J.-F. Yang, B. Wang, High-performance epoxy composites reinforced with three-dimensional $Al_2O_3$ ceramic framework, Compos. A: Appl. Sci. Manuf. 127 (2019) 105648.

[113] S. Bhavsar, G.B. Patel, N.L. Singh, Investigation of optical properties of aluminium oxide doped polystyrene polymer nanocomposite films, Phys. B Condens. Matter 533 (2018) 12–16.
[114] E. Ritzhaupt-Kleissl, J. Boehm, J. Hausselt, T. Hanemann, Thermoplastic polymer nanocomposites for applications in optical devices, Mater. Sci. Eng. C 26 (5-7) (2006) 1067–1071.
[115] T. Hirvikorpi, M. Vähä-Nissi, A. Harlin, M. Salomäki, S. Areva, J.T. Korhonen, M. Karppinen, Enhanced water vapor barrier properties for biopolymer films by polyelectrolyte multilayer and atomic layer deposited Al2O3 double-coating, Appl. Surf. Sci. 257 (22) (2011) 9451–9454.
[116] M.T. Ramesan, Dynamic mechanical properties, magnetic and electrical behavior of iron oxide/ethylene vinyl acetate nanocomposites, Polym. Compos. 35 (10) (2014) 1989–1996.
[117] A. Laachachi, E. Leroy, M. Cochez, M. Ferriol, J.M. Lopez Cuesta, Use of oxide nanoparticles and organoclays to improve thermal stability and fire retardancy of poly(methyl methacrylate), Polym. Degrad. Stab. 89 (2) (2005) 344–352.
[118] H. Donya, T.A. Taha, A. Alruwaili, I.B.I. Tomsah, M. Ibrahim, Micro-structure and optical spectroscopy of PVA/iron oxide polymer nanocomposites, J. Mater. Res. Technol. 9 (4) (2020) 9189–9194.
[119] K.R. Reddy, W. Park, B.C. Sin, J. Noh, Y. Lee, Synthesis of electrically conductive and superparamagnetic monodispersed iron oxide-conjugated polymer composite nanoparticles by in situ chemical oxidative polymerization, J. Colloid Interface Sci. 335 (1) (2009) 34–39.
[120] K.R. Reddy, K.P. Lee, A.I. Gopalan, Self-assembly approach for the synthesis of electromagnetic functionalized $Fe_3O_4$/polyaniline nanocomposites: effect of dopant on the properties, Colloids Surf. A Physicochem. Eng. Asp. 320 (1–3) (2008) 49–56.
[121] A.S. Bhatt, D.K. Bhat, M.S. Santosh, Crystallinity, conductivity, and magnetic properties of PVDF-$Fe_3O_4$ composite films, J. Appl. Polym. Sci. 119 (2) (2011) 968–972.
[122] L.A. Ramajo, A.A. Cristóbal, P.M. Botta, J.M. Porto López, M.M. Reboredo, M.S. Castro, Dielectric and magnetic response of $Fe_3O_4$/epoxy composites, Compos. A: Appl. Sci. Manuf. 40 (4) (2009) 388–393.
[123] C.-C.M. Ma, Y.-J. Chen, H.-C. Kuan, Polystyrene nanocomposite materials: preparation, morphology, and mechanical, electrical, and thermal properties, J. Appl. Polym. Sci. 98 (5) (2005) 2266–2273.
[124] S. Tavassoli Hojati, H. Alaghemand, F. Hamze, F. Ahmadian Babaki, R. Rajab-Nia, M.B. Rezvani, M. Kaviani, M. Atai, Antibacterial, physical and mechanical properties of flowable resin composites containing zinc oxide nanoparticles, Dent. Mater. 29 (5) (2013) 495–505.
[125] J. Ahmed, Y.A. Arfat, H. Al-Attar, R. Auras, M. Ejaz, Rheological, structural, ultraviolet protection and oxygen barrier properties of linear low- density polyethylene films reinforced with zinc oxide (ZnO) nanoparticles, Food Packag. Shelf Life 13 (2017) 20–26.
[126] S.-C. Li, Y.-N. Li, Mechanical and antibacterial properties of modified nano-ZnO/high-density polyethylene composite films with a low doped content of nano-ZnO, J. Appl. Polym. Sci. (2010). NA-NA.
[127] Y.A. Arfat, J. Ahmed, A. Al Hazza, H. Jacob, A. Joseph, Comparative effects of untreated and 3-methacryloxypropyltrimethoxysilane treated ZnO nanoparticle

reinforcement on properties of polylactide-based nanocomposite films, Int. J. Biol. Macromol. 101 (2017) 1041–1050.
[128] A. Marra, C. Silvestre, D. Duraccio, S. Cimmino, Polylactic acid/zinc oxide biocomposite films for food packaging application, Int. J. Biol. Macromol. 88 (2016) 254–262.
[129] A.M. Diez-Pascual, A.L. Diez-Vicente, ZnO-reinforced poly(3-hydroxybutyrate-co-3-hydroxyvalerate) bionanocomposites with antimicrobial function for food packaging, ACS Appl. Mater. Interfaces 6 (12) (2014) 9822–9834.
[130] A.M. Diez-Pascual, A.L. Diez-Vicente, Poly(3-hydroxybutyrate)/ZnO bionanocomposites with improved mechanical, barrier and antibacterial properties, Int. J. Mol. Sci. 15 (6) (2014) 10950–10973.
[131] I.S. Elashmawi, N.A. Hakeem, L.K. Marei, F.F. Hanna, Structure and performance of ZnO/PVC nanocomposites, Phys. B Condens. Matter 405 (19) (2010) 4163–4169.
[132] M. Hajibeygi, M. Maleki, M. Shabanian, F. Ducos, H. Vahabi, New polyvinyl chloride (PVC) nanocomposite consisting of aromatic polyamide and chitosan modified ZnO nanoparticles with enhanced thermal stability, low heat release rate and improved mechanical properties, Appl. Surf. Sci. 439 (2018) 1163–1179.
[133] A. Batool, F. Kanwal, M. Imran, T. Jamil, S.A. Siddiqi, Synthesis of polypyrrole/zinc oxide composites and study of their structural, thermal and electrical properties, Synth. Met. 161 (23–24) (2012) 2753–2758.
[134] F. Ram, P. Kaviraj, R. Pramanik, A. Krishnan, K. Shanmuganathan, A. Arockiarajan, PVDF/$BaTiO_3$ films with nanocellulose impregnation: investigation of structural, morphological and mechanical properties, J. Alloys Compd. 823 (2020) 153701.
[135] F. Beigmohammadi, S.H. Peighambardoust, J. Hesari, S. Azadmard-Damirchi, S.J. Peighambardoust, N.K. Khosrowshahi, Antibacterial properties of LDPE nanocomposite films in packaging of UF cheese, LWT- Food Sci. Technol. 65 (2016) 106–111.
[136] Y. Li, X.S. Sun, Preparation and characterization of polymer – inorganic nanocomposites by in situ melt polycondensation of l-lactic acid and surface-hydroxylated MgO, Biomacromolecules 11 (7) (2010) 1847–1855.
[137] T.A. Taha, N. Hendawy, S. El-Rabaie, A. Esmat, M.K. El-Mansy, Effect of NiO NPs doping on the structure and optical properties of PVC polymer films, Polym. Bull. 76 (9) (2018) 4769–4784.
[138] F.H.A. El-kader, N.A. Hakeem, R.S. Hafez, A.M. Ismail, Effect of $Li_4Ti_5O_{12}$ nanoparticles on structural, optical and thermal properties of PVDF/PEO blend, J. Inorg. Organomet. Polym. Mater. 28 (3) (2017) 1037–1048.
[139] M.B. Mohamed, M.H. Abdel-Kader, Effect of excess oxygen content within different nano-oxide additives on the structural and optical properties of PVA/PEG blend, Appl. Phys. A 125 (3) (2019).
[140] K. Thanigai Arul, E. Manikandan, P.P. Murmu, J. Kennedy, M. Henini, Enhanced magnetic properties of polymer-magnetic nanostructures synthesized by ultrasonication, J. Alloys Compd. 720 (2017) 395–400.
[141] M. Khairy, Synthesis, characterization, magnetic and electrical properties of polyaniline/ $NiFe_2O_4$ nanocomposite, Synth. Met. 189 (2014) 34–41.
[142] M.H. Naveen, N.G. Gurudatt, Y.-B. Shim, Applications of conducting polymer composites to electrochemical sensors: a review, Appl. Mater. Today 9 (2017) 419–433.
[143] M.A. Kashfipour, N. Mehra, J. Zhu, A review on the role of interface in mechanical, thermal, and electrical properties of polymer composites, Adv. Compos. Hybrid Mater. 1 (3) (2018) 415–439.

[144] X.-J. Huang, X.-F. Zeng, J.-X. Wang, J.-F. Chen, Transparent dispersions of monodispersed ZnO nanoparticles with ultrahigh content and stability for polymer nanocomposite film with excellent optical properties, Ind. Eng. Chem. Res. 57 (12) (2018) 4253–4260.

[145] J. Loste, J.-M. Lopez-Cuesta, L. Billon, H. Garay, M. Save, Transparent polymer nanocomposites: an overview on their synthesis and advanced properties, Prog. Polym. Sci. 89 (2019) 133–158.

[146] J. Kruenate, R. Tongpool, T. Panyathanmaporn, P. Kongrat, Optical and mechanical properties of polypropylene modified by metal oxides, Surf. Interface Anal. 36 (8) (2004) 1044–1047.

# Medical applications of polymer/functionalized nanoparticle composite systems, renewable polymers, and polymer–metal oxide composites

5

*Muhammad Umar Aslam Khan*[a,b,c,g], *Mohsin Ali Raza*[d], *Sajjad Haider*[e], *Saqlain A. Shah*[f], *Muhammad Arshed*[g], *Saiful Izwan Abd Razak*[b,h], *and Adnan Haider*[i]
[a]Department of Polymer Engineering and Technology, University of the Punjab, Lahore, Pakistan, [b]BioInspired Device and Tissue Engineering Research Group, School of Biomedical Engineering and Health Sciences, Faculty of Engineering, Universiti Teknologi Malaysia, Skudai, Johor, Malaysia, [c]Institute for Personalized Medicine, School of Biomedical Engineering, Shanghai Jiao Tong University, Shanghai, China, [d]Institute of Metallurgy and Materials Engineering, University of the Punjab, Lahore, Pakistan, [e]Department of Chemical Engineering, College of Engineering, King Saud University, Riyadh, Saudi Arabia, [f]Materials Science Lab, Department of Physics, Forman Christian College (University), Lahore, Pakistan, [g]Nanoscience and Technology Department (NS & TD), National Center for Physics, Islamabad, Pakistan, [h]Centre for Advanced Composite Materials, Universiti Teknologi Malaysia, Skudai, Johor, Malaysia, [i]Department of Biological Sciences, National University of Medical Sciences, Rawalpindi, Punjab, Pakistan

## 1 Introduction

Bone tissue engineering (BTE) is an advanced approach that combines engineering technology and biological principles for repairing and regeneration of fractured or damaged bones [1–3]. The techniques of BTE focus on autogenous cell/tissue transplant in contrast to conventional autograph and allograft processes. These techniques would eliminate donor shortages, restrictions, and pathogen transmission and immune rejection [4–8]. Hence, BTE has drawn the attention of researchers in the last decade due to the development of advanced fabrication approaches. The natural bone is a highly porous structure that communicates the ECM of the host bone for osteogenesis by providing biochemical and mechanical properties. Biodegradable scaffold systems are three-dimensional (3D) cell substratum that carries cells into a defect and releases growth factors during their degradation [9–12]. Scaffold porosity plays an integral role in the seeding, replication, and growth of cells.

The main characteristics of a suitable structural scaffold for BTE are biocompatibility, biodegradability, cell adhesion, and proliferation. Different materials have

**Renewable Polymers and Polymer-Metal Oxide Composites. https://doi.org/10.1016/B978-0-323-85155-8.00006-6**

been used to replace or repair damaged or traumatized bone tissue. These materials include plastics and ceramics, which have poor processability and biodegradability [13–15]. Biopolymers are an attractive choice because they are highly flexible in design. They can adopt any shape and structure to design scaffolds, and further incorporation of ceramic nanofillers in biopolymers can impart their properties similar to natural bone [16,17]. Nanosized materials can intensively change the characteristics of the polymer matrix, making it possible to produce biomaterials that individual elements or conventional microsized materials cannot produce.

In BTE, polymeric composite materials offer a multidisciplinary scientific approach [8,18] based on a combination of physicochemical and biomechanical analysis of basic science and engineering of composite materials to develop, create, and regenerate new tissues and organs. It involves cell proliferation and adequate biochemical signals that encourage new tissues and organs to regenerate [19,20]. Tissue engineering restores fractured bones and removes defects to facilitate osteogenesis. However, the disease, infection, immunogenicity, high manufacturing costs, and patient specificity are the significant drawbacks of a typical BTE process [21–23]. Hereafter, new structural scaffolding technologies have been developed for the regeneration of injured bone tissue with biodegradable, biopolymer matrix, and bioactive/resorbable nanomaterials facilitating the in vivo bioreactor approach [24–26]. Nanocomposite biomaterials can imitate natural bone tissue's structural and biological properties and have gained importance in BTE.

Biopolymers provide an excellent biological environment that facilitates cell adhesion, proliferation, differentiation, migration, and protein adsorption to create new tissues. Biopolymers obtained from natural sources like polysaccharides, glycosaminoglycans (GAGs), and proteins provide substantial similarities with the extracellular matrix (ECM) [27–30]. The resemblance with ECM makes these biopolymers chemically versatile to achieve excellent biological response without causing any host reaction, toxicity, and immunological response. Ceramic materials/bioresorbable fillers such as calcium phosphate and bioglass have promising biocompatibility, osteoconductivity, and osteogenesis resorbability properties [31–33]. The process of developing a biopolymer nanocomposite material for bone-implant is extensively challenging (Fig. 1). Biopolymer and bioresorbable materials with desired properties synthesize the composites materials suitable for the bone scaffolding. The developed biomaterials are subjected to characterization and tested in small animal models for their healing properties. However, load-bearing and graft functionality should be fully accessed in large animal models before the clinical trials [15]. Herein, we report a comprehensive development of biopolymeric composites with potential applications in BTE and regeneration.

## 2 Properties of ceramic and biopolymers

In the production of polymeric nanocomposite products for BTE, different materials perform an essential role. These organic, bio-polymer, and ceramic materials connect to the natural bone matrix, activate the crucial bone biomarkers, and regenerate the

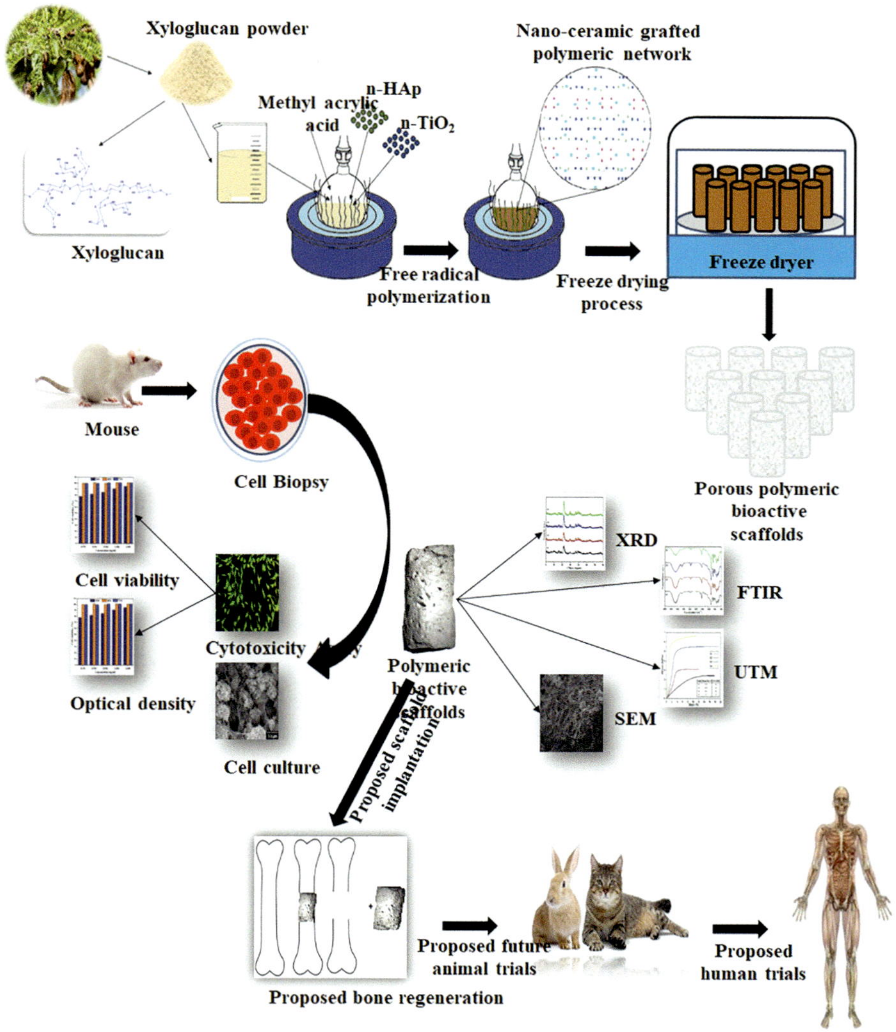


**Fig. 1** Method implantation of BTE scaffold composite for bone tissue engineering applications [8].

host tissue over the scaffold surface. The process continues till the implanted bone scaffold leads to the generation of a healthy bone [34].

## 2.1 Bioceramic materials

Bioceramic materials are fundamental to the manufacture of bone tissue scaffolds and provide structural and biological properties. The biochemical interaction of the bone and the scaffold is bioactivity. Bioceramics such as hydroxyapatite have primarily contributed to the evolution of BTE. Bioceramics interacts with the cells, resorbs, and initiates new bone formation at the bioceramic–tissue interface [35]. Substituting carbonate for hydroxyl groups in such biominerals adds weight to the fundamental bone and when solidifies forms a bone mineral [36].

### 2.1.1 Hydroxyapatite

Hydroxyapatite (HAp) is a ceramic product with the formula Ca10 $(PO_4)_6(OH)_2$. HAp is well known for bone cement and applied in BTE due to structural and crystallographic similarities with the vertebrates' bones and teeth. It is suitable for processing composite materials for bone tissue applications and dental implants [27,37]. The polymeric composite materials of HAp have several biocompatible features to support bone structure by supporting skin, soft organs, muscles, and hard material. HAp cannot be used directly for bone tissue or dental implants due to its brittle nature, but it provides desirable mechanical properties [37]. However, its polymeric composite makes it potential material for bone tissue and dental implants. Moreover, by controlling parameters of agglomeration, shape, particle distribution, and particle size, the properties of Hap ceramics can be improved. Nanocrystalline HAp ceramic has a larger surface area that can be densified and sintered for BTE applications [38].

### 2.1.2 Calcium phosphate

Calcium phosphate (CaP) is a mammal calcification inorganic factor. It is essential to form bone tissue because it has unusual biological behaviors such as osteoconductivity, biocompatibility, and noncytotoxicity [39,40]. CaP is a tunable bioactive material for the restoration and augmentation of bone tissue. The synthesis of calcium phosphate can be achieved at high temperature due to the thermal decomposition of calcium minerals at 800 °C or above [41].

### 2.1.3 Bioactive glass

The bioglass (45S5 Bioglass®), together with other components, is a successful candidate for BTE. The HAp-coated bioglass scaffolds bind with the recipient's bone to stabilize gene expression via osteogenesis enhancement. Silicon is a core element of bioglass, which stimulates gene expression for osteogenesis, works well instead of Ca [42,43]. Bioglass (45S5 Bioglass®) promoted vascular endothelial growth factor secretion for rapid bone vascularization. It promotes enzyme activity, cell adhesion, proliferation, and the regulation of biodegradation by altering structural and chemical composition at the molecular level [44].

## 2.2 Biopolymers

Biopolymers are produced by living organisms and are also called biological macromolecules. The considerable amount of biopolymers produces biodegradable and biodegradable polymeric composite with structural integrity to support bone regeneration. Since biopolymers are multifunctional, this multifunctional character makes them excellent candidates to fabricate different biomaterials with desirable properties. Multifunctional biopolymers include polysaccharides, GAGs, and proteins. Most water-based biopolymers form a viscous liquid or gel. Biopolymers also play an important role in structural polymer scaffolds; among those, chitosan (CS), silk, fibrin, alginate, collagen, and hyaluronic acid are also used in BTE. These biopolymers are biocompatible, biodegradable, drug carrier, and noncytotoxic.

### 2.2.1 Polysaccharides

Polysaccharides are macromolecules of carbohydrate biopolymers produced mainly by the 1,4 or 1,6 glycosidic connections of monosaccharides and have linear or well-connected structures. The polysaccharides are the homopolymers such as cellulose or the heteropolymers such as arabinoxylan (ARX), gellan gum, alginate, CS, and carrageenan (CRG) (Fig. 2).

#### 2.2.1.1 Gellan gum

Gellan gum (GG) has a high molecular weight, anionic structure, and ECM structural polysaccharides produced by *Sphingomonas paucimobilis*. It contains approximately 20% rhamnose, 20% glucuronic acid, and 60% fructose with multiple units. There are two acyl, glycerate, and acetate groups associated with glucose [45,46]. The GG is a highly versatile biopolymer for bio-fabrication, adhesion, biodegradation, cell biocompatibility, drug delivery, easy functionalization, and tunable mechanical properties. Such properties entail GG as capable material for tissue engineering and regenerative medicine. Though, GG alone has poor physiochemical properties such as mechanical strength and stability. However, GG composites with other polymers like agar, starch, CS, sodium alginate, pullulan, cellulose, xanthan gum, or metal nanoparticles have displayed excellent physicochemical properties [47]. GG composites have several applications in different areas of cellular and acellular tissue engineering. Because of its superior mechanical strength, biological, and hydrogel properties, the composite scaffold of GG with HAp was suggested for BTE applications [48]. Two approaches for cartilage and bony parts are mixed with and without HAp (20, w/w) or GG (2%, w/v) and bilayered biomaterials [49]. Bilayer's chemical properties suggest that it is suitable for BTE with a pore size of $278.3 \pm 35.6\,\mu m$ and $83.4 \pm 0.9\%$ porosity and $61.9 \pm 4.99\%$ connectivity [50,51]. Gantar et al. and coworkers developed a new hydrophilic gellan-gum bioactive-glass-reinforced spongy scaffold for BTE. These hydrophilic scaffolds have an enhanced mechanical and biocompatible microstructure and are promising osteogenic biomaterials [52]. For soft tissue regeneration, GG-based halloysite nanotubes (HNTs) were synthesized by Bonifacio and co-workers. They found that HNT was biocompatible with controllable mechanical properties, and Young's modulus was found to be in the range of 20–75 kPa, which is appropriate for BTE [53].

#### 2.2.1.2 Alginate

Alginates are a group of polysaccharides derived from algae. It consists of two monomers, α-L-guluronate (G), and (1,4)-linked β-D-mannuronate (M), and can also be obtained from different sources with different compositions and sequences of M

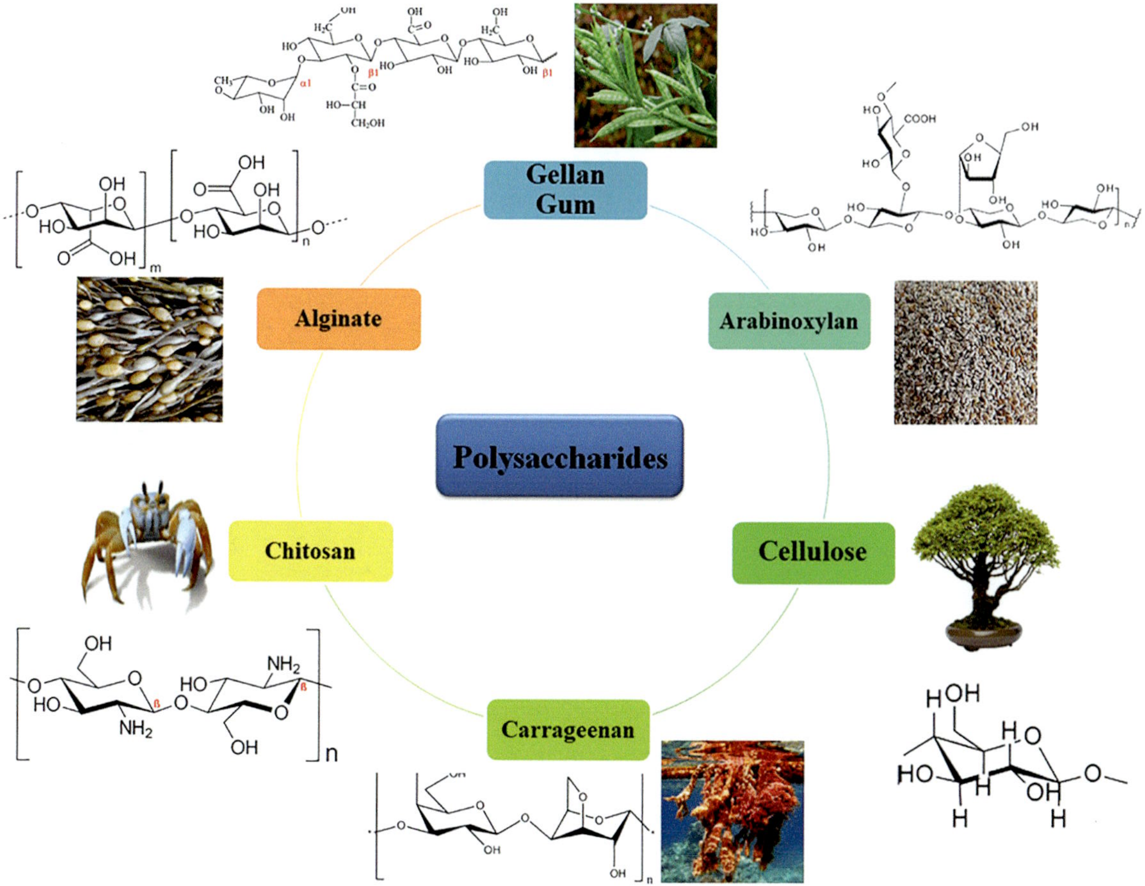


**Fig. 2** A range of different types of polysaccharides is primarily used in BTE.

and G blocks. M-content is more immunogenic and capable of producing cytokines than G-content. At the same time, G-content and alginate molecular weight are critical for enhanced mechanical characteristics. Alginate has been identified as noncytotoxic, biocompatible, and nonimmunogenic, making it an attractive BTE material [54]. Scaffolds are prepared by Schiff base reaction from biphasic gelatin and alginate and have satisfactory biodegradation, gelation, excellent biocompatibility, and mechanical stability [55]. Besides, composite scaffolds were synthesized with alginate and calcium phosphate for osteoblast production and found high potential for osteogenic differentiation and successful proliferation [56]. Alginate-HAp based scaffolds support human dental pulp stem cells for osteogenesis, biomineralization, and calcium deposition in vitro [57–59]. Luo et al. reported core structure alginate/n-HAp bone scaffolds and found enhanced mineral (particle size, crystallinity, morphology, etc.) and mechanical properties with controlled protein delivery [60]. Shaheen and coworkers have synthesized CS/alginate/HAp/nanocrystalline cellulose-based scaffolds via the freeze-drying method. They reported suitable swelling with enhanced compressive strength, increased pore size, cell growth, adherence, and viability, suitable for bone regeneration [61].

#### 2.2.1.3 Chitosan

CS is produced by deacetylation of chitin, a regional polysaccharide found in ladybug, butterfly, medley, lobster, shrimp, crabs, etc. The length, sequence, and structure of the molecular chain are randomly distributed by D-glucosamine and *N*-acetyl glucosamine [62,63]. Moreover, it can have an electrostatic relationship with the negative-loaded bio-molecule and interact with the cell surface, an essential requirement of BTE. CS is structurally cationic with the unique chemical composition of D-glucosamine and *N*-acetyl-glucosamine, which attributes negative-charged to CS with electrostatic activity and makes it compatible with the cell membrane. CS in various forms such as hydrogel, particles, films, fibers, and foams has been used to treat cartilage, bone, tissue, and wound [64,65]. The high mechanical strength of CS enables its use in load-bearing applications. CS being nonosteoconductive limits and its polymeric composites can be used as bioactive scaffolds for bone tissue. However, these polymeric bioactive scaffolds' degradation rate takes a long time and sometimes months. Being osteoinductive, CS supports the osteoblast cells for proliferation and differentiation, leading to forming a mineralized bone matrix [66].

Nevertheless, CaPs coating on CS improved the bone-bonding capability and osteoconductivity [66]. Zhang et al. and coworkers fabricated porous composite bone scaffolds that helped MC3T3-E1 cell line proliferation, alkaline-phosphate displacement, and cell adhesion properties [47]. Sadeghi and coworkers fabricated scaffolds based on poly(ε-caprolactone) (PCL), CS, and polypyrrole (PPy) through electrospinning and analyzed surface morphology, hydrophilicity, and bioactivity [67]. This study showed that CS-base nanocomposite plays a vital role in hydrophilicity (water contact angle = 66 degree) with suitable porosity 30–180 μm. It helped cell adherence, growth with increasing proliferation up to 356% compared to pure PCL and neurite extension of PC12 [68].

#### 2.2.1.4 Arabinoxylan

ARX is a biopolymer having biomedical and pharmaceutical applications due to hydrophilic, water-soluble, biodegradable, and biocompatible characteristics. ARX provides interconnectivity with cross-linkers and other polymeric materials [69]. Functional ARX's hydrogels with biopolymers and reinforcement agents respond to temperature and pH changes and swell to release drugs [70,71]. ARX has high molecular mass, load density, and adhesive attributes. Stimulant-responsive ARX-based hydrogels can control the release of drugs for wound healing [72–75]. Aduba Jr. et al. reported gelatin-ARX-based electrospun composite fibrous mats, and the results showed desirable morphology, pore size, and porosity biocompatibility for chronic wound healing [76].

#### 2.2.1.5 Bacterial cellulose

Cellulose is a fundamental component of a primary cell wall structure and is a polysaccharide that contains a linear chain of an enormous number of $\beta(1 \rightarrow 4)$ linked D-glucose units. Cellulose is the most available biomacromolecule on earth in the plant. It has a wide range of medicinal applications and is used as a potential material in the pharmaceutical industry. The cellulose extraction process from the plant is not easy and involves a series of chemical or enzymatic methods. However, bacterial cellulose (BC) is produced by Gram-negative bacteria as an extracellular section that can be easily extracted. BS has been used to fabricate scaffolds for BTE applications [77–79]. Müller et al. fabricated a scaffold based on non-woven cellulose II for bone tissue repair and found cell growth, adherence, and chondrocyte cell proliferation for cartilage regeneration [80]. Courtenay et al. reported the fabrication of BC sheets and cell adhesion by 70% compared to unmodified cellulose. The cell attachment analysis was found to have the same patterns as the adhesion studies, and on the cationic surfaces, the mean cell area and aspect ratio were the highest [81]. Zhang et al. and coworkers have fabricated scaffolds based on BC and investigated their effects on stem cell growth, adhesion, and differentiation. They used the rabbit model in vivo, and histological analysis has shown that these scaffolds are suitable for BTE [82].

#### 2.2.1.6 Carrageenan

CRGs are linear sulfated polysaccharides that are derived from red marine algae. It shows new attributes for establishing a controlled release technique because of its gelling capability. Traditionally, kappa, iota, and lambda are the most common types of CRG and only vary in sulfate per disaccharide. Because of its biocompatible behavior, it has multiple cosmetic and pharmaceutical applications [83–85]. CRG's cross-linking hydrogels are biocompatible and raise the stability of the encapsulated protein. Platelet-dependent growth factor exists in five isoforms and plays a vital role in stimulating angiogenesis in vivo. While it does not stimulate endothelial cell proliferation, it acts as an active mitogen to produce distinctive angiogenic factors in smooth muscle and fibroblasts cells [86]. The recruitment of pericytes and the separation of endothelial precursor cells are also assumed. CRGs are ideal for scaffold preparation and even for the construction of new chemical structures. Different polysaccharides, ceramic materials, or cross-linkers can also load and control the release of different types of drugs and proteins [87,88]. Araujo et al. prepared pH stimuli porous scaffolds based on CS/CRG for bone regeneration, and CSCRG1 scaffolds were more hydrophilic with higher water absorption capacity. The improvement in the CS:CRG molar ratios strengthened the mechanical properties of the scaffold, with CSCRG3 providing the lower compressive strength under wet conditions [89].

### 2.2.2 *Glycosaminoglycan*

GAGs consist of dual sugar (disaccharide) repeating units, and because of polarity, they are readily soluble. Their various types of GAGs are shown in Fig. 3 and are discussed in the following subsections.

#### 2.2.2.1 Hyaluronic acid

Hyaluronic acid (HA) is a natural anionic, regular D-glucuronic acid-glucuronic acid-β-1,3-*N*-acetyl-D-glucosamine-β-1,4 units. Nonimmunogenic, water absorption, biocompatible, and viscoelastic are the main characteristics of HA, which make it attractive to be used as an ECM for regenerative medicine [90,91]. HA has been modified for use in various forms such as foams, fibers, mesh, and hydrogels with proven biological and physicochemical properties. Composite materials with HAp/HA were tested for fracture strength, structural stability, and hardness. Nevertheless, the HAp composites also demonstrated a pronounced capacity for damaged cartilage treatment [92–95]. Leach et al. reported biodegradable hydrogel for tissue engineering and found that these have good biocompatibility [96]. Kenar et al. have prepared microfibrous scaffolds based on poly(L-lactide-*co*-ε-caprolactone) and HA for vascularization and observed higher water uptake capacity (103% vs 66%), which decreased Young's modulus (from 1.31 to 0.89 MPa) [97].

#### 2.2.2.2 Chondroitin sulfate

Chondroitin sulfate, a form of GAG, is composed of an alternating sugar chain (glucuronic acid and *N*-acetylgalactosamine) usually connected as a proteoglycan to proteins [98,99]. The chondroitin chain has many sugars, and the variable position and quantity of each sugar have been sulfated. It is an integral part of the cartilage bone and displays compression resistance [100–102]. Chondroitin sulfate was used as a nutritional supplement for osteoarthritis together with glucosamine. It is mainly extracted from cow and pig cartilage to extract chondroitin, and some other sources are sharks, fish, and birds [103–105]. Chondroitin is nonuniform composition and varies from source to source substance. Chondroitin sulfate porous scaffolds with HAp have sufficient distinctive porosity, proliferation, and osteoblast migration [106,107]. Chen et al. fabricated chondroitin-6-sulfate/dermatan sulfate/CS-based composite scaffolds for cartilage tissue engineering and found enhanced cell morphology [108]. Chang et al. prepared a gelatin-chondroitin-hyaluronan scaffold for cartilage tissue engineering and reported that the scaffold has a uniform pore size with adequate porosity of 75% [109].

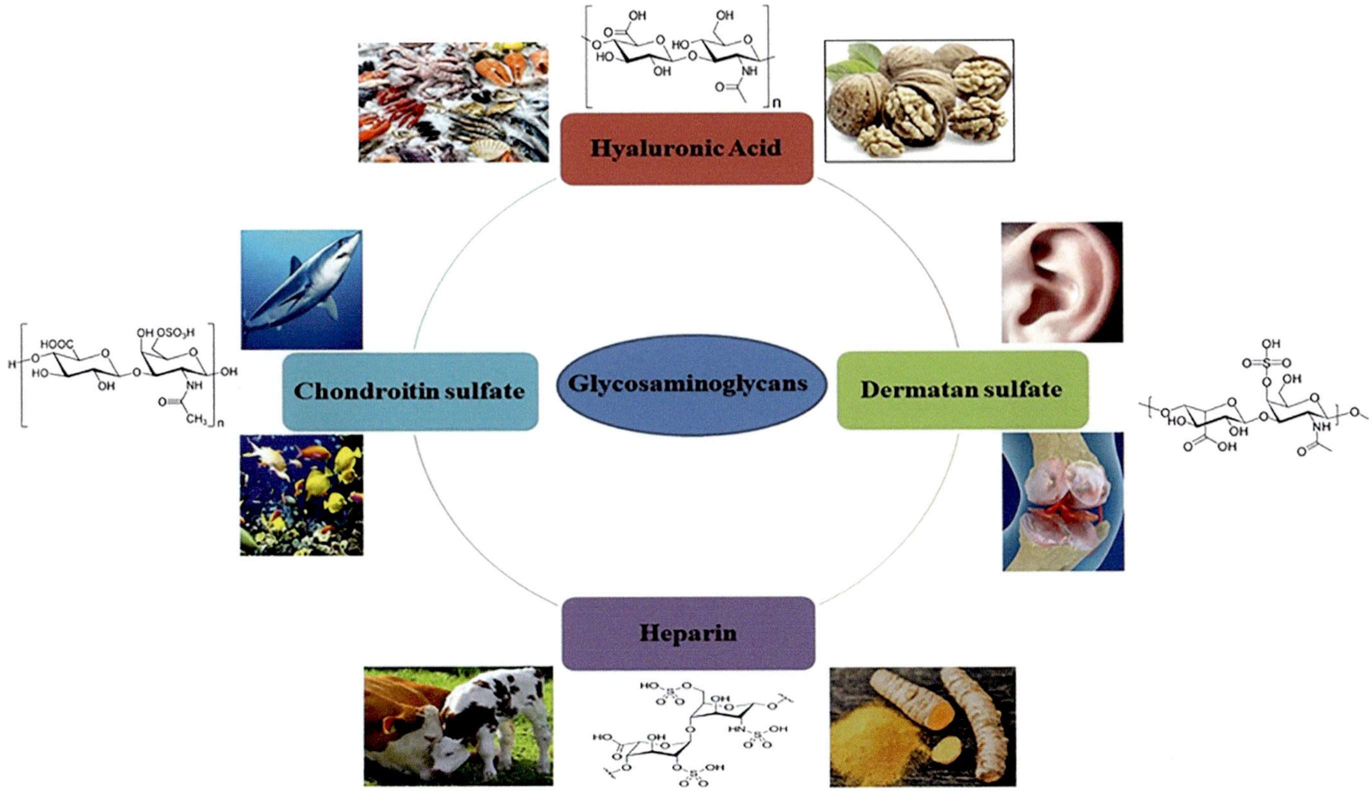


**Fig. 3** The chemical structure and natural source of GAGs.

#### 2.2.2.3 Dermatan sulfate

Dermatan sulfate is from the GAG class and is also known as chondroitin sulfate B [101,110,111]. It can be found in the lungs, tendons, heart valves, and blood vessels. It plays a vital role in medical sciences such as cardiovascular disease, infection, repair of wounds, skin repair, and fibrosis. Dermatan sulfate composite scaffolds help cell proliferation [112,113]. It supports osteoblast differentiation and excellent osteogenesis [113]. Its polymeric matrix is analogous to an ECM which plays a crucial role in controlling differentiation and migration of osteoblasts [114–116]. Mouw et al. fabricated GAG scaffolds for cartilage tissue engineering and found significant cell proliferation (20–40 days) and ECM accumulation. The lowest proportion of unsulfated residues with a disaccharide ratio of 6-sulfated/4-sulfated is very close to that of natural articular cartilage [117].

#### 2.2.2.4 Heparin

As an anticoagulant, unfractionated heparin is the most commonly used natural medicine and belongs to the GAG group. It has been used in hemodialysis to administer the patients with unstable angina or heart issues to prevent thrombosis propagation [101,118–120]. But in other biological processes, it has a unique role [121,122]. It can be used as a carrier because of its interaction with the protein, which helps cell proliferation. The heparin composite is biocompatible with other polymers and helps osteogenesis. Cell differentiation and cell migration are responsible for the porosity of heparin composite [123,124]. Lu et al. reported cytocompatible and blood compatible multifunctional fibroin/collagen/heparin scaffolds for tissue engineering and found that these fibroin-based scaffolds have excellent microstructures. Besides, the scaffolds retain collagen bioactivity and become structures that slowly release heparin and allow the scaffolds to flush with water. Compared to fibroin/collagen scaffolds, heparin-containing scaffolds promoted the development of HepG2 cells, as heparin facilitated a more nuanced, dynamic microenvironment to sustain cell growth and viability [125].

### *2.2.3 Proteins*

Proteins include large chains of amino acid residues. Proteins are essential biomolecules that function in humans and perform catalytic reactions, DNA replication, stimulus tolerance, molecular structure, and motion. The essential sources of protein have been demonstrated in Fig. 4. Proteins are integral biopolymers as they can provide structural and mechanical functions for cells and organisms [126–128]. Joerg et al. prepared three-dimensional scaffolds to support cell growth for protein delivery in tissue engineering. These scaffolds interacted biochemically through active sites with cells and helped cell adhesion, growth, and proliferation by controlled release of bioactive proteins such as growth factors and cytokines [126]. The significant proteins used in BTE are silk fibroin, collagen, and gelatin (Fig. 4).

#### 2.2.3.1 Silk fibroin

Silk fibroin is derived from spiders, worms, and insects with 70%–80% fibroin, 20%–30% sericin, and a water-based fibroin-like glue protein. As a biomedical material, the silk fibroin extracted from the silkworm is of great interest. It is obtained by boiling the

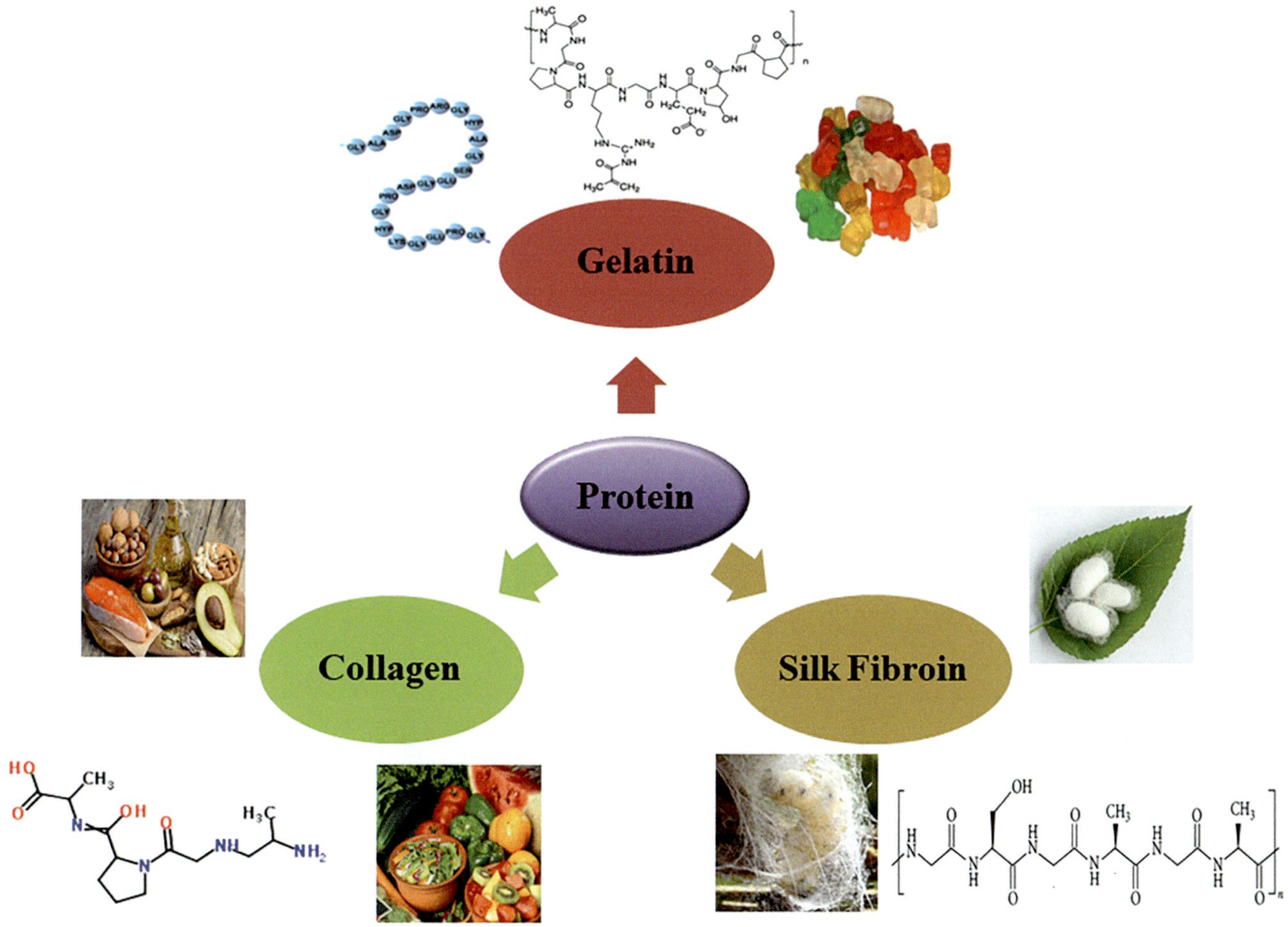


**Fig. 4** Chemical structure and natural source of protein-based biomolecules or biological macromolecules.

alkaline solution from the silkworm cocoon [129,130]. Scaffolds of silk fibroin have poor mechanical strength, elastic and swelling behavior, along with biocompatibility, noncytotoxic, and biodegradability characteristics. The mechanical properties can be enhanced by adding other materials to support the physical and biological characteristics of silk fibroin-based scaffolds [131,132]. Silk/CaPs scaffold provides elastic nanostructure, osteogenic differentiation, and bone formation in vivo. HAp-based scaffolds have improved mechanical strength and osteoconductivity. MSCs and silk scaffolds are manufactured for connective tissue technology [133–137]. Tellado et al. fabricated silk fibroin scaffolds for tendon/ligament of bone tissue and found that these scaffolds had porosity ranging from 50 to 80%, and average pore sizes (80–90%) were ]<100–300 μm. The mechanical properties varied from 689 to 1322 kPa and helped cell attachment and proliferation [138].

### 2.2.3.2 Collagen and gelatin

Collagen is a popular biopolymer that resembles a fibrous protein ECM and is biocompatible. It has several potential applications in BTE. It increases the response and migration of osteoblast cells to the bone matrix [139,140]. Collagen is an ideal biopolymer to fabricate scaffolds for various body tissues due to its fibrous nature and ability to interact with the biological environment. The scaffolds of collagen are well architecturally biocompatible, biodegradable, and noncytotoxic, but with poor mechanical behavior and insufficient swelling properties at the isoelectric point. This biocompatibility has made collagen a favorite and potential candidate for tissue engineering [141–145].

Moreover, many strategies have been developed to enhance the mechanical stability of collagen scaffolds for BTE. Tangsadthakun et al. prepared a scaffold from collagen/CS for skin tissue engineering and found enhanced mechanical properties with significantly developed biodegradability [146]. Collagen and CaP nanocomposites are developed using a technique such as electro-spinning. Kikuchi and co-workers developed porous scaffolds from HAp-collagen for BTE. They found enhanced osteoblast cell differentiation when scaffolds cocultivated with osteoblast cells MG-63 and bone marrow cells [147].

Gelatin is a biopolymer that is noncytotoxic, biocompatible, and bioresorbable but with weak mechanical properties to be used in bone tissue synthesis. Gelatin has excellent cell adherence and increased crystal/nanoparticulate microstructures and mechanical integrity. Gelatin has been widely used in the biomedical field to generate porous nanocomposite with improved structural rigidity and exceptional cell adherence [148–150]. Gentile et al. have synthesized scaffolds from CS/gelatin and engrafted them with simvastatin-loaded PLGA-microparticles for BTE. The increasing amount of PLGA in composite materials enhanced mechanical properties by reducing swelling from $1245 \pm 56\%$ to $570 \pm 35\%$ while increasing compressive modulus to $23.0 \pm 1.0$ kPa. The hFOB cell viability, proliferation, and osteoblastic differentiation were observed after 11 days [151]. Gelatin/CaP-based nanocomposite revealed improved mechanical strength and cell adhesion [152,153]. Ko et al. investigated gelatin and HAp microstructure. Researchers showed that a decrease in gelatin content makes the material more compact and durable due to increased fracture strength [154]. Synthesized nanocomposite with porous bone-like structure displayed extraordinary mechanical strength, structure migration, penetration, and cell adhesion [155–159].

# 3 Requirements for BTE

Biopolymer-based bone tissue scaffolds provide significant advantages over metallic implants for cell adhesion, movement, proliferation, and differentiation. The performance of a porous biomaterial in BTE applications depends on mechanical, chemical, and biological properties. Targeted scaffolds should have various features for BTE, as discussed below, such as (i) swelling, (ii) porosity, (iii) mechanical strength, (iv) biocompatibility, (v) protein adherence, (vi) biodegradability, and (vii) biomineralization. Such characteristics can be regulated by different concentrations of polymer and ceramic materials to grow bone tissue scaffolds.

## 3.1 Swelling

Swelling plays a vital role in reinforcing BTE scaffolds because of their contact with the biological environment. This quality is known as the water retention capability of the scaffold. The scaffold absorbs water from surrounding tissue during implantation. On swelling, the pore size of the scaffold increases for penetration of the cells into the deep internal structure of the scaffold [160–162]. Reducing water swelling minimizes the contact with the implant site and reduces deep cell penetration. However, increased water retention leads to implant behavior analogous to natural soft tissue, increased biocompatibility, enhanced oxygen permeability, and mobilization of nutrients. The scaffold's swelling potential must be so developed that the implant should have suitable mechanical strength and can interact well with the neighboring cells [163–165]. Swelling depends on the nature of the biopolymers and the media available for biopolymers scaffolds. The functional groups play a crucial role in water retention and hence increases or decreases swelling. Depending on the nature of the functional groups in biopolymers, the swelling of the scaffold may change [166–169].

## 3.2 Porosity

Porosity also plays an important role in ECM adhesion, adaptation, invasion, proliferation, differentiation, and secretion. The pore size of scaffolds is essential as narrow pore size allows a minimal number of cells can penetrate the scaffold. The sufficient cell adhesion, penetration, and proliferation on scaffold surface, similarly the flow of oxygen and nutrients into and toxin out of the scaffold simply depends on the pore sizes [170–173]. .The scaffold with a large pore size results in reduced cell binding. The pore size from 100 to 300 μm can induce neovascularization and supports endochondral ossification. At the same time, pores exceeding 300 μm can increase osteogenesis [174–176]. Decreasing and increasing the number of pores in conjunction with the scaffold size affects cell proliferation and differentiation. Various ceramic materials can be used to regulate the pore size and density of the scaffold for BTE [172,176–180].

## 3.3 Mechanical strength

Mechanical strength is an essential behavior of any biopolymeric scaffold for load-bearing in BTE applications. The load-bearing behavior of the scaffold at the implanted site might be similar to the host bone. The compressive modulus of natural cancellous bone is 0.1–2 GPa and for the cortical bone is 15–20 GPa, respectively [42,181–183]. Depending on the structural nature of biopolymers, the mechanical properties of their scaffolds vary from soft to moderate, and they can be modified for load-bearing applications. However, the combination of different polymers with ceramic substrates has been found to increase the mechanical properties of scaffolds with desired load-bearing applications [184–186]. Several studies have reported that biopolymeric composite scaffolds have several reasons to be considered for BTE with controlled and tailored mechanical properties similar to the host bone. Therefore, the biopolymer-based scaffolds are mechanically appropriate at the bone implantation site [187,188].

## 3.4 Biodegradation

Biodegradation of biopolymeric scaffolds is a slow chemical breakdown and decay. This process continues without causing any cell, tissue reaction, or skin damage within a biological system. Biodegradation of scaffolds starts with many valuable materials under the physiological influence in the biological system [189,190]. The biodegradation of scaffolds can be controlled to transport essential biomolecules for osteogenesis. Biodegradation of scaffolds is a critical chemical process before implantation, while some of the biomaterials subjected to serum or fluid irregularly deteriorated [144,191]. Chemical degradation should be nontoxic and harmful to the biological immune system. The biodegraded scaffolds are smart enough that body fluids can be absorbed and excreted in the metabolic process [14,184,192–194].

## 3.5 Protein adhesion

A perfect scaffold tends to attach to protein when communicating with the biological system's physiological environment as cell contact with the implant. Protein adherence controls the adhesion of tissue implanted scaffolds, cell proliferation, and other physiological interactions. Biological interaction is determined by the amino acid content of the protein [4,171,195–197]. Research reports indicate that peptide-REDV fibronectin supports endothelial cell adhesion while not promoting fibroblast and muscle cell adhesion. During cellular processes, proteins are regulated by controlling signaling mechanism that results in cell adhesion, migration, proliferation, and differentiation [198–200]. These adsorbed proteins monitor cellular functions and maintain transcription factors in the cell configuration. Hence, functional scaffolds have a particular affinity for protein adsorption [201,202].

## 3.6 Biomineralization

When scaffolds are subjected to body fluids, the deposition of an appetite layer on scaffolds is called biomineralization. The formation of the appetite layer is due to the deposition of minerals present in body fluids. Biomineralization helps cell adhesion to facilitate osteoblast activity. Bones contain inorganic crystals (60%–70%), most of which comprise HAp. These play a crucial role in optimizing scaffold binding with the normal bone for osteogenesis [203]. The behavior of the scaffold is considered well when it produces an appetite-like material in situ. Simulated body fluid can be used to test biomineralizing activities of scaffold [204]. The scaffolds serve as a template for biomineralization due to CO and $NH_2$ on many biopolymers. Such scaffolds are potential candidates for biomineralization in BTE applications [205–207]. The biomineralization can be enhanced by increasing the incorporation of ceramic materials or fillers into the biopolymeric matrix of the polymeric composite material. Nanoparticles-based scaffolds raise the substratum's surface area and create an active or nucleation site for appetite deposition [208,209]. In a scaffold, the reactive system improves bone-based apatite with an acceptable combination of Ca and P, similar to natural bone. The biopolymers like CS have several functional groups ($NH_2$, OH, and $COO^-$) to encourage appetite deposition due to nucleation sites [210–212].

## 3.7 Manufacturing technology

The structure of scaffolds plays a significant role in BTE as these provide pores for cell penetration which are interconnected and engineered to produce an ECM. Biopolymers are degraded naturally through a gradual process, and by-products are evacuated from the body without damaging the environment of the organs and tissues [213–216]. The material's degradation is a significant concern for tissue regeneration. Cell adhesion is a phenomenon attributable to chemical groups over the scaffolding surface. The optimum porosity encourages the mobility of cells throughout the scaffold and forms associations with efficient scaffolding groups (ligands). The scaffold structure is an essential factor in BTE, and scaffolds of the cell type should be built [171,217–219].

## 3.8 Scaffold architecture

Scaffold's mechanical integrity, clinically cost-effective, and scale-up design are vital to manufacturing effectively with clinical approaches to BTE. The distribution and accessibility of scaffolds for patients are also an essential element in BTE. The preference of the clinician is the shelf availability of scaffolds without any surgical requirements [220,221]. A scaffold analysis is sometimes needed before implantation by an in vitro engineer. The required scaffolds in BTE must follow said conditions for manufacturing targeted biomaterials [221–224].

# 4 Fabrication composite biomaterials scaffold

Scaffolds can be fabricated by various methods. These are discussed in the following subsections.

## 4.1 Phase separation

The phase separation method consists of separating different phases from more than one stage to reduce the free energy of the cycle. The polymer solution divides a lean polymer or rich polymer mixture into two phases during this phase separation process. This approach helps isolate and filter porous membranes with a back draw scattering pores in a nonuniform manner [225–227]. Phase separation is also used to distinguish between heat-induced phases and thermally induced phase separation. It is a beneficial methodology for the construction of the scaffolding. The phase separation optimization can give up to 68%–91% interconnected porosity, while the overall porosity is 77%–93% [228–230]. Desired morphology of scaffolds can be achieved by adjusting parameters like polymer concentration, freezing temperature, and the use of various porogen [231]. Different types of scaffolds with altered shapes can be prepared by this method as per demands by keeping consistency between polymer and porogen. Phase separation is a laboratory-scale method, and only a few polymers can be used to synthesize scaffolds with potential applications in bone engineering [227].

## 4.2 Solvent casting and particulate-leaching

A combination of solvent casting and particulate leaching methods can be used for active 3D porous scaffolding. Salt is spread evenly into solution with dissolved salt into polymeric composite formation such as sodium chloride, ammonium bicarbonate, etc., solidifying the polymers by removing the solvent [232–234]. Treating with water leaches salt that causes pore formation. This process can control the number of pores to determine the size of the salt crystal. One can obtain a pore quality of 93% with an average pore diameter of 500 μm, although this approach cannot manage interconnectivity and pore shape [234,235]. A polymer is dissolved in an organic solvent and ceramic and primarily salts for scaffold synthesis of particular geometry, dimensions, and shape. The organic solvent was evaporated with rubber, and the concrete substance is then put into a water bath to produce a porous scaffold [232].

## 4.3 Foam replica method

This technique can be used to fill an aqueous suspension of flexible synthetic polymer foam and resin pores. The water-suspended impregnated foam is either centrifuged or passed through rollers after drying the foam at 300–800 °C/min to remove the excess suspension and allow it to dry gradually at 1 °C/min [236,237]. The gradual decomposition and diffusion leave behind a porous surface and sinter between 1100 and 1700 °C to increase its density. This method creates a macroporous structure with a profoundly entangled pore size range between 200 μm and 3 mm and a total porosity between 40% and 95%. Whereas, sintering of scaffold's structural stability may degrade during foam decay due to crack formation [230,238–240]. The pore size greater than 100 μm cannot be acquired and the porosity of the scaffolds from 80% to 90% is obtained by this technique. These porosity characteristics are essential in vivo vascularization. But this high porosity results in low mechanical strength [241].

## 4.4 Gas foaming

High-pressure $CO_2$ gas is purged into a polymer salt solution, like sodium chloride, and the solubility of $CO_2$ reduces with the environment, and nucleation and gas bubbles increase. The porogen is extracted when the process is complete by interconnecting porous structures [171,242]. The functional, structural, and mechanical characteristics of the surface are highly dependent on the state of the system (e.g., time, temperature, and pressure venting) [243–245]. The construction of scaffolds in this technique does not require high temperatures or organic solvents. Scientists have been paying more attention to this approach in recent years to apply tissue engineering scaffold synthesis. Research findings of scaffolds show that microstructures play a significant role in differentiating cell seeding, adherence, migration, proliferation, and postproduction [246–249]. Porous scaffolds are prepared through this technique without using any solvent by creating gas bubbles into a biopolymer. Typically, $CO_2$ is pressurized into polymeric molds until the polymer is saturated to get porous scaffolds on gas removal.

## 4.5 Freeze-drying

Lyophilizer is also the name of this strategy that refers to the idea of dehydration, i.e., the vacuum drying of a frozen product. A polymeric substance is initially dissolved into a solvent and frozen at various temperatures. The solvent molecules are forced to settle and crystallize in the freezing material's microscopic pores [250,251]. Frozen product is purposefully and gradually vacuum dried to remove accumulated solvents at low temperatures [196,252,253]. The freezing time can be adjusted to control pore size. However, porous scaffolds provided by this method are not mechanically stable, which is the main backdrop of this technique. This approach is suitable for synthesizing polymeric porous composite scaffolds for BTE [8,12]. It has a porosity level of up to 91% with interconnectivity of up to 97% and involves the composition and distribution of polymers. In this method, the polymeric mixture is transferred into molds of desired shapes for prefreezing for 12 h at −20 °C and then frozen for 48 h at −80 °C. The prepared scaffolds through this method have low or poor mechanical strength but excellent porosity that helps adherence and proliferation for regeneration of the bone tissue [17,254].

## 4.6 Rapid prototyping

This method is also referred to as additive development and solid-free manufacturing process. In this process, CAD (computer-aided design), MRI (magnetic resonance imaging), and CT (computer tomography) data are used to create an outline of the object [255–257]. This approach deals directly with the 3D design layer by layer. This method is ideal for controlling the geometry, pore size, and interconnected porosity that facilitates penetration of tuning cell scaffolds and tissue engineering activities. This approach is more desirable for BTE owing to its pore size and material shape control [258–260].

## 4.7 Electrospinning

Electrospinning has attracted considerable interest in tissue engineering as it can be used to fabricate nanofibers from polymeric materials. Nanofibers are membranous scaffolds with structural integrity, mechanical strength, and surface area [261–263]. Electrospinning is carried out by applying a substantial potential difference between two electrodes attached to a polymer-filled syringe and deposited over a collector. Because of the high electrical potential difference between the two electrodes, the polymer solutions cause a jet and generate nanofiber on the collector after liquid evaporation [264]. The preparation of electrospinning fibers depends on the solution's features. It, therefore, provides suitable pores for scaffolding, including structural integrity and mechanical strength for tissue engineering [265–267]. The electrospun scaffolds are fabricated by pumping polymeric solution from the syringe. The polymeric solution is collected over the collector due to the high direct voltage (5–50 kV) difference between the syringe and collector. The shape, size, and morphology of the nanofiber scaffolds can be controlled by the size and diameter of the syringe needle, viscosity of the polymeric solution, voltage difference, and steady feed pressure.

The synthesis of polymeric nanomaterial from well-organized nanostructured polymers is also known as nanocomposites. Several methods have been developed to adequately synthesize and produce new biomaterials to address cell activity [268–271] ***[272]. New solutions are required to fix the problems for restoration with sufficient cell connectivity with specific cellular and biological characteristics and gene activation. These solutions should support cell adhesion, proliferation, differentiation, and migration through ECM production to boost tissue regeneration [76,268,269]. The bone scaffolds based on nanocomposites have three-dimensional, hierarchical structures for BTE. They can be different from macroscopic tissue configurations to macromolecular protein arrangements to multiple organizational levels. The mechanical behavior of polymeric nanocomposite materials can be enhanced by choosing an appropriate ratio of polymers and nanomaterials to optimize mechanical integrity to the cellular level [20,269–271]. The required osteogenesis can be achieved using nanocomposite as biomaterials in BTE.

# 5 Conclusion and future directions

Research on biopolymeric scaffolds has opened new horizons and new gateways in BTE in recent decades by introducing new technology, manufacturing techniques, and production methods. An integral system designed to assist the osteogenesis and angiogenesis for bone scaffold preparation can help BTE. The porosity of scaffolds is critical for cell adhesion, cell migration, proliferation, and differentiation, thus developing a fully functional bone matrix. Porous scaffolds have low structural strength, and porosity is evenly distributed around scaffold boundaries. Since scaffold is a possible treatment for fractured or damaged bone, there is no need for uniform porosity because the natural bone has no uniform porosity. Natural bone is denser and thinner from the outside, whereas the core is porous. These issues require new biomaterials to be developed with different combinations of polymers and ceramics for BTE. Biopolymer scaffolds have ideal properties as their biodegradability is faster and leave ceramic content behind after biodegradation to address osteolysis. Biosorption of the ceramic polymer scaffold must be sufficient to resolve the problem of biodegradation. Alternative calcium phosphate degrades faster and creates a good calcium environment for the accelerated deposition of the appetite. Hence, creating a model of natural bone is challenging, but cutting-edge technologies, advanced techniques, and accessible bone scaffolds have empowered scientists fully for BTE applications. Such biomaterials should be chosen for biopolymeric scaffold synthesis, producing desirable pores and strong mechanical scaffolds for BTE. These scaffolds should have a structural design, and biomolecular release, and controlled biodegradation. It will play a crucial role in the future development of scaffolds for bone tissue. The precise biomechanical and physicochemical behaviors of the biopolymeric composite scaffolds and their biological properties and interaction with bioactive materials are highlighted. Biopolymeric composite scaffolds in animal models are also reviewed and discussed.

# References

[1] J.R. Xavier, T. Thakur, P. Desai, M.K. Jaiswal, N. Sears, E. Cosgriff-Hernandez, R. Kaunas, A.K. Gaharwar, Bioactive nanoengineered hydrogels for bone tissue engineering: a growth-factor-free approach, ACS Nano 9 (3) (2015) 3109–3118.

[2] A.S. Mistry, A.G. Mikos, Tissue engineering strategies for bone regeneration, in: Regenerative Medicine II, Springer, 2005, pp. 1–22.

[3] M.U.A. Khan, S. Haider, A. Haider, S.I. Abd Razak, M.R.A. Kadir, S.A. Shah, A. Javed, I. Shakir, A.A. Al-Zahrani, Development of porous, antibacterial and biocompatible GO/n-HAp/bacterial cellulose/β-glucan biocomposite scaffold for bone tissue engineering, Arab. J. Chem. 14 (2) (2021) 102924.

[4] G. Wei, P.X. Ma, Structure and properties of nano-hydroxyapatite/polymer composite scaffolds for bone tissue engineering, Biomaterials 25 (19) (2004) 4749–4757.

[5] H.J. Haugen, S.P. Lyngstadaas, F. Rossi, G. Perale, Bone grafts: which is the ideal biomaterial? J. Clin. Periodontol. 46 (2019) 92–102.

[6] R. Ferracini, A. Bistolfi, R. Garibaldi, V. Furfaro, A. Battista, G. Perale, Composite xenohybrid bovine bone-derived scaffold as bone substitute for the treatment of tibial plateau fractures, Appl. Sci. 9 (13) (2019) 2675.

[7] W.S. Al-Arjan, M.U. Aslam Khan, S. Nazir, S.I. Abd Razak, M.R. Abdul Kadir, Development of Arabinoxylan-reinforced apple pectin/graphene oxide/nano-hydroxyapatite based nanocomposite scaffolds with controlled release of drug for bone tissue engineering: in-vitro evaluation of biocompatibility and cytotoxicity against MC3T3-E1, Coatings 10 (11) (2020) 1120.
[8] M.U.A. Khan, S. Haider, S.A. Shah, S.I. Abd Razak, S.A. Hassan, M.R.A. Kadir, A. Haider, Arabinoxylan-co-AA/HAp/TiO2 nanocomposite scaffold a potential material for bone tissue engineering: an in vitro study, Int. J. Biol. Macromol. 151 (2020) 584–594.
[9] M. Kouhi, M. Fathi, M.P. Prabhakaran, M. Shamanian, S. Ramakrishna, Poly L lysine-modified PHBV based nanofibrous scaffolds for bone cell mineralization and osteogenic differentiation, Appl. Surf. Sci. 457 (2018) 616–625.
[10] A. Baird, N. Dominguez-Falcon, A. Saeed, D. Guest, Biocompatible 3D printed thermoplastic scaffolds for osteoblast differentiation of equine iPS cells, Tissue Eng. Part C Methods 25 (5) (2019) 253–261, https://doi.org/10.1089/ten.tec.2018.0343.
[11] M.U.A. Khan, M.A. Raza, H. Mehboob, M.R.A. Kadir, S.I. Abd Razak, S.A. Shah, M.Z. Iqbal, R. Amin, Development and in vitro evaluation of κ-carrageenan based polymeric hybrid nanocomposite scaffolds for bone tissue engineering, RSC Adv. 10 (66) (2020) 40529–40542.
[12] M.U.A. Khan, M.A. Al-Thebaiti, M.U. Hashmi, S. Aftab, S.I. Abd Razak, S. Abu Hassan, M.R. Abdul Kadir, R. Amin, Synthesis of silver-coated bioactive nanocomposite scaffolds based on grafted beta-glucan/hydroxyapatite via freeze-drying method: anti-microbial and biocompatibility evaluation for bone tissue engineering, Materials 13 (4) (2020) 971.
[13] C.Y. Goh, S.S. Lim, K.Y. Tshai, A.W.Z.Z. El Azab, H.-S. Loh, Fabrication and in vitro biocompatibility of sodium tripolyphosphate-crosslinked chitosan–hydroxyapatite scaffolds for bone regeneration, J. Mater. Sci. 54 (4) (2019) 3403–3420.
[14] T. Ghassemi, A. Shahroodi, M.H. Ebrahimzadeh, A. Mousavian, J. Movaffagh, A. Moradi, Current concepts in scaffolding for bone tissue engineering, Arch. Bone Jt. Surg. 6 (2) (2018) 90.
[15] M.F.M.A. Zamri, R. Bahru, R. Amin, M.U.A. Khan, S.I. Abd Razak, S.A. Hassan, M.R. A. Kadir, N.H.M. Nayan, Waste to health: a review of waste derived materials for tissue engineering, J. Clean. Prod. 125792 (2021).
[16] M.U.A. Khan, S.I. Abd Razak, H. Mehboob, M.R. Abdul Kadir, T.J.S. Anand, F. Inam, S. A. Shah, M.E. Abdel-Haliem, R. Amin, Synthesis and characterization of silver-coated polymeric scaffolds for bone tissue engineering: antibacterial and in vitro evaluation of cytotoxicity and biocompatibility, ACS Omega 6 (6) (2021) 4335–4346.
[17] M.U. Aslam Khan, H. Mehboob, S.I. Abd Razak, M.Y. Yahya, A.H. Mohd Yusof, M.H. Ramlee, T.J. Sahaya Anand, R. Hassan, A. Aziz, R. Amin, Development of polymeric nanocomposite (xyloglucan-co-methacrylic acid/hydroxyapatite/SiO2) scaffold for bone tissue engineering applications—in-vitro antibacterial, cytotoxicity and cell culture evaluation, Polymers 12 (6) (2020) 1238.
[18] L. Bedian, A.M. Villalba-Rodriguez, G. Hernandez-Vargas, R. Parra-Saldivar, H.M. Iqbal, Bio-based materials with novel characteristics for tissue engineering applications—a review, Int. J. Biol. Macromol. 98 (2017) 837–846.
[19] R. Fradique, T.R. Correia, S. Miguel, K. De Sa, D. Figueira, A. Mendonça, I. Correia, Production of new 3D scaffolds for bone tissue regeneration by rapid prototyping, J. Mater. Sci. Mater. Med. 27 (4) (2016) 69.
[20] M.U.A. Khan, S. Haider, A. Haider, M.R.A. Kadir, S.I. Abd Razak, S.A. Shah, A. Javad, I. Shakir, A.A. Al-Zahrani, Development of porous, antibacterial and biocompatible GO/n-HAp/bacterial cellulose/β-glucan biocomposite scaffold for bone tissue engineering, Arab. J. Chem. 102924 (2020).

[21] A.W. Glaudemans, P.C. Jutte, M.A. Cataldo, V. Cassar-Pullicino, O. Gheysens, O. Borens, A. Trampuz, K. Wörtler, N. Petrosillo, H. Winkler, Consensus document for the diagnosis of peripheral bone infection in adults: a joint paper by the EANM, EBJIS, and ESR (with ESCMID endorsement), Eur. J. Nucl. Med. Mol. Imaging 46 (4) (2019) 957–970.
[22] W. Lin, L. Xu, S. Zwingenberger, E. Gibon, S.B. Goodman, G. Li, Mesenchymal stem cells homing to improve bone healing, J. Orthop. Transl. 9 (2017) 19–27.
[23] M.U. Aslam Khan, S.I. Abd Razak, W.S. Al Arjan, S. Nazir, T.J. Sahaya Anand, H. Mehboob, R. Amin, Recent advances in biopolymeric composite materials for tissue engineering and regenerative medicines: a review, Molecules 26 (3) (2021) 619.
[24] H. Hajiali, S. Karbasi, M. Hosseinalipour, H.R. Rezaie, Preparation of a novel biodegradable nanocomposite scaffold based on poly (3-hydroxybutyrate)/bioglass nanoparticles for bone tissue engineering, J. Mater. Sci. Mater. Med. 21 (7) (2010) 2125–2132.
[25] M.U. Aslam Khan, A. Haider, S.I. Abd Razak, M.R. Abdul Kadir, S. Haider, S.A. Shah, A. Hasan, R. Khan, S.U.D. Khan, Arabinoxylan/graphene-oxide/nHAp-NPs/PVA bio-nano composite scaffolds for fractured bone healing, J. Tissue Eng. Regen. Med. 15 (4) (2021) 322–335.
[26] M.U.A. Khan, M.A. Raza, H. Mehboob, M.R.A. Kadir, S.I. Abd Razak, S.A. Shah, M.Z. Iqbal, R. Amin, Correction: development and in vitro evaluation of κ-carrageenan based polymeric hybrid nanocomposite scaffolds for bone tissue engineering, RSC Adv. 11 (30) (2021) 18615–18616.
[27] H. Cao, N. Kuboyama, A biodegradable porous composite scaffold of PGA/β-TCP for bone tissue engineering, Bone 46 (2) (2010) 386–395.
[28] C.R. Nuttelman, D.J. Mortisen, S.M. Henry, K.S. Anseth, Attachment of fibronectin to poly (vinyl alcohol) hydrogels promotes NIH3T3 cell adhesion, proliferation, and migration, J. Biomed. Mater. Res. 57 (2) (2001) 217–223.
[29] S. Chen, Y. Guo, R. Liu, S. Wu, J. Fang, B. Huang, Z. Li, Z. Chen, Z. Chen, Tuning surface properties of bone biomaterials to manipulate osteoblastic cell adhesion and the signaling pathways for the enhancement of early osseointegration, Colloids Surf. B: Biointerfaces 164 (2018) 58–69.
[30] C. Heinemann, R. Brünler, C. Kreschel, B. Kruppke, R. Bernhardt, D. Aibibu, C. Cherif, H.-P. Wiesmann, T. Hanke, Bioinspired calcium phosphate mineralization on net-shape-nonwoven chitosan scaffolds stimulates human bone marrow stromal cell differentiation, Biomed. Mater. 25 (20) (2019), https://doi.org/10.3390/molecules25204785.
[31] O.H. Jeon, L.M. Panicker, Q. Lu, J.J. Chae, R.A. Feldman, J.H. Elisseeff, Human iPSC-derived osteoblasts and osteoclasts together promote bone regeneration in 3D biomaterials, Sci. Rep. 6 (2016) 26761.
[32] P. Zarrintaj, I. Rezaeian, B. Bakhshandeh, B. Heshmatian, M.R. Ganjali, Bio-conductive scaffold based on agarose-polyaniline for tissue engineering, J. Skin Stem Cell 4 (2) (2017).
[33] Q.Z. Chen, I.D. Thompson, A.R. Boccaccini, 45S5 bioglass®-derived glass–ceramic scaffolds for bone tissue engineering, Biomaterials 27 (11) (2006) 2414–2425.
[34] B. Lowe, J.G. Hardy, L.J. Walsh, Optimizing nanohydroxyapatite nanocomposites for bone tissue engineering, ACS Omega 5 (1) (2019) 1–9.
[35] I.V. Antoniac, Handbook of Bioceramics and Biocomposites, Springer, Berlin, Germany, 2016.
[36] S. Li, W. Yu, W. Zhang, G. Zhang, L. Yu, E. Lu, Evaluation of highly carbonated hydroxyapatite bioceramic implant coatings with hierarchical micro-/nanorod topography optimized for osseointegration, Int. J. Nanomedicine 13 (2018) 3643.

[37] Y. Zhang, J.R. Venugopal, A. El-Turki, S. Ramakrishna, B. Su, C.T. Lim, Electrospun biomimetic nanocomposite nanofibers of hydroxyapatite/chitosan for bone tissue engineering, Biomaterials 29 (32) (2008) 4314–4322.
[38] M. Azami, A. Samadikuchaksaraei, S.A. Poursamar, Synthesis and characterization of a laminated hydroxyapatite/gelatin nanocomposite scaffold with controlled pore structure for bone tissue engineering, Int. J. Artif. Organs 33 (2) (2010) 86–95.
[39] K. Sajesh, R. Jayakumar, S.V. Nair, K. Chennazhi, Biocompatible conducting chitosan/polypyrrole–alginate composite scaffold for bone tissue engineering, Int. J. Biol. Macromol. 62 (2013) 465–471.
[40] S. Saravanan, D. Sameera, A. Moorthi, N. Selvamurugan, Chitosan scaffolds containing chicken feather keratin nanoparticles for bone tissue engineering, Int. J. Biol. Macromol. 62 (2013) 481–486.
[41] C. Liewhiran, S. Seraphin, S. Phanichphant, Synthesis of nano-sized ZnO powders by thermal decomposition of zinc acetate using Broussonetia papyrifera (L.) vent pulp as a dispersant, Curr. Appl. Phys. 6 (3) (2006) 499–502.
[42] L.-C. Gerhardt, A.R. Boccaccini, Bioactive glass and glass-ceramic scaffolds for bone tissue engineering, Materials 3 (7) (2010) 3867–3910.
[43] C. Vitale-Brovarone, E. Verné, L. Robiglio, P. Appendino, F. Bassi, G. Martinasso, G. Muzio, R. Canuto, Development of glass–ceramic scaffolds for bone tissue engineering: characterisation, proliferation of human osteoblasts and nodule formation, Acta Biomater. 3 (2) (2007) 199–208.
[44] C. Xu, P. Su, X. Chen, Y. Meng, W. Yu, A.P. Xiang, Y. Wang, Biocompatibility and osteogenesis of biomimetic bioglass-collagen-phosphatidylserine composite scaffolds for bone tissue engineering, Biomaterials 32 (4) (2011) 1051–1058.
[45] C.M. Haslauer, A.K. Moghe, J.A. Osborne, B.S. Gupta, E.G. Loboa, Collagen–PCL sheath–core bicomponent electrospun scaffolds increase osteogenic differentiation and calcium accretion of human adipose-derived stem cells, J. Biomater. Sci. Polym. Ed. 22 (13) (2011) 1695–1712.
[46] R. Chen, C. Huang, Q. Ke, C. He, H. Wang, X. Mo, Preparation and characterization of coaxial electrospun thermoplastic polyurethane/collagen compound nanofibers for tissue engineering applications, Colloids Surf. B: Biointerfaces 79 (2) (2010) 315–325.
[47] P.J. Wood, J.T. Braaten, F.W. Scott, D. Riedel, L.M. Poste, Comparisons of viscous properties of oat and guar gum and the effects of these and oat bran on glycemic index, J. Agric. Food Chem. 38 (3) (1990) 753–757.
[48] M.G. Manda, L.P. da Silva, M.T. Cerqueira, D.R. Pereira, M.B. Oliveira, J.F. Mano, A.P. Marques, J.M. Oliveira, V.M. Correlo, R.L. Reis, Gellan gum-hydroxyapatite composite spongy-like hydrogels for bone tissue engineering, J. Biomed. Mater. Res. A 106 (2) (2018) 479–490.
[49] B.P. Chan, T. Hui, C. Yeung, J. Li, I. Mo, G. Chan, Self-assembled collagen–human mesenchymal stem cell microspheres for regenerative medicine, Biomaterials 28 (31) (2007) 4652–4666.
[50] S. Eshghi, D.V. Schaffer, Engineering microenvironments to control stem cell fate and function, in: StemBook [Internet], Harvard Stem Cell Institute, 2008.
[51] D. Wahl, J. Czernuszka, Collagen-hydroxyapatite composites for hard tissue repair, Eur. Cell. Mater. 11 (2006) 43–56.
[52] A. Gantar, L.P. da Silva, J.M. Oliveira, A.P. Marques, V.M. Correlo, S. Novak, R.L. Reis, Nanoparticulate bioactive-glass-reinforced gellan-gum hydrogels for bone-tissue engineering, Mater. Sci. Eng. C 43 (2014) 27–36.

[53] M.A. Bonifacio, P. Gentile, A.M. Ferreira, S. Cometa, E. De Giglio, Insight into halloysite nanotubes-loaded gellan gum hydrogels for soft tissue engineering applications, Carbohydr. Polym. 163 (2017) 280–291.
[54] M. Szekalska, A. Puciłowska, E. Szymańska, P. Ciosek, K. Winnicka, Alginate: current use and future perspectives in pharmaceutical and biomedical applications, Int. J. Polym. Sci. 2016 (2016).
[55] A. Pettignano, M. Häring, L. Bernardi, N. Tanchoux, F. Quignard, D.D. Díaz, Self-healing alginate–gelatin biohydrogels based on dynamic covalent chemistry: elucidation of key parameters, Mater. Chem. Front. 1 (1) (2017) 73–79.
[56] W. Thein-Han, H.H. Xu, Collagen-calcium phosphate cement scaffolds seeded with umbilical cord stem cells for bone tissue engineering, Tissue Eng. A 17 (23–24) (2011) 2943–2954.
[57] S.I. Jeong, M.D. Krebs, C.A. Bonino, J.E. Samorezov, S.A. Khan, E. Alsberg, Electrospun chitosan–alginate nanofibers with in situ polyelectrolyte complexation for use as tissue engineering scaffolds, Tissue Eng. A 17 (1–2) (2010) 59–70.
[58] L. Zhao, M.D. Weir, H.H. Xu, An injectable calcium phosphate-alginate hydrogel-umbilical cord mesenchymal stem cell paste for bone tissue engineering, Biomaterials 31 (25) (2010) 6502–6510.
[59] N.T. Khanarian, N.M. Haney, R.A. Burga, H.H. Lu, A functional agarose-hydroxyapatite scaffold for osteochondral interface regeneration, Biomaterials 33 (21) (2012) 5247–5258.
[60] Y. Luo, A. Lode, C. Wu, J. Chang, M. Gelinsky, Alginate/nanohydroxyapatite scaffolds with designed core/shell structures fabricated by 3D plotting and in situ mineralization for bone tissue engineering, ACS Appl. Mater. Interfaces 7 (12) (2015) 6541–6549.
[61] T.I. Shaheen, A. Montaser, S. Li, Effect of cellulose nanocrystals on scaffolds comprising chitosan, alginate and hydroxyapatite for bone tissue engineering, Int. J. Biol. Macromol. 121 (2019) 814–821.
[62] T. Chandy, C.P. Sharma, Chitosan—as a biomaterial, Biomater. Artif. Cells Artif. Organs 18 (1) (1990) 1–24.
[63] I. Younes, M. Rinaudo, Chitin and chitosan preparation from marine sources. Structure, properties and applications, Mar. Drugs 13 (3) (2015) 1133–1174.
[64] J. Berger, M. Reist, J.M. Mayer, O. Felt, R. Gurny, Structure and interactions in chitosan hydrogels formed by complexation or aggregation for biomedical applications, Eur. J. Pharm. Biopharm. 57 (1) (2004) 35–52.
[65] P.K. Dutta, J. Dutta, V. Tripathi, Chitin and chitosan: chemistry, properties and applications, J. Sci. Ind. Res. (2004).
[66] S.K.L. Levengood, M. Zhang, Chitosan-based scaffolds for bone tissue engineering, J. Mater. Chem. B 2 (21) (2014) 3161–3184.
[67] M. Abbasian, B. Massoumi, R. Mohammad-Rezaei, H. Samadian, M. Jaymand, Scaffolding polymeric biomaterials: are naturally occurring biological macromolecules more appropriate for tissue engineering? Int. J. Biol. Macromol. 134 (2019) 673–694, https://doi.org/10.1016/j.ijbiomac.2019.04.197.
[68] A. Sadeghi, F. Moztarzadeh, J.A. Mohandesi, Investigating the effect of chitosan on hydrophilicity and bioactivity of conductive electrospun composite scaffold for neural tissue engineering, Int. J. Biol. Macromol. 121 (2019) 625–632.
[69] S. Nazir, M.U.A. Khan, W.S. Al-Arjan, S.I. Abd Razak, A. Javed, M.R.A. Kadir, Nanocomposite hydrogels for melanoma skin cancer care and treatment: in-vitro drug delivery, drug release kinetics and anti-cancer activities, Arab. J. Chem. 14 (5) (2021) 103120.
[70] H. Ijaz, U.R. Tulain, F. Azam, J. Qureshi, Thiolation of arabinoxylan and its application in the fabrication of pH-sensitive thiolated arabinoxylan grafted acrylic acid copolymer, Drug Dev. Ind. Pharm. 45 (5) (2019) 754–766.

[71] M.U.A. Khan, M.A. Raza, S.I.A. Razak, M.R. Abdul Kadir, A. Haider, S.A. Shah, A.H. Mohd Yusof, S. Haider, I. Shakir, S. Aftab, Novel functional antimicrobial and biocompatible arabinoxylan/guar gum hydrogel for skin wound dressing applications, J. Tissue Eng. Regen. Med. 14 (10) (2020) 1488–1501.
[72] C. Péroval, F. Debeaufort, D. Despré, A. Voilley, Edible arabinoxylan-based films. 1. Effects of lipid type on water vapor permeability, film structure, and other physical characteristics, J. Agric. Food Chem. 50 (14) (2002) 3977–3983.
[73] S. Xin, X. Li, Q. Wang, R. Huang, X. Xu, Z. Lei, H. Deng, Novel layer-by-layer structured nanofibrous mats coated by protein films for dermal regeneration, J. Biomed. Nanotechnol. 10 (5) (2014) 803–810.
[74] S. Atta, S. Khaliq, A. Islam, I. Javeria, T. Jamil, M.M. Athar, M.I. Shafiq, A. Ghaffar, Injectable biopolymer based hydrogels for drug delivery applications, Int. J. Biol. Macromol. 80 (2015) 240–245.
[75] D. Archana, J. Dutta, P. Dutta, Evaluation of chitosan nano dressing for wound healing: characterization, in vitro and in vivo studies, Int. J. Biol. Macromol. 57 (2013) 193–203.
[76] D.C. Aduba Jr., S.-S. An, G.S. Selders, W.A. Yeudall, G.L. Bowlin, T. Kitten, H. Yang, Electrospun gelatin–arabinoxylan ferulate composite fibers for diabetic chronic wound dressing application, Int. J. Polym. Mater. Polym. Biomater. 68 (11) (2019) 660–668.
[77] S. Yamanaka, K. Watanabe, N. Kitamura, M. Iguchi, S. Mitsuhashi, Y. Nishi, M. Uryu, The structure and mechanical properties of sheets prepared from bacterial cellulose, J. Mater. Sci. 24 (9) (1989) 3141–3145.
[78] A. Sarko, R. Muggli, Packing analysis of carbohydrates and polysaccharides. III. Valonia cellulose and cellulose II, Macromolecules 7 (4) (1974) 486–494.
[79] M. Stroescu, G. Isopencu, C. Busuioc, A. Stoica-Guzun, Antimicrobial food pads containing bacterial cellulose and polysaccharides, in: Cellulose-Based Superabsorbent Hydrogels, 2019, pp. 1303–1338.
[80] F.A. Müller, L. Müller, I. Hofmann, P. Greil, M.M. Wenzel, R. Staudenmaier, Cellulose-based scaffold materials for cartilage tissue engineering, Biomaterials 27 (21) (2006) 3955–3963.
[81] J.C. Courtenay, M.A. Johns, F. Galembeck, C. Deneke, E.M. Lanzoni, C.A. Costa, J.L. Scott, R.I. Sharma, Surface modified cellulose scaffolds for tissue engineering, Cellulose 24 (1) (2017) 253–267.
[82] X. Zhang, C. Wang, M. Liao, L. Dai, Y. Tang, H. Zhang, P. Coates, F. Sefat, L. Zheng, J. Song, Aligned electrospun cellulose scaffolds coated with rhBMP-2 for both in vitro and in vivo bone tissue engineering, Carbohydr. Polym. 213 (2019) 27–38.
[83] V.D. Prajapati, P.M. Maheriya, G.K. Jani, H.K. Solanki, Carrageenan: a natural seaweed polysaccharide and its applications, Carbohydr. Polym. 105 (2014) 97–112.
[84] N. Anderson, J. Campbell, M. Harding, D. Rees, J. Samuel, X-ray diffraction studies of polysaccharide sulphates: double helix models for κ- and ι-carrageenans, J. Mol. Biol. 45 (1) (1969) 85–97.
[85] Y.A. Shchipunov, Sol–gel-derived biomaterials of silica and carrageenans, J. Colloid Interface Sci. 268 (1) (2003) 68–76.
[86] A. Kalsoom Khan, A.U. Saba, S. Nawazish, F. Akhtar, R. Rashid, S. Mir, B. Nasir, F. Iqbal, S. Afzal, F. Pervaiz, Carrageenan based bionanocomposites as drug delivery tool with special emphasis on the influence of ferromagnetic nanoparticles, Oxidative Med. Cell. Longev. 2017 (2017).
[87] W.J. Li, C.T. Laurencin, E.J. Caterson, R.S. Tuan, F.K. Ko, Electrospun nanofibrous structure: a novel scaffold for tissue engineering, J. Biomed. Mater. Res. 60 (4) (2002) 613–621.
[88] S. Rodrigues, A.M.R. da Costa, A. Grenha, Chitosan/carrageenan nanoparticles: effect of cross-linking with tripolyphosphate and charge ratios, Carbohydr. Polym. 89 (1) (2012) 282–289.

[89] J. Araujo, N. Davidenko, M. Danner, R. Cameron, S. Best, Novel porous scaffolds of pH responsive chitosan/carrageenan-based polyelectrolyte complexes for tissue engineering, J. Biomed. Mater. Res. A 102 (12) (2014) 4415–4426.
[90] A. Ogston, J. Stanier, The physiological function of hyaluronic acid in synovial fluid; viscous, elastic and lubricant properties, J. Physiol. 119 (2–3) (1953) 244–252.
[91] J. Necas, L. Bartosikova, P. Brauner, J. Kolar, Hyaluronic acid (hyaluronan): a review, Vet. Med. 53 (8) (2008) 397–411.
[92] M. Swetha, K. Sahithi, A. Moorthi, N. Srinivasan, K. Ramasamy, N. Selvamurugan, Biocomposites containing natural polymers and hydroxyapatite for bone tissue engineering, Int. J. Biol. Macromol. 47 (1) (2010) 1–4.
[93] Y. Hu, J. Chen, T. Fan, Y. Zhang, Y. Zhao, X. Shi, Q. Zhang, Biomimetic mineralized hierarchical hybrid scaffolds based on in situ synthesis of nano-hydroxyapatite/chitosan/chondroitin sulfate/hyaluronic acid for bone tissue engineering, Colloids Surf. B: Biointerfaces 157 (2017) 93–100.
[94] A.E. Erickson, J. Sun, S.K.L. Levengood, S. Swanson, F.-C. Chang, C.T. Tsao, M. Zhang, Chitosan-based composite bilayer scaffold as an in vitro osteochondral defect regeneration model, Biomed. Microdevices 21 (2) (2019) 34.
[95] C. Huang, G. Fang, Y. Zhao, S. Bhagia, X. Meng, Q. Yong, A.J. Ragauskas, Bio-inspired nanocomposite by layer-by-layer coating of chitosan/hyaluronic acid multilayers on a hard nanocellulose-hydroxyapatite matrix, Carbohydr. Polym. 222 (2019) 115036.
[96] J. Baier Leach, K.A. Bivens, C.W. Patrick Jr., C.E. Schmidt, Photocrosslinked hyaluronic acid hydrogels: natural, biodegradable tissue engineering scaffolds, Biotechnol. Bioeng. 82 (5) (2003) 578–589.
[97] H. Kenar, C.Y. Ozdogan, C. Dumlu, E. Doger, G.T. Kose, V. Hasirci, Microfibrous scaffolds from poly (l-lactide-co-ε-caprolactone) blended with xeno-free collagen/hyaluronic acid for improvement of vascularization in tissue engineering applications, Mater. Sci. Eng. C 97 (2019) 31–44.
[98] U. Häcker, K. Nybakken, N. Perrimon, Developmental cell biology: heparan sulphate proteoglycans: the sweet side of development, Nat. Rev. Mol. Cell Biol. 6 (7) (2005) 530.
[99] K. Prydz, K.T. Dalen, Synthesis and sorting of proteoglycans, J. Cell Sci. 113 (2) (2000) 193–205.
[100] J.E. Silbert, G. Sugumaran, Biosynthesis of chondroitin/dermatan sulfate, IUBMB Life 54 (4) (2002) 177–186.
[101] J.D. Esko, K. Kimata, U. Lindahl, Proteoglycans and sulfated glycosaminoglycans, in: Essentials of Glycobiology, second ed., Cold Spring Harbor Laboratory Press, 2009.
[102] V. Profant, C. Johannessen, E.W. Blanch, P. Bouř, V. Baumruk, Effects of sulfation and the environment on the structure of chondroitin sulfate studied via Raman optical activity, Phys. Chem. Chem. Phys. 21 (14) (2019) 7367–7377.
[103] D.M. Arenson, S.L. Friedman, D.M. Bissell, Formation of extracellular matrix in normal rat liver: lipocytes as a major source of proteoglycan, Gastroenterology 95 (2) (1988) 441–447.
[104] R. Leach Jr., A.-M. Muenster, E.M. Wien, Studies on the role of manganese in bone formation: II. Effect upon chondroitin sulfate synthesis in chick epiphyseal cartilage, Arch. Biochem. Biophys. 133 (1) (1969) 22–28.
[105] A. Gressner, M. Bachem, Cellular sources of noncollagenous matrix proteins: role of fat-storing cells in fibrogenesis, Semin. Liver Dis. (1990) 30–46 (© 1990 by Thieme Medical Publishers, Inc).
[106] H. Omi, S. Itoh, T. Ikoma, Y. Asou, S. Nishikawa, M. Tanaka, K. Shinomiya, S. Toh, Biocompatibility and osteoconductivity of hydroxyapatite/polysaccharide nanocomposite microparticles, Key Eng. Mater. (2006) 561–564 (Trans Tech Publ).

[107] A.N. Renth, M.S. Detamore, Leveraging "raw materials" as building blocks and bioactive signals in regenerative medicine, Tissue Eng. B Rev. 18 (5) (2012) 341–362.
[108] Y.-L. Chen, H.-P. Lee, H.-Y. Chan, L.-Y. Sung, H.-C. Chen, Y.-C. Hu, Composite chondroitin-6-sulfate/dermatan sulfate/chitosan scaffolds for cartilage tissue engineering, Biomaterials 28 (14) (2007) 2294–2305.
[109] C.-H. Chang, H.-C. Liu, C.-C. Lin, C.-H. Chou, F.-H. Lin, Gelatin–chondroitin–hyaluronan tri-copolymer scaffold for cartilage tissue engineering, Biomaterials 24 (26) (2003) 4853–4858.
[110] J.E. Silbert, M. Bernfield, R. Kokenyesi, Proteoglycans: a special class of glycoproteins, in: Glycoproteins II, vol. 29, 1997, p. 1.
[111] K. Sugahara, T. Mikami, T. Uyama, S. Mizuguchi, K. Nomura, H. Kitagawa, Recent advances in the structural biology of chondroitin sulfate and dermatan sulfate, Curr. Opin. Struct. Biol. 13 (5) (2003) 612–620.
[112] S. Enoch, D.J. Leaper, Basic science of wound healing, Surgery (Oxford) 23 (2) (2005) 37–42.
[113] T. Manon-Jensen, Y. Itoh, J.R. Couchman, Proteoglycans in health and disease: the multiple roles of syndecan shedding, FEBS J. 277 (19) (2010) 3876–3889.
[114] M. Kawano, W. Ariyoshi, K. Iwanaga, T. Okinaga, M. Habu, I. Yoshioka, K. Tominaga, T. Nishihara, Mechanism involved in enhancement of osteoblast differentiation by hyaluronic acid, Biochem. Biophys. Res. Commun. 405 (4) (2011) 575–580.
[115] A.B. Roberts, B.K. McCune, M.B. Sporn, TGF-β: regulation of extracellular matrix, Kidney Int. 41 (3) (1992) 557–559.
[116] K.Y. Tsang, M.C. Cheung, D. Chan, K.S. Cheah, The developmental roles of the extracellular matrix: beyond structure to regulation, Cell Tissue Res. 339 (1) (2010) 93.
[117] J. Mouw, N. Case, R. Guldberg, A. Plaas, M. Levenston, Variations in matrix composition and GAG fine structure among scaffolds for cartilage tissue engineering, Osteoarthr. Cartil. 13 (9) (2005) 828–836.
[118] A. Koenig, K. Norgard-Sumnicht, R. Linhardt, A. Varki, Differential interactions of heparin and heparan sulfate glycosaminoglycans with the selectins. Implications for the use of unfractionated and low molecular weight heparins as therapeutic agents, J. Clin. Invest. 101 (4) (1998) 877–889.
[119] Conrad, H. E.; Guo, Y., Non-anticoagulant heparin derivatives. Google Patents: 1994.
[120] Y. Nakayama, R. Sakata, M. Ura, T.-A. Miyamoto, Coronary artery bypass grafting in dialysis patients, Ann. Thorac. Surg. 68 (4) (1999) 1257–1261.
[121] T.N. Wight, M.G. Kinsella, E.E. Qwarnström, The role of proteoglycans in cell adhesion, migration and proliferation, Curr. Opin. Cell Biol. 4 (5) (1992) 793–801.
[122] D.S. Benoit, A.R. Durney, K.S. Anseth, The effect of heparin-functionalized PEG hydrogels on three-dimensional human mesenchymal stem cell osteogenic differentiation, Biomaterials 28 (1) (2007) 66–77.
[123] S. Blotnick, G.E. Peoples, M.R. Freeman, T.J. Eberlein, M. Klagsbrun, T lymphocytes synthesize and export heparin-binding epidermal growth factor-like growth factor and basic fibroblast growth factor, mitogens for vascular cells and fibroblasts: differential production and release by CD4+ and CD8+ T cells, Proc. Natl. Acad. Sci. 91 (8) (1994) 2890–2894.
[124] K.M. Malinda, L. Ponce, H.K. Kleinman, L.M. Shackelton, A.J. Millis, Gp38k, a protein synthesized by vascular smooth muscle cells, stimulates directional migration of human umbilical vein endothelial cells, Exp. Cell Res. 250 (1) (1999) 168–173.
[125] Q. Lu, S. Zhang, K. Hu, Q. Feng, C. Cao, F. Cui, Cytocompatibility and blood compatibility of multifunctional fibroin/collagen/heparin scaffolds, Biomaterials 28 (14) (2007) 2306–2313.

[126] J.K. Tessmar, A.M. Göpferich, Matrices and scaffolds for protein delivery in tissue engineering, Adv. Drug Deliv. Rev. 59 (4–5) (2007) 274–291.
[127] S. Gräslund, P. Nordlund, J. Weigelt, B.M. Hallberg, J. Bray, O. Gileadi, S. Knapp, U. Oppermann, C. Arrowsmith, R. Hui, Protein production and purification, Nat. Methods 5 (2) (2008) 135.
[128] H.J. Kim, U.-J. Kim, G. Vunjak-Novakovic, B.-H. Min, D.L. Kaplan, Influence of macroporous protein scaffolds on bone tissue engineering from bone marrow stem cells, Biomaterials 26 (21) (2005) 4442–4452.
[129] N. Minoura, M. Tsukada, M. Nagura, Physico-chemical properties of silk fibroin membrane as a biomaterial, Biomaterials 11 (6) (1990) 430–434.
[130] M. Mondal, The silk proteins, sericin and fibroin in silkworm, Bombyx mori Linn.—a review, Caspian J. Environ. Sci. 5 (2) (2007) 63–76.
[131] S. Aznar-Cervantes, M.I. Roca, J.G. Martinez, L. Meseguer-Olmo, J.L. Cenis, J.M. Moraleda, T.F. Otero, Fabrication of conductive electrospun silk fibroin scaffolds by coating with polypyrrole for biomedical applications, Bioelectrochemistry 85 (2012) 36–43.
[132] W. Shao, J. He, F. Sang, B. Ding, L. Chen, S. Cui, K. Li, Q. Han, W. Tan, Coaxial electrospun aligned tussah silk fibroin nanostructured fiber scaffolds embedded with hydroxyapatite–tussah silk fibroin nanoparticles for bone tissue engineering, Mater. Sci. Eng. C 58 (2016) 342–351.
[133] H.J. Kim, U.-J. Kim, H.S. Kim, C. Li, M. Wada, G.G. Leisk, D.L. Kaplan, Bone tissue engineering with premineralized silk scaffolds, Bone 42 (6) (2008) 1226–1234.
[134] B. Niu, B. Li, Y. Gu, X. Shen, Y. Liu, L. Chen, In vitro evaluation of electrospun silk fibroin/nano-hydroxyapatite/BMP-2 scaffolds for bone regeneration, J. Biomater. Sci. Polym. Ed. 28 (3) (2017) 257–270.
[135] C. Xu, R. Inai, M. Kotaki, S. Ramakrishna, Electrospun nanofiber fabrication as synthetic extracellular matrix and its potential for vascular tissue engineering, Tissue Eng. 10 (7–8) (2004) 1160–1168.
[136] H.J. Park, O.J. Lee, M.C. Lee, B.M. Moon, H.W. Ju, J. min Lee, J.-H. Kim, D.W. Kim, C. H. Park, Fabrication of 3D porous silk scaffolds by particulate (salt/sucrose) leaching for bone tissue reconstruction, Int. J. Biol. Macromol. 78 (2015) 215–223.
[137] Y.H. Kang, H.C. Kim, S.H. Shin, H.S. Kim, K.C. Kim, S.H. Lee, Osteoconductive effect of chitosan/hydroxyapatite composite matrix on rat skull defect, Tissue Eng. Regen. Med. 8 (1) (2011) 23–31.
[138] S. Font Tellado, W. Bonani, E.R. Balmayor, P. Foehr, A. Motta, C. Migliaresi, M. van Griensven, Fabrication and characterization of biphasic silk fibroin scaffolds for tendon/ligament-to-bone tissue engineering, Tissue Eng. A 23 (15–16) (2017) 859–872.
[139] D.R. Eyre, S. Apon, J.-J. Wu, L.H. Ericsson, K.A. Walsh, Collagen type IX: evidence for covalent linkages to type II collagen in cartilage, FEBS Lett. 220 (2) (1987) 337–341.
[140] S.P. Ho, S.J. Marshall, M.I. Ryder, G.W. Marshall, The tooth attachment mechanism defined by structure, chemical composition and mechanical properties of collagen fibers in the periodontium, Biomaterials 28 (35) (2007) 5238–5245.
[141] C.H. Lee, A. Singla, Y. Lee, Biomedical applications of collagen, Int. J. Pharm. 221 (1–2) (2001) 1–22.
[142] Y.-C. Lin, F.-J. Tan, K.G. Marra, S.-S. Jan, D.-C. Liu, Synthesis and characterization of collagen/hyaluronan/chitosan composite sponges for potential biomedical applications, Acta Biomater. 5 (7) (2009) 2591–2600.
[143] C. Rodrigues, P. Serricella, A. Linhares, R. Guerdes, R. Borojevic, M. Rossi, M. Duarte, M. Farina, Characterization of a bovine collagen–hydroxyapatite composite scaffold for bone tissue engineering, Biomaterials 24 (27) (2003) 4987–4997.

[144] A. Oryan, A. Kamali, A. Moshiri, H. Baharvand, H. Daemi, Chemical crosslinking of biopolymeric scaffolds: current knowledge and future directions of crosslinked engineered bone scaffolds, Int. J. Biol. Macromol. 107 (2018) 678–688.
[145] C.-F. Liu, S.-J. Li, W.-T. Hou, Y.-L. Hao, H.-H. Huang, Enhancing corrosion resistance and biocompatibility of interconnected porous β-type Ti-24Nb-4Zr-8Sn alloy scaffold through alkaline treatment and type I collagen immobilization, Appl. Surf. Sci. 476 (2019) 325–334.
[146] C. Tangsadthakun, S. Kanokpanont, N. Sanchavanakit, T. Banaprasert, S. Damrongsakkul, Properties of collagen/chitosan scaffolds for skin tissue engineering, J. Met. Mater. Miner. 16 (1) (2017).
[147] S.C. Leeuwenburgh, J. Jo, H. Wang, M. Yamamoto, J.A. Jansen, Y. Tabata, Mineralization, biodegradation, and drug release behavior of gelatin/apatite composite microspheres for bone regeneration, Biomacromolecules 11 (10) (2010) 2653–2659.
[148] Y. Tabata, Y. Ikada, Protein release from gelatin matrices, Adv. Drug Deliv. Rev. 31 (3) (1998) 287–301.
[149] S.-C. Wu, W.-H. Chang, G.-C. Dong, K.-Y. Chen, Y.-S. Chen, C.-H. Yao, Cell adhesion and proliferation enhancement by gelatin nanofiber scaffolds, J. Bioact. Compat. Polym. 26 (6) (2011) 565–577.
[150] X. Liu, L.A. Smith, J. Hu, P.X. Ma, Biomimetic nanofibrous gelatin/apatite composite scaffolds for bone tissue engineering, Biomaterials 30 (12) (2009) 2252–2258.
[151] P. Gentile, V.K. Nandagiri, J. Daly, V. Chiono, C. Mattu, C. Tonda-Turo, G. Ciardelli, Z. Ramtoola, Localised controlled release of simvastatin from porous chitosan–gelatin scaffolds engrafted with simvastatin loaded PLGA-microparticles for bone tissue engineering application, Mater. Sci. Eng. C 59 (2016) 249–257.
[152] D. Nguyen, J. McCanless, M. Mecwan, A. Noblett, W. Haggard, R. Smith, J. Bumgardner, Balancing mechanical strength with bioactivity in chitosan–calcium phosphate 3D microsphere scaffolds for bone tissue engineering: air- vs. freeze-drying processes, J. Biomater. Sci. Polym. Ed. 24 (9) (2013) 1071–1083.
[153] B.S. Kim, J.S. Kim, Y.S. Chung, Y.W. Sin, K.H. Ryu, J. Lee, H.K. You, Growth and osteogenic differentiation of alveolar human bone marrow-derived mesenchymal stem cells on chitosan/hydroxyapatite composite fabric, J. Biomed. Mater. Res. A 101 (6) (2013) 1550–1558.
[154] G. Zuo, C. Liu, H. Luo, F. He, H. Liang, J. Wang, Y. Wan, Synthesis of intercalated lamellar hydroxyapatite/gelatin nanocomposite for bone substitute application, J. Appl. Polym. Sci. 113 (5) (2009) 3089–3094.
[155] R. Murugan, S. Ramakrishna, Development of nanocomposites for bone grafting, Compos. Sci. Technol. 65 (15–16) (2005) 2385–2406.
[156] P. Vashisth, J.R. Bellare, Development of hybrid scaffold with biomimetic 3D architecture for bone regeneration, Nanomedicine 14 (4) (2018) 1325–1336.
[157] K. Maji, S. Dasgupta, K. Pramanik, A. Bissoyi, Preparation and characterization of gelatin-chitosan-nanoβ-TCP based scaffold for orthopaedic application, Mater. Sci. Eng. C 86 (2018) 83–94.
[158] C.C. Lin, K.B. Chen, C.H. Tsai, F.J. Tsai, C.Y. Huang, C.H. Tang, J.S. Yang, Y.M. Hsu, S. F. Peng, J.G. Chung, Casticin inhibits human prostate cancer DU 145 cell migration and invasion via Ras/Akt/NF-κB signaling pathways, J. Food Biochem. (2019), e12902.
[159] S. Yu, Z. Mao, C. Gao, Preparation of gelatin density gradient on poly (ε-caprolactone) membrane and its influence on adhesion and migration of endothelial cells, J. Colloid Interface Sci. 451 (2015) 177–183.
[160] M.S. Chapekar, Tissue engineering: challenges and opportunities, J. Biomed. Mater. Res. 53 (6) (2000) 617–620.

[161] J.P. Kumar, L. Lakshmi, V. Jyothsna, D. Balaji, S. Saravanan, A. Moorthi, N. Selvamurugan, Synthesis and characterization of diopside particles and their suitability along with chitosan matrix for bone tissue engineering in vitro and in vivo, J. Biomed. Nanotechnol. 10 (6) (2014) 970–981.
[162] N. Davidenko, J.J. Campbell, E. Thian, C.J. Watson, R.E. Cameron, Collagen–hyaluronic acid scaffolds for adipose tissue engineering, Acta Biomater. 6 (10) (2010) 3957–3968.
[163] R. Jayakumar, K.P. Chennazhi, S. Srinivasan, S.V. Nair, T. Furuike, H. Tamura, Chitin scaffolds in tissue engineering, Int. J. Mol. Sci. 12 (3) (2011) 1876–1887.
[164] R. Sainitya, M. Sriram, V. Kalyanaraman, S. Dhivya, S. Saravanan, M. Vairamani, T. Sastry, N. Selvamurugan, Scaffolds containing chitosan/carboxymethyl cellulose/mesoporous wollastonite for bone tissue engineering, Int. J. Biol. Macromol. 80 (2015) 481–488.
[165] R. Khajavi, M. Abbasipour, A. Bahador, Electrospun biodegradable nanofibers scaffolds for bone tissue engineering, J. Appl. Polym. Sci. 133 (3) (2016).
[166] K. Madhumathi, P.S. Kumar, K. Kavya, T. Furuike, H. Tamura, S. Nair, R. Jayakumar, Novel chitin/nanosilica composite scaffolds for bone tissue engineering applications, Int. J. Biol. Macromol. 45 (3) (2009) 289–292.
[167] M. Barbosa, P. Granja, C. Barrias, I. Amaral, Polysaccharides as scaffolds for bone regeneration, ITBM-RBM 26 (3) (2005) 212–217.
[168] Y. Qin, S. Zhang, J. Yu, J. Yang, L. Xiong, Q. Sun, Effects of chitin nano-whiskers on the antibacterial and physicochemical properties of maize starch films, Carbohydr. Polym. 147 (2016) 372–378.
[169] T. Rasheed, M. Bilal, Y. Zhao, A. Raza, S.Z.H. Shah, H.M. Iqbal, Physiochemical characteristics and bone/cartilage tissue engineering potentialities of protein-based macromolecules—a review, Int. J. Biol. Macromol. 121 (2019) 13–22.
[170] A. Cheng, Z. Schwartz, A. Kahn, X. Li, Z. Shao, M. Sun, Y. Ao, B.D. Boyan, H. Chen, Advances in porous scaffold design for bone and cartilage tissue engineering and regeneration, Tissue Eng. B Rev. 25 (1) (2019) 14–29.
[171] X. Liu, P.X. Ma, Polymeric scaffolds for bone tissue engineering, Ann. Biomed. Eng. 32 (3) (2004) 477–486.
[172] P.X. Ma, J.-W. Choi, Biodegradable polymer scaffolds with well-defined interconnected spherical pore network, Tissue Eng. 7 (1) (2001) 23–33.
[173] Y. Lai, H. Cao, X. Wang, S. Chen, M. Zhang, N. Wang, Z. Yao, Y. Dai, X. Xie, P. Zhang, Porous composite scaffold incorporating osteogenic phytomolecule icariin for promoting skeletal regeneration in challenging osteonecrotic bone in rabbits, Biomaterials 153 (2018) 1–13.
[174] M. Borden, M. Attawia, Y. Khan, C.T. Laurencin, Tissue engineered microsphere-based matrices for bone repair: design and evaluation, Biomaterials 23 (2) (2002) 551–559.
[175] F. Dehghani, N. Annabi, Engineering porous scaffolds using gas-based techniques, Curr. Opin. Biotechnol. 22 (5) (2011) 661–666.
[176] R.A. Perez, G. Mestres, Role of pore size and morphology in musculo-skeletal tissue regeneration, Mater. Sci. Eng. C 61 (2016) 922–939.
[177] M.J. Gupte, W.B. Swanson, J. Hu, X. Jin, H. Ma, Z. Zhang, Z. Liu, K. Feng, G. Feng, G. Xiao, Pore size directs bone marrow stromal cell fate and tissue regeneration in nanofibrous macroporous scaffolds by mediating vascularization, Acta Biomater. 82 (2018) 1–11.
[178] N. Celikkin, S. Mastrogiacomo, J. Jaroszewicz, X.F. Walboomers, W. Swieszkowski, Gelatin methacrylate scaffold for bone tissue engineering: the influence of polymer concentration, J. Biomed. Mater. Res. A 106 (1) (2018) 201–209.

[179] A. Grémare, V. Guduric, R. Bareille, V. Heroguez, S. Latour, N. L'heureux, J.C. Fricain, S. Catros, D. Le Nihouannen, Characterization of printed PLA scaffolds for bone tissue engineering, J. Biomed. Mater. Res. A 106 (4) (2018) 887–894.
[180] A. Rogina, M. Antunović, D. Milovac, Biomimetic design of bone substitutes based on cuttlefish bone-derived hydroxyapatite and biodegradable polymers, J. Biomed. Mater. Res. B Appl. Biomater. 107 (1) (2019) 197–204.
[181] A.J.W. Johnson, B.A. Herschler, A review of the mechanical behavior of CaP and CaP/polymer composites for applications in bone replacement and repair, Acta Biomater. 7 (1) (2011) 16–30.
[182] H. Zhuang, Y. Han, A. Feng, Preparation, mechanical properties and in vitro biodegradation of porous magnesium scaffolds, Mater. Sci. Eng. C 28 (8) (2008) 1462–1466.
[183] V. Guarino, F. Causa, L. Ambrosio, Bioactive scaffolds for bone and ligament tissue, Expert Rev. Med. Devices 4 (3) (2007) 405–418.
[184] D.W. Hutmacher, Scaffolds in tissue engineering bone and cartilage, in: The Biomaterials: Silver Jubilee Compendium, Elsevier, 2000, pp. 175–189.
[185] Y. Kang, A. Scully, D.A. Young, S. Kim, H. Tsao, M. Sen, Y. Yang, Enhanced mechanical performance and biological evaluation of a PLGA coated β-TCP composite scaffold for load-bearing applications, Eur. Polym. J. 47 (8) (2011) 1569–1577.
[186] S. Prasadh, R.C.W. Wong, Unraveling the mechanical strength of biomaterials used as a bone scaffold in oral and maxillofacial defects, Oral Sci. Int. 15 (2) (2018) 48–55.
[187] R.C. Thomson, M.J. Yaszemski, J.M. Powers, A.G. Mikos, Fabrication of biodegradable polymer scaffolds to engineer trabecular bone, J. Biomater. Sci. Polym. Ed. 7 (1) (1996) 23–38.
[188] C. Shuai, W. Guo, P. Wu, W. Yang, S. Hu, Y. Xia, P. Feng, A graphene oxide-Ag co-dispersing nanosystem: dual synergistic effects on antibacterial activities and mechanical properties of polymer scaffolds, Chem. Eng. J. 347 (2018) 322–333.
[189] C.M. Agrawal, R.B. Ray, Biodegradable polymeric scaffolds for musculoskeletal tissue engineering, J. Biomed. Mater. Res. 55 (2) (2001) 141–150.
[190] M.I. Sabir, X. Xu, L. Li, A review on biodegradable polymeric materials for bone tissue engineering applications, J. Mater. Sci. 44 (21) (2009) 5713–5724.
[191] J.D. Kretlow, A.G. Mikos, Mineralization of synthetic polymer scaffolds for bone tissue engineering, Tissue Eng. 13 (5) (2007) 927–938.
[192] K. Rezwan, Q. Chen, J. Blaker, A.R. Boccaccini, Biodegradable and bioactive porous polymer/inorganic composite scaffolds for bone tissue engineering, Biomaterials 27 (18) (2006) 3413–3431.
[193] M. Bouyer, R. Guillot, J. Lavaud, C. Plettinx, C. Olivier, V. Curry, J. Boutonnat, J.-L. Coll, F. Peyrin, V. Josserand, Surface delivery of tunable doses of BMP-2 from an adaptable polymeric scaffold induces volumetric bone regeneration, Biomaterials 104 (2016) 168–181.
[194] J. Aragón, S. Salerno, L. De Bartolo, S. Irusta, G. Mendoza, Polymeric electrospun scaffolds for bone morphogenetic protein 2 delivery in bone tissue engineering, J. Colloid Interface Sci. 531 (2018) 126–137.
[195] I. Smith, X. Liu, L. Smith, P. Ma, Nanostructured polymer scaffolds for tissue engineering and regenerative medicine, Wiley Interdiscip. Rev. Nanomed. Nanobiotechnol. 1 (2) (2009) 226–236.
[196] A. Hasan, G. Waibhaw, V. Saxena, L.M. Pandey, Nano-biocomposite scaffolds of chitosan, carboxymethyl cellulose and silver nanoparticle modified cellulose nanowhiskers for bone tissue engineering applications, Int. J. Biol. Macromol. 111 (2018) 923–934.

[197] K. Joshy, S. Snigdha, S. Thomas, Plasma modified polymeric materials for scaffolding of bone tissue engineering, in: Non-Thermal Plasma Technology for Polymeric Materials, Elsevier, 2019, pp. 439–458.
[198] W. Wang, L. Guo, Y. Yu, Z. Chen, R. Zhou, Z. Yuan, Peptide REDV-modified polysaccharide hydrogel with endothelial cell selectivity for the promotion of angiogenesis, J. Biomed. Mater. Res. A 103 (5) (2015) 1703–1712.
[199] S.A. DeLong, A.S. Gobin, J.L. West, Covalent immobilization of RGDS on hydrogel surfaces to direct cell alignment and migration, J. Control. Release 109 (1–3) (2005) 139–148.
[200] H. Zhang, X. Zheng, W. Ahmed, Y. Yao, J. Bai, Y. Chen, C. Gao, Design and applications of cell-selective surfaces and interfaces, Biomacromolecules 19 (6) (2018) 1746–1763.
[201] R.G. LeBaron, K.A. Athanasiou, Extracellular matrix cell adhesion peptides: functional applications in orthopedic materials, Tissue Eng. 6 (2) (2000) 85–103.
[202] O. Guillaume, M. Geven, C. Sprecher, V. Stadelmann, D. Grijpma, T. Tang, L. Qin, Y. Lai, M. Alini, J. de Bruijn, Surface-enrichment with hydroxyapatite nanoparticles in stereolithography-fabricated composite polymer scaffolds promotes bone repair, Acta Biomater. 54 (2017) 386–398.
[203] N. Reznikov, M. Bilton, L. Lari, M.M. Stevens, R. Kröger, Fractal-like hierarchical organization of bone begins at the nanoscale, Science 360 (6388) (2018), eaao2189.
[204] X. Guo, D. Li, Synthesis of hydroxyapatite containing some trace amounts elements in simulated body fluids, Iran. J. Chem. Chem. Eng. 38 (1) (2019) 83–91.
[205] A.H. Ambre, D.R. Katti, K.S. Katti, Biomineralized hydroxyapatite nanoclay composite scaffolds with polycaprolactone for stem cell-based bone tissue engineering, J. Biomed. Mater. Res. A 103 (6) (2015) 2077–2101.
[206] M.C. Nerantzaki, I.G. Koliakou, M.G. Kaloyianni, Z.N. Terzopoulou, E.K. Siska, M.A. Karakassides, A.R. Boccaccini, D.N. Bikiaris, New N-(2-carboxybenzyl) chitosan composite scaffolds containing nanoTiO2 or bioactive glass with enhanced cell proliferation for bone-tissue engineering applications, Int. J. Polym. Mater. Polym. Biomater. 66 (2) (2017) 71–81.
[207] Z. Hong, R.L. Reis, J.F. Mano, Preparation and in vitro characterization of scaffolds of poly (L-lactic acid) containing bioactive glass ceramic nanoparticles, Acta Biomater. 4 (5) (2008) 1297–1306.
[208] V.V. Seregin, J.L. Coffer, Biomineralization of calcium disilicide in porous polycaprolactone scaffolds, Biomaterials 27 (27) (2006) 4745–4754.
[209] R. Zhang, P.X. Ma, Biomimetic polymer/apatite composite scaffolds for mineralized tissue engineering, Macromol. Biosci. 4 (2) (2004) 100–111.
[210] F. Farshi Azhar, A. Olad, R. Salehi, Fabrication and characterization of chitosan–gelatin/nanohydroxyapatite–polyaniline composite with potential application in tissue engineering scaffolds, Des. Monomers Polym. 17 (7) (2014) 654–667.
[211] P.E. Mikael, S.P. Nukavarapu, Functionalized carbon nanotube composite scaffolds for bone tissue engineering: prospects and progress, J. Biomater. Tissue Eng. 1 (1) (2011) 76–85.
[212] A.S. Mesgar, Z. Mohammadi, S. Khosrovan, Improvement of mechanical properties and in vitro bioactivity of freeze-dried gelatin/chitosan scaffolds by functionalized carbon nanotubes, Int. J. Polym. Mater. Polym. Biomater. 67 (5) (2018) 267–276.
[213] B. Huang, G. Caetano, C. Vyas, J. Blaker, C. Diver, P. Bártolo, Polymer-ceramic composite scaffolds: the effect of hydroxyapatite and β-tri-calcium phosphate, Materials 11 (1) (2018) 129.

[214] S.J. Kalita, S. Bose, H.L. Hosick, A. Bandyopadhyay, Development of controlled porosity polymer-ceramic composite scaffolds via fused deposition modeling, Mater. Sci. Eng. C 23 (5) (2003) 611–620.
[215] X. Wang, M. Jiang, Z. Zhou, J. Gou, D. Hui, 3D printing of polymer matrix composites: a review and prospective, Compos. Part B 110 (2017) 442–458.
[216] C. Shuai, Y. Xu, P. Feng, G. Wang, S. Xiong, S. Peng, Antibacterial polymer scaffold based on mesoporous bioactive glass loaded with in situ grown silver, Chem. Eng. J. 374 (2019) 304–315.
[217] D.A.L.V.d. Cunha, P. Inforçatti Neto, K.C. Micocci, C.F. Bellani, H.S. Selistre-de-Araujo, Z.C. Silveira, M.C. Branciforti, Fabrication and characterization of scaffolds of poly (ε-caprolactone)/Biosilicate® biocomposites prepared by generative manufacturing process, Int. J. Biomater. 2019 (2019).
[218] U. Kalsoom, P.N. Nesterenko, B. Paull, Recent developments in 3D printable composite materials, RSC Adv. 6 (65) (2016) 60355–60371.
[219] Y. Wu, X. Chen, G. Zhao, R. Chen, Y. Liu, H. Ren, X. Qu, Y. Liu, β-Tricalcium phosphate/ε-polycaprolactone composite scaffolds with a controllable gradient: fabrication and characterization, Ceram. Int. 45 (13) (2019) 16188–16194, https://doi.org/10.1016/j.ceramint.2019.05.140.
[220] A. Berner, M. Woodruff, C. Lam, M. Arafat, S. Saifzadeh, R. Steck, J. Ren, M. Nerlich, A. Ekaputra, I. Gibson, Effects of scaffold architecture on cranial bone healing, Int. J. Oral Maxillofac. Surg. 43 (4) (2014) 506–513.
[221] J.M. Williams, A. Adewunmi, R.M. Schek, C.L. Flanagan, P.H. Krebsbach, S.E. Feinberg, S.J. Hollister, S. Das, Bone tissue engineering using polycaprolactone scaffolds fabricated via selective laser sintering, Biomaterials 26 (23) (2005) 4817–4827.
[222] L.M. Mathieu, T.L. Mueller, P.-E. Bourban, D.P. Pioletti, R. Müller, J.-A.E. Månson, Architecture and properties of anisotropic polymer composite scaffolds for bone tissue engineering, Biomaterials 27 (6) (2006) 905–916.
[223] M. Schieker, H. Seitz, I. Drosse, S. Seitz, W. Mutschler, Biomaterials as scaffold for bone tissue engineering, Eur. J. Trauma 32 (2) (2006) 114–124.
[224] C.N. Kelly, A.T. Miller, S.J. Hollister, R.E. Guldberg, K. Gall, Design and structure–function characterization of 3D printed synthetic porous biomaterials for tissue engineering, Adv. Healthc. Mater. 7 (7) (2018).
[225] J.M. Holzwarth, P.X. Ma, Biomimetic nanofibrous scaffolds for bone tissue engineering, Biomaterials 32 (36) (2011) 9622–9629.
[226] V.J. Chen, L.A. Smith, P.X. Ma, Bone regeneration on computer-designed nano-fibrous scaffolds, Biomaterials 27 (21) (2006) 3973–3979.
[227] R. Zhang, P.X. Ma, Poly (α-hydroxyl acids)/hydroxyapatite porous composites for bone-tissue engineering. I. Preparation and morphology, J. Biomed. Mater. Res. 44 (4) (1999) 446–455.
[228] B. Yuan, M. Zhu, C.Y. Chung, Biomedical porous shape memory alloys for hard-tissue replacement materials, Materials 11 (9) (2018) 1716.
[229] J. Richard, F.S. Deschamps, Supercritical fluid processes for polymer particle engineering, in: Colloidal Biomolecules, Biomaterials, and Biomedical Applications, Marcel Dekker, New York, 2003, pp. 429–487.
[230] S. Pina, J.M. Oliveira, R.L. Reis, Natural-based nanocomposites for bone tissue engineering and regenerative medicine: a review, Adv. Mater. 27 (7) (2015) 1143–1169.
[231] M.A. Ghalia, Y. Dahman, Advanced nanobiomaterials in tissue engineering: synthesis, properties, and applications, in: Nanobiomaterials in Soft Tissue Engineering, Elsevier, 2016, pp. 141–172.

[232] C.J. Liao, C.F. Chen, J.H. Chen, S.F. Chiang, Y.J. Lin, K.Y. Chang, Fabrication of porous biodegradable polymer scaffolds using a solvent merging/particulate leaching method, J. Biomed. Mater. Res. 59 (4) (2002) 676–681.
[233] A. Prasad, M.R. Sankar, V. Katiyar, State of art on solvent casting particulate leaching method for orthopedic scaffolds fabrication, Mater. Today Proc. 4 (2) (2017) 898–907.
[234] A. Sola, J. Bertacchini, D. D'Avella, L. Anselmi, T. Maraldi, S. Marmiroli, M. Messori, Development of solvent-casting particulate leaching (SCPL) polymer scaffolds as improved three-dimensional supports to mimic the bone marrow niche, Mater. Sci. Eng. C 96 (2019) 153–165.
[235] M.N. Abdallah, S. Abdollahi, M. Laurenti, D. Fang, S.D. Tran, M. Cerruti, F. Tamimi, Scaffolds for epithelial tissue engineering customized in elastomeric molds, J. Biomed. Mater. Res. B Appl. Biomater. 106 (2) (2018) 880–890.
[236] A.H. Aghajanian, B.A. Khazaei, M. Khodaei, M. Rafienia, Fabrication of porous mg-Zn scaffold through modified replica method for bone tissue engineering, J. Bionic Eng. 15 (5) (2018) 907–913.
[237] I. Denry, O.M. Goudouri, J. Harless, J.A. Holloway, Rapid vacuum sintering: a novel technique for fabricating fluorapatite ceramic scaffolds for bone tissue engineering, J. Biomed. Mater. Res. B Appl. Biomater. 106 (1) (2018) 291–299.
[238] Q. Chen, D. Mohn, W.J. Stark, Optimization of bioglass® scaffold fabrication process, J. Am. Ceram. Soc. 94 (12) (2011) 4184–4190.
[239] J. Hum, A. Boccaccini, Collagen as coating material for 45S5 bioactive glass-based scaffolds for bone tissue engineering, Int. J. Mol. Sci. 19 (6) (2018) 1807.
[240] M. Kopeć, M. Lamson, R. Yuan, C. Tang, M. Kruk, M. Zhong, K. Matyjaszewski, T. Kowalewski, Polyacrylonitrile-derived nanostructured carbon materials, Prog. Polym. Sci. (2019).
[241] S. Bose, M. Roy, A. Bandyopadhyay, Recent advances in bone tissue engineering scaffolds, Trends Biotechnol. 30 (10) (2012) 546–554.
[242] Y.S. Nam, J.J. Yoon, T.G. Park, A novel fabrication method of macroporous biodegradable polymer scaffolds using gas foaming salt as a porogen additive, J. Biomed. Mater. Res. 53 (1) (2000) 1–7.
[243] M. Floren, S. Spilimbergo, A. Motta, C. Migliaresi, Porous poly (D,L-lactic acid) foams with tunable structure and mechanical anisotropy prepared by supercritical carbon dioxide, J. Biomed. Mater. Res. B Appl. Biomater. 99 (2) (2011) 338–349.
[244] P. Zhao, H. Gu, H. Mi, C. Rao, J. Fu, L.-S. Turng, Fabrication of scaffolds in tissue engineering: a review, Front. Mech. Eng. 13 (1) (2018) 107–119.
[245] J. Ju, X. Peng, K. Huang, L. Li, X. Liu, C. Chitrakar, L. Chang, Z. Gu, T. Kuang, High-performance porous PLLA-based scaffolds for bone tissue engineering: preparation, characterization, and in vitro and in vivo evaluation, Polymer 180 (2019) 121707.
[246] K. Whang, K. Healy, D. Elenz, E. Nam, D. Tsai, C. Thomas, G. Nuber, F. Glorieux, R. Travers, S. Sprague, Engineering bone regeneration with bioabsorbable scaffolds with novel microarchitecture, Tissue Eng. 5 (1) (1999) 35–51.
[247] C. Zhou, K. Yang, K. Wang, X. Pei, Z. Dong, Y. Hong, X. Zhang, Combination of fused deposition modeling and gas foaming technique to fabricated hierarchical macro/microporous polymer scaffolds, Mater. Des. 109 (2016) 415–424.
[248] R.M. Duarte, J. Correia-Pinto, R.L. Reis, A.R.C. Duarte, Subcritical carbon dioxide foaming of polycaprolactone for bone tissue regeneration, J. Supercrit. Fluids 140 (2018) 1–10.

[249] I. Manavitehrani, T.Y. Le, S. Daly, Y. Wang, P.K. Maitz, A. Schindeler, F. Dehghani, Formation of porous biodegradable scaffolds based on poly (propylene carbonate) using gas foaming technology, Mater. Sci. Eng. C 96 (2019) 824–830.
[250] A. Hottot, S. Vessot, J. Andrieu, A direct characterization method of the ice morphology. Relationship between mean crystals size and primary drying times of freeze-drying processes, Dry. Technol. 22 (8) (2004) 2009–2021.
[251] Z. Ge, Z. Jin, T. Cao, Manufacture of degradable polymeric scaffolds for bone regeneration, Biomed. Mater. 3 (2) (2008), 022001.
[252] L.-P. Yan, J.M. Oliveira, A.L. Oliveira, S.G. Caridade, J.F. Mano, R.L. Reis, Macro/microporous silk fibroin scaffolds with potential for articular cartilage and meniscus tissue engineering applications, Acta Biomater. 8 (1) (2012) 289–301.
[253] H. Radhouani, D. Bicho, C. Gonçalves, F.R. Maia, R.L. Reis, J.M. Oliveira, Kefiran cryogels as potential scaffolds for drug delivery and tissue engineering applications, Mater. Today Commun. 100554 (2019).
[254] B. Sarker, W. Li, K. Zheng, R. Detsch, A.R. Boccaccini, Designing porous bone tissue engineering scaffolds with enhanced mechanical properties from composite hydrogels composed of modified alginate, gelatin, and bioactive glass, ACS Biomater. Sci. Eng. 2 (12) (2016) 2240–2254.
[255] D.J. Hoelzle, A.G. Alleyne, A.J.W. Johnson, Micro-robotic deposition guidelines by a design of experiments approach to maximize fabrication reliability for the bone scaffold application, Acta Biomater. 4 (4) (2008) 897–912.
[256] H.S. Tuan, D.W. Hutmacher, Application of micro CT and computation modeling in bone tissue engineering, Comput. Aided Des. 37 (11) (2005) 1151–1161.
[257] S.S. Kumar, R. Chhibber, R. Mehta, PEEK composite scaffold preparation for load bearing bone implants, Mater. Sci. Forum (2018) 77–82 (Trans Tech Publ).
[258] S. Agarwal, J.H. Wendorff, A. Greiner, Use of electrospinning technique for biomedical applications, Polymer 49 (26) (2008) 5603–5621.
[259] A. Martins, R. Reis, N. Neves, Electrospinning: processing technique for tissue engineering scaffolding, Int. Mater. Rev. 53 (5) (2008) 257–274.
[260] S. Yang, K.-F. Leong, Z. Du, C.-K. Chua, The design of scaffolds for use in tissue engineering. Part II. Rapid prototyping techniques, Tissue Eng. 8 (1) (2002) 1–11.
[261] P. Miranda, E. Saiz, K. Gryn, A.P. Tomsia, Sintering and robocasting of β-tricalcium phosphate scaffolds for orthopaedic applications, Acta Biomater. 2 (4) (2006) 457–466.
[262] T. Chae, H. Yang, V. Leung, F. Ko, T. Troczynski, Novel biomimetic hydroxyapatite/alginate nanocomposite fibrous scaffolds for bone tissue regeneration, J. Mater. Sci. Mater. Med. 24 (8) (2013) 1885–1894.
[263] T. Hemamalini, V.R.G. Dev, Comprehensive review on electrospinning of starch polymer for biomedical applications, Int. J. Biol. Macromol. 106 (2018) 712–718.
[264] W.-E. Teo, R. Inai, S. Ramakrishna, Technological advances in electrospinning of nanofibers, Sci. Technol. Adv. Mater. 12 (1) (2011), 013002.
[265] C.P. Barnes, C.W. Pemble IV, D.D. Brand, D.G. Simpson, G.L. Bowlin, Cross-linking electrospun type II collagen tissue engineering scaffolds with carbodiimide in ethanol, Tissue Eng. 13 (7) (2007) 1593–1605.
[266] M. Shin, H. Yoshimoto, J.P. Vacanti, In vivo bone tissue engineering using mesenchymal stem cells on a novel electrospun nanofibrous scaffold, Tissue Eng. 10 (1–2) (2004) 33–41.
[267] B. Singh, N. Panda, R. Mund, K. Pramanik, Carboxymethyl cellulose enables silk fibroin nanofibrous scaffold with enhanced biomimetic potential for bone tissue engineering application, Carbohydr. Polym. 151 (2016) 335–347.

[268] B.K. Mann, J.L. West, Cell adhesion peptides alter smooth muscle cell adhesion, proliferation, migration, and matrix protein synthesis on modified surfaces and in polymer scaffolds, J. Biomed. Mater. Res. 60 (1) (2002) 86–93.

[269] S. Bose, S. Vahabzadeh, A. Bandyopadhyay, Bone tissue engineering using 3D printing, Mater. Today 16 (12) (2013) 496–504.

[270] B. Duan, M. Wang, W.Y. Zhou, W.L. Cheung, Z.Y. Li, W.W. Lu, Three-dimensional nanocomposite scaffolds fabricated via selective laser sintering for bone tissue engineering, Acta Biomater. 6 (12) (2010) 4495–4505.

[271] A.G. Mikos, S.W. Herring, P. Ochareon, J. Elisseeff, H.H. Lu, R. Kandel, F.J. Schoen, M. Toner, D. Mooney, A. Atala, Engineering complex tissues, Tissue Eng. 12 (12) (2006) 3307–3339.

# Polymer-$MoS_2$-metal oxide composite: An eco-friendly material for wastewater treatment

6

*Selvaraj Mohana Roopan*[a] *and Mohammad Ahmed Khan*[b]
[a]Chemistry of Heterocycles & Natural Research Laboratory, Department of Chemistry, School of Advances Sciences, Vellore Institute of Technology, Vellore, Tamilnadu, India,
[b]School of Chemical Engineering, Vellore Institute of Technology, Vellore, Tamilnadu, India

## 1 Introduction

Water is considered one of the most limited essential commodities that humans depend upon. According to a World Health Organization (WHO) report, 62% of people will be affected due to water scarcity by 2030 [1]. It is often argued that World War III would be fought over water resources. More than 80% of the wastewater is discharged into water bodies without proper treatment. Global water demand has increased by 600% over the past 100 years [2]. Experts believe that over 70% of the available freshwater supply would be used by 2025. The global water demand is expected to increase by 20%–30% by 2050 [3]. Water-borne pathogens are frequently detected in water sources which pose serious health risks to humans [4,5]. A large amount of money is invested by corporations to tackle the problem of water shortage. To avoid water shortage crisis, water from wastewater must be reused. The environmental wastewater problem cannot be resolved solely by cutting down wastewater production from industries; we need to develop ways to improve the quality of existing wastewater. There are several methods employed for water purification. Since water purification and wastewater treatment are evergreen hot topics, new methods are developed consistently, and advancements are made in older methods. It is the need of the hour to address problems related to wastewater treatment. Heavy metals, pharmaceuticals, microorganisms, dyes, and pesticides are the common pollutants present in wastewater globally. Moreover, accidental oil spills and leakage of service tanks are also responsible for water contamination. Plastics and plastic products are increasingly seen in wastewater in recent times.

Wastewater treatment research needs to be focused on simple, economical, and easy-to-use methods. In this context, advanced oxidation processes (AOPs) have gained significant attention in recent years. Hydroxyl radical produced in AOPs is highly reactive and reacts with volatile organic compounds to break them into non-polluting compounds [6]. Hydroxyl radical is the second-highest potent oxidizing agent (first being fluorine) and can react with recalcitrant compounds [6]. Moreover, hydroxyl radical has a short lifespan (nanoseconds in water) and therefore can be itself eliminated from the treatment system [7]. The most commonly researched AOP

**Renewable Polymers and Polymer-Metal Oxide Composites. https://doi.org/10.1016/B978-0-323-85155-8.00002-9**

system is of titanium oxide-based system ($TiO_2$/UV) [8]. Since the demonstration of photocatalytic activity of $TiO_2$ by Fujishima and Honda [9], there has been a tremendous surge in the publications dealing with photocatalytic systems [10,11]. There is abundant literature on $TiO_2$ [8], reduced graphene oxide [12], and graphitic carbon nitrile [13–15]. Doping of these with other photocatalysts such as ZnO [16], $SnO_2$ [17], $Fe_3O_4$ [18], etc., makes plenty of permutations and combinations of photocatalytic systems [19–22]. However, most of the research has been focused on dye-degradation systems. Further, photocatalysis is considered as one of the most environmentally friendly and economical methods [23].

There is a lot of research interest in the field of transition metal dichalcogenides system-molybdenum disulfide ($MoS_2$) because of its easy synthesis, low toxicity, good thermal and chemical stability, super adsorption capacity, and lubrication property. Moreover, band gap for $MoS_2$ is not wide as in the case of $TiO_2$. Due to this, it can be utilized as a photocatalyst under the exposure of visible light (no need of UV light). Therefore, photocatalysis using $MoS_2$ is interesting due to its potential in utilizing freely available solar energy [24]. Hydrogen evolution reaction by the electrolysis of water is also a highly researched photocatalytic application of $MoS_2$-based compounds [25]. The extensive research on $MoS_2$ for photocatalytic degradation shows optimism. Similarly, there are many reports on $MoS_2$-based polymeric (g-$C_3N_4$, poly-polyethyleneglycol (PEG), PPM, etc.) in recent years (2019 and 2020).

The g-$C_3N_4$ is one of the most effective organic polymeric semiconductor materials. Due to their appropriate band gap (2.7 eV), the nonmetallic polymer g-$C_3N_4$ has recently gained a lot of interest in photocatalytic degradation of contaminant systems [26]. As a result, materials based on g-$C_3N_4$ are commonly used for photocatalytic degradation of various contaminants in the atmosphere, including dyes, antibiotics, and toxic pollutants [27]. PVDF (polyvinylidene fluoride) has a wide range of properties and is cost-effective at the same time. PVDF is lightweight and versatile, is a good adsorbent, solvent-resistant, and stable under high electric fields. Depending on crystallization conditions, the PVDF crystalline structure exhibits five distinct polymorphs, namely, $\alpha$, $\beta$, $\gamma$, $\delta$, and $\varepsilon$ phases [28]. PVDF can be used as a photocatalyst carrier to separate and recover photocatalysts from reaction solutions. Different conductive polymers such as polyaniline (PANI), chitosan, polypyrrole (PPy), polyvinylpyrrolidone (PVP), PEG, and polyethyleneimine (PEI) are also used to prepare nanocomposite photocatalysts. Because of their ability to provide aligned band structures with other semiconductor materials, they have shown enhanced photocatalytic activity powered by visible light. Increasing electron and hole separation, extending the light absorption ability, increasing reactant adsorption, suppressing photocorrosion, and assisting in forming particles aggregates [29].

In this era of booming study on environmental remediation and advancement in advanced oxidative processes, this chapter emphasizes the importance of degradation of organic pollutants from wastewater using $MoS_2$-metal oxide composite and $MoS_2$-based polymer metal oxide composite. This is the first report of its kind in this field. We have summarized the recent literature in photocatalytic water treatment with $MoS_2$-based polymer metal oxide composite. Recent trends and forthcoming challenges regarding $MoS_2$-metal oxide-polymer composite for wastewater treatment are highlighted.

# 2 Structure and mechanism

## 2.1 Structure

The structure of $MoS_2$ is quite similar to that of graphene. $MoS_2$ is a layered structure material of S-Mo-S layers, where Mo is squeezed in-between a pair of S atoms. Covalent bonds hold these atoms together while the Vander Waal forces hold the stack of layers [30]. Due to weak Van der Waals forces between layers, it can be exfoliated into single- or few- layered sheets (two-dimensional sheets). Therefore, due to 2D structure of $MoS_2$, it possesses extraordinary properties such as a large surface area, short carrier diffusion distance, and superior electronic conductivity. Superior photocatalytic activity of $MoS_2$ is ascribed to these properties.

## 2.2 Mechanism

Due to the effect of weak Van der Waals forces, the layers of $MoS_2$ are coupled weakly and offer a direct bandgap of approximately 1.8 eV. However, in bulk form, the indirect bandgap for $MoS_2$ is about 1.2 eV and sufficient for photocatalysis reactions to take place. $MoS_2$, when used as a semiconductor, can execute considerable absorption in the visible region of the spectrum and is an appropriate material for preparing hetero photocatalysts (composites). These composites have proved to be promising photocatalysts.

Photocatalytic degradation means breaking down pollutants in the presence of light. In almost every literature on photocatalysis system, the mechanism has been described. Photocatalysis using $MoS_2$/visible light is based upon adsorption of photon with energy higher than 1.2 eV (wavelengths greater than 420 nm) which results in initiating excitation related to charge separation event. In a typical semiconductor, high-energy electrons and hole pairs are generated when irradiated with energy higher than their bandgap energy. As a result, an electron gets excited into the conduction band (CB), which results in the formation of a positive hole in the valence band. These CB electrons ($e-$) and VB holes ($h+$) are both potent reducing and oxidizing agents [11,31]. These $h+$ can oxidize organic contaminants, which causes the formation of $CO_2$ and $H_2O$ as the final products. These $h+$ can also produce hydroxyl radicals (•OH). These hydroxyl ions formed are powerful oxidizing agents because of their electrophilic nature and can oxidize almost all electron-rich organic compounds to $CO_2$ and $H_2O$. It is to be noted that harmful organic contaminants are broken into simple and nonpolluting compounds.

In heterogeneous photocatalysts, two or more semiconductors are synthesized with proper bandgap structures to ensure the mobility of photogenerated charge carriers.

# 3 Applications

## 3.1 Degradation of dyes

The molecules of dyes are chemically very stable and resistant to physiochemical methods. Removal of dyes from the wastewater system is essential because dyes can be carcinogenic even in small amounts [32]. Dye degradation is possible using biodegradable techniques (which is economical), but this process is slow and time-consuming [33]. Dye degradation using photocatalysts is easy and is very well researched. Herein, we have discussed some $MoS_2$-metal oxide composites for dye degradation.

Recently, Gopal et al. fabricated the 2D layered $MoS_2$ on $TiO_2$ [34]. The heterojunction's photocatalytic activity was tested using CR dye. After 2 h, 97% degradation efficiency was achieved under visible light irradiation. Experiments showed that various factors such as pH, amount of catalyst, and initial dye concentration affect the degradation rate. It was observed that dye degradation happened at a faster rate in an acidic environment (pH 3). This is mainly because anionic dye decomposed faster in the presence of protons and decomposed slower in the presence of hydroxyl groups. This heterojunction photocatalyst showed outstanding recyclability as degradation efficiency even after 4 cycles of experiments was 95%, and the XRD pattern showed no noticeable change confirming the stability of the photocatalyst.

Phung et al. designed the $MoS_2$/V, N co-doped $TiO_2$ heterostructure [35]. The heterojunction was utilized in photo-degrading MB dye under visible light exposure for 150 min. The degradation efficiency of the heterostructure was 99% compared to 6%, 26%, and 86% for $TiO_2$, $MoS_2$, and V,N co-doped $TiO_2$, respectively. A positive synergetic effect between V, N co-doped $TiO_2$ and $MoS_2$ is believed to effectively suppress charge recombination and improve interfacial transfer, thus improving photoactivity of the composite.

$TiO_2$@$MoS_2$ hollow microtubes were synthesized by He et al. [36]. The photocatalytic activity of the composite was tested against RhB dye for 1 h. The degradation efficiency of the composite was 81.31% compared to 3.90% and 43.77% for $TiO_2$ microtube and $MoS_2$ nanosheets, respectively. Improved adsorption on $MoS_2$ is because of its negatively charged surface, indicating abundant sites for adsorption of positively charged dye. Excellent dispersion of $MoS_2$ on $TiO_2$ resulted in synergistic adsorption, which is proved using the Langmuir adsorption model.

Recently, Islam et al. developed a C-ZnO/$MoS_2$ nanocomposite anchored on 3D mesoporous carbon framework [37]. The photocatalytic efficiency of the composite was tested against MB and MO dyes under visible light irradiation for 2 h. The degradation efficiency of MB was 92%. This nanocomposite showed outstanding recyclability as there was no apparent change in degradation efficiency even after 4 cycles of experiments, and the XRD pattern and SEM image showed no noticeable difference confirming the stability of the photocatalyst. This composite not only degraded dyes but was also very efficient in degrading other organic pollutants such as nitrobenzene and phenol.

0D/2D binary heterostructure of $Bi_2O_3$/$MoS_2$ was synthesized by Goud et al. [38]. This composite showed extraordinary photocatalytic activity by degrading 100% of CV in 70 min under visible light irradiation. 100% of the dye was also degraded with pristine $MoS_2$ and pristine $Bi_2O_3$ in 100 min and 130 min, respectively. The improved photocatalytic performance of $Bi_2O_3$/$MoS_2$ composite was associated with the interfacial interaction between $Bi_2O_3$ and $MoS_2$. The scavenger tapping experiments proposed the formation of *Z*-schemed heterojunction.

Lejbini and Sangpour [39] synthesized $\alpha$-$Fe_2O_3$-decorated $MoS_2$ nanosheets. The photocatalytic efficiency of the composite was tested against RhB dye under visible light irradiation for 75 min. The degradation efficiency of the composite was 98%. From the kinetic rate constant, it is evident that the composites' degradation process is 3.56 times faster than compared to pristine $\alpha$-$Fe_2O_3$ nanoparticles. The synthesized composites are of increased surface area and engineered band gap for better photocatalytic properties than $\alpha$-$Fe_2O_3$ and $MoS_2$.

Kebede et al. synthesized the p-n heterojunction of Sb doped $MoS_2/WO_3$ [40]. The photocatalytic efficiency of the heterojunction was tested against MB dye degradation. Matt-Schottky analysis was performed for the confirmation of p-n heterojunction. The heterojunction degraded 99.4% of the dye in 14 min in dark conditions. Dye degradation under dark conditions is not commonly observed as most of the research utilizes visible light or UV light. Degradation in the dark happens via self-supporting charge–carrier transfer mechanism. This heterojunction showed outstanding recyclability as degradation efficiency even after 4 cycles of experiments was reported to be 96%.

p-n heterostructured hollow tubes of $WO_3@MoS_2$ were synthesized by Zeng et al. [41]. The photocatalytic efficiency of this heterojunction was evaluated using CR, MB, MO, and RhB dyes under visible light irradiation for 35 min. The degradation efficiency of this heterojunction was 98.8% for degradation of CR, while the degradation efficiency of pure MoS2 was only 35.6%. In the case of MB, RhB, and MO, the degradation efficiencies of the heterojunction are 89.3%, 85.2%, and 57.6%, respectively. The scavenger tapping experiments suggested that $h^+$ was the main active specie involved in photocatalytic reaction.

Magnetic nanocomposites of $Fe_3O_4/MoS_2$ were fabricated by Song et al. [42]. The photocatalytic efficiency of this composite was tested against CR, MB, MG, RhB, and EY dyes. The results show that the adsorption capacity of CR was much higher than MB, MG, RhB, and EY. The degradation efficiency for CR was more than 99% within 2 min. Ultrafast adsorption for dye removal is not much reported in literature.

Fabrication of $MoS_2@CuO$ heterogenous structured nanoflowers was done by Li et al. [43]. The heterojunction's photocatalytic efficiency was tested against MB dye under visible light irradiation for 100 min. The degradation efficiency of this heterojunction was 95.7%, while the degradation efficiency of pristine MoS2 was only 72.5%. Humidity sensing mechanisms showing the effect of water molecule have also been described. $MoS_2@CuO$ heterogenous showed better water adsorption characteristics than CuO due to high specific surface area and increased polarization. Humidity sensing mechanism is very rarely reported in photocatalysis literature.

$ZnO/MoS_2$-deposited photocatalytic membrane was fabricated by Rameshkumar et al. [44]. The photocatalytic efficiency of this membrane was tested against MB dye under UV irradiation. The degradation efficiency of $ZnO/MoS_2$ composite (not membrane) was 97.21%, while the degradation efficiency of pristine ZnO was 89% after 3 h. Maximum degradation of a photocatalytic membrane deposited with $ZnO/MoS_2$ was approximately 99.95% in 2 h. It is believed that active sites of $MoS_2$ produce the radicals that bind to the pollutants and mineralize them. The reusability of the membrane was nearly 99.45%. The process of removal of MB dye was easier and better when compared with powdered composite since centrifugation and drying were not required (as in the case of composite) and only backwashing was sufficient.

Rahimi et al. fabricated ZnO nanorods in $MoS_2$ nanosheets [45]. The photocatalytic efficiency of this composite was tested against MB dye. Since ZnO has limited photocatalytic activity in the visible light spectrum, the addition of $MoS_2$ could enhance its photocatalytic activity by 74%. It was concluded that the optical bandgap of the composite was lower than pure ZnO and the recombination rate of the composite was also lower than pure ZnO. The authors have clearly explained the activity of ZnO and $MoS_2$ in different wavelength regions.

P-doped ZnO/$MoS_2$ hybrid photocatalyst was fabricated by Liu et al. [46]. The photocatalytic efficiency of this photocatalyst was tested against MB and RhB dyes under visible light exposure. The degradation efficiencies for MB and RhB were 83% in 2 min and 95% in 6 min, respectively. Both of the dyes were completely degraded after 10 min of visible light exposure. The enhanced photocatalytic activity was ascribed to the 2D-2D heterojunction, which showed higher performance than other metal oxide nanoparticle photocatalysts.

$MoS_2$@ZnO nano-heterojunctions were fabricated by Tan et al. [47]. The photocatalytic efficiency of this heterojunction was tested using MB dye in 100 min. The degradation efficiency of this heterojunction was 92.7% whereas the degradation efficiency of pure MoS2 was 76.3%. Increase in surface area was the primary reason for improved photocatalytic activity. The photocatalytic activity of $MoS_2$ sheets was improved by the use of small ZnO nanoparticles with a wide band gap.

Composite of $MoS_2$ nanosheets with $SnO_2$ nanoparticles were prepared by Vattikuti et al. [48]. The photocatalytic efficiency of these composites was tested with RhB dye in 50 min under UV light. The order of degradation was $MoS_2$/$SnO_2$ composite > pure $MoS_2$ > pure SnO, in which the degradation efficiency for the composite was around 99.8%. This composite showed outstanding recyclability as degradation efficiency even after 4 cycles of experiments was around 98.4%.

## 3.2 Removal of heavy metals

Heavy metals in wastewater streams are a critical issue for both environment and health because they are nonbiodegradable and highly toxic. They can enter the human body through the food chain [49]. Industrial processes such as electroplating, mineral extraction, pigment manufacture, and leather tanning are few examples that cause environmental contamination of heavy metals [50]. Herein, we have discussed some $MoS_2$-metal oxide composites for heavy metal removal.

Gang et al. fabricated the $MoS_2$ quantum dots/ZnO nanosheet 0D/2D heterojunction [51]. The efficiency of the heterojunction was tested against mercury. 99.8% removal efficiency was achieved using the composite, whereas 78.2% removal occurred while using pristine ZnO in 1 h under UV light irradiation. This heterojunction photocatalyst showed outstanding recyclability as degradation efficiency even after 6 cycles of experiments was 96%, and the XRD pattern showed no obvious change confirming the stability of the photocatalyst. XPS analysis suggested the oxidation of $Hg^0$ to HgO. This composite not only removed heavy metal but was also very efficient in the degradation of Rh B dye.

$Fe_3O_4$@$MoS_2$ core-shell particles were synthesized by Wang and Zeng [52]. The efficiency of these core-shell structures was evaluated using mercury [Hg(II)] ions. Adsorption kinetics indicate that these core-shell particles removed one-half of Hg(II) ions in 1 h. The overall removal efficiency of Hg(II) ions was 99.2%. Adsorption capacity was compared with other metal ions and decreasing order of interaction found was Hg(II), Cu(II), Pb(II), and Cd(II). The results conclude that there are strong chemical interactions between $Fe_3O_4$@$MoS_2$ core-shell particles and Hg(II) ions. These core-shell particles showed outstanding recyclability and degradation efficiency of 89.6% even after 5 cycles of experiments.

$TiO_2$/Ag/$MoS_2$/Ag composites were synthesized by Jiang et al. [53]. The efficiency of this composite was tested against Cr(VI). Photoreduction of Cr(VI) was around 100% by $TiO_2$/Ag/$MoS_2$/Ag composites, while about 80% was observed in the case of $TiO_2$/Ag/$MoS_2$ composites. If more Ag nanoparticles are incorporated in $TiO_2$/Ag/$MoS_2$/Ag composites, photoreduction is reduced, maybe because too much nanoparticles loading could lead to aggregation. The localized surface plasmon resonance due to silver is ascribed as one of the primary factors for improved photocatalytic efficiency of the composites. This composite not only removed heavy metals but was also very efficient in the degradation of RhB dye.

Synthesis of defective $MoS_2$ nanosheets with $Fe_3O_4$ was described by Song et al. [54]. The heavy metal removal efficiency of these nanohybrids was evaluated using mercury [$Hg^{2+}$] ions. These nanohybrids showed a rapid adsorption rate as around 99% of mercury ions were removed in a time interval of 5 min. After 20 min of adsorption, the concentration of mercury met the drinking water standard. It was also shown that nondefective $MoS_2$/$Fe_3O_4$ hybrids were not as effective as defective $MoS_2$/$Fe_3O_4$ hybrids because the concentration of mercury ions remained high even after 2 h. These nanohybrids showed good recyclability as removal efficiency even after 5 cycles of experiments decreased by around 15% compared to its original adsorption capacity. It was also observed that defective $MoS_2$ sheets possessed poor recyclability since restacking occurred in nanosheets which were not observed in hybrids due to $Fe_3O_4$ nanoparticles, which served as nanospacer inhibiting the restacking of nanosheets.

Porous composition aerogel of Au/$Fe_3O_4$/$MoS_2$ was fabricated by Zhiet et al. [55]. The efficiency of this composition aerogel was tested against mercury [$Hg^{2+}$] ions. These aerogels showed a rapid adsorption rate as more than 99% of mercury ions were removed in a time interval of 10 min. This aerogel showed an excellent adsorption rate as around 99.99% of mercury ions were removed after 30 min. The 3D porous assembly was the reason for the efficient removal of mercury ions. It is worth mentioning that the concentration of mercury met the drinking water standards. These aerogels showed outstanding recyclability as removal efficiency even after 10 cycles of experiments remained 95%, and SEM and TEM images of used catalyst were similar to that of fresh catalyst.

Core-shell composites of $Fe_3O_4$@$MoS_2$ and $MoS_2$@$Fe_3O_4$ were synthesized by Yang et al. [56]. The efficiency of these core-shell composites was tested against

Cr(VI). The removal efficiency of Cr(VI) for $Fe_3O_4@MoS_2$ was 99.5%, and $MoS_2@Fe_3O_4$ was 34.8%. These core-shell composites showed outstanding recyclability as removal efficiency even after 5 cycles of experiments were 95.3% in $Fe_3O_4@MoS_2$ and 30.5% in the case of $MoS_2@Fe_3O_4$. Both $MoS_2$ and $Fe_3O_4$ possessed good sorption abilities and reduction potential. The effect of pH was also observed, and it was concluded that the removal efficiency decreased with the increase in pH. This might be because the anion exchange reaction proceeds well in acidic conditions.

Core-shell nanospheres of $Fe_3O_4$@Polydopamine-$MoS_2$ were synthesized by Wang et al. [57]. The efficiency of these core-shell nanospheres was tested against $Pb^{2+}$. The maximum adsorption capacity was around 508.9 mg/g at pH around 5.5 and it was attributed to the increased surface area and binding of $Pb^{2+}$ on the surface of $MoS_2$. These nanospheres showed good recyclability as there was only a tiny decrease in the adsorption capacity even after 10 cycles.

$CeO_2/MoS_2$ nanocomposite system was fabricated by Wang et al. [58]. The nanocomposite's photocatalytic efficiency was tested against Cr(VI) under visible light exposure. The removal efficiency of Cr(VI) for the composite was 99.6%. This nanocomposite showed outstanding recyclability as removal efficiency even after 3 cycles of experiments was 92.1%. The photocatalytic activity of this composite was significantly better than commercial $TiO_2$ P25.

## 3.3 Degradation of pharmaceutical contaminants

More than 200 different pharmaceuticals have been reported in river waters globally. The maximum concentration found is that of antibiotic ciproflaxacin [59]. Traditional wastewater plans are not capable of eliminating antibiotics effectively [60]. This might be because the degradation of pharmaceutical contaminants is difficult compared with that of dyes [61,62]. Herein, we have discussed some $MoS_2$-metal oxide composites for the degradation of pharmaceutical pollutants.

Heterojunction photocatalyst of $MoS_2/Co_3O_4$ was synthesized for the first time by Ji et al. [63]. The designed photocatalyst was tested against 2-mercaptobenzothiazole (MBT) and tetracycline under visible light irradiation for 2 h. Degradation efficiency of this heterojunction photocatalyst was 87.41% for 2-mercaptobenzothiazole and 85.05% for tetracycline, which are higher than that of pure $MoS_2$ and $Co_3O_4$. This heterojunction photocatalyst showed outstanding recyclability as degradation efficiency even after 5 cycles of experiments was around 80% and the XRD pattern showed no obvious change confirming the stability of the photocatalyst.

Heterostructured photocatalyst of $Bi_2O_3/MoS_2$ was synthesized by Ma et al. [64]. The photocatalytic efficiency of the photocatalyst was tested against MB dye and TC antibiotic under visible light irradiation in 100 min. At 23.81% molar ratio of MoS2, MB was completely degraded while degradation of TC was 97%. Trapping experiments show that hydroxyl radical was the main reactive oxidation specie in the degradation of dye. This photocatalyst showed excellent recyclability in case of both pollutants.

N-doped $ZnO/MoS_2$ semiconductor heterojunction was synthesized by Kumar et al. [65]. The heterojunction's photocatalytic efficiency of this heterojunction was tested against TC antibiotic under visible light irradiation for 2 h. Degradation efficiency of this heterojunction photocatalyst was 84%, while the degradation efficiency of N-doped ZnO was only 32%. This heterojunction photocatalyst showed good recyclability as there was a negligible loss in photocatalytic activity after 3 cycles of experiments and the XRD pattern showed no obvious change, confirming the stability of the photocatalyst. The scavenger tapping experiments suggested that holes ($h^+$) and hydroxyl radicals were the most active species in photocatalytic reaction.

## 3.4 Degradation of organic pollutants

Farmers use organic pollutants such as pesticides/herbicides/insecticides to save crops. However, an overdose of these can have an impact on human health and the environment. In developing nations, pesticide poisoning is a severe public health issue [66]. Photodegradation of pesticides/herbicides/insecticides has been widely studied [67]. Herein, we have discussed some $MoS_2$-metal oxide composites in degrading organic pollutants.

Ji et al. synthesized the *Z*-scheme heterojunction of $MoS_2/Bi_2O_3$ [68]. The photocatalytic efficiency of the heterojunction was tested against MBT, TC, and RhB under visible irradiation for 2 h. Degradation efficiency of this heterojunction photocatalyst was 89.6% for MBT, while the degradation efficiencies of bare $MoS_2$ and $Bi_2O_3$ were 13.5% and 37%, respectively. Degradation efficiency for heterojunction was 79.3% for tetracycline, which is 4.65 and 1.94 times higher than that of pure $MoS_2$ and $Bi_2O_3$, respectively. Degradation efficiency for heterojunction in the case of RhB was 90% which was again higher than that of pristine $MoS_2$ and $Bi_2O_3$. This heterojunction photocatalyst showed outstanding recyclability as there was no distinct inactivation in the performance of photocatalyst even after 5 cycles of experiments and the XRD pattern showed no obvious change, confirming the stability of the photocatalyst.

Li et al. fabricated the novel attapulgite (ATP)-$CeO_2/MoS_2$ [69]. Photocatalytic desulfurization of model oil was done by the dissolution of DBT in acetonitrile with 200 ppm sulfur. The degradation efficiency of the composite in desulfurization of DBT was 95% under visible light irradiation for 3 h. $CeO_2$ and $MoS_2$ constitute heterogeneous junction at the interface, which may be the reason for improved charge separation and efficient transfer of photogenerated electrons between the two components. The composite's increase in photocatalytic efficiency was credited to the ATP skeleton, which maintained stability and maintained interacting area to promote adsorption efficiency. The composite showed excellent reusability as desulfurization of DBT even after 10 consecutive cycles was 87%. No change in morphology and structure was observed before and after the desulfurization processes, suggesting that the ternary composite was highly stable.

## 3.5 Photocatalytic disinfection

Inactivation of bacteria using photocatalysis is an essential application for environmental remediation. Traditional wastewater disinfection methods have some shortcomings which can be treated with photocatalytic disinfection [21]. However, antimicrobial activity in photocatalysis is not researched as much as photocatalytic degradation. Herein, we have discussed some antibacterial activities of $MoS_2$-metal oxide composites.

Awasti et al. synthesized the 2D sheet of ZnO nanoflowers on the surface of $MoS_2$ [70]. The antibacterial activity of the 2D composite was studied via zone inhibition test and was tested against Gram-negative bacteria—*Escherichia coli*. 45% of the bacteria were destroyed by ZnO/$MoS_2$ composite, while pristine ZnO destroyed only 25%. The reason for increased antibacterial activity is the polar surface existing on the composite, which can provide more reactive oxygen species than ZnO. Moreover, small particle size can easily penetrate the cell wall of the bacteria. This composite was also very efficient in the degradation of PR dye.

Yan et al. fabricated $MoS_2$/$TiO_2$ p-n heterojunction nanotube arrays [71]. Water disinfection was studied by *E. coli* and *Staphylococcus aureus* under visible light irradiation for 150 min. In the case of ternary heterojunction, disinfection efficiency was 98.5% and it reached up to 92.1% after 3 recycles. The agar plates used were cultured for 48 and 72 h (after the disinfection process) to observe if colonies are formed again and it was concluded that the bacteria did not repair itself and hence formed no new colonies. The mechanism of action was that bacterial cell membrane was damaged by reactive oxygen species (ROS). Superoxide radical played a major role in ROS and is believed to have started the destruction of bacterial cells, which was found via electron spin resonance.

## 3.6 Polymer-supported $MoS_2$ composites and their applications in water treatment

Zhang et al. synthesized fibrous heterostructured core@shell structure of $TiO_2$/PVDF@$MoS_2$ [72]. These composites were in the shape of cauliflowers and were synthesized using electro-spun tetrabutyl orthotitanate treated using one-step hydrothermal method. It showed 58.4% degradation of RhB in 120 min under visible light irradiation. $TiO_2$/PVDF@$MoS_2$ core@shell heterostructured fibers preserved photocatalytic properties and high adsorption even after five cycles of experiments. Most of the traditional photocatalysts use powdered composites and hence loss of composites means loss in photocatalytic activity. But here in the case of fibrous composites, loss of catalyst is minimized. Self-cleaning makes this composite unique and reduces its maintenance coast.

Efficient, stable, mesoporous, and eco-friendly photocatalyst of g-$C_3N_4$/$MoS_2$ was synthesized and studied for degradation of RhB [73]. It showed 97% degradation within 120 min. No significant loss in photocatalytic efficiency was observed after 3 cycles indicating the stable nature of these heterojunction catalysts. The integration of $MoS_2$ denotes charge separation and improved photocatalytic performance. However, excessive loading of $MoS_2$ decreases the photocatalytic activity because of the shielding effect which inhibits the generation of charge carriers. This study found that the loading of 1.25% $MoS_2$ into the g-$C_3N_4$ shows ideal photocatalytic performance.

Li et al. fabricated chitosan-glyoxal/polyvinylpyrrolidone/$MoS_2$ nanocomposite via ultrasonic-hydrothermal route [74]. This composite was employed as photocatalyst in the degradation of diclofenac sodium under UV light in 50 min. The composite had coarse and semi-flake structure. The synthesized nanocomposite exhibited broader and stronger photo-absorption as observed red shift in the UV region in comparison with pristine $MoS_2$. Also, it was observed that with the high ratio of $MoS_2$, maximum absorption takes place.

Ahamad et al. developed g-$C_3N_4$/$MoS_2$/PANI (polyaniline) by In-situ polymerization [75]. Photocatalytic degradation studies show that under visible light, 92.66% degradation of bisphenol A occurred in 60 min. This composite showed outstanding reusability with excellent catalytic activity after six cycles of experiments. This study showed that the presence of inorganic ions such as $NO^{3-}$, $Cl^-$, $SO_4^{2-}$, and $HCO_3^-$ in wastewater affects the catalytic efficiency of the photocatalysts which may be attributed to decrease in light absorption. The enhanced photodegradation activity of $MoS_2$/g-$C_3N_4$/PANI was due to conductive nature of PANI and good structural morphology of the polymer.

$TiO_2$/g-$C_3N_4$/$MoS_2$ is formed by liquid-exfoliation and solvothermal methods. This composite showed 96.5% of degradation of MO within 60 min, under simulated sun-light study [76]. The g-$C_3N_4$/$MoS_2$ hybrid composite acted as an electron donor for interfacial electron transfer thus enhancing the photocatalytic activity. Density functional theory (DFT) analysis specifies the solid-solid interfaces of contact between $TiO_2$ and g-$C_3N_4$/$MoS_2$ which were not connected by physical absorption but connected via chemical bonds. The heterojunction makes the catalyst effective in photodegradation. Photoluminescence (PL) intensity of $TiO_2$/g-$C_3N_4$/$MoS_2$ composite decreased compared to pure catalysts and binary composites. The studies suggest that the $TiO_2$/g-$C_3N_4$/$MoS_2$ nanocomposites have lower rate of recombination of photo-generated electron/hole pairs compared to that of pure g-$C_3N_4$.

Shi et al. fabricated novel heterojunction of $MoS_2$ QDs/g-$C_3N_4$ composite through one-step facile polymerization technique [77]. The photocatalytic efficiency of $MoS_2$ QDs /g-$C_3N_4$ composite and pristine g-$C_3N_4$ toward organic pollutants degradation such as methyl orange (MO) and phenol on visible light exposure was performed and compared. The as-prepared $MoS_2$ QDs/g-$C_3N_4$ composite displayed improved the visible light-driven photocatalytic activity and degraded 96.3% of MO in 40 min and 74.8% of phenol in 90 min with the catalyst loading of 10 mg. This improved visible light-driven photoactivity could be ascribed to the band alignment and powerful coupling between nanosheets of g-$C_3N_4$ and $MoS_2$ QDs, allowing enhanced photo-induced charge carriers generation and separation.

Zhang et al. designed $MoS_2$/g-$C_3N_4$ heterojunction composite that resembled flower-like arrangement [78]. The prepared catalyst was utilized in the photoreduction of hexavalent uranium U(VI) under the catalyst dosage of 50 mg. The composite containing 5% $MoS_2$ showed the highest photoreduction efficiency for U(VI) and the removal rate reached 86.3% in 75 min, which was 2.24 times faster than in nanosheets of g-$C_3N_4$. The photoreduction of U(VI) to U(IV) happens via electron-hole pair charge transfer mechanism. This composite showed excellent recyclability as photocatalytic reduction efficiency was 83.07% even after five runs.

Shi et al. prepared ultrathin positively charged g-$C_3N_4$/$MoS_2$ heterojunction composite using electrostatic self-assembly preceded by hydrothermal treatment [79]. The as-prepared catalyst with 2 wt% $MoS_2$ loading displayed excellent activity and degraded 98% of MO dye within 50 min and 82% of phenol within 120 min, which is approximately 3.5 times and 8 times higher than those of positively charged ultrathin g-$C_3N_4$ and bulk g-$C_3N_4$, respectively. The synergistic effect of this positively charged g-$C_3N_4$ and $MoS_2$ drives this degradation by the suppression of charge recombination. The ultrathin layered positively charged g-$C_3N_4$ provides new insights for organic pollutants degradation.

Bai et al. fabricated $MoS_2$/S-doped porous g-$C_3N_4$ composite through calcination and ultrasound technique [80]. The design was an excellent catalyst in the photodegradation of rhodamine (RhB) dye and degraded 91.1% of RhB within 20 min of visible light exposure. The designed composite outperformed S-doped porous g-$C_3N_4$ and $MoS_2$/bulk g-$C_3N_4$ because of the higher visible light absorption, suppression of electron-hole pair recombination and enriched active sites. Furthermore, the composite showed better chemical stability and excellent recyclability.

Kang et al. prepared heterojunction $MoS_2$/g-$C_3N_4$/$Bi_{24}O_{31}Cl_{10}$ ternary composite through impregnation-calcination technique [81]. Furthermore, the fabricated composite was employed as heterojunction photocatalyst to decompose tetracycline (TC) on visible light exposure. The optimal $MoS_2$/g-$C_3N_4$/$Bi_{24}O_{31}Cl_{10}$ ternary composite removes 97.5% of TC within 50 min, which was 5.38, 1.96 and 2.51 times higher than that of the $Bi_{24}O_{31}Cl_{10}$, g-$C_3N_4$/$Bi_{24}O_{31}Cl_{10}$ and $MoS_2$/$Bi_{24}O_{31}Cl_{10}$ composites under visible light exposure. This improved TC degradation by $MoS_2$/g-$C_3N_4$/$Bi_{24}O_{31}Cl_{10}$ composite was mainly due to its stimulated transportation of charge pairs, suppression of electron-hole pair recombination and significant visible light absorption ability.

In 2020, Liu et al. investigated the $MoS_2$/g-$C_3N_4$ polymeric nanocomposites, using the synthesis route of ultrasonic dispersion and annealing [82]. Strong interface electrostatic interaction shown by the X-ray photoelectron spectroscopy attractions between the $MoS_2$-based g-$C_3N_4$ polymeric nanocomposites. Furthermore, they are confirmed by the structure, electronic, and optical properties. Also, configured $MoS_2$/g-$C_3N_4$ heterostructures obtained a greater photo-efficiency against bisphenol A (BPA) degeneration. Due to the creation of an intimate Z-scheme surface heterostructure around $MoS_2$ and g-$C_3N_4$, photocatalytic activity of $MoS_2$/g-$C_3N_4$ improved its light-harvesting ability. Quick load separation was also facilitated, and more active sites were formed. In photocatalytic degradation, loading of 20 mg of catalyst under visible light irradiation caused 96% degradation of organic pollutants in 150 min.

Core@shell material of $Fe_2O_3$@$MoS_2$/polyethyleneimine was synthesized by Ran et al. [83]. A polymeric material of polyethyleneimine coated with the 2-D materials of $MoS_2$-based metal oxide $Fe_3O_4$ surface functions as a spacer arm and assists enzyme immobilization. It has stable catalytic activity and can be used. In this, organic pollutants degradation was performed using various organic pollutants like Malachite green (MB) 15 mg/L, bisphenol A (BPA) 25 mg/L, and bisphenol F (BPF) 25 mg/L. Higher efficiencies and immobilized laccases degrade (MB 82.7%), (BPA 87.6%), (BPF 70.6%) for in h. The built material is expected to be considered a green polymer in biomacromolecules for mobility.

In 2020, new composite of cyclodextrins/$MoS_2$/g-$C_3N_4$ was synthesized using photo-catalysts for removing the glyphosates and Cr(VI) [84]. In this study, grafting of three cyclodextrins (α-CD, β-CD, and μ-CD) was done to make several photo-catalysis materials and then their efficiencies were compared. CDs/$MoS_2$/g-$C_3N_4$ significantly improves photocatalyst activity (for both glyphosates and Cr(VI)) in comparison to pure $MoS_2$ and g-$C_3N_4$. In the photocatalytic degradation, 70% of glyphosates was degraded in 170 min and photoreduction of 81% of Cr(VI) was observed in 50 min under the sunlight radiation at the catalyst loading 20 mg. This study concludes that the cyclodextrin with grafted *co*-polymer is an effective strategy for enhancing the photocatalytic efficiency of glyphosate and Cr(VI) removal composites $MoS_2$/g-$C_3N_4$.

Ma et al. fabricated E-$MoS_2$/polyvinylidene fluoride (PVDF)/electrospunfiber membranes (EFMs) using electro spinning method [85]. The photocatalytic efficiency of the EEMs was studied using degradation of oxytetracycline (OTC) antibiotic. The 10.0 wt% E-$MoS_2$/PVDF in EFMs showed greater efficiency in degrading 93.08% OTC in 24 min with a kinetic constant of pseudo-first-order. It was due to the electrostatic interaction of the incorporated E-$MoS_2$ nanosheets for the increase in the electronic effect assets of composite. This analysis is significant in the case of polymer surface piezocatalysts for large-scale use in wastewater treatment.

Peng et al. developed the $MoS_2$/g-$C_3N_4$ polymeric nanocomposites synthesized via low-temperature hydrothermal method [86]. $MoS_2$ significantly improved the photocatalysis on g-$C_3N_4$ nanosheets. Methyl orange (MO) was degraded under virtual solar light environment with the catalyst loading of 50 mg. 63% of degradation was observed after 300 min.

Tomer et al. used the nano-casting method to fabricate composite of Ag/$MoS_2$/g-$C_3N_4$ [87]. The photocatalytic performance of the synthesized composite was evaluated using AOPs against RhB under visible light. At pH 8, the Ag/$MoS_2$/g-$C_3N_4$ nanocomposite had 85% degradation efficacy, compared to 70% for Ag/$MoS_2$/g-$C_3N_4$, 50% for Ag/$MoS_2$, and 35% for pure g-$C_3N_4$. The presence of Ag NPs decreases the recombination rate of electron and holes, allowing additional dopant nanoparticles to participate in the redox reaction and form active oxygen species, resulting in high degradation efficiency. Because of the catalytic action of incorporated Ag and $MoS_2$ nanoparticles and the intense absorption of g-$C_3N_4$ in the visible-light, Ag/$MoS_2$/g-$C_3N_4$ has the highest degradation efficacy compared to other materials. Since the negatively charged surface of the Ag/$MoS_2$/g-$C_3N_4$ nanocomposite will absorb the positively charged RhB dye molecules by electrostatic interaction at pH = 12, the composite's degradation efficiency increases to 97%.

Ntakadzeni et al. fabricated a PEGylated $MoS_2$ nanocomposite and tested its photocatalytic activity against the dyes Rhodamine B (RhB) and Methylene Blue (MB) under visible light irradiation [88]. The removal percentages were 97.3% of RhB and 98.05% of MB in 90 min. The results showed that the PEGylated $MoS_2$ had higher efficiency compared to MoS2 in all processes. This study suggests that in addition to functioning as a capping agent, PEG 400 improved the photodegradation performance of $MoS_2$. Furthermore, the catalyst is involved in the photocatalytic reduction of Cr(VI). In 75 min, Cr(VI) was then reduced to Cr(III) by 91.05%.

Wang et al. developed a ternary nanocomposite $Fe^0$/g-$C_3N_4$/$MoS_2$ using hydrothermal method [89]. The synthesized composite was examined for its photocatalytic activity against the dye RhB and Cr(VI) under visible light irradiation. Compared with the pure g-$C_3N_4$, g-$C_3N_4$/$MoS_2$, and $Fe^0$/g-$C_3N_4$, the photodegradation efficiency of the $Fe^0$/g-$C_3N_4$/$MoS_2$ toward the RhB and Cr(VI) under visible light was 98.2% for RhB and 91.4% for Cr(VI).

Qiu et al. created a flower-like $MoS_2$/PANI/PAN nanocomposite and tested its Cr(VI) disposal capability in polluted water [90]. The Cr(VI) removal potential of the $MoS_2$/PANI/PAN nanocomposite enhanced quickly at pH 3 after 30 min. The quick removal of Cr(VI) was due to improved properties in nanocomposites such as ion exchange, electrostatic interaction, and redox reaction. The $MoS_2$/PANI/PAN nanocomposites had outstanding mechanical and hydrophilic properties and exemplary regeneration and reusability. It has a high potential for removing Cr(VI), even in the existence of organic substances and inorganic anions. This nanocomposite was found to be an excellent material for the emergency disposal of Cr(VI)-contaminated water.

Gogoi et al. investigated the fabrication of a hierarchical $Cd_{0.5}Zn_{0.5}S$/g-$C_3N_4$/$MoS_2$ nanocomposite and evaluated its photocatalytic activity against the organic pollutant RhB as well as the production of $H_2$ [91]. 91 mol/h/0.02 g hydrogen was produced by synthesized nanocomposite from water. In addition, the $Cd_{0.5}Zn_{0.5}S$/g-$C_3N_4$/$MoS_2$ nanocomposite degraded 86% of RhB, while the $Cd_{0.5}Zn_{0.5}S$ and $Cd_{0.5}Zn_{0.5}S$/g-$C_3N_4$ nanocomposite degraded 33% and 65% of RhB after 30 min, respectively. The $e^-$-$h^+$ recombination rate is diminished in the fabricated composite since it was attributed to the electrons with greater interfacial charge transfer. This results in the availability of more electrons for reaction at the active sites, which enhances the photocatalytic performance of the synthesized nanocomposite.

Recently, A. Beyhaqi et al. constructed the polymeric g-$C_3N_4$/$WO_3$/$MoS_2$ nanosheets, which were prepared using cocalcination and hydrothermal methods [20]. The composite was employed as photocatalyst in degrading colored model pollutants RhB, MO, MB, and $AO_7$ dyes corresponding to 50, 20, 20, and 20 ppm concentrations, respectively. Furthermore, colorless organic liquid phase pollutants in a particular concentration 10 ppm of Bisphenol A, Atrazine, ciprofloxacin, and 2-chlorophenol were tested by similar photocatalytic experiment. The photocatalytic degradation efficiency of the organic pollutants using g-$C_3N_4$/$WO_3$/$MoS_2$ catalyst is provided in Table 1. Due to the synergistic effect between the ternary composites, which improves charge carrier lifetime, g-$C_3N_4$/$WO_4$/$MoS_2$ has a higher visible light absorption than g-$C_3N_4$, g-$C_3N_4$/$WO_3$ and g-$C_3N_4$/$MoS_2$. In addition to being determined by the low-energy bandgap semiconductors like $WO_3$ and $MoS_2$, the improvement in ternary composite catalytic activity was also affected by the polymeric g-$C_3N_4$, which supports separate photo-induced pairs $e^-/h^+$.

$MoS_2$/S-g-$C_3N_4$ photocatalyst was fabricated by Chegeni et al. [92]. The nanocomposite was investigated as a potential adsorbent for removing MB dye. The adsorption mechanism indicated Π-Π stacking relationship between the nanocomposite's delocalized Π-Π electron network and the aromatic ring structure of dye molecules. The photocatalytic efficiency of the $MoS_2$/S-g-$C_3N_4$ composite

**Table 1** Study on $MoS_2$ based polymeric composites and their applications in photocatalytic degradation.

| S. no. | Polymer composite | Synthesis method | Catalyst characterization | Pollutant and its concentration | Degradation procedure, catalyst dosage, and light source | Degradation (%) and time | Reference |
|---|---|---|---|---|---|---|---|
| 1. | $TiO_2$/PVDF@$MoS_2$ core-shell | Hydrothermal | XRD, XPS, SEM, PL, TEM, BET, UV-DRS | RhB, 15 mg/L | Photocatalytic degradation, photocatalysts (1 g $L^{-1}$), and 9 W white light LED | 58.4%, 120 min | [72] |
| 2. | Mesoporous g-$C_3N_4$/$MoS_2$ | Directly heating | XRD, FT-IR, SEM, TEM, BET, UV-DRS, PL, XPS, | RhB, $2 \times 10^{-5}$ mol $L^{-1}$ | Photocatalytic degradation, photocatalysts (30 mg) and light irradiation (500 W Xe lamp) | 97%, 120 min | [73] |
| 3. | Chitosan-glyoxal/PVP/$MoS_2$ | Hydrothermal-ultrasonic | XRD, UV-DRS, FT-IR, SEM, EDS, XPS | Diclofenac sodium, 100 mg/L | Adsorption and photo-degradation, Dosage (0.1 g $L^{-1}$), UV light (100 W, $\lambda = 254$ nm) | 94.5%, 50 min | [74] |
| 4. | g-$C_3N_4$/$MoS_2$-PANI (polyaniline) | In-situ polymerization | UV-vis DRS, FTIR, TGA, SEM, XRD, TEM, XPS, BET | BPA, 20 mg/L | Photocatalytic degradation, visible light | 92.66%, 60 min | [75] |

*Continued*

**Table 1** Continued

| S. no. | Polymer composite | Synthesis method | Catalyst characterization | Pollutant and its concentration | Degradation procedure, catalyst dosage, and light source | Degradation (%) and time | Reference |
|---|---|---|---|---|---|---|---|
| 5. | $TiO_2$/g-$C_3N_4$/$MoS_2$ | Liquid-exfoliation and solvothermal | SEM, HRTEM, XPS, XRD, FT-IR, PL, UV-vis DRS, BET | MO, 20 mg/L | Photocatalytic degradation, dosage (100 mg), Sun light and 500 xenon long-arc lamp | 96.5%, 60 min | [76] |
| 6. | $MoS_2$ QDs/g-$C_3N_4$ | Polymerization | XRD, SEM, EDS, TEM, UV-DRS, PL, XPS, ESR | MO and phenol, 20 mg/L | Photocatalytic degradation, dosage (10 mg), Visible light (300 W Xe lamp) | 96.3% of MO in 40 min and 74.8% of phenol in 90 min | [77] |
| 7. | $MoS_2$/g-$C_3N_4$ | Hydrothermal | XRD, PL, FT-IR, BET, XPS, EIS, TEM, UV–Vis | U(VI), 40 mg/L $UO_2^{2+}$ solution | Photocatalytic reduction (1.25 mL of methanol as the hole sacrificial agent.), 50 mg, xenon lamp | 86.8%, 90 min | [78] |
| 8. | Ultrathin g-$C_3N_4$/$MoS_2$ | Ultrasonic and hydrothermal | FT-IR, XRD, SEM,, UV-DRS, PL, XPS, TEM, Zeta potential | MO and phenol, 20 mg/L | Photocatalytic degradation, 10 mg (dosage), visible light (300 W Xe lamp) | 98% of MO in 50 min and 82% of phenol in 120 min | [79] |
| 9. | $MoS_2$/S-doped porous g-$C_3N_4$ | Calcination and ultrasound assembly | XRD, TEM, XPS, UV-DRS, PL, BET | RhB, 10 mg/L | Photocatalytic degradation, visible light irradiation (500 W Xe) | 91.1%, 20 min | [80] |

| | | | | | | | |
|---|---|---|---|---|---|---|---|
| 10. | $MoS_2$/g-$C_3N_4$/ $Bi_{24}O_{31}Cl_{10}$ | Impregnation-calcination | XRD, ESR, BET, EIS, SEM, EDS, TEM, Life time-PL UV-DRS, PL, XPS | TC, 20mg/L | Photocatalysis, 0.01 g of photocatalyst, 300W Xe lamp | ~96%, 50min | [81] |
| 11. | $MoS_2$/g-$C_3N_4$ | Ultrasonic dispersion and annealing | XRD, FESEM, HRTEM, UV-DRS, PL, BET, ESR, EIS | BPA, 10mg/L | Photocatalytic degradation, 20mg catalyst, visible light irradiation (500W Xe) | 96%, 150min | [82] |
| 12. | $Fe_3O_4$/$MoS_2$/ PEI | Laccase immobilization | TEM, XPS, BET, EDS, FT-IR, TGA, VSM | MG (15mg/L), BPA (25mg/L), BPF (25mg/L) | Laccase exhibits the removal, 5mg dosage, shaken at 180rpm at 25°C overnight | MG, BPA, and BPF is 82.7%, 87.6%, and 70.6% for 24h | [83] |
| 13. | Cyclodextrins grafted $MoS_2$/g-$C_3N_4$ | Hydrothermal, thermal polymerization, functionalization | XRD, SEM, XPS, FT-IR, AFM, UV-DRS, PL, ESR | Glyphosate (9 g/L) and Cr (VI) (10mg/L) | Photocatalytic degradation and reduction, 20mg catalyst, sunlight irradiation | Degradation of Glyphosate is 79% (170min) and Reduction of Cr (VI) is 81% in 50min | [84] |
| 14. | E-$MoS_2$/ polyvinylidene fluoride (PVDF)/ electrospun fiber membranes (EFMs) | Electrospinning | FESEM, HRTEM, XRD, ESR, FTIR, EIS, XPS, UV-DRS, | OTC, 20mg/L | Piezocatalysis, fixed membrane ~100mg, could send ultrasonic waves at a frequency of | 93.08%, 24min | [85] |

*Continued*

**Table 1** Continued

| S. no. | Polymer composite | Synthesis method | Catalyst characterization | Pollutant and its concentration | Degradation procedure, catalyst dosage, and light source | Degradation (%) and time | Reference |
|---|---|---|---|---|---|---|---|
| | | | | | 20 kHz and with a power output ranging from 100 W to 650 W | | |
| 15. | $MoS_2$/g-$C_3N_4$ | Low temperature hydrothermal | XRD, FT-IR, UV-DRS, TEM, BET | MO, 20 mg/L | Photocatalytic degradation, 50 mg of catalyst, sun light | ~63%, 300 min | [86] |
| 16. | Ag/$MoS_2$/g-$C_3N_4$ | Nanocasting | XRD, BET, TEM, FESEM, EDX | RhB, 10 mg/L | Photocatalysis, 10 mg photocatalyst, visible light (300 W Xe) | 78%, 30 min | [87] |
| 17. | PEGylated $MoS_2$ | Hydrothermal | XRD, TEM, FESEM, EDS, FTIR, TGA, UV-DRS, | RhB, 10 mg/L MB, 10 mg/L | Visible light with UV filter to a solar simulator (250 W) | 97.30% (RhB) in 75 min and 98.05% (MB) in 90 min | [88] |
| 18. | $Fe^0$/g-$C_3N_4$/$MoS_2$ | Hydrothermal | XRD, FT-IR, UV-DRS, SEM, XPS, PL, EIS, BET | RhB, (20 mg/L) Cr(VI), (20 mg/L) | Photocatalytic activity, photocatalyst 30 mg, Sun light | 98.2% of RhB in 120 min and 91.4% Cr (VI) in 120 min | [89] |
| 19. | $MoS_2$/PANI/PAN | Electrospinning, hydrothermal | FESEM, FTIR, XRD, BET, AFM | Cr (VI) (2 mmol/L) | Batch removal, 50 mg dosage, centrifugation speed of 180 rpm | 94. 2%, 1 h | [90] |

| | | | | | | | |
|---|---|---|---|---|---|---|---|
| 20. | $Cd_{0.5}Zn_{0.5}S/g\text{-}C_3N_4/MoS_2$ | Ultrasonication | XRD, XPS, FESEM, TEM, UV, PL, Time-resolved PL, FTIR, BET, EIS | RhB, $1 \times 10^{-5}$ M | Photocatalytic degradation, 50 mg catalyst, 500 W tungsten-halogen lamp | 97%, 60 min | [91] |
| 21. | $g\text{-}C_3N_4/WO_3/MoS_2$ | Co-Calcination and hydrothermal | XRD, XPS, SEM, TEM, UV-DRS | RhB (50 ppm), MO (20 ppm), MB (20 ppm), $AO_7$ (20 ppm) Bisphenol A (10 ppm), Atrazine (10 ppm), ciprofloxacin (10 ppm), and 2-chlorophenol (10 ppm) | Photocatalysis, 100 mg photocatalyst, visible-light (300 W of xenon arc lamp) | RhB (99.9%), MO (91.8%), MB (83.4%), $AO_7$ (94.2%) in 60 min. Bisphenol A (32%), Atrazine (46%), ciprofloxacin (100%), and chlorophenol (28%) | [20] |
| 22. | $MoS_2/S\text{-}g\text{-}C_3N_4$ | Calcination | FTIR, XRD, SEM, TGA, BET | MB, 8 mg/L | Adsorption and degradation condition, 50 mg dosage, LED | 98%, 30 min | [92] |
| 23. | $MoS_2/C_3N_4$ | Ultrasonic chemical | XRD, FESEM, EDX, TEM, UV, PL, FTIR, BET | RhB, 5 mg/L MO, 5 mg/L | Photocatalysis, Catalyst, Visible light (300 W Xe lamp) | 99% RhB in 20 min and ~82% MO for 3 h | [93] |
| 24. | $MoS_2/Al_2O_3/g\text{-}C_3N_4$ | Hydrothermal | XRD, FT-IR, UV-DRS, SEM, XPS, PL, TGA | CV, 100 mg/L | Photocatalytic degradation, 5 mg catalyst, 150 W tungsten-halogen lamp | 97.3%, 90 min | [94] |

*Continued*

**Table 1** Continued

| S. no. | Polymer composite | Synthesis method | Catalyst characterization | Pollutant and its concentration | Degradation procedure, catalyst dosage, and light source | Degradation (%) and time | Reference |
|---|---|---|---|---|---|---|---|
| 25. | $MoS_2$/$TiO_2$/poly (methyl methacrylate) (PMMA) | Hydrothermal | FESEM, TEM, XRD, BET, Raman, XPS | MO, 10 mg/L | Photodegradation, 100 mg of photocatalyst, UV (254 nm, 2 W high-pressure mercury lamp) | 95.7%, 50 min | [95] |

Methylene blue (MB); methyl orange (MO); rhodamine B (RhB); congo red (CR); tetracycline (TC); acid orange-7 (AO-7); phenol red (PR); dibenzothiophene (DBT); 2-mercaptobenzothiazole (MBT), bisphenol A (BPA).

was tested against MB dye for 30 min under visible light irradiation. The degradation efficiency of the composite was noted to be 98%.

The preparation of nanocomposite $MoS_2/C_3N_4$ using ultrasonic chemical method was reported by Li et al. [93]. The $MoS_2/C_3N_4$ nanocomposites were investigated for their efficacy in removing RhB and MO from wastewater under visible light irradiation for 3 h. The $MoS_2/C_3N_4$ heterostructures showed better photocatalytic degradation assets than that of $C_3N_4$, in which 0.05% loading of $MoS_2$ on $C_3N_4$ showed the superior photocatalytic performance. The photocatalytic enrichment of $MoS_2/C_3N_4$ was attributed to its high photoelectrical conversion efficiency, good light absorption, and inhibition of charge carrier recombination.

Vattikuti and Byon synthesized the heterostructure of $MoS_2/Al_2O_3$/g-$C_3N_4$ using the hydrothermal method [94]. The photocatalytic efficiency of this composite was tested against CV dye (100 mg/L). This finding suggests that the synergetic effects of visible light absorption and the favorable heterojunction structure of $MoS_2/Al_2O_3$/g-$C_3N_4$ may explain this increase in photocatalytic capacity. Excess $MoS_2$ in the $MoS_2/Al_2O_3$/g-$C_3N_4$ nanocomposite reduced photocatalytic capacity because agglomeration reduces charge transfer capability of $MoS_2$ and high $MoS_2$ content can have a significant impact on surface chemistry as well as a negative effect on visible absorption.

$MoS_2/TiO_2$/poly(methyl methacrylate) (PMMA) was fabricated by Li et al. [95]. The photocatalytic efficiency of the nanocomposites (100 mg) was investigated by measuring MO (10 mg/L) photodegradation in the presence of UV light. After 50 min of the UV illumination, photodegradation efficiency exceeds 95.7, 83.8, and 77.5%, respectively, for $MoS_2/TiO_2$/PMMA, $TiO_2$/PMMA, and $MoS_2$/PMMA. Apart from introducing $MoS_2$ nanosheets, $MoS_2/TiO_2$/PMMA seems to exhibit higher efficiency as photocatalyst than $TiO_2$/PMMA, which had a strong photocatalytic activity and a high specific surface area. Furthermore, the composite could be easily retrieved and recycled while still maintaining high photocatalytic efficiency. The research implies to provide a novel approach to the development of high-performance $MoS_2$ based polymer photocatalysts.

Advantages of polymer supported on composites:

- Enhanced the adsorption property of the materials.
- Decrease the electron-hole pair recombination rate of the photocatalyst.
- In photocatalysis, polymers are enhancing the light absorption ability.
- Easily available and is a less-weight material.
- Mainly, stability, recovery, and reusability.

# 4 Reliability

Reliability in wastewater treatment can be defined as the probability of success in wastewater treatment that should meet the effluent quality standards [96]. In other words, it can be defined as the probability of achieving satisfactory performance for a specified period under specified conditions [97]. However, in case of photocatalytic systems, no concrete work is done to explore reliability. Since researchers combine two or more systems (nanoparticles/nanosheets) to form composite, there is no reasonable data on bandgap energy and there is no fixed way to find the reliability of the system yet. Therefore, not much can be said for the term "reliability" in photocatalytic systems. But if seen from a practical viewpoint, operation conditions and design are fairly reliable for small-scale applications [98]. In the application part, we have seen that most of the composites could be re-used for 3–5 cycles of experiments. This stability of the system also accounts for the reliability. More research is needed to explore the reliability of photocatalytic systems.

# 5 Conclusion and outlook

In summary, this chapter highlights the photocatalytic applications of $MoS_2$-metal oxide composites and polymer-$MoS_2$-metal oxide composites. $MoS_2$ has proved to be a promising candidate for fabricating and designing advanced composite photocatalysts for wastewater treatment. As evident from the reviewed literature, $MoS_2$-metal oxide-based polymer composites have been tested for the removal of pollutants for water treatment. A recent surge in the literature of photocatalysis (and $MoS_2$) reflects optimism. This shows the success of photocatalysis in lab-scale applications. However, more research is needed for the application of photocatalytic systems for large-scale water treatment plants. It has been more than three decades since the discovery of photocatalysis, and it is time we reduced the gap between academic interest and practical applications.

Overall, this chapter is a comprehensive study of the current progress of $MoS_2$-metal oxide-polymer composites in the field of wastewater treatment. The promising results ensure further development of sustainable environmental remediation techniques using freely available solar energy as a renewable source of energy.

# References

[1] WHO, WHO's Annual World Health Statistics Report: World Health Statics, Available at: *http://apps.who.int/iris/bitstream/handle/10665/272596/9789241565585-eng.pdf2018*.

[2] Y. Wada, M. Flörke, N. Hanasaki, S. Eisner, G. Fischer, S. Tramberend, Y. Satoh, M.T.H. van Vliet, P. Yillia, C. Ringler, P. Burek, D. Wiberg, Modeling global water use for the 21st century: the water futures and solutions (WFaS) initiative and its approaches, Geosci. Model Dev. 9 (1) (2016) 175–222, https://doi.org/10.5194/gmd-9-175-2016.

[3] A. Boretti, L. Rosa, Reassessing the projections of the world water development report, NPJ Clean Water 2 (1) (2019) 15, https://doi.org/10.1038/s41545-019-0039-9.

[4] S.-H. Lee, S.-J. Kim, Detection of infectious enteroviruses and adenoviruses in tap water in urban areas in Korea, Water Res. 36 (1) (2002) 248–256, https://doi.org/10.1016/S0043-1354(01)00199-3.

[5] E.T. Gensberger, T. Kostić, Novel tools for environmental virology, Curr. Opin. Virol. 3 (1) (2013) 61–68, https://doi.org/10.1016/j.coviro.2012.11.005.

[6] M.A. Oturan, J.-J. Aaron, Advanced oxidation processes in water/wastewater treatment: principles and applications. A review, Crit. Rev. Environ. Sci. Technol. 44 (23) (2014) 2577–2641, https://doi.org/10.1080/10643389.2013.829765.

[7] H. Suty, C.D. Traversay, M. Cost, Applications of advanced oxidation processes: present and future, Water Sci. Technol. 49 (4) (2004) 227–233, https://doi.org/10.2166/wst.2004.0270.
[8] M. Pelaez, N.T. Nolan, S.C. Pillai, M.K. Seery, P. Falaras, A.G. Kontos, P.S.M. Dunlop, J. W.J. Hamilton, J.A. Byrne, K. O'Shea, M.H. Entezari, D.D. Dionysiou, A review on the visible light active titanium dioxide photocatalysts for environmental applications, Appl. Catal. B Environ. 125 (2012) 331–349, https://doi.org/10.1016/j.apcatb.2012.05.036.
[9] A. Fujishima, K.A. Honda, Electrochemical photolysis of water at a semiconductor electrode, Nature 238 (1972) 37–38, https://doi.org/10.1038/238037a0.
[10] S.K. Loeb, P.J.J. Alvarez, J.A. Brame, E.L. Cates, W. Choi, J. Crittenden, D.D. Dionysiou, Q. Li, G. Li-Puma, X. Quan, D.L. Sedlak, T. David Waite, P. Westerhoff, J.-H. Kim, The technology horizon for photocatalytic water treatment: sunrise or sunset? Environ. Sci. Technol. 53 (6) (2019) 2937–2947, https://doi.org/10.1021/acs.est.8b05041.
[11] A.B. Djurišić, Y. He, A.M.C. Ng, Visible-light photocatalysts: prospects and challenges, APL Mater. 8 (3) (2020), https://doi.org/10.1063/1.5140497, 030903.
[12] K. Krishnamoorthy, R. Mohan, S.-J. Kim, Graphene oxide as a photocatalytic material, Appl. Phys. Lett. 98 (24) (2011) 244101, https://doi.org/10.1063/1.3599453.
[13] D. Masih, Y. Ma, S. Rohani, Graphitic C3N4 based noble-metal-free photocatalyst systems: a review, Appl. Catal. B Environ. 206 (2017) 556–588, https://doi.org/10.1016/j.apcatb.2017.01.061.
[14] J. Wen, J. Xie, X. Chen, X. Li, A review on G-$C_3N_4$ g-$C_3N_4$ -based photocatalysts, Appl. Surf. Sci. 391 (2017) 72–123, https://doi.org/10.1016/j.apsusc.2016.07.030.
[15] W.-J. Ong, L.-L. Tan, Y.H. Ng, S.-T. Yong, S.-P. Chai, Graphitic carbon nitride (g-$C_3N_4$)-based photocatalysts for artificial photosynthesis and environmental remediation: are we a step closer to achieving sustainability? Chem. Rev. 116 (12) (2016) 7159–7329, https://doi.org/10.1021/acs.chemrev.6b00075.
[16] N. Daneshvar, D. Salari, A.R. Khataee, Photocatalytic degradation of azo dye acid red 14 in water on ZnO as an alternative catalyst to $TiO_2$, J. Photochem. Photobiol. A Chem. 162 (2–3) (2004) 317–322, https://doi.org/10.1016/S1010-6030(03)00378-2.
[17] G. Elango, S.M. Roopan, Efficacy of $SnO_2$ nanoparticles toward photocatalytic degradation of methylene blue dye, J. Photochem. Photobiol. B Biol. 155 (2016) 34–38, https://doi.org/10.1016/j.jphotobiol.2015.12.010.
[18] L. Xu, J. Wang, Magnetic nanoscaled $Fe_3O_4/CeO_2$ composite as an efficient Fenton-like heterogeneous catalyst for degradation of 4-chlorophenol, Environ. Sci. Technol. 46 (18) (2012) 10145–10153, https://doi.org/10.1021/es300303f.
[19] S. Shanavas, S. Mohana Roopan, A. Priyadharsan, D. Devipriya, S. Jayapandi, R. Acevedo, P.M. Anbarasan, Computationally guided synthesis of (2D/3D/2D) RGO/$Fe_2O_3$/g-$C_3N_4$ nanostructure with improved charge separation and transportation efficiency for degradation of pharmaceutical molecules, Appl. Catal. B Environ. 255 (2019) 117758, https://doi.org/10.1016/j.apcatb.2019.117758.
[20] A. Beyhaqi, Q. Zeng, S. Chang, M. Wang, S.M. Taghi Azimi, C. Hu, Construction of g-$C_3N_4/WO_3/MoS_2$ ternary nanocomposite with enhanced charge separation and collection for efficient wastewater treatment under visible light, Chemosphere 247 (2020) 125784, https://doi.org/10.1016/j.chemosphere.2019.125784.
[21] M. Wu, L. Li, Y. Xue, G. Xu, L. Tang, N. Liu, W. Huang, Fabrication of ternary GO/g-$C_3N_4/MoS_2$ flower-like heterojunctions with enhanced photocatalytic activity for water remediation, Appl. Catal. B Environ. 228 (2018) 103–112, https://doi.org/10.1016/j.apcatb.2018.01.063.
[22] H. Jiang, Z. Xing, T. Zhao, Z. Yang, K. Wang, Z. Li, S. Yang, L. Xie, W. Zhou, Plasmon ag nanoparticle/Bi2S3 ultrathin nanobelt/oxygen-doped flower-like $MoS_2$ nanosphere

ternary heterojunctions for promoting charge separation and enhancing solar-driven photothermal and photocatalytic performances, Appl. Catal. B Environ. 274 (2020) 118947, https://doi.org/10.1016/j.apcatb.2020.118947.
[23] M.N. Chong, B. Jin, C.W.K. Chow, C. Saint, Recent developments in photocatalytic water treatment technology: a review, Water Res. 44 (10) (2010) 2997–3027, https://doi.org/10.1016/j.watres.2010.02.039.
[24] M. Wu, L. Li, N. Liu, D. Wang, Y. Xue, L. Tang, Molybdenum disulfide ($MoS_2$) as a coco-catalyst for photocatalytic degradation of organic contaminants: a review, Process Saf. Environ. Prot. 118 (2018) 40–58, https://doi.org/10.1016/j.psep.2018.06.025.
[25] U. Gupta, C.N.R. Rao, Hydrogen generation by water splitting using $MoS_2$ and other transition metal dichalcogenides, Nano Energy 41 (2017) 49–65, https://doi.org/10.1016/j.nanoen.2017.08.021.
[26] Q. Liang, X. Liu, J. Wang, Y. Liu, Z. Liu, L. Tang, B. Shao, W. Zhang, S. Gong, M. Cheng, Q. He, C. Feng, In-situ self-assembly construction of hollow tubular g-$C_3N_4$ isotype heterojunction for enhanced visible-light photocatalysis: experiments and theories, J. Hazard. Mater. 401 (2021) 123355, https://doi.org/10.1016/j.jhazmat.2020.123355.
[27] R. Malik, V.K. Tomer, State-of-the-art review of morphological advancements in graphitic carbon nitride (g-CN) for sustainable hydrogen production, Renew. Sust. Energ. Rev. 135 (August 2020) (2021) 110235, https://doi.org/10.1016/j.rser.2020.110235.
[28] Z. He, F. Rault, M. Lewandowski, E. Mohsenzadeh, F. Salaün, Electrospun PVDF nanofibers for piezoelectric applications: a review of the influence of electrospinning parameters on the β phase and crystallinity enhancement, Polymers (Basel) 13 (2) (2021) 1–23, https://doi.org/10.3390/polym13020174.
[29] A.M. Mohammed, S.S. Mohtar, F. Aziz, M. Aziz, A. Ul-Hamid, $Cu_2O$/ZnO-PANI ternary nanocomposite as an efficient photocatalyst for the photodegradation of Congo red dye, J. Chem. Eng. 9 (2) (2021) 105065, https://doi.org/10.1016/j.jece.2021.105065.
[30] A. Molina-Sánchez, K. Hummer, L. Wirtz, Vibrational and optical properties of $MoS_2$: from monolayer to bulk, Surf. Sci. Rep. 70 (4) (2015) 554–586, https://doi.org/10.1016/j.surfrep.2015.10.001. J. Low, J. Yu, M. Jaroniec, S. Wageh, A.A. Al-Ghamdi, Heterojunction photocatalysts, Adv. Mater. 29 (20) (2017) 1601694, https://doi.org/10.1002/adma.201601694.
[31] Z. Wang, B. Mi, Environmental applications of 2D molybdenum disulfide ($MoS_2$) nanosheets, Environ. Sci. Technol. 51 (15) (2017) 8229–8244, https://doi.org/10.1021/acs.est.7b01466.
[32] N. Muhd Julkapli, S. Bagheri, S. Bee Abd Hamid, Recent advances in heterogeneous photocatalytic decolorization of synthetic dyes, Sci. World J. 2014 (2014) 1–25, https://doi.org/10.1155/2014/692307.
[33] S.M. Ghoreishi, R. Haghighi, Chemical catalytic reaction and biological oxidation for treatment of non-biodegradable textile effluent, Chem. Eng. J. 95 (1–3) (2003) 163–169,-https://doi.org/10.1016/S1385-8947(03)00100-1.
[34] R. Gopal, M.M. Chinnapan, A.K. Bojarajan, N.K. Rotte, J.S. Ponraj, R. Ganesan, I. Atanas, M. Nadarajah, R.K. Manavalan, J. Gaspar, Facile synthesis and defect optimization of 2D-layered $MoS_2$ on $TiO_2$ heterostructure for industrial effluent, wastewater treatments, Sci. Rep. 10 (1) (2020) 21625, https://doi.org/10.1038/s41598-020-78268-4.
[35] H.N.T. Phung, N.D. Truong, P.A. Duong, L.V. Tuan Hung, Influence of $MoS_2$ deposition time on the photocatalytic activity of $MoS_2$/V, N Co-doped $TiO_2$ heterostructure thin film in the visible light region, Curr. Appl. Phys. 18 (6) (2018) 737–743, https://doi.org/10.1016/j.cap.2018.02.004.
[36] S. He, Y. Zhang, J. Ren, B. Wang, Z. Zhang, M. Zhang, Facile synthesis of $TiO_2$@$MoS_2$ hollow microtubes for removal of organic pollutants in water treatment,

Colloids Surf. A Physicochem. Eng. Asp. 600 (2020) 124900, https://doi.org/10.1016/j.colsurfa.2020.124900.

[37] S.E. Islam, D.-R. Hang, C.-H. Chen, M.M.C. Chou, C.-T. Liang, K.H. Sharma, Rational design of hetero-dimensional C-ZnO/$MoS_2$ nanocomposite anchored on 3D mesoporous carbon framework towards synergistically enhanced stability and efficient visible-light-driven photocatalytic activity, Chemosphere 266 (2021) 129148, https://doi.org/10.1016/j.chemosphere.2020.129148.

[38] B.S. Goud, G. Koyyada, J.H. Jung, G.R. Reddy, J. Shim, N.D. Nam, S.V.P. Vattikuti, Surface oxygen vacancy facilitated Z-scheme $MoS_2/Bi_2O_3$ heterojunction for enhanced visible-light driven photocatalysis-pollutant degradation and hydrogen production, Int. J. Hydrog. Energy 45 (38) (2020) 18961–18975, https://doi.org/10.1016/j.ijhydene.2020.05.073.

[39] M. Behtaj Lejbini, P. Sangpour, Hydrothermal synthesis of $\alpha$-$Fe_2O_3$-decorated $MoS_2$ nanosheets with enhanced photocatalytic activity, Optik 177 (2019) 112–117, https://doi.org/10.1016/j.ijleo.2018.09.019.

[40] W.L. Kebede, D.-H. Kuo, K.E. Ahmed, H. Abdullah, Dye degradation over the multivalent charge- and solid solution-type n-$MoS_2$/p-$WO_3$ based diode catalyst under dark condition with a self-supporting charge carrier transfer mechanism, Adv. Powder Technol. 31 (7) (2020) 2629–2640, https://doi.org/10.1016/j.apt.2020.04.028.

[41] Y. Zeng, N. Guo, H. Li, Q. Wang, X. Xu, Y. Yu, X. Han, H. Yu, A novel route to manufacture $WO_3$@$MoS_2$ p–n heterostructure hollow tubes with enhanced photocatalytic activity, Chem. Commun. 55 (5) (2019) 683–686, https://doi.org/10.1039/C8CC08614A.

[42] H.J. Song, S. You, X.H. Jia, J. Yang, $MoS_2$ nanosheets decorated with magnetic $Fe_3O_4$ nanoparticles and their ultrafast adsorption for wastewater treatment, Ceram. Int. 41 (10) (2015) 13896–13902, https://doi.org/10.1016/j.ceramint.2015.08.023.

[43] H. Li, K. Yu, X. Lei, B. Guo, C. Li, H. Fu, Z. Zhu, Synthesis of the $MoS_2$@CuO heterogeneous structure with improved photocatalysis performance and $H_2O$ adsorption analysis, Dalton Trans. 44 (22) (2015) 10438–10447, https://doi.org/10.1039/C5DT01125F.

[44] S. Rameshkumar, R. Henderson, R.B. Padamati, Improved surface functional and photocatalytic properties of hybrid ZnO-$MoS_2$-deposited membrane for photocatalysis-assisted dye filtration, Membranes 10 (5) (2020) 106, https://doi.org/10.3390/membranes10050106.

[45] K. Rahimi, M. Moradi, R. Dehghan, A. Yazdani, Enhancement of sunlight-induced photocatalytic activity of ZnO nanorods by few-layer $MoS_2$ nanosheets, Mater. Lett. 234 (2019) 134–137, https://doi.org/10.1016/j.matlet.2018.09.103.

[46] Y. Liu, S. Xie, H. Li, X. Wang, A highly efficient sunlight driven ZnO nanosheet photocatalyst: synergetic effect of P-doping and $MoS_2$ atomic layer loading, ChemCatChem 6 (9) (2014) 2522–2526, https://doi.org/10.1002/cctc.201402191.

[47] Y.-H. Tan, K. Yu, J.-Z. Li, H. Fu, Z.-Q. Zhu, $MoS_2$@ZnO Nano-heterojunctions with enhanced photocatalysis and field emission properties, J. Appl. Phys. 116 (6) (2014), https://doi.org/10.1063/1.4893020, 064305.

[48] S.V.P. Vattikuti, C. Byon, C.V. Reddy, R.V.S.S.N. Ravikumar, Improved photocatalytic activity of $MoS_2$ nanosheets decorated with $SnO_2$ nanoparticles, RSC Adv. 5 (105) (2015) 86675–86684, https://doi.org/10.1039/C5RA15159G.

[49] L. Zhang, M. Fang, Nanomaterials in pollution trace detection and environmental improvement, Nano Today 5 (2) (2010) 128–142, https://doi.org/10.1016/j.nantod.2010.03.002.

[50] Y. Wang, B. Zou, T. Gao, X. Wu, S. Lou, S. Zhou, Synthesis of orange-like $Fe_3O_4$/PPy composite microspheres and their excellent Cr(viVI) ion removal properties, J. Mater. Chem. 22 (18) (2012) 9034, https://doi.org/10.1039/c2jm30440f.

[51] R. Gang, L. Xu, Y. Xia, J. Cai, L. Zhang, S. Wang, R. Li, Fabrication of $MoS_2$ QDs/ZnO nanosheet 0D/2D heterojunction photocatalysts for organic dyes and gaseous heavy metal removal, J. Colloid Interface Sci. 579 (2020) 853–861, https://doi.org/10.1016/j.jcis.2020.06.116.

[52] L. Wang, L. Hehua, H. Zeng, $Fe_3O_4$@$MoS_2$ core-shell magnetism nanocomposite for water remediation through highly efficient and selective removal of Hg(II) ions, chemistry letters, Chem. Lett. 47 (12) (2018) 1515–1518, https://doi.org/10.1246/cl.180793.

[53] N. Jiang, Y. Du, P. Ji, S. Liu, B. He, J. Qu, J. Wang, X. Sun, Y. Liu, H. Li, Enhanced photocatalytic activity of novel $TiO_2$/Ag/$MoS_2$/Ag nanocomposites for water-treatment, Ceram. Int. 46 (4) (2020) 4889–4896, https://doi.org/10.1016/j.ceramint.2019.10.225.

[54] Y. Song, M. Lu, B. Huang, D. Wang, G. Wang, L. Zhou, Decoration of defective $MoS_2$ nanosheets with $Fe_3O_4$ nanoparticles as superior magnetic adsorbent for highly selective and efficient mercury ions ($Hg^{2+}$) removal, J. Alloys Compd. 737 (2018) 113–121, https://doi.org/10.1016/j.jallcom.2017.12.087.

[55] L. Zhi, W. Zuo, F. Chen, B. Wang, 3D $MoS_2$ composition aerogels as chemosensors and adsorbents for colorimetric detection and high-capacity adsorption of $Hg^{2+}$, ACS Sustain. Chem. Eng. 4 (6) (2016) 3398–3408, https://doi.org/10.1021/acssuschemeng.6b00409.

[56] S. Yang, Q. Li, L. Chen, Z. Chen, Z. Pu, H. Wang, S. Yu, B. Hu, J. Chen, X. Wang, Ultra-high sorption and reduction of Cr(VI) by two novel core-shell composites combined with $Fe_3O_4$ and $MoS_2$, J. Hazard. Mater. 379 (2019) 120797, https://doi.org/10.1016/j.jhazmat.2019.120797.

[57] Q. Wang, L. Peng, Y. Gong, F. Jia, S. Song, Y. Li, Mussel-inspired $Fe_3O_4$@Polydopamine (PDA)-$MoS_2$ core–shell nanosphere as a promising adsorbent for removal of $Pb^{2+}$ from water, J. Mol. Liq. 282 (2019) 598–605, https://doi.org/10.1016/j.molliq.2019.03.052.

[58] H. Wang, F. Wen, X. Li, X. Gan, Y. Yang, P. Chen, Y. Zhang, Cerium-doped $MoS_2$ nanostructures: efficient visible photocatalysis for Cr(VI) removal, Sep. Purif. Technol. 170 (2016) 190–198, https://doi.org/10.1016/j.seppur.2016.06.049.

[59] B. Petrie, R. Barden, B. Kasprzyk-Hordern, A review on emerging contaminants in wastewaters and the environment: current knowledge, understudied areas and recommendations for future monitoring, Water Res. 72 (2015) 3–27, https://doi.org/10.1016/j.watres.2014.08.053.

[60] V. Homem, L. Santos, Degradation and removal methods of antibiotics from aqueous matrices—a review, J. Environ. Manag. 92 (10) (2011) 2304–2347, https://doi.org/10.1016/j.jenvman.2011.05.023.

[61] E. Topkaya, M. Konyar, H.C. Yatmaz, K. Öztürk, Pure ZnO and composite ZnO/$TiO_2$ catalyst plates: a comparative study for the degradation of azo dye, pesticide and antibiotic in aqueous solutions, J. Colloid Interface Sci. 430 (2014) 6–11, https://doi.org/10.1016/j.jcis.2014.05.022.

[62] W.S. Koe, J.W. Lee, W.C. Chong, Y.L. Pang, L.C. Sim, An overview of photocatalytic degradation: photocatalysts, mechanisms, and development of photocatalytic membrane, Environ. Sci. Pollut. Res. 27 (3) (2020) 2522–2565, https://doi.org/10.1007/s11356-019-07193-5.

[63] R. Ji, M. Zhang, W. Ma, Z. Zhu, C. Ma, P. Huo, Y. Yan, Y. Liu, C. Li, Heterojunction photocatalyst fabricated by deposition $Co_3O_4$ nanoparticles on $MoS_2$ nanosheets with enhancing photocatalytic performance and mechanism insight, J. Taiwan Inst. Chem. Eng. 97 (2019) 158–169, https://doi.org/10.1016/j.jtice.2019.01.011.

[64] Z. Ma, L. Hu, X. Li, L. Deng, G. Fan, Y. He, A novel nano-sized $MoS_2$ decorated $Bi_2O_3$ heterojunction with enhanced photocatalytic performance for methylene blue

and tetracycline degradation, Ceram. Int. 45 (13) (2019) 15824–15833, https://doi.org/10.1016/j.ceramint.2019.05.085.
[65] S. Kumar, V. Sharma, K. Bhattacharyya, V. Krishnan, N-doped ZnO–MoS $_2$ binary heterojunctions: the dual role of 2D $MoS_2$ in the enhancement of photostability and photocatalytic activity under visible light irradiation for tetracycline degradation, Mater. Chem. Front. 1 (6) (2017) 1093–1106, https://doi.org/10.1039/C6QM00274A.
[66] S. Bougarrani, K. Skadell, R. Arndt, M. El Azzouzi, R. Gläser, Novel CaxMnOy/$TiO_2$ composites for efficient photocatalytic degradation of methylene blue and the herbicide imazapyr in aqueous solution under visible light irradiation, J. Environ. Chem. Eng. 6 (2) (2018) 1934–1942, https://doi.org/10.1016/j.jece.2018.02.026.
[67] H.D. Burrows, L.M. Canle, J.A. Santaballa, S. Steenken, Reaction pathways and mechanisms of photodegradation of pesticides, J. Photochem. Photobiol. B Biol. 67 (2) (2002) 71–108, https://doi.org/10.1016/S1011-1344(02)00277-4.
[68] R. Ji, C. Ma, W. Ma, Y. Liu, Z. Zhu, Y. Yan, Z-scheme $MoS_2/Bi_2O_3$ heterojunctions: enhanced photocatalytic degradation performance and mechanistic insight, New J. Chem. 43 (30) (2019) 11876–11886, https://doi.org/10.1039/C9NJ02521A.
[69] X. Li, Z. Zhang, C. Yao, X. Lu, X. Zhao, C. Ni, Attapulgite-$CeO_2/MoS_2$ ternary nanocomposite for photocatalytic oxidative desulfurization, Appl. Surf. Sci. 364 (2016) 589–596, https://doi.org/10.1016/j.apsusc.2015.12.196.
[70] G.P. Awasthi, S.P. Adhikari, S. Ko, H.J. Kim, C.H. Park, C.S. Kim, Facile synthesis of ZnO flowers modified graphene like $MoS_2$ sheets for enhanced visible-light-driven photocatalytic activity and antibacterial properties, J. Alloys Compd. 682 (2016) 208–215,-https://doi.org/10.1016/j.jallcom.2016.04.267.
[71] H. Yan, L. Liu, R. Wang, W. Zhu, X. Ren, L. Luo, X. Zhang, S. Luo, X. Ai, J. Wang, Binary composite $MoS_2/TiO_2$ nanotube arrays as a recyclable and efficient photocatalyst for solar water disinfection, Chem. Eng. J. 401 (2020) 126052, https://doi.org/10.1016/j.cej.2020.126052.
[72] Z.G. Zhang, H. Liu, X.X. Wang, J. Zhang, M. Yu, S. Ramakrishna, Y.Z. Long, One-step low temperature hydrothermal synthesis of flexible $TiO_2$/PVDF@$MoS_2$ core-shell heterostructured fibers for visible-light-driven photocatalysis and self-cleaning, Nanomaterials 9 (3) (2019) 1–22, https://doi.org/10.3390/nano9030431.
[73] Y.R. Qi, Q.H. Liang, R.T. Lv, W.C. Shen, F.Y. Kang, Z.H. Huang, Subject category: subject areas: synthesis and photocatalytic activity of mesoporous g-$C_3N_4/MoS_2$ hybrid catalysts, R. Soc. 2 (2018).
[74] X. Li, Z. Zhang, A. Fakhri, V.K. Gupta, S. Agarwal, Adsorption and photocatalysis assisted optimization for drug removal by chitosan-glyoxal/polyvinylpyrrolidone/$MoS_2$ nanocomposites, Int. J. Biol. Macromol. 136 (2019) 469–475, https://doi.org/10.1016/j.ijbiomac.2019.06.003.–10.
[75] T. Ahamad, M. Naushad, Y. Alzaharani, S.M. Alshehri, Photocatalytic degradation of bisphenol-a with g-$C_3N_4/MoS_2$-PANI nanocomposite: kinetics, main active species, intermediates and pathways, Mol. Liq. 311 (2020) 113339, https://doi.org/10.1016/j.molliq.2020.113339.
[76] W. Zhang, X. Xiao, Y. Li, X. Zeng, L. Zheng, C. Wan, Liquid-exfoliation of layered $MoS_2$ for enhancing photocatalytic activity of $TiO_2$/g-$C_3N_4$ photocatalyst and DFT study, Appl. Surf. Sci. 389 (2016) 496–506, https://doi.org/10.1016/j.apsusc.2016.07.154.
[77] L. Shi, Z. He, S. Liu, $MoS_2$ quantum dots embedded in G-$C_3N_4$ frameworks: a hybrid 0D-2D heterojunction as an efficient visible-light driven photocatalyst, Appl. Surf. Sci. 457 (June) (2018) 30–40, https://doi.org/10.1016/j.apsusc.2018.06.132.

[78] Z. Zhang, C. Liu, Z. Dong, Y. Dai, G. Xiong, Y. Liu, Y. Wang, Y. Wang, Y. Liu, Synthesis of flower-like $MoS_2$/g-$C_3N_4$ nanosheet heterojunctions with enhanced photocatalytic reduction activity of uranium(VI), Appl. Surf. Sci. 520 (February) (2020) 146352, https://doi.org/10.1016/j.apsusc.2020.146352.
[79] L. Shi, W. Ding, S. Yang, Z. He, S. Liu, Rationally designed $MoS_2$/protonated g-$C_3N_4$ nanosheet composites as photocatalysts with an excellent synergistic effect toward photocatalytic degradation of organic pollutants, J. Hazard. Mater. 347 (2018) 431–441,-https://doi.org/10.1016/j.jhazmat.2018.01.010.
[80] J. Bai, W. Lv, Z. Ni, Z. Wang, G. Chen, H. Xu, H. Qin, Z. Zheng, X. Li, Integrating $MoS_2$ on sulfur-doped porous g-$C_3N_4$ isotype heterojunction hybrids enhances visible-light photocatalytic performance, J. Alloys Compd. 768 (2018) 766–774, https://doi.org/10.1016/j.jallcom.2018.07.286.
[81] J. Kang, C. Jin, Z. Li, M. Wang, Z. Chen, Y. Wang, Dual Z-scheme $MoS_2$/g-$C_3N_4$/$Bi_{24}O_{31}Cl_{10}$ ternary heterojunction photocatalysts for enhanced visible-light photodegradation of antibiotic, J. Alloys Compd. 825 (2020) 153975, https://doi.org/10.1016/j.jallcom.2020.153975.
[82] H. Liu, J. Liang, L. Shao, J. Du, Q. Gao, S. Fu, L. Li, M. Hu, F. Zhao, J. Zhou, Promoting charge separation in dual defect mediated Z-scheme $MoS_2$/g-$C_3N_4$ photocatalysts for enhanced photocatalytic degradation activity: synergistic effect insight, Colloids Surf. A Physicochem. Eng. Asp. 594 (2020), https://doi.org/10.1016/j.colsurfa.2020.124668.
[83] F. Ran, Y. Zou, Y. Xu, X. Liu, H. Zhang, $Fe_3O_4$@$MoS_2$@PEI-facilitated enzyme tethering for efficient removal of persistent organic pollutants in water, Chem. Eng. J. (2019) 375, https://doi.org/10.1016/j.cej.2019.121947.
[84] Z. Wu, X. He, Y. Xue, X. Yang, Y. Li, Q. Li, B. Yu, Cyclodextrins grafted $MoS_2$/g-$C_3N_4$ as high-performance photocatalysts for the removal of glyphosate and Cr(VI) from simulated agricultural runoff, Chem. Eng. J. 399 (March) (2020) 125747, https://doi.org/10.1016/j.cej.2020.125747.
[85] W. Ma, B. Yao, W. Zhang, Y. He, Y. Yu, J. Niu, Fabrication of PVDF-based piezocatalytic active membrane with enhanced oxytetracycline degradation efficiency through embedding few-layer E-$MoS_2$ nanosheets, Chem. Eng. J. 415 (February) (2021) 129000, https://doi.org/10.1016/j.cej.2021.129000.
[86] W.C. Peng, X.Y. Li, Synthesis of $MoS_2$/g-$C_3N_4$ as a solar light-responsive photocatalyst for organic degradation, Catal. Commun. 49 (2014) 63–67, https://doi.org/10.1016/j.catcom.2014.02.008.
[87] V.K. Tomer, R. Malik, V. Chaudhary, Y.K. Mishra, L. Kienle, R. Ahuja, L. Lin, Superior visible light photocatalysis and Low-operating temperature VOCs sensor using cubic $ag^{(0)}$-$MoS_2$ loaded g-$C_3N_4$ 3D porous hybrid, Appl. Mater. Today 16 (2019) 193–203,-https://doi.org/10.1016/j.apmt.2019.05.010.
[88] M. Ntakadzeni, W.W. Anku, N. Kumar, P.P. Govender, L. Reddy, PeGylated $MoS_2$ nanosheets: a dual functional photocatalyst for photodegradation of organic dyes and photoreduction of chromium from aqueous solution, Bull. Chem. React. Eng. Catal. 14 (1) (2019) 142–152, https://doi.org/10.9767/bcrec.14.1.2258.142-152.
[89] X. Wang, M. Hong, F. Zhang, Z. Zhuang, Y. Yu, Recyclable nanoscale zero valent iron doped G-$C_3N_4$/$MoS_2$ for efficient photocatalysis of RhB and Cr(VI) driven by visible light, ACS Sustain. Chem. Eng. 4 (7) (2016) 4055–4063, https://doi.org/10.1021/acssuschemeng.6b01024.
[90] J. Qiu, F. Liu, S. Cheng, L. Zong, C. Zhu, C. Ling, A. Li, Recyclable nanocomposite of flowerlike $MoS_2$@hybrid acid-doped PANI immobilized on porous PAN nanofibers for

the efficient removal of Cr(VI), ACS Sustain. Chem. Eng. 6 (1) (2018) 447–456, https://doi.org/10.1021/acssuschemeng.7b02738.
[91] G. Gogoi, S. Keene, A.S. Patra, T.K. Sahu, S. Ardo, M. Qureshi, Hybrid of G-$C_3N_4$ and $MoS_2$ integrated onto $Cd_{0.5}Zn_{0.5}S$: rational design with efficient charge transfer for enhanced photocatalytic activity, ACS Sustain. Chem. Eng. 6 (5) (2018) 6718–6729,-https://doi.org/10.1021/acssuschemeng.8b00512.
[92] M. Chegeni, M. Mehri, M. Hosseini, The $MoS_2$/S-doped graphitic carbon nitride: synthesis and application as a composite for removing organic pollutant, J. Water Environ. Nanotechnol. 5 (4) (2020) 331–341, https://doi.org/10.22090/jwent.2020.04.004.
[93] Q. Li, N. Zhang, Y. Yang, G. Wang, D.H.L. Ng, High efficiency photocatalysis for pollutant degradation with $MoS_2/C_3N_4$ heterostructures, Langmuir 30 (29) (2014) 8965–8972, https://doi.org/10.1021/la502033t.
[94] S.V.P. Vattikuti, C. Byon, Hydrothermally synthesized ternary heterostructured $MoS_2$/$Al_2O_3$/g-$C_3N_4$ photocatalyst, Mater. Res. Bull. 96 (2017) 233–245, https://doi.org/10.1016/j.materresbull.2017.03.008.
[95] Y. Li, Z. Wang, H. Zhao, X. Huang, M. Yang, 3D $MoS_2$@$TiO_2$@poly(methyl methacrylate) nanocomposite with enhanced photocatalytic activity, J. Colloid Interface Sci. 557 (2019) 709–721, https://doi.org/10.1016/j.jcis.2019.09.
[96] M. Taheriyoun, S. Moradinejad, Reliability analysis of a wastewater treatment plant using fault tree analysis and Monte Carlo simulation, Environ. Monit. Assess. 187 (1) (2015) 4186, https://doi.org/10.1007/s10661-014-4186-7.
[97] S.C. Oliveira, M. Von Sperling, Reliability analysis of wastewater treatment plants, Water Res. 42 (4–5) (2008) 1182–1194, https://doi.org/10.1016/j.watres.2007.09.001.
[98] L. Matoh, B. Žener, R.C. Korošec, U.L. Štangar, Chapter-27 — Photocatalytic water treatment, 2019, pp. 675–702, https://doi.org/10.1016/B978-0-08-102641-0.00027-X.

# Metal oxide-conducting polymer-based composite electrodes for energy storage applications

7

*Mohsin Ali Raza, Zaeem Ur Rehman, Muhammad Gulraiz Tanvir, and Muhammad Faheem Maqsood*
Institute of Metallurgy and Materials Engineering, University of the Punjab, Lahore, Pakistan

## 1 Introduction

Fossil fuels are on the verge of extinction and cannot cope with the increasing energy demand due to modern industrialism and infrastructure. These indeed have a major share in providing energy globally, but their harmful impacts have endangered the lives of living creatures. Emission of $CO_2$ alone is the largest contributor to global climate changes that have intensified global warming and contributed to air pollution resulting in millions of deaths every year. According to the World Energy Outlook, global electricity demand will grow at 1.3% each year [1]. To fulfill future energy demands, alternative and renewable energy resources have emerged over the last five decades as a potential replacement for fossil fuels including wind energy, geothermal energy, solar energy, bioenergy, etc. [2]. One can imagine the importance of such green energy sources by the fact that at the end of 2019, 629 GW of solar power stations were installed throughout the world [3].

Clean energy sources like solar and wind energies are important sustainable energy resources of the future, and their efficient usage is critical for their constant energy production. To acquire energy as per demand, these energy resources are merged with some efficient energy storage technologies such as batteries and supercapacitors. Recently, electrochemical energy storage (EES) has emerged with features like high energy density, smart size, and easy configuration which have given them popularity among technologists, scientists, and manufacturers. Such EES technologies include fuel cells, lithium-ion batteries, sodium-ion batteries, and supercapacitors.

## 2 Supercapacitor

Supercapacitors have emerged as a promising technology for energy storage. This field is extensively researched in recent era [4]. Supercapacitors are also known as electrochemical capacitors [5], electric double-layer capacitors (EDLCs) [6], or ultracapacitors [7]. Supercapacitor is a device consisting of two porous electrodes separated by an ion-conductive membrane immersed in an electrolyte. The electric field

**Renewable Polymers and Polymer-Metal Oxide Composites. https://doi.org/10.1016/B978-0-323-85155-8.00008-X**

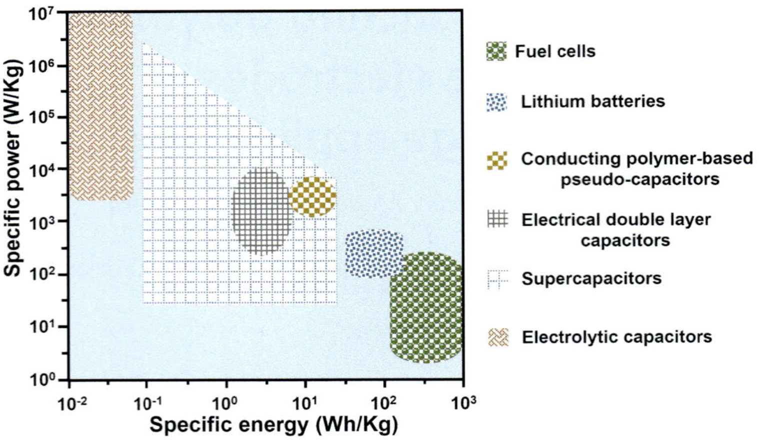


**Fig. 1** Ragone plot represents energy vs. power density for different energy storage devices [8].

created between the electrodes and electrolyte facilitates energy storage in the electrodes. The charge storage capacity and mechanism depend on the nature of electrode materials, electrolyte porosity, morphology of the electrode, and interactions between electrode and electrolyte.

Supercapacitors bridge a gap between conventional capacitors and batteries with regard to energy and power densities. This can be well understood by Ragone plot (Fig. 1) which shows supercapacitors have a greater energy density than solid-state or conventional capacitors while a higher power density than batteries. One of the best features of supercapacitors is fast charging and discharging time that makes them suitable for many applications requiring rapid charge/discharge cycles and high-power bursts such as automobiles, buses, trains, cranes, and elevators.

# 3 Types of supercapacitors

On the basis of energy storage mechanism, supercapacitors are classified into three main categories, i.e., EDLCs, pseudocapacitors, and hybrid supercapacitors [9] as shown in Fig. 2. Each type is based on different electrode materials and charge storage mechanisms that are discussed in the following sections.

## *3.1 Electrical double-layer capacitors (EDLCs)*

Electrical double-layer capacitors' charge storage mechanism involves electrostatic charge adsorption at the electrode surface without a charge transfer between electrode and electrolyte, i.e., without any faradaic reaction [10,11]. Such a capacitor consists of two porous electrodes immersed in a conductive electrolyte mechanically separated by a separator [12] as shown in Fig. 3. When a potential is applied, one electrode gets negatively charged and the other positively charged. Due to this polarization, ions

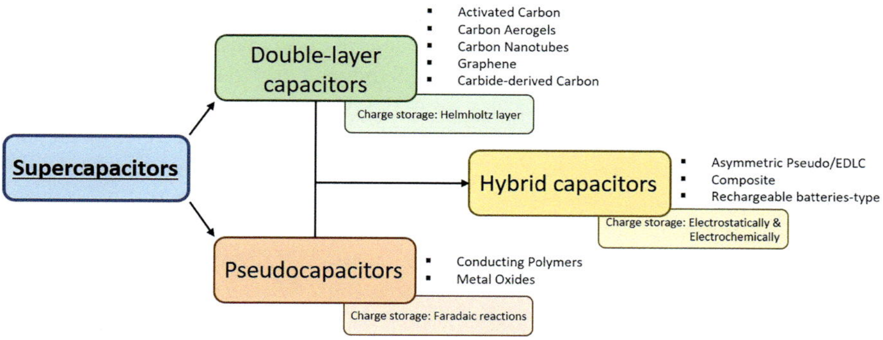


**Fig. 2** Various types of supercapacitors on the basis of charge storage mechanism [9].

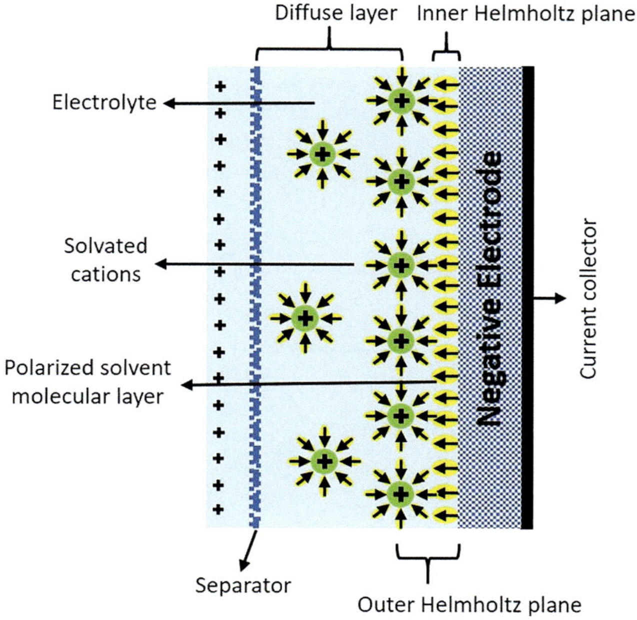


**Fig. 3** Schematic of charge storage mechanism of EDLCs [14].

move toward their respective oppositely charged electrodes forming a layer of ions known as double layers or inner Helmholtz layer [13] which is produced as a result of the separation of ions. The higher the concentration of ions, the narrower will be the inner layer.

The porous separator facilitates the migration of ions and prevents remixing. The large surface area of electrodes about 1000 $m^2/g$ and a reduced distance between electrodes due to the formation of inner Helmholtz layer produce high capacitance and high energy density [15]. The movement of ions through a solution makes EDLCs capable of displaying thousands of charge/discharge cycles without a significant loss of capacity. As the distance for ions to travel is reduced due to the formation of inner layers, so do their charging/discharging time, ranging between 1 and 5 s. The rate of charging/discharging time of electrodes can be tuned by the conductivity of the electrodes [16]. The double-layer capacitance depends on the applied voltage. In EDLC, the only electrochemical interaction between electrode and electrolyte interface involves adsorption and desorption of cations and anions. The electrodes are designed to maximize the electrode-electrolyte interface area. EDLC electrodes are produced with porous carbon conducting powders, fibers, or felts. The higher graphitic character of the carbon electrode results in higher conductivity of the electrode [17].

## 3.2 Pseudocapacitors

This type of supercapacitor stores electrical energy by reversible faradic redox reactions on the surface of the electrodes. Pseudocapacitance arises from an electron transfer between electrode and electrolyte coming from a desolvated and adsorbed ion (Fig. 4). The adsorbed ion has no chemical interaction with the electrode's atom and so no chemical bond occurs [18]. The faradic current involved in pseudocapacitors made them to attain higher capacitance and energy densities than EDLCs [19]. However, pseudocapacitors possess the lack of power due to slow faradic processes involved [20]. Electrodes exhibiting pseudocapacitance are liable to swelling and shrinkage upon charging/discharging resulting in poor cyclic stability and a low

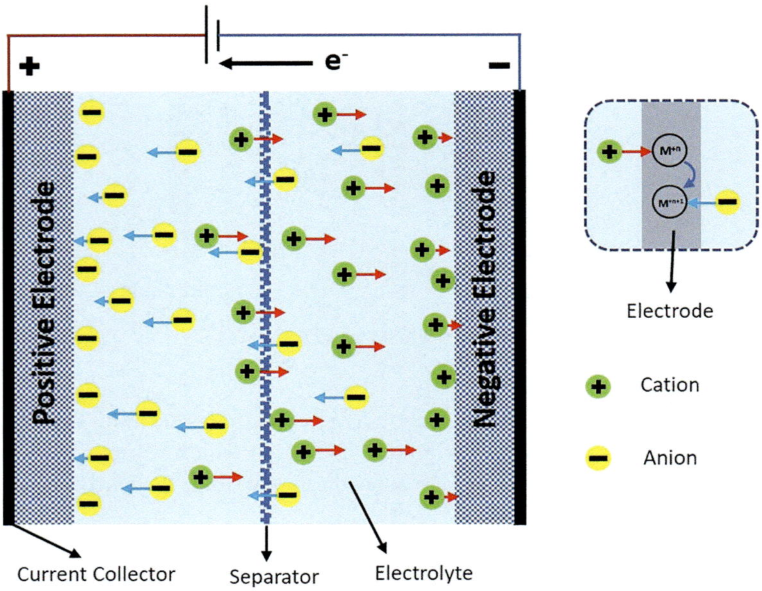


**Fig. 4** Schematic of energy storage mechanism of pseudocapacitors [24].

life cycle [21]. The value of pseudocapacitance is usually higher than EDLCs by a factor of 100 [16]. The faradic reactions involved in pseudocapacitors are much faster than batteries because adsorbed ion has no chemical reaction with the atoms of electrode since the electrons are transferred to or from the valence band of the redox reagent of the electrode [17]. Pseudocapacitive phenomenon can occur through three types of behaviors including underpotential deposition, reduction-oxidation reactions (redox reactions), and intercalation [17,22]. Electrodes involved in pseudocapacitors are typically transition metal oxides like $MnO_2$ and $RuO_2$ and conducting polymers like polyaniline (PANI) and polypyrrole (PPy) [23].

## 3.2.1 *Types of pseudocapacitive behavior*

### 3.2.1.1 Underpotential deposition (UPD)

The UPD is analogous to electrodeposition typically involves the reduction of metal cations to solid metal on the substrate at a potential less negative than the equilibrium potential (Fig. 5A) [25,26]. The deposition of Pb on Au is a typical example [27,28].

### 3.2.1.2 Redox reactions

The pseudocapacitance occurs as a result of charge transfer at the electrode surface [30]. Redox pseudocapacitance is accompanied with faradic charge transfer when ions chemically adsorb onto the surface or near the surface of the electrode (Fig. 5B) [31].

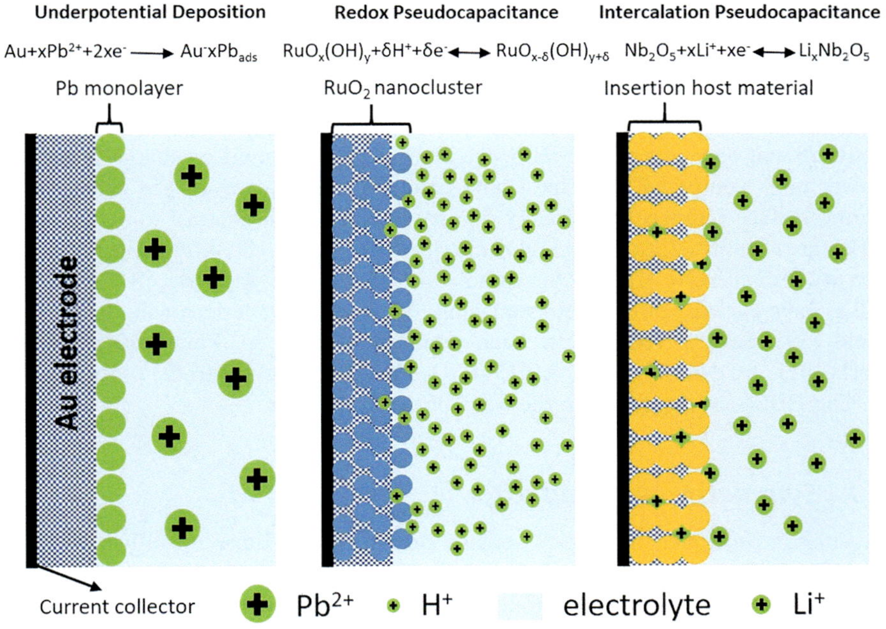


**Fig. 5** Schematic representation of different types of pseudocapacitance behaviors [29].

Materials exhibiting such redox electrochemical behavior are also referred to as intrinsic pseudocapacitive materials [31]. Examples of such materials include $RuO_2$, $MnO_2$, and $Fe_3O_4$ [30].

#### 3.2.1.3 Intercalation pseudocapacitance

Intercalation of ions into galleries or layers of redox-active materials along with faradic charge transfer results in intercalation pseudocapacitance. This process does not affect the crystallographic structure of electrode material. Some 2D materials and few-layered materials including $TiO_2$, $Nb_2O_5$, and $MoO_3$ [30] display fast and reversible charge storage rates approaching or even exceeding traditional surface-redox pseudocapacitor material (Fig. 5C). [32]. The electrochemical features of intercalation pseudocapacitance include [30]:

**(a)** Current increases with an increase of scan rate.
**(b)** Charging time does not affect charge storage capacity.
**(c)** Change in scan rate does not affect peak potential.

# 4 Hybrid supercapacitor (HBS)

The EDLCs offer good cyclic stability and high power density, while pseudocapacitors provide greater specific capacitance (SC) and energy density. HBS combines both the features by utilizing two different electrodes, i.e., EDLC-type electrode and pseudocapacitor-type electrode [33]. One-half of the hybrid supercapacitor acts as EDLC and the other as pseudocapacitor. HBSs have higher energies and power densities than EDLC and pseudocapacitors. Similarly, they have higher power densities than fuel cells and batteries, but their energy densities are lower than conventional capacitors [34]. HBS can store a large amount of charge at elevated power rates compared to rechargeable batteries [35]. However, the charge stored on the surface restricts their energy density or capacity in contrast to batteries [34]. HBSs are capable of achieving higher energy density up to 100 Wh/kg [36,37]. Energy density prevailed by HBS is due to active materials' SC and net cell voltage [38]. HBSs have a strong prospect for replacing rechargeable batteries owing to their high yield [39]. Several HBSs have been reported such as AC/$PbO_2$ [40], graphene/$PbO_2$ [41], AC/$Ni(OH)_2$ [42], AC/$Li_4Ti_5O_{12}$ [43], and AC/$LiMn_2O_4$ [44].

The HBSs are classified on the basis of their assembly.

## *4.1 Symmetric supercapacitor*

Symmetric supercapacitor (SSC) consists of two similar supercapacitive electrodes, i.e., EDLC [34,45]. Commercially available SSCs are comprised of binary electrodes of activated carbon (AC) inside organic electrolyte with an operational potential up to 2.7 V [46]. Researchers are utilizing the advantages of most commonly employed materials, for instance, carbon nanotubes [47], carbon aerogels [48], and graphene

[49] for development of SSCs. SSCs have relatively lower energy density, around 5 Wh/kg, but their power density can be as high as 9 kW/kg [50].

### 4.2 Asymmetric supercapacitor

Asymmetric supercapacitor consists of two different types of electrodes, i.e., one is comprised of EDLC material and the other of pseudocapacitive material [45,50]. The most notable asymmetric supercapacitor electrodes being used are AC and $MnO_2$ along with AC-$Ni(OH)_2$ [34]. AC/$MnO_2$ is one of the most promising materials that have been widely reported for energy storage [45]. Asymmetric supercapacitors can achieve high energy and power density [45]. These types of supercapacitors are very expensive and possess lower charge storage efficiency and operational potential. Furthermore, they possess poor cyclic stability [51]). Asymmetric supercapacitors can be utilized in those applications where compromise between energy and power density and cost is not a major issue.

## 5 Metal oxide/polymer materials for supercapacitor applications

Metal oxides offer high SC because of which they can become potential electrode materials for supercapacitor applications, in particular, when high energy and power density are required. Transition metal oxides have gained wider interest due to their layered structure and multioxidation states. Metal oxides are suitable for pseudocapacitors due to highly surface induced reversible chemical reactions or swift reversible lattice intercalation. Metal oxides can deliver higher energy density than traditional carbon materials and better chemical stability than polymeric materials as their mechanism of energy storage is similar to carbon materials; however, they also exhibit faradic reactions between electrode and electrolyte within a certain potential window. Many metal oxides like CuO, $MnO_2$, NiO, $Co_3O_4$, and $Fe_3O_4$ have been widely investigated [52,53].

For supercapacitor applications, metal oxides should possess the following characteristics [54]:

**i.** Metal oxides should have good electrical conductivity.
**ii.** Metal oxides should have multiple valence states and there should not be any phase change involved.
**iii.** Ions should be free to move into the structure for redox reactions to occur.

### 5.1 Conducting polymers (CPs)

There are some limitations of metal oxides like lower power density and poor cycling stability. Lack of power density is ascribed to metal oxide's poor electrical conductance that limits the charge transfer rate [55]. Their morphology is liable to degradation caused by charging/discharging which will result in poor cyclic stability [55]. To tackle this issue, intrinsically conducting polymers (ICPs) have been suggested [56]. CPs being active materials not only offer enhanced conductance, but these also act as a binder for metal oxide [57]. PANI, poly(3,4-ethylenedixoythiophene) (PEDOT), and polypyrrole (PPy) are widely used conducting polymers that can be combined with metal oxides to form composite electrodes for supercapacitors [58]. CPs are "bridging the gap" between batteries and EDLCs with superior energy densities compared to supercapacitors made up of carbon-based materials. The first pseudocapacitor was reported in 1990 [59]. A brief introduction of different CPs used as supercapacitors is as follows:

### 5.1.1 Polyaniline (PANI)

First discovered back in 1834 by Runge [60], polyaniline has become a focal among the intrinsically conductive polymers (ICPs) [61]. The general formula of polyaniline is $[(\text{—B—NH—B—NH—})_n(\text{—B—N=Q=N—})_{1-n}]_m$, where B and Q denote the rings in the benzenoid and quinoid forms, respectively [62]. PANI has chains consisting of repeatedly ordered phenyl rings and nitrogen groups as shown in Fig. 6. The structure helps in conjugation with backbone chain forming a zigzag lying in one plain while π-electronic clouds overlapping above and below this plain. Lone pair of nitrogen also offers a role in polyconjugation just like π-electrons. This polyconjugated system is responsible for giving charge a pathway resulting in conduction [63].

The PANI possesses excellent physiochemical properties and has unique ability to conduct both electronic and ionic charges [65]. PANI applications are notable among electrochemical devices including batteries, supercapacitors (estimated capacitance is 750 F/g), pH sensing, biosensing, electronic displays, frequency converters, modulators, dielectric amplifiers, sensors, electromagnetic shielding, and photovoltaic and light-emitting devices [66,67]. PANI can store high energy because of multiredox transitions and high electrical conductivity. It offers good stability, ecofriendliness, and cost-effectiveness [68,69]. The performance of PANI for energy storage is strongly relevant to its structure, especially specific surface area. Tweaking PANI can result in its many applications [70] including electromagnetic shielding and microwave absorption [71], antistatic materials, heating elements [72], membrane materials [73], coatings for anticorrosion [74], welding of plastics, electrochromic display devices, electronic devices, gas sensor materials [75], smart conductive textiles, and electronic textiles for the operation of capacitive touch screen displays [76].

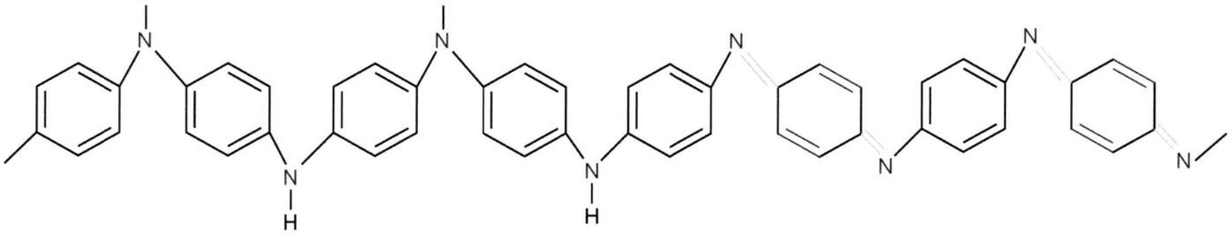

**Fig. 6** Chemical structure of PANI [64].

**Table 1** Various synthesis techniques of PANI.

| Synthesis methods | Refs. |
|---|---|
| Chemical/electrochemical methods | [78,79] |
| Electroactive oxidative polymerization | [80] |
| Interfacial polymerization | [81] |
| Radical polymerization | [82] |
| Seeding polymerization | [83] |
| Microwave-assisted polymerization | [83] |
| Chemical oxidative polymerization | [84] |
| In situ chemical polymerization | [85] |
| Bulk and slurry oxidative polymerization | [86] |
| Reflux method | [87] |
| Melt process, emulsion, inversion emulsion, and interfacial polymerization methods | [88,89] |
| Electrochemical synthesis | [90] |
| Enzymatic synthesis | [91] |

There are some limitations of PANI like difficulty in processing and dispersion due to phenyl rings that make its backbone very stiff [77]. In the perspective of energy storage, PANI displays poor capacitance retention at higher current density or longer term cycling stability [77]. Various synthesis methods of PANI are presented in Table 1.

## 5.1.2 Polypyrrole (PPy)

The PPy is an amorphous, highly branched, and possibly cross-linked polymer having moderately low molecular weight [92]. The chemical structure of PPy is shown in Fig. 7.

The PPy is one of the attractive electroconductive polymers that integrate metallic and semiconductor properties including characteristics such as flexibility, strength, elasticity, and ease of synthesis [94]. The most notable feature of PPy is superior ease of electrochemical processing over other conducting polymers. Furthermore, its properties can be better tuned depending on the nature of synthesis methods [95]. PPy can retain its conductivity and mechanical properties as function of temperature fluctuation ranging from 150°C to 280°C [96]. Among conducting polymers, PPY is being widely studied because of the unique features of pyrrole monomer including ease of

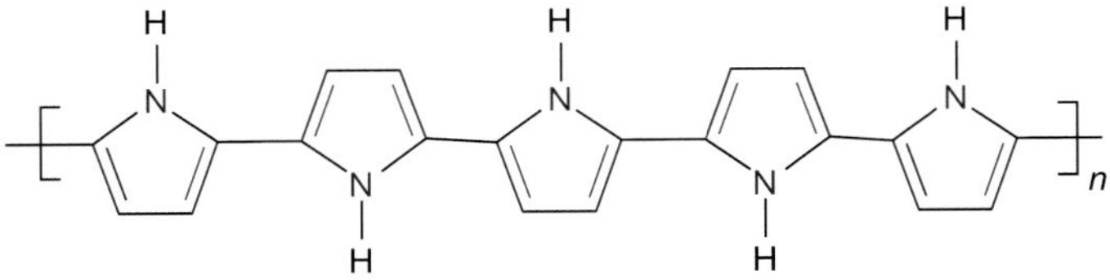


**Fig. 7** Chemical structure of PPy [93].

oxidation, water solubility, ease of availability, environmental stability, and good redox properties [97]. PPy-based materials can be a part of a variety of display members, antistatic coatings, anticorrosion coatings, electronic devices, batteries, sensors, microactuators, and biomedical [98]. PPy has been a focus of attention among researchers for batteries and supercapacitor's electrodes due to its excellent theoretical capacitance of 620 F/g [99,100].

One of the limiting properties of PPy is its poor mechanical strength and poor processability [101]. It cannot be n-doped restricting its use only for cathode material [102]. PPy has high density resulting in significantly a high capacitance of 400–500 F/cm$^3$ [103]. But the negative aspect of high density is inaccessibility of dopant ions to the internal sites of the polymer that results in low capacitance per gram, especially when electrodes with thicker coating are employed [104].

In PPy, pyrrole rings generally link through α-position to give polymer [105]. There are a range of synthesis techniques such as electrochemical [106], vapor-phase polymerization [107], chemical oxidation pyrrole monomer [108] in various organic solvents and aqueous media, photochemistry, metathesis, concentrated emulsion, and soluble precursor polymer polymerization [109].

### *5.1.3 Poly(3,4-ethylenedioxythiophene) (PEDOT)*

The PEDOT was first reported by Bayer AG in 1989 and is based on 3,4-ethylenedioxythiophene (EDOT) [110]. The polymer is generated by oxidation of EDOT monomer (resulting in radical cation which is added in neutral EDOT causing deprotonation) in the presence of peroxydisulfate as shown as follows:

$$nC_2H_4O_2C_4H_2S + n\,(OSO_3)^{-2} \rightarrow [C_2H_4O_2C_4S]_n + 2nHOSO_3^{-}$$

Chemical structure of PEDOT is shown in Fig. 8. Reports on PEDOT have increased significantly over the years due to its attractive properties including higher conductivity and greater stability over other polymers [67,111–113]. Its extraordinary conductivity has been reported to increase up to 6259 S/cm for thin films and 8797 S/cm for single crystals, which is an order of magnitude less than copper and silver [114,115]. Its electron surplus and low oxidation potential accompanied with

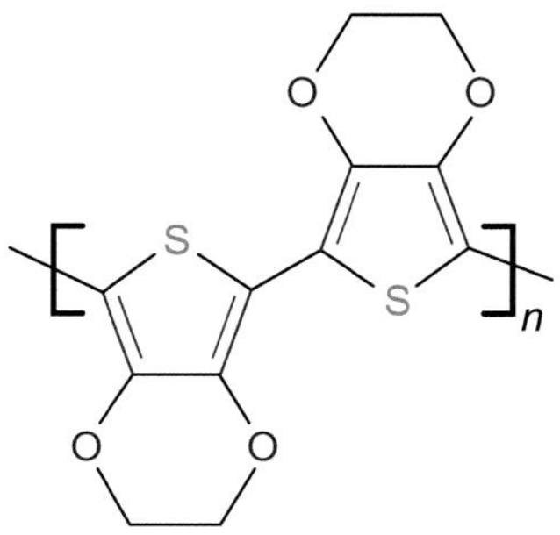


**Fig. 8** Chemical structure of PEDOT [110].

a wide potential window (1.2–1.5 V) result in high capacitance [116]. The PEDOT has low band gap of 1–3 eV and is highly conductive in p-doped state (300–500 S/cm), has good thermal and chemical stability and high charge mobility, resulting in fast electrochemical kinetics [67]. It also possesses good film-forming properties [116]. Due to these remarkable features of PEDOT, many researchers have reported its synthesis, integration in devices and applications such as electrochromic devices, antistatics, electrostatic coatings, metallization of insulators, electronic circuits for cars and satellites, and electrodes for capacitors or photodiodes [117].

The PEDOT limitations include low theoretical capacitance of 210 F/g, and the practical one is even lower, i.e., 90 F/g [67,100]. It has high molecular weight, doping level of about 0.33 [67], insolubility in aqueous solvents [102], air sensitivity, and poor processability [118]. It can be synthesized by three main polymerization processes [119]:

**i.** Transition metal-mediated coupling of dihalogen derivatives of EDOT
**ii.** Electrochemical polymerization of EDOT-based monomers
**iii.** Oxidative chemical polymerization of EDOT-based monomers

Comparison of conductivities of various CPs is presented in Table 2.

The following sections discuss the overview of various metal oxide/polymer-based composites studies for supercapacitor applications.

## 5.2 Manganese oxide/polymer-based supercapacitors

Being one of the most stable oxides, manganese oxide exhibits excellent physical and chemical properties under ambient conditions along with polymorphism and structural variety being widely used in catalysis, biosensors, and energy storage [126,127]. It is notable for its pseudocapacitive property, high SC, environmental benignity, and has low cost. It has theoretical capacitance of 1370 F/g [78]. The charge accumulation process of $MnO_2$ mainly occurs near the surface area of electrode, and its capacitive performance degrades rapidly with increased thickness suggesting surface area is an indicator of device efficiency [128]. Various $MnO_2$ electrodes have been suggested based on nanostructures like nanoparticles [129], nanowires [130],

**Table 2** Comparison of electrical conductivity of various polymers [67,120–125].

| Conducting polymers | Conductivity (S/cm) |
|---|---|
| Polyaniline (PANI) | 30–200 |
| Polypyrrole (PPy) | 10–7500 |
| Poly(3,4-ethylenedioxythiophene) (PEDOT) | 0.4–400 |
| Polythiophene (PT) | 10–1000 |
| Polyacetylene (PA) | 200–1000 |
| Polyparaphenylene (PPP) | 500 |
| Polyparaphenylene sulfide (PPS) | 3–300 |
| Polyparavinylene (PPv) | 1–1000 |
| Polyisothionapthene (PITN) | 1–50 |

nanotubes [131], and hierarchical nanostructures [132]. The limitation of $MnO_2$ is its low electrical conductivity, dense morphology of oxide, higher electrolyte resistance in acidic medium, and poor full mass utilization along with poor stability at high proton-conducting mediums (which are favorable for supercapacitors) [133]. To overcome this issue, many reports [134,135] suggested adding conductive materials to $MnO_2$ including conducting polymers to form a composite electrode for supercapacitor applications.

Staiti et al. [133] synthesized hybrid supercapacitor based on $MnO_2$, activated carbon and solid polymer electrolyte Nafion. The specific capacitance (SC) obtained was 48 F/g in which one of positive electrode, $MnO_2$ showed 384 F/g, while negative electrode, carbon had 117 F/g at 5 mA/g current density. SEM micrograph, cyclic voltammetry (CV), and galvanostatic charge/discharge (GCD) curves are shown in Fig. 9. Jin et al. [136] developed composite of polypyrrole (PPy)/$MnO_2$/polypropylene fibrous films (PPf) by in situ chemical oxidation polymerization in the PPy vapors at room temperature. The SC of the cells assembled by the optimum PPy/$MnO_2$/PPf obtained was 110 F/g based on the active material. Jafri et al. [137] reported a novel binary nanocomposite based on polyaniline (PANI) and α-$MnO_2$ nanotubes (MNTs) synthesized by in situ polymerization without using any additional oxidizing agent. The hybrid nanocomposite showed a SC value of 626 F/g and energy density of 17.8 Wh/kg at a current density of 2 A/g. Lv et al. [138] synthesized 3D porous ternary composite electrode (CF-ACNT/$MnO_2$/PEDOT) prepared by electrodeposition of

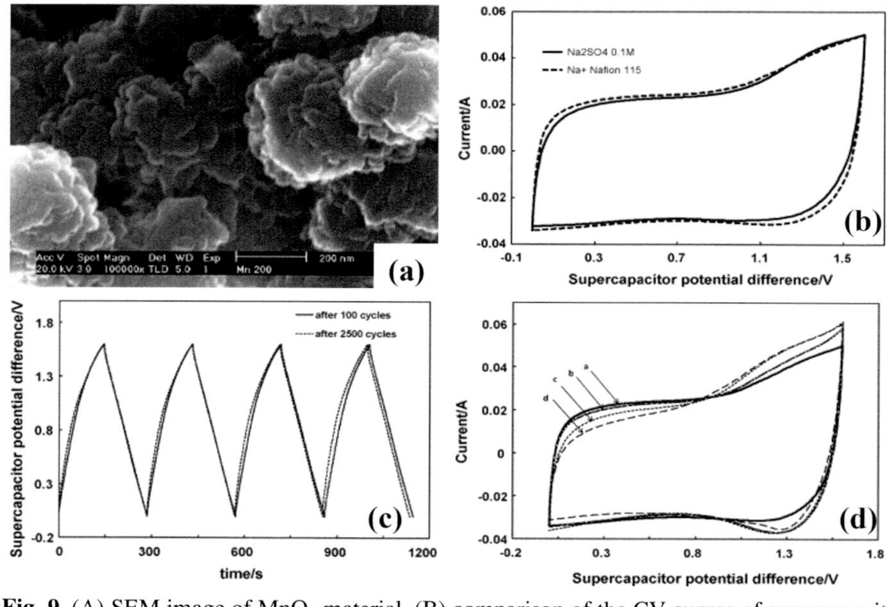


**Fig. 9** (A) SEM image of $MnO_2$ material, (B) comparison of the CV curves of supercapacitor derived at 20 mV/s in 0.1 M $Na_2SO_4$ and Nafion electrolyte, (C) GCD curves recorded after 100 and 2500 cycles, and (D) CV recorded at 20 mV/s after 100, 1000, 1700, and 2500 cycles [133].

$MnO_2$ and poly(3,4-ethylenedioxythiophene) (PEDOT) sequentially on carbon-fabric-aligned carbon nanotube (CF-ACNT) hybrids for supercapacitors. $MnO_2$ served as high charge storage medium and PEDOT as a conductive bridge. The ternary composite morphology enabled high $MnO_2$ utilization of 36% with $MnO_2$ mass loading up to 3.11 mg/cm$^{-2}$. Duay et al. [139] synthesized supercapacitor electrodes that consist of PEDOT nanowires and $MnO_2$-based materials. These electrodes showed enhanced capacitance and charge retention compared to pure materials. The flexible asymmetric supercapacitor cell outperformed symmetric cell and exhibited fast charge/discharge with superior charge retention of ca. 86%. The electrode material exhibited an energy density of 9.8 Wh/kg and power density of 850 Wh/kg. PEDOT as anode material offers a large electrochemically reversible voltage window (up to 1.7 V) due to which high energy density of the device was attained. Xiao et al. [140] synthesized two types of composites that include PEDOT-$MnO_2$ and PANI-$MnO_2$ using aqueous dispersion polymerization technique which exhibited nanorods like morphology. PEDOT-$MnO_2$ and PANI-$MnO_2$ displayed SC of 315 and 221 F/g, respectively, which was much superior to neat materials alone.

Yang et al. [141] prepared nanoparticles of PEDOT/$MnO_2$ using chemical vapor polymerization that showed high SC of 321.4 F/g at a 0.5A/g current density and retained SC of 163 F/g at much higher current density of 7 A/g along with more than 90% of charge retention after 4000 charge/discharge cycles. Choi et al. [142] synthesized stretchable electrodes for the solid-state supercapacitors using giant inserted twist to coil a nylon sewing thread that was helically wrapped with a carbon nanotube sheet on which pseudocapacitive $MnO_2$ nanofibers were electrochemically deposited. The maximum linear, areal capacitances, areal energy storage, and power densities were 5.4 mF/cm, 40.9 mF/cm$^2$, 2.6 mWh/cm$^2$, and 66.9 mW/cm$^2$, respectively, despite the engineered superelasticity of the fiber supercapacitor. He et al. [143] utilized polypyrrole nanowires (NWs) as a core and K-Birnessite-type $MnO_2$ nanofibers as a shell to construct a novel hierarchically porous $MnO_2$ NFs@PPy NW core/shell nanostructures. $MnO_2$ NFs@PPy NW core/shell nanostructures demonstrated not only high SC and good cycle stability, but also low and stable internal resistance. Moreover, the capacitances retained were 72.5% (276–200 F/g), when the current densities ranged from 2 to 20 A/g, indicating the material being suitable for working under high current attributed to synergetic effect between $MnO_2$ and PPy. PPy NWs build a reliable conductive network for fast electron transport throughout the electrode, while core/shell nanostructure significantly enhanced the active materials contact area with the electrolyte. The charge storage mechanism of these $MnO_2$@PPy involves surface adsorption/desorption and insertion/extraction of electrolyte cations and reversible doping/dedoping of anions in PPy. The assembled asymmetrical flexible supercapacitors (AFSCs) ($MnO_2$/PPy/AC) gave rise to a maximum energy density of 25.8 Wh/kg at a power density of around 901.7 W/kg and maintained 17 Wh/kg at an extremely high power density of 9000 W/kg with an excellent cycling performance. Roy et al. [134] synthesized polyaniline-manganese dioxide (PANI-$MnO_2$) composites of varying composition prepared by in situ chemical oxidative polymerization of aniline monomer in aqueous acidic medium. The SC increased with an increase of PANI wt% in the composite reaching up to 242 F/g for composite consisting of

35 wt% PANI with current density of 0.10 A/g. This increase was attributed to the contribution of conducting PANI in facilitating charge transport and energy storage. He et al. [144] devised a facile two-step strategy to construct $MnO_2$/PANI hybrid nanostructures on carbon cloth (CC) for high-performance supercapacitor applications. The vertically aligned PANI nanofiber arrays grow on CC helped in reducing the ionic diffusion path and improved utilization of electrode materials. $MnO_2$/PANI/CC-based symmetric supercapacitor displayed maximum energy density of 39.9 Wh/kg at a power density of 642 W/kg.

Relekar et al. [145] synthesized $MnO_2$/PANI composite thin films using electrochemical route for the fabrication of supercapacitor electrodes. The thin film with composition (80% PANI-20% $MnO_2$) showed better supercapacitance with SC of 409 F/g along with 92% capacitance retention after 1000 cycles. Table 3 summarizes the composition, SC, energy density, power density, and capacitance retention of $MnO_2$/polymer-based composites developed for supercapacitor applications.

## 5.3 Ruthenium oxide/polymer-based supercapacitors

Ruthenium oxide has been realized as a prime material for supercapacitor applications due to its high SC of about 1000 F/g, highly reversible redox reactions, wide potential window, high proton conductivity, low resistivity, good thermal stability, long cycle life, metallic-type conductivity, and high-rate capability [146]. The highly redox nature of ruthenium oxide arises due to the ability of $RuO_2$ to undergo various oxidation states like $Ru^{+2}$, $Ru^{+3}$, and $Ru^{+4}$ which are possible within 0–1.2 V window [147,148], where pseudocapacitance comes into play contributing capacitive current. However, the limitations of ruthenium oxide such as high cost, toxicity, and rarity pose problems for its commercial use. Moreover, a very thin layer of $RuO_2$ takes part in charge storage process, while the underlying active material largely remains inactive [149]. $RuO_2$ has also been reported to delaminate from the electrodes in acidic medium [150]. One major issue of ruthenium is that in hydrous form $RuO_2{\cdot}xH_2O$, it has low electronic conductivity about 89 S/cm which makes it difficult to balance the electronic and protonic channels for achieving a high-power supercapacitor electrode [151]. To address such issues, conducting polymers like PANI, PPy, PEDOT, polyethylene oxide, polyethylene glycol, poly(3-octylthiophene-2,5-diyl), and polymer gels have been used with $RuO_2$ to form composite electrode to enhance SC and cycling life [152].

Sopčić et al. [153] manufactured composite electrodes with amorphous and hydrous $RuO_2$ with polymeric binder Nafion or poly vinylidene fluoride (PVDF). Among the composite electrodes, the highest SC of 526 F/g was found for the electrode prepared with PVDF as a binder. Physiochemical properties of polymeric binders influence the overall behavior of composite electrodes differently. Nafion consisting of perfluorinated chains having hydrophilic sulfonic groups attached on side chains tends to block the surface of $RuO_2$, preventing deeper layers charging/discharging resulting in decreased overall capacitance of the $RuO_2$. PVDF forms separate domains within the composite electrode leaving the surface of $RuO_2$ free for the ionic transport accompanied with charging/discharging reactions. As a result, higher

**Table 3** Comparison of published literature on $MnO_2$/polymer-based supercapacitors.

| Research group | Composite detail | Method | Electrolyte | Specific capacitance (F/g) | Areal-specific capacitance (F/cm$^2$) | Energy density (Wh/kg)/power density (W/kg) | Capacitance retention (%) |
|---|---|---|---|---|---|---|---|
| He et al. [144] | $MnO_2$/PANI/CC | Chemical oxidative polymerization | 1 M $Na_2SO_4$ | 642 (at 10 mV/s) | 1.56 (at 10 mV/s) | 39.9/642 | 88.1 |
| Jaidev et al. [137] | PANI/$MnO_2$ nanotube hybrid composite | In situ polymerization | 1 M $H_2SO_4$ | 626 at a current density of 2 A/g | – | 17.8 | – |
| Yang et al. [141] | PEDOT/$MnO_2$ nanoparticles | Chemical vapor polymerization | 1 M $H_2SO_4$ | 321.4 at 0.5 A/g | – | – | 90 |
| He et al. [143] | $MnO_2$ NFs@PPy NW core/shell nanostructures | –Chemical oxidative polymerization | 1 M $Na_2SO_4$ | 271 at 2 A/g | – | 25.8/901.7 | 93.4 |
| Duay et al. [139] | PEDOT nanowires/$MnO_2$ | Co-electrodeposition | 1 M $LiClO_4$ | 114 at 0.5 A/g | – | 9.8/850 | 86 |
| Roy et al. [134] | PANI/$MnO_2$ | In situ chemical oxidative polymerization | 0.5 $Na_2SO_4$ | 242 at 0.25 A/g | – | – | – |
| Relekar et al. [145] | PANI/$MnO_2$ | Potentiostatic electrode position | 1 M $Na_2SO_4$ | 409 | – | – | 92 |

amount of $RuO_2$ material is available for charge storage, and consequently, higher energy electrodes can be constructed, but with limited ability for higher power generation. Sellam et al. [154] reported the studies on all-solid-state flexible pseudocapacitor based on poly(3,4-ethylenedioxythiophene)-poly (styrene sulfonate) (PEDOT-PSS)/$RuO_2 \cdot xH_2O$ electrode sand protic ionic liquid-based polymer. Pseudocapacitor with symmetrical configuration using PEDOT-PSS/$RuO_2 \cdot xHO$ electrodes showed enhanced SC (33 F/g) over pure PEDOT-PSS, indicating the effective role of protonation-deprotonation of $RuO_2 \cdot xH_2O$ entrapped in conducting network of PEDOT-PSS. This enhancement is attributed to the compatibility of hydrous $RuO_2$ with the proton-conducting electrolyte. It was found that incorporation of $RuO_2 \cdot xH_2O$ in PEDOT-PSS network deteriorated its high rate performance; however, the composite electrode showed its rate capability higher than many capacitive interfaces of the $RuO_2 \cdot xH_2O$ composites with carbonaceous or oxide materials. Stable capacitance of 70 F/g was found with PEDOT-PSS electrode for 1000 charge/discharge cycles after initial improvement in the SC.

Deshmukh et al. [155] constructed porous PANI-$RuO_2$ composite that was synthesized using a successive ionic layer adsorption and reaction (SILAR) method. The PANI-$RuO_2$ composite electrode exhibited a considerable SC of 664 F/g with better rate capability and stable cyclability. The PANI-$RuO_2$ composite electrode showed 89% stability after 5000th cycles. It was observed that PANI-$RuO_2$ showed less SC compared to PANI (1078 F/g), but the composite electrode showed higher stability than the PANI electrode alone. SEM micrographs and CV curves are shown in Fig. 10. Nam et al. [156] devised a composite electrode by electrodeposition of PANI on hydrous $RuO_2$ electrode and evaluated its supercapacitive properties. The surface morphology confirmed the porous structure of $RuO_2$ and the deposition of PANI on the hydrous $RuO_2$. The SC of PANI/$RuO_2$ and hydrous $RuO_2$ electrodes decreased from 708 to 295 F/g and from 517 to 240 F/g, respectively, as the scan rate increased from 5 to 200 mV/s. The results showed a promise of PANI/$RuO_2$ electrode for supercapacitor applications. CV curves of various electrodes are shown in Fig. 12A–D. Li et al. [157] reported polypyrrole (PPy)-embedded $RuO_2$ electrodes prepared by the electrodeposition of PPy particles on $RuO_2$ films. Cyclic voltammetry curves of PPy-$RuO_2$ electrodes with different deposition time of PPy particles (10, 20, 25, and 30 min) showed excellent capacitive behaviors. The SCs of 657, 553, 471, and 396 F/g were obtained at deposition time of 10, 20, 25, and 30 min, respectively. After 1000 cycles, decay of nearly 3.4%, 4.5%, 5.9%, and 7.5% in capacity values was observed and composite electrode was found to be very stable. Table 4 summarizes the composition, SC, energy density, power density, and capacitance retention of $RuO_2$/polymer-based composites developed for supercapacitor applications.

## 5.4 Copper oxide/polymer-based supercapacitors

In the last few years, copper oxides have emerged as strong prospect for energy storage devices. Though it has been exploited as positive electrode for batteries, their utilization for pseudocapacitive energy storage mechanism is something new [158]. CuO has good physiochemical properties, affordability, high abundance, and nontoxic

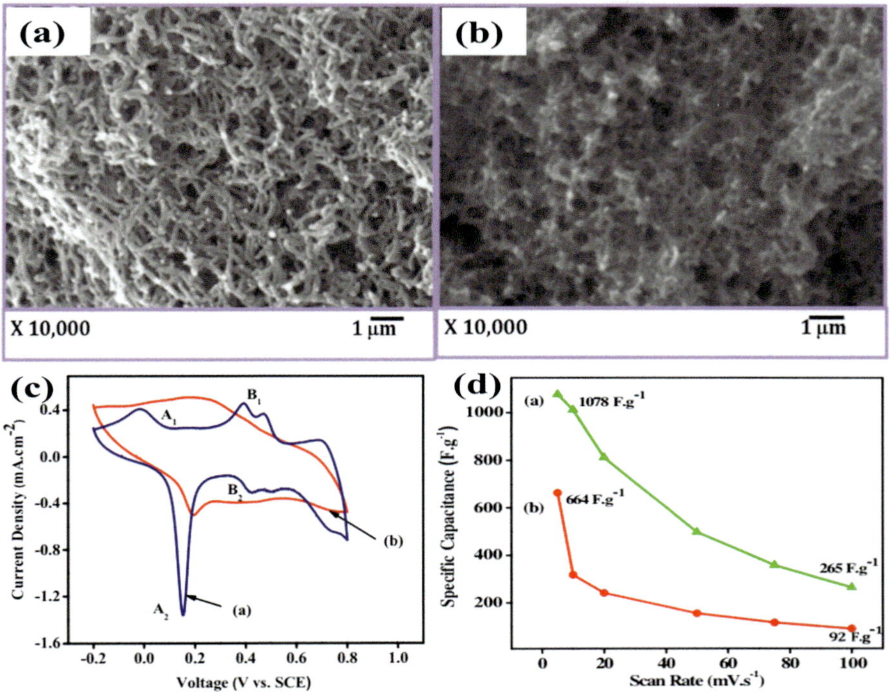


**Fig. 10** SEM images of (A) PANI and (B) PANI-$RuO_2$ composite thin films, (C) CV curves of PANI and PANI-$RuO_2$ composite thin-film electrodes in 1 M $H_2SO_4$ electrolyte, and (D) scan rate effect on SC of PANI and PANI-$RuO_2$ composite thin-film electrodes [155].

nature. Particularly, cupric oxide (CuO) and cuprous oxide ($Cu_2O$) are effective due to their semiconductor nature, especially CuO, which has narrow band gap ($E_g = 1.2\,eV$) and p-type semiconductor nature that enables it to be used for solar cells, photocatalysis, supercapacitors, and lithium-ion batteries. They are suitable for energy storage devices because of their faradic nature that occurs in both lithium-ion batteries and pseudocapacitors due to their various oxidation states (form 0 to +3) [159]. $Cu_2O_3$ has large practical SC of (3560 F/g), while CuO could deliver capacitance of 1800 F/g, a value close to the widely studied hydrous ruthenium oxide (2200 F/g) [160].

The $CuO_x$ nanoparticles have been synthesized by using various techniques like electrodeposition [161], template-free growth [135], precipitation method [162], least square-support vector [163], impregnation method [164], SILAR method [165], and chemical bath deposition [166] for supercapacitor applications. Various morphologies of CuO nanostructures have been reported such as nanoparticles (NPs) [167], nanosheets [161], nanoflowers [168], nanowires [160,169], nanoribbons [170], nanorods [168], nanoleaves [171], grasslike [172], cauliflower, and nanobelt [173], which maximizes the surface area for easy access of ions from the electrolyte.

**Table 4** Comparison of published literature of $RuO_2$/polymer-based supercapacitors.

| Research group | Composite detail | Method | Electrolyte | Specific capacitance (F/g) | Energy density (Wh/kg)/power density (W/kg) | Capacitance retention (%) |
|---|---|---|---|---|---|---|
| Li et al. [157] | PPy-$RuO_2$ | Electrodeposition of PPy on $RuO_2$ films | 0.5 M $H_2SO_4$ | 657 | – | 96.6 |
| Deshmukh et al. [155] | PANI-$RuO_2$ | SILAR | 1 M $H_2SO_4$ | 664 | 432/357 | 89 |
| Hashmi and Upadhyaya [154] | | – | PVA:PVP blend (1:4) | 57 | 2/846 | – |
| Sopčić et al. [153] | $RuO_2$/PANI | Electrochemical deposition of PANI | 0.5 M $H_2SO_4$ | 526 | – | – |
| Nam et al. [156] | $RuO_2$/PANI | Electrochemical deposition of PANI on $RuO_2$ electrode | 1 M $H_2SO_4$ | 708 | – | – |

Dubal et al. [173] synthesized cauliflower-shaped CuO particles for supercapacitor application with SC of 179 F/g in 1 M $Na_2SO_4$ electrolytic solution. Vidhyadharan et al. [169] reported that nanowire-shaped CuO acts as anodic material for supercapacitor application with 620 F/g in aqueous electrolyte. Xue et al. [174] synthesized $Cu_2O$ with a hollow octahedral structure with capacitive value of 58 F/g and $Cu_2O$ with core/shell morphology with capacitance value of 88 F/g. Patake et al. [161] reported thin-film fabrication of the $Cu_2O$ and studied as a supercapacitive electrode with the maximum SC value of 36 F/g.

There are few studies on the application of CuO as a supercapacitive material, ascribing to low conductivity and rapid capacitance reduction. These disadvantages of CuO occur due to degradation of crystal structure during the ion-extraction process [175]. For CuO-based supercapacitors, problem arises during continuous redox reaction where it shows lack of stability during cycling, volumetric shrinkage during ejection of ions (doped ions) and low conductance at dedoped state. PANI creates high ohmic polarization in supercapacitors [176]. The best way to rectify this is to use polymer-metal oxide nanocomposite. It will provide two benefits, i.e., polymer will provide flexibility, toughness, coat ability, while metal oxides offer hardness and durability. This technique also prevents the chemical swelling and shrinking whenever charge diffusion takes place from electrode-electrolyte-electrode during the cyclic process [177].

Ashokkumar et al. [176] synthesized PANI/CuO nanocomposite (PCN) using electrochemical deposition method and achieved a long cyclic stability of 4000 cycles with a SC value of 294 F/g at highest concentration of copper nanoparticles in PANI matrix. The morphology of PANI/copper nanoparticle composite changed substantially from granular shape to rodlike structure with increasing copper concentration. PANI was responsible for enhanced cyclic stability for supercapacitor applications. Chakraborty et al. [178] combined multiwalled carbon nanotubes (MWCNTs), CuO as transition metal oxide, NiO, and a conducting polymer PANI to make electrode for supercapacitor. The highest value of SC was observed for CuO@NiO/PANI/MWCNTs quaternary nanocomposite (1372 F/g at 5 mV/s scan rate) compared to all other electrodes prepared in this work. Further, the CuO@NiO/PANI/MWCNT electrode showed up to 83% capacitance retention (highest among others) after 1500 charge/discharge cycles, showing good cyclic stability, thus making it a promising electrode material for supercapacitor application [178].

Viswanathan et al. [179] synthesized reduced graphene oxide, copper oxide, and PANI (GCP) nanocomposites by in situ single-step chemical method by varying the weight composition of each of the constituents. The composite having weight percentage of G12%: $Cu_2O$/CuO40%/P48% (G12CP) displayed SC of 684.93 F/g at a current density of 0.25 A/g, energy density of 136.98 Wh/kg, power density of 1315.76 W/kg, and cyclic stability of 84% up to 5000 cycles at a scan rate of 700 mV/s. Ates et al. [180] developed composites of PANI/CuO, poly (3,4-ethylenedioxythiophene) (PEDOT)/CuO, and polypyrrole (PPy)/CuO electrochemically on glassy carbon electrodes. The SCs were determined in 0.2 M $H_2SO_4$ solution at various scan rates (5, 10, 20, 30, 40, 50, 60, 70, 80, 90, and 100 mV/s) by CV method. The highest capacitance 286.35 F/g was obtained for

PANI/CuO composite at 20 mV/s compared to 198.89 F/g for PEDOT/CuO at 5 mV/s and 20.78 F/g for PPy/CuO at 5 mV/s. The stability of these electrodes was confirmed for (500 cycles). Gholivand et al. [181] employed in situ polymerization method to develop PANI/CuO nanocomposite. The CV analysis of PANI/CuO nanocomposite and CuO was carried in 1 M $Na_2SO_4$ electrolytic solution and found SC of 185 F/g for PANI/CuO nanocomposite compared to 76 F/g of CuO at a scan rate of 5 mV/s. The rate capability of PANI/CuO nanocomposite was better than CuO alone. Majumder et al. [182] studied the impact of europium oxide ($Eu_2O_3$) on the SC of ternary composites of PPy/CuO/$Eu_2O_3$. They found that the incorporation of $Eu_2O_3$ in PPy/CuO matrix can promote charge transportation in the ternary composites by increasing porosity. Table 5 summarizes the composition, SC, energy density, power density, and capacitance retention of CuO/polymer-based composites developed for supercapacitor applications.

## 5.5 Cobalt oxide/polymer-based supercapacitors

Cobalt oxide is another naturally abundant transition metal oxide that possesses excellent reversible redox behavior, large surface area, low toxicity, stability, controllable size and shape, ecofriendliness, high conductivity, corrosion resistance, chemical durability, solar energy absorbance, diverse morphology, high surface area, high theoretical SC (3560 F/g), low cost compared to ruthenium, and durability [183]. CoO is a p-type semiconductor that exists in three different forms (CoO, $Co_2O_3$, and $Co_3O_4$) that have cubic spinel structure and has direct and indirect band gaps of 2.10 and 1.60 eV, respectively [184]. Its high capacitance owes to both pseudocapacitive-type and battery-type behaviors [185]. Due to these characteristics, $Co_3O_4$ is considered as a potential candidate for supercapacitor applications [186]. However, its practical capacitance is significantly lower than its theoretical value attributed to its slow kinetics, fast capacity decay, and impaired morphologies during its redox electrochemical reactions [187]. Its applications in supercapacitors are prevented by poor conductivity [188] and unstable solid electrolyte interface [189].

The SC of cobalt oxide heavily depends on morphology, surface area, and pore size distribution [190]. Many CoO nanostructures' (nanowalls, nanograss, nanoflowers, nanoneedles, nanoplatelets, nanocages, nanotubes, and nanosheets [191]) synthesis methods have been developed [192] like solvothermal [193], spray pyrolysis [194], electrodeposition [195], combustion synthesis method, microwave-assisted synthesis method, hydrothermal, sol-gel, and chemical precipitation [196]. Yan et al. [197] synthesized composite of $Co_3O_4$ and graphene nanosheet (GNS)/$Co_3O_4$ that showed SC of 243 F/g. Wang et al. [198] obtained $Co_3O_4$ nanorods that exhibited SC of 280 F/g. Tummala et al. [199] utilized plasma spray route to obtain electrode films with SC of 166 F/g. Gao et al. [200] fabricated aligned $Co_3O_4$ nanowires array on nickel foam via hydrothermal process and attained SC of 746 F/g. Yuan et al. [201] reported hierarchically porous $Co_3O_4$ film on nickel foam by the cathodic electrodeposition using liquid crystalline material as a template and obtained SC of 443 F/g. Zhen et al. [202] prepared $Co_3O_4$ nanoparticle-based electrode for a pseudosupercapacitor that exhibited SC of 304 F/g in 1 M KOH solution. Xia and coworkers [203] synthesized

**Table 5** Comparison of published literature of CuO/polymer-based supercapacitors.

| Research group | Composite detail | Method | Electrolyte | Specific capacitance (F/g) | Energy density (Wh/kg)/power density (W/kg) | Capacitance retention (%) |
|---|---|---|---|---|---|---|
| Ashokkumar et al. [176] | PANI/CuO | Electrochemical deposition | 1 M KCl | 424.52 | – | – |
| Viswanathan and Shetty [179] | Reduced graphene oxide/CuO/PANI | In situ single-step chemical method | 0.4 M $H_2SO_4$ | 684.93 | 136.98/1315.76 | 84 |
| Chakraborty et al. [178] | CuO@NiO/PANI/MWCNT | Mixing method | 3 M NaOH | 1372 | – | 83 |
| Majumder et al. [182] | PPY/CuO/$Eu_2O_3$ | In situ chemical oxidative polymerization | 1 M $H_2SO_4$ | 320 | 11.1/3147 | 99.86 |
| Gholivand et al. [181] | CuO/PANI | In situ polymerization | 1 M $Na_2SO_4$ | 185 | – | 75 |
| Ates et al. [180] | CuO/PANI, CuO/PEDOT, CuO/PPy | Electrochemical synthesis | 0.2 M $H_2SO_4$ | 286.35 for CuO/PANI, 198 for CuO/PEDOT, and 20.78 for CuO/PPy | – | 81.82(CuO/PANI) |

hollow mesoporous $Co_3O_4$ spheres on nickel foil through electrochemical method and fabricated supercapacitor that displayed SC of 358 F/g. Lin et al. [204] reported a maximum SC of 291 F/g using $CoO_x$ xerogel calcined at 150°C.

The low conductivity of cobalt oxide and limitation of ion diffusion are the reason for low theoretical capacitance, therefore, researchers looked up for different strategies to address this issue. One effective remedy is to combine cobalt oxide with conducting polymers like PPy, PEDOT, and PANI [205,206]. Sandhya et al. [207] designed highly oriented well-aligned nanoshrubs of polyaniline decorated $Co_3O_4$ (COP) electrode for supercapacitor application. The as-synthesized nanoshrubs offered low charge transfer resistance, high conductivity, and good electrolyte accessibility. COP with 70% PANI and 30% $Co_3O_4$ (COP 70) showed excellent performance with a SC value of 1151 F/g at 3 A/g, high rate performance of 70% at the current density ranging from 3 to 20 A/g and excellent cyclic stability (92% after 5000 charging/discharging cycles) using two symmetric electrode systems. Bashir et al. [208] prepared hybrid poly *N,N*-dimethylacrylamide (PDMA) hydrogels cross-linked by sodium montmorillonite through free radical mechanism. Magnesium trifluoromethane sulfonate ($MgTf_2$) and cobalt oxide ($Co_3O_4$) nanoparticles were added to provide the conduction pathway. The hydrogel containing $MgTf_2$ and $Co_3O_4$ nanoparticles (NPs) displayed maximum SC of 26.1 F/g at 3 mV/s and 29.48 F/g at 30 mA/g. It was able to retain 97.4% of its initial capacitance over 8000 cycles at a current density of 200 mA/g. Babu et al. [209] reported ultrahigh energy and power density of asymmetric supercapacitor (ASC) consisting of a novel porous carbon nanofiber derived from hyper-cross-linked polymers (HCP-CNF) and two-dimensional copper cobalt oxide nanosheets (CCO-NS) as the negative and positive electrodes, respectively. The fabricated CCO-NS/HCP-CNF ASC exhibited a SC of 244 F/g at a current density of 1 A/g owing to the unique porous architecture of CCO-NS and the interconnected microporous carbon skeleton with a high surface area of HCP-CNF. The ASC displayed an energy density of 25.1 Wh/kg and power density of 400 Wh/kg with capacitance retention of 91.1% after 5000 cycles in 3 M aqueous KOH solution.

Wei et al. [210] synthesized PPy/$Co_3O_4$/carbon paper (CP) ternary composites by electropolymerization of pyrrole monomers onto flower-like $Co_3O_4$ (f-$Co_3O_4$) nanoparticles coated on CP to be used as electrode material for supercapacitor application. The CV and GCD curves in 2 M KOH aqueous solution revealed that the PPy/f-$Co_3O_4$/CP ternary composite exhibited superior supercapacitive performance (398.4 F/g) than f-$Co_3O_4$/CP alone. PPy/f-$Co_3O_4$/CP also displayed excellent cycling ability with negligible loss after 1000 cycles. Abidin et al. [211] fabricated highly conductive nanofiber composite by coating polyvinyl alcohol graphene quantum dot-cobalt oxide (PVA-GQD-$Co_3O_4$) with PEDOT for supercapacitor application. The presence of cauliflower-like structure PVA-GQD-$Co_3O_4$ studied by field-emission scanning electron microscope (FESEM) revealed that PEDOT was uniformly coated on PVA-GQD-$Co_3O_4$ electrospun nanofibers. The (PVA-GQD-$Co_3O_4$-PEDOT) nanofiber composite exhibited a SC of 361.97 F/g with high energy density ranging from 16.51 to 19.98 Wh/kg and excellent power density ranging from 496.10 to 2396.99 W/kg as current density increased from 1.0 to 5.0 A/g.

Sulaiman et al. [212] synthesized PEDOT/graphene oxide/$Co_3O_4$ (PEDOT/GO/$Co_3O_4$) nanocomposite prepared using a single-step electropolymerization technique via chronoamperometry at 1.2 V. The ternary nanocomposite electrode exhibited a SC of 536.50 F/g and excellent cycle stability with charge retention of 92.69% after 2000 cycles. SEM images of ternary composites along with CV and GCD curves are shown in Fig. 11A–F.

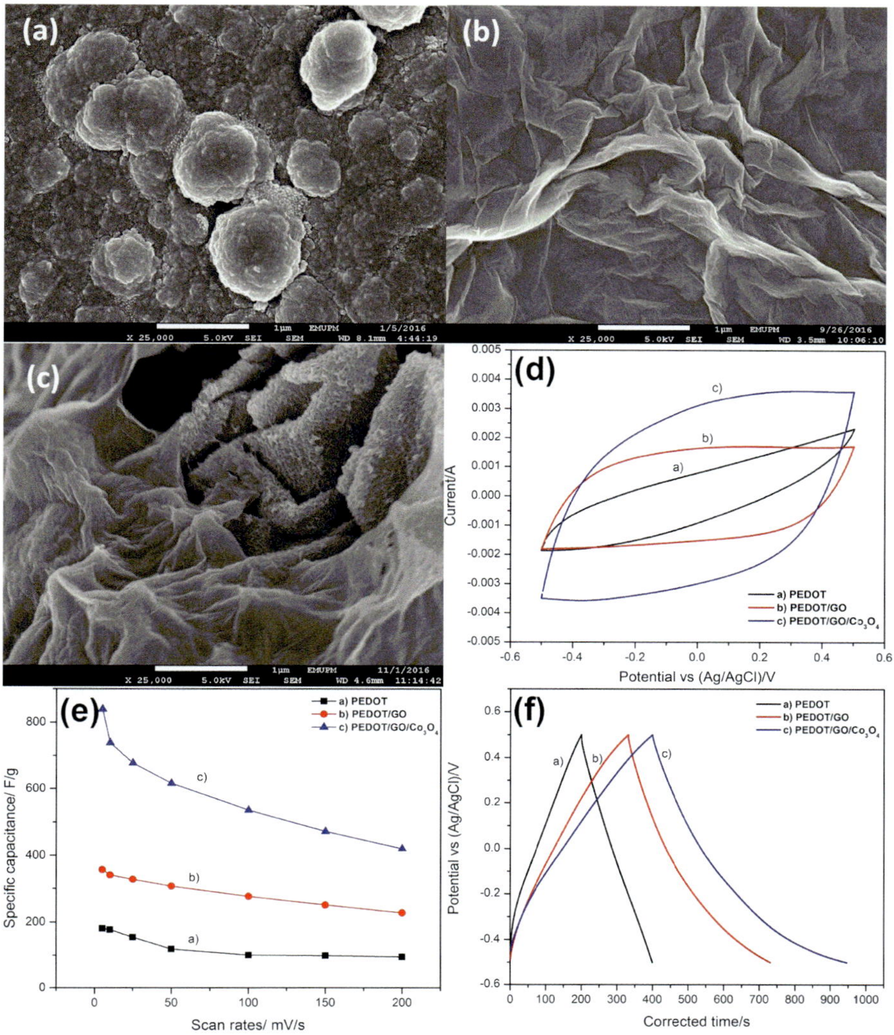


**Fig. 11** FESEM images of (A) PEDOT, (B) PEDOT/GO, and (C) PEDOT/GO/$Co_3O_4$, and (D) CV curves, (E) SC and (F) GCD curves of PEDOT, PEDOT/GO, and PEDOT/GO/$Co_3O_4$ [212].

Yang et al. [213] reported two nanocomposites with different morphologies of polypyrrole nanofibers/cobalt oxide (PPyNF/$Co_3O_4$) and polypyrrole nanotubes/$Co_3O_4$ (PPyNT/$CoO_x$) were successfully prepared via rapid and cost-efficient microwave approach. PPyNF/$Co_3O_4$ nanocomposite showed much higher SC (270.6 F/g) than PPyNT/$CoO_x$ (167.5 F/g) at the same current density of 1 A/g. Nevertheless, the PPyNT/$CoO_x$ nanocomposite exhibited excellent cycling stability (105% retention after 4000 cycles) than PPyNF/$Co_3O_4$ (100% retention after 1000 cycles) at a current density of 1 A/g. It was noteworthy that the capacitance of PPyNT/$CoO_x$ could be improved further to 278.4 F/g at 1 A/g in the initial 1200 cycles. The ASC was also prepared using activated carbon (AC) as a negative electrode and the PPy/$CoO_x$ nanocomposite as the positive electrode (PPy/$CoO_x$//AC). The CV curves PPyNF/$CoO_x$//AC and PPyNT/$CoO_x$//AC were also evaluated at a scan rate of 5–100 mV/s. The SC of PPyNT/$CoO_x$/AC was 79.2 F/g at a current density of 1 A/g with a charge retention of 60.8% as the current density improved to 10 A/g. Furthermore, 100% of initial capacitance could be maintained over 4000 cycles at 1 A/g. The ASC of PPyNF/$CoO_x$//AC presented a better capacitance of 109 F/g at 1 A/g, but that downgraded to 69.1 F/g when current density increased to 10 A/g. After 4000 successive GCD cycles at 1 A/g, 92.7% of initial capacitance was maintained. Table 6 summarizes the composition, SC, energy density, power density, and capacitance retention of cobalt oxide/polymer-based composites developed for supercapacitor applications.

## 5.6 Molybdenum oxide/polymer-based supercapacitors

In recent years, molybdenum oxide has been a center of attention among researchers for the electrochemical capacitors [214]. $MoO_3$ is a notable example of a potential prospect for anode and cathode materials for electrochemical capacitors due to its stable layered (2D) structure, rich polymorphism, excellent electrical conductivity (> $10^4$ S/cm), and high theoretical capacitance of (1117 mA h/g). It also exhibits high SC of (2700 F/g) and fast faradic redox reaction kinetics. It is also an important battery electrode material candidate due to high discharge capacity of around 300 mAh/g. $MoO_3$ has three polymorph forms such as orthorhombic $MoO_3$ ($\alpha$-$MoO_3$) which has 2D structure favorable for $Li^+$ intercalation, monoclinic ($\beta$-$MoO_3$), and hexagonal (h-$MoO_3$) [215,216]. $MoO_3$ has good gas-sensitive properties, photosensitive properties, photochromic, and electrochemical properties [217]. It has some important uses such as gas sensors, optical switching devices, energy devices, information storage, variable reflectivity lens, and efficient smart windows [218].

Currently, the synthesis of $MoO_2$/$MoO_3$ is limited to a "bottom-up method," for example, pulsed layer deposition (PLD) method, gas phase method, chemical precipitation method, sol-gel method, and hydrothermal method, which are based on $MoO_3$ powder, $Na_2MoO_4$, and $(NH_4)_2MoO_4$ [217]. All of these methods require high temperature and pressure which made the cost of production much higher. Moreover, the poor stability and surface properties of $MoO_3$ such as fast recombination of photogenerated electron–hole pairs limit photocatalytic activity [219]. Many reports on $MoO_3$ electrode showed sluggish faradic redox kinetics, rapid degradation of cycling performance, and low volumetric capacitance due to its poor intrinsic

**Table 6** Comparison of published literature of cobalt oxide/polymer-based supercapacitors.

| Research group | Composite detail | Method | Electrolyte | Specific capacitance (F/g) | Energy density (Wh/kg)/ power density (W/kg) | Capacitance retention (%) |
|---|---|---|---|---|---|---|
| Wei et al. [210] | PPy/$Co_3O_4$/carbon paper (CP) | Potentiodynamic electropolymerization | 2 M KOH | 398.4 | 18.4/2.9 | 100 |
| Sulaiman et al. [212] | PEDOT/GO/$Co_3O_4$ | Single-step electropolymerization | 1 M KCl | 535.60 | 75.5/497 | 92.69 |
| Abidin et al. [211] | PVA-GQD-$Co_3O_4$-PEDOT | Two-step electrospinning and electropolymerization | 3 M KCl | 361.97 | 19.98/ 2396.99 | 96 |
| Yang et al. [213] | PPyNF/$Co_3O_4$ and PPyNT/ $CoO_x$ | Microwave synthesis | 6 M KOH | 270.2 | 24.22/6.8 | 100 |
| Sandhya et al. [207] | Aligned nanoshrubs of PANI decorated with $Co_3O_4$ | Sol-gel route | 1 M $Na_2SO_4$ | 1151 | 54/12 | 92 |
| Bashir et al. [208] | PDMA hydrogel/$MgTF_2$/ $Co_3O_4$ | Hydrothermal method and free radical mechanism | Poly (*N*, *N*-dimethylacrylamide) composite hydrogel electrolyte | 29.48 | 2.62/24.04 | 97.4 |

conductivity ($10^{-5}$ S/cm) [220]. The performance of $MoO_3$ can be enhanced by various methods, such as forming a core/shell structure or mixing $MoO_3$ with carbon materials like graphene and CNTs or enhancing the conductivity of $MoO_3$ through ethanol reduction, ion-intercalation, or oxygen/nitrogen doping [220,221]. Guan et al. [222] coated $MoO_3$ on $TiO_2$ nanotubes exhibited a SC up to 74.9 F/g at 5 mV/s in 1 M KCl solution. Shakir et al. [223] reported $MoO_3$ nanowires by dissolving $MoO_2$ in $H_2O_2$ and mixing $SnCl_2$ with $MoO_3$ to fabricate $MoO_3$ nanowires decorated with $SnCl_2$ in order to get better performance. Kumar et al. [224] demonstrated that maximum SC of 250 F/g at a current density of 0.25 A/g could be obtained with hexagonal-$MoO_3$ rods. Dhanabal et al. [225] reported RGO/h-$MoO_3$ nanocomposite for SC application which exhibited SC of 134 F/g at a current density of 1 A/g with outstanding cyclic stability of 2000 cycles. Zhang et al. [226] synthesized 3D porous graphene sheets/$MoO_2$ nanohybrids that displayed SC of 356 F/g at the current density of 0.1 A/g in KOH solution.

To improve the electrochemical performance of $MoO_3$, one effective way is to coat it with conducting polymers [227]. Xia et al. synthesized reduced graphene oxide/$MoO_3$/PANI ternary composite with a SC of 553 F/g in 1 M $H_2SO_4$ at a scan rate of 1 mV/s and found good cycling stability after 200 cycles [228]. Jiang et al. [229] fabricated a large scale of $MoO_3$/PANI coaxial heterostructure nanobelts via a simple and green approach without any surfactant. The assembled PANI conducting layer on the surface of the well-crystallized $\alpha$-$MoO_3$ nanobelts (prepared by hydrothermal method [230]) was carried out using ammonium persulfate (APS) as an oxidant by in situ polymerization at room temperature. The as-synthesized coaxial heterostructure nanobelts demonstrated high SC of 714 F/g at a scan rate of 1 mV/s and 632 F/g at a current density of 1 A/g with capacity retention of 76.7% after 3000 cycles showing good cycling stability. CV, CGD, and EIS of $MoO_3$/PANI and $\alpha$-$MoO_3$ nanobelts are presented in Fig. 12. Das et al. [231] fabricated PANI/$MoO_3$/graphene nanoplatelets (GNP) (PMG) composite electrode for the next-generation supercapacitor through in situ polymerization of aniline in the presence of $MoO_3$ and graphene nanoplatelets. They obtained a fiber-like morphology of PANI coated on GNP/$MoO_3$ composite and also intercalated within the graphene layers, thereby restricting the restacking of GNP. The PMG possessed SC of 734 F/g at 10 mV/s, 593 F/g at 1 A/g, and 92.4% charge retention after 1000 charging/discharging cycles.

Bian et al. [232] synthesized a novel ternary graphene/molybdenum oxide/poly(-*p*-phenylenediamine) nanocomposite (GMP) via two-step process including the generation of binary graphene/$MoO_3$ composite through a hydrothermal method and chemical polymerization of p-phenylenediamine monomer. The GMP electrode exhibited SC of 1042.6 F/g at 1 A/g and 418 F/g at 1 A/g in two electrode systems, while energy density reached 24.56 and 6.8 Wh/kg at a power density of 325 and 3263 W/kg, respectively. It showed negligible decay of 13.3% of initial capacitance after 3000 cycles revealing excellent cycle stability. Hsu et al. [233] reported synthesis of core/shell structure of carbon-coated $MoO_3$ (C/$MoO_3$) by simple hydrothermal approach and used in situ method to fabricate high-performance nanocomposite with polypyrrole (PPy) and graphene nanoribbon (GNR). The PPy/C/$MoO_3$ nanoparticle/

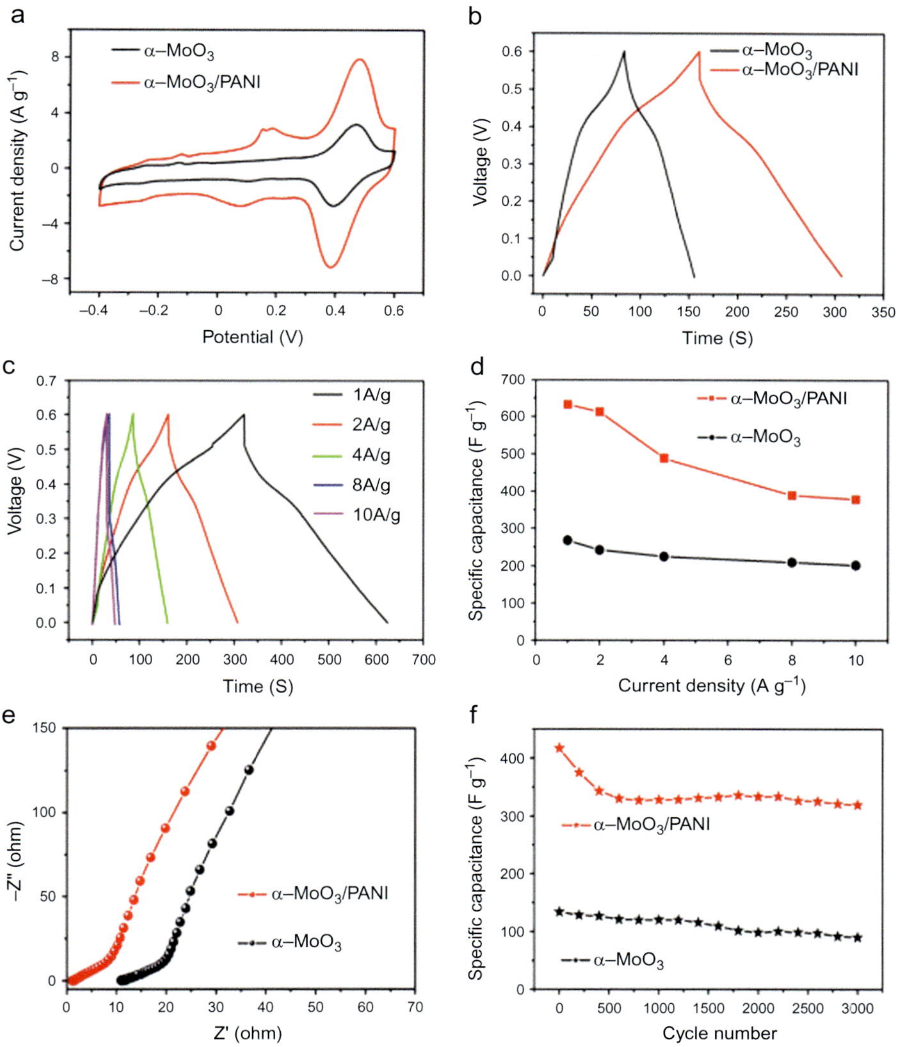


**Fig. 12** (A) CV curves showing comparison of the synthesized $MoO_3$/PANI coaxial heterostructure nanobelts and α-$MoO_3$ nanobelts at a scan rate of 5 mV/s, (B) comparative GCD curves at a current density of 2 A/g, (C) GCD curves of the $MoO_3$/PANI coaxial heterostructure nanobelts at different current densities, (D) SC of $MoO_3$/PANI and α-$MoO_3$ materials at different current densities, (E) comparative EIS spectra of two materials, and (F) cycling performance of the two materials for 3000 cycles at a scan rate of 50 mV/s [229].

GNR nanocomposite exhibited the highest SC of about 991 F/g at 5 mV/s scan rate in a 1 M $H_2SO_4$ electrolyte and also retained a high capacitance of 92.1% after 1000 charge/discharge cycles. Liu et al. [234] synthesized molybdenum trioxide (α-$MoO_3$) nanobelts coated with PPy fabricated by in situ polymerization route. The energy density of SC based on virgin $MoO_3$ was 24 Wh/kg at the power density of 150 W/kg and faded to 4.5 Wh/kg at 1.6 W/kg. In contrast, energy density of $MoO_3$/PPy nanocomposite was 20 Wh/kg at the power density of 75 W/kg. However, supercapacitor based on PPy-$MoO_3$ nanocomposite as the anode, activated carbon as the cathode, and 0.5 M $K_2SO_4$ aqueous solution as electrolyte exhibited better rate capability as well as cycling stability than virginal $MoO_3$. Table 7 summarizes the composition, SC, energy density, power density, and capacitance retention of molybdenum oxide/polymer-based composites developed for supercapacitor applications.

## 5.7 Strontium oxide/polymer-based supercapacitors

Due to high electronic conductivity, strontium oxide has gained wide attention among transition metal oxides. References [235, 236] reported in-depth research about electrical, thermal, and electrochemical properties of stoichiometric strontium.

To the best of our knowledge, a very few researchers integrated conducting polymers with strontium oxide (SrO) for supercapacitors application. Iqbal et al. [237] synthesized SrO using sonochemical method followed by calcination, and the device was fabricated by combining SrO and activated carbon (AC). SrO was integrated with PANI and polyaniline/graphene (PANI/Gr) for making composite electrodes. Three symmetric electrodes (SrO/AC, SrO/PANI/AC, and SrO/PANI/Gr/AC) were analyzed for electrochemical properties. The SC of SrO, SrO/PANI, and SrO/PANI/Gr were found to be 206, 393, and 417 F/g, respectively, and the SrO/PANI/Gr showed the highest capacitance.

For symmetric devices, SrO/PANI/Gr/AC exhibited the highest SC of 111 F/g whereas SrO/AC showed 62 F/g, while SrO/PANI/AC possessed 70 F/g at 0.4 A/g current density. SrO/PANI/Gr//AC exhibited the highest energy density of 40 Wh/kg followed by SrO/PANI/AC with 25 Wh/kg and SrO/AC with 22 Wh/kg at a specific power of 319, 321, and 320 Wh/kg, respectively, at a current density of 0.4 A/g. After the 2500 galvanic charge/discharge cycles, the SrO/PANI/Gr/AC displayed the highest charge retention of 97%. They suggested SrO/PANI/Gr//AC electrode as a strong candidate for SC applications. Iqbal et al. [238] reported an innovative route for the synthesis of SrO nanorods integrated with PANI (SrO/PANI) by using a physical blending method and fabricated high-performance asymmetric supercapacitor. The SrO/PANI nanocomposite formed possessed maximum specific capacity of 258 F/g at a current density of 0.8 A/g. The supercapattery device exhibited an energy density of 24 Wh/kg and maximum power density of 2240 W/kg. Moreover, the device displayed excellent cyclic stability with charge retention of 114% after continuous 3000 galvanostatic charge/discharge cycles. CV curves of various electrodes are shown in Fig. 13.

Table 8 summarizes the composition, SC, energy density, power density, and capacitance retention of SrO/polymer-based composites developed for supercapacitor applications.

## 5.8 Titanium oxide/polymer-based supercapacitors

Among transition metal oxides, $TiO_2$ is one of the most promising support materials for the next-generation supercapacitors because of its characteristics: good dielectric property, ideal capacitive response (rectangular CV curve), high surface area with

**Table 7** Comparison of published literature of molybdenum oxide/polymer-based supercapacitors.

| Research group | Composite detail | Method | Electrolyte | Specific capacitance (F/g) | Energy density (Wh/kg)/power density (W/kg) | Capacitance retention (%) |
|---|---|---|---|---|---|---|
| Liu et al. [234] | ($\alpha$-$MoO_3$) nanobelts coated with PPy | In situ polymerization | 0.5 M $K_2SO_4$ | 110 | 20/75 | 83 |
| Jiang et al. [229] | $MoO_3$/PANI coaxial heterostructure nanobelts | In situ polymerization | 1 M $H_2SO_4$ | 714 | – | 76.7 |
| Das et al. [231] | PANI/$MoO_3$/graphene nanoplatelets | In situ polymerization | 1 M $H_2SO_4$ | 593 | 99.68/550.4 | 92.4 |
| Li et al. [232] | ternary graphene/molybdenum oxide/poly(p-phenylenediamine) nanocomposite | Hydrothermal method and chemical polymerization | 1 M $H_2SO_4$ | 1042.6 | 24.56/325 | 86.7 |
| Hsu et al. [233] | PPy/C/$MoO_3$ nanoparticle/GNR | Hydrothermal and In situ synthesis | 1 M $H_2SO_4$ | 991 | – | 92.1 |

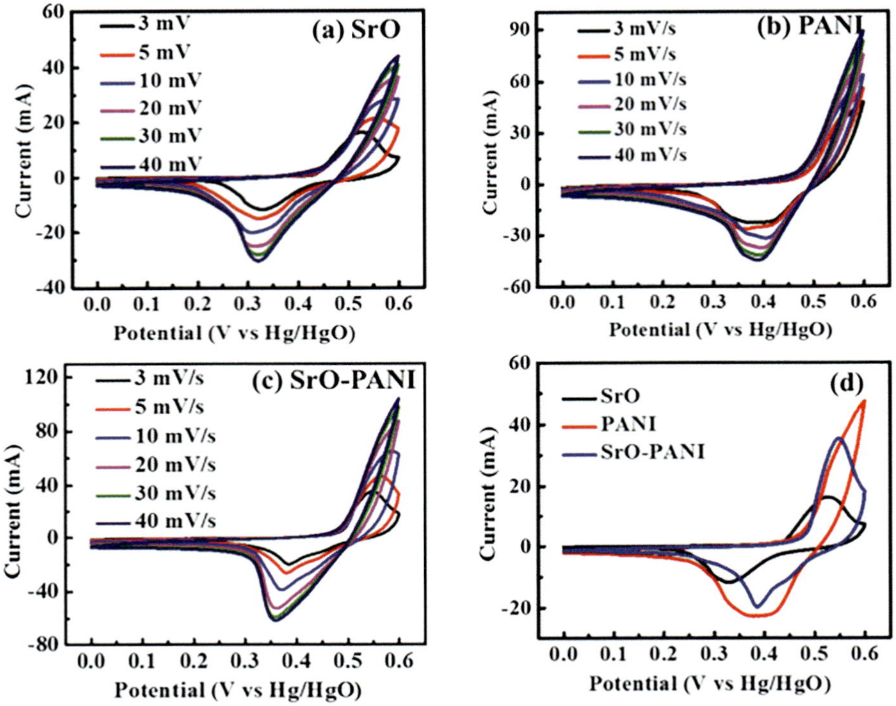


**Fig. 13** CV curves of (A) SrO NPs, (B) PANI flakes, (C) SrO-PANI composites in 1 M KOH, and (D) comparative CV curves at 3 mV/s [238].

**Table 8** Comparison of published literature of strontium oxide/polymer-based supercapacitors.

| Research group | Composite detail | Method | Electrolyte | Specific capacitance (F/g) | Energy density (Wh/kg)/power density (W/kg) | Capacitance retention (%) |
|---|---|---|---|---|---|---|
| Iqbal et al. [237] | SrO/PANI/Gr | Sonochemical method | 1 M KOH | 417 | 40/319 | 97 |
| Iqbal al. [238] | SrO/PANI/Gr/AC | Physical blending | 1 M KOH | – | 24/2240 | 114 |

normalized areal capacitance of 90–120 μF/cm$^2$, high theoretical capacitance (1206 F/g) due to multivalent vacancies, abundancy, nontoxicity, physical and chemical stability, excellent optoelectronic property, biocompatibility, and small volume change during charging/discharging process. $TiO_2$ is an n-type semiconductor with high charge density on the surface. It has wide applications in solar cells, photocatalysis, photo-/electrochromic devices, sensors, and lithium-ion batteries [239].

During voltage cycling in air or aqueous electrolyte, Ti-ion oxidation states switch between $Ti^{+3}$ and $Ti^{+4}$ resulting in an outward diffusion of oxygen in the native titanium oxide surface layer and thus an increase of oxygen vacancies resulting in high supercapacitive behavior [240].

The $TiO_2$ is a widely investigated supercapacitor material as single or comaterial with other metal oxides such as activated carbon [241]. $TiO_2$ has several polymorphs like $TiO_2$ (monoclinic), anatase (tetragonal), brookite (orthorhombic), and rutile (tetragonal) in which the first two exhibit high electrochemical activity over the other two polymorphs [242]. As one-dimensional nanostructures, nanotubes, nanorods, and nanowires provide unidirectional and rapid electron transfer pathways [243]. Many researchers have tried to manipulate $TiO_2$ surface morphology by doping and studied other methods to develop binder-free and core/shell hybrid electrodes to improve its energy and power density [244,245]. Sol-gel and hydrothermal methods were used to fabricate $TiO_2$ nanoparticles, while solvothermal, spin coating, sputtering, chemical vapor deposition, and electrodeposition have been employed for the immobilization of $TiO_2$ nanoparticles on substrates [246].

$TiO_2$ has low electrical conductivity ($>10^{-12}$ S/cm), low theoretical capacity, wide electronic band gap (3.2 eV), and high resistivity that would lead to strong internal resistance of charge storage that hinders its application for supercapacitor devices [247]. The SC of $TiO_2$ along with other metal oxides decreases with increasing scan rate [248]. $TiO_2$ conductivity can be increased through introducing metallic [249] and nonmetallic impurities [250] into oxide that can generate donor or acceptor states in the band, hence increases charge carrier concentration and hydrogenation [251]. $TiO_2$ has been mixed with various conductive agents like metals, graphene, carbon frameworks, CNTs, and conducting polymers to improve its conductance and charge storage capacity [252].

Mujawar et al. [253] synthesized vertically aligned PANI nanotubes using electrochemical polymerization. They demonstrated that by optimizing electropolymerization conditions of PANI within nanospores of titania nanotube (TNT) template, one can produce tailored PANI nanotubes. The key parameter to get nanotubular structure was the polymer growth rate vs monomer diffusion flux. PANI nanotubes having diameter of c. 200 nm with low disparity were prepared. SC of 740 F/g was obtained for PANI nanotube structures (measured at charge/discharge rate of 3 A/g) with 87% of original capacitance that was retained over 1100 cycles. A specific power of 3000 W/kg was obtained at the specific energy of 220 Wh/kg, while specific energy of 194 Wh/kg at a specific power of 7000 W/kg. Kim et al. [254] explored a composite based on the incorporation of titanium dioxide/single-walled carbon nanotube (SWCNT) complex during polymerization of hollow nanotubes of polypyrrole. The ternary composite showed SC of 282 F/g and specific energy density of 1 Wh/kg that was the result of synergistic interaction of components during polymerization of polymeric hallow nanotubes and core/shell structures of SWCNTs/polypyrrole decorated with titanium dioxide nanoparticles. Gottam et al. [255] made a stable electrode material for supercapacitor by selecting polyaniline and titanium dioxide ($TiO_2$) hybrid material. The composite material in cell configuration exhibited a SC of 320 F/g at 0.33 A/g charge current density. They found equivalent series resistance and efficiency independent of the cycle number. The composite showed a capacitive retention of 83.5%. Singu et al. [256] reported PANI sulfate salt titanium dioxide composite (PANI-$H_2SO_4$·$TiO_2$) by chemical in situ polymerization of aniline in the presence of $TiO_2$. The effect of anionic surfactant (sodium lauryl sulfate) was observed which, during polymerization, converted to dodecyl hydrogen sulfate (DHS) in the presence of acidic medium and got doped onto polyaniline along with sulfuric acid dopant, i.e., formation of polyaniline-sulfate-dodecyl hydrogen sulfate-titanium dioxide composite (PANI-$H_2SO_4$-DHS.$TiO_2$). In PANI-$H_2SO_4$-DHS.$TiO_2$ system, the nanoparticles of $TiO_2$ (10–20 nm) were embedded on nanofibers (20–60 nm) of PANI-$H_2SO_4$-DHS, and some part of PANI-$H_2SO_4$-DHS formed core/shell morphology, where $TiO_2$ was in core and PANI-$H_2SO_4$-DHS as shell. The SCs at low and high current densities were found to be 280 and 205 F/g, respectively, and after 1700 charge/discharge cycles, their retention in supercapacitor was found to be the same (65%–66%) with coulombic efficiency of (98%–100%). Azizi et al. [257] reported a novel reduced graphene oxide/poly(1,5-dihydroxynaphthalene)/$TiO_2$ (RGO/PDHN/$TiO_2$) ternary nanocomposite electrochemically synthesized on gold electrodes for supercapacitor applications. The RGO/PDHN/$TiO_2$ exhibited a much higher SC of 556 F/g than RGO/PDHN (432 F/g) and PDHN (223 F/g) at 2.4 A/g current density. Capacitance retention of 73% after 1700 cycles was obtained for RGO/PDHN/$TiO_2$.

Thakur et al. [258] reported a ternary composite of polythiophene (PTP)/PANI/$TiO_2$ synthesized using in situ oxidative polymerization. Material design involved the blending of two polymers named PANI and PTP and further incorporating $TiO_2$ particles in blended polymer matrix. The ternary composite exhibited SC of 265 F/g at 1 A/g in 1 M $H_2SO_4$ solution. The hybrid composite showed energy density of 9.09 Wh/kg at 1 A/g, power density of 3770 Wh/kg at a current density of 10 A/g, and charge retention of 92.3%. Xiao et al. [259] synthesized $TiO_2$ nanoarray (TNTA) which was first impregnated with ethanol solution containing aniline monomer; then, PANI-$TiO_2$ nanotube array (PANI-TNTA) composite electrode was prepared through polymerization of aniline monomer using ammonium persulfate (APS) as the initiator. The composite electrode exhibited SC of 816.7 F/g and a maximum energy density of 1410 J/g. The capacitive retention of the sample was over 95% as the current density increased from 0.1 to 5 mA/cm$^2$ and the charge retention was over 97% after 5000 cycles when tested in $H_2SO_4$ solution. Kim et al. [254] reported vertically oriented titanium oxide nanotube/polypyrrole (PPy) nanocomposites to increase SC of $TiO_2$-based energy storage devices. To increase its electrical storage capacity, titanium oxide nanotube was coated with PPy which elevated the SC of titanium oxide nanotube-based supercapacitor system (247 F/g) due to increased surface area and additional pseudocapacitance. Deshmukh et al. [260] developed a unique and cost-effective chemical route for synthesis of a polyaniline-titanium dioxide (PANI-$TiO_2$) composite thin film at room temperature. The PANI-$TiO_2$ composite showed a SC of 783 F/g at the scan rate of 5 mV/s. It also showed the stability of 78% after 5000 cycles. The specific power, specific energy, and coulombic efficiency were found to be 1.83 kW/kg, 391.5 Wh/kg, and 96%, respectively. Wang et al. [261] developed reduced graphene oxide (RGO)/PANI/urchin-like mesoporous $TiO_2$ sphere (UMTS) composite assembled by an in situ polymerization of PANI with UMTS (synthesized by hydrothermal method) as template/inset into the GO layer/reduction of GO and redoping process. The maximum SC of the composite was as high as 464 F/g at a current density of 0.62 A/g. In addition, the material has a high energy density of 34 Wh/kg at a high power density of 3720 W/kg with excellent cyclic stability (13.5% capacitance loss after 2000 cycles at a current density of 12.4 A/g).

Yu et al. [262] reported that PANI/$TiO_2$ nanoweb composite fabricated through electrochemical deposition and electrospinning techniques for supercapacitor application. The PANI/$TiO_2$ composite film displayed three-dimensional hierarchical micro-/nanoarchitecture. The composite electrode showed a SC of 306.5 F/g at a scan rate of 20 mV/s with a capacitive retention being 103% after 500 cycles at a scan rate of 50 mV/s. Thakur et al. [263] developed polythiophene (PTP)-$TiO_2$ composite prepared via oxidative polymerization process using varying ratios of $TiO_2$ content. UV-Vis study revealed that there was a successive decrease in the band gap (2.7–2.7 eV) of PTP-$TiO_2$ with increasing $TiO_2$ content. The highest SC recorded was 250 F/g for PTP-$TiO_2$ (10:2) composite at the current density of 1 A/g, while the energy recorded was 5.54 Wh/kg at a power density of 263.8 W/kg.

Xiao et al. [264] synthesized PANI/$TiO_2$ nanotube arrays composite electrodes by depositing PANI onto $TiO_2$ nanotube arrays through cyclic voltammetry method. The composite fabricated in ethanol solvent exhibited an excellent electrochemical performance in $Na_2SO_4$ aqueous solution and its areal capacitance was 6.25 mF/cm$^2$ at the scan rate of 50 mV/s. The maximum energy density of active material obtained was 727.3 J/g with a capacitive retention of 85% at a current density of 5 mA/cm$^2$ and capacitive retention was greater than 95% after 5000 cycle. Singh et al. [265] reported improved specific capacity of PANI by incorporating minimum amount of $TiO_2$ nanoparticles where the structure could facilitate an efficient access to electrolyte ions to the electrode surface and shorten the ion diffusion path. They used very low temperature (0–5°C) synthesis of PANI-n$TiO_2$ composite by in situ chemical oxidative chemical polymerization. Electron microscopy showed a well-dispersed network of PANI-n$TiO_2$ nanofibers and their presence in the composite was confirmed by the EDX analysis (Fig. 14A–G). The PT5 (PANI with 5 wt% $TiO_2$ nanoparticles composite) possessed a highest SC of 813 F/g at 1 mA/cm$^2$. It displayed specific energy of 16.8 Wh/kg at maximum power of 9.75 W/kg. Aslan et al. [266] reported that $TiO_2$ nanoparticles electrospunned with polyacrylonitrile (PAN) polymer solution onto the discharged battery coal (DBC) electrode and produced (PAN+$TiO_2$/DBC) composite electrode. It showed SC of 156 F/g while also exhibited stability performance over 2000 cycles with a high energy density of 21.66 Wh/kg.

Table 9 summarizes the composition, SC, energy density, power density, and capacitance retention of titanium oxide/polymer-based composites developed for supercapacitor applications.

## 5.9 Vanadium oxide/polymer-based supercapacitors

Vanadium-based materials are suitable candidates for energy storage applications due to their high energy/power density, high SC and long cycle life, high conductivity, and good electrochemical reversibility [267]. Vanadium has five valence electrons that can participate in bonding. Due to various oxidation states (+5, +4, +3, and +2),

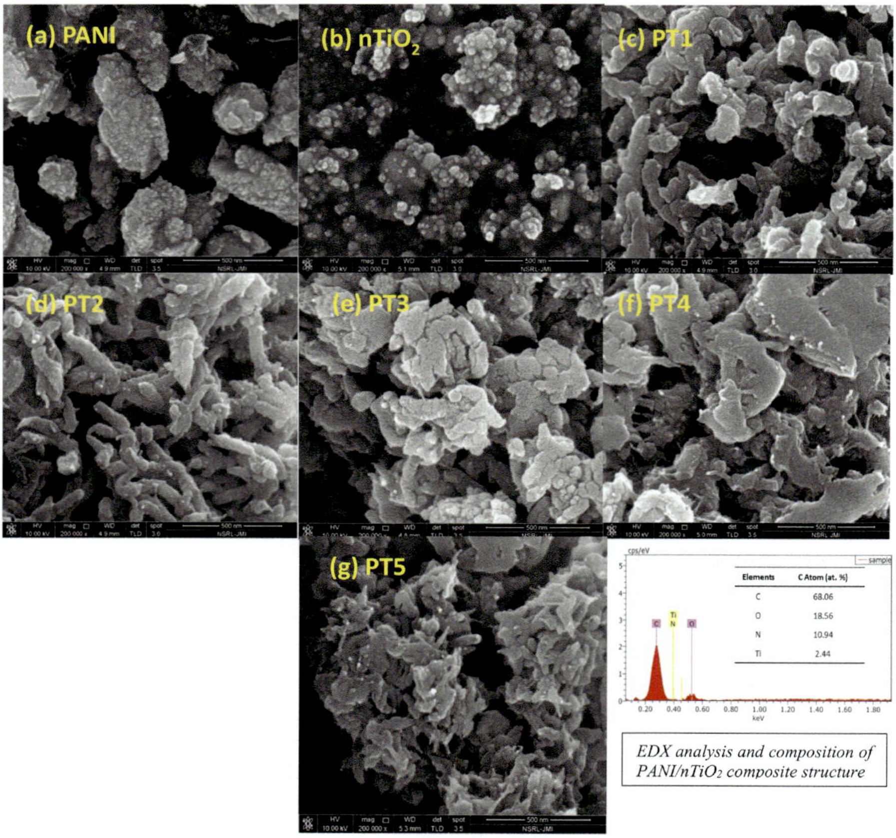


**Fig. 14** FESEM images of as-synthesized (A) PANI, (B) $nTiO_2$, (C) PT1, (D) PT2, (E) PT3, (F) PT4, and (G) PT5 composite nanostructures [265].

vanadium oxide can provide excellent pseudocapacitance. However, the stability of vanadium compounds depends on the valence state, i.e., compounds having +5 valence state are the most stable, while having +2 and +4 have poor stability [268]. Vanadium oxide has 13 different phases varying stoichiometrically including structures forming compounds like $VO_2$, $V_2O_5$, $V_2O_3$, $V_6O_{11}$, and $V_7O_{13}$ making this material a perfect choice for energy storage materials [269]. Vanadium oxides are very promising for lithium-ion battery electrodes. They are comprised of distinct surface morphology with a consecutively layered structure that gives rise to pseudocapacitance as it gives room to $Li^+$ ions for reversible intercalation as well as good catalytic activity for energy conversion [54,270].

Among vanadium's different compounds, $V_2O_5$ has been widely investigated as electrode material for supercapacitor material owing to its multiple valences ($V^{+2}$, $V^{+3}$, $V^{+4}$, and $V^{+5}$), low cost, low toxicity, abundancy, ease of synthesis, environmental friendliness, stable crystal structure, 3D architecture, having a wide layered

**Table 9** Comparison of published literature of titanium oxide/polymer-based supercapacitors.

| Research group | Composite details | Method | Electrolyte | Specific capacitance (F/g) | Energy density (Wh/kg)/power density (W/kg) | Capacitance retention (%) |
|---|---|---|---|---|---|---|
| Aslan et al. [266] | PAN + $TiO_2$ nanocomposite fibers | Sol-gel and electrospinning | 6 M KOH | 156 | 21.66 | – |
| Deshmukh et al. [260] | PANI-$TiO_2$ thin film | In situ polymerization | 1 M $H_2SO_4$ | 783 | 391.5/183 | 78 |
| Wang et al. [261] | (RGO)/PANI/urchin-like mesoporous $TiO_2$ spheres | In situ polymerization | 1 M HCl | 464 | 34/3720 | 86.5 |
| Yu et al. [262] | PANI/$TiO_2$ nanoweb | Electrochemical deposition and electrospinning | 1 M $H_2SO_4$ | 306.5 | 51.5/1430 | 103 |
| Thakur et al. [263] | Polythiophene (PTP)-$TiO_2$ | In situ oxidative polymerization | 1 M $H_2SO_4$ | 250 | 5.54/263.8 | – |
| Thakur et al. [258] | polythiophene (PTP)/PANI/ $TiO_2$ | In situ oxidative polymerization | 1 M $H_2SO_4$ | 265 | 9.09/3770 | 92.3 |
| Xiao et al. [264] | PANI/$TiO_2$ nanotube arrays | Cyclic voltammetry | 0.5 M $Na_2SO_4$ | – | 727.3 | 95 |
| Azizi et al. [257] | graphene oxide /poly (1,5-dihydroxynaphthalene)/ $TiO_2$ | Electrochemical | 1 M $HClO_4$ | 556 | – | 74 |
| Xiao et al. [259] | PANI-$TiO_2$ nanotube arrays | In situ microcavity polymerization | 0.5 M $H_2SO_4$ | 816.7 | 1410.8 | 98.6 |
| Singh et al. [265] | PANI-$TiO_2$ nanoparticle composite | In situ chemical oxidative polymerization | 0.1 M $H_2SO_4$ | 813 | 16.8/9.75 | 91.39 |

structure of $V_2O_5$ stacking perpendicular to c-axis through van der Waals interaction, wide potential window, high faradic activity, high SC in basic electrolytes and has high theoretical SC (883.3 F/g working in a potential range of 1.2 V, 2120 F/g at a potential window of 1 V) [271–273]. Different physical forms of vanadium pentoxide such as fine amorphous, crystalline, and porous structure for electrochemical capacitors have been investigated [274,275]. Pseudocapacitive behavior of vanadium oxide can be understood as a reaction where monovalent electrolyte cation ($M^+$) intercalates into the $V_2O_5$ involving the reduction of V(V) to V(IV) while preserving the oxide structure [276,277]:

$$(V_2O_5)_{surface} + 2M^+ + 2e^- \leftrightarrow (M_2V_2O_5)_{surface}$$

A wide range of 0D, 1D, and 3D nanostructured $V_2O_5$ such as nanotubes, nanofibers, nanorods, nanoflowers, nanoarrays, nanobelts, hollow/microspheres, quantum dots, and nanowires have been synthesized for faster ion transportation using hydrothermal/solvothermal reactions, thermal decomposition, chemical vapor deposition, microwave-assisted synthesis, template-assisted growth, electrospinning, and sol-gel reactions, and many other studies reported these as electrode material [269,278–280]. According to literature, $V_2O_5$ showed a better ion diffusion rate and improved energy and power densities compared to bulk material [281]. It has wide applications that include solar cells, sensors, electrical and optical switching devices, electrochromic and thermochromic devices, actuators, sensors, and photocatalyst and electrochemical capacitors [282–284].

$V_2O_5$ has slow rate stability and poor cycling performance, owing to low electronic conductivity ($10^{-2}$ to $10^{-3}$ S/cm), slow electrochemical kinetics and structural instability, which are considered as a huge obstacle in its utilization as electrode material for supercapacitor application [285,286]. To address these obstacles, composites of vanadium oxides with highly conductive materials including graphene, reduced graphene oxide, carbon nanofiber, carbon nanotube, carbonaceous materials, metal fibers, and conducting polymers were reported. Such hybrid composite electrodes (e.g., $V_2O_5$/CNT, $V_2O_5$/RGO, and $V_2O_5$/PPy) could resolve the conductivity issue and enhance SC [287,288].

Mak et al. [289] synthesized vanadium pentoxide nanofibers (VNF) and polyaniline nanofibers (PANF) using electrospinning and rapid chemical polymerization routes, respectively. The electrochemical performance was analyzed in symmetrical (VNF-VNF, PANF-PANF) and asymmetrical (PANF-VANF) configuration. The asymmetrical device exhibited a high working potential of 2 V resulting in a high energy density of 21.7 Wh/kg at a power density of 0.22 kW/kg. It also possessed capacitive retention of 73% up to 2000 cycles. Bai et al. [290] reported electrochemical codeposition of vanadium pentoxide and polypyrrole (PPy) conducted from vanadyl sulfate ($VOSO_4$) and pyrrole in their aqueous solution to get $V_2O_5$-PPy (VPy) composite electrode, in which one-dimensional growth of polypyrrole was directed. They tested the composite electrode using symmetrical configuration VPy//VPy and found high SC of 412 F/g at 4.5 mA/cm$^2$ and high energy density of 82 Wh/kg at a power density of 800 W/kg with 2 V working voltage. It also retained

80% of capacitive retention over 5000 galvanostatic charge/discharge cycles. Bai et al. [291] synthesized a high-performance negative $V_2O_5$-Polyaniline composite electrode by electro-codeposition. It displayed a wide charge storage potential window of 1.6 V and a high capacitance of 443 F/g (664.5 mF/cm$^2$) at a current density of 0.5 mA/cm$^2$. The high performance of composite electrode was believed to be due to the formation of dense 1D nanowires without aggregation. In symmetric configuration, it displayed a maximum energy density of 69.2 Wh/kg at a current density of 0.5 mA/cm$^2$ with power density of 0.72 kW/kg at a current density of 5 mA/cm$^2$. It also exhibited charge retention of 92% after 5000 cycles. Karaca et al. [292] reported one-step electrochemical synthesis of polypyrrole-vanadium oxide (PPy-VO$_x$) composite on the vanadium-intercalated pencil graphite (PG) surface in an acrylonitrile solution with the presence of carboxymethyl cellulose (CMC). Owing to the intercalation of vanadium on PG surface PPy-VO$_x$-CMC composite coatings provided higher SC of 800 F/g (5.3 A/g or 1 mA/cm$^2$) at mass loading of 0.19 mg/cm$^2$ (excluding the capacitance of bare graphite). The SC decreased with an increase in mass loading. The asymmetric configuration delivered energy density of 18 Wh/kg at a power density of 0.43 kW/kg at 0.5 A/g in the potential range of 1.2 V for a mass loading of 20 mg/cm$^2$. SEM images of composite electrodes along with CV and GCD curves of various electrodes are shown in Fig. 15A–D.

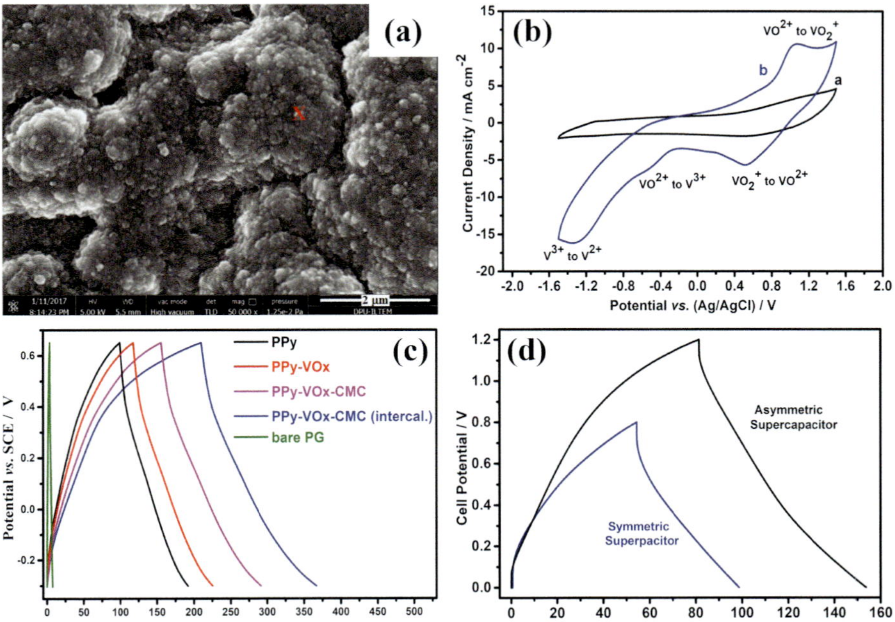


**Fig. 15** SEM images of (A) PPy-VOx-CMC on V-intercalated PG surface, (B) CV of PG electrode in an acetonitrile solution containing TBABF$_4$ and VO(BF$_4$)$_2$, (C) GCD curves recorded in 100 mM $H_2SO_4$ solution at 1 mA/cm$^2$ for various electrodes, and (D) galvanostatic charge/discharge curves recorded at 1.5 A/g for symmetric and asymmetric supercapacitor. Karaca, Pekmez [292].

Yasoda et al. [293] used $V_3O_7$ as electrode material for supercapacitor applications because of its multiple oxidation states. $V_3O_7$ nanoparticles were coated on reduced graphene oxide (rGO) to form a peroxovanadate-rGO composite by the thermal micro-explosive decomposition method. By coating conducting polymer polyaniline on $V_3O_7$ decorated rGO using in situ polymerization, a nanocomposite electrode ($V_3O_7$-rGO-PANI with HCl dopant) was developed which gave brush-like 1D nanostructure. It exhibited SC of 579 F/g at 0.2 A/g current density. It showed 95% charge retention over 2500 cycles. They believed that the synergistic effect of polyaniline conducting fibers helped in enhancing the capacitance, while $V_3O_7$-rGO was responsible for stability. Bi et al. [294] demonstrated that oxygen vacancies (Vö) played a crucial role in energy storage materials. Oxygen vacancy-enriched vanadium pentoxide/poly (3,4-ethylenedioxythiophene) (Vö-$V_2O_5$/PEDOT) nanocables were prepared through the one-pot oxidative polymerization of PEDOT. PEDOT was used to tune the concentrations of Vö in $V_2O_5$ that was confirmed by X-ray absorption near-edge structure (XANES) and X-ray photoelectron spectroscopy (XPS). The Vö-$V_2O_5$/PEDOT nanocables with an overall concentration of 1.3% Vö displayed energy density of 85 Wh/kg with a power density of 250 W/kg and 105% capacitive retention after 10,000 cycles at 5 A/g which was ascribed to synergy between Vö and PEDOT. Bi et al. [295] investigated different kinds of conducting polymers (CPs) including PEDOT-, PPy-, PANI-coated $V_2O_5$ nanofibers with Vö generated at the interface during polymerization process and Vö induced a local electric field at the interface between $V_2O_5$-NFs and CP coating. The SC of Vö-$V_2O_5$/PEDOT, Vö-$V_2O_5$/PANI, and Vö-$V_2O_5$/PPy was found to be 614, 523, and 437 F/g, respectively, at 0.5 A/g, while at a power density of 2500 W/kg, the energy density of Vö-$V_2O_5$/PEDOT, Vö-$V_2O_5$/PPy, and Vö-$V_2O_5$/PANI was 85, 73, and 61 Wh/kg, respectively. The Vö-$V_2O_5$/CP-based supercapacitors possessed excellent cycling stability over 15,000 cycles.

Bi et al. [296] synthesized $V_2O_5$/poly(3,4-ethylenedioxythiophene) nanocables with oxygen vacancies gradually decreasing from the surface to core (G-$V_2O_5$/PEDOT nanocables) and studied as electrode materials for supercapacitors. The G-$V_2O_5$/PEDOT nanocable-based supercapacitor displayed high SC of 614 F/g and energy density of 85 Wh/kg at a power density of 250 W/kg in neutral aqueous electrolyte. It also exhibited long cycling life with 122% capacitance retention after 50,000 cycles as oxygen vacancies migrated and redistributed during charge/discharge process. Asen et al. [297] reported a ternary composite comprising graphene oxide (GO), polypyrrole (PPy), and vanadium pentoxide ($V_2O_5$) obtained via one-step electrochemical deposition process on a stainless steel substrate conducted from an aqueous solution containing vanadyl acetate, pyrrole, and GO to get $V_2O_5$/PPy/GO nanocomposite. The $V_2O_5$/PPy/GO nanocomposite exhibited SC of 750 F/g at a current density of 5 A/g and retained 83% of its initial capacitance after 3000 cycles. The symmetric configuration of $V_2O_5$/PPy/GO nanocomposite electrode possessed energy density of 27.6 and 22.8 Wh/kg at a power density of 3600 and 13,680 W/kg, respectively.

Table 10 summarizes the composition, SC, energy density, power density, and capacitance retention of vanadium oxide/polymer-based composites developed for supercapacitor applications.

**Table 10** Comparison of published literature of vanadium oxide/polymer-based supercapacitors.

| Research group | Composite detail | Method | Electrolyte | Specific capacitance (F/g) | Energy density (Wh/kg)/power density (W/kg) | Capacitance retention (%) |
|---|---|---|---|---|---|---|
| Karaca et al. [292] | polypyrrole-vanadium oxide (PPy-$VO_x$) composite | One-step electrochemical synthesis | $HBF_4$/ACN/Adiponitrile | 800 | 18/430 | 87 |
| Yasoda et al. [293] | $V_3O_7$-rGO-PANI | Thermal microexplosive decomposition and in situ polymerization | 1 M $H_2SO_4$ | 579 | – | 94 |
| Bi et al. [294] | Vö-$V_2O_5$/PEDOT | One-pot oxidative polymerization | 1 M $Na_2SO_4$ | 614 | 85/250 | 105 |
| Bi, W., et al. [295] | Vö-$V_2O_5$/PEDOT, Vö-$V_2O_5$/PANI and Vö-$V_2O_5$/PPy | Oxidative polymerization | 1 M $Na_2SO_4$ | 614 (Vö-$V_2O_5$/PEDOT), 523 (Vö-$V_2O_5$/PANI) and 437 (Vö-$V_2O_5$/PPy) | 85, 73 and 61 at 2500 W/kg | – |
| Bi et al. [296] | G-$V_2O_5$/PEDOT nanocables | In situ oxidative polymerization | 1 M $Na_2SO_4$ | 614 | 85/250 | 122 |
| Asen et al. [297] | $V_2O_5$/PPy/GO | Electrochemical deposition | 0.5 M $Na_2SO_4$ | 750 | 27.6/3600 | 83 |

# 6 Comparison of different metal oxide/polymer-based composites

Comparison of best performing metal oxide/polymer composites is presented in Table 11. Among $MnO_2$/polymer composites, the best performance was displayed by $MnO_2$/PANI/CC ternary composite. The performance of this ternary composite was attributed to nanosized array of PANI fibers that helped in increasing the electrolyte/electrode interface area, shortening the diffusion path of electrolyte and preventing the agglomeration of $MnO_2$ nanoparticles. The deposition of $MnO_2$/PANI on carbon cloth (CC) eased electron transport to current collector without polymer binder resulting in decreased ohmic polarization and increased rate capability. The combination of MnO, PANI, and CC in the composite electrode enhanced electrical conductivity and reduced the internal resistance of the composite electrode. Similar to ternary composite, Jaidev et al. [137] reported PANI/$MnO_2$ nanotube hybrid composite that gave an almost similar performance as $MnO_2$/PANI/CC. In the former composite, the $MnO_2$ nanotube (MNT) incorporation in PANI matrix improved the dissolution of MNT in acid electrolyte and promoted electrochemical activity. The study suggests that $MnO_2$-based electrodes' stability in acidic electrolyte can be enhanced by developing composites with PANI or conducting polymers. Furthermore, $MnO_2$/polymer-based electrodes or $MnO_2$ nanofibers/PPy nanowire-based electrodes showed excellent charge storage capacity which was reported by He et al. [143]. This enhanced charge storage capacity of electrodes is mainly attributed to PPy nanowires, which provided conducting network for fast electron transfer while $MnO_2$ nanostructure provided high surface area and good electrode-electrolyte interaction. Assembled supercapacitor device showed maximum energy density of 25.8 Wh/kg at 901.7 W/kg power density along with excellent cyclic stability.

$RuO_2$-based electrodes despite their high cost have shown a significant promise for energy storage applications. The most promising composite electrode was found to be $RuO_2$/PANI. Electrodeposited PANI on $RuO_2$ electrode showed maximum SC of 708 F/g which is superior among other reported electrodes. This higher charge storage capacity is because of highly porous structure of $RuO_2$ which is verified by SEM analysis [156]. Chakraborty et al. [178] prepared quaternary composites based on CuO@NiO/PANI/MWCNTs as electrode material which showed maximum SC of 1372 F/g. This higher SC is attributed to highly porous structure and synergetic effect of various components that provided lower charge transfer resistance as verified by EIS analysis. Furthermore, $Co_3O_4$ nanoshrubs coated with PANI showed enhanced SC of 1151 F/g at 3 A/g current density. Higher charge storage capacity is mainly attributed due to high conductivity, low charge transfer resistance and excellent electrode-electrolyte interaction of nanoshrubs that were developed due to well-aligned structure of composite [207]. $MoO_3$ is an important metal oxide for supercapacitor applications due to its excellent electrical conductivity and high theoretical SC of c. 2700 F/g. Bian et al. [232] developed ternary GMP composites using hydrothermal technique. It showed excellent charge storage capacity. Superior charge storage properties of these composites are attributed to pseudocapacitive behavior of

**Table 11** Comparison of published literature of various metal oxide/polymer-based supercapacitors.

| Research group | Composite detail | Method | Electrolyte | Specific capacitance (F/g) | Energy density (Wh/kg)/power density (W/kg) | Capacitance retention (%) |
|---|---|---|---|---|---|---|
| He et al. [144] | $MnO_2$/PANI/CC | Chemical oxidative polymerization | 1 M $Na_2SO_4$ | 642 (at 10 mV/s) | 39.9/642 | 88.1 |
| Nam et al. [156] | $RuO_2$/PANI | Electrochemical deposition of PANI on $RuO_2$ electrode | 1 M $H_2SO_4$ | 708 | – | – |
| Chakraborty et al. [178] | CuO@NiO/PANI/MWCNTs | Mixing method | 3 M NaOH | 1372 | – | 83 |
| Sandhya et al. [207] | Aligned nanoshrubs of PANI decorated with $Co_3O4$ | Sol-gel route | 1 M $Na_2SO_4$ | 1151 | 54/12 | 92 |
| Li et al. [232] | Ternary graphene/molybdenum oxide/poly(p-phenylenediamine) nanocomposite | Hydrothermal method and chemical polymerization | 1 M $H_2SO_4$ | 1042.6 | 24.56/325 | 86.7 |
| Iqbal et al. [237] | SrO/PANI/Gr | Sonochemical method | 1 M KOH | 417 | 40/319 | 97 |
| Xiao et al. [259] | PANI-$TiO_2$ nanotube arrays | In situ microcavity polymerization | 0.5 M $H_2SO_4$ | 816.7 | 1410.8 | 98.6 |
| Karaca et al. [292] | polypyrrole-vanadium oxide (PPy-$VO_x$) composite | One-step electrochemical synthesis | $HBF_4$/ACN/Adiponitrile | 800 | 18/430 | 87 |

$MoO_3$, binding capacity of poly(*p*-phenylenediamine) that enhanced capacitance retention of composite electrodes and high electrical conductivity imparted by graphene. Among various composite of SrO, SrO/PANI/Gr composites showed maximum SC of 417 F/g which is excellent among other SrO composites [237]. High performance of SrO/PANI/Gr composites is attributed to fast faradic reactions of composites. $TiO_2$ is another candidate of supercapacitor electrode due to high surface area and availability of multivalent vacancies. Xiao et al. [264] developed PANI/$TiO_2$-based electrodes by depositing PANI on $TiO_2$ using CV. Concentration of PANI had a significant effect on charge storage capacity of electrodes. These electrodes also exhibited a wide potential window of −0.2 to 1.8 V. PANI not only provided binding to the electrodes but also imparted excellent conductivity to composites that were responsible for high charge storage capacity of electrodes. Vanadium oxide, due to its multiple oxidation states, can provide excellent pseudocapacitive behavior along with much higher energy and power density [267]. Among $V_2O_5$ composites, PPy-$VO_x$-CMC composite showed excellent charge storage capacity, which is attributed to intercalation of V to composite structure while carboxymethyl improved charge retention of electrodes. Furthermore, improvement in capacitance is attributed to synergetic effects between PPy and $VO_x$.

# 7 Conclusions

This chapter gives a comprehensive overview of the supercapacitive performance of various metal oxide/conducting polymer-based composites. The presented studies showed that drawbacks of metal oxides such as low electrical conductivity and low cyclic stability can be well compensated by developing their composites with conducting polymers. Conducting polymers not only provide binding but also impart excellent electrical conductivity to composite electrodes required for energy storage. Among all the reported composite electrodes, a ternary composite of CuO@NiO/PANI/MWCNTs showed excellent charge storage capacity via pseudocapacitive mechanism due to multiple valence state of CuO. CuO imparted high surface area to composite electrodes; PANI provided both binding ability and electrical conductivity and MWCNTs offered both electrical conductivity and higher surface area for charge storage. The superior supercapacitive performance of composite electrodes is attributed to the synergetic effects between polymers and metal oxides.

This overview showed that a combination of metal oxides along with conducting polymers can provide a good combination of high pseudocapacitive behavior, excellent binding ability, and enhanced electrical conductivity which is suitable for charge storage applications.

# References

[1] W.E. Outlook, Electricity, 2019, Available from: *https://www.iea.org/reports/world-energy-outlook-2019*.

[2] M.S. Dresselhaus, I.L. Thomas, Alternative energy technologies, Nature 414 (6861) (2001) 332–337.

[3] J. Smeets, A Realistic Future of Solar Fuels Assessing Possibilities of Solar Liquid Hydrocarbon Fuels Via Photovoltaics and Electrolysis and their Potential Application in the Transportation Sector, Utrecht University, 2020.
[4] A. Berrueta, Supercapacitors: electrical characteristics, modeling, applications, and future trends, IEEE 7 (2019) 50869–50896.
[5] J. Li, et al., Facile preparation of nitrogen/sulfur co-doped and hierarchical porous graphene hydrogel for high-performance electrochemical capacitor, J. Power Sources 345 (2017) 146–155.
[6] K. Parida, et al., Fast charging self-powered electric double layer capacitor, J. Power Sources 342 (2017) 70–78.
[7] L. Zhang, et al., A comparative study of equivalent circuit models of ultracapacitors for electric vehicles, J. Power Sources 274 (2015) 899–906.
[8] I. Shown, et al., Conducting polymer-based flexible supercapacitor, Energy Sci. Eng. 3 (1) (2015) 2–26.
[9] J. Libich, et al., Supercapacitors: properties and applications, J. Energy Storage 17 (2018) 224–227.
[10] M.V. Kiamahalleh, et al., Multiwalled carbon nanotubes based nanocomposites for supercapacitors: a review of electrode materials, Nano 7 (02) (2012) 1230002.
[11] M.S. Halper, J.C. Ellenbogen, Supercapacitors: A Brief Overview, The MITRE Corporation, McLean, Virginia, USA, 2006, pp. 1–34.
[12] Z. Stojek, The electrical double layer and its structure, in: Electroanalytical Methods, Springer, 2010, pp. 3–9.
[13] H. Helmholtz, Uber einige Gesetze der Vertheilung elektrischer Strome in korperlichen Leitem, mit Anwendung auf die thierisch-elektrischen Versuche, Ann. Phys. Chem. 89 (1853) 21.
[14] R. Sudhakar, M. Krishnappa, Mesoporous materials for high-performance electrochemical supercapacitors, in: Mesoporous Materials-Properties and Applications, IntechOpen, 2019.
[15] P. Sharma, T. Bhatti, A review on electrochemical double-layer capacitors, Energy Convers. Manag. 51 (12) (2010) 2901–2912.
[16] M.G. Bakker, et al., Perspectives on supercapacitors, pseudocapacitors and batteries, Nanomater. and Energy 1 (3) (2012) 136–158.
[17] B.E. Conway, Electrochemical Supercapacitors: Scientific Fundamentals and Technological Applications, Springer Science & Business Media, 2013.
[18] J. Garthwaite, How Ultracapacitors Work (and Why They Fall Short), Earth2Tech, GigaOM Network, 2011.
[19] M.A. Scibioh, B. Viswanathan, Materials for Supercapacitor Applications, Elsevier, 2020, p. 400.
[20] C.-M. Chuang, et al., Effects of carbon nanotube grafting on the performance of electric double layer capacitors, Energy Fuels 24 (12) (2010) 6476–6482.
[21] Z.-z. Zhu, et al., Fabrication and electrochemical characterization of polyaniline nanorods modified with sulfonated carbon nanotubes for supercapacitor applications, Electrochim. Acta 56 (3) (2011) 1366–1372.
[22] E. Frackowiak, et al., Nanotubular materials for supercapacitors, J. Power Sources 97 (2001) 822–825.
[23] Y. Jiang, J. Liu, Definitions of pseudocapacitive materials: a brief review, Energy Environ. Mater. 2 (1) (2019) 30–37.
[24] N.S.A. Manaf, M.S.A. Bistamam, M.A. Azam, Development of high performance electrochemical capacitor: a systematic review of electrode fabrication technique based on different carbon materials, ECS J. Solid State Sci. Technol. 2 (10) (2013) M3101.

[25] K. Engelsmann, W. Lorenz, E. Schmidt, Underpotential deposition of lead on polycrystalline and single-crystal gold surfaces: part I. thermodynamics, J. Electroanal. Chem. Interfacial Electrochem. 114 (1) (1980) 1–10.

[26] F. Will, C. Knorr, Investigation of formation and removal of hydrogen and oxygen coverage on platinum by a new, nonstationary method, Z. Elektrochem. 64 (1960) 258–259.

[27] B.E. Conway, Electrochemical oxide film formation at noble metals as a surface-chemical process, Prog. Surf. Sci. 49 (4) (1995) 331–452.

[28] B. Conway, Two-dimensional and quasi-two-dimensional isotherms for Li intercalation and UPD processes at surfaces, Electrochim. Acta 38 (9) (1993) 1249–1258.

[29] J. Liu, et al., Advanced energy storage devices: basic principles, analytical methods, and rational materials design, Adv. Sci. 5 (1) (2018) 1700322.

[30] N.R. Chodankar, et al., True meaning of pseudocapacitors and their performance metrics: asymmetric versus hybrid supercapacitors, Small 16 (37) (2020) 2002806.

[31] V. Augustyn, P. Simon, B. Dunn, Pseudocapacitive oxide materials for high-rate electrochemical energy storage, Energy Environ. Sci. 7 (5) (2014) 1597–1614.

[32] V. Augustyn, et al., High-rate electrochemical energy storage through Li+ intercalation pseudocapacitance, Nat. Mater. 12 (6) (2013) 518–522.

[33] F.-X. Ma, et al., Self-supported formation of hierarchical $NiCo_2O_4$ tetragonal microtubes with enhanced electrochemical properties, Energy Environ. Sci. 9 (3) (2016) 862–866.

[34] A. Muzaffar, et al., A review on recent advances in hybrid supercapacitors: design, fabrication and applications, Renew. Sust. Energ. Rev. 101 (2019) 123–145.

[35] H.-W. Wang, et al., Design and synthesis of $NiCo_2O_4$-reduced graphene oxide composites for high performance supercapacitors, J. Mater. Chem. 21 (28) (2011) 10504–10511.

[36] C. Liu, et al., Exploiting high-performance anode through tuning the character of chemical bonds for Li-ion batteries and capacitors, Adv. Energy Mater. 7 (1) (2017) 1601127.

[37] F. Zhang, et al., A high-performance supercapacitor-battery hybrid energy storage device based on graphene-enhanced electrode materials with ultrahigh energy density, Energy Environ. Sci. 6 (5) (2013) 1623–1632.

[38] I. Plitz, et al., The design of alternative nonaqueous high power chemistries, Appl. Phys. A 82 (4) (2006) 615–626.

[39] A.G. Pandolfo, A.F. Hollenkamp, Carbon properties and their role in supercapacitors, J. Power Sources 157 (1) (2006) 11–27.

[40] W. Zhang, Y.H. Qu, L.J. Gao, Performance of PbO2/activated carbon hybrid supercapacitor with carbon foam substrate, Chin. Chem. Lett. 23 (5) (2012) 623–626.

[41] M. Soumya, et al., Electrochemical performance of PbO2 and PbO2–CNT composite electrodes for energy storage devices, J. Nanosci. Nanotechnol. 15 (1) (2015) 703–708.

[42] W. Sun, et al., Few-layered $Ni(OH)_2$ nanosheets for high-performance supercapacitors, J. Power Sources 295 (2015) 323–328.

[43] J. Zhu, et al., A novel electrochemical supercapacitor based on $Li_4Ti_5O_{12}$ and LiNi1/3Co1/3Mn1/3O2, Mater. Lett. 115 (2014) 237–240.

[44] J. Li, et al., Asymmetric supercapacitors with high energy and power density fabricated using $LiMn_2O_4$ nano-rods and activated carbon electrodes, Int. J. Electrochem. Sci. 12 (2) (2017) 1157–1166.

[45] Y. Wang, Y. Song, Y. Xia, Electrochemical capacitors: mechanism, materials, systems, characterization and applications, Chem. Soc. Rev. 45 (21) (2016) 5925–5950.

[46] M. Mastragostino, C. Arbizzani, F. Soavi, Conducting polymers as electrode materials in supercapacitors, Solid State Ionics 148 (3–4) (2002) 493–498.

[47] C. Lei, C. Lekakou, Carbon-based nanocomposite EDLC supercapacitors, in: Nanotechnology: Advanced Materials, CNTs, Particles, Films Composites, vol. 1, TechConnect Briefs, 2010, pp. 176–179.

[48] S.H. Kwon, et al., Preparation of nano-porous activated carbon aerogel using a single-step activation method for use as high-power EDLC electrode in organic electrolyte, J. Nanosci. Nanotechnol. 16 (5) (2016) 4598–4604.
[49] S. Ye, et al., Characterization of expanded graphene nanosheet as additional material and improved performances for electric double layer capacitors, J. Ind. Eng. Chem. 43 (2016) 53–60.
[50] Y. Maletin, et al., Electrochemical double layer capacitors and hybrid devices for green energy applications, Green 4 (1–6) (2014) 9–17.
[51] Z.-S. Wu, et al., High-energy MnO2 nanowire/graphene and graphene asymmetric electrochemical capacitors, ACS Nano 4 (10) (2010) 5835–5842.
[52] N. Parveen, et al., Intercalated reduced graphene oxide and its content effect on the supercapacitance performance of the three dimensional flower-like $\beta$-$Ni(OH)_2$ architecture, New J. Chem. 41 (18) (2017) 10467–10475.
[53] Z. Wu, et al., Transition metal oxides as supercapacitor materials, in: Nanomaterials in Advanced Batteries and Supercapacitors, Springer, 2016, pp. 317–344.
[54] M. Aulice Scibioh, B. Viswanathan, Electrode Materials for Supercapacitors, Elsevier, 2020, pp. 35–204 (Chapter 3).
[55] F. Shi, et al., Metal oxide/hydroxide-based materials for supercapacitors, RSC Adv. 4 (79) (2014) 41910–41921.
[56] R. Holze, Spectroelectrochemistry of conducting polymers, in: Handbook of Advanced Electronic and Photonic Materials and Devices, Academic Press, 2001, pp. 171–222.
[57] R. Holze, Metal Oxides in Supercapacitors, Technische Universität Chemnitz, Chemnitz, Germany, 2017, p. 219.
[58] R. Holze, Metal oxide/conducting polymer hybrids for application in supercapacitors, in: Metal Oxides in Supercapacitors, Elsevier, 2017, pp. 219–245.
[59] M. Mastragostino, et al., in: B. Scrosati, W. van Schalkwi (Eds.), Advances in Lithium-Ion Batteries, first ed., Springer US, 2002.
[60] H. Letheby, XXIX.—On the production of a blue substance by the electrolysis of sulphate of aniline, J. Chem. Soc. 15 (1862) 161–163.
[61] H. Wang, J. Lin, Z.X. Shen, Polyaniline (PANi) based electrode materials for energy storage and conversion, J. Sci.: Adv. Mater. Devices 1 (3) (2016) 225–255.
[62] C. Oueiny, S. Berlioz, F.-X. Perrin, Carbon nanotube–polyaniline composites, Prog. Polym. Sci. 39 (4) (2014) 707–748.
[63] I.Y. Sapurina, M. Shishov, Oxidative polymerization of aniline: molecular synthesis of polyaniline and the formation of supramolecular structures, in: New Polymers for Special Applications, vol. 740, IntecOpen Book Series, 2012, p. 272.
[64] V. Ivanova, et al., The sorption of influenza viruses and antibiotics on carbon nanotubes and polyaniline nanocomposites, J. Phys. Conf. Ser. 291 (2011) 012004. (IOP Publishing). In press.
[65] S.J. Park, C.S. Park, H. Yoon, Chemo-electrical gas sensors based on conducting polymer hybrids, Polymers 9 (5) (2017) 155.
[66] R.J. Tseng, et al., Polyaniline nanofiber/gold nanoparticle nonvolatile memory, Nano Lett. 5 (6) (2005) 1077–1080.
[67] K. Lota, V. Khomenko, E. Frackowiak, Capacitance properties of poly (3, 4-ethylenedioxythiophene)/carbon nanotubes composites, J. Phys. Chem. Solids 65 (2–3) (2004) 295–301.
[68] H. Li, et al., Theoretical and experimental specific capacitance of polyaniline in sulfuric acid, J. Power Sources 190 (2) (2009) 578–586.
[69] Y. Lee, et al., Grafting of polyaniline onto the radiation crosslinked chitosan, Macromolecules 33 (2000) 7341.

[70] A. Khan, et al., Synthesis by in situ chemical oxidative polymerization and characterization of polyaniline/iron oxide nanoparticle composite, Polym. Int. 59 (12) (2010) 1690–1694.
[71] T. Taka, EMI shielding measurements on poly (3-octyl thiophene) blends, Synth. Met. 41 (3) (1991) 1177–1180.
[72] F. Jonas, G. Heywang, Technical applications for conductive polymers, Electrochim. Acta 39 (8–9) (1994) 1345–1347.
[73] V. Misoska, et al., Polypyrrole membranes containing chelating ligands: synthesis, characterisation and transport studies, Polymer 42 (21) (2001) 8571–8579.
[74] B. Wessling, J. Posdorfer, Corrosion prevention with an organic metal (polyaniline): corrosion test results, Electrochim. Acta 44 (12) (1999) 2139–2147.
[75] R. Gangopadhyay, A. De, Conducting polymer composites: novel materials for gas sensing, Sensors Actuators B Chem. 77 (1–2) (2001) 326–329.
[76] N.Y. Abu-Thabit, Chemical oxidative polymerization of polyaniline: a practical approach for preparation of smart conductive textiles, J. Chem. Educ. 93 (9) (2016) 1606–1611.
[77] P. Röse, U. Krewer, S. Bilal, An amazingly simple, fast and green synthesis route to polyaniline nanofibers for efficient energy storage, Polymers 12 (10) (2020) 2212.
[78] S. Bhadra, N.K. Singha, D. Khastgir, Electrochemical synthesis of polyaniline and its comparison with chemically synthesized polyaniline, J. Appl. Polym. Sci. 104 (3) (2007) 1900–1904.
[79] G.-R. Li, et al., Electrochemical synthesis of polyaniline nanobelts with predominant electrochemical performances, Macromolecules 43 (5) (2010) 2178–2183.
[80] H. Wei, et al., Electropolymerized polyaniline nanocomposites from multi-walled carbon nanotubes with tuned surface functionalities for electrochemical energy storage, J. Electrochem. Soc. 160 (7) (2013) G3038.
[81] M. Fahim, S. Bilal, Highly stable and efficient performance of binder-free symmetric supercapacitor fabricated with electroactive polymer synthesized via interfacial polymerization, Materials 12 (10) (2019) 1626.
[82] H.J. Kim, et al., Phytic acid doped polyaniline nanofibers for enhanced aqueous copper (II) adsorption capability, ACS Sustain. Chem. Eng. 5 (8) (2017) 6654–6664.
[83] S. Xing, et al., Morphology and conductivity of polyaniline nanofibers prepared by 'seeding' polymerization, Polymer 47 (7) (2006) 2305–2313.
[84] K. Wang, J. Huang, Z. Wei, Conducting polyaniline nanowire arrays for high performance supercapacitors, J. Phys. Chem. C 114 (17) (2010) 8062–8067.
[85] W. Hu, et al., Flexible electrically conductive nanocomposite membrane based on bacterial cellulose and polyaniline, J. Phys. Chem. B 115 (26) (2011) 8453–8457.
[86] J.A. Marins, et al., Hybrid polyaniline-coated sepiolite nanofibers for electrorheological fluid applications, Synth. Met. 185 (2013) 9–16.
[87] E. Lahiff, et al., Synthesis and characterisation of controllably functionalised polyaniline nanofibres, Synth. Met. 159 (7–8) (2009) 741–748.
[88] O. Ikkala, et al., Counter-ion induced processibility of polyaniline: conducting melt processible polymer blends, Synth. Met. 69 (1–3) (1995) 97–100.
[89] J. Huang, R.B. Kaner, A general chemical route to polyaniline nanofibers, J. Am. Chem. Soc. 126 (3) (2004) 851–855.
[90] E.A. Gizzie, et al., Electrochemical preparation of photosystem I–polyaniline composite films for biohybrid solar energy conversion, ACS Appl. Mater. Interfaces 7 (18) (2015) 9328–9335.

[91] A.F. Naves, A.M. Carmona-Ribeiro, D.F. Petri, Immobilized horseradish peroxidase as a reusable catalyst for emulsion polymerization, Langmuir 23 (4) (2007) 1981–1987.
[92] F. Bradner, et al., Some insights into the microstructure of polypyrrole, Polym. Bull. 30 (5) (1989) 914–917.
[93] S.A. Popli, U.D. Patel, Electrochemical decolorization of reactive black 5 in an undivided cell using Ti and graphite anodes: effect of polypyrrole coating on anodes, J. Electrochem. Sci. Eng. 5 (2) (2015) 145–156.
[94] H. Eisazadeh, Studying the characteristics of polypyrrole and its composites, World J. Chem. 2 (2) (2007) 67–74.
[95] H. Naarmann, Polymers, electrically conducting, in: Ullmann's Encyclopedia of Industrial Chemistry, Wiley, 2000.
[96] A. Diaz, B. Hall, Mechanical properties of electrochemically prepared polypyrrole films, IBM J. Res. Dev. 27 (4) (1983) 342–347.
[97] A. Kassim, Z.B. Basar, H.E. Mahmud, Effects of preparation temperature on the conductivity of polypyrrole conducting polymer, J. Chem. Sci. 114 (2) (2002) 155–162.
[98] M. De la Plaza, et al., Electrosynthesis, electrochemical behavior and structure of poly [bis (phenoxyphosphazene)]-polypyrrole doped composite film, Synth. Met. 106 (2) (1999) 121–127.
[99] G.A. Snook, et al., Studies of deposition of and charge storage in polypyrrole–chloride and polypyrrole–carbon nanotube composites with an electrochemical quartz crystal microbalance, J. Electroanal. Chem. 568 (2004) 135–142.
[100] G.A. Snook, G.Z. Chen, The measurement of specific capacitances of conducting polymers using the quartz crystal microbalance, J. Electroanal. Chem. 612 (1) (2008) 140–146.
[101] A.B. Slimane, M.M. Chehimi, M.-J. Vaulay, Polypyrrole-coated poly (vinyl chloride) powder particles: surface chemical and morphological characterisation by means of X-ray photoelectron spectroscopy and scanning electron microscopy, Colloid Polym. Sci. 282 (4) (2004) 314–323.
[102] G.A. Snook, P. Kao, A.S. Best, Conducting-polymer-based supercapacitor devices and electrodes, J. Power Sources 196 (1) (2011) 1–12.
[103] F.M. Delnick, Proceedings of the Symposium on Electrochemical Capacitors II, The Electrochemical Society, 1997.
[104] G.A. Snook, et al., Achieving high electrode specific capacitance with materials of low mass specific capacitance: potentiostatically grown thick micro-nanoporous PEDOT films, Electrochem. Commun. 9 (1) (2007) 83–88.
[105] L.A. Samuelson, M.A. Druy, Kinetics of the degradation of electrical conductivity in polypyrrole, Macromolecules 19 (3) (1986) 824–828.
[106] M. Wysocka-Żołopa, K. Winkler, Electrochemical synthesis and properties of conical polypyrrole structures, Electrochim. Acta 258 (2017) 1421–1434.
[107] M.R. Gandhi, Morphological Studies and Structure-Property Relationship of Polypyrrole (Master's thesis), Department of Materials Engineering, University of Wollongong, 1993. In press.
[108] Z.D. Kojabad, S.A. Shojaosadati, Chemical synthesis of polypyrrole nanostructures: Optimization and applications for neural microelectrodes, Mater. Des. 96 (2016) 378–384.
[109] D. Kumar, R. Sharma, Advances in conductive polymers, Eur. Polym. J. 34 (8) (1998) 1053–1060.
[110] F. Jonas, et al., Polythiophenes, Process for Their Preparation and Their Use, 1990. Google Patents.

[111] D. Villers, et al., The influence of the range of electroactivity and capacitance of conducting polymers on the performance of carbon conducting polymer hybrid supercapacitor, J. Electrochem. Soc. 150 (6) (2003) A747.

[112] S. Hashmi, H.M. Upadhyaya, Polypyrrole and poly (3-methyl thiophene)-based solid state redox supercapacitors using ion conducting polymer electrolyte, Solid State Ionics 152 (2002) 883–889.

[113] L.-M. Huang, T.-C. Wen, A. Gopalan, Electrochemical and spectroelectrochemical monitoring of supercapacitance and electrochromic properties of hydrous ruthenium oxide embedded poly (3, 4-ethylenedioxythiophene)–poly (styrene sulfonic acid) composite, Electrochim. Acta 51 (17) (2006) 3469–3476.

[114] M.N. Gueye, et al., Progress in understanding structure and transport properties of PEDOT-based materials: a critical review, Prog. Mater. Sci. 108 (2020) 100616.

[115] X. Wang, et al., High electrical conductivity and carrier mobility in oCVD PEDOT thin films by engineered crystallization and acid treatment, Sci. Adv. 4 (9) (2018), eaat5780.

[116] J.D. Stenger-Smith, et al., Poly (3, 4-alkylenedioxythiophene)-based supercapacitors using ionic liquids as supporting electrolytes, J. Electrochem. Soc. 149 (8) (2002) A973.

[117] F. Jonas, J. Morrison, 3, 4-polyethylenedioxythiophene (PEDT): conductive coatings technical applications and properties, Synth. Met. 85 (1–3) (1997) 1397–1398.

[118] D. Waldow, et al., Molecular weight dependence of local segmental dynamics in dilute solution, Polym. Commun. 29 (10) (1988) 296–299.

[119] A. Elschner, S. Kirchmeyer, W. Lovenich, U. Merker, Pedot: Principles and Applications of an Intrinsically Conductive Polymer, CRC Press, 2010.

[120] G. Kaur, et al., Electrically conductive polymers and composites for biomedical applications, RSC Adv. 5 (47) (2015) 37553–37567.

[121] N.K. Guimard, N. Gomez, C.E. Schmidt, Conducting polymers in biomedical engineering, Prog. Polym. Sci. 32 (8–9) (2007) 876–921.

[122] K.S. Ryu, et al., Symmetric redox supercapacitor with conducting polyaniline electrodes, J. Power Sources 103 (2) (2002) 305–309.

[123] K.S. Ryu, et al., Redox supercapacitor using polyaniline doped with Li salt as electrode, Solid State Ionics 152 (2002) 861–866.

[124] M.E. Abdelhamid, A.P. O'Mullane, G.A. Snook, Storing energy in plastics: a review on conducting polymers & their role in electrochemical energy storage, RSC Adv. 5 (15) (2015) 11611–11626.

[125] F. Faverolle, et al., Highly conducting and strongly adhering polypyrrole coating layers deposited on glass substrates by a chemical process, Chem. Mater. 10 (3) (1998) 740–752.

[126] H. Zhang, et al., Growth of manganese oxide nanoflowers on vertically-aligned carbon nanotube arrays for high-rate electrochemical capacitive energy storage, Nano Lett. 8 (9) (2008) 2664–2668.

[127] Z.U. Rehman, et al., Effect of morphology of manganese oxide on the capacitive behavior of electrodes, Mater. Res. Express 6 (11) (2019) 115552.

[128] M. Toupin, T. Brousse, D. Bélanger, Charge storage mechanism of MnO2 electrode used in aqueous electrochemical capacitor, Chem. Mater. 16 (16) (2004) 3184–3190.

[129] S. Devaraj, N. Munichandraiah, The effect of nonionic surfactant triton X-100 during electrochemical deposition of MnO2 on its capacitance properties, J. Electrochem. Soc. 154 (10) (2007) A901.

[130] M.-S. Wu, Electrochemical capacitance from manganese oxide nanowire structure synthesized by cyclic voltammetric electrodeposition, Appl. Phys. Lett. 87 (15) (2005) 153102.

[131] W. Xiao, et al., Growth of single-crystal α-MnO2 nanotubes prepared by a hydrothermal route and their electrochemical properties, J. Power Sources 193 (2) (2009) 935–938.

[132] I. Ryu, et al., Hierarchically nanostructured $MnO_2$ electrodes for pseudocapacitor application, RSC Adv. 6 (104) (2016) 102814–102820.

[133] P. Staiti, F. Lufrano, Investigation of polymer electrolyte hybrid supercapacitor based on manganese oxide–carbon electrodes, Electrochim. Acta 55 (25) (2010) 7436–7442.

[134] H.S. Roy, et al., Polyaniline-MnO2 composites prepared in-situ during oxidative polymerization of aniline for supercapacitor applications, Mater. Today: Proc. 29 (2020) 1013–1019.

[135] G. Wang, et al., Preparation and supercapacitance of CuO nanosheet arrays grown on nickel foam, J. Power Sources 196 (13) (2011) 5756–5760.

[136] M. Jin, et al., Flexible electrodes based on polypyrrole/manganese dioxide/polypropylene fibrous membrane composite for supercapacitor, Electrochim. Acta 56 (27) (2011) 9838–9845.

[137] R.I. Jafri, A.K. Mishra, S. Ramaprabhu, Polyaniline–$MnO_2$ nanotube hybrid nanocomposite as supercapacitor electrode material in acidic electrolyte, J. Mater. Chem. 21 (44) (2011) 17601–17605.

[138] P. Lv, et al., Carbon fabric-aligned carbon nanotube/MnO2/conducting polymers ternary composite electrodes with high utilization and mass loading of MnO2 for supercapacitors, J. Power Sources 220 (2012) 160–168.

[139] J. Duay, et al., Highly flexible pseudocapacitor based on freestanding heterogeneous MnO 2/conductive polymer nanowire arrays, Phys. Chem. Chem. Phys. 14 (10) (2012) 3329–3337.

[140] P. Sen, et al., Conducting polymer based manganese dioxide nanocomposite as supercapacitor, Electrochim. Acta 108 (2013) 265–273.

[141] Y. Yang, et al., Manganese dioxide nanoparticle enrichment in porous conducting polymer as high performance supercapacitor electrode materials, Electrochim. Acta 165 (2015) 323–329.

[142] C. Choi, et al., Stretchable, weavable coiled carbon nanotube/MnO 2/polymer fiber solid-state supercapacitors, Sci. Rep. 5 (1) (2015) 1–6.

[143] W. He, et al., Flexible and high energy density asymmetrical supercapacitors based on core/shell conducting polymer nanowires/manganese dioxide nanoflakes, Nano Energy 35 (2017) 242–250.

[144] Y. He, et al., MnO 2/polyaniline hybrid nanostructures on carbon cloth for supercapacitor electrodes, J. Solid State Electrochem. 20 (5) (2016) 1459–1467.

[145] B. Relekar, et al., Development of porous manganese oxide/polyaniline composite using electrochemical route for electrochemical supercapacitor, J. Electron. Mater. 48 (4) (2019) 2449–2455.

[146] H. Lee, et al., RuOx/polypyrrole nanocomposite electrode for electrochemical capacitors, Synth. Met. 160 (9–10) (2010) 1055–1059.

[147] P. Simon, Y. Gogotsi, Materials for electrochemical capacitors, in: Nanoscience and Technology: A Collection of Reviews From Nature Journals, World Scientific, 2010, pp. 320–329.

[148] J. Zheng, P. Cygan, T. Jow, Hydrous ruthenium oxide as an electrode material for electrochemical capacitors, J. Electrochem. Soc. 142 (8) (1995) 2699v–2703.

[149] I.-H. Kim, K.-B. Kim, Electrochemical characterization of hydrous ruthenium oxide thin-film electrodes for electrochemical capacitor applications, J. Electrochem. Soc. 153 (2) (2006) A383.

[150] C. Lokhande, D. Dubal, O.-S. Joo, Metal oxide thin film based supercapacitors, Curr. Appl. Phys. 11 (3) (2011) 255–270.

[151] E. Seo, et al., Versatile double hydrophilic block copolymer: dual role as synthetic nanoreactor and ionic and electronic conduction layer for ruthenium oxide nanoparticle supercapacitors, J. Mater. Chem. 22 (23) (2012) 11598–11604.
[152] T.M. Ali, N. Padmanathan, S. Selladurai, Effect of nanofiller $CeO_2$ on structural, conductivity, and dielectric behaviors of plasticized blend nanocomposite polymer electrolyte, Ionics 21 (3) (2015) 829–840.
[153] S. Sopčić, M. Kraljić Roković, Z. Mandić, Preparation and characterization of RuO2/polyaniline/polymer binder composite electrodes for supercapacitor applications, J. Electrochem. Sci. Eng. 2 (1) (2012) 41–52.
[154] S. Hashmi, High rate performance of flexible pseudocapacitors fabricated using ionic-liquid-based proton conducting polymer electrolyte with poly (3, 4-ethylenedioxythiophene): poly (styrene sulfonate) and its hydrous ruthenium oxide composite electrodes, ACS Appl. Mater. Interfaces 5 (9) (2013) 3875–3883.
[155] P. Deshmukh, et al., Inexpensive synthesis route of porous polyaniline–ruthenium oxide composite for supercapacitor application, Chem. Eng. J. 257 (2014) 82–89.
[156] H.-S. Nam, et al., Supercapacitive properties of polyaniline/hydrous $RuO_2$ composite electrode, Polym. Bull. 68 (2) (2012) 553–560.
[157] X. Li, et al., Preparation and characterization of RuO2/polypyrrole electrodes for supercapacitors, Solid State Commun. 197 (2014) 57–60.
[158] J. Xu, et al., Copper-based nanomaterials for high-performance lithium-ion batteries, Part. Part. Syst. Charact. 33 (11) (2016) 784–810.
[159] Z. Zang, A. Nakamura, J. Temmyo, Nitrogen doping in cuprous oxide films synthesized by radical oxidation at low temperature, Mater. Lett. 92 (2013) 188–191.
[160] B. Vidhyadharan, et al., Superior supercapacitive performance in electrospun copper oxide nanowire electrodes, J. Mater. Chem. A 2 (18) (2014) 6578–6588.
[161] V. Patake, et al., Electrodeposited porous and amorphous copper oxide film for application in supercapacitor, Mater. Chem. Phys. 114 (1) (2009) 6–9.
[162] H. Zhang, J. Feng, M. Zhang, Preparation of flower-like CuO by a simple chemical precipitation method and their application as electrode materials for capacitor, Mater. Res. Bull. 43 (12) (2008) 3221–3226.
[163] M. Ghaedi, et al., Least square-support vector (LS-SVM) method for modeling of methylene blue dye adsorption using copper oxide loaded on activated carbon: kinetic and isotherm study, J. Ind. Eng. Chem. 20 (4) (2014) 1641–1649.
[164] O.O. Fasanya, et al., Copper zinc oxide nanocatalysts grown on cordierite substrate for hydrogen production using methanol steam reforming, Int. J. Hydrog. Energy 44 (41) (2019) 22936–22946.
[165] S.K. Shinde, et al., Influence of Mn incorporation on the supercapacitive properties of hybrid CuO/Cu $(OH)_2$ electrodes, RSC Adv. 5 (39) (2015) 30478–30484.
[166] D. Dubal, et al., Fabrication of copper oxide multilayer nanosheets for supercapacitor application, J. Alloys Compd. 492 (1–2) (2010) 26–30.
[167] S.E. Moosavifard, et al., Designing 3D highly ordered nanoporous CuO electrodes for high-performance asymmetric supercapacitors, ACS Appl. Mater. Interfaces 7 (8) (2015) 4851–4860.
[168] M. Lugo-Ruelas, et al., Synthesis, microstructural characterization and optical properties of CuO nanorods and nanowires obtained by aerosol assisted CVD, J. Alloys Compd. 643 (2015) S46–S50.
[169] B. Vidyadharan, et al., High performance asymmetric supercapacitors using electrospun copper oxide nanowires anode, J. Alloys Compd. 633 (2015) 22–30.

[170] S. Wu, et al., Enhanced field emission properties of CuO nanoribbons decorated with Ag nanoparticles, Mater. Lett. 171 (2016) 220–223.
[171] Y. Yu, J. Zhang, Solution-phase synthesis of rose-like CuO, Mater. Lett. 63 (21) (2009) 1840–1843.
[172] P. Gao, et al., Non-enzymatic amperometric detection of hydrogen peroxide using grass-like copper oxide nanostructures calcined in nitrogen atmosphere, Electrochim. Acta 173 (2015) 31–39.
[173] G. Navathe, et al., Rapid synthesis of nanostructured copper oxide for electrochemical supercapacitor based on novel [HPMIM][Cl] ionic liquid, J. Electroanal. Chem. 738 (2015) 170–175.
[174] K. Chen, S. Song, D. Xue, Chemical reaction controlled synthesis of $Cu_2O$ hollow octahedra and core–shell structures, CrystEngComm 15 (46) (2013) 10028–10033.
[175] A. Pendashteh, M.F. Mousavi, M.S. Rahmanifar, Fabrication of anchored copper oxide nanoparticles on graphene oxide nanosheets via an electrostatic coprecipitation and its application as supercapacitor, Electrochim. Acta 88 (2013) 347–357.
[176] S. Ashokkumar, et al., Electrochemically synthesized polyaniline/copper oxide nano composites: to study optical band gap and electrochemical performance for energy storage devices, Inorg. Chem. Commun. (2020) 107865.
[177] C. Zhao, et al., Facile synthesis of $Co(OH)_2/Al(OH)_3$ nanosheets with improved electrochemical properties for asymmetric supercapacitor, J. Phys. Chem. Solids 112 (2018) 54–60.
[178] I. Chakraborty, et al., CuO@ NiO/polyaniline/MWCNT nanocomposite as high-performance electrode for supercapacitor, J. Phys. Chem. C 122 (48) (2018) 27180–27190.
[179] A. Viswanathan, A.N. Shetty, Single step synthesis of rGO, copper oxide and polyaniline nanocomposites for high energy supercapacitors, Electrochim. Acta 289 (2018) 204–217.
[180] M. Ates, et al., Supercapacitor behaviors of polyaniline/CuO, polypyrrole/CuO and PEDOT/CuO nanocomposites, Polym. Bull. 72 (10) (2015) 2573–2589.
[181] M.B. Gholivand, et al., Nanostructured CuO/PANI composite as supercapacitor electrode material, Mater. Sci. Semicond. Process. 30 (2015) 157–161.
[182] M. Majumder, et al., Impact of rare-earth metal oxide ($Eu_2O_3$) on the electrochemical properties of a polypyrrole/CuO polymeric composite for supercapacitor applications, RSC Adv. 7 (32) (2017) 20037–20048.
[183] J. Lang, X. Yan, Q. Xue, Facile preparation and electrochemical characterization of cobalt oxide/multi-walled carbon nanotube composites for supercapacitors, J. Power Sources 196 (18) (2011) 7841–7846.
[184] M.K. Lima-Tenório, et al., Pseudocapacitance properties of Co3O4 nanoparticles synthesized using a modified sol-gel method, Mater. Res. 21 (2) (2018).
[185] P.J. Kulesza, et al., Spectroelectrochemical characterization of cobalt hexacyanoferrate films in potassium salt electrolyte, Electrochim. Acta 43 (8) (1998) 919–923.
[186] G. Sun, et al., Incorporation of homogeneous $Co_3O_4$ into a nitrogen-doped carbon aerogel via a facile in situ synthesis method: implications for high performance asymmetric supercapacitors, J. Mater. Chem. A 4 (24) (2016) 9542–9554.
[187] Y. Huang, J. Liang, Y. Chen, An overview of the applications of graphene-based materials in supercapacitors, Small 8 (12) (2012) 1805–1834.
[188] C. Xiang, et al., A reduced graphene oxide/Co3O4 composite for supercapacitor electrode, J. Power Sources 226 (2013) 65–70.
[189] L. Tao, et al., Co3O4 nanorods/graphene nanosheets nanocomposites for lithium ion batteries with improved reversible capacity and cycle stability, J. Power Sources 202 (2012) 230–235.

[190] S.J. Uke, et al., Recent advancements in the cobalt oxides, manganese oxides, and their composite as an electrode material for supercapacitor: a review, Front. Mater. 4 (2017) 21.

[191] N. Yan, et al., Co3O4 nanocages for high-performance anode material in lithium-ion batteries, J. Phys. Chem. C 116 (12) (2012) 7227–7235.

[192] V. Shinde, et al., Supercapacitive cobalt oxide (Co3O4) thin films by spray pyrolysis, Appl. Surf. Sci. 252 (20) (2006) 7487–7492.

[193] X. Hou, et al., Facile synthesis of Co3O4 hierarchical microspheres with improved lithium storage performances, Appl. Surf. Sci. 383 (2016) 159–164.

[194] C. Niveditha, et al., Feather like highly active Co3O4 electrode for supercapacitor application: a potentiodynamic approach, Mater. Res. Express 5 (6) (2018), 065501.

[195] A. Jagadale, V. Kumbhar, C. Lokhande, Supercapacitive activities of potentiodynamically deposited nanoflakes of cobalt oxide (Co3O4) thin film electrode, J. Colloid Interface Sci. 406 (2013) 225–230.

[196] S. Navale, et al., Solution-processed rapid synthesis strategy of Co3O4 for the sensitive and selective detection of H2S, Sensors Actuators B Chem. 245 (2017) 524–532.

[197] J. Yan, et al., Rapid microwave-assisted synthesis of graphene nanosheet/Co3O4 composite for supercapacitors, Electrochim. Acta 55 (23) (2010) 6973–6978.

[198] G. Wang, et al., Hydrothermal synthesis and optical, magnetic, and supercapacitance properties of nanoporous cobalt oxide nanorods, J. Phys. Chem. C 113 (11) (2009) 4357–4361.

[199] R. Tummala, R.K. Guduru, P.S. Mohanty, Nanostructured Co3O4 electrodes for supercapacitor applications from plasma spray technique, J. Power Sources 209 (2012) 44–51.

[200] J. Deng, et al., Solution combustion synthesis of cobalt oxides (Co3O4 and Co3O4/CoO) nanoparticles as supercapacitor electrode materials, Electrochim. Acta 132 (2014) 127–135.

[201] Y. Yuan, et al., Hierarchically porous Co3O4 film with mesoporous walls prepared via liquid crystalline template for supercapacitor application, Electrochem. Commun. 13 (10) (2011) 1123–1126.

[202] Z.-Y. Li, et al., Enhanced electrochemical activity of low temperature solution process synthesized Co3O4 nanoparticles for pseudo-supercapacitors applications, Ceram. Int. 42 (1) (2016) 1879–1885.

[203] X.-H. Xia, et al., Mesoporous Co3O4 monolayer hollow-sphere array as electrochemical pseudocapacitor material, Chem. Commun. 47 (20) (2011) 5786–5788.

[204] C. Lin, J.A. Ritter, B.N. Popov, Characterization of sol-gel-derived cobalt oxide xerogels as electrochemical capacitors, J. Electrochem. Soc. 145 (12) (1998) 4097.

[205] L. Zhang, et al., Recent progress on nanostructured conducting polymers and composites: synthesis, application and future aspects, Sci. China Mater. 61 (3) (2018) 303–352.

[206] Y. Xie, D. Wang, Supercapacitance performance of polypyrrole/titanium nitride/polyaniline coaxial nanotube hybrid, J. Alloys Compd. 665 (2016) 323–332.

[207] C. Sandhya, et al., Polyaniline-cobalt oxide nano shrubs based electrodes for supercapacitors with enhanced electrochemical performance, Electrochim. Acta 324 (2019) 134876.

[208] S. Bashir, et al., Synthesis and characterization of hybrid poly (N, N-dimethylacrylamide) composite hydrogel electrolytes and their performance in supercapacitor, Electrochim. Acta 332 (2020) 135438.

[209] R.S. Babu, et al., Asymmetric supercapacitor based on carbon nanofibers as the anode and two-dimensional copper cobalt oxide nanosheets as the cathode, Chem. Eng. J. 366 (2019) 390–403.

[210] H. Wei, et al., Electropolymerized polypyrrole nanocomposites with cobalt oxide coated on carbon paper for electrochemical energy storage, Polymer 67 (2015) 192–199.
[211] S.N.J.S.Z. Abidin, et al., Electropolymerization of poly (3, 4-ethylenedioxythiophene) onto polyvinyl alcohol-graphene quantum dot-cobalt oxide nanofiber composite for high-performance supercapacitor, Electrochim. Acta 261 (2018) 548–556.
[212] Y. Sulaiman, et al., One step electrodeposition of poly-(3, 4-ethylenedioxythiophene)/graphene oxide/cobalt oxide ternary nanocomposite for high performance supercapacitor, Electrochim. Acta 253 (2017) 581–588.
[213] X. Yang, et al., Comparative evaluation of PPyNF/CoOx and PPyNT/CoOx nanocomposites as battery-type supercapacitor materials via a facile and low-cost microwave synthesis approach, Electrochim. Acta 311 (2019) 230–243.
[214] H. Xu, et al., Flexible fiber-shaped supercapacitors based on hierarchically nanostructured composite electrodes, Nano Res. 8 (4) (2015) 1148–1158.
[215] X. Hu, et al., Nanostructured Mo-based electrode materials for electrochemical energy storage, Chem. Soc. Rev. 44 (8) (2015) 2376–2404.
[216] M. Sarfraz, M.F. Aboud, I. Shakir, Molybdenum oxide nanowires based supercapacitors with enhanced capacitance and energy density in ethylammonium nitrate electrolyte, J. Alloys Compd. 650 (2015) 123–126.
[217] D. Ban, et al., Low-temperature synthesis of large-area films of molybdenum trioxide microbelts in air and the dependence of their field emission performance on growth conditions, J. Mater. Sci. Technol. 26 (7) (2010) 584–588.
[218] S.K. Pradhan, et al., Growth of TiO2 nanorods by metalorganic chemical vapor deposition, J. Cryst. Growth 256 (1–2) (2003) 83–88.
[219] I. Shakir, et al., $MoO_3$-MWCNT nanocomposite photocatalyst with control of light-harvesting under visible light and natural sunlight irradiation, J. Mater. Chem. 22 (38) (2012) 20549–20553.
[220] M. Yu, et al., Dual-doped molybdenum trioxide nanowires: a bifunctional anode for fiber-shaped asymmetric supercapacitors and microbial fuel cells, Angew. Chem. 128 (23) (2016) 6874–6878.
[221] T. Li, et al., Ethanol reduced molybdenum trioxide for Li-ion capacitors, Nano Energy 26 (2016) 100–107.
[222] D. Guan, et al., Enhanced capacitive performance of TiO2 nanotubes with molybdenum oxide coating, Appl. Surf. Sci. 300 (2014) 165–170.
[223] I. Shakir, et al., Tin oxide coating on molybdenum oxide nanowires for high performance supercapacitor devices, Electrochim. Acta 72 (2012) 134–137.
[224] V. Kumar, X. Wang, P.S. Lee, Formation of hexagonal-molybdenum trioxide (h-$MoO_3$) nanostructures and their pseudocapacitive behavior, Nanoscale 7 (27) (2015) 11777–11786.
[225] R. Dhanabal, et al., Reduced graphene oxide supported molybdenum oxide hybrid nanocomposites: high performance electrode material for supercapacitor and photocatalytic applications, J. Nanosci. Nanotechnol. 20 (7) (2020) 4035–4046.
[226] L. Zhang, et al., Enhanced supercapacitor performance based on 3D porous graphene with MoO2 nanoparticles, J. Mater. Res. 32 (2) (2017) 292.
[227] S. Li, et al., Rational synthesis of coaxial MoO3/PTh nanowires with improved electrochemical cyclability, Int. J. Electrochem. Sci. 6 (4) (2011) 4504–4513.
[228] X. Xia, et al., Reduced-graphene oxide/molybdenum oxide/polyaniline ternary composite for high energy density supercapacitors: synthesis and properties, J. Mater. Chem. 22 (17) (2012) 8314–8320.

[229] F. Jiang, et al., MoO3/PANI coaxial heterostructure nanobelts by in situ polymerization for high performance supercapacitors, Nano Energy 7 (2014) 72–79.
[230] W. Tang, et al., Aqueous supercapacitors of high energy density based on $MoO_3$ nanoplates as anode material, Chem. Commun. 47 (36) (2011) 10058–10060.
[231] A.K. Das, S.K. Karan, B. Khatua, High energy density ternary composite electrode material based on polyaniline (PANI), molybdenum trioxide (MoO3) and graphene nanoplatelets (GNP) prepared by sono-chemical method and their synergistic contributions in superior supercapacitive performance, Electrochim. Acta 180 (2015) 1–15.
[232] L.-J. Bian, H.-L. He, X.-X. Liu, Self-doped polyaniline/molybdenum oxide composite nanorods for supercapacitors, RSC Adv. 5 (92) (2015) 75374–75379.
[233] F.-H. Hsu, T.-M. Wu, Facile synthesis of polypyrrole/carbon-coated $MoO_3$ nanoparticle/graphene nanoribbon nanocomposite with high-capacitance applied in supercapacitor electrode, J. Mater. Sci. Mater. Electron. 29 (1) (2018) 382–391.
[234] Y. Liu, et al., Polypyrrole-coated $\alpha$-$MoO_3$ nanobelts with good electrochemical performance as anode materials for aqueous supercapacitors, J. Mater. Chem. A 1 (43) (2013) 13582–13587.
[235] X. Li, et al., Electrical conduction behavior of La, Co co-doped SrTiO3 perovskite as anode material for solid oxide fuel cells, Int. J. Hydrog. Energy 34 (15) (2009) 6407–6414.
[236] M. Søgaard, P.V. Hendriksen, M. Mogensen, Oxygen nonstoichiometry and transport properties of strontium substituted lanthanum ferrite, J. Solid State Chem. 180 (4) (2007) 1489–1503.
[237] M.Z. Iqbal, et al., Scrutinizing the charge storage mechanism in SrO based composites for asymmetric supercapacitors by diffusion-controlled process, Appl. Nanosci. 10 (11) (2020) 3999–4011.
[238] M.Z. Iqbal, et al., Binary composites of strontium oxide/polyaniline for high performance supercapattery devices, Solid State Ionics 347 (2020) 115276.
[239] M. Jayalakshmi, K. Balasubramanian, Simple capacitors to supercapacitors-an overview, Int. J. Electrochem. Sci. 3 (11) (2008) 1196–1217.
[240] B. Avasarala, P. Haldar, Electrochemical oxidation behavior of titanium nitride based electrocatalysts under PEM fuel cell conditions, Electrochim. Acta 55 (28) (2010) 9024–9034.
[241] H. Liang, et al., Electrochemical study of activated carbon-semiconducting oxide composites as electrode materials of double-layer capacitors, Electrochim. Acta 49 (21) (2004) 3463–3467.
[242] A. Ramadoss, S.J. Kim, Improved activity of a graphene–TiO2 hybrid electrode in an electrochemical supercapacitor, Carbon 63 (2013) 434–445.
[243] K. Lee, A. Mazare, P. Schmuki, One-dimensional titanium dioxide nanomaterials: nanotubes, Chem. Rev. 114 (19) (2014) 9385–9454.
[244] A. Ramadoss, S.J. Kim, Enhanced supercapacitor performance using hierarchical TiO2 nanorod/Co $(OH)_2$ nanowall array electrodes, Electrochim. Acta 136 (2014) 105–111.
[245] N.S. Peighambardoust, et al., Band-gap narrowing and electrochemical properties in N-doped and reduced anodic TiO2 nanotube arrays, Electrochim. Acta 270 (2018) 245–255.
[246] I. Heng, et al., Low-temperature synthesis of TiO2 nanocrystals for high performance electrochemical supercapacitors, Ceram. Int. 45 (4) (2019) 4990–5000.
[247] A. Elmouwahidi, et al., Carbon–$TiO_2$ composites as high-performance supercapacitor electrodes: synergistic effect between carbon and metal oxide phases, J. Mater. Chem. A 6 (2) (2018) 633–644.

[248] D. Ponnamma, M.A.A. Al-Maadeed, 3D architectures of titania nanotubes and graphene with efficient nanosynergy for supercapacitors, Mater. Des. 117 (2017) 203–212.
[249] Z. Zheng, et al., Facile in situ synthesis of visible-light plasmonic photocatalysts M@TiO2 (M = Au, Pt, Ag) and evaluation of their photocatalytic oxidation of benzene to phenol, J. Mater. Chem. 21 (25) (2011) 9079–9087.
[250] X. Chen, S.S. Mao, Titanium dioxide nanomaterials: synthesis, properties, modifications, and applications, Chem. Rev. 107 (7) (2007) 2891–2959.
[251] Y. Xie, et al., Improved performance of dye-sensitized solar cells by trace amount Cr-doped TiO2 photoelectrodes, J. Power Sources 224 (2013) 168–173.
[252] X. Yang, et al., Anatase $TiO_2$ nanocubes for fast and durable sodium ion battery anodes, J. Mater. Chem. A 3 (16) (2015) 8800–8807.
[253] S.H. Mujawar, et al., Electropolymerization of polyaniline on titanium oxide nanotubes for supercapacitor application, Electrochim. Acta 56 (12) (2011) 4462–4466.
[254] M.S. Kim, J.H. Park, Polypyrrole/titanium oxide nanotube arrays composites as an active material for supercapacitors, J. Nanosci. Nanotechnol. 11 (5) (2011) 4522–4526.
[255] R. Gottam, P. Srinivasan, One-step oxidation of aniline by peroxotitanium acid to polyaniline–titanium dioxide: a highly stable electrode for a supercapacitor, J. Appl. Polym. Sci. 132 (13) (2015).
[256] B.S. Singu, et al., Use of surfactant in aniline polymerization with $TiO_2$ to PANI-$TiO_2$ for supercapacitor performance, J. Solid State Electrochem. 18 (7) (2014) 1995–2003.
[257] E. Azizi, J. Arjomandi, J.Y. Lee, Reduced graphene oxide/poly (1, 5 dihydroxynaphthalene)/TiO2 nanocomposite conducting polymer coated on gold as a supercapacitor electrode, Electrochim. Acta 298 (2019) 726–734.
[258] A.K. Thakur, et al., Fairly improved pseudocapacitance of PTP/PANI/$TiO_2$ nanohybrid composite electrode material for supercapacitor applications, Ionics 24 (1) (2018) 257–268.
[259] T. Xiao, et al., Preparation and performance of PANI-TiO2 nanotube arrays composite electrode by in-situ microcavity polymerization, Mater. Chem. Phys. 240 (2020) 122179.
[260] P. Deshmukh, et al., Chemical synthesis of PANI–$TiO_2$ composite thin film for supercapacitor application, RSC Adv. 5 (84) (2015) 68939–68946.
[261] H. Wang, et al., Design and assembly of reduced graphene oxide/polyaniline/urchin-like mesoporous TiO2 spheres ternary composite and its application in supercapacitors, Compos. Part B 92 (2016) 405–412.
[262] H. Yu, et al., Fabrication and electrochemical characterization of polyaniline/titanium oxide nanoweb composite electrode for supercapacitor application, J. Nanosci. Nanotechnol. 16 (3) (2016) 2937–2943.
[263] A.K. Thakur, R.B. Choudhary, High-performance supercapacitors based on polymeric binary composites of polythiophene (PTP)–titanium dioxide (TiO2), Synth. Met. 220 (2016) 25–33.
[264] T. Xiao, et al., Effects of monomer solvent on the supercapacitance performance of PANI/TiO2 nanotube arrays composite electrode, Mater. Lett. 230 (2018) 245–248.
[265] G. Singh, Y. Kumar, S. Husain, High charge retention and optimization of polyaniline–titanium dioxide nanoparticles composite nanostructures for dominantly stable pseudocapacitive nature, J. Energy Storage 31 (2020) 101660.
[266] S. Aslan, et al., Electrochemical evaluation of titanium (IV) oxide/polyacrylonitrile electrospun discharged battery coals as supercapacitor electrodes, Electroanalysis 33 (2020) 120–128.
[267] S. Surnev, M. Ramsey, F. Netzer, Vanadium oxide surface studies, Prog. Surf. Sci. 73 (4–8) (2003) 117–165.

[268] B. Saravanakumar, et al., Interconnected V2O5 nanoporous network for high-performance supercapacitors, ACS Appl. Mater. 4 (9) (2012) 4484–4490.
[269] Y.L. Cheah, et al., Morphology, structure and electrochemical properties of single phase electrospun vanadium pentoxide nanofibers for lithium ion batteries, J. Power Sources 196 (15) (2011) 6465–6472.
[270] C.F. Armer, et al., Electrospun vanadium-based oxides as electrode materials, J. Power Sources 395 (2018) 414–429.
[271] D. Shu, et al., Enhanced capacitance and rate capability of nanocrystalline VN as electrode materials for supercapacitors, Int. J. Electrochem. Sci. 8 (1) (2013) 1209–1225.
[272] E.A. de Souza, et al., Ethanol electro-oxidation on partially alloyed Pt-Sn-Rh/C catalysts, Electrochim. Acta 147 (2014) 483–489.
[273] N.A. Chernova, et al., Layered vanadium and molybdenum oxides: batteries and electrochromics, J. Mater. Chem. 19 (17) (2009) 2526–2552.
[274] C.K. Chan, et al., Fast, completely reversible Li insertion in vanadium pentoxide nanoribbons, Nano Lett. 7 (2) (2007) 490–495.
[275] A. Pan, et al., Facile synthesized nanorod structured vanadium pentoxide for high-rate lithium batteries, J. Mater. Chem. 20 (41) (2010) 9193–9199.
[276] A. Engstrom, F. Doyle, Exploring the cycle behavior of electrodeposited vanadium oxide electrochemical capacitor electrodes in various aqueous environments, J. Power Sources 228 (2013) 120–131.
[277] D.H. Nagaraju, et al., Two-dimensional heterostructures of $V_2O_5$ and reduced graphene oxide as electrodes for high energy density asymmetric supercapacitors, J. Mater. Chem. A 2 (40) (2014) 17146–17152.
[278] J. Zhu, et al., Building 3D structures of vanadium pentoxide nanosheets and application as electrodes in supercapacitors, Nano Lett. 13 (11) (2013) 5408–5413.
[279] H.B. Wu, et al., Template-assisted formation of rattle-type V2O5 hollow microspheres with enhanced Lithium storage properties, Adv. Funct. Mater. 23 (45) (2013) 5669–5674.
[280] M. Lee, et al., One-step hydrothermal synthesis of graphene decorated $V_2O_5$ nanobelts for enhanced electrochemical energy storage, Sci. Rep. 5 (2015) 8151.
[281] M. Malta, R.M. Torresi, Electrochemical and kinetic studies of lithium intercalation in composite nanofibers of vanadium oxide/polyaniline, Electrochim. Acta 50 (25–26) (2005) 5009–5014.
[282] J.-S. Huang, et al., Solution-processed vanadium oxide as an anode interlayer for inverted polymer solar cells hybridized with ZnO nanorods, Org. Electron. 10 (6) (2009) 1060–1065.
[283] M. Benmoussa, et al., Electrochromism in sputtered V2O5 thin films: structural and optical studies, Thin Solid Films 405 (1–2) (2002) 11–16.
[284] A. Jin, et al., Multi-electrochromism behavior and electrochromic mechanism of electrodeposited molybdenum doped vanadium pentoxide films, Electrochim. Acta 55 (22) (2010) 6408–6414.
[285] H. Liu, et al., Porous V2O5 nanorods/reduced graphene oxide composites for high performance symmetric supercapacitors, Appl. Surf. Sci. 478 (2019) 383–392.
[286] Z. Liu, et al., Graphene/V2O5 hybrid electrode for an asymmetric supercapacitor with high energy density in an organic electrolyte, Electrochim. Acta 287 (2018) 149–157.
[287] M. Sathiya, et al., V2O5-anchored carbon nanotubes for enhanced electrochemical energy storage, J. Am. Chem. Soc. 133 (40) (2011) 16291–16299.
[288] B. Yan, et al., Crumpled reduced graphene oxide conformally encapsulated hollow V2O5 nano/microsphere achieving brilliant lithium storage performance, Nano Energy 24 (2016) 32–44.

[289] W.F. Mak, et al., High-energy density asymmetric supercapacitor based on electrospun vanadium pentoxide and polyaniline nanofibers in aqueous electrolyte, J. Electrochem. Soc. 159 (9) (2012) A1481.

[290] M.-H. Bai, et al., Electrochemical codeposition of vanadium oxide and polypyrrole for high-performance supercapacitor with high working voltage, ACS Appl. Mater. Interfaces 6 (15) (2014) 12656–12664.

[291] M.-H. Bai, et al., Electrodeposition of vanadium oxide–polyaniline composite nanowire electrodes for high energy density supercapacitors, J. Mater. Chem. A 2 (28) (2014) 10882–10888.

[292] E. Karaca, K. Pekmez, N.Ö. Pekmez, Electrosynthesis of polypyrrole-vanadium oxide composites on graphite electrode in acetonitrile in the presence of carboxymethyl cellulose for electrochemical supercapacitors, Electrochim. Acta 273 (2018) 379–391.

[293] K.Y. Yasoda, et al., Brush like polyaniline on vanadium oxide decorated reduced graphene oxide: efficient electrode materials for supercapacitor, J. Energy Storage 22 (2019) 188–193.

[294] W. Bi, et al., Tailoring energy and power density through controlling the concentration of oxygen vacancies in V2O5/PEDOT Nanocable-based supercapacitors, ACS Appl. Mater. Interfaces 11 (18) (2019) 16647–16655.

[295] W. Bi, et al., $V_2O_5$–conductive polymer nanocables with built-in local electric field derived from interfacial oxygen vacancies for high energy density supercapacitors, J. Mater. Chem. A 7 (30) (2019) 17966–17973.

[296] W. Bi, et al., Gradient oxygen vacancies in V2O5/PEDOT nanocables for high-performance supercapacitors, ACS Appl. Energy Mater. 2 (1) (2018) 668–677.

[297] P. Asen, S. Shahrokhian, One step electrodeposition of V2O5/polypyrrole/graphene oxide ternary nanocomposite for preparation of a high performance supercapacitor, Int. J. Hydrog. Energy 42 (33) (2017) 21073–21085.

# Synthesis and properties of percolative metal oxide-polymer composites

# 8

*Srikanta Moharana[a], Bibhuti B. Sahu[b], Rozalin Nayak[c], and Ram Naresh Mahaling[c]*
[a]School of Applied Sciences, Centurion University of Technology and Management, Odisha, India, [b]Department of Physics, Veer Surendra Sai University of Technology, Odisha, India, [c]Laboratory of Polymeric and Materials Chemistry, School of Chemistry, Sambalpur University, Odisha, India

## 1 Introduction

During the past decade, the preparation of nanostructured metal oxide and their composites are one of the most promising materials for the fabrication of energy-saving and energy storage devices including lithium-ion battery [1–3], fuel cells [4–6], solar cells [7–9], transistors, and light-emitting devices (LEDs) [10,11] for the reason that of their higher carrier drifting, better transparency, tremendous consistency, and realistic electrical stability [7–11]. The combination of various metal oxides into a polymer matrix to produce composite material with novel physical and chemical properties leads to comparatively better performances in the field of various technological applications [12–16]. The metal oxides exhibit a variety of functional properties including ferromagnetism, ferroelectricity, multiferroelectricity, and superconductivity with superior carrier mobility even in the amorphous state and suitable environmental stability [16–20]. Most importantly, the amorphous phase of metal oxides is more suitable for using in flexible devices as compared to the crystalline phase but when it is folded, the crystalline materials tend to crack. However, the metals have isolated units that get their independent properties which are modified by interaction between metal–metal and metal-oxygen-metal [21,22]. These interactions of various constituents of materials have been significantly improved the efficiency of charge separation, which increases charge carrier lifetime and also simultaneously improves the properties of nanostructured system. Most of the mixed oxide-based monophase materials are united with properties of the constituent oxides [21–24]. The two metal oxide-based multiphase nanocomposites (e.g., $CeO_2$-MnOx [25], CuO-ZnO [26], ZnO-NiO [27], $TiO_2$-$WO_3$ [28], and $Co_3O_4$-ZnO [29]) have been synthesized for various prospective applications in the domain of photocatalysis, antibacterial activity, etc.

Besides, the metal oxide-based nanocomposites are having beyond two phases (ZnO-$ZnWO_3$-$WO_3$ [23,30], Mo-V-W mixed oxide [31], La-Mo-V oxide solid solution [32], and pseudoquaternary system CaO-CoO-$SiO_2$ [33] with enhanced properties and applicability in energy storage devices [12–16]. Moreover, carbon-based

**Renewable Polymers and Polymer-Metal Oxide Composites. https://doi.org/10.1016/B978-0-323-85155-8.00001-7**

nanofillers extensively applicable while preparing the nanocomposites with polymer matrix show homogeneous dispersion and enhanced properties (electrical, mechanical, and optical). The nanocomposites with various carbon-based fillers are carbon black (CB), expanded graphite (EG), carbon nanofiber (CNF), carbon nanotube (CNT), and graphene, etc. [34–38]. In the past few decades, graphene is the one atom thick $sp^2$ hybridized 2D carbon-based materials with shape of honeycomb lattice [39]. Each carbon atom is bonded in the four outer shell electrons and out of them only three are used for chemical bonding and the rest is highly mobile and accessible for electrical conduction. The high electrical conductivity of graphene is due to the existence of these mobile electrons. The monolayer structure of graphene provides a very low aspect ratio [39–41]. However, recently studies have reported that the metal-based filler particles particularly with low and high aspect ratios will lead to very low percolation threshold [42,43]. Therefore, the carbon-based filler particles especially graphene or carbon nanotubes (CNTs) are incorporated into the polymer matrix and they lead to the formation of conductive networks at very low concentration of filler contents. For instance, percolation threshold of graphene at low concentration (0.1%) compared with CNTs but in general, the graphene is easily prepared with lesser cost for mass production [44]. Thus, graphene is a potentially applicable material than that of the CNTs. Many studies have reported that low percolation threshold and increment of electrical conductivity of the composites are due to the increase in the filler contents [43–46]. The nanocomposites show various properties at particular concentration of the nanoparticles to the percolation threshold. These types of properties are changed considerably for those of bulk materials as well as the composite obtains these during its development [47].

The controlled synthesis and possible applications of metal oxide- and polymer-based composites for specific concentration at percolation threshold have increased to an amazing scale, which open new challenges and opportunities for the promotion of individual property of the materials. This archetypal chapter is mainly alienated into two main segments. The first segment covers the properties and various synthesis techniques of metal oxides. The second segment covers the overview of polymer-metal oxide composite at percolation threshold and also discussed the dielectric and electrical properties of the resultant composite systems in the field of energy storage devices. Moreover, we conclude this chapter with a few words on this new and exciting research area of the metal oxide-based advanced polymer composites followed by summary and perspective.

## 2 Properties of metal oxide nanostructures

In recent years, nanostructured metal oxides (NMOs) are the extensively investigated materials in the field of materials science due to their multitude of technological and device applications in optoelectronics, magnetoelectric sensors, and transducers [48,49]. Moreover, with various interesting exclusive properties like electrical, optical, sensing, ferroelectricity, magnetism, and superconductivity, etc. are exhibited by nanostructured metal oxides [50,51]. Numerous metal oxide nanostructured materials have also widespread utility in the field of sensors, synchronized drug delivery, pigments, etc., and become competent material due to their minor bandgap, nontoxicity, cost effectiveness, and elevated thermal and chemical stability [52,53]. In the past few decades, many metal oxide nanostructured materials have been reported; among these, nanostructured $BaTiO_3$ (BT) is one of the interesting and most studied perovskite-type metal oxides, due to its high dielectric constant and ferroelectricity with numerous potential applications in the field of high energy storage devices and optical memories [54]. Zheng et al. [55] have investigated the temperature-dependent magnetic and ferroelectric coupling parameters of nanostructured $BaTiO_3$-$CoFe_2O_4$ metal oxide composite. Moreover, they have prepared the composite by self-assembled technique as well as produced a spinel-perovskite system and reported about the variation of magnetizing behavior at the ferroelectric transition temperature ($T_c$). The X-ray diffraction (XRD) analysis revealed the superior degree of crystallographic orientation in the $BaTiO_3$-$CoFe_2O_4$ nanostructured system. The microstructural analysis has unfolded the heteroepitaxy among $BaTiO_3$ and $CoFe_2O_4$. The ferroelectric and piezoelectric ($d_{33}$) measurements exhibited ferroelectric hysteresis and archetypal piezoelectric perovskite behavior, respectively. Bao et al. have successfully synthesized $BaTiO_3$ nanostructures using $Na_2Ti_3O_7$ nanowires and nanotubes as synthetic antecedents obtained from transmission electron microscopy (TEM) and high-resolution transmission electron microscopy (HRTEM) as shown in the Fig. 1. They have processed nanostructured $BaTiO_3$, which demonstrates various shapes like nanowires, nanosheets, nanocubes, and hexagonal nanoparticles or ordered structural design of coral-like nanostructures with collection of nanorods, starfish, and sword-like nanostructures. This different shape of product of $BaTiO_3$ depends on the barium hydroxide concentration as well as nature of the precursors and temperature [56].

Corral-Flores et al. have prepared nanostructures of $CoFe_2O_4$-$BaTiO_3$ with a variation of cobalt ferrite from 20 to 60 wt% via wet chemical method combining coprecipitation as well as sol-gel techniques. Meanwhile, the cobalt ferrite- and barium titanate-based materials show spinel and perovskite phase, respectively, and were confirmed from X-ray diffraction analysis with an average crystallite size in the range of 15 to 28 nm. Due to crystallographic deformation, there is a significant enhancement of tetragonality of $BaTiO_3$ and lattice parameters of $CoFe_2O_4$. Cobalt ferrite as core with shell-like covering of $BaTiO_3$ was revealed from TEM data. The room temperature ferromagnetic behavior of cobalt ferrite and the nanostructures were confirmed from the ferromagnetic hysteresis loop with a significant coercivity value. This multiferroic nanostructured $CoFe_2O_4$-$BaTiO_3$ composite increased the mechanical coupling with the phases and signify magnetostrictive-piezoelectric core-shell nanostructures [57]. These multifunctional nanostructured composites have the prospective for producing ceramic nanofibers to use in practical applications.

The synthesis of nickel-zinc ferrite-barium titanate via chemical combustion and hydrothermal synthesis techniques was reported by Verma et al. The composites prepared with chemical combustion produce nanoparticles of average size of 4 nm, whereas nanowires were obtained via hydrothermal technique with a diameter of 3 nm and about 150 nm length. The nanowires were obtained due to the involvement of Ostwald ripening process and variation in the ionic radii of metal ions. Further, the

**Fig. 1** TEM and HRTEM images of synthesized $BaTiO_3$ nanostructures using $Na_2Ti_3O_7$ nanotubes: ($A_1$–$A_3$) hexagonal $BaTiO_3$ nanoparticles ($B_1$–$B_3$) coral-like nanostructures of assembled nanorods. TEM and HRTEM images of $BaTiO_3$ nanostructures prepared from $Na_2Ti_3O_7$ nanowires: (C1–C3) starfish-like nanoparticles and ($D_1$–$D_3$) sword-like nanostructures.
Reprinted with permission from reference N. Bao, L. Shen, G. Srinivasan, K. Yanagisawa, A. Gupta, J. Phys. Chem. C 112 (2008) 8634–8642.

XRD analysis confirmed the crystalline nature of the metal oxide nanostructure with cubic spinel and tetragonal structure of nickel-zinc ferrite (NZF) and BT, respectively. Magnetoelectric coupling is more prominently increasing in the case of hydrothermally produced $Ni_{0.6}Zn_{0.4}Fe_2O_4$ (NZF)-BT nanowires than that of the NZF-BT nanoparticles prepared by chemical combustion. The saturation magnetization of the nanostructured composite is increased from 12.26 to 18.35 emu/g for hydrothermally prepared as compared to chemical combustion prepared (NZF-BT) samples [58]. Liu et al. have prepared three different types of $BaTiO_3$ in the form of nanofibers (NFs), nanocubes (NCs), and nanoparticles (NPs) via electrospun, hydrothermal, and sol-gel routes for piezocatalyzing the dye degradation. The nanofibers have superior degree of catalyzing degradation performance in case of nanofibers (NFs) as a result of greater surface area, fine crystal size, and ease of deformation structure compared to NCs and NPs. Moreover, the piezoelectric effect-induced dye degradation occurs due to production of the OH, $O_2^-$ radicals, which may have potential practical application in water remediation [59]. Yang et al. have reported synthesis of monocrystal barium titanate hollow nanoparticles by sol-gel-hydrothermal technique using acetylacetone as modifier with an average diameter of less than 100 nm. This approach will lead to development and design of shell-type structure of perovskite oxides [60]. Huang et al. [61] reported on a microstructural study of $BaTiO_3$ nanostructures and their synthesis by molten salt method. They have successfully prepared various shapes such as cubical-, rod-, and spherical-shaped nanostructured $BaTiO_3$. In addition, the shapes of the precursors play an important role in developing the various shapes of barium titanate nanostructures. In the morphological study, the SEM images signify that the shapes of the products have resulted as the shape and pace of barium and titanium antecedents. The spherical and rod shape of $TiO_2$ are reflected in the form of spherical- and rod-shaped final product as barium titanate. Therefore, it can be concluded that by varying the shape of the precursors, the product form can be changed effectively [62] and his research group investigated the structure and electronic properties of Ti-W metal oxide nanostructure. They have synthesized these mixed metal oxides by microemulsion route. The structural study by XRD and Raman spectroscopy confirmed the presence of titania content and the crystallinity diminished with the increasing W content and formation of Ti-W mixed metal oxide. Interestingly, it is observed that strain, particle size, and energy bandgap of Ti-W metal oxide as a function of W percentage in the material represent qualitatively the structural properties and electronic behavior of the metal oxide. The diminished bandgap as a result of W content is justified by density functional theory (DFT).

# 3 Synthesis techniques of metal oxides

For the multifunctional device applications, sample preparation techniques with optimizing physical properties play an important role in the field of materials science [63]. Recently, many superior techniques are obtainable in the literature for the synthesis of high purity and homogeneous metal oxide nanostructured composites. The routes with ease of processing, economically viable with high purity and performance are generally adopted.

In the recent past decade, the academic and industrial researchers are most widely using synthesis methods of different metal oxides such as solid-state reaction, sol-gel, combustion, and hydrothermal methods. In the solid-state reaction, technique is simple and easy processing, cost-effective due to the use of carbonates and oxide precursors. On the contrary, the chemical routes require solution comprising acetate, nitrate, chloride, etc., with some reducing agents including ammonium hydroxide in the presence of stabilizing agent. These synthesis techniques are discussed separately in the subsections in detail as follows.

## 3.1 Solid-state reaction technique

The solid-state reaction is a distinguished processing method for achieving stable phases via solid-state diffusion thermodynamically at high range of temperature [64]. The solid-state reaction method has been employed to prepare polycrystalline samples by using oxides and carbonates of high purity, which can be shown by the flowchart as in the Scheme 1. It is a suitable technique for low production costs and easy adaptability for mass production of the materials. The systematic steps toward the synthesis of the materials are comprehended as the following:

### 3.1.1 Step-I: Raw precursor materials

To process the required solid solution, raw materials such as oxides and carbonates of different particle sizes of different elements are selected with high purity. These focused to gain chemical equilibrium and a phase pure solid solution. The presence of impurities in the reactants influences the reactivity as well as the structure correlated properties of the solid solution. The knowledge of volatile materials within the raw material is essential to maintain proper stoichiometric proportions. First of all the

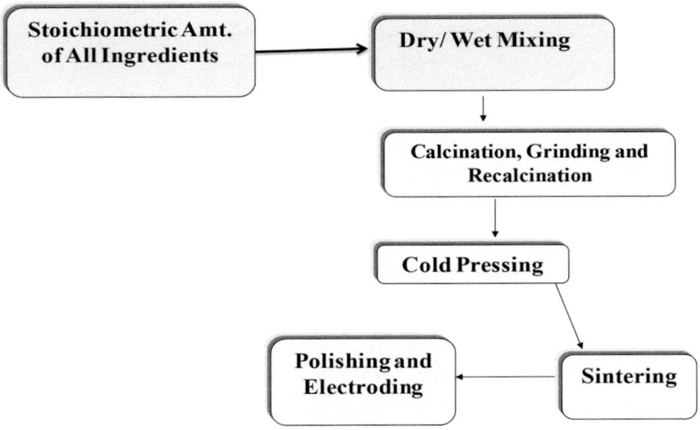


**Scheme 1** Flowchart of Solid-state reaction route.

required amounts of precursors were calculated and weighed according to desired stoichiometry. The calculation of different chemicals can be done in the following way. If the total molecular weight of the reactants of a balanced equation is $M$, the amount of ceramics to be prepared is $X$, the molecular weight of $n$th metallic carbonate/oxide utilized in the formation of ceramic material is $M_n$, and the fraction of $n$th ion is $Z$ (say), then the required amount of $n$th ion is i.e., $M_n = ZM_n/M \times X$ gm. An agate mortar and pestle mix all the measured chemicals (oxides/carbonates). The grinding mixture is obtained by dry grinding in the agate mortar including wet grinding with methyl alcohol medium for several hours for attaining homogeneity.

### 3.1.2 Step-II: Calcination

Calcination is the heating process used to change the physical or chemical constitution of substances without fusion. During calcinations, phase transition, removal of water (which may be absorbed by the materials during the grinding process), volume shrinkage of the mixture, control of the particle size, and removal of volatile fractions occur from the substance entirely or partially along with other thermal decomposition processes. In the thermal decomposition process, some solid solution characteristics such as particle size and their distribution, agglomeration, voids, and microstructure appear. The temperature for calcination aims at reducing Gibb's free energy to zero [65]. This temperature is generally less than the sintering temperature and melting point of the final solid solution. A suitable crucible takes the mixture sample, and the same can be placed in a furnace for a certain interval of time. The calcination temperature is optimized with the completion of the reaction and prevention of volatile oxides. In some cases, repeated calcination is required to obtain single-phase ceramics.

### 3.1.3 Step-III: Grinding and pelletization

The lump formed during calcination is to be grinded using the agate mortar. The ceramics generally exhibit intergranular voids and low density in case of rough grinding. To obtain the desired material, the process of calcinations followed by grinding is repeated. Then a few drops of polyvinyl alcohol (PVA) [66] are added to power of solid solution, acting as a binder to compact the granules of the desired material. By the performance of cold press of materials, various shapes like cylindrical, rectangular, or circular are formed with the help of a die-punch and hydraulic press to compact it.

### 3.1.4 Step-IV: Sintering

Sintering is a thermal process of heating the pellets at an optimized high temperature that is selected between calcination temperature and melting point. During the process, the pellets are densified with the reduction of pores by the approach of grain centers. Crystal growth occurs in sintering due to mass transport that occurs by volume and grain boundary diffusions. This sintering process may occur as constant heating rate sintering, multistage sintering, microwave sintering, isothermal sintering, pressure imparted sintering, plasma-assisted sintering, etc. It is pertinent to mention that PVA (binder) is burnt out during this process. Based on different densification mechanisms, sintering processes may be of the following types:

**(i) Solid-phase sintering**: The densification is carried out by a change in the grains' shape, and all the components remain solid throughout the sintering. Mass transport occurs by volume and grain boundary diffusion. **(ii) Liquid-phase sintering**: The densification occurs mainly by dissolution and reprecipitation of the solid, allowing rapid mass transport. The formation of a viscous liquid fills the pore spaces of the initial green body. **(iii) Reactive sintering**: The formation of a new compound carries out the densification. In this case, two or more constituents react during sintering.

When the particles merge, the reduction of surface energy occurs, this acts as the driving force for sintering. Generally, to achieve the liquid-phase free composition, the sintering was carried out below the melting point by atomic motions that eliminate the high surface energy. The actual mechanisms involved during the sintering process are to increase the interparticle contact area as a function of sintering time and temperature, to achieve maximum physical densification with respect to its theoretical density values, to minimize the volume of interconnected pores of the sintered products, to enable pore migration to the surface, to control grain growth, and to control shrinkage without geometrical deformation.

**Step-V: Electroding**

For electrical characterization, the electrode of the samples is essential. In this process, fine emery paper, and then a thin layer of conducting material such as gold, silver, or platinum, are polished with the opposite faces of the sample. After the electroding, the sintering pellet becomes a capacitor and is used for different electrical measurements.

## 3.2 Sol-gel method

Sol-gel technique in material processing is one of the versatile routes to synthesize and understand the materials as well [67]. The materials prepared via sol-gel method acquire the benefits of organic as well as inorganic characteristics. This processing technique is advantageous due to less expensiveness, ease of processing at low temperature, and permitting a good control on chemical composition of the developed material [68]. Moreover, this technique allows us to have a control over the chemical properties, morphology, structure, and texture of the product. It provides a better-quality chemical reaction with the molecules, which becomes favorable for enhancing the chemical homogeneity [69]. However, this technique is a creative route for developing metal oxide nanostructure by offering mixture of the precursors with homogeneity on molecular level. Generally, organometallic or inorganic salts are taken as starting precursors, and then the ingredients undergo a sequence of hydrolysis as well as polymerization reactions for producing "sol" in the form of colloidal suspension (Scheme 2). Moreover, the sol is subjected to further processing to obtain various forms like fibers, films, monolith, and monosized powders [70]. The process of sol-gel engrosses the transformation of a solution phase from a colloidal liquid system

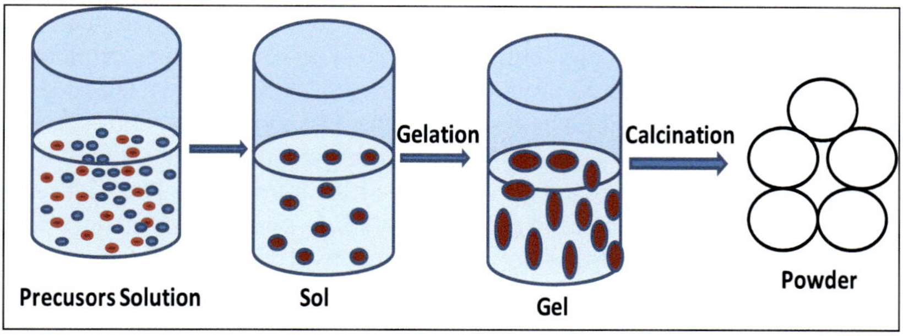


**Scheme 2** Schematic flow diagram of sol-gel synthesis route.

i.e., "sol" into a solid "gel" system. There are several literatures based on sol-gel technique, which have been reported during the past decade: Gao et al. [71] have synthesized zinc oxide (ZnO) monolith via sol-gel route using solution of zinc nitrate with an initiator. In this route, the zinc nitrates act as a solvent and are then constantly stirred in the presence of propylene oxide. The prepared solution was kept still to produce ZnO monolith. Similarly, Li et al. [72] and his group have synthesized zinc oxide nanostructure by sol-gel technique. In this typical route, zinc nitrate was taken as precursor with deionized water that acts as a solvent and stabilizing agent as methenamine. Both the precursors were taken in equimolar concentration and then homogenized with the help of magnetic stirring for several hours with a temperature of 60°C. Further, the solution was kept in air environment for 24 h to obtain homogeneous neutral solution for preparation of zinc oxide nanostructured materials.

## 3.3 Combustion method

Combustion synthesis technique is one of the well-known techniques for processing a broad range of materials because of its simplicity and wide range of applicability, ease of producing with preferred composition [73]. This process involves a series of exothermic reactions among fuel and an oxidant to maintain a chemical reaction. The fuels which are used in this technique are mostly hydrocarbons that may be in liquid, solid, or gaseous states. On the basis of the physical state of the fuel, it can be broadly classified into solid-phase, solution-phase, and gas-phase combustion synthesis. This synthesis technique allows a wide variety of materials such as oxides, spinel ferrites, metals, alloys, and sulfides. Jayalaxmi et al. [74] have prepared ZnO nanopowder via combustion technique by taking zinc nitrate and aqueous dextrose solution as starting materials. These solutions were then heated for few minutes to obtain a gel and followed by heating at high temperature of 400°C for 5 min in a furnace to produce a highly amorphous powder.

## 3.4 Hydrothermal method

Hydrothermal synthesis is generally aqueous solution-based wet chemical route for preparing various metal oxide nanoparticles [75–79]. Generally in this process, the aqueous solution-based metal precursors were placed in an autoclave (Teflon lined stainless steel closed chamber as thermal reactor). The various types of metal oxide nanostructures can be produced by inducing different temperatures and time during the treatment of hydrothermal process. The dimension, morphology, and crystal structure of the nanoparticles can be controlled depending on the use of various solvents, nature and concentration of metal precursors, surfactant/polymer reaction temperature, duration of time, and pH of the precursor solution. Yang et al. [80] have reported $Fe_3O_4$@C hybrid nanoparticles were prepared via hydrothermal synthesis route. In this synthesis route, metal nitrate, ammonium hydrogen phosphate, and glucose were first dissolved in distilled water and mix homogeneously by magnetic stirring. Then the solution transfer into Teflon lined stainless steel autoclave by using temperature 180°C for 48 h. After completion of the reaction, the solution was cooled to room temperature and washed several times with deionized water and dried overnight followed by calcinations at 450°C to get core-shell hybrid nanoparticles.

# 4 Percolation theory

The percolation threshold is the simplest fundamental model based on the percolation theory. This typical theory shows the presence of metal-insulator-semiconductor transition for conductive networks in the polymer matrix of composites with specific concentrations at percolation threshold [81–83]. It also revealed that each conductive composite at a particular concentration shows the presence of critical filler concentration (recognized as percolation threshold). This percolation theory was established in the year of 1957 by Broadbent and Hammersley [84], and it is used to explain the problems of fluid flow via static medium. The percolation threshold was initially determined with the progress of 2D and further was established in 3D geometries. Generally, the percolation has been classified into two types, i.e., site percolation and bond percolation [85]. In site percolation, the sites are either filled or vacant (uniform distribution or nonhomogeneity) in a matrix; on the contrary, in case of bond percolation, the sites in a matrix are filled completely but are either interconnected between particles in polymer matrix or not.

In addition, percolation theory verifies the performance of composite materials comprising conducting and nonconducting phases. These materials exhibit abrupt transition from insulator to conductor at percolation threshold. The relationship between electrical conductivity of the composite and concentration of conductive filler near the percolation threshold is signified by the simple universal power law [86–88] given by the equation.

$$\sigma_c = \sigma_o \left(\Phi_f - \Phi_c\right) t \tag{1}$$

where $\sigma_c$ is the conductivity of the composite, $\sigma_o$ is the electrical conductivity of filler content, $\Phi_f$ is the concentration of conductive fillers (wt%), $\mathit{\Phi}_c$ is the percolation threshold, and $t$ is the fitting parameter (critical exponent).

From the Eq. (1), it is clear that the concentration above percolation threshold (when $\phi_f > \phi_c$) with the value of the fitting parameter "$t$" dependent upon the lattice dimension with aspect ratio of filler content or quality dispersion and orientation of filler particles. For instance, the superior aspect ratios of the fiber-reinforced composites have larger value than that of the other filler-based composites. The calculations from the universal power law equation depend on the critical exponents of the $\phi_c$ and $t$ values. Moreover, the $\phi_c$ value can be determined experimentally and $t$ value is obtained by fitting the experimental data. However, the pristine polymer and filler particles have the same values for the density; mass fraction is equal to the volume fraction of filler contents applied in this equation. This equation can be described by the percolation theory and most of the previous literatures have reported of determining the value of percolation threshold by the experimental fitted data. The value of the percolation threshold is normally determined by plotting a graph between electrical conductivity of the composites with volume fraction of filler content fitted with a power law [86–90]. There are various reports which have been demonstrated that the percolation theory depends on the type of polymeric materials, preparative technique of the composites, aspect ratio of conductive materials with aggregation and agglomeration, homogeneous distribution of conductive materials, and their degree of orientation [86,88,89].

# 5 Polymer-metal oxide nanocomposites

The polymer-based hybrid nanocomposites have attractive characteristics for practical application in the field of academics and industrial research through the combination of matrix materials and metal oxides. The polymer-based nanocomposites exhibit outstanding dielectric and electrical properties, sensing, catalytic, and excellent optical performance, which have an immense area of interest in this present time. However, the effect of other parameters including microstructural properties could be determined, controlled, and enhanced accurately by tailoring carefully during the preparation of the composite materials [12–15,91,92]. In this respect, the core-shell nanocomposites were prepared by using one-step template-assisted oxidative polymerization reaction through various oxidative agents. These nanocomposites exhibit unique size, higher catalytic properties, and different morphology of metal nanoparticles [e.g., gold (Au) [93–97], palladium (Pd), silver (Ag) [98–100], and copper (Cu) [101–103] can be encapsulated with conducting polymers. There are several strategies that have been made based on the elevated dielectric constant metal oxide/ceramic-based fillers [e.g., barium titanate ($BaTiO_3$; BT)] incorporated into the polymer matrix composites, which exhibit excellent physical properties (dielectric and conductivity), easy processing, good flexibility in the field of dielectric-based energy storage system [12–16,89,104–106]. Further, the large dielectric constant values have changed nearer to the percolation threshold but the value of the dielectric constant of the composite simultaneously reduced; on the contrary, the higher volume fraction of ceramic fillers that are incorporated into the polymer matrix may cause some undesirable effects on the composites due to their high surface energy, poor processability, mechanical brittleness, high dielectric loss, and high defect density [89,105,107,108]. Dang et al. [109] have fabricated barium titanate ($BaTiO_3$; BT)-poly(vinylidene fluoride) (PVDF) nanocomposites with different volume fractions of filler contents by using solution processing technique, which revealed that the particle dimension of BT is reduced to 100 nm as compared to from 500 nm. The low-frequency dielectric constant and dielectric loss of these composites have significantly increased with decreased particle size of BT. Fan et al. have reported that the dielectric performance of BT-PVDF nanocomposites film has enhanced with particle size decrease in the range from 100–150 nm to 30–50 nm [110].

Besides, the percolation behavior of different parameters occurred in the conductive filler-based polymer composites i.e., high dielectric insulating filler particles, size of metal particles, role of polymer matrix, semiconducting fillers, and conducting polymers. For instance, Thakur et al. [111] have stated that the synthesis and percolation behavior of polymer-based metal composites. It is observed that the composites have larger critical exponents and percolation threshold (0.30) with effective dielectric constant (300) and simultaneously achieved low loss (0.1) at 100 Hz. Besides, the nanocomposites ($Fe_2O_3 + TiO_2 + NiFe_2O_4$) were synthesized via sol and coprecipitation technique with encapsulation into the unsaturated polyester resin [112]. The synthesized nanocomposites show high electrical conductivity, low resistivity, and better dispersion into the polymer matrix. The nanocomposites with maximum tensile strength (21.62%) and Young's modulus (6.56%) might be potentially used in both industrial and light shielding applications [112,113]. He et al. [81] have synthesized the three-phase syndiotactic polystyrene (sPS)-$BaTiO_3$-graphite nanosheeets (GNs) composites via solution blending technique. The composites exhibit high dielectric permittivity of 51.8 at $10^2$ Hz when the concentration of GNs is increased at low percolation threshold ($f_c = 1.22$ vol%), which is nearly 18 and 7 times larger than that of the pristine polymer matrix and two-phase PS-$BaTiO_3$ composite systems. For the improvement of compatibility between filler particles and polymer matrix, polyethylene glycol (PEG) is used as surface modifier in the ferroferric oxide ($Fe_3O_4$)-based poly(vinylidene fluoride) (PVDF) composites [114]. The PEG-modified filler particles are in the good contact with the interfaces in the polymer matrix, so that the increase in the interfacial polarization results in larger value of the dielectric constant above percolation threshold.

# 6 Properties of conductive filler-based percolative polymer composites

The dielectric and electrical performances of polymer composites are extremely dependent on the presence of filler particles and their potential application in energy storage devices [81]. The most common technique of preparing 0–3 type composites is to add high dielectric constant ceramic filler particles into the polymer matrix. However, the higher volume fraction of filler particles (>40%) is essential to achieve high dielectric constant but simultaneously affects the mechanical performance of such polymer composites [81–83]. In order to overcome the shortcomings, addition of conductive particles into the matrix as the third phase is required in which total amount of filler particles is much lesser. According to percolation theory, the electrical (conductivity) and dielectric properties (dielectric constant) are enhanced considerably when the quantities of conductive filler particles come closer to the critical concentration (i.e., percolation threshold) [82,83]. Percolation is a distinguished phenomenon observed in the filler particle-polymer matrix system. In literatures, it is reported that there is considerable change in certain physical properties within a quite narrow concentration range noticed including composites comprising of insulating polymer matrix and conductive particles. Moreover, the change in electrical properties (conductivity) is clarified by the formation of conductive network during the sample when adequate contacts are formed between conductive particles corresponding to the percolation threshold [81,115,116]. The percolation is a type of geometric phase transition which is scale invariant. In this phase of transition, the single cluster forms a percolation cluster, which changes abruptly the state of a percolation system. Further, the percolation concentration depends on various factors including physical nature of the filler (surface area, shape, and aggregation), chemical composition of the surface, and the processing conditions [115,116].

## 6.1 Dielectric properties

The fabrication of three-phase percolative composite systems comprising of $Fe_3O_4$@ $BaTiO_3$ core-shell nanoparticles was well embedded into the poly(vinylidene fluoride) (PVDF) matrix. The experimental results of the $Fe_3O_4$@$BaTiO_3$-PVDF composites show higher value of the dielectric constant (893) and simultaneously achieved reduced dielectric loss of 0.9. As a result, these three-phase composites could be applicable in the field of superior dielectric application [107,117]. Li et al. have demonstrated a polydopamine (PDA)-coated barium titanate particles were then incorporated into the polymer PVDF matrix. These composites exhibit excellent enhancement of the dielectric constant ($\approx$70) with substantially low loss ($<$1), which is due to the presence of PDA coating layer. Moreover, the prepared composites possess high dielectric constant and ultralow loss making them a significant material for the industrial and electronics applications, especially embedded capacitors [118]. Lee et al. have synthesized dopamine-functionalized $BaTiO_3$ nanoparticles under the influence of reflux method and were successfully embedded into the poly(vinylidene fluoride) (PVDF) matrix to form unmodified (Fig. 2A and B) and dopamine-modified $BaTiO_3$-PVDF (Fig. 2C and D) nanocomposites film [119]. The resultant modified nanocomposites at 50 wt% of barium titanate filler contents showed high dielectric constant of 56.8 and suppressed dielectric loss up to 0.04 at $10^3$ Hz due to the presence of organic surface layer on the nanoparticles (Fig. 1). Moreover, the dielectric constant of the modified nanocomposites is significantly enhanced as compared to the composites with unmodified $BaTiO_3$-PVDF composites [119,120].

Phatharapeetranun et al. have designed new type of composites by using barium titanate nanofibers based on poly(vinylidene fluoride) nanohybrid (BT NFs-PVDF) via 3D-printing-fused deposition modeling (FDM) technique. The resultant barium titanate nanofibers (BTNFs)-based PVDF nanohybrid with 20 wt% of filler contents

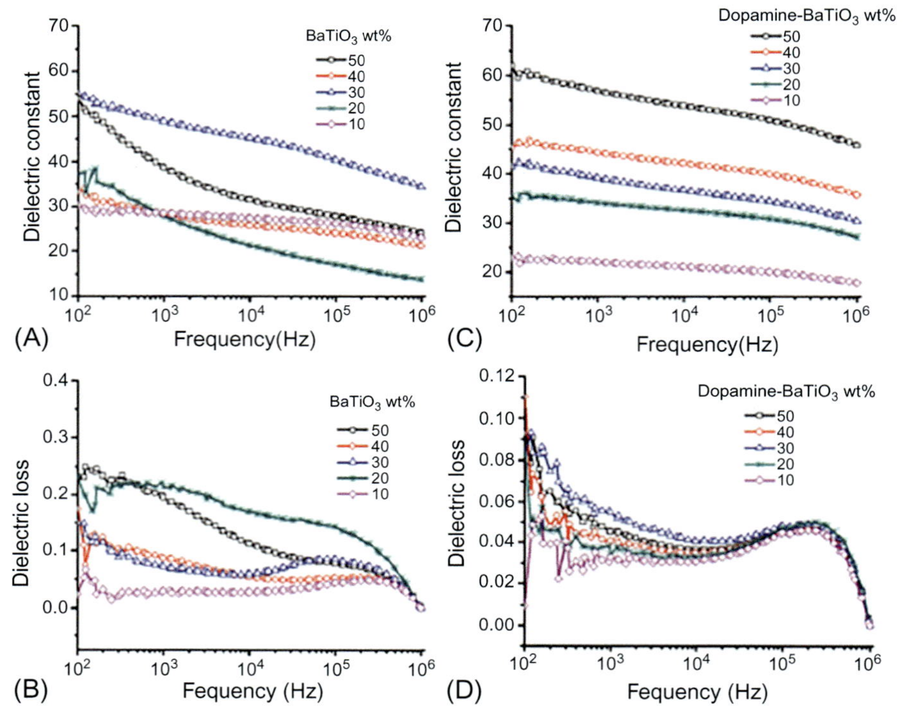


**Fig. 2** Frequency dependence of (A, C) dielectric constant and (B, D) dielectric loss of unmodified $BaTiO_3$-PVDF and dopamine-modified $BaTiO_3$-PVDF composites measured at room temperature.
Reprinted with permission from reference M.F. Lin, V.K. Thakur, E.J. Tan, P.S. Lee, RSC Adv. 1 (2011) 576–578.

revealed a large improvement in dielectric constant about 200 at 1 kHz, which was much higher than that of the pristine PVDF matrix. These excellent enhancements of dielectric constant of the anisotropic flexible 3D nanohybrid composites could have potential applications for complex-shaped embedded capacitors and electric energy storage devices [120,121]. Xie et al. [122] have successfully synthesized polyimide-based $BaTiO_3$ composites via colloidal process. In this technique, the ultrafine $BaTiO_3$ particles are incorporated into the poly(amic acid) (PAA) solution, followed by film casting and imidization. The resultant composites showed high dielectric constant (35) (50 wt%) which is 10 times greater than that of the neat polyimide matrix. Meanwhile, it could be concluded that the synthesized composites should be used as the most promising candidates for dielectric applications [15,19,122]. Li et al. have reported a facile and effective method of the calcined barium titanate powder based on PVDF composites. The obtained barium titanate-PVDF composites unfold that the barium titanate filler loading with suitable temperature at 950° C could enhance the dielectric property. However, the enhancement of the

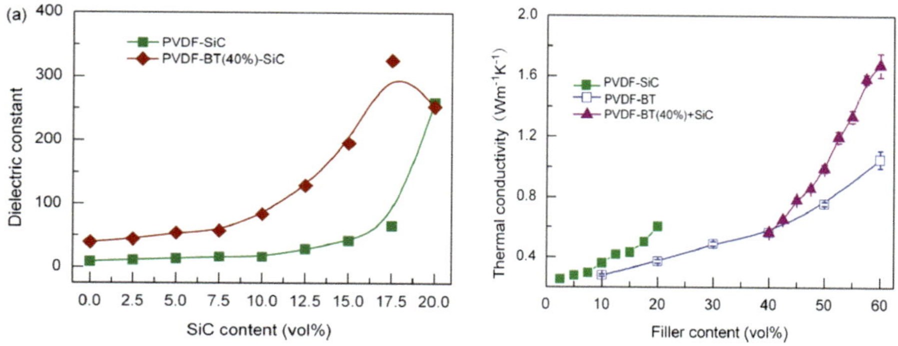


**Fig. 3** Dielectric properties of PVDF-$BaTiO_3$-β-SiC composites at $10^3$ Hz and (b) thermal conductivity of PVDF-BT, PVDF-β-SiC, and PVDF-$BaTiO_3$-β-SiC composites at 25°C. Reprinted with permission from reference Y. Li, X. Huang, Z. Hu, P. Jiang, S. Li, T. Tanaka, ACS Appl. Mater. Interfaces 3 (2011) 4396–4403.

dielectric constant was up to 166.38 and negligible loss with improved piezoelectric strain constant (25 pC/N) at $10^4$ Hz [123]. The three-phase composite system containing poly(vinylidene fluoride), barium titanate (BT), and β-silicon carbide (β-SiC) whiskers act as fillers were prepared by Li et al. and his group. The three-phase composite with 17.5 vol% of β-SiC whiskers showed large enhancement of dielectric constant from 39 to 325 (Fig. 3A) at $10^3$ Hz. On the contrary, the addition of 20 vol% of β-SiC whiskers improved the thermal conductivity of PVDF-BT nanocomposites (1.05–1.68 $W m^{-1} K^{-1}$ at 25°C) than that of the PVDF-BT (0.37 $W m^{-1} K^{-1}$) and PVDF-β-SiC (0.6 $W m^{-1} K^{-1}$) composites. Thus, this synthesized composite system could provide a better dielectric material for the potential utility in the field of modern electronics and electrical industries [124].

Fu et al. [125] have successfully fabricated the barium titanate ($BaTiO_3$)-PVDF composites and the surface-modifying agent polyvinylpyrrolidone (PVP) were used on top of the facade of barium titanate particles through simple molten salt method [125]. The resultant PVP-modified BT-PVDF composites exposed that the BT particles of the dimension of 600 nm acquired the tough polarization, attributed to the ferroelectric size effect, which was significantly enhanced by the dielectric constant of 65 at 25°C. Although the energy storage density of the composite with 40 wt% of BT loading was increased up to $30 \times 10^{-3}$ J/cm$^3$ at 10 kV/mm, which is 4.5 times greater than that of the pristine PVDF matrix, the synthesis of functionalized PVP-BT-PVDF composites with excellent dielectric properties could be considered as better promising candidates for high-technology field applications [81,104]. Sheng et al. [126] have provided an effective and facile approach toward the synthesis of the core-shell polydopamine (PDA), encapsulated into the barium titanate (D-BT) particles and then successfully grafted into the poly(vinylidene fluoride-*co*-trifluoroethylene) [P (VDF-TrFE)] matrix to improve the dielectric properties and energy storage density through a convenient solution casting method [Fig. 4]. The results showed that elevated dielectric constant of 46.4 and reduced tangent loss of 0.07 with 100 Hz in

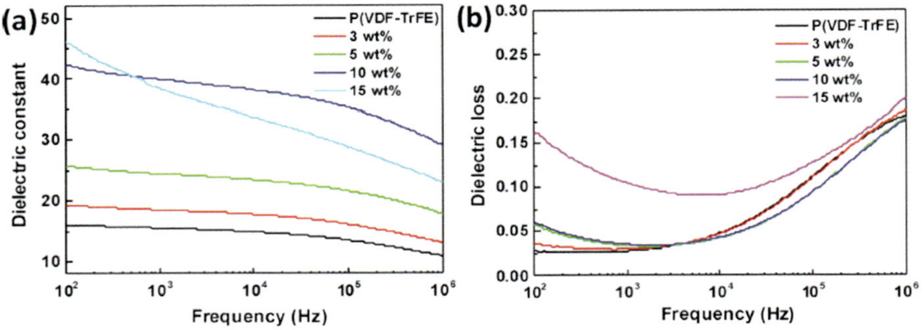


**Fig. 4** Variation of frequency of (A) dielectric constant and (B) dielectric loss of P(VDF-TrFE) nanocomposites film.
Reprinted with permission from reference Y. Sheng, X. Zhang, H. Ye, Y. Liang, L. Xu, H. Wu, Colloids Surf. A Physicochem. Eng. Asp. 585 (2020) 124091.

15 wt% of D-BT-P(VDF-TrFE) composites in Fig. 4A and B. This may be attributed to the effect of the interfacial polarization and also the presence of strong hydrogen bond between PDA layers and macromolecular chains, so that the value of the dielectric constant increased and also achieved negligible dielectric loss (<1). Consequently, the synthesized D-BT-P(VDF-TrFE) composite provides green routes to accomplish largely enhanced dielectric properties, which could be applicable in the polymer dielectric nanocomposites for film capacitors application [126,127].

Wang et al. have demonstrated barium titanate-polyimide nanocomposites film via in situ polymerization process followed by the surface modification by the two dispersants such as 2-phosphonobutane-1,2,4-tricarboxylic acid (PBTCA) and TH-615 acrylic-acrylate-amide copolymers [128]. Moreover, the use of modifying agent in the nanocomposites would give better dispersion and uniformity of the filler particles in the polymer matrix, also the interfacial compatibility between two phases was improved with 8% of PBTCA contents. Besides, the dielectric constant of $BaTiO_3$-polyimide composite film was 23.5 with loss tangent 0.00942 at $10^{-3}$ Hz, breakdown strength $80 MVm^{-1,}$ and energy storage density $0.67 J cm^{-3}$. On the contrary, $BaTiO_3$ was modified by 6% of TH-615, with the value of the dielectric constant 20.3, loss tangent 0.00571, breakdown strength $73 MV m^{-1,}$ and energy storage density is $0.68 J cm^{-3}$. Feng et al. have illustrated a PVDF-based barium titanate (BT) ternary nanocomposites system (Fig. 5) through the introduction of MXene nanoparticles via solution casting process [129]. The results show that the $BaTiO_3$-MXene-PVDF nanocomposites (Fig. 5A and B) with low filler loading (8 wt% of BT and 2 wt% MXene) achieved high dielectric constant of 77, which is nearly 7.7 times higher than that of the neat PVDF and low loss of 0.15 at 100 Hz. Thus, this approach might open up the way for large-scale production of dielectric materials in the field of electronic industries.

A novel core-shell type composite comprising of barium titanate (BT) and polypyrrole (PPy) was modified with graphene oxide (GO) via in situ polymerization method by

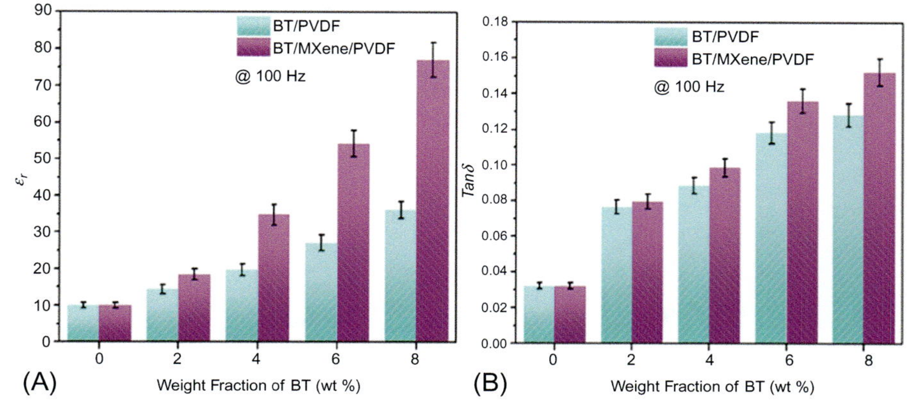


**Fig. 5** Frequency dependence of (A) dielectric constant and (B) dielectric loss of $BaTiO_3$-MXene-PVDF nanocomposites film.
Reprinted with permission from reference Y. Feng, Q. Deng, C. Peng, Q. Wu, Ceram. Int. 45 (2019) 7923–7930.

Wang et al. [130]. In this report, there are two types of composites such as PVDF-$BaTiO_3$@PPy and PVDF-$BaTiO_3$@PPy-GO, which have fabricated and measured dielectric properties. Although the PVDF-BT@PPy-GO composites show lower dielectric constant as well as lower dielectric loss and higher breakdown strength compared with the unmodified PVDF-BT@PPy composites, this provides the PVDF-BT@PPy-GO composites that might have better applicability in the field of microelectronic devices. However, the nano-barium titanate (NBT)-based PVDF composites were prepared via hydrothermal processing technique through the introduction of $Ni(OH)_2$ into the NBT for better dispersion of particles within the polymer matrix. Therefore, Yang et al. [131] and his research groups have synthesized NBT-PVDF composites and it exhibits effectively enhanced dielectric constant at low (2.5 wt%) filler contents and also better functional properties of the resultant composites. Furthermore, this method lay out better promising candidates for the preparation of conducting high dielectric materials. Zha et al. [132] have demonstrated the BT@$SnO_2$ fillers with nanoscale semiconductive $SnO_2$ particles coated on the BT surface were prepared by a simple chemical treatment comprising the PVDF matrix. Meanwhile, the experimental results show that with lower content of $SnO_2$ nanodots effectively enhanced the dielectric properties of BT-PVDF nanocomposites. Feng et al. have synthesized the BT-nanowires (NWs) successfully through one-step hydrothermal method by incorporating poly(vinylidene fluoride)-*co*-hexafluoropropylene [P(VDF-HFP)] matrix to form P(VDF-HFP)-BT-NWs composite (Fig. 6). As a result, the obtained composites (10 vol%) show significant improvement (9603, 391, and 49 at 0.01, 1, and 100 Hz) in the dielectric properties (Fig. 6). These improved dielectric performances of P(VDF-HFP)-BT-nanowires composite were accredited to the greater interfacial polarization and elevated aspect ratio of BT-nanowires [133].

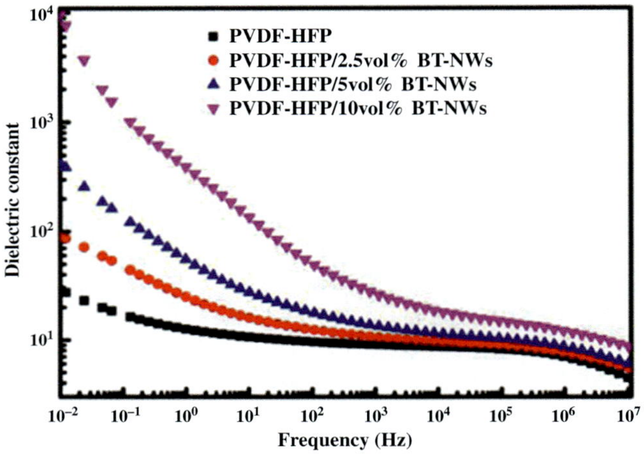


**Fig. 6** Variation of frequency with dielectric constant and of P(VDF-HFP)-$BaTiO_3$-NWs composites film.
Reprinted with permission from reference Y. Feng, W.L. Li, Y.F. Hou, Y. Yu, W.P. Cao, T.D. Zhang, W.D. Fei, J. Mater. Chem. C 3 (2015) 1250–1260.

Kim et al. [134] have fabricated a novel composite system of barium titanate ($BaTiO_3$)-epoxy resin composites where barium titanate ($BaTiO_3$) particles were directionally aligned in the epoxy resin via ice-templating method. It is observed that there is an effective increase in the dielectric permittivity with low loss as compared to the traditional composites structure, which could be used as a good dielectric material in the field of high energy density capacitors. The satellite core-shell-structured $Fe_2O_3$@$BaTiO_3$ ($Fe_2O_3$@BT) nanoparticles were fabricated and then incorporated into the PVDF-HFP matrix by Zhang et al. [135]. The satellite core structure prevents $Fe_2O_3$ particles from direct contact with each other, to increase the dielectric constant of 31.7 at 1 kHz, which was nearly 1.8 and 3.0 times higher than that of the 20 wt% BT and neat P (VDF-HFP), respectively, with a suppressed loss of 0.05. In addition, this composite-processing technique may be a green synthesis route for the potential applications in the flexible electronics. Niu et al. [136] have reported on the preparation of BT-PVDF composite in the form of thin films by ball-milling technique with enhancement in the dielectric performance compared to that of the composites prepared via stirring method. The results show high dielectric permittivity at 30 wt% of BT particles with high breakdown strength, which might be valuable in terms of application in the realm of future energy storage devices [81,92].

Shen et al. [137] have prepared dopamine-modified $BaTiO_3$ nanofibers that act as filler with a large aspect ratio comprising PVDF-TrFE matrix-based nanocomposites through electrospinning method. The composites possess high dielectric performance with low volume fraction of $BaTiO_3$ contents into the matrix. This synthesis strategy for modifying $BaTiO_3$ nanofibers could be applied for the improvement of both dielectric and ferroelectric properties in the polymeric nanocomposites. The barium titanate was successfully embedded into the polydimethylsiloxane-α, ω- diols matrix-based electroactive elastomeric composites by Bele et al. and his coworkers [138]. The results revealed significant enhancement of dielectric permittivity of 4.41 at 100 Hz and 20 °C temperature. This may provide a good approach for potential future application of such materials in energy harvesting capability, electrochemical sensitivity, and electrochemical device applications [122,123]. Upadhyay et al. have prepared the composite films of barium titanate as filler with β-phase PVDF via solution casting method where Ti (IV) triethanolaminato isopropoxide and hydrated barium hydroxide as precursors and tetramethylammonium hydroxide (TMAH) as base are used [139]. The dielectric constant and dissipation factors of the obtained nanocomposites were found to be about 7 and 0.03 respectively, which may be potential candidates for novel capacitors application [114,117]. Similarly, Choudhury et al. [140] and his groups have synthesized polyetherimide-encapsulated barium titanate (PEI-$BaTiO_3$) nanocomposites film through mixing of fine $BaTiO_3$ particles into poly(amic acid) solution under ultrasonication followed by film casting and thermal imidization. These nanocomposites exhibit high dielectric constant with 50 wt% $BaTiO_3$ filler loading, which was nearly about 12 times greater as compared to the neat polymer matrix in one hand and the dielectric tangent loss only improved by five times on the other hand.

## 6.2 AC electrical conductivity

The composites of BT-$Fe_3O_4$-PVDF on deposition of magnetic iron oxide ($Fe_3O_4$) nanoparticles on the surface of barium titanate were successfully grafted onto PVDF matrix followed by the introduction of external magnetic field by Zhang et al. [141]. The experimental results showed the significant enhancement of the dielectric permittivity value (385 at 100 Hz) with low dielectric loss 0.3 and low electrical conductivity $4.9 \times 10^{-9}$ S/cm in the composites. This work provides a green synthetic route for the synthesis of polymer composites with high dielectric properties, which might be used as promising candidate in the field of electronic industries [92,104]. Yang et al. [142] have prepared barium titanate ($BaTiO_3$) as fillers of size 100 nm and 200 nm, respectively, which were incorporated into the epoxy composites. Moreover, the $BaTiO_3$-based epoxy composites exhibit high electrical conductivities at lower frequency region. Silakaew et al. have studied the dielectric and electrical properties of multiwalled carbon nanotube (CNT)-deposited $BaTiO_3$ -PVDF composites with variation of $BaTiO_3$ particles sizes about 100 nm (nBT) and 0.5–1.0 μm (μBT). So, the resultant composite systems at 0.019 wt% of CNT contents revealed the excellent enhancement of dielectric performance ($\approx 155$ at $10^3$ Hz) and low electrical conductivity (Fig. 7) about $6.82 \times 10^{-9}$ S/cm. The outstanding improvement of the dielectric performance and low electrical conductivity of the CNT-μBT-PVDF composites are attributed to the interfacial polarization and also the tetragonal phase of μBT [143].

Xie et al. have reported double core-shell-structured barium titanate ($BaTiO_3$) encapsulated with hyperbranched aromatic polyamide (HBP) nanoparticles were embedded into the poly(methylmethacrylate) (PMMA) matrix to form BT@HBP-PVDF composites via two-step polymerization such as ATRP and conventional blending technique. It is observed that the prepared BT@HBP-PVDF composites possess

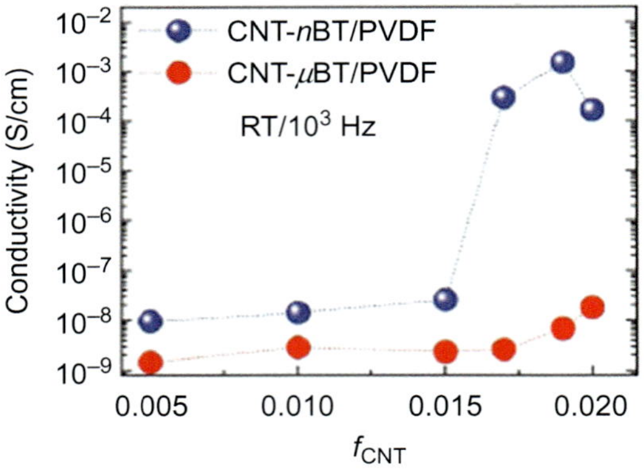


**Fig. 7** Dependence of electrical conductivity of CNT-μBT-PVDF composites as a function of filler contents.
Reprinted with permission from reference K. Silakaew, P. Thongbai, RSC Adv. 9 (2019) 23498.

lower electrical conductivity at lower frequency region due to the transfer of charge carrier into the dielectric material [144]. Ren et al. have synthesized BT@Ag-PVDF composites consisting of barium titanate ($BaTiO_3$, BT) particles that were surface modified by silver (Ag) nanoparticles with an average diameter of 5–10 nm via electroless plating and then grafted into the PVDF matrix by using solution casting technique [145]. Meanwhile, the BT@Ag-PVDF composites [Fig. 8] with 35 vol% of BT@Ag content show that the DC conductivity was retained to $10^{-12}$ S/cm, instead of increases rapidly owing to the coulomb block effect of Ag nanoparticles. For the temperature dependence of electrical conductivity, there are some potential differences between BT@Ag-PVDF and BT-PVDF composites in high-temperature range. Therefore, this work could shed some light on better optimization of the polymer composites with unique dielectric and electrical properties.

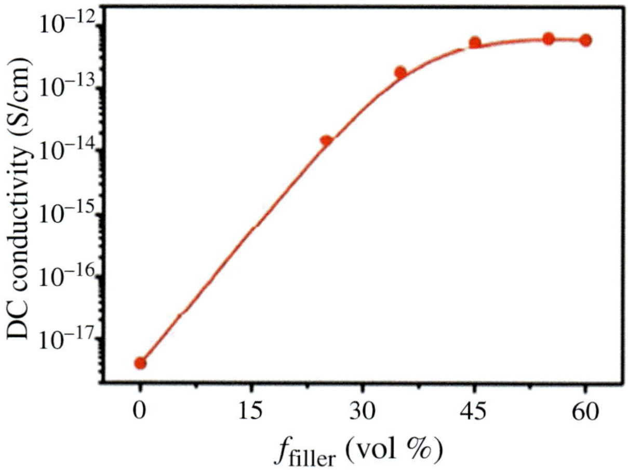


**Fig. 8** Dependence of DC electrical conductivity of BT@Ag-PVDF composites as a function of filler contents.
Reprinted with permission from reference L. Ren, X. Meng, J.W. Zha, Z.M. Dang, RSC Adv. 5 (2015) 65167–65174.

The preparation of composites with surface-modified barium titanate particles by polyethylene glycol (PEG) was successfully rooted into the poly(vinylidene fluoride) (PVDF) matrix to form PEG-BT-PVDF composites. The AC electrical conductivity of the prepared composites enhances up to $5.4 \times 10^{-6}$ S/cm with 7 vol% of filler content followed by the Jonscher's universal power law. Thus, this significant increase in the electrical properties could be potential applications for ideal capacitor and microscopic development of the composite [146]. The same group has also demonstrated a three-phase percolative composite system comprising of polystyrene 2% divinylbenzene (PDB)-modified barium titanate-based PVDF matrix composites via solution casting technique. It is observed that the AC electrical conductivity is significantly increased with the increase of filler loading and owing to the addition of PDB, which suppressed the effective energy separation of charge carriers. Moreover, the obtained ceramic-based polymer (PDB-BT-PVDF) composites may be attributed to the development of high-performance dielectric materials for energy storage devices [147].

Similarly, Li et al. [148] have investigated the nanographite (GN)-doped ternary nanocomposites system consisting of binary polyvinylidene fluoride and barium titanate (PVDF-$BaTiO_3$) via simple solution casting method followed by compression molding technique. In this respect, the electrical behavior of ternary PVDF-$BaTiO_3$-GN hybrids can be explained on the basis of percolation theory. On the contrary, the electrical conductivity of hybrids was found to be strongly frequency and temperature dependent with the increase in the filler loading. Wang et al. [149] have reported the deposition of Ag nanoparticles on the surface of barium titanate particles ($BaTiO_3$) which were well embedded into the polymer PVDF matrix to form Ag@BT-PVDF composites by using coaxial electrospinning technology followed by the calcinations. The resulting Ag@BT-PVDF composite shows that higher AC electrical conductivity is nearly about $10^{-10}$ S/cm at lower frequency, which indicates the electrically insulate composites.

# 7 Conclusions

The synthesis techniques of various metal oxides and their composites have brought new properties and potential applications in various ranges with improved performance; there are still many confrontations in this field. Since the high dielectric constant, metal oxide-based polymer composites show huge potential in the field of dielectric applications due to their ease of processing, flexibility, low cost, and good chemical stability.

A lot of research studies have been undertaken for making better dielectric with electrical properties of the metal oxide-based polymer composites. However, a larger aspect ratio and lesser percolation threshold of conductive filler-encapsulated polymer matrix composites is a remarkable subject. The percolation models are mainly simple, and their application with attractive mathematical structure and fundamental aspects of materials science in terms of the composite systems is very important for practical applications. The high dielectric constant and reduced dielectric loss can be achieved for metal oxide-polymer composites with very low filler concentration at percolation threshold. Thus, the progress in metal oxide-based polymer composites is significantly dependent on valuable and interdisciplinary collaboration with diverse fields of chemistry, physics, and materials science.

## Acknowledgment

The author, RN Mahaling, gratefully acknowledged DST-SERB for financial support obtained through project grant of (CRG/2018/004101), New Delhi, Government of India.

## References

[1] Y. Zhao, X. Li, B. Yan, D. Xiong, D. Li, S. Lawes, Recent developments and understanding of novel mixed transition-metal oxides as anodes in lithium ion batteries, Adv. Energy Mater. 6 (2016) 1502175.

[2] Z.S. Wu, G. Zhou, G.Z. Yin, W. Ren, F. Li, H.M. Cheng, Graphene/metal oxide composite electrode materials for energy storage, Nano Energy 1 (2012) 107–131.

[3] H.B. Wu, G. Zhang, L. Yu, X. Wen, One-dimensional metal oxide–carbon hybrid nanostructures for electrochemical energy storage, Nanoscale Horiz. 1 (2016) 27–40.

[4] K.T. Adjemian, R. Dominey, L. Krishnan, H. Ota, P. Majsztrik, T. Zhang, J. Mann, B. Kirby, L. Gatto, M.V. Simpson, J. Leahy, S. Srinivasan, J.B. Benziger, A.B. Bocarsly, Function and characterization of metal oxide−nafion composite membranes for elevated-temperature $H_2/O_2$ PEM fuel cells, Chem. Mater. 18 (2006) 2238–2248.

[5] W.Y. Wong, W.R.M. Daud, A.B. Mohamad, A.A.H. Kadhum, K.S. Loh, E.H. Majlan, Recent progress in nitrogen-doped carbon and its composites as electrocatalysts for fuel cell applications, Int. J. Hydrogen Energy 38 (2013) 9370–9386.

[6] K. Kannan, D. Radhika, K.K. Sadasivuni, K.R. Reddy, A.V. Raghu, Nanostructured metal oxides and its hybrids for photocatalytic and biomedical applications, Adv. Colloid Interface Sci. 281 (2020) 10178.

[7] C.E. Small, S. Chen, J. Subbiah, C.M. Amb, S.W. Tsang, T.H. Lai, J.R. Reynolds, F. So, High-efficiency inverted dithienogermole–thienopyrrolodione-based polymer solar cells, Nat. Photonics 6 (2012) 115–120.

[8] N.M. Saidi, F.S. Omar, A. Numan, D.C. Apperley, M.M. Algaradah, R. Kasi, A.J. Avestro, R.T. Subramaniam, Enhancing the efficiency of a dye-sensitized solar cell based on a metal oxide nanocomposite gel polymer electrolyte, ACS Appl. Mater. Interfaces 11 (2019) 30185–30196.

[9] S.V. Kuppu, A.R. Jeyaraman, P.K. Guruviah, S. Thambusamy, Preparation and characterizations of PMMA-PVDF based polymer composite electrolyte materials for dye sensitized solar cell, Curr. Appl. Phys. 18 (2018) 619–625.

[10] W. Cai, Z. Chen, Z. Li, Y. Lei, D. Zhang, L. Liu, Q.H. Xu, Y. Ma, F. Huang, H.L. Yip, Y. Cao, Polymer-assisted in situ growth of all-inorganic perovskite nanocrystal film for efficient and stable pure-red light-emitting devices, ACS Appl. Mater. Interfaces 10 (2018) 42564–42572.

[11] J. Liu, D.B. Buchholz, J.W. Hennek, R.P. Chang, A. Facchetti, T.J. Marks, All-amorphous-oxide transparent, flexible thin-film transistors. Efficacy of bilayer gate dielectrics, J. Am. Chem. Soc. 132 (2010) 11934–11942.

[12] I.Y. Lee, H.Y. Park, J.H. Park, G. Yoo, M.H. Lim, J. Park, S. Rathi, W.S. Jung, J. Kim, S. W. Kim, Poly-4-vinylphenol and poly (melamine-co-formaldehyde)-based graphene

passivation method for flexible, wearable and transparent electronics, Nanoscale 6 (2014) 3830–3836.
[13] J.H. Kim, B.U. Hwang, D.I. Kim, J.S. Kim, Y.G. Seol, T.W. Kim, N.E. Lee, Nanocomposites of polyimide and mixed oxide nanoparticles for high performance nanohybrid gate dielectrics in flexible thin film transistors, Electron. Mater. Lett. 13 (2017) 214–221.
[14] N. Madusanka, S.G. Shivareddy, P. Hiralal, M.D. Eddleston, Y. Choi, R.A. Oliver, G.A.J. Amaratunga, Nanocomposites of $TiO_2$/cyanoethylated cellulose with ultra high dielectric constants, Nanotechnology 27 (2016) 195402.
[15] Z.M. Dang, J.K. Yuan, J.W. Zha, T. Zhou, S.T. Li, G.H. Hu, Fundamentals, processes and applications of high-permittivity polymer–matrix composites, Prog. Mater. Sci. 57 (2012) 660–723.
[16] H. Althues, J. Henle, S. Kaskel, Functional inorganic nanofillers for transparent polymers, Chem. Soc. Rev. 36 (2007) 1454–1465.
[17] K.K. Banger, R.L. Peterson, K. Mori, Y. Yamashita, T. Leedham, H. Sirringhaus, High performance, low temperature solution-processed barium and strontium doped oxide thin film transistors, Chem. Mater. 26 (2014) 1195–1203.
[18] K. Nomura, H. Ohta, A. Takagi, T. Kamiya, M. Hirano, H. Hosono, Room-temperature fabrication of transparent flexible thin-film transistors using amorphous oxide semiconductors, Nature 432 (2004) 488–492.
[19] W. Li, J. Shi, K.H.L. Zhang, D. JLM, Defects in complex oxide thin films for electronics and energy applications: challenges and opportunities, Mater. Horiz. 7 (2020) 2832–2859.
[20] D.B. Buchholz, Q. Ma, D. Alducin, A. Ponce, M. Jose Yacaman, R. Khanal, J.E. Medvedeva, R.P.H. Chang, The structure and properties of amorphous indium oxide, Chem. Mater. 26 (2014) 5401–5411.
[21] W. Yang, K. Song, Y. Jung, S. Jeong, J. Moon, Solution-deposited Zr-doped AlOx gate dielectrics enabling high-performance flexible transparent thin film transistors, J. Mater. Chem. C 1 (2013) 4275–4282.
[22] Y.S. Min, Y.J. Cho, C.S. Hwang, Atomic layer deposition of $Al_2O_3$ thin films from a 1-methoxy-2-methyl-2-propoxide complex of aluminum and water, Chem. Mater. 17 (2005) 626–631.
[23] Y. Wang, L. Cai, Y. Li, Y. Tang, C. Xie, Structural and photoelectrocatalytic characteristic of $ZnO/ZnWO_4/WO_3$ nanocomposites with double heterojunctions, Phys. E Low-Dimens. Syst. Nanostruct. 43 (2010) 503–509.
[24] J.A. Rodriguez, X. Wang, J.C. Hanson, G. Liu, A. Iglesias-Juez, M.J. Fernandez-Garcia, The behavior of mixed-metal oxides: structural and electronic properties of $Ce_{1--x}CaxO_2$ and $Ce_{1-x}CaxO_{2-x}$, Chem. Phys. 119 (2003) 5659–5669.
[25] X. Jiang, X. Zhao, L. Duan, H. Shen, H. Liu, T. Hou, F. Wang, Enhanced photoluminescence and photocatalytic activity of ZnO-$ZnWO_4$ nanocomposites synthesized by a precipitation method, Ceram. Int. 42 (2016) 15160–15165.
[26] P.P. Dorneanu, A. Airinei, N. Olaru, M. Homocianu, V. Nica, F. Doroftei, Preparation and characterization of NiO, ZnO and NiO–ZnO composite nanofibers by electrospinning method, Mater. Chem. Phys. 148 (2014) 1029–1035.
[27] J. Vidic, S. Stankic, F. Haque, D. Ciric, R. Le Goffic, A. Vidy, J. Jupille, B. Delmas, Selective antibacterial effects of mixed ZnMgO nanoparticles, J. Nanopart. Res. 15 (2013) 1595.
[28] M.V. Dozzi, S. Marzorati, M. Longhi, M. Coduri, L. Artiglia, E. Selli, Photocatalytic activity of $TiO_2$-$WO_3$ mixed oxides in relation to electron transfer efficiency, Appl. Catal. Environ. 186 (2016) 157–165.

[29] R.K. Sharma, R. Ghose, Synthesis of $Co_3O_4$-ZnO mixed metal oxide nanoparticles by homogeneous precipitation method, J. Alloys Compd. 686 (2016) 64–73.
[30] K. Santhi, C. Rani, R.D. Kumar, S. Karuppuchamy, Synthesis of nanoporous Zn-$WO_3$ by microwave irradiation method for photocatalytic applications, J. Mater. Sci. Mater. Electron. 26 (2015) 10068–10074.
[31] B. Ramachandra, J.S. Choi, K.S. Kim, K.Y. Choo, J.S. Sung, T.H. Kim, MoVW-mixed oxide as a partial oxidation catalyst for methanol to formaldehyde, Stud. Surf. Sci. Catal. 159 (2006) 273–276.
[32] H.R. Arandiyan, M. Parvari, Studies on mixed metal oxides solid solutions as heterogeneous catalysts, Braz. J. Chem. Eng. 26 (2009) 63–74.
[33] C. Barrios, M.A. Baltanas, R. Bolmaro, A.L. Bonivardi, Preparation and structural characterization of ZnO and $CeO_2$ nanocomposite powders as 'active catalytic supports', Powder Technol. 267 (2014) 180–192.
[34] Z. Spitalsky, D. Tasis, K. Papagelis, C. Galiotis, Carbon nanotube–polymer composites: chemistry, processing, mechanical and electrical properties, Prog. Polym. Sci. 35 (2010) 357–401.
[35] B.A. Rozenberg, R. Tenne, Polymer-assisted fabrication of nanoparticles and nanocomposites, Prog. Polym. Sci. 33 (2008) 40–112.
[36] S.K. Srivastava, Y.K. Mishra, Nanocarbon reinforced rubber nanocomposites: detailed insights about mechanical, dynamical mechanical properties, payne, and mullin effects, Nanomaterials 8 (2018) 945.
[37] A. Kasgoz, D. Akin, A. Durmus, Effects of size and shape originated synergism of carbon nano fillers on the electrical and mechanical properties of conductive polymer composites, J. Appl. Polym. Sci. 132 (2015) 42313.
[38] J.E.Q. Quinsaat, L. Budra, R. Kramer, D. Hafliger, F.A. Nuesch, M. Dascalu, D.M. Opris, Conductive silicone elastomers electrodes process able by screen printing, Sci. Rep. 9 (2019) 13331.
[39] M.J. Allen, V.C. Tung, R.B. Kaner, Honeycomb carbon: a review of graphene, Chem. Rev. 1 (2010) 132–145.
[40] T.K. Das, S. Prusty, Graphene based polymer composites and their applications, Polym. Plast. Technol. Eng. 52 (4) (2013) 319–331.
[41] S. Moharana, R.N. Mahaling, Silver (Ag)-graphene oxide (GO)–poly (vinylidene fluoride-co-hexafluoropropylene) (PVDF-HFP) nanostructured composites with high dielectric constant and low dielectric loss, Chem. Phys. Lett. 680 (2017) 31–36.
[42] Y. Wang, G.J. Weng, S.A. Meguid, A.M. Hamouda, A continuum model with a percolation threshold and tunneling-assisted interfacial conductivity for carbon nanotube-based nanocomposites, J. Appl. Phys. 115 (2014) 193706.
[43] S. Mondal, S. Ganguly, P. Das, D. Khastgir, N.C. Das, Low percolation threshold and electromagnetic shielding effectiveness of nano-structured carbon based ethylene methyl acrylate nanocomposites, Compos. Part B Eng. 119 (2017) 41–56.
[44] H. Chen, M.B. Muller, K.J. Gilmore, G.G. Wallace, D. Li, Mechanically strong, electrically conductive, and biocompatible graphene paper, Adv. Mater. 20 (18) (2008) 3557–3561.
[45] H. Pang, T. Chen, G.M. Zhang, B. Zeng, Z.M. Li, An electrically conducting polymer/graphene composite with a very low percolation threshold, Mater. Lett. 64 (20) (2010) 2226–2229.
[46] L. He, S.C. Tjong, Low percolation threshold of graphene/polymer composites prepared by solvothermal reduction of graphene oxide in the polymer solution, Nanoscale Res. Lett. 132 (2013) 132.

[47] S.A. Zavyalov, A.N. Pivkina, J. Schoonman, Formation and characterization of metal-polymer nanostructured composites, Solid State Ion. 147 (2002) 415–419.
[48] X. Yu, T.J. Marks, A. Facchetti, Metal oxides for optoelectronic applications, Nat. Mater. 15 (2016) 383–396.
[49] J. Zhang, Z.D. Qin, C. Xie, Metal-oxide-semiconductor based gas sensors: screening, preparation, and integration, Phys. Chem. Chem. Phys. 19 (2017) 6313–6329.
[50] S.A. Mir, M. Ikram, V. Asokan, Structural, optical and dielectric properties of Ni substituted $NdFeO_3$, Optik 125 (2014) 6903–6908.
[51] S. Xun, G. LeClair, J. Zhang, X. Chen, J. Gao, Z. Wang, Tuning the electrical and optical properties of dinuclear ruthenium complexes for near infrared optical sensing, Org. Lett. 8 (8) (2006) 1697–1700.
[52] C. Zhu, G. Yang, H. Li, D. Du, L. Lin, Electrochemical sensors and biosensors based on nanomaterials and nanostructures, Anal. Chem. 87 (2015) 230–249.
[53] M. Kulkarni, R.A. John, M. Rajput, Transparent flexible multifunctional nanostructured architectures for non-optical readout, proximity, and pressure sensing, ACS Appl. Mater. Interfaces 9 (2017) 17.
[54] T.F. Zhang, X.G. Tang, Q.X. Liu, Y.P. Jiang, Y.P. Jiang, L. Luo, Optical and dielectric properties of $PbZrO_3$ thin films prepared by a sol–gel process for energy-storage application, Mater. Des. 90 (2016) 410–415.
[55] H. Zheng, J. Wang, S.E. Lofland, Z. Ma, L. Mohaddes-Ardabili, T. Zhao, L. Salamanca-Riba, S.R. Shinde, S.B. Ogale, F. Bai, D. Viehland, Y. Jia, D.G. Schlom, M. Wuttig, A. Roytburd, R. Ramesh, Multiferroic $BaTiO_3$-$CoFe_2O_4$ nanostructures, Science 303 (2004) 661.
[56] N. Bao, L. Shen, G. Srinivasan, K. Yanagisawa, A. Gupta, Shape-controlled monocrystalline ferroelectric barium titanate nanostructures: from nanotubes and nanowires to ordered nanostructures, J. Phys. Chem. C 112 (2008) 8634–8642.
[57] V. Corral-Flores, D. Bueno-Baque's, R.F. Ziolo, Shape-controlled monocrystalline ferroelectric barium titanate nanostructures: from nanotubes and nanowires to ordered nanostructures, Acta Mater. 58 (2010) 764–769.
[58] K.C. Verma, S. Singh, S.K. Tripathi, R.K. Kotnala, Multiferroic $Ni_{0.6}Zn_{0.4}Fe_2O_4$-$BaTiO_3$ nanostructures: magnetoelectric coupling, dielectric, and fluorescence, J. Appl. Phys. 116 (2014) 124103.
[59] D. Liu, C. Jin, F. Shan, J. He, F. Wang, Synthesizing $BaTiO_3$ nanostructures to explore morphological influence, kinetics, and mechanism of piezocatalytic dye degradation, ACS Appl. Mater. Interfaces 12 (2020) 17443–17451.
[60] X. Yang, Z. Ren, G. Xu, C. Chao, S. Jiang, S. Deng, G. Shen, X. Wei, G. Hann, Monodisperse hollow perovskite $BaTiO_3$ nanostructures prepared by a sol–gel–hydrothermal method, Ceram. Int. 140 (2014) 9663–9670.
[61] K. Huang, T. Huang, W. Hsieh, Monodisperse hollow perovskite $BaTiO_3$ nanostructures prepared by a sol–gel–hydrothermal method, Inorg. Chem. 48 (2009) 9180–9184.
[62] G.A. Fernández, A.A. Martínez, J.C. Fuerte, Nanostructured Ti−W mixed-metal oxides: structural and electronic properties, J. Phys. Chem. B 109 (2005) 6075–6083.
[63] E. Fortunato, A. Gonçalves, A. Pimentel, P. Barquinha, G. Gonçalves, L.I. Pereira, I. Ferreira, R. Martins, Zinc oxide, a multifunctional material: from material to device applications, Appl. Phys. A 96 (2009) 197–205.
[64] S.J. Han, T.H. Jang, Y.B. Kim, B.G. Park, J.H. Park, Y.H. Jeong, Magnetism in Mn-doped ZnO bulk samples prepared by solid state reaction, Appl. Phys. Lett. 83 (2003) 920.

[65] K.M. Dermenci, S. Turan, Achieving high performance for aluminum stabilized $Li_7La_3Zr_2O_{12}$ solid electrolytes for all solid-state Li-ion batteries: a thermodynamic point of view, Int. J. Energy Res. 43 (2018) 141–149.
[66] C. Yang, S. Hsu, W. Chien, All solid-state electric double-layer capacitors based on alkaline polyvinyl alcohol polymer electrolytes, J. Power Sources 152 (2005) 303–310.
[67] J. Fan, S. Boettcher, G.D. Stucky, Nanoparticle assembly of ordered multicomponent mesostructured metal oxides via a versatile sol−gel process, Chem. Mater. 18 (2006) 6391–6396.
[68] S. Shen, X. Zhang, Y. Zhou, H. Li, Preparation and characterization of nanocrystalline $Li_4Ti_5O_{12}$ by sol–gel method, Mater. Chem. Phys. 78 (2003) 437–441.
[69] S.C. Luciana, M.B.C. Tiago, A.R. Liana, D.B. Deborah, P.T. Gilmar, Review of mullite synthesis routes by sol–gel method, J. Sol-Gel Sci. Technol. 55 (2010) 111–125.
[70] P. Colomban, Gel technology in ceramics, glass-ceramics and ceramic-ceramic composites, Ceram. Int. 15 (1989) 23–50.
[71] Y.P. Gao, C.N. Sisk, W.L.J. Hope, A sol–gel route to synthesize monolithic zinc oxide aerogels, Chem. Mater. 19 (2007) 6007.
[72] J. Li, S. Srinivasan, G.N. He, J.Y. Kang, S.T. Wu, F.A. Ponce, Synthesis and luminescence properties of ZnO nanostructures produced by the sol–gel method, J. Cryst. Growth 310 (2008) 599.
[73] T. Aruna, M.S. Alexander, Combustion synthesis and nanomaterials, Curr. Opin. Solid State Mater. Sci. 12 (2008) 44–50.
[74] M. Jayalakshmi, M. Palaniappa, K. Balasubramanian, Single step solution combustion synthesis of ZnO/carbon composite and its electrochemical characterization for supercapacitor application, Int. J. Electrochem. Sci. 3 (2008) 96.
[75] A.C. Lawrence, C. Bor-Rong, M.K. Robert, W. Jianguo, R.P. Kenneth, J.B. Michael, D. M. Laurence, All roads lead to $TiO_2$: $TiO_2$-rich surfaces of barium and strontium titanate prepared by hydrothermal synthesis, Chem. Mater. 30 (2018) 841–846.
[76] Q. Yang, Z. Lu, J. Liun, X. Lei, Z. Chang, L. Luo, X. Sun, Metal oxide and hydroxide nanoarrays: Hydrothermal synthesis and applications as supercapacitors and nanocatalysts, Prog. Nat. Sci.: Mater. Int. 23 (2013) 351–366.
[77] M.A. Lim, D.H. Kim, C.O. Park, A new route toward ultrasensitive, flexible chemical sensors: metal nanotubes by wet-chemical synthesis along sacrificial nanowire templates, ACS Nano 6 (2012) 598–608.
[78] Y. Zhang, K. Fugane, T. Mori, L. Niu, J. Yea, Wet chemical synthesis of nitrogen-doped graphene towards oxygen reduction electrocatalysts without high-temperature pyrolysis, J. Mater. Chem. 22 (2012) 6575.
[79] M. Pal, N.R. Mathew, E. Sanchez-Mora, U. Pal, F. Paraguay-Delgado, X. Mathew, Synthesis of CuS nanoparticles by a wet chemical route and their photocatalytic activity, J. Nanopart. Res. 17 (2015) 301.
[80] H. Cao, X. Qian, J. Zai, J. Yin, Z. Zhu, Synthesis of 3-D hierarchical dendrites of lead chalcogenides in large scale via microwave-assistant method, Cryst. Growth Des. 7 (2007) 425.
[81] F.A. He, K.H. Lam, J.T. Fan, L.W. Chan, Novel syndiotactic polystyrene/ $BaTiO_3$-graphite nanosheets three-phase composites with high dielectric permittivity, Polym. Test. 32 (2013) 927–931.
[82] N. Yousefi, X.Y. Sun, X.Y. Lin, X. Shen, J.J. Jia, B. Zhang, B.Z. Tang, M.S. Chan, J.K. Kim, Highly aligned graphene/polymer nanocomposites with excellent dielectric properties for high-performance electromagnetic interference shielding, Adv. Mater. 26 (2014) 5480–5487.

[83] Z. Wang, K. Sun, P. Xie, Y. Liu, Q. Gu, R. Fan, Permittivity transition from positive to negative in acrylic polyurethane-aluminum composites, Compos. Sci. Technol. 188 (2020) 107969.
[84] S.R. Broadbent, J.M. Hammersley, Percolation processes, Math. Proc. Camb. Philos. Soc. 53 (3) (1957) 629–641.
[85] A.A. Saberi, Recent advances in percolation theory and its applications, Phys. Rep. 578 (2015) 1–32.
[86] G. Pandey, E.T. Thostenson, Carbon nanotube-based multifunctional polymer nanocomposites, Polym. Rev. 52 (2012) 355–416.
[87] R.H. Cruz-Estrada, M.J. Folkes, Structure formation and modelling of the electrical conductivity in SBS-polyaniline blends, Part II: generalized effective media theories approach, J. Mater. Sci. Lett. 21 (2002) 1431–1434.
[88] M. Weber, M.R. Kamal, Estimation of the volume resistivity of electrically conductive composites, Polym. Compos. 18 (1997) 711–725.
[89] Q. Guo, Q. Xue, J. Sun, M. Dong, F. Xia, Z. Zhang, Gigantic enhancement in the dielectric properties of polymer-based composites using core/shell MWCNT/amorphous carbon nanohybrids, Nanoscale 7 (2015) 3660–3667.
[90] H. Wang, Q. Fu, J. Luo, D. Zhao, L. Luo, W. Li, Three-phase $Fe_3O_4$/MWNT/PVDF nanocomposites with high dielectric constant for embedded capacitor, Appl. Phys. Lett. 110 (2017) 242902.
[91] M.E. Rhazi, S. Majid, M. Elbasri, F.E. Salih, L. Oularbi, K. Lafdi, Recent progress in nanocomposites based on conducting polymer: application as electrochemical sensors, Int. Nano Lett. 8 (2018) 79–99.
[92] G. Shimoga, S.Y. Kim, High-*k* polymer nanocomposite materials for technological applications, Appl. Sci. 10 (12) (2020) 4249.
[93] Z. Liu, S. Poyraz, Y. Liu, X. Zhang, Seeding approach to noble metal decorated conducting polymer nanofiber network, Nanoscale 4 (2012) 106–109.
[94] X. Feng, H. Huang, Q. Ye, J.J. Zhu, W. Hou, Ag/polypyrrole core-shell nanostructures: interface polymerization, characterization, and modification by gold nanoparticles, J. Phys. Chem. C 111 (2007) 8463–8468.
[95] H. Huang, X. Feng, J.J. Zhu, Synthesis, characterization and application in electrocatalysis of polyaniline/Au composite nanotubes, Nanotechnology 19 (2008) 145607.
[96] M. Magnozzi, Y. Brasse, T.A.F. König, F. Bisio, E. Bittrich, A. Fery, M. Canepa, Plasmonics of Au/polymer core/shell nanocomposites for thermoresponsive hybrid metasurfaces, ACS Appl. Nano Mater. 3 (2020) 1674–1682.
[97] W.H. Eisa, E. Ashkar, S.M.E. Mossalamy, S.S.M. Ali, PVP induce self-seeding process for growth of Au@Ag core@shell nanocomposites, Chem. Phys. Lett. 651 (2016) 28–33.
[98] Y. Wang, G. Zhao, G. Zhang, Y. Zhang, H. Wang, W. Cao, T. Li, Q. Wei, An electrochemical aptasensor based on gold-modified $MoS_2$/rGO nanocomposite and gold-palladium-modified Fe-MOFs for sensitive detection of lead ions, Sens. Actuators B 319 (2020) 128313.
[99] P.M. Anjana, M.R. Bindhu, M. Umadevi, R.B. Rakhi, Antibacterial and electrochemical activities of silver, gold, and palladium nanoparticles dispersed amorphous carbon composites, Appl. Surf. Sci. 479 (2019) 96–104.
[100] P. Singh, M. Halder, S. Ray, A. Bose, K. Sen, Green synthesis of silver and palladium nanocomposites: a study of catalytic activity towards etherification reaction, Mater. Adv. 1 (2020) 2937–2952.

[101] Y. Liu, Z. Liu, N. Lu, E. Preiss, S. Poyraz, M.J. Kim, X. Zhang, Facile synthesis of polypyrrole coated copper nanowires: a new concept to engineered core–shell structures, Chem. Commun. 48 (2012) 2621–2623.

[102] U. Khalil, S. Haider, M.S. Khan, A. Haider, R. Khan, A.A. Alghyamah, W.A. Almasry, M. Bououdina, Synthesis of novel copper nanoparticles/ternary polymer blend nanocomposites and their structural, thermal and rheological properties and AC impedance, Polym. Int. 66 (8) (2017) 1182–1189.

[103] I. Aktitiz, R. Varol, N. Akkurt, M.F. Saraç, In-situ synthesis of 3D printable mono- and Bi-metallic (Cu/Ag) nanoparticles embedded polymeric structures with enhanced electromechanical properties, Polym. Test. 90 (2020) 106724.

[104] A.S. Zeraati, M. Arjmand, U. Sundararaj, Silver nanowire/$MnO_2$ nanowire hybrid polymer nanocomposites: materials with high dielectric permittivity and low dielectric loss, ACS Appl. Mater. Interfaces 9 (2017) 14328–14336.

[105] Y. Li, Y. Zhou, Y. Zhu, S. Cheng, C. Yuan, J. Hu, J. He, Q. Li, Polymer nanocomposites with high energy density and improved charge–discharge efficiency utilizing hierarchically-structured nanofillers, J. Mater. Chem. A 8 (2020) 6576–6585.

[106] J. Yang, X. Zhu, H. Wang, X. Wang, C.C. Hao, R. Fan, D. Dastan, Z. Shi, Achieving excellent dielectric performance in polymer composites with ultralow filler loadings via constructing hollow-structured filler frameworks, Compos. Part A Appl. Sci. Manuf. 131 (2020) 105814.

[107] Z. Qiang, G. Liang, A. Gu, L. Yuan, The dielectric behavior and origin of high-k composites with very low percolation threshold based on unique multi-branched polyaniline/carbon nanotube hybrids and epoxy resin, Compos. Part A Appl. Sci. Manuf. 64 (2014) 1–10.

[108] K. Yu, Y. Niu, Y. Zhou, Y. Bai, H. Wang, Nanocomposites of surface-modified $batio_3$ nanoparticles filled ferroelectric polymer with enhanced energy density, J. Am. Ceram. Soc. 96 (2013) 2519–2524.

[109] Z.M. Dang, H.P. Xu, H.Y. Wang, Significantly enhanced low-frequency dielectric permittivity in the $BaTiO_3$/poly(vinylidene fluoride) nanocomposites, Appl. Phys. Lett. 90 (2007), 012901.

[110] B.H. Fan, J.W. Zha, D. Wang, J. Zhao, Z.M. Dang, Size-dependent low-frequency dielectric properties in the $BaTiO_3$/poly(vinylidene fluoride) nanocomposite films, Appl. Phys. Lett. 100 (2012), 012903.

[111] M. Panda, V. Srinivas, A.K. Thakur, Percolation behavior of polymer/metal composites on modification of filler, Mod. Phys. Lett. B 28 (2014) 1450055–1450056.

[112] M.T. Rahmana, M.A. Hoquea, G.T. Rahmana, M.A. Gafurb, R.A. Khanc, M.K. Hossain, Study on the mechanical, electrical and optical properties of metal-oxide nanoparticles dispersed unsaturated polyester resin nanocomposites, Results Phys. 13 (2019) 102264–102268.

[113] R. Baskaran, M. Sarojadevi, C.T. Vijayakumar, Vijayakumar, polyester nanocomposites filled with nano alumina, J. Mater. Sci. 46 (2011) 4864–4871.

[114] J. Zhu, W. Li, X. Huo, L. Li, Y. Li, L. Luo, Y. Zhu, An ultrahigh dielectric constant composite based on polyvinylidene fluoride and polyethylene glycol modified ferroferric oxide, J. Phys. D Appl. Phys. 48 (2015) 355301–355305.

[115] A.J. Marsden, D.G. Papageorgious, C. Valles, A. Liscio, V. Palermo, M.A. Bissett, R.J. Young, I.A. Kinloch, Electrical percolation in graphene–polymer composites, 2D Mater. 5 (2018), 032003.

[116] C.J.R. Verbeek, Effect of percolation on the mechanical properties of sand-filled polyethylene composites, J. Thermoplast. Compos. Mater. 20 (2007) 137–149.

[117] X. Huo, W. Li, J. Zhu, L. Li, Y. Li, L. Luo, Y. Zhu, A composite based on $Fe_3O_4$@$BaTiO_3$ particles and PVDF with excellent dielectric properties and high energy density, J. Phys. Chem. C 119 (2015) 25786–25791.
[118] L. Yuhan, J. Yuan, J. Xue, F. Cai, F. Chen, Q. Fu, Towards suppressing loss tangent: effect of polydopamine coating layers on dielectric properties of core-shell barium titanate filled polyvinylidene fluoride composites, Compos. Sci. Technol. 118 (2015) 198–206.
[119] M.F. Lin, V.K. Thakur, E.J. Tan, P.S. Lee, Surface functionalization of $BaTiO_3$ nanoparticles and improved electrical properties of $BaTiO_3$/polyvinylidene fluoride composite, RSC Adv. 1 (2011) 576–578.
[120] N. Phatharapeetranun, B. Ksapabutr, D. Marani, J.R. Bowen, V. Esposito, 3D-printed barium titanate/poly-(vinylidene fluoride) nano-hybrids with anisotropic dielectric properties, J. Mater. Chem. C 5 (2017) 12430–12440.
[121] D. Majumdar, Recent progress in copper sulfide based nanomaterials for high energy supercapacitor applications, J. Electroanal. Chem. 880 (2021) 114825.
[122] S.H. Xie, B.K. Zhu, X.Z. Wei, Z.K. Xu, Y.Y. Xu, Polyimide/$BaTiO_3$ composites with controllable dielectric properties, Compos. Part A Appl. Sci. Manuf. 36 (2005) 1152–1157.
[123] R. Li, Z. Zhao, Z. Chen, J. Pei, Novel $BaTiO_3$/PVDF composites with enhanced electrical properties modified by calcined $BaTiO_3$ ceramic powders, Mater. Express 7 (2017) 2158–5849.
[124] Y. Li, X. Huang, Z. Hu, P. Jiang, S. Li, T. Tanaka, Large dielectric constant and high thermal conductivity in poly(vinylidene fluoride)/barium titanate/silicon carbide three-phase nanocomposites, ACS Appl. Mater. Interfaces 3 (2011) 4396–4403.
[125] J. Fu, Y. Hou, M. Zheng, W. Wei, M. Zhu, H. Yan, Improving dielectric properties of PVDF composites by employing surface modified strong polarized $BaTiO_3$ particles derived by molten salt method, ACS Appl. Mater. Interfaces 7 (2015) 24480–24491.
[126] Y. Sheng, X. Zhang, H. Ye, Y. Liang, L. Xu, H. Wu, Improved energy density in core–shell poly(dopamine) coated barium titanate/poly(fluorovinylidene-co-trifluoroethylene) nanocomposite with interfacial polarization, Colloids Surf. A Physicochem. Eng. Asp. 585 (2020) 124091.
[127] X. Dou, X. Liu, Y. Zhang, H. Feng, J.F. Chen, S. Du, Improved dielectric strength of barium titanate-polyvinylidene fluoride nanocomposite, Appl. Phys. Lett. 95 (2009) 132904.
[128] Y. Wang, X. Wu, C. Feng, Q. Zeng, Improved dielectric properties of surface modified $BaTiO_3$/polyimide composite films, Microelectron. Eng. 154 (2016) 17–21.
[129] Y. Feng, Q. Deng, C. Peng, Q. Wu, High dielectric and breakdown properties achieved in ternary BaTiO3/ MXene/PVDF nanocomposites with low-concentration fillers from enhanced interface polarization, Ceram. Int. 45 (2019) 7923–7930.
[130] J.H. Yang, X. Xie, Z. He, Y. Lu, X. Qi, Y. Wang, Graphene oxide-tailored dispersion of hybrid barium titanate@polypyrrole particles and the dielectric composites, Chem. Eng. J. 355 (2019) 137–149.
[131] Y. Yang, Z. Li, W. Ji, C. Sun, H. Deng, Q. Fu, Enhanced dielectric properties through using mixed fillers consisting of nano-barium titanate/nickel hydroxide for polyvinylidene fluoride based composites, Compos. Part A Appl. Sci. Manuf. 104 (2018) 24–31.
[132] J.W. Zha, X. Meng, D. Wang, Z.M. Dang, R.K.Y. Li, Dielectric properties of poly(vinylidene fluoride) nanocomposites filled with surface coated $BaTiO_3$ by $SnO_2$ nanodot, Appl. Phys. Lett. 104 (2014), 072906.

[133] Y. Feng, W.L. Li, Y.F. Hou, Y. Yu, W.P. Cao, T.D. Zhang, W.D. Fei, Enhanced dielectric properties of PVDFHFP/$BaTiO_3$-nanowires composites induced by interfacial polarization and wire-shape, J. Mater. Chem. C 3 (2015) 1250–1260.
[134] D.S. Kim, C. Baek, H.J. Ma, D.K. Kim, Enhanced dielectric permittivity of $BaTiO_3$/epoxy resin composites by particle alignment, Ceram. Int. 42 (2016) 7141–7147.
[135] Y. Jiang, Z. Zhang, Z. Zhou, H. Yang, Q. Zhang, Enhanced dielectric performance of P (VDF-HFP) composites with satellite–core-structured $Fe_2O_3$@$BaTiO_3$ nanofillers, Polymers 11 (2019) 1541.
[136] Y. Niu, K. Yu, Y. Bai, H. Wang, Enhanced dielectric performance of $BaTiO_3$/ PVDF composites prepared by modified process for energy storage applications, IEEE Trans. Ultrason. Ferroelectr. Freq. Control 62 (2015) 108–115.
[137] Y. Song, Y. Shen, H. Liu, Y. Lin, M. Li, C.W. Nan, Enhanced dielectric and ferroelectric properties induced by dopamine-modified $BaTiO_3$ nanofibers in flexible poly(vinylidene fluoride-trifluoroethylene) nanocomposites, J. Mater. Chem. 22 (2012) 8063.
[138] A. Bele, M. Cazacu, G. Stiubianu, S. Viad, M. Ignat, Polydimethylsiloxane–barium titanate composites: preparation and evaluation of the morphology, moisture, thermal, mechanical and dielectric behaviour, Compos. Part B Eng. 68 (2015) 237–245.
[139] R.H. Upadhyay, R.R. Deshmukh, Investigation of dielectric properties of newly prepared β-phase polyvinylidene fluoride-barium titanate nanocomposite films, J. Electrostat. 71 (2013) 945–950.
[140] A. Choudhury, Dielectric and piezoelectric properties of polyetherimide/$BaTiO_3$ nanocomposites, Mater. Chem. Phys. 121 (2010) 280–285.
[141] C. Zhang, Q. Chi, J. Dong, Y. Cui, X. Wang, L. Liu, Q. Lei, Enhanced dielectric properties of poly(vinylidene fluoride) composites filled with nano iron oxide-deposited barium titanate hybrid particles, Sci. Rep. 6 (2016) 33508.
[142] W. Yang, S. Yu, S. Luo, R. Sun, W.H. Liao, C.P. Wong, A systematic study on electrical properties of the $BaTiO_3$–epoxy composite with different sized $BaTiO_3$ as fillers, J. Alloys Compd. 620 (2015) 315–323.
[143] K. Silakaew, P. Thongbai, Significantly improved dielectric properties of multiwall carbon nanotube-$BaTiO_3$/PVDF polymer composites by tuning the particle size of the ceramic filler, RSC Adv. 9 (2019) 23498.
[144] L. Xie, X. Huang, Y. Huang, K. Yang, P. Jiang, Core@double-shell structured $BaTiO_3$–-polymer nanocomposite with high dielectric constant and low dielectric loss for energy storage application, J. Phys. Chem. C 117 (2013) 22525–22537.
[145] L. Ren, X. Meng, J.W. Zha, Z.M. Dang, Coulomb block effect inducing distinctive dielectric properties in electroless plated barium titanate@silver/poly(vinylidene fluoride) nanocomposites, RSC Adv. 5 (2015) 65167–65174.
[146] S. Moharana, M.K. Mishra, B. Behera, R.N. Mahaling, Enhanced dielectric properties of polyethylene glycol (PEG) modified $BaTiO_3$ (BT)-poly(vinylidene fluoride) (PVDF) composites, Polym. Sci., Ser. A 59 (2017) 53–63.
[147] S. Moharana, M.K. Mishra, R.N. Mahaling, Preparation of divinyl benzene modified barium titanate-poly (vinylidene fluoride) composites with high dielectric constant, J. Chin. Adv. Mater. Soc. 5 (2017) 269–282.
[148] S.C. Li, S.C. Tjong, R.K.Y. Li, Dielectric properties of binary polyvinylidene fluoride/barium titanate nanocomposites and their nanographite doped hybrids, Express Polym. Lett. 5 (2011) 526–534.
[149] M. Wang, X. Pan, X. Qi, N. Zhang, T. Huang, J.H. Yang, Y. Wang, Fabrication of Ag@$BaTiO_3$ hybrid nanofibers via coaxial electrospinning toward polymeric composites with highly enhanced dielectric performances, Compos. Commun. 21 (2020) 100411.

# Polymer-metal oxide composite as sensors

# 9

*Manuel Palencia*[a,b], *Jorge A. Ramírez-Rincón*[a], *and Diego F. Restrepo-Holguín*[a]
[a]Research Group in Science with Technological Applications (GI-CAT), Department of Chemistry, Universidad del Valle, Cali, Colombia, [b]Mindtech Research Group (Mindtech-RG), Mindtech s.a.s., Cali, Colombia

## 1 Introduction

Analytical technology is one of the essential tools for the acquisition of useful information, and its underlying foundations are based on the combination of multiple disciplines ranging from mathematics, chemistry, engineering, and productive sectors, including social environments, biomedical applications, and communication systems [1]. On the contrary, analysis and interpretation of the analytical information allow us to know and describe any system in its present state, in order to understand its evolution in the future; also, through regression techniques, it is possible to obtain information from a previous state. Thus, all analytical information is a key for the control and design of processes, decision-making, and detection and correction of faults. In this point, it is important to clarify that the term "system" is understood to be any object under study, whereas "analytical technology" is defined to be a set of tools or devices used for the qualitative or quantitative acquisition of information about the target system. In this context, the acquisition of relevant information is where sensors are functionally circumscribed. In a wide meaning, a sensor is defined as any device capable of detecting a change and generating a producible response that is directly related to the respective change. From a general point of view, sensors are an extrapolation of our senses, like eye (i.e., optic sensor), skin (i.e., contacting sensor), tongue (i.e., chemosensor), and ears (i.e., acoustic sensors), since these are biological devices acting like sensors for the acquisition of information from the environment [1–5].

Keeping in mind the intuitive definition of sensor, we can approach a more formal definition. As previously it was stated, a sensor is a material device capable of detecting a change; however, in precise terms, a sensor is a material system capable of experiencing a change in some property, in a reproducible way, as a result of a stimulus or change in environment. Note that we place particular emphasis on the reproducibility of the property change. The existence of this characteristic, i.e., a reproducible change, is what makes its use as a sensor technologically viable. If the answer is not reproducible, the uncertainty is too high, and consequently, the information obtained will have no practical value. Currently, there are a number of sensors for the measurement or monitoring of a large number of parameters, in a large number of sectors such as medicine, food industry, and telecommunications. With the evolution of telecommunications technology, cell phones are commonly used and more

**Renewable Polymers and Polymer-Metal Oxide Composites. https://doi.org/10.1016/B978-0-323-85155-8.00003-0**

popular devices implementing multiple sensors in their operation. From a more specialized approach, we can exemplify the versatility of sensing strategies by referring to variables, or parameters, such as pressure [6], temperature [7], humidity [8], the presence of a certain gas [5,6], metal ion [9], microorganism [4], or the effect of a biotic or abiotic entity on a target system [4]. Spectral sensors are known as photodiodes, these are arrays of semiconductor materials able to transform light into an electrical response. Some examples of these materials are silicon, germanium, lead sulfide, and indium gallium arsenide. As sensors, these are used in the making of charge-coupled devices that are used as image sensors. Among specific analytical applications are the quantification of proteins for food industry [10], monitoring of membrane's fouling in filtration processes [11], and the determination of cellular concentration in microbiological analysis [12]. An illustration of optical sensor used in the previously mentioned works is shown in Fig. 1.

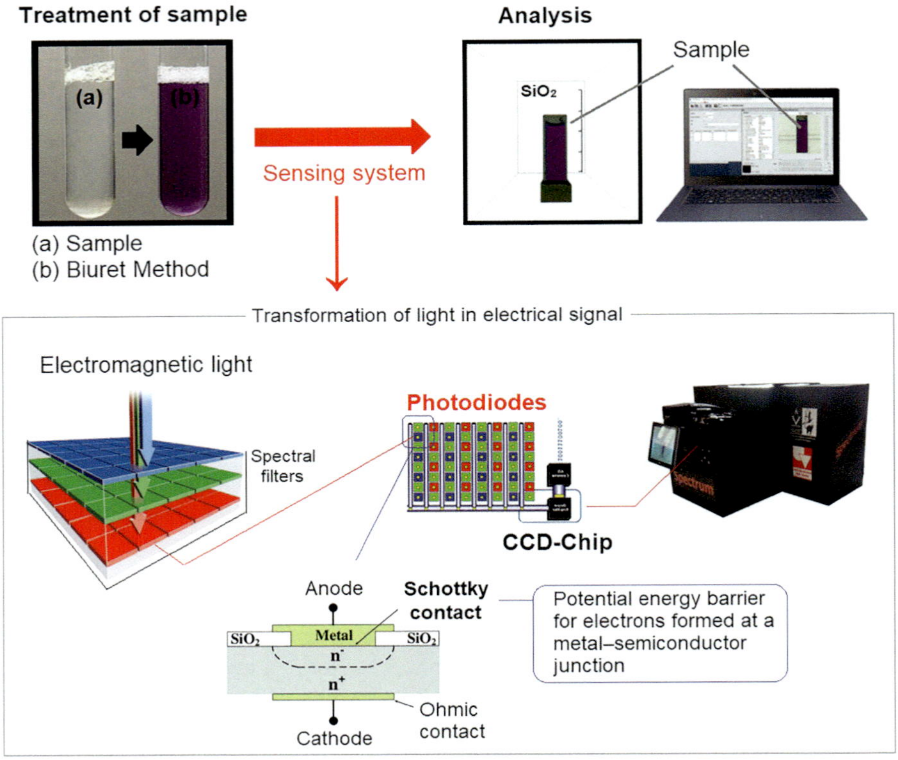


**Fig. 1** Illustration of optical sensors: (A) treatment of sample, which is an in-lab or out-lab procedure, (B) information of analysis by algorithms and computational tools, and (C) sensing systems (component of process where sensor is acting, in this case, by capturing of spectral information and converting it to electrical signal).
Adapted from T. Lerma, J. Martinez, E. Combatt, Determination of total protein content by digital image analysis: an approach for food quality remote analysis. J. Sci. Technol. Appl. 6 (2019) 53–64. 10.34294/j.jsta.19.6.41. Copyright © 2019 CC BY-NC-SA 4.0.

Sensors are usually designed under three approaches: (i) contact-based sensors in which contact is required between the material acting as a sensor and the target system; (ii) sensors without contact, or sensors for remote sensing, in which no contact is required between the sensor and the target system, and in this case, the change is usually observed at the level of the electromagnetic spectrum and is characterized by a minimal disturbance of the system; and (iii) sensors based on sample extraction, under this approach the sensing is invasive and requires a sample to be extracted [1]. Depending on the response, the sensors are classified into mechanical sensors (e.g., pressure sensors, accelerometers, and potentiometers) [13], optical sensors (e.g., photodetectors, and infrared-sensitive, fiber optic, interferometers) [14], sensors based on semiconductors (e.g., gas, temperature, magnetic, optical, and ion-sensitive field sensors) [15–17], electrochemical sensors (e.g., potentiometric, amperometric, coulometric, and dielectric sensors) [2,3], biosensors (e.g., microbiological, enzymatic, and biobased) [4], and chemical sensors (e.g., sensors based on specific chemical reactions) [5,9].

According to the foregoing, sensors can be made of different materials, such as metals, metalloids, nanostructured metals, conglomerates of low molecular weight molecules, synthetic polymers, natural polymers, porous materials, composites, and nanocomposites [1]. In particular, this chapter shows and discusses fundamental and applied information related to sensors based on polymer-metal oxide composites, including the physical properties associated with detection applications, manufacturing methods, and analytical principles of the detection.

# 2 Sensing by materials based on polymer and metal oxides

## *2.1 Sensors from overview perspective*

Previously, it was indicated that sensors are material systems, in other words, they are formed by matter and energy. Under this initial point, it is well known that changes in the systems are associated with the change in their internal energy. When a material system is in equilibrium, its properties can be classified as intensive or extensive properties, and they can be defined by an initial value which is in many cases a reference value. Thus, the first does not depend on the amount of matter (e.g., pressure, temperature, density, and chemical potential) whereas the magnitude of the second is directly related to the mass (e.g., internal energy, and entropy). In sensors, external stimuli are commonly used to produce a change in thermodynamic equilibrium of materials; for instance, changes of one intensive property can be easily monitored by changes of extensive or intensive property. A typical example is a thermosensitive sensor in which changes in temperature, produced by heat flow from surroundings, are measured through changes in electrical resistance [1,18,19].

In many studies, in the field of sensors, the mechanism underlying the sensor response remains hidden, or they do not acquire a leading role. Usually, the physical phenomena behind sensing processes often involve great complexity, or in some cases, they are well known or are treated as reproducible empirical relationships under controlled conditions. The aforementioned should not be seen as a negative something since sensors are beyond fundamental research; they are essentially technological forms of application of materials. That is, they do not seek to understand the physical phenomenon as a primary purpose, but rather their objective is focused on the use of said phenomenon for specific technological applications. In conclusion, a sensor is a useful information acquisition device, which can be based on well-established physical laws, but it can also be based on empirical relationships, usually linear or approximately linear, within the operating range [1,18,19].

To conceptualize the potential application of material as a sensor, it is necessary at least: (i) the identification of an easily measurable property, (ii) the identification of a stimulus that produces reproducible changes in the target property, (iii) the characterization of the change in the property as a result of the stimulus (i.e., operating range which is defined by the minimum value and maximum value in which the stimulus produces a useful change, uncertainty associated with the variability of the response, the definition of operating conditions, among others), and (iv) the detection and processing of the signals in order to facilitate the acquisition and interpretation of the information obtained [18,19]. An illustration of general functioning of sensors is shown in Fig. 2.

## 2.2 Sensors based on polymer-metal oxide composites

A metal oxide is defined to be a chemical compound resulting from metal oxidation. They are characterized to show a wide number of characteristics useful for functional applications, including optical, electrical, magnetic, chemical, and catalytic properties [20–23]. In addition, these properties are strongly related to the nature of atoms forming the oxide, crystal structure, morphology, intrinsic defects and impurities, among others structural and compositional aspects, which determine their optical, electrical, chemical, and catalytic properties [20,21]. In particular, energy gap and electronic structure of metal oxides depend on the size of oxide particles, and consequently, these properties can be controlled by the control of size and shape of metal-oxide nanometric particles. Among typical applications of metal oxide and metal-oxide-based materials are adsorption of dyes (e.g., ZnO nanosheets) [24], adsorption of ions (e.g., core-shell materials based on $Fe_3O_4$) [25], degradation of dyes (e.g., nanoparticles of ZnO-$TiO_2$ supported on clays) [26], removal of CO from reformed fuel (e.g., Cu/ZnO impregnated on $Al_2O_3$) [27], photodegradation of organic molecules used in microbiology like carbolfuchsin (e.g., nanoparticles of $Fe_3O_4$ doped with $Co^{2+}$ and $Ni^{2+}$) [28], photoreduction of $CO_2$ (e.g., perovskite titanates ($SrTiO_3$, $CaTiO_3$, and $PbTiO_3$) [29], making of supercapacitors based on $TiO_2$-poly(pyrrole) [30], and photodegradation of organic substances (e.g., core-shell nanostructure based on $Fe_3O_4$/$WO_3$) [31].

On the contrary, polymer-metal oxide composites are materials characterized by two phases, a continuous polymer phase containing metal-oxide nanoparticles. Usually, for sensor-like applications conductive polymers are used, however, conductivity is not an indispensable requirement. Many times, sensing properties are exclusively a

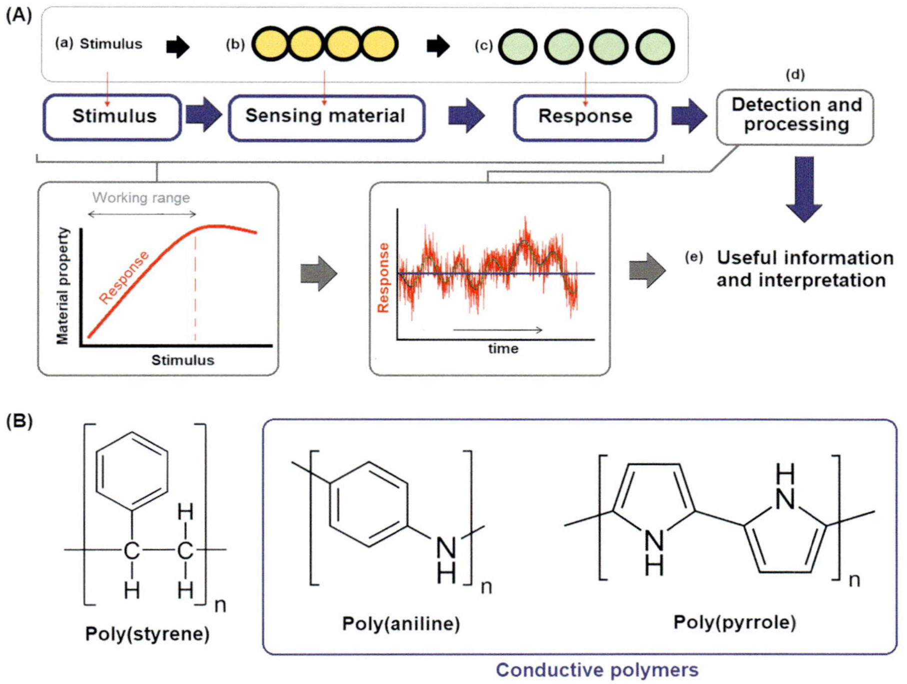


**Fig. 2** (A) Illustration of general functioning of sensors: (a) generation of stimulus, (b) change of material property, (c) generation of reproducible response, (d) detection and processing of signals, (e) obtaining of useful information and interpretation. (B) Example of polymers used for the making of sensors based on polymer-metal oxide nanocomposites (poly(styrene)) is used in mixing with some conductive polymer like poly(aniline)).


result of the presence of metal-oxide nanoparticles. The polymers used for this type of sensing materials are poly(styrene)/poly(aniline), poly(aniline), and poly(pyrrole) (structure are shown in Fig. 2B) [32]. Some examples are shown in Table 1.

# 3 Fabrication methods

Manufacture of polymer-metal oxide composites can be described, in a simple way, to be the intrinsic and correct combination of metal-oxide nanomaterials being dispersed in a continuous polymeric phase. This simplification evidences the nature (i.e., metal oxide and polymer) and characteristics of the main phases of this kind of materials (i.e., nanostructured inorganic phase corresponding to metal oxide and an organic phase acting as dispersion medium). Also, it is emphasized that the formation of polymer-metal oxide composites implies an intrinsic and correct combination. These concepts have been introduced to maintain the coherence with the target application: sensing. Note that an important characteristic of sensing is the reproducibility;

**Table 1** Examples of polymer-metal oxide nanocomposites used as sensors.

| Nanoparticle | Polymer | Application | Reference |
|---|---|---|---|
| $SnO_2$ | Poly(styrene)/poly (aniline) | Gas sensor for CO | [33] |
| | Poly(aniline) | Sensor for ethanol and acetone | [34] |
| | Poly(thiophene) | Gas sensor for $NO_2$ | [35] |
| | Poly(aniline) | Gas sensor for $NO_2$ | [36] |
| | Poly(aniline) | Gas sensor for CO | [37] |
| $SnO_2+Ag$ | Poly(pyrrole) | Gas sensor for $NH_3$ | [38] |
| $SnO_2+Cu^{2+}$ | Poly(pyrrole) | Gas sensor for $H_2S$ | [39] |
| $SnO_2+ZnO$ | Poly(aniline) | Gas sensor for triethanolamine | [40] |
| $SiO_2+CuO+TiO_2$ | Poy(aniline) | Gas sensor for $NH_3$ | [41] |
| $TiO_2$ | Poly(aniline) | Gas sensor for trimethylamine | [42] |
| | Poly(pyrrole) | Humidity sensor | [43] |
| $Fe_2O_3$ | Poly(pyrrole) | Gas sensor for $CO_2$, $N_2$, and $CH_4$ | [44] |
| ZnO | Poly(aniline) | Gas sensor for $NH_3$ | [45] |
| ZnO/GQD | Poly(aniline) | Gas sensor for acetone | [46] |
| $WO_3$ | Poly(pyrrole) | Gas sensor for acetone | [47] |
| | Poly(pyrrole) | Gas sensor for $NO_2$ | [48] |

*GQD*, graphene quantum dots.

therefore, a high degree of homogeneity should be achieved during the making of composite. Key aspects which should be taken into account are as follows: (i) uniformity of metal-oxide particles in terms of size and shape, (ii) high degree of isotropy into final material, (iii) high degree of stability between contact phases, (iv) adequate dispersion of particles avoiding agglomerates, and (v) high reproducibility of manufacture procedure. Obviously, other aspects are highly important like low cost, low production of wastes, and eco-friendly and green processes [49,50].

The making of polymer-metal oxide composites can be described to be a two-stage process: (i) obtaining of components (i.e., polymer and metal-oxide nanoparticles) and (ii) formation of nanocomposites. In the first case, synthesis or obtaining of polymer materials are not described here, since this topic is away from the focus of chapter objective. On the contrary, metal-oxide nanoparticles can be obtained using chemical, physical, and biological methods. A summary of the main aspects of these methods is shown in Table 2.

Three general techniques are usually described for the manufacture of polymer-metal oxide composites; however, we identified at least five different techniques. The first technique is based on the in situ monomer polymerization in the presence of metal-oxide nanoparticles. The second is the direct mixing of polymer and

**Table 2** Manufacture method of metal-oxide particles.

| Strategy | Method | Description | References |
|---|---|---|---|
| Chemical | Sol-gel method | Obtaining of colloidal system ("sol"), gelation of the dispersion, and subsequent removal of the solvent. Low-temperature and low-cost process.<br>When obtaining the metal oxide, the following can occur: processes of precipitation, shrinkage, formation of pores, and residual hydroxyl groups could remain in the product. To carry out this method, salt like nitrates, chlorides, and acetates of target metals, and oxidant compounds like alkoxides are used. In general, salt should be highly soluble in the working solvent.<br>Typical metal oxides obtained by this method are ZnO, $TiO_2$, $SnO_2$, $Al_2O_3$, among others. | [51–55] |
| | Hydrothermal method | It is based on the metal-oxide crystal growth, via a chemical reaction in heterogeneous phase, using water or aqueous solvent. It implies the dissolution precursors and recrystallization of metal oxides. Different morphologies are obtained, with a good control of size. However, factors like pH, concentration, auxiliary reagents (e.g., surfactants), impurities, temperature, time, among other parameters can affect the features of metal oxide obtained. In consequence, a strict control of these aspects should be performed.<br>Typical metal oxides obtained by this method are ZnO, $TiO_2$, CuO, among others. | [56,57] |
| | Chemical vapor deposition | It consists of deposition of solid material on some surface; in this method, crystal growth is carried out from gaseous phase previous to chemical decomposition of precursor. Metal oxide is obtained with high purity and quality.<br>In general, this method can be used in a wide range of temperatures, its substrates have a passive role, and its operation is simple and adaptable to several configurations.<br>Main disadvantage is the deposition of metal oxide on reactor walls, and therefore, a large consumption of precursors is produced.<br>Typical metal oxides obtained by this method are $TiO_x$, ZnO, $In_2O_3$, $Al_2O_3$, among others. | [58–61] |

*Continued*

**Table 2** Continued

| Strategy | Method | Description | References |
|---|---|---|---|
| | Thermal decomposition processing | It is based on the chemical decomposition of precursor, usually an organometallic compound, in presence of ligands and some adequate solvent; it is carried out at high temperature using a coordinating or noncoordinating solvent. Ligands are required to stabilize the metal-oxide nanocrystals and correspond commonly to surfactants. Since it is a process at a high temperature, solvents with a high boiling point are used.<br>Typical metal oxides obtained by this method are $TiO_x$ and ZnO. | [62–64] |
| | Sonochemical method | It is based on the use of high-intensity ultrasonic energy which is used for the generation, growth, and collapse in the liquid phase. It can be mediated through free radicals (homogeneous sonication) or by cavitation (heterogeneous sonication). In addition, it can be carried out under ambient conditions. However, it should be performed using inert atmosphere. The main disadvantage is that irregular and random morphologies of metal-oxide nanoparticles are obtained.<br>Typical metal oxides obtained by this method are CuO, $Cu_2O$, $TiO_2$, NiO, and $BaTiO_3$. | [64–68] |
| Physical | Comminution | It is based on crushing, grinding, and milling of precursors. In general terms, it is a mechanical processing method in which physical degradation is promoted in order to obtain particulate material of small size. This method can be carried out in dry or wet conditions.<br>Typical metal oxides obtained by this method are $B_2O_3$, $TiO_2$, nanoparticles of perovskite oxides from (Bi, Na)$TiO_3$ and (K, Na)$NbO_3$, $Fe_2TiO_5$, and $CaTiO_3$. | [69–73] |
| | Spray drying | It is based on the atomization of a solution, slurry, suspension, or emulsion containing one or more precursors of target oxide. Thus, small droplets are sprayed and subsequently solvent is evaporated. Finally, calcination is carried out to produce the metal-oxide powders. Aspects like slid content, viscosity, and density must be optimized in order to achieve an adequate homogeneity of particles.<br>Typical oxides obtained by this method are ZnO, $MgAl_2O_4$, $Al_2O_3$, $FeAl_2O_4$, among others. | [74–76] |

| | | | |
|---|---|---|---|
| | Spray pyrolysis | It is based on the drying of the atomized droplets of metal-oxide dispersion into a solid surface. It is carried out in three stages: (i) dispersion in a liquid feedstock, (ii) evaporation, and (iii) drying. It has the advantage that the operating conditions can be easily changed, and furthermore, the properties of metallic oxides can be controlled by mixing processes and changes in drying conditions.<br>Typical oxides obtained by this method are $ZnMn_2O_4$, $CoAl_2O_4$, NiO-CuO-MgO, among others. | [77–79] |
| | Freeze drying | It is based on the sublimation process for the removal of solvent. It is carried out by the formation of particulate material from salt solution containing the target metal. Later, metal oxides are obtained by calcination of salt particles. It has the advantage that can produce metal-oxide nanoparticles with low polydispersity of particle size.<br>Typical oxides obtained by this method are $Al_2O_3$, MgO, $ZrO_2$, $CeO_2$, $SiO_2$, MgO, $MgAl_2O_4$, among others. | [80–82] |
| | Pulsed laser ablation | It is based on the use of pulsed laser beam to produce the fragmentation of material from target surface. It is influenced by the nature of raw material, environmental operation conditions (i.e., vacuum or nature of gaseous atmosphere), pressure, and laser characteristics. Chemical reaction also can be involved in this process. It is a simple, low-cost, and fast method. In some cases, nanopowders tend to the agglomeration. Also, industrial applications are limited by the low amount of metal oxide produced.<br>Typical oxides obtained by this method are $\alpha$-$Fe_2O_3$, CdO, $Y_2O_3$, among others. | [83–85] |
| | Vaporization-condensation | It is based on the heating and vaporization of precursor material and its subsequent and fast condensation of the vapor. Depending if chemical reaction occurs, then it can be named to be chemical or physical method. The aforementioned is directly related to the changes on nature experimented by the vapor phase respect to condensed phase. When change is observed, the process is chemical process whereas if no change is observed then the process occurred through a physical route. However, its application is limited by the melting point which in some cases can be very high. In addition, a strict control of heating should be maintained during all processes to avoid | [86–88] |

*Continued*

**Table 2** Continued

| Strategy | Method | Description | References |
|---|---|---|---|
| | | a high heterogeneity of particles.<br>Typical oxides obtained by this method are $WO_3$, ZnO, α-$Fe_2O_3$, $SiO_2$, among others. | |
| Biological | Biosynthesis based on microorganisms | Fungi, bacteria, and yeast are able to accumulate and biotransform heavy metals which can be utilized to produce metal-oxide nanoparticles. Some microorganisms used for these biosynthetic processes are *Actinobacter sp.* (e.g., $Fe_2O_3$ nanoparticles), *Aeromonas hydrophila* (e.g., ZnO nanoparticles), *Hypocrea lixii* (e.g., NiO nanoparticles), *and Aspergillus niger* (e.g., ZnO nanoparticles). | [89–91] |
| | Biobased synthesis based on plant extracts | It is based on the use of plant extracts to promote the chemical modification of precursors. In general, a wide number of extracts with different origins have been described in the field of nanotechnology. Depending on solvent, different biomolecules can be present in the extracts, for example, water (anthocyanins, tannins, lectins, among others), ethanol (tannins, polyphenols, flavanols, sterols, alkaloids, among others), methanol (anthocyanins, terpenoids, tannins, lactones, flavones, among others), chloroform (terpenoids, flavonoids, among others), ether (alkaloids, terpenoids, coumarins, fatty acids, among others), and acetone (phenols, flavanols, and substance with relatively low polarity).<br>Some examples are CuO nanoparticles that have been obtained using *Aloe barbadensis*, *Bifurcaria bifurcata*, and *Carica papaya* as reducing agents; $Fe_3O_4$ that has been obtained using extracts from *Sargassum muticum*, *Hordeum vulgare*, and *Rumex acetosa*; $TiO_2$ that has been prepared from *Jatropha curcas L.*, *Psidium guajava*, and *Ageratina altissima*; and ZnO that has been obtained from *Solanum nigrum*, *Pongamia pinnata*, and *Cassia fistula*. | [92] |

metal-oxide nanoparticles, and the third one is based on the inclusion of particles via melt or solution mixing (i.e., sol-gel processes). Also, other techniques are the in situ crystal growth and direct surface deposition of metal-oxide nanoparticles on the polymer phase [49].

The methods to generate and deposit thin films are mainly divided into two groups, both differentiated by the physicochemical processes followed by the material when going from source (feedstock or precursors) to substrate. The method in which the material to deposit begins at liquid or solid state goes to vapor phase to be transported until a substrate and finally, becomes a solid thin film is known as physical vapor deposition (PVD). These processes take place in vacuum environments to manage oxygen presence and avoid uncontrollable changes in stoichiometry of original material. On the contrary, chemical vapor deposition (CVD) refers to a chemical process to synthesize and deposit thin film from volatile precursors that react with a substrate to obtain the material required, i.e., the desired composition is generated at the same time it is deposited. A reaction chamber is used to eliminate secondary gas-volatile products and control the atmosphere of reaction [93–95]. Main PVD and CVD methods are presented later.

## 3.1 Spray coating methods

Spray is a physical technique allowing the deposit of particles of specific material through its spraying and shooting at high velocity toward a surface. Depending on the particles' temperature and the shooting method, spray coating can be classified as thermal or cold. Short explanation of both processes is developed following.

Thermal spray coating process is based on the use of electric or chemical energy to create molten particles (droplets of diameter $<100\,\mu m$) from a material initially in solid state (wire, rod, or powder). By using a spray gun (arc or plasma), the particles are propelled by an inert gas at high temperature toward a surface named substrate. Two mechanisms are usually referred to as follows: (i) arc spray gun uses an electric arc to melt two solid conductive wires that contain the material into molten droplets (see Fig. 3), and the particles obtained are propelled toward the substrate by a previously heated up inert gas jet (temperature $>500\,K$) [96,97]; and (ii) plasma spray gun employs an ionized conductive gas (usually N, He or H) to melt the material to deposit (powder in this case), and then shoot it toward the substrate. The expansion of the gas inside the chamber induces propel supersonic velocity in the particles ($>1000\,m/s$) at the escape nozzle, improving the bond between material and substrate. After impact, the droplets solidify at cooling rates up to $10^8\,K/s$ forming individual microstructural layers [97,98]. High temperature of plasma allows several materials (ceramics, polymers, glasses, and metals) to be melted and deposited under a controllable atmosphere in order to avoid oxidation and porosity.

In contrast to thermal spray, in *cold spray coating method,* powder particles (diameter $<50\,\mu m$) are exposed briefly to heating, and consequently, the material does not present state change impacting on the substrate in solid phase. During the powder deposition, the propelling inert gas is heated up and expanded inside a chamber, then powder gets into the chamber and is driven by the gas toward the substrate. If gas velocity is high enough (up to $1200\,m/s$) and mechanical properties are adequate, after

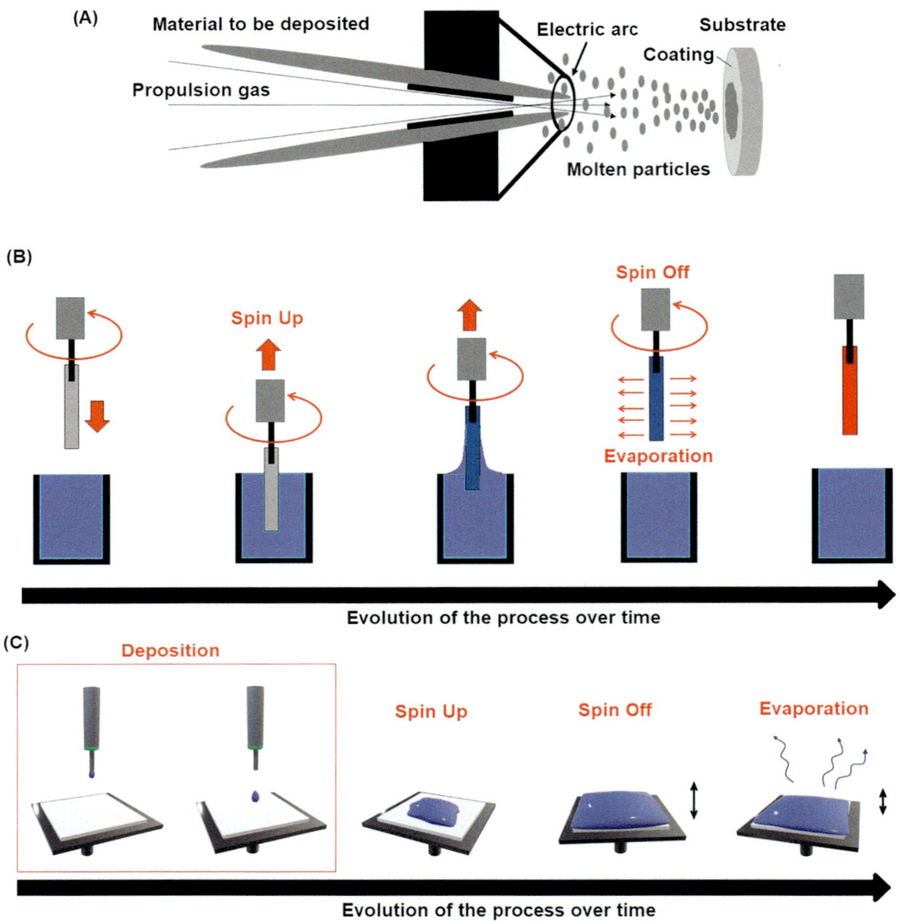


**Fig. 3** (A) Scheme of a basic arc spray gun used in thermal spray deposition method, (B) illustration of dip coating technique, and (C) illustration of spin coating technique. 

impact solid particles deform plastically creating a hydrodynamic stability with substrate [99,100]. Despite the versatility of spray methods to deposit many types of materials on different substrate conditions, high temperature of processes could induce evaporation of volatile species that modifies stoichiometry and shifts the composition of the coating. Furthermore, the size of the acquired particles limits the deposit of layers with thickness of less than 5 μm [100,101].

## 3.2 Dip coating

Dip coating is a simple deposition process widely used in the industry to coat from sol-gel and liquids solutions, nonflat and nonrigid materials with protective, antireflective, or selective thin films. The method consists of the immersion at constant velocity of a substrate into a liquid or sol-gel bath with enough time (proportional to liquid viscosity) to adhere to the surface. Coated material is slowly withdrawn to avoid vibrations that affect film quality and thermally treated to eliminate volatile solvents or water excesses. Coating thickness is proportional to the velocity of the process and the liquid viscosity, and inversely proportional to its density [102,103]. Illustration is shown in Fig. 3B.

### 3.3 Spin coating

Spin coating is a method to deposit thin films on flat and rigid surfaces from liquids or sol-gel materials. In this process, a substrate impregnated with drops of liquid rotates in a range from 300 to 1000 rpm to form the film. During rotation, liquid in the surface flows to the edge radially due to centrifugal force, while the viscous force acts radially inward. The balance between these two forces induces a uniform coating. The film deposited becomes thinner as viscosity and rotation time decrease, while angular velocity increases [93,104]. Illustration is shown in Fig. 3C.

## 4 Specific sensing applications by polymer-metal oxide composites

### 4.1 Gas sensors

The semiconductor properties offered by many metallic oxides have allowed their application in gas sensing; generally, they are electrochemical devices with a gas-sensitive solid layer corresponding to the recognition layer. This layer has a high degree of affinity to the gas and interacts with it by chemical or physical phenomena; however, in most cases, oxidation-reduction reactions are processes resulting from interaction of gas with the recognition layer [105]. In general, gas sensors require a high reactivity sensing material to increase its sensibility and time response in the presence of one or more types of gases in the environment. Polymeric sensors use a mixed metal oxide as sensing layer due to the high sensitivity in the presence of gases, which is mainly generated by the interaction between different electronic structures that modifies the nanocrystals [106]. The stability of the electrical signal requires the building of strong conductive networks capable of increasing tunneling effect between electrical domains, which can be achieved by (i) high particle concentration or (ii) nonrandom particle distribution inside the matrix.

Some metal oxides used as semiconductors in gas sensing are $SnO_2$, ZnO, $TiO_2$, CoO, NiO, and $WO_3$. These show a great detection capacity in terms of sensitivity in most of the applications; however, current research objectives are focused on these types of sensors, and they are directed toward the improvement of three fundamental characteristics: sensitivity, selectivity, and stability. The first characteristic, sensitivity, is strongly related to the nature and properties of metallic oxides, whereas the other two, selectivity and stability, can be improved by the incorporating of matrices based on conductive polymers. Conducting polymers offer improved mechanical properties as well as these ease the manufacture of sensors with different morphologies, transmission of electrical signals, and even, they can act as an active component into the sensing process [16,17,107].

A wide number of works for the gas detection are available (see Table 2). In some cases, results must be carefully analyzed since small differences can limit or enhance one or more applications. For example, detection of $NH_3$ has been described using composites based on ZnO/poly(aniline) as well as composites based on $TiO_2$/poly(aniline). In consequence, at first glance, it is not possible to identify a difference between the different sensors, and it could even be inferred that the active phase of the sensing process is the poly(aniline). In fact, sensors based on poly(aniline) without the use of metal-oxide nanoparticles have been recently described [108], as well as, sensors based on $TiO_2$/poly(aniline) also have been described for the sensing of trimethylamine [42]. In consequence, it is not only one important reproducible change, but also selectivity and differentiation play a very important role in establishing the practical usefulness of a sensor.

## 4.2 Humidity sensors

Humidity sensing is a key factor in various industries like agriculture, food, and fuels. Thus, polymer-metal oxide composites have played an enormous role in recent technological advances in the manufacture of humidity sensors with high response speed, high stability, and high sensitivity. Again, the nanostructures of metal oxides are based mainly on Cu, Zn, Sn, Sr, Ni, Cd, and Ti because they usually exhibit optical and electronic properties which can be altered by adsorption phenomena, exactly, water adsorption or surface hydration [16,17].

Polymer-based sensitive materials for humidity sensing present faster water desorption than ceramic or solid-state sensors since they do not require drying assistance, which optimize their time response. The polymer matrix used must be insoluble in water, hydrophilic to absorb enough molecules that generate ionic conductive paths, and, in turn, porous to improve the contact with the atmosphere [109]. According to the electrical conductivity of the material, polymeric-based humidity sensors can be divided into two main groups: (i) resistive humidity sensor that uses a high electrical conductivity material as transductor, which under vapor water presence (by adsorption) increases its electrical resistivity consequence of the low conductivity of water [110] and (ii) capacitive humidity sensor which employs a dielectric material that modifies its electric permittivity in wet atmospheres as a result of the dynamic equilibrium between liquid water on the material and vapor in the air. By using a parallel layer stack (electrode/polymer/electrode), changes in the capacitance of the sensor are monitored and associated with humidity [111]. Typical humidity detection range for these sensors goes from 20% to 80% where the electric signal presents linearity and stability [111]. For higher or lower humidity values, novel materials such as graphene oxide, ZnO, ZnS doped with $SiO_2$ encapsulated in porous polymers have been developed.

For this type of sensor, the introduction of polymeric matrices, generally conductive polymers, facilitates and promotes adsorption processes through new interactions, improvement of permeability, and incorporation of improved mechanical properties. These aspects ease the manufacture of more compact sensors, with greater sensitivity and durability [16,17,112]. For instance, composites based on $SnO_2$/poly(o-anisidine) achieve an increase in sensitivity of up to 10 times greater than that of pure components for the moisture detection; in addition, an excellent linearity was observed ($R^2 = 0.9992$) in a range between 20% and 100% of relative humidity [113]. Likewise, humidity sensors based on $TiO_2$/poly(aniline) composite show a high sensitivity in a range between 25% and 95% of relative humidity, with a response speed of 60 s [114].

## 4.3 Temperature sensor

Temperature sensors based on polymer-metal oxide composites use changes in the resistivity of the polymers as transductor to measure thermal information of the environment. The relationship between the temperature and the resistivity of the materials is given by a coefficient known as temperature coefficient, such that it takes a negative (positive) value when the resistivity of the sensing layer decreases (increases) as the temperature rises. A negative temperature coefficient indicates that electrons of valence band undergo thermal agitation, which maximizes the number of active charges, i.e., electric conduction increases with temperature. For materials with positive temperature coefficient, thermal lattice agitation associated with impurities and defects in nanocrystals is induced with the increment of temperature, which in turn increases (decreases) its electrical resistivity (conductivity) [115].

Temperature sensors usually use NTC materials to obtain high sensitivity and strong electrical response under low-temperature stimuli [116]. Additionally, conductive polymers, with negative and positive temperature coefficients, have been employed to develop current limiters, over-temperature protectors, self-regulators, and thermistors [117].

## 4.4 Organic molecules sensors

Organic molecules are of great interest in chemical analysis as contaminants, quality control, monitoring of health state of patients, biological system characterizations, etc. In this context, sensors based on $CeO_2$/poly(pyrrole-dodecylbenzene) with good linearity of the calibration curve ($R^2 = 0.9922$) have been designed for the determination of peroxide. Thus, when peroxide is used as oxidant agent or organic matter, peroxide remaining in the system can be detected and indirectly correlated the amount consumed with the organic matter contents [118]. In addition, sensors based on $Bi_2O_3$/poly(pyrrole) were developed for the determination of clofazimine, which is a drug used for the treatment of leprosy [119]. Other examples are the determination of glucose, lactate, ascorbic acid, dopamine, and xanthine using sensors made of $TiO_2$/poly(aniline) and ZnO/Nafion [120,121]. Other examples were previously shown in Table 2.

# 5 Conclusions and remarks

Polymer-metal oxides are a promissory strategy for the making of sensors. At present, despite the notable advances made, it can be concluded that this field of technological development is in an intermediate stage of development. The foregoing is based on the availability in the market of sensing technologies based on this type of hybrid materials. However, many aspects or potential lines of work remain unexplored and are therefore a niche for new developments in this field. Some of these are as follows. (i) To advance in the development of polymer-metal oxide composite sensors based on nonlinear response. At present, the great advances in data processing, portability, and web connectivity for remote analysis make the premise of requiring a linear response model unnecessary. (ii) The developments of polymer-metal oxide composite sensors are strongly focused on the same group of raw materials. The use of conductive polymers is strongly limited to 2 or 3 (i.e., poly(aniline), poly(pyrrole), and to a lesser extent, poly(thiophene)). Usually, modifications of these are not considered for copolymerization strategies. Some alternative of conductive polymers are poly(phenylene)s, poly(pyrene)s, poly(fluorene)s, poly(*p*-phenylene vinylene), poly(carbazole)s, poly(thiophene)s, poly(*p*-phenylene sulfide), among others, which are completely absent in the researches. Similar situation occurs with inorganic polymers, metal-polymer composites, and nonconductive polymers, which could be useful if dielectric behavior is used as an analytical response. In addition, small advances have been performed using mixed oxides. (iii) Applications of this type of sensor are strongly limited to gases, humidity, temperature, and some biomolecules. Consequently, the stock of alternative analytes of technological interest remains extensive. (iv) Multifunctional strategies of sensing are practically unexplored. Composites based on polymer-metal oxide are usually made using one single oxide.

# Acknowledgments

The authors thank for financial support to Universidad del Valle (C.I. 71263) and the Ministry of Science (Minciencias) of Colombia (Call 848 of 2019 "Program of Postdoctoral stays in entities of the SNCTeI 2019").

# References

[1] M.J. McGrath, C.N. Scanaill, Sensing and sensor fundamentals, in: Sensor Technologies, Apress, Berkeley, CA, 2013, pp. 15–50, https://doi.org/10.1007/978-1-4302-6014-1_2.

[2] H. Beitollahi, S.Z. Mohammadi, M. Safarei, S. Tajik, Applications of electrochemical sensors and biosensors based on modified screen-printed electrodes: a review, Anal. Methods 12 (2020) 1547–1560, https://doi.org/10.1039/C9AY02598G.

[3] O. Isildak, O. Ozbek, Application of potentiometric sensors in real samples, Crit. Rev. Anal. Chem. (2020) 1–14, https://doi.org/10.1080/10408347.2019.1711013.

[4] X. Su, L. Sutarlie, X. Jun, Sensors, biosensors, and analytical technologies for aquaculture water quality, Research (2020) 1–15, https://doi.org/10.34133/2020/8272705.

[5] X. Zhou, S. Lee, Z. Xu, J. Yoon, Recent progress on the development of chemosensors for gases, Chem. Rev. 115 (2015) 7944–8000, https://doi.org/10.1021/cr500567r.

[6] S. Sikarwar, S. Satyedra, S. Singh, B.C. Yadav, Review on pressure sensors for structural health monitoring, Photonic Sens. 7 (2017) 1–11, https://doi.org/10.1007/s13320-017-0419-z.
[7] B.A. Kuzubasoglu, S.K. Bahadir, Flexible temperature sensors: a review, Sensors Actuators A Phys. 315 (2020) 112282, https://doi.org/10.1016/j.sna.2020.112282.
[8] C.Y. Lee, G.B. Lee, Humidity sensors: a review, Sens. Lett. 3 (2004) 1–15, https://doi.org/10.1166/sl.2005.001.
[9] N.S. Patil, R.B. Dhake, M.I. Ahamed, U. Fegade, A mini review on organic chemosensors for cation recognition, J. Fluoresc. 30 (2020) 1295–1330, https://doi.org/10.1007/s10895-020-02554-7.
[10] T. Lerma, J. Martinez, E. Combatt, Determination of total protein content by digital image analysis: an approach for food quality remote analysis, J. Sci. Technol. Appl. 6 (2019) 53–64, https://doi.org/10.34294/j.jsta.19.6.41.
[11] M. Palencia, T. Lerma, V. Palencia, Description of fouling, surface changes and heterogeneity of membranes by color-based digital analysis, J. Membr. Sci. 510 (2016) 229–237, https://doi.org/10.1016/j.memsci.2016.02.057.
[12] N. Cajiao, H. Mora, M. Palencia, Determination of microorganism cellular concentration by digital image quantitative analysis (DIQA), J. Sci. Technol. Appl. 1 (2016) 77–86,-https://doi.org/10.34294/j.jsta.16.1.7.
[13] A.C. Peixoto, A.F. Silva, Smart devices: micro- and nanosensors, in: L. Rodrigues, M. Mota (Eds.), Bioinspired Materials for Medical Applications, Science Direct, 2017, pp. 297–329, https://doi.org/10.1016/B978-0-08-100741-9.00011-5.
[14] N. Sabri, S.A. Aljunid, M. Sabri, R. Badlishah, R. Kamaruddin, Toward optical sensors: review and applications, J. Phys. Conf. Ser. 423 (2020) 2064, https://doi.org/10.1088/1742-6596/423/1/012064.
[15] H. Gu, Z. Wang, Y. Hu, Hydrogen gas sensors based on semiconductor oxide nanostructures, Sensors 12 (5) (2012) 5517–5550, https://doi.org/10.3390/s120505517.
[16] Y. Wang, Q. Gong, Q. Miao, Structured and functionalized organic semiconductors for chemical and biological sensors based on organic field effect transistors, Mater. Chem. Front. 4 (2020) 3505–3520, https://doi.org/10.1039/D0QM00202J.
[17] Y. Wang, A. Liu, Y. Han, T. Li, Sensors based on conductive polymers and their composites: a review, Polym. Int. 69 (2020) 7–17, https://doi.org/10.1002/pi.5907.
[18] C.W. de Silva, Sensor Systems: Fundamentals and Applications, CRC Press, 2016, p. 746.
[19] P.F. Dunn, Fundamentals of Sensors for Engineering and Science: Measurement and Data Analysis for Engineering and Science, CRC Press, 2019, p. 614, https://doi.org/10.1201/9781315275390.
[20] D. Nunes, A. Pimentel, L. Santos, P. Barquinha, L. Pereira, E. Fortunato, R. Martins, Structural, optical, and electronic properties of metal oxide nanostructures, in: Metal Oxide Nanostructes, 2019, pp. 59–102, https://doi.org/10.1016/B978-0-12-811512-1.00003-5.
[21] M.S. Shah, A. Bhattacharya, D. Stepanova, A. Mikhaylov, M.L. Grilli, et al., A systematic review of metal oxide applications for energy and environmental sustainability, Metals 10 (2020) 1604, https://doi.org/10.3390/met10121604.
[22] R. Villalobos, C. Medina, C. Canales, M. Melendrez, P. Flores, Development of nanocomposite materials with thermostable matrix from nanoreinforcement of titanium dioxide (TNPS and TNTS) and zinc oxide (ZNPS and ZNBS) on epoxy resin matrix, J. Sci. Technol. Appl. 6 (2019) 65–78, https://doi.org/10.34294/j.jsta.19.6.42.

[23] R. Villalobos, D. Rojas, A. Diaz, J. Ramirez, M. Melendrez, P. Flores, Synthesis and characterization of 1d metal oxide nano-structures: titanium dioxide nanotubes (TNTS) and zinc oxide nanobars (ZNBS), J. Sci. Technol. Appl. 6 (2019) 79–95, https://doi.org/10.34294/j.jsta.19.6.43.
[24] C. Pei, G. Han, Y. Zhao, H. Zhao, B. Liu, L. Cheng, H. Yang, S. Liu, Superior adsorption performance for triphenylmethane dyes on 3D architectures assembled by ZnO nanosheets as thin as ~1.5 nm, J. Hazard. Mater. 318 (2016) 732–741, https://doi.org/10.1016/j.jhazmat.2016.07.066.
[25] J. Fan, D. Chen, N. Li, Q. Xu, H. Li, J. He, J. Lu, Adsorption and biodegradation of dye in wastewater with Fe3O4@MIL-100 (Fe) core–shell bio-nanocomposites, Chemosphere 191 (2018) 315–323, https://doi.org/10.1016/j.chemosphere.2017.10.042.
[26] H.B. Hadjltaief, M. Ben-Zina, M.E. Galvez, P. Da Costa, Photocatalytic degradation of methyl green dye in aqueous solution over natural clay-supported ZnO–TiO2 catalysts, J. Photochem. Photobiol. A Chem. 315 (2016) 25–33, https://doi.org/10.1016/j.jphotochem.2015.09.008.
[27] Y. Tanaka, T. Utaka, R. Kikuchi, K. Sasaki, K. Eguchi, CO removal from reformed fuel over Cu/ZnO/Al2O3 catalysts prepared by impregnation and coprecipitation methods, Appl. Catal. A Gen. 238 (2003) 11–18, https://doi.org/10.1016/S0926-860X(02)00095-9.
[28] P.B. Koli, K.H. Kapadnis, U.G. Deshpande, Transition metal decorated Ferrosoferric oxide (Fe3O4): an expeditious catalyst for photodegradation of Carbol Fuchsin in environmental remediation, J. Environ. Chem. Eng. 7 (2019) 103373, https://doi.org/10.1016/j.jece.2019.103373.
[29] H. Zhou, J. Guo, P. Li, T. Fan, D. Zhang, J. Ye, Leaf-architectured 3D hierarchical artificial photosynthetic system of perovskite titanates towards CO2 photoreduction into hydrocarbon fuels, Sci. Rep. 3 (2013), https://doi.org/10.1038/srep01667, srep01667.
[30] M. Yu, Y. Zeng, C. Zhang, X. Lu, C. Zeng, C. Yao, Y. Yang, Y. Tong, Titanium dioxide polypyrrole core-shell nanowires for all solid-state flexible supercapacitors, Nanoscale 5 (2013) 10806–10810, https://doi.org/10.1039/C3NR03578F.
[31] G. Xi, B. Yue, J. Cao, J. Ye, Fe3O4/WO3 hierarchical core-shell structure: high-performance and recyclable visible-light photocatalysis, Chem. A Eur. J. 17 (2011) 5145–5154, https://doi.org/10.1002/chem.201002229.
[32] X. Zhao, L. Lv, B. Pan, W. Zhang, S. Zhang, Q. Zhang, Polymer-supported nanocomposites for environmental application: a review, Chem. Eng. J. 170 (2011) 381–394, https://doi.org/10.1016/j.cej.2011.02.071.
[33] M.K. Ram, Ö. Yavuz, V. Lahsangah, M. Aldissi, CO gas sensing from ultrathin nanocomposite conducting polymer film, Sens. Actuators B 106 (2005) 750–757, https://doi.org/10.1016/j.snb.2004.09.027.
[34] L. Geng, Y.Q. Zhao, X.L. Huang, S.R. Wang, S.M. Zhang, S.H. Wu, Characterization and gas sensitivity study of polyaniline/SnO2 hybrid material prepared by hydrothermal route, Sens. Actuators B 120 (2007) 568–572, https://doi.org/10.1016/j.snb.2006.03.009.
[35] M. Xu, J. Zhang, S. Wang, X. Guo, H. Xia, Y. Wang, et al., Gas sensing properties of SnO2 hollow spheres/polythiophene inorganic–organic hybrids, Sens. Actuators B 146 (2010) 8–13, https://doi.org/10.1016/j.snb.2010.01.053.
[36] H. Xu, D. Ju, W. Li, H. Gong, J. Zhang, et al., Low-working-temperature, fast-response-speed NO2 sensor with nanoporous-SnO2/polyaniline double-layered film, Sens. Actuators B 224 (2016) 654–660, https://doi.org/10.1016/j.snb.2015.10.076.

[37] K.S. Jian, C.J. Chang, J.J. Wu, Y.C. Chang, C.Y. Tsay, et al., High response CO sensor based on a polyaniline/SnO2 nanocomposite, Polymers (Basel) 11 (2019) 184, https://doi.org/10.3390/polym11010184.

[38] T. Jiang, Z. Wang, Z. Li, W. Wang, X. Xu, J. Wang, C. Wang, Synergic effect within n-type inorganic–p-type organic nano-hybrids in gas sensors, J. Mater. Chem. C 1 (2013) 3017, https://doi.org/10.1039/C3TC00370A.

[39] J. Shu, Z. Qiu, S. Lv, K. Zhang, D. Tang, Cu2+-doped SnO2 nanograin/polypyrrole nanospheres with synergic enhanced properties for ultrasensitive room-temperature H2S gas sensing, Anal. Chem. 89 (2017) 11135–11142, https://doi.org/10.1021/acs.analchem.7b03491.

[40] L. Quan, J. Sun, S. Bai, et al., A flexible sensor based on polyaniline hybrid using ZnO as template and sensing properties to triethylamine at room temperature, Appl. Surf. Sci. 399 (2017) 583–591, https://doi.org/10.1016/j.apsusc.2016.12.133.

[41] Z. Pang, Q. Nie, P. Lv, J. Yu, F. Huang, Q. Wei, Design of flexible PANI-coated CuO-TiO2-SiO2 heterostructure nanofibers with high ammonia sensing response values, Nanotechnology 28 (2017) 225501, https://doi.org/10.1088/1361-6528/aa6dd5.

[42] J.B. Zheng, G. Li, X.F. Ma, Y.M. Wang, G. Wu, Y.N. Cheng, Polyaniline–TiO2 nano-composite-based trimethylamine QCM sensor and its thermal behavior studies, Sens. Actuators B 133 (2008) 374–380, https://doi.org/10.1016/j.snb.2008.02.037.

[43] P.G. Su, L.N. Huang, Humidity sensors based on TiO2 nanoparticles/polypyrrole composite thin films, Sens. Actuators B 123 (2007) 501–507, https://doi.org/10.1016/j.snb.2006.09.052.

[44] K. Suri, S. Annapoorni, A.K. Sarkar, R.P. Tandon, Gas and humidity sensors based on iron oxide–polypyrrole nanocomposites, Sens. Actuators B 81 (1–2) (2002) 277–282,-https://doi.org/10.1016/S0925-4005(01)00966-2.

[45] M. Das, D. Sarkar, One-pot synthesis of zinc oxide—polyaniline nanocomposite for fabrication of efficient room temperature ammonia gas sensor, Ceram. Int. 43 (2017) 11123–11131, https://doi.org/10.1016/j.ceramint.2017.05.159.

[46] Y. Yan, G. Yang, J.L. Xu, M. Zhang, C.C. Kuo, S.D. Wang, Conducting polymer-inorganic nanocomposite-based gas sensors: a review, Sci. Technol. Adv. Mater. 21 (2020) 768–786, https://doi.org/10.1080/14686996.2020.1820845.

[47] H. Jamalabadi, N. Alizadeh, Enhanced low-temperature response of PPy-WO3 hybrid nanocomposite based gas sensor deposited by electrospinning method for selective and sensitive acetone detection, IEEE Sens. J. 17 (2017) 2322–2328, https://doi.org/10.1109/JSEN.2017.2662716.

[48] A.T. Mane, S.T. Navale, V.B. Patil, Room temperature NO2 gas sensing properties of DBSA doped PPy–WO3 hybrid nanocomposite sensor, Org. Electron. 19 (2015) 15–25,-https://doi.org/10.1016/j.orgel.2015.01.018.

[49] S.R. Prasanna, K. Balaji, S. Pandey, S. Rana, Metal oxide based nanomaterials and their polymer nanocomposites, in: Nanomaterials and Polymer Nanocomposites, 2019, pp. 123–144, https://doi.org/10.1016/b978-0-12-814615-6.00004-7.

[50] S. Zhuiykov, Semiconductor nanocrystals in environmental sensors, in: S. Zhuiykov (Ed.), Nanostructured Semiconductor Oxides for the Next Generation of Electronics and Functional Devices, Woodhead Publishing, 2014, pp. 374–426, https://doi.org/10.1533/9781782422242.374.

[51] S.K. Omanwar, S.R. Jaiswal, V.B. Bhatkar, K.A. Koparkar, Comparative study of nano-sized Al2O3 powder synthesized by sol-gel (citric and stearic acid) and aldo-keto gel method, Optik 158 (2018) 1248–1254, https://doi.org/10.1016/j.ijleo.2017.12.068.

[52] R. Peña-Garcia, Y. Guerra, B.V.M. Farias, D.M. Buitrago, A. Franco, E. Padron-Hernández, Effects of temperature and atomic disorder on the magnetic phase transitions in ZnO nanoparticles obtained by sol-gel method, Mater. Lett. 233 (2018) 146–148,-https://doi.org/10.1016/j.matlet.2018.08.148.

[53] K. Sakthiraj, M. Hema, K.K. Balachandra, The effect of reaction temperature on the room temperature ferromagnetic property of sol-gel derived tin oxide nanocrystal, Phys. B Condens. Matter 538 (2018) 109–115, https://doi.org/10.1016/j.physb.2018.03.023.

[54] S. Javed, M. Islam, M. Mujahid, Synthesis and characterization of TiO2 quantum dots by sol gel reflux condensation method, Ceram. Int. 45 (2019) 2676–2679, https://doi.org/10.1016/j.ceramint.2018.10.163.

[55] K. Mahendraprabhu, S.A. Selva, P. Elumalai, CO sensing performances of YSZ-based sensor attached with sol-gel derived ZnO nanospheres, Sens. Actuators B 283 (2019) 842–847, https://doi.org/10.1016/j.snb.2018.11.164.

[56] Q. Yang, Z. Lu, J. Liu, X. Lei, Z. Chang, L. Luo, et al., Metal oxide and hydroxide nanoarrays: hydrothermal synthesis and applications as supercapacitors and nanocatalysts, Prog. Nat. Sci.: Mater. Int. 23 (2013) 351–366, https://doi.org/10.1016/j.pnsc.2013.06.015.

[57] S.M. Saleh, ZnO nanospheres based simple hydrothermal route for photocatalytic degradation of azo dye, Spectrochim. Acta A Mol. Biomol. Spectrosc. 211 (2019) 141–147,-https://doi.org/10.1016/j.saa.2018.11.065.

[58] S. Lukic, I. Stijepovic, S. Ognjanovic, V.V. Srdic, Chemical vapour synthesis and characterisation of Al2O3 nanopowders, Ceram. Int. 41 (2015) 3653–3658, https://doi.org/10.1016/j.ceramint.2014.11.034.

[59] D.W. Sheel, J.M. Gaskell, Deposition of fluorine doped indium oxide by atmospheric pressure chemical vapour deposition, Thin Solid Films 520 (2011) 1242–1245, https://doi.org/10.1016/j.tsf.2011.04.206.

[60] J. Sung, M. Shin, P.R. Deshmukh, H.S. Hyun, Y. Sohn, W.G. Shin, Preparation of ultrathin TiO2 coating on boron particles by thermal chemical vapor deposition and their oxidation-resistance performance, J. Alloys Compd. 767 (2018) 924–931, https://doi.org/10.1016/j.jallcom.2018.07.152.

[61] J. Tatebayashi, G. Yoshii, T. Nakajima, M. Mishina, Y. Fujiwara, Formation and optical properties of Tm, Yb-codoped ZnO nanowires grown by sputtering-assisted metalorganic chemical vapor deposition, J. Cryst. Growth 503 (2018) 13–19, https://doi.org/10.1016/j.jcrysgro.2018.09.006.

[62] M.M. Lencka, R.E. Riman, Thermodynamic modeling of hydrothermal synthesis of ceramic powders, Chem. Mater. 5 (1993) 61–70, https://doi.org/10.1021/cm00025a014.

[63] H. Dong, C. Feldmann, Porous ZnO platelets via controlled thermal decomposition of zinc glycerolate, J. Alloys Compd. 513 (2012) 125–129, https://doi.org/10.1016/j.jallcom.2011.10.004.

[64] H. Xu, B.W. Zeiger, K.S. Suslick, Sonochemical synthesis of nanomaterials, Chem. Soc. Rev. 42 (2013) 2555–2567, https://doi.org/10.1039/C2CS35282F.

[65] M. Xu, Y.N. Lu, Y.F. Liu, S.Z. Shi, T.S. Qian, D.Y. Lu, Sonochemical synthesis of monosized spherical BaTiO3 particles, Powder Technol. 161 (2006) 185–189, https://doi.org/10.1016/j.powtec.2005.10.001.

[66] W. Zhu, A. Shui, L. Xu, X. Cheng, P. Liu, H. Wang, Template-free sonochemical synthesis of hierarchically porous NiO microsphere, Ultrason. Sonochem. 21 (2014) 1707–1713, https://doi.org/10.1016/j.ultsonch.2014.02.026.

[67] M.A. Bhosale, B.M. Bhanage, A simple approach for sonochemical synthesis of Cu2O nanoparticles with high catalytic properties, Adv. Powder Technol. 27 (2016) 238–244,- https://doi.org/10.1016/j.apt.2015.12.00.
[68] D.S. Kim, J.C. Kim, B.K. Kim, D.W. Kim, One-pot low-temperature sonochemical synthesis of CuO nanostructures and their electrochemical properties, Ceram. Int. 42 (2016) 19454–19460, https://doi.org/10.1016/j.ceramint.2016.09.044.
[69] M. Alizadeh, F. Sharifianjazi, E. Haghshenasjazi, M. Aghakhani, L. Rajabi, Production of nanosized boron oxide powder by high-energy ball milling, Synth. React. Inorg., Met.-Org., Nano-Met. Chem. 45 (2015) 11–14, https://doi.org/10.1080/15533174.2013.797438.
[70] T. Nguyen, J.L. He, Preparation of titanium monoxide nanopowder by lowenergy wet ball-milling, Adv. Powder Technol. 27 (2016) 1868–1873, https://doi.org/10.1016/j.apt.2016.04.022.
[71] G.J. Lee, E.K. Park, S.A. Yang, J.J. Park, S.D. Bu, M.K. Lee, Rapid and direct synthesis of complex perovskite oxides through a highly energetic planetary milling, Sci. Rep. 7 (2017) 46241, https://doi.org/10.1038/srep46241.
[72] D.T. Franca, B.F. Amorim, A.M. de Morais Araujo, M.A. Morales, F. Bohn, S.N. de Medeiros, Structural and magnetic properties of Fe2TiO5 nanopowders prepared by ball-milling and post annealing, Mater. Lett. 236 (2019) 526–529, https://doi.org/10.1016/j.matlet.2018.10.149.
[73] D. Stoyanova, I. Stambolova, V. Blaskov, K. Zaharieva, I. Avramova, O. Dimitrov, et al., Mechanical milling of hydrothermally obtained CaTiO3 powders—morphology and photocatalytic activity, Nano-Struct. Nano-Objects 18 (2019) 100301, https://doi.org/10.1016/j.nanoso.2019.100301.
[74] B.D. Ehrhart, B.J. Ward, B.M. Richardson, K.S. Anseth, A.W. Weimer, Partial flocculation for spray drying of spherical mixed metal oxide particles, J. Am. Ceram. Soc. 101 (2018) 4452–4457. Available from *https://doi.org/10.1111/jace.15727*.
[75] O. Yıldız, A.M. Soydan, Parameters for spray drying ZnO nanopowders as spherical granules, J. Am. Ceram. Soc. 101 (2018) 103–115, https://doi.org/10.1111/jace.15191.
[76] O. Yıldız, A.M. Soydan, Synthesis of zirconia toughened alumina nanopowders as soft spherical granules by combining co-precipitation with spray drying, Ceram. Int. 45 (2019) 17521–17528, https://doi.org/10.1016/j.ceramint.2019.05.314.
[77] T. Tani, A. Kato, H. Morisaka, Effects of solvent on powder characteristics of zinc oxide and magnesia prepared by flame spray pyrolysis, J. Cerma. Soc. Jpn. 113 (2005) 255–258, https://doi.org/10.2109/jcersj.113.255.
[78] J. Yu, D. Kim, The preparation of nano size nickel oxide powder by spray pyrolysis process, Powder Technol. 235 (2013) 1030–1037, https://doi.org/10.1016/j.powtec.2012.11.031.
[79] I.V. Krasnikova, I.V. Mishakov, Y.I. Bauman, T.M. Karnaukhov, A.A. Vedyagin, Preparation of NiO-CuO-MgO fine powders by ultrasonic spray pyrolysis for carbon nanofibers synthesis, Chem. Phys. Lett. 684 (2017) 36–38, https://doi.org/10.1016/j.cplett.2017.06.036.
[80] C. Tallon, R. Moreno, M.I. Nieto, Synthesis of ZrO2 nanoparticles by freeze drying, Int. J. Appl. Ceram. Technol. 6 (2009) 324–334, https://doi.org/10.1111/j.1744-7402.2008.02279.x.
[81] B. Liu, Y. You, H. Zhang, H. Wu, J. Jin, H. Liu, Synthesis of ZnO nano-powders via a novel PVA-assisted freeze-drying process, RSC Adv. 6 (2016) 110349–110355, https://doi.org/10.1039/C6RA24154A.

[82] L.B. Chiriac, M. Todea, A. Vulpoi, M. Muresan-Pop, R.V.F. Turcu, S. Simon, Free-zedrying assisted sol-gel-derived silica-based particles embedding iron: synthesis and characterization, J. Sol-Gel Sci. Technol. 87 (2018) 195–203, https://doi.org/10.1007/s10971-018-4702-2.
[83] S.J. Henley, S. Mollah, C.E. Giusca, S.R.P. Silva, Laser-induced self-assembly of iron oxide nanostructures with controllable dimensionality, J. Appl. Phys. 106 (2009), https://doi.org/10.1063/1.3224854, 064309.
[84] K. Siraj, Y. Sohail, A. Tabassum, Metals and metal oxides particles produced by pulsed laser ablation under high vacuum, Turk. J. Phys. 35 (2011) 179–183.
[85] V.A. Svetlichnyi, A.V. Shabalina, I.N. Lapin, Structure and properties of nanocrystalline iron oxide powder prepared by the method of pulsed laser ablation, Russ. Phys. J. 59 (2017) 2012–2016, https://doi.org/10.1007/s11182-017-1008-8.
[86] M.S. El-Shall, W. Slack, W. Vann, D. Kane, D. Hanley, Synthesis of nanoscale metal oxide particles using laser vaporization/condensation in a diffusion cloud chamber, J. Phys. Chem. 98 (1994) 3067–3070, https://doi.org/10.1021/j100063a001.
[87] C. Stotzel, H.D. Kurland, J. Grabow, F.A. Muller, Gas phase condensation of super-paramagnetic iron oxide-silica nanoparticles-control of the intraparticle phase distribution, Nanoscale 7 (2015) 7734–7744, https://doi.org/10.1039/C5NR00845J.
[88] A.V. Vodopyanov, A.V. Samokhin, N.V. Alexeev, M.A. Sinayskiy, A.I. Tsvetkov, M.Y. Glyavin, et al., Application of the 263 GHz/1 kW gyrotron setup to produce a metal oxide nanopowder by the evaporation-condensation technique, Vacuum 145 (2017) 340–346,-https://doi.org/10.1016/j.vacuum.2017.09.018.
[89] A.A. Bharde, R.Y. Parikh, M. Baidakova, S. Jouen, B. Hannoyer, T. Enoki, et al., Bacteria-mediated precursor-dependent biosynthesis of superparamagnetic iron oxide and iron sulfide nanoparticles, Langmuir 24 (2008) 5787–5794, https://doi.org/10.1021/la704019p.
[90] H. Almoammar, M. Rai, E. Said-Galiev, K.A. Abd-Elsalam, Myconanoparticles: synthesis and their role in phytopathogens management, Biotechnol. Biotechnol. Equip. 29 (2015) 221–236, https://doi.org/10.1080/13102818.2015.1008194.
[91] M.R. Salvadori, R.A. Ando, C.A. Oller Nascimento, B. Correa, Extra and intracellular synthesis of nickel oxide nanoparticles mediated by dead fungal biomass, PLoS ONE 10 (2015), https://doi.org/10.1371/journal.pone.0129799, e0129799.
[92] J. Varghese, R. Zikalala, N. Sakho, O.S. Oluwafemi, Green synthesis protocol on metal oxide nanoparticles using plant extracts, in: Colloidal Metal Oxide Nanoparticles, 2020, pp. 67–82, https://doi.org/10.1016/b978-0-12-813357-6.00006-1.
[93] J. Danglad-Flores, S. Eickelmann, H. Riegler, Deposition of polymer films by spin casting: a quantitative analysis, Chem. Eng. Sci. 179 (2018) 257–264, https://doi.org/10.1016/j.ces.2018.01.012.
[94] B. Fotovvati, N. Namdari, A. Dehghanghadikolaei, On coating techniques for surface protection: a review, J. Manuf. Mater. Process. 3 (2019) 1–22, https://doi.org/10.3390/jmmp3010028.
[95] F. Lévy, Film growth and epitaxy: methods, in: Reference Module in Materials Science and Materials Engineering, Elsevier, 2016, pp. 210–222, https://doi.org/10.1016/B978-0-12-803581-8.01012-2.
[96] M.F. Smith, J.E. Brockmann, R.C. Dykhuizen, D.L. Gilmore, R.A. Neiser, et al., Cold spray direct fabrication-high rate solid state, Mater. Res. Soc. Proc. 542 (1999) 65–76,-https://doi.org/10.1557/PROC-542-65.

[97] D. Tejero-Martin, M. Rezvani, A. McDonald, T. Hussain, Beyond traditional coatings: a review on thermal-sprayed functional and smart coatings, J. Therm. Spray Technol. 28 (2019) 598–644, https://doi.org/10.1007/s11666-019-00857-1.

[98] M.F. Smith, Comparing cold spray with thermal spray coating technologies, in: V.K. Champagne (Ed.), The Cold Spray Materials Deposition Process, Woodhead Publishing, 2007, pp. 43–61, https://doi.org/10.1533/9781845693787.1.43.

[99] A.C. Hall, D.J. Cook, R.A. Neiser, T.J. Roemer, D.A. Hirschfeld, The effect of a simple annealing heat treatment on the mechanical properties of cold-sprayed aluminum, J. Therm. Spray Technol. 15 (2006) 233–238, https://doi.org/10.1361/105996306X108138.

[100] A. Srikanth, G. Mohammed, B. Venkateshwarlu, A brief review on cold spray coating process, Mater. Today Proc. 22 (2020) 1390–1397, https://doi.org/10.1016/j.matpr.2020.01.482.

[101] R.C. Dykhuizen, M.F. Smith, D.L. Gilmore, R.A. Neiser, X. Jiang, et al., Impact of high velocity cold spray particles, J. Therm. Spray Technol. 8 (1999) 559–564, https://doi.org/10.1361/105996399770350250.

[102] P.C. Innocenzi, M. Guglielmi, M. Gobbin, P. Colombo, Coating of metals by the sol-gel dip-coating method, J. Eur. Ceram. Soc. 10 (1992) 4331–4436, https://doi.org/10.1016/0955-2219(92)90018-9.

[103] X. Tang, X. Yan, Dip-coating for fibrous materials: mechanism, methods and applications, J. Sol-Gel Sci. Technol. 81 (2017) 378–404, https://doi.org/10.1007/s10971-016-4197-7.

[104] N. Sahu, B. Parija, S. Panigrahi, Fundamental understanding and modeling of spin coating process: a review, Indian J. Phys. 83 (2009) 493–502, https://doi.org/10.1007/s12648-009-0009-z.

[105] G. Korotcenkov, B.K. Cho, Metal oxide composites in conductometric gas sensors: achievements and challenges, Sens. Actuators B 244 (2017) 182–210, https://doi.org/10.1016/j.snb.2016.12.117.

[106] K.C. Persaud, Polymers for chemical sensing, Mater. Today 8 (2005) 38–44, https://doi.org/10.1016/S1369-7021(05)00793-5.

[107] S.J. Park, C.S. Park, H. Yoon, Chemo-electrical gas sensors based on conducting polymer hybrids, Polymers 9 (2017) 1–24, https://doi.org/10.3390/polym9050155.

[108] L. Kumar, I. Rawal, A. Kaur, S. Annpoorni, Flexible room temperature ammonia sensor based on polyaniline, Sens. Actuators B 240 (2017) 408–416, https://doi.org/10.1016/j.snb.2016.08.173.

[109] Y. Sakai, Y. Sadaoka, M. Matsuguchi, Humidity sensors based on polymer thin films, Sens. Actuators B 35 (1996) 85–90, https://doi.org/10.1016/S0925-4005(96)02019-9.

[110] M. Packirisamy, I. Stiharu, X. Li, G. Rinaldi, A polyimide based resistive humidity sensor, Sens. Rev. 25 (2005) 271–276, https://doi.org/10.1108/02602280510620123.

[111] J. Majewski, Low humidity characteristics of polymer-based capacitive humidity sensors, Metrol. Meas. Syst. 24 (2017) 607–616, https://doi.org/10.1515/mms-2017-0048.

[112] K. Shaheen, Z. Shah, B. Khan, M. Adnan, M. Alamzeb, et al., Electrical, photocatalytic, and humidity sensing applications of mixed metal oxide nanocomposites, ACS Omega 5 (2020) 7271–7279, https://doi.org/10.1021/acsomega.9b04074.

[113] D. Patil, P. Patil, Y.K. Seo, Y.K. Hwang, Poly(o-anisidine)–tin oxide nanocomposite: synthesis, characterization and application to humidity sensing, Sens. Actuators B 148 (2010) (2010) 41–48, https://doi.org/10.1016/j.snb.2010.04.046.

[114] S.C. Nagaraju, S.R. Aashis, K. Prasanna, R.A. Koppalkar, G. Ramagopal, Moisture detection properties of modified polyaniline metal oxide compounds on the surface, J. Eng. (2014) 1–8, https://doi.org/10.1155/2014/925020.
[115] H. Pang, L. Xu, D.X. Yan, Z.M. Li, Conductive polymer composites with segregated structures, Prog. Polym. Sci. 39 (2014) 1908–1933, https://doi.org/10.1016/j.progpolymsci.2014.07.007.
[116] C. Zhang, C.A. Ma, P. Wang, M. Sumita, Temperature dependence of electrical resistivity for carbon black filled ultrahigh molecular weight polyethylene composites prepared by hot compaction, Carbon 43 (2005) 2544–2553, https://doi.org/10.1016/j.carbon.2005.05.006.
[117] H.P. Xu, Z.M. Dang, D.H. Shi, J.B. Bai, Remarkable selective localization of modified nanoscale carbon black and positive temperature coefficient effect in binary–polymer matrix composites, J. Mater. Chem. 18 (2008) 2685–2690, https://doi.org/10.1039/b717591d.
[118] A. Karimi, S.W. Husain, M. Hosseini, P.A. Azar, M.R. Ganjali, Rapid and sensitive detection of hydrogen peroxide in milk by enzyme-free electrochemiluminescence sensor based on a polypyrrole-cerium oxide nanocomposite, Sensors Actuators B 271 (2018) 90–96, https://doi.org/10.1016/j.snb.2018.05.066.
[119] A.L. Khan, A. Sinha, R. Jain, Design, fabrication, and optimization of polypyrrole/bismuth oxide nanocomposite as voltammetric sensor for the electroanalysis of clofazimine, J. Electrochem. Soc. 165 (2018) (2018) H979–H990, https://doi.org/10.1149/2.1121814jes.
[120] M. Shukla, T. Dixit, R. Prakash, I. Palani, V. Singh, Influence of aspect ratio and surface defect density on hydrothermally grown ZnO nanorods towards amperometric glucose biosensing applications, Appl. Surf. Sci. 422 (2017) 798–808, https://doi.org/10.1016/j.apsusc.2017.06.119.
[121] J. Zhu, X. Huo, X. Liu, H. Ju, Gold nanoparticles deposited polyaniline–TiO2 nanotube for surface plasmon resonance enhanced photoelectrochemical biosensing, ACS Appl. Mater. Interfaces 8 (2015) 341–349, https://doi.org/10.1021/acsami.5b08837.

# Production of bio-cellulose from renewable resources: Properties and applications

10

*Mazhar Ul-Islam[a], Shaukat Khan[a,b], Atiya Fatima[a], Md. Wasi Ahmad[a], Mohd Shariq Khan[a], Salman Ul Islam[c], Sehrish Manan[d], and Muhammad Wajid Ullah[d]*
[a]Department of Chemical Engineering, College of Engineering, Dhofar University, Salalah, Sultanate of Oman, [b]School of Chemical Engineering, Yeungnam University, Gyeongsan, South Korea, [c]School of Life Sciences, College of Natural Sciences, Kyungpook National University, Daegu, Republic of Korea, [d]Department of Biomedical Engineering, Huazhong University of Science and Technology, Wuhan, PR China

## 1 Introduction

The modern age industrialized lifestyle is producing tons of waste materials every day globally and adding them to the environment, thus creating serious environmental and economic threats [1]. One particular type is the food waste produced by food processors and manufacturers, most of which is either stoppable or recyclable. A global estimate shows that 1/3rd of food produced is either wasted or lost [2]. Besides direct loss, the food wastes from various processing industries are rich sources of nutrients, e.g., carbohydrates. This wastage not only leads to high cost but also provides a food reservoir for microbial growth leading to environmental and health risks. Therefore, it is desirable to seek ways to reduce the waste production and also devise plausible techniques for their recycling or conversion into valuable products.

Cellulose, the most abundant biopolymer on earth, is mainly produced by plants and cell walls. Besides plants, some microbes such as *Acetobacter* bacteria also produce cellulose from simple carbon sources like sugar. The produced cellulose is called microbial cellulose or bacterial cellulose (BC) [3,4]. BC possesses the same chemical structure as plant cellulose; however, its morphology and physicochemical properties are different from plant-derived cellulose. It shows superiority to plant cellulose due to its high purity, nanofiber arrangement, better crystallinity, moldability, 3D fibrous structure, and better mechanical properties [5]. These distinguishing features lead to exploration of BC for various important applications in tissue engineering [6–8], wound dressing [9–11], drug delivery [12], cancer diagnosis [13,14], biosensing [15,16], cosmetics [17,18], environment [19,20], energy [21], bio- and optoelectronics [8,22], additive manufacturing [23–25], and several other fields [26]. Currently, BC is mainly produced from fructose-based media; however, exploration of cheap carbon sources for BC production is under investigation [27].

**Renewable Polymers and Polymer-Metal Oxide Composites. https://doi.org/10.1016/B978-0-323-85155-8.00009-1**

Recycling waste materials for the production of important bioproducts such as biofuels (biogas and bioethanol) and biopolymers (bioplastic, biofilms, and bio-cellulose) is a hot research topic currently [28]. The wastes from agriculture, forestry, industrial, municipal, and food sectors have been routinely recycled through various approaches. Most food wastes are nutrient-rich, and thus could be utilized as the cheap microbial growth media or carbon source for the production of various important bioproducts. Our group has reported the production of bioethanol from the waste of beer fermentation broth (WBFB), which served as a sole source of the required nutrients and microbes [29,30]. Also, WBFB contained the required sugar content for BC production [31]. Besides the food waste, several other commonly available and low-cost carbon sources could also serve as alternates to the chemically defined synthetic media.

This chapter mainly deals with the overview of utilization of waste materials and low-cost resources as alternates to the synthetic media for BC production. We have summarized the individual reports of utilizing different wastes for BC production. The sources have been comparatively analyzed based on yield, productivity, and structural features of BC. Furthermore, we have highlighted the BC synthetic process, its important features, multiple applications of BC and BC-based composites, and future directions of BC research and application areas.

## 2 Biosynthesis of bacterial cellulose

The first report of biosynthesis of BC by microbial cells was reported by A. G. Brown who demonstrated its synthesis as an extracellular pellicle in a study [32]. Later on, several bacterial genera of acetic acid bacteria, including *Acetobacter*, *Rhizobium*, *Agrobacterium*, *Aerobacter*, *Achromobacter*, *Azotobacter*, *Salmonella*, *Escherichia*, and *Sarcina*, were reported to produce BC by utilizing different carbon sources. Among the different strains, *Gluconacetobacter xylinum* and *Gluconacetobacter hansenii* are the most explored for BC production [33]. Besides the in vivo microbial synthesis, bio-cellulose or BC is also synthesized in vitro by the cell-free enzyme systems [34–36]. By any method, glucose is the primary carbon source for BC production; nevertheless, other carbon sources like sucrose, fructose, and galactose have also been utilized [27]. The BC synthesis pathway is very complex, involving several specific enzymes and cofactors which mediate different biochemical reactions. The different steps involved in the conversion of glucose and other carbon sources to various intermediates and the polymerization into cellulose take place inside the microbial cells and within the growth medium, respectively. During these reactions, the UDP glucose, produced intracellularly, serves as a BC precursor [37]. The process of microbial synthesis of BC consists of four steps: (1) production of glucose nucleotides, (2) polymerization of glucose to produce cellulose chains, (3) addition of the acyl groups to glucose units, and (4) the excretion of $\beta$-1 $\rightarrow$ 4 glucan chains through terminal complexes (TCs) to the extracellular medium [38]. The polymerization process involving the production of glucan chains occurs in the cytoplasm. These individual chains combine through intermolecular forces and produce ribbons, which in turn combine to produce fibrils. The self-assembly of fibrils leads to the formation of BC pellicles at the air-medium interface [39]. The thickness of the membrane increases with the addition of more pellicles until the growth media is depleted of the essential nutrients or the death of the microbial cells due to oxygen deficiency [37]. A typical process of biosynthesis of BC in the microbial cells, and the extracellular transport of cellulose fibrils and their in vitro aggregation into highly ordered structures are illustrated in Fig. 1.

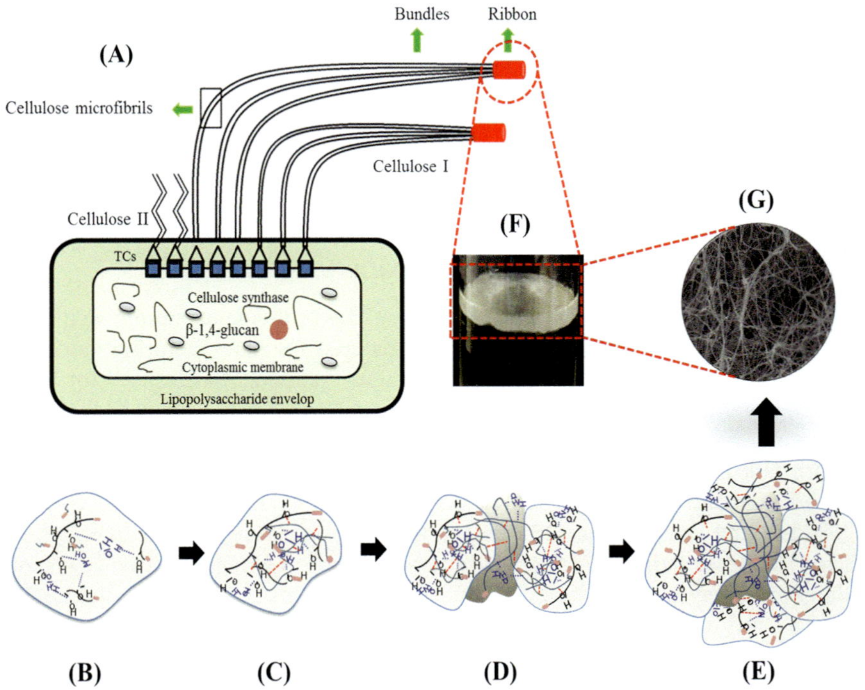


**Fig. 1** Schematic illustration of (A) synthesis of β-1,4-glucan chains and their excretion from bacterial cells across the cell wall through TCs, involving (B) synthesis and aggregation of fibrils, (C) formation of pellicles, (D) movement of pellicles toward the air-medium interface due to the density gradient (in a cell-free system), and (E) formation of the BC sheet at the air-medium interface in the form of a (F) hydrogel, which is seen as a (G) reticulated fibrous structure forming a network of cellulose fibers.
Figure reproduced from M.U. Islam, S. Khan, W.A. Khattak, J.K. Park, Synthesis, chemistry, and medical application of bacterial cellulose nanocomposite, in: Eco-friendly Polymer Nanocomposites, 2015, pp. 399–437. https://doi.org/10.1007/978-81-322-2473-0 and Y. Kim, M.W. Ullah, M. Ul-Islam, S. Khan, J.H. Jang, J.K. Park, Self-assembly of bio-cellulose nanofibrils through intermediate phase in a cell-free enzyme system, Biochem. Eng. J. 142 (2019) 135–144.

# 3 BC production from renewable resources

Despite having tremendous physicochemical, mechanical, thermal, and biological features, the large-scale production and commercialization of BC-based products are yet to reach the marketplace due to the high production cost of BC. The fermentation media used for BC production accounts for 30% of its production cost, thus limiting its large-scale economic production. Therefore, researchers have shown a great deal of interest in the exploration of low-cost fermentation media from various waste materials. The utilization of low-cost materials not only minimizes the production cost but also contributed to a great deal in addressing the environmental issues associated with the disposal of such industrial wastes [40]. However, the conversion of different wastes into value-added BC products requires pretreatment such as solid waste separation and sterilization for contaminant removal, acid hydrolysis, or super/subcritical water hydrolysis. The latter is the most advanced method involving pressure hydrolysis of cellulose and lignocellulose into smaller molecules at 100–374°C [41]. Among many, the agro-industrial, municipal, bakery, brewery, and textile wastes are extensively utilized as alternative sources of BC production.

## 3.1 BC production from agricultural waste materials

The agricultural waste, although a huge source of worldwide economic significance, only about 10% of it is utilized as the raw material for manufacturing value-added products [42]. Table 1 shows a summary of agro-industrial wastes used for BC production. Herein, some examples of agro-industrial waste materials used for BC production are discussed.

Cornstalk hydrolysate is known for its high sugar content as it contains glucose (3.87 g/L), mannose (1.84 g/L), xylose (29.61 g/L), lignin (4.01 g/L), furfural (2.95 g/L), and acetic acid (18.71 g/L). Cheng et al. reported the production of 2.86 g/L BC by utilizing the acetic acid prehydrolysate of cornstalk as the sugar source with optimum pretreatment and detoxification conditions [58]. The obtained BC possessed a fibril diameter of 20–70 nm and length in the 300 nm to micrometer range. Besides, rice barks have also been used as the carbon source for BC production. In a study, rice barks were used as a medium for BC production after enzymatic hydrolysis which produced up to 2.42 and 1.57 g/L BC under static and shaking conditions, respectively [59]. These examples and many more demonstrate the potential of different crop wastes as the cost-effective feedstock for BC production.

Wheat straw, another abundant agro-waste, is important biomass that is usually wasted by setting fire in it which then causes immense air pollution. The acid or enzymatic hydrolysis treatment can turn the wheat straw into the BC production medium. The acid hydrolyzed wheat straw has been utilized as a carbon source for BC production [60]. The detoxification with various alkalis such as sodium and calcium hydroxide and ammonia combined with laccase or activated charcoal was carried out to remove the bacterial growth inhibitors. The results indicated a 50% increased BC production by utilizing the calcium hydroxide and activated charcoal-treated wheat straw

**Table 1** Agro-industrial wastes utilized as a feedstock for producing BC.

| Agro-industrial wastes | Additional nutrients | Microbial strain | Maximum BC production | Ref. |
|---|---|---|---|---|
| ***Waste as a complex medium without any additional nutrients*** | | | | |
| Citrus peels (lemon, mandarin, orange, and grapefruit) | – | *Komagataeibacter hansenii* GA2016 | 3.92 BC/100 g peel | [43] |
| Sugar cane juice and pineapple residues | – | *Gluconacetobacter medellinensis* | 3.24 g/L | [44] |
| Discarded waste durian shell | – | *G. xylinus* CH001 | 2.67 g/L | [45] |
| ***Waste as a carbon source with additional nutrients*** | | | | |
| Enzymatic hydrolysate of wheat straw | Other components are the same as of HS medium | *G. xylinus* ATCC 23770 | 8.3 g/L | [46] |
| Coffee cherry husk | Urea and corn steep liquor | *Gluconacetobacter hansenii* UAC09 | 8.2 g/L | [47] |
| Juice samples watermelon, pineapple, and paw | Other components are the same as that of HS medium | *Gluconacetobacter pasteurianus* PW1 | 7.7 g/L | [48] |
| Cashew tree exudates | Other components are the same as of HS medium | *Komagataeibacter rhaeticus* | 6.0 g/L | [49] |
| Date-extracted syrup | Other components are the same as of HS medium | *G. xylinus* 0416 MARDI | 5.8 g/L | [50] |
| Pineapple waste medium and pawpaw waste medium | Other components are the same as of HS medium | *G. pasteurianus PW1* | 3.9 g/L | [48] |
| Orange peel fluid and orange peel hydrolysate | Acetate buffer, peptone, and yeast extract | *G. xylinus* BCRC 12334 | 3.40 g/L | [51] |
| Cheap agricultural product konjac powder | Yeast extract and tryptone | *G. aceti* ATCC 23770 | 2.12 g/L | [52] |

*Continued*

**Table 1** Continued

| Agro-industrial wastes | Additional nutrients | Microbial strain | Maximum BC production | Ref. |
|---|---|---|---|---|
| Sago byproduct | Other components are the same as of HS medium | *Beijerinkia fluminensis* WAUPM53 and *G. xylinus* 0416 (Reference strain) | 0.47 g/L and 1.55 g/L for the reference strain | [53] |
| Grape skins aqueous extract, cheese whey, crude glycerol, and sulfite pulping liquor | Organic or inorganic nitrogen | *Gluconacetobacter sacchari* | 0.1 g/L | [54] |
| ***Waste as a nitrogen source*** | | | | |
| Pineapple peel and sugar cane juice | Glucose, fructose, and sucrose | *Gluconacetobacter swingsii* | 2.8 g/L | [55] |
| ***Others*** | | | | |
| Poor quality apple residues in combination with glycerol | Apple glucose equivalents, glycerol, ammonium sulfate, and citric acid | *G. xylinus* DSMZ-2004 | 8.6 g/L | [56] |
| Pineapple and watermelon peels | Sucrose, ammonium sulfate, and cycloheximide | *G. hansenii* MCM B-967 | 125 g/L (on a wet weight basis) | [57] |

Table reproduced from M. Ul-Islam, M.W. Ullah, S. Khan, J.K. Park, Production of bacterial cellulose from alternative cheap and waste resources: a step for cost reduction with positive environmental aspects, Korean J. Chem. Eng. 37 (2020) 925–937. https://doi.org/10.1007/s11814-020-0524-3.

hydrolysate compared to the chemically defined medium under the same experimental conditions [60]. Another group of researchers utilized the chemically and thermally-treated wheat straw for BC production, which contained up to 52.12 g/L sugars. The pretreated wheat straw was hydrolyzed enzymatically through cellulose, xylanase, and β-glucosidase. The hydrolyzed medium produced as high as 10.6 g/L BC upon microbial fermentation [61].

Peels, which account for 5%–40% of fruit or vegetable weight, are the inedible part and are discarded. These peels are a rich source of different sugar, vitamins, and other important nutrients, and thus could be utilized as the carbon sources for the cost-effective production of different value-added products, including BC. Several studies have reported BC production from peels of different fruits and vegetables. For example, the orange peels from the juice-producing industry have been used as a BC production medium. The orange peels consists of 10% moisture, 30%–40% sugar, 8%–10% cellulose, 5%–7% hemicellulose, and 15%–25% pectin. The enzymatic hydrolysis of orange peels with cellulose and pectinase increased the sugar content up to 60–80 g/L that led to the production of up to 4.2–6.32 g/L BC, a much high amount compared to the chemically defined medium under the same conditions [51].

Oat hulls are also cheap and renewable carbon sources. These account for 28% of grain weight and contain 45 weight percent cellulose [62]. The oat processing industries producing oat cereals, cookies, and slices accumulate tons of them. Oat hulls possess the same chemical composition and mechanical features (similar size, elasticity, and strength) even when grown in different climates, thus making them industrially sustainable and globally standardized waste. Thus, a standardized medium for BC production can be prepared using the oat hulls. As the BC producing bacteria are unable to directly utilize the oat hulls as the carbon source, various pretreatment procedures are followed to hydrolyze them into fermentable sugars. Recently, BC production from oat hulls has been reported involving a four-step procedure: (1) chemical treatment of oat hulls with 2%–6% $HNO_3$ under atmospheric pressure, (2) enzymatic saccharification resulting in 79.5% sugars, (3) BC production using *Medusomyces gisevii* in mixed-culture fermentation, and (4) purification yielding BC accounting for 10% of the reducing sugars with high purity and crystallinity index of BC up to 93% [63].

## 3.2 Brewery industrial wastes

The fermentation wastewater from ethanol, butanol, and acetone industries is a rich source of sugars (glucose and xylose), organic acids (butyric acid and acetic acid), and butanol and ethanol, and thus could be used as the BC production media. In a study, Huang et al. reported the production of 1.34 g/L BC by *G. xylinus* CH001 after 7 days by utilizing the fermentation wastewater. The produced BC resembled in structure and properties to the one produced using HS medium, thus highlighting the application of the fermentation wastewater as a cheap medium for economical BC production [64].

Also, different sludges like thin stillage (TS), Makgeolli sludge (MS), vinasse, waste from beer fermentation broth (WBFB), waste beer yeast (WBY), sugarcane jaggery (gur), and lipid fermentation wastewater are rich in carbohydrates, proteins, vitamins, and minerals, and thus used as the cheap carbon sources for BC production. TS and vinasse are byproducts produced in the ethanol industry during the ethanol distillation from molasses fermentation obtained from sugar beet, sugar cane, or corn starch, are rich in carbohydrates and useful organic acids as well as inorganic salts of Na, K, Ca, and Mg in the form of sulfates and phosphates. TS has been used as a supplement to BC fermentation medium for *G. xylinum* growth and BC production. Additionally, the replacement of TS with water in the preparation of HS medium further increased the BC production by 2.5-fold to a concentration of 10.38 g/L [65]. Another study reported the utilization of TS from wheat and whey for BC production up to 6.19 and 2.14 g/L after 3 days, respectively. The BC produced from two media showed variant morphology and crystallinity despite their same chemical structure [66]. MS, a discarded waste, is produced in rice wine distilleries. It is a rich source of glucose (10.24 g/L), organic acids (1.15 g/L), nitrogen sources (0.81 g/L), alcohol (0.93 v/v %), and also metal ions, and thus considered a suitable growth medium for *G. xylinus* and BC production. The utilization of MS as the BC production medium produced a dense fibrous network and cellulose I crystalline form [67].

Besides sludges, different beer industry wastes are also considered as the sources of rich nutrients for microbial growth and thus show the potential to replace the high-cost synthetic media for BC production. For example, WBY, produced during the fermentation of cereals, contains carbohydrates (23%–28%), proteins (48%–55%), vitamins (2%), RNA (6%–8%), and glutathione (around 1%) in addition to minerals (Mingyuan [68]). However, despite its rich nutrient composition, WBY cannot be utilized directly for BC production due to the presence of large chain carbohydrates and proteins; and thus, pretreatment through acid or alkali hydrolysis and ultrasonication is required to convert these large chain molecules into small consumable molecules. A study reported the utilization of WBY as a nutrient source for *G. hansenii* CGMCC 3917 growth and showed 1.21 g/L and 7.02 g/L BC production from untreated and ultrasonicated WBY, respectively. Similarly, WBFB has also been utilized as a BC production medium. WBFB consists of semi-solid waste produced from the beer fermentation process containing a reasonable amount of sugars. These sugars might be simple ones like glucose or starch-like polysaccharides, depending on the initial sources. WBFB needs sterilization prior to its usage as the BC production medium as it contains a high density of yeast cells. A comparatively high amount of BC was produced in shaking cultivation compared to the static method by utilizing WBFB as the sole nutrient source [29].

The biodiesel industry produces lipid fermentation wastes such as residual waters and glycerol, which are the rich sources of sugars, including glucose, arabinose, and xylose, as well as lipids, and exopolysaccharides, and are thus considered a suitable BC fermentation media. BC fermentation medium, when supplemented with glycerol and grape bagasse, led to BC production as high as 10 g/L. The produced BC possessed several micrometer long microfibrils with a rectangular cross-section with 35–70 μm width and 13–24 μm thickness and 79% crystallinity [69]. Another study reported the application of wastewater from lipid fermentation as the carbon source for the growth of *G. xylinus*, which produced 0.659 g/L BC after 5 days [70]. This study shows the application of low-value wastewater from lipid fermentation for the production of the value-added BC polymer.

Corn wet milling produces a byproduct called corn steep liquor (CSL), which is a rich source of carbon, nitrogen, and vitamins, and thus could be utilized as the supporting microbial growth and fermentation medium [71]. BC production was carried out from CSL with molasses as the additional carbon source in the shaking cultivation method, which produced 3.12 g/L BC by the *Acetobacter* sp. V6. The yield was twofold higher compared to the HS medium and possessed better crystallinity compared to the BC produced in the conventional medium [72]. Another study reported a comparative study of BC production from CSL supplemented with a mixture of carbon and nitrogen sources and the synthetic HS medium using *G. hansenii* UCP1619. The results indicated the highest yield achieved with the medium containing 2.5% CSL and 1.5% glucose produced BC yield equal to 73% achieved with the HS medium [73].

Bakery wastes like dough, flour, and bread are generally thrown into wastage due to spoilage, expiration, or fungal attack and are also suitable for BC production. A pretreatment with boiling $H_2SO_4$ or HCl hydrolyzes them to glucose leading to high-quality BC production. A study reported 13 g/L BC production from hydrolyzed bakery wastes using *Komagataeibacter sucrofermentans* DSM15973. The obtained BC showed high water holding capacity (102–138 g water/ g of dry BC), stress at break (72–139 MPa), and Young's modulus (0.97–1.64 GPa). These properties were comparable to the BC produced using the high-cost HS medium [74].

Industrial wastewater is normally rich in nutrients, wastage of which causes environmental issues and economic loss. Utilization of this waste as the medium for the production of high-value products such as BC can add to environmental and economic perspectives. Li et al. utilized the wastewater from the Candied Jujube industry for BC production. The acid-treated and untreated wastewater was used as the carbon source using *G. xylinus* CGMCC 295. The treated medium produced 2.25 g/L BC, which was 1.5 times higher compared to the untreated one. The acid-treated wastewater produced BC possessed an average diameter of 5.9 nm, however, possessed a lower crystallinity compared to the BC produced from the untreated wastewater [75].

Tons of waste fibers and textiles from the textile industries and consumers is a big environmental challenge, and demands their sustainable management, recycling, and utilization. One possible solution could be their use for the production of value-added products as these are rich in cellulose content. A pretreatment such as hydrolysis and detoxification is required prior to their use. For example, enzymatic and ionic liquid-based hydrolysis of cotton textile has led to a hydrolysate with 17 g/L sugar content. The utilization of this hydrolysate produced 10.8 g/L BC, which was about 83% of the yield obtained from the HS medium [52]. Another study reported the pretreatment of cellulose-based textiles with 85% concentrated phosphoric acid, *N*-methylmorpholine oxide monohydrate, ionic liquid 1-butyl-3-methylimidazolium chloride, and NaOH/urea solution followed by enzymatic hydrolysis, which produced as high as 1.88 g/L and 1.59 g/L BC for discolored hydrolysate and colored hydrolysate, respectively [76].

The above-mentioned examples demonstrate that, if treated wisely, the different wastes from the brewery industry could be utilized as the growth medium for BC-producing microorganisms. Thus, these wastes could be converted into valuable polymer like BC.

## 3.3 BC production from cheap sources

As already stated, the high production cost of BC owing to the expensive fermentation medium limits its large-scale production and broad-spectrum applications, the utilization of different waste materials could minimize the production cost. Besides different waste materials, the utilization of low-cost carbon sources can also lead to minimizing the overall production cost of this valuable polymer, BC. This has led to extensive efforts made to explore cheap carbon and nitrogen sources [33]. Fig. 2 illustrates the possibilities of BC production from several commonly available cheap resources. Furthermore, several efforts made to produce BC from cheap resources are summarized in Table 1. The following sections describe BC production from low-cost carbon sources.

The sugar-producing and refining industries produce a viscous liquid called molasses, which is widely used as the fermentation medium for microbes and is highly rich

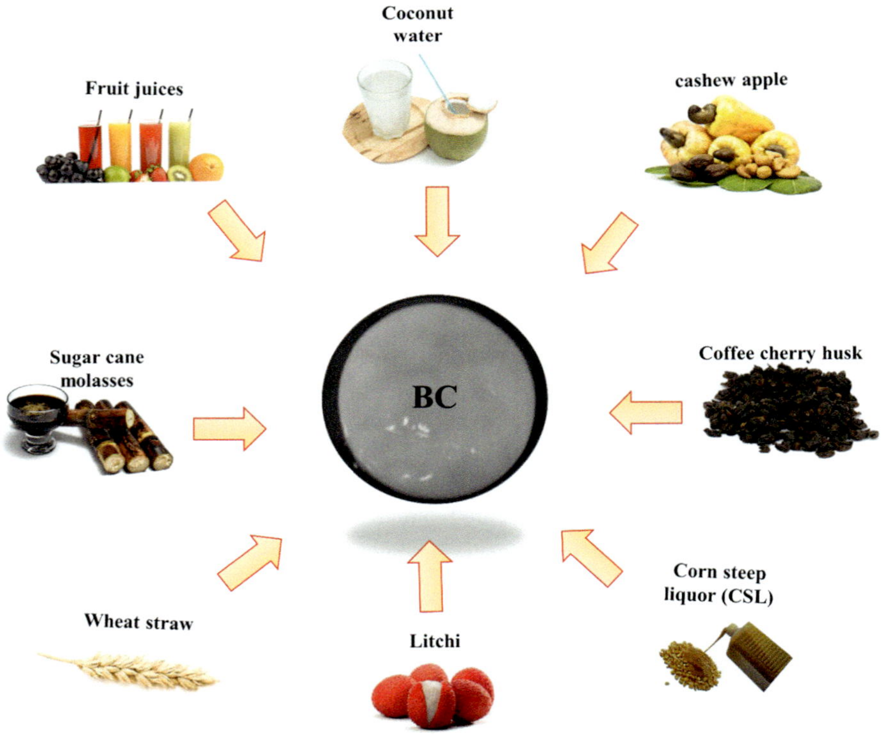


**Fig. 2** Bacterial cellulose production from various cheap resources.
Figure reproduced from M. Ul-Islam, M.W. Ullah, S. Khan, J.K. Park, Production of bacterial cellulose from alternative cheap and waste resources: a step for cost reduction with positive environmental aspects, Korean J. Chem. Eng. 37 (2020) 925–937. https://doi.org/10.1007/s11814-020-0524-3.

in sugars, including glucose, sucrose, and fructose, as well as vitamins and minerals [77]. Its low cost and high sugar and mineral contents make molasses a suitable substrate for low-cost BC production [78,79]. Molasses after pretreatment with $H_2SO_4$ and dilution (1:4 v/v) has shown the production of 12.6 g/L BC by *Gluconacetobacter intermedius* SNT-1 under static conditions. The yield was comparable to that when yeast extract or CSL is used as the nitrogen source [79]. Another detailed study compared 36 alternate culture media using acid and high temperature pretreated and rawhide molasses as the substrate combined with various nutrients such as yeast extract, peptone, glucose, citric acid, and $Na_2HPO_4$ for BC production potential. The results showed that the medium formulated with 15 g/L molasses, 5 g/L glucose, 1.5 g/L acid citric, and 2.7 g/L $Na_2HPO_4$, and no nitrogen source showed the lowest production cost compared to the HS medium and yielded 52% dry and 59% hydrated mass and 65% BC yield [78]. These examples elucidate the potential of sugarcane molasses for low-cost BC production and additionally pave a way to address the environmental issues associated with the mall-handling of molasses. Similarly, the scum obtained during the preparation of sugarcane jaggery from brown sugar industry contains a high amount of glucose and other fermentable sugars. A study by Khattak et al. reported the production of 2.51 and 2.13 g/L BC in shaking and static cultures, respectively, after 10 days, by *G. xylinum* (ATCC 23768). The produced BC showed structural homology with the BC produced by the chemically defined medium and demonstrated better physical and biological properties [80].

Wastes from food processing industries, such as palm oil mill waste, pineapple juice waste, rotten apples, and apple juice industry waste and sweet potato pulp, are rich sources of valuable nutrients, including glucose, sucrose, vitamins, and proteins, and are therefore, candidate materials for low-cost BC production. For example, a BC production medium consisting of grapes, pear, orange, apple, and pineapple led to high BC production using *G. xylinus* NBRC 13693, with further increase in the yield when supplemented with a nitrogen source [81]. The orange pulp and peel have also been used for BC production, and 0.65 g dry BC weight was obtained using a 17.2 g solid orange peel and squeezed residues [81]. Another study reported the utilization of citrus fruit juice supplemented with 10% sucrose, 1% ethanol, and 1% acetic acid as the medium using *Gluconacetobacter* sp. gel_SEA623-2 at pH 3.5 and 30°C, which produced BC of high tensile strength and water uptake [82]. Sisal juice, supplemented with 15 sugars and 7.5 g/L yeast extract, when used as the medium produced 3.38 g/L BC by the *G. hansenii* ATCC 23769 at pH5 under static cultivation method for 10 days, indicating a threefold greater yield than the HS medium [83]. Another study reported the utilization of watermelon juice (70%, v/v) and mandarin (80%, v/v) as the medium for *G. xylinus* CICC10529 and BC production. The medium was externally supplemented with 1% ethanol (v/v), 1.5% $MgSO_4{\cdot}7H_2O$ (w/v), and 0.1% $K_2HPO_4$ (w/v). BC was produced from this medium through static and shaking cultivation methods for 7–10 days at 30°C. The width of the ribbons after 2 days was higher for the static method (40–50μm) as compared to the shaking cultivation (25–37 μm) [84]. Coconut juice is also a rich source of carbon and nitrogen, and thus could be used as the BC fermentation medium. Hungund reported that in the presence of 1.6% sugars in coconut water supplemented with 2% peptone, 0.5% yeast extract, and 0.115% citric acid, *Gluconacetobacter persimmonis* produced 6.18 g/L at pH 6 in 14 days, which was comparable with that obtained using the orange juice [85]. Another study compared the BC production by *G. xylinus* from coconut water and pineapple juice using static and shaking methods. A high yield and conversion rates were achieved for coconut water compared to the pineapple juice [86]. Another study optimized the percentage of an external carbon source, fermentation period, and pH for maximum BC production from coconut water by *G. xylinus*. The amount of supplemented sugar varied at 3%, 5%, and 7%, cultivation period 3, 5, and 7 days, and pH 3, 5, and 7. An optimum BC production was recorded with 5% sugar added and cultivation carried out at pH 5 for 7 days [87].

### 3.4 BC production from vegetables

Vegetables are rich sources of carbon, nitrogen, and minerals, and, are therefore, considered cheap nutrient sources for the production of value-added products such as BC. For example, dry processing of coffee cherry produces coffee cherry husk (CCH), which is a rich source of carbohydrates, proteins, polyphenols, and minerals. According to an estimate, 0.18 tons of CCH is produced from the processing of a ton of coffee [47]. The contaminants such as caffeine and other polyphenols and tannins restrict the application in agriculture, while their disposal leads to environmental issues. Rani and Appaiah reported BC production from CCH-supplemented medium by *G. hansenii* UAC09 by varying its concentration along with other added supplements such as CSL (10%), alcohol (0.5%), and acetic acid (1.13%). After 14 days of cultivation at pH 6.64, a 6.24 g/L BC was produced [88]. The same group of researchers reported a production of 5.6–8.2 g/L BC, threefold high compared to the control medium, through the addition of the following nutrients in their specified ratio: CCH extract 1:1 (w/v), CSL 8% (v/v), ethanol 1.5%, acetic acid 1.0% (v/v), and urea 0.2% (w/v) [47]. These studies show the potential of CCH as a cheap carbon source for low-cost BC production, while the process also minimizes the environmental issues concerning the safe disposal of CCH. Similarly, litchi has high edible value and highly rich in nutrients; however, its short shelf-life limits its market value [89]. In a study, a litchi extract was used as a primary carbon source for BC production under static cultivation using *G. xylinus*, which produced 2.53 g/L BC after 14 days. The produced BC showed 94% crystallinity, better than the one produced through the HS medium. Interestingly, the produced BC was doped with sodium and magnesium elements, indicating an in situ composite synthesis [90]. Another study evaluated BC production using four different types of tea and carbon sources and comparing the yield and fiber morphology of the produced BC. The combination of green tea and sucrose as nitrogen and carbon sources produced the best results. A 0.213 mm thick BC sheet was produced with 74% crystallinity and high mechanical strength [91]. Kombucha (the black tea broth) has also been used as a nitrogen source for BC production, evaluating the effect of the fermentation period and sucrose concentration. The results showed a 67% yield when 90 g/L sucrose was used as the carbon source while the yield increased with the extension of fermentation time [92].

# 4 Application of bacterial cellulose

## 4.1 Medical and pharmaceutical applications

Lack of functional groups, except hydroxyl group, and inadequate surface loading limit the direct application of BC in biomedical field. Moreover, the insolubility of BC in common aqueous/organic solvents facilitates, maintaining the biocompatible 3D structure and restricting any functionalization of active surface chemical groups [93]. Such characteristics hinder the adherence of bioactive compounds utilized in tissue regeneration such as proteins, polyelectrolytes, and drugs under normal conditions, which presents a major challenge in exploiting BC to its optimum potential therapeutic role [94]. Several functionalization techniques have been developed to overcome these limitations and add biorecognition, conductivity or electrostatic potential, and charged interfacial groups to BC for widespread applications in regenerative medicine [7,95,96]. Synthesis of BC composites with two or more components to incorporate the desired antimicrobial, biological, chemical, and physical characteristics to BC has gained considerable attention [97–99].

Despite the inertness of BC toward pH variations and ionic strength, there have been several reports on the production of BC composites displaying a broad range of applications in diagnostic [15,100] and biomedical applications [101]. The BC microstructure undergoes adhesion and adsorption with proteins as well as polysaccharides of different polarities, permitting its sequential functionalization [102]. Several other conjugation processes have also been reported with alginate, chitosan, and gelatin [103–105]. These modifications tend to increase the hydration capacity or bonding efficiency of BC. In a study, the BC/pectin composite displayed a 20-fold increase in compression modulus as well as an increased resistance to compression and stress. The development of a BC composite with cross-linked carboxymethyl cellulose was reported, which showed enhanced entrapment capacity toward ibuprofen as compared to the unmodified BC, thus offering a good drug delivery potential [106]. Other modifications are targeted to increase the porosity of BC for enhanced drug diffusion and cell communication. Therefore, the studies are targeted to acquire BC composites as novel biomaterials for bioengineering applications with the added advantage of biologically active components incorporated into the BC microstructure [105]. BC composite synthesis is generally aimed to make BC composites with better wound healing capacity, antibacterial properties, and diagnostic applications. Some more common applications of BC-based scaffolds are shown in Fig. 3.

### 4.1.1 Wound healing

BC is considered to be an excellent wound dressing material as it can absorb wound exudates, reduce pain, and prevent infections [107]. However, BC-based composites are synthesized for specific target applications such as better tissue regeneration or improved cell adhesion by incorporating different biomolecules delivering potential candidates for the next generation of wound healing devices. The BC/Ag nanocomposites have proved to be the most important composite material in wound

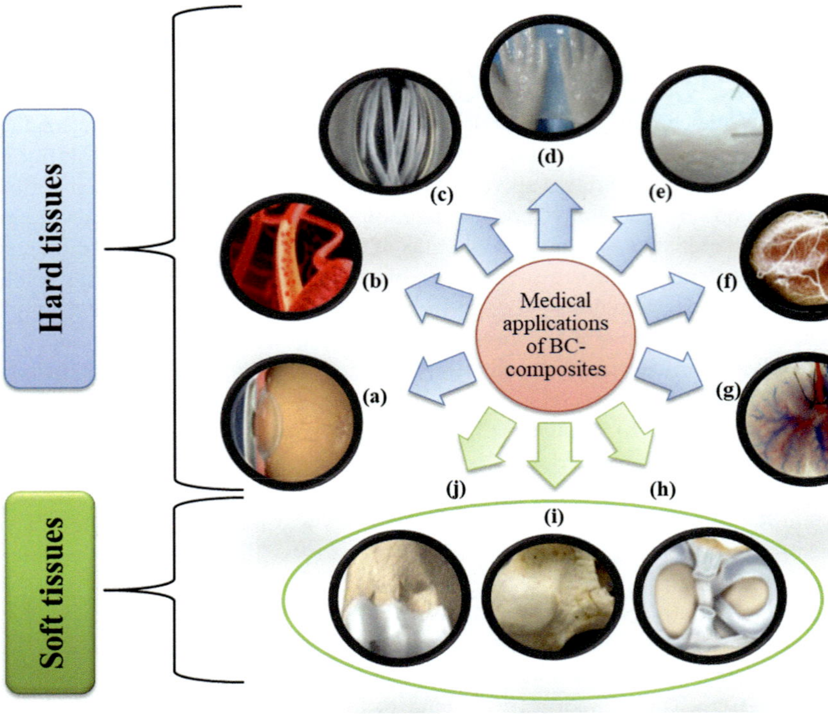


**Fig. 3** Prospects of various biomedical applications of BC-based scaffolds: hard tissues (A) cornea, (B) blood vessels, (C) urethra, (D) skin, (E) scaffold, (F) heart valves, (G) liver, and soft tissues (H) menisci, (I) bone, (J) cartilage.
Figure reproduced from M.U. Islam, S. Khan, W.A. Khattak, J.K. Park, Synthesis, chemistry, and medical application of bacterial cellulose nanocomposite, in: Eco-friendly Polymer Nanocomposites, 2015, pp. 399–437. https://doi.org/10.1007/978-81-322-2473-0.

dressing due to the bactericidal and bacteriostatic activity of Ag nanoparticles (NPs). In these composites, BC prevents the nucleation and aggregation of nanoparticles by acting as a stabilizing agent, thus leading to uniform distribution of Ag NPs in the nanocomposites [108]. The hydrophilicity and surface porosity of BC play a crucial role in the synthesis and stabilization of Ag NPs. Although Ag NPs are found to be cytotoxic in certain cases [109], the BC/Ag nanocomposites display antibacterial effects against various kinds of bacterial strains, including *Saccharomyces aureus*, *Escherichia coli*, *Bacillus subtilis*, *Klebsiella pneumoniae*, and *Pseudomonas aeruginosa* [110,111], thus establishing the BC/Ag nanocomposites as the efficient wound dressing materials.

The macromolecular structure, transparency, and flexibility of BC make it a suitable candidate to be exploited as a drug reservoir and administration system for tracking the wound regeneration process. This could be achieved by in situ transformation of the 3D BC network with drugs and polyelectrolytes for controlled drug delivery.

Several studies reported the efficacy of BC composites with hydroxyapatite [112], AgNPs-curcumin-cyclodextrin [113], benzalkonium chloride, methylglyoxal [114], vaccarine [115], and Aloe vera [116,117], as well as and plant-extracts [118], as the wound dressings.

Although there have been several reports describing the functionalization of BC interfacial fibers containing positively charged [96] or negatively charged groups [119], there are still few reports related to their application as a drug delivery system or biomedical device. In a study by Picheth et al., the oxidized BC membrane containing chitosan and alginate layers, which demonstrated a controlled release of epidermal growth factor, thus could be modulated in case of bacterial infection on the skin [120]. Such modifications inside the BC matrix were facilitated by the presence of the negatively charged carboxyl groups existing at the material interface, permitting spray assisted "layer-by-layer" coating [121]. This technique helps in preserving the properties and integrity of BC while allowing the adsorption of interfacial layers for controlled drug delivery [122]. BC composites with several drugs, including diclofenac [123], ciprofloxacin [124], doxorubicin (L. [16,125]), tetracycline [126], octenidine [127], and others have been reported for local or transdermal delivery in treating superficial skin diseases to cancerous tissues. To achieve a controlled drug release, the use of different materials such as methoxylated pectin (HMP) [128], sodium alginate [125] as well as light irradiation of BC [129] or differential drug content [126] has been reported in few studies. These BC-based composites act as the potential sources of drug release and wound healing materials and provide an adequate environment for tissue repair and epithelialization through combined drug release and healing processes.

### 4.1.2 Ophthalmic scaffolds and contact lenses

BC and its composites have emerged as important biomaterials capable of adherence and proliferation of retinal pigment epithelium (RPE) and keratinocyte cells. The BC-based composites support neovascularization and minimize the side effects, and at the same time lower the intervals of surgical recovery. Thus, these can greatly reduce the rejection rate of corneas and improve the treatment of ocular diseases. To date, different BC-based composites have been developed as the materials required for ocular therapy, such as BC/polyvinyl alcohol (BC/PVA) composites displayed increased transmittance and UV absorbance [130]. Similarly, the BC-carboxymethyl cellulose-chitosan composites exhibited amplified RPE proliferation through an increase in hydrophilicity [131]. The high liquid holding capacity of BC, flexibility, biocompatibility, and higher mechanical resistance to sutures make it a suitable material for ocular surface bandage [132]. Moreover, the BC-based composites have also been reported to relieve glaucoma or artificial cornea [99,103]. Therefore, these composites not only promote the growth of corneal stromal cells in patients but also restore their complete view. These composites have the potential to be used as ocular scaffolds instead of lesser biocompatible hydroxyapatite or poly (methyl) methacrylate (PMMA) systems, which are currently in clinical use [133].

Moldability of BC has found its application in the manufacturing of BC-based convex-shaped stable contact lenses for correcting astigmatism, presbyopia, myopia, and hyperopia [134]. These contact lenses are also capable of drug delivery by maintaining a controlled drug release during eye infections. The development of BC-based contact lenses containing cyclodextrin/ciprofloxacin complexes has already been developed by Cavicchioli and co-workers [135]. Therefore, it can be deduced that BC-based contact lenses can prove to be a great alternative in wound dressing applications after eye surgery to improving the ocular burns recovery and replacing the antibiotics in eye drops.

### *4.1.3 Bone, cartilage, and connective tissue repair*

Bone tissues are composed of osteocytes, osteoblasts, and lining cells with the solid matrix made of calcium phosphates, i.e., hydroxyapatite and tricalcium phosphate [136,137]. The patients suffering from fractures and bone diseases require biocompatible grafts for filling the defective area and tissue regeneration. Generally, the transplant of allogeneic or autologous bone faces the limitation of limited sizes and shapes along with the risk of rejection and transmission of pathogens. The BC-based scaffolds with well-defined porous nanostructures have considerable potential to act as suitable tissue scaffolds for bone tissue engineering. In a study, bone tissue scaffolds were fabricated comprising PVA/hexagonal boron nitride (hBN)/BC composites using 3D printing technology, which showed a significant increase in human osteoblast cell viability [138].

The development of BC-based composites with a variety of materials has been reported for improving bone regeneration, which leads to better osteogenic potential and faster healing. Particularly, the surface phosphorylated BC with high porosity has been extensively researched due to its ability to form complexes with calcium causing better mineralization rates during the regeneration process [139,140]. The morphological similarity of the BC network with collagen facilitates bone regeneration, which thus allows mineralization with hydroxyapatite to mimic the bone in various reconstitutes or fillings [141]. The BC-based bone implants can be utilized to fill the defects and provide long-lasting support in case of injuries without replacement. Recent studies have reported new strategies for therapeutic natural bone regeneration. Hu and coworkers synthesized cellulase enzymes loaded BC-calcium phosphate composites capable of local release of calcium phosphate, which thus helps in maintaining a viable scaffold in mouse embryo preosteoblasts (MC3T3-E1) before biodegradation by the cellulase enzyme, thus acting as a temporary bone substitute [142].

BC presents an excellent option and is currently used in its pristine form as an artificial replacement of Dura mater in patients requiring surgical procedures. In vivo studies suggest low inflammatory response and restricted adhesion to brain tissue leading to repair in Dura mater defect in case of a BC membrane delivering results superior to a commercially available Dura Mater substituent (NormalGEN) [143]. In addition to this, BC can be easily stitched in neurosurgery, thus improving the post-surgery healing period [144].

## 4.2 *Industrial applications of BC composites*

Besides its great applicability in biomedical and pharmaceutical fields, BC and its composites have depicted considerable applications in various other fields. Indeed, a compilation of all its application fields in a single report is difficult. Herein, we have summarized a few common industrial, environmental, and food applications of BC. Fig. 4 lists some important applications of BC in different industrial and biomedical areas.

### 4.2.1 *Food and food packaging applications*

The rheological properties of BC make it a good candidate to be used in the food industry. It is possible to add color, flavor, texture, and shape by modifying the BC production media. BC, being a dietary fiber, offers valuable benefits for food digestion and reduces the risk of chronic diseases such as diabetes, cardiovascular diseases, obesity, and constipation [145,146]. It is also used as an additive for low-calorie food production. Lin et al. synthesized a compound by treating BC with an alkaline medium, which increased the water retention property of BC. This compound was added to surimi as a fat replacement and a dietary source [147]. Chau et al. reported that BC administration in hamsters decreased the level of triglycerides and total cholesterol in serum while at the same time decreased the cholesterol in the liver, making

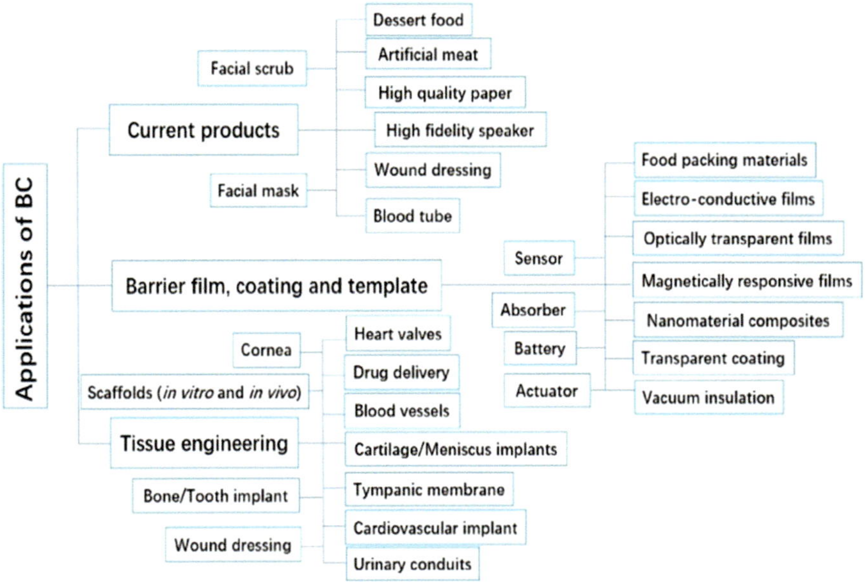


**Fig. 4** Applications of BC in different industrial and biomedical areas.
The figure reproduced from J. Wang, J. Tavakoli, Y. Tang, Bacterial cellulose production, properties and applications with different culture methods—a review, Carbohydr. Polym. 219 (2019) 63–76. https://doi.org/10.1016/j.carbpol.2019.05.008.

it a suitable alternative for gelatin and similar products [148]. Similarly, Okiyama et al. found that BC acts as an exceptional suspending agent that prevents the cocoa precipitation of chocolate drinks by entrapping it in its fibrillar mesh [149]. It also improves the mechanical properties of pasty condiments by reducing the adhesion and can also replace the stabilizer gum in ice creams owing to similar properties. It is also utilized as the body reinforcement in tofu as it significantly improves its mechanical properties with a minimal percentage of BC.

Edible BC antimicrobial membranes are synthesized using lactoferrin as an active compound, thus promoting its applications in developing active and smart food packaging materials [150]. These antimicrobial films could be used as the packaging material for highly perishable foods by hindering the growth of different bacteria such as *S. aureus* and *E. coli*. Jipa et al. synthesized BC composites by incorporating sorbic acid (SA) as an antimicrobial agent into BC. In their study, powdered BC (PBC) and PVA were composited to form a monolayer film, which was further coated with BC to produce multilayer films. It was found that SA and BCP concentration influenced the release rate, water sensitivity, and antimicrobial capacity. In addition, the concentration and water solubility of SA affected the antimicrobial efficacy of the composite films [151].

BC nanowhiskers (BCNW) incorporated with gelatin-PVA form a blend film matrix. Studies were performed to monitor the effect of BCNW content (1–10 wt% of biopolymer) on the mechanical, microstructural, optical, and water barrier properties of bionanocomposites. The incorporation of BCNW up to 7.5% into the GL/PVA blend caused a reduction in the water vapor transmission rate and water vapor permeability by about 22% and 14%, respectively. The tensile strength, elongation at break, and elastic modulus were found to increase up to 21.5%, 41%, and 19%, respectively ($P < .05$). The addition of BCNW ($P > .05$) did not affect the film transparency, suggesting the uniform dispersion of BCNW in the film [152].

Kuswandi et al. developed a new label sensor by immobilizing red methyl in the BC film through absorption. A linear relationship was observed between the sensor color and chicken storage time, suggesting its possible application for evaluating the freshness of the chicken. In another report, the BC/curcumin film was developed as a color indicator to volatile amines released during the fish spoilage [153]. Similarly, BC/bromophenol blue responded through a color change to indicate the freshness of guava fruit. Pourjavaher et al. synthesized a pH indicator by immersing BC into an anthocyanin solution extracted from Brassica oleracea, which gave a good response in case of pH change from 2 to 10 [154].

### 4.2.2 Sensors

BC-based sensors are synthesized by taking the advantage of their surface hydroxyl groups, large surface area, and high water retention and absorption capacity [15,155]. The BC nanofibers are used as low-cost and sensitive humidity sensors by developing the BC fiber coating and applying it as the cover to quartz crystal microbalance (QCM) [155]. In another study, a formaldehyde sensor was developed by using the QCM and BC covering with PEI, where the surface hydroxyl groups of BC and PEI surface amide groups formed hydrogen bonding. The developed BC/PEI fibrous nanocomposite was used as a formaldehyde sensor, where PEI acted as the detector owing to its good reproducibility and reversibility and gave a linear relationship with the increasing formaldehyde concentration (1–100 ppm) at ambient temperature [156]. In a similar approach, Farooq et al. developed a PEI-modified BC nanocomposite with conductive carbon nanotubes as a platform for the immobilization of high density phage particles. The developed BC/phage interface effectively detected as low as 3 and 5 CFU/mL *S. aureus* in phosphate buffer saline and milk, respectively, in mixed bacterial culture, thus demonstrating ultra-sensitive and selective electrochemical detection [15,157,158].

Wang et al. developed a BC/Au sensor where Au NPs were synthesized in the presence of a BC suspension. The surface of the BC/Au nanocomposite served as the substrate for immobilization of heme proteins like hemoglobin (Hb), horseradish peroxide, and myoglobin (Mb). The heme proteins displayed electrocatalytic activity toward $H_2O_2$ reduction in the presence of hydroquinone (HQ) acting as a mediating agent. The $H_2O_2$ biosensor demonstrated biocatalytic activity with high sensitivity, rapid response, and low detection limit [159]. They used the same approach for the development of BC/Au biosensors for glucose detection [160,161].

A hybrid thin film of BC loaded with AgNPs and alginate-molybdenum trioxide nanoparticles ($MoO_3$NPs) was synthesized as a hydrogen sulfide ($H_2S$) gas sensor. The film successfully detected $H_2S$ gas where the film changed its color with the shift of oxidation number of $MoO_3$NPs [162].

### 4.2.3 Separation membranes

The nanofibrous structure of BC makes it a suitable candidate to be used as an ion-exchange membrane for its application in element separation, dialysis, or filtration. In a study, BC/silica composites illustrated recyclable compressibility and super-elasticity with high hydrophobicity about 10 times their weight and oil binding capacity [163]. Dubey et al. synthesized a BC/chitosan composite for the separation of ethanol from the ethanol/azeotrope mixture in water (EtOH/$H_2O$). The BC/chitosan composite was found to be more stable than the PVA-chitosan mixture, which dissolves in highly concentrated water. It also yielded better dimensional and thermal stability and high mechanical properties [164]. In another study, BC/acrylic acid composites were also explored as ion exchange membranes [165].

In a recent study, the plant cellulose needle-leaf bleached kraft pulp (NBKP) was incorporated into BC to prepare a superhydrophobic/superoleophilic membrane for oil/water separation. The surface of the modified BC membrane displayed a petal-like micro-structure, a water contact angle (WCA) of 162.3 degrees, and an oil contact angle of 0 degree. The membrane displayed resistance toward both pH and salt, and showed oil/water separation under the gravitational pull, recyclability, and separation efficiency (>95%) [166].

### 4.2.4 Optical materials

The optical materials must demonstrate high transparency, high stability, and flexibility, and should be free of dispersion. The BC-based composites with fiber diameter of approximately 40–70 nm can exhibit excellent optical properties, good flexibility, and dimensional stability [167]. In contrast, the nanostructures with components size <1/10 wavelength size does not show any dispersion [168]. Yano et al. obtained nanocomposites by impregnating dry BC films with acrylic resins, which exhibited a fiber content of 70% with less than 10% loss of transparency. These composites displayed a very low coefficient of thermal expansion and improved mechanical properties as compared to the plastics [169]. Similarly, Nogi et al. synthesized BC composites with dimethanoldimethacrylate and tricyclodecan by varying the cellulose content, and found that varying the weight of cellulose up to 7.4% caused a light transmission impairment of only 2.4%; however, it reduced the thermal expansion coefficient of the acrylic resin from $86 \times 10^{-6}$ to $38 \times 10^{-6}\,K^{-1}$ [170]. The BC nanofibers are also acetylated for better transparency of acrylic resins reinforced with BC nanofibers. A reduction in the refractive index was found with acetylation and loss of regular transmission of the material with 63% of fiber content from 13% to 3.4%. Also, the coefficient of thermal expansion obtained was around $1 \times 10^{-6}\,K^{-1}$ [171]. Fernandes et al. developed BC/chitosan composites that showed a transmittance of 90% in the range of 400–700 nm. The BC composite with uncured chitosan displayed high Young's modulus and low elongation at fracture with good thermal stability and low $O_2$ permeability [172].

### 4.2.5 Energy storage

The fascinating features of BC are ideal for fabricating highly versatile three-dimensional carbon nanomaterials [22], and its conductive composites can be utilized in energy storage applications. For example, the BC/Pd materials have been explored for energy storage applications as BC can catalyze palladium (Pd) precipitation-producing structures with very high surface area [173]. BC has found its applications in the development of polyelectrolyte membranes and fuel cells. Yang et al. developed BC/Pt composites through in situ reduction, which showed high electrocatalytic activity. BC films doped with protonic acids illustrated high proton conductivity, suggesting the usability of BC as a fuel cell membrane [174].

The materials used for energy storage in modern-day technological applications should possess flexibility along with good mechanical strength. Therefore, there is a need to produce stretchable and conductive membranes. The pyrolyzed BC, possessing interwoven carbon nanofibers, is an excellent alternative for a mechanically robust 3D conducting network [175]. The pyrolyzed BC/polydimethylsiloxane composites prepared using BC as the precursor were found to exhibit electrical conductivity of 0.2–0.41 S/cm and retained their electromechanical properties under high tension [176]. BC has also been utilized in the development of supercapacitors. Chen et al. synthesized a supercapacitor by coating $MnO_2$ on BC (BC/$MnO_2$), which formed a positive electrode and BC coated with nitrogen as the negative electrode. The device showed high charge/discharge ability, energy density (32.91 Wh/kg), and cycling stability [177].

In another report, Shu et al. fabricated excellent supercapacitor hierarchical composite porous carbon (HPC) materials by one-step carbonization and activation from polysaccharides carboxymethyl cellulose, BC, and citric acid. The resultant HPC possessed high oxygen content showing a high specific capacitance of $350\,F\,g^{-1}$, good rate performance, and excellent cycling stability [178].

# 5 Conclusions and future recommendations

Among the different other directions of BC-based research like structural modification, functional characterization, molecular characterization of biosynthesis, development of novel materials and scaffolds for new technological applications, and others, many research groups are determined to minimize its production cost. The utilization of wastes and low-cost substrates for BC production is aimed to (1) reduce the overall production cost, (2) enhance the structural features of BC, (3) ensure the availability of cost-effective BC for value-added applications, and (4) address the environmental issues caused by the release of different industrial and nonindustrial wastes. Presently, the low yield, high capital investment, and several other challenges are the major limitations for the large-scale production and industrialization of BC-based products. Some efforts like isolation and identification of new strains as well as genetic and metabolic engineering of the existing strains, developing advanced bioreactors, and utilization of different industrial and food wastes have greatly contributed to minimizing the BC production cost. Despite the high production cost, the application areas of BC are quite fascinating. To date, both pristine BC and BC-based composites have found application in different areas, mainly in the biomedical field, where these are used in tissue engineering, development of scaffolds, drug delivery systems, artificial organs (e.g., blood vessels, cornea, bones, skin, wound dressings, and others), and several others. Besides, different BC-based scaffolds find applications in the environment (e.g., developing filter membranes), energy (e.g., supercapacitors, fuel cells, solar cells, and others), optoelectronics (sensors, light-emitting diodes, flexible displays, and other), food (e.g., additive, emulsifier, packaging, and others), and several other areas.

The isolation and identification of novel BC-producing strains from the environment, mainly the industrial and food wastes, could greatly contribute to in-site degradation of biological wastes and their conversion to useful products, thereby addressing some environmental issues. The improvement of pretreatment approaches and process optimization during the utilization of different wastes as alternative carbon sources could lead to improved yield. To this end, the development of an automated process could largely contribute to the large-scale production of BC. Altogether, these efforts would lead to the large-scale production of BC and the industrialization and commercialization of BC-based products at affordable prices. The development of value-added products like medical implants, biosensors, energy storage devices, cancer diagnosis models, and food products, and their commercialization, could largely address the concerns associated with the high production cost of BC.

## Acknowledgments

This work was supported by the "The Research Council (TRC)" of Oman through Block Research Funding Program (BFP/RGP/EBR/20/261) and Undergraduate research grant (BFP/URG/EBR/20/055).

# References


[1] A. Demirbas, Waste management, waste resource facilities and waste conversion processes, Energ. Conver. Manage. 52 (2011) 1280–1287, https://doi.org/10.1016/j.enconman.2010.09.025.

[2] Food Loss and Waste Database. (2021). Available from: *https://www.fao.org/platform-food-loss-waste/resources/detail/en/c/1378978/*.

[3] M. Ul-Islam, S. Khan, M.W. Ullah, J.K. Park, Bacterial cellulose composites: synthetic strategies and multiple applications in bio-medical and electro-conductive fields, Biotechnol. J. 10 (2015) 1847–1861, https://doi.org/10.1002/biot.201500106.

[4] M. Ul-Islam, W.A. Khattak, M.W. Ullah, S. Khan, J.K. Park, Synthesis of regenerated bacterial cellulose-zinc oxide nanocomposite films for biomedical applications, Cellul. 21 (2014) 433–447, https://doi.org/10.1007/s10570-013-0109-y.

[5] M. Ul-Islam, S. Khan, M.W. Ullah, J.K. Park, Comparative study of plant and bacterial cellulose pellicles regenerated from dissolved states, Int. J. Biol. Macromol. 137 (2019) 247–252.

[6] S. Khan, M. Ul-Islam, M. Ikram, S.U. Islam, M.W. Ullah, M. Israr, et al., Preparation and structural characterization of surface modified microporous bacterial cellulose scaffolds: a potential material for skin regeneration applications in vitro and in vivo, Int. J. Biol. Macromol. 117 (2018) 1200–1210, https://doi.org/10.1016/j.ijbiomac.2018.06.044.

[7] S. Khan, M. Ul-Islam, W.A. Khattak, M.W. Ullah, J.K. Park, Bacterial cellulose-poly(3,4-ethylenedioxythiophene)-poly(styrenesulfonate) composites for optoelectronic applications, Carbohydr. Polym. 127 (2015) 86–93, https://doi.org/10.1016/j.carbpol.2015.03.055.

[8] S. Khan, M. Ul-Islam, M.W. Ullah, Y. Kim, J.K. Park, Synthesis and characterization of a novel bacterial cellulose–poly(3,4-ethylenedioxythiophene)–poly(styrene sulfonate) composite for use in biomedical applications, Cellul. 22 (2015) 2141–2148, https://doi.org/10.1007/s10570-015-0683-2.

[9] Z. Di, Z. Shi, M.W. Ullah, S. Li, G. Yang, A transparent wound dressing based on bacterial cellulose whisker and poly(2-hydroxyethyl methacrylate), Int. J. Biol. Macromol. 105 (2017) 638–644, https://doi.org/10.1016/j.ijbiomac.2017.07.075.

[10] W. Sajjad, F. He, M.W. Ullah, M. Ikram, S.M. Shah, R. Khan, et al., Fabrication of bacterial cellulose-curcumin nanocomposite as a novel dressing for partial thickness skin burn, Front. Bioeng. Biotechnol. 8 (2020), https://doi.org/10.3389/fbioe.2020.553037.

[11] L. Wang, S. Hu, M.W. Ullah, X. Li, Z. Shi, G. Yang, Enhanced cell proliferation by electrical stimulation based on electroactive regenerated bacterial cellulose hydrogels, Carbohydr. Polym. 249 (2020) 116829, https://doi.org/10.1016/j.carbpol.2020.116829.

[12] S. Li, A. Jasim, W. Zhao, L. Fu, M.W. Ullah, Z. Shi, et al., Fabrication of pH-electroactive bacterial cellulose/polyaniline hydrogel for the development of a controlled drug release system, ES Mater. Manuf. (2018) 41–49, https://doi.org/10.30919/esmm5f120.

[13] M.Y. Ahmad, M.W. Ahmad, H. Cha, I.-T. Oh, T. Tegafaw, X. Miao, et al., Cyclic RGD-coated ultrasmall $Gd_2O_3$ nanoparticles as tumor-targeting positive magnetic resonance imaging contrast agents, Eur. J. Inorg. Chem. 2018 (2018) 3070–3079, https://doi.org/10.1002/ejic.201800023.

[14] M. Ul-Islam, F. Subhan, S.U. Islam, S. Khan, N. Shah, S. Manan, et al., Development of three-dimensional bacterial cellulose/chitosan scaffolds: analysis of cell-scaffold interaction for potential application in the diagnosis of ovarian cancer, Int. J. Biol. Macromol. 137 (2019) 1050–1059.

[15] U. Farooq, M.W. Ullah, Q. Yang, A. Aziz, J. Xu, L. Zhou, et al., High-density phage particles immobilization in surface-modified bacterial cellulose for ultra-sensitive and selective electrochemical detection of *Staphylococcus aureus*, Biosens. Bioelectron. 157 (2020) 112163, https://doi.org/10.1016/j.bios.2020.112163.
[16] A. Jasim, M.W. Ullah, Z. Shi, X. Lin, G. Yang, Fabrication of bacterial cellulose/polyaniline/single-walled carbon nanotubes membrane for potential application as biosensor, Carbohydr. Polym. 163 (2017) 62–69, https://doi.org/10.1016/j.carbpol.2017.01.056.
[17] P. Boonme, T. Amnuaikit, T. Chusuit, P. Raknam, Effects of a cellulose mask synthesized by a bacterium on facial skin characteristics and user satisfaction, Med Devices (Auckl) 77 (2011), https://doi.org/10.2147/MDER.S20935.
[18] B. Mbituyimana, L. Mao, S. Hu, M. Wajid, Bacterial cellulose/glycolic acid/glycerol composite membrane as a system to deliver glycolic acid for anti-aging treatment, J. Bioresour. Bioprod. (2021) 1–26, https://doi.org/10.1016/j.jobab.2021.02.003.
[19] A. Shoukat, F. Wahid, T. Khan, M. Siddique, S. Nasreen, G. Yang, et al., Titanium oxide-bacterial cellulose bioadsorbent for the removal of lead ions from aqueous solution, Int. J. Biol. Macromol. 129 (2019) 965–971, https://doi.org/10.1016/j.ijbiomac.2019.02.032.
[20] M. Ul-Islam, S. Ul-Islam, S. Yasir, A. Fatima, M.W. Ahmed, Y.S. Lee, et al., Potential applications of bacterial cellulose in environmental and pharmaceutical sectors, Curr. Pharm. Des. 26 (2020) 5793–5806, https://doi.org/10.2174/1381612826666201008165241.
[21] S. Li, D. Huang, J. Yang, B. Zhang, X. Zhang, G. Yang, et al., Freestanding bacterial cellulose-polypyrrole nanofibres paper electrodes for advanced energy storage devices, Nano Energy 9 (2014) 309–317, https://doi.org/10.1016/j.nanoen.2014.08.004.
[22] M. Ul-Islam, S. Yasir, L. Mombasawala, S.M. Ullah, W. Muhammad, Bacterial cellulose: a versatile material for fabrication of conducting nanomaterials, Curr. Nanosci. 16 (2020) 1–13, https://doi.org/10.2174/1573413716999201005214832.
[23] W. Aljohani, M.W. Ullah, X. Zhang, G. Yang, Bioprinting and its applications in tissue engineering and regenerative medicine, Int. J. Biol. Macromol. 107 (2018) 261–275,-https://doi.org/10.1016/j.ijbiomac.2017.08.171.
[24] R.R. McCarthy, M.W. Ullah, P. Booth, E. Pei, G. Yang, The use of bacterial polysaccharides in bioprinting, Biotechnol. Adv. 37 (2019), https://doi.org/10.1016/j.biotechadv.2019.107448.
[25] R.R. McCarthy, M.W. Ullah, E. Pei, G. Yang, Antimicrobial inks: the anti-infective applications of bioprinted bacterial polysaccharides, Trends Biotechnol. 37 (2019) 1153–1155, https://doi.org/10.1016/j.tibtech.2019.05.004.
[26] N. Shah, M. Ul-Islam, W.A. Khattak, J.K. Park, Overview of bacterial cellulose composites: a multipurpose advanced material, Carbohydr. Polym. 98 (2013) 1585–1598.
[27] M. Ul-Islam, Comparative synthesis and characterization of bio-cellulose from local waste and cheap resources, Curr. Pharm. Des. 25 (2019) 3664–3671.
[28] M. Ul-Islam, M.W. Ullah, S. Khan, J.K. Park, Production of bacterial cellulose from alternative cheap and waste resources: a step for cost reduction with positive environmental aspects, Korean J. Chem. Eng. 37 (2020) 925–937, https://doi.org/10.1007/s11814-020-0524-3.
[29] J.H. Ha, O. Shehzad, S. Khan, S.Y. Lee, J.W. Park, T. Khan, et al., Production of bacterial cellulose by a static cultivation using the waste from beer culture broth, Korean J. Chem. Eng. 25 (2008) 812–815, https://doi.org/10.1007/s11814-008-0134-y.
[30] S. Khan, M. Ul-Islam, W.A. Khattak, M.W. Ullah, B. Yu, J.K. Park, Enhanced bio-ethanol production via simultaneous saccharification and fermentation through a cell free enzyme system prepared by disintegration of waste of beer fermentation broth, Korean J. Chem. Eng. 32 (2015) 694–701, https://doi.org/10.1007/s11814-014-0242-9.

[31] O. Shezad, S. Khan, T. Khan, J.K. Park, Physicochemical and mechanical characterization of bacterial cellulose produced with an excellent productivity in static conditions using a simple fed-batch cultivation strategy, Carbohydr. Polym. 82 (2010) 173–180.
[32] A.J. Brown, XLIII.—on an acetic ferment which forms cellulose, J. Chem. Soc. Trans. 49 (1886) 432–439.
[33] M.U. Islam, M.W. Ullah, S. Khan, N. Shah, J.K. Park, Strategies for cost-effective and enhanced production of bacterial cellulose, Int. J. Biol. Macromol. 102 (2017) 1166–1173.
[34] Y. Kim, M.W. Ullah, M. Ul-Islam, S. Khan, J.H. Jang, J.K. Park, Self-assembly of bio-cellulose nanofibrils through intermediate phase in a cell-free enzyme system, Biochem. Eng. J. 142 (2019) 135–144, https://doi.org/10.1016/j.bej.2018.11.017.
[35] M.W. Ullah, W.A. Khattak, M. Ul-Islam, S. Khan, J.K. Park, Metabolic engineering of synthetic cell-free systems: strategies and applications, Biochem. Eng. J. 105 (2016) 391–405.
[36] M.W. Ullah, M. Ul-Islam, S. Khan, Y. Kim, J.K. Park, Innovative production of bio-cellulose using a cell-free system derived from a single cell line, Carbohydr. Polym. 132 (2015) 286–294, https://doi.org/10.1016/j.carbpol.2015.06.037.
[37] M. Ul-Islam, M.W. Ullah, T. Khan, J.K. Park, Bacterial cellulose: trends in synthesis, characterization, and applications, in: Handbook of Hydrocolloids, Elsevier, 2021, pp. 923–974.
[38] I.W. Sutherland, Microbial polysaccharides from gram-negative bacteria, Int. Dairy J. 11 (2001) 663–674.
[39] M.W. Ullah, M. Ul-Islam, S. Khan, Y. Kim, J.K. Park, Structural and physico-mechanical characterization of bio-cellulose produced by a cell-free system, Carbohydr. Polym. 136 (2016) 908–916.
[40] M. Ul-Islam, M.W. Ullah, S. Khan, T. Kamal, S. Ul-Islam, N. Shah, et al., Recent advancement in cellulose based nanocomposite for addressing environmental challenges, Recent Pat. Nanotechnol. 10 (2016) 169–180, https://doi.org/10.2174/1872210510666160429144916.
[41] P. Krammer, H. Vogel, Hydrolysis of esters in subcritical and supercritical water, J. Supercrit. Fluids 16 (2000) 189–206, https://doi.org/10.1016/S0896-8446(99)00032-7.
[42] Z. Hussain, W. Sajjad, T. Khan, F. Wahid, Production of bacterial cellulose from industrial wastes: a review, Cellul. 26 (2019) 2895–2911.
[43] M. Güzel, Ö. Akpınar, Production and characterization of bacterial cellulose from citrus peels, Waste Biomass Valoriz. 10 (2019) 2165–2175, https://doi.org/10.1007/s12649-018-0241-x.
[44] I. Algar, S.C.M. Fernandes, G. Mondragon, C. Castro, C. Garcia-Astrain, N. Gabilondo, et al., Pineapple agroindustrial residues for the production of high value bacterial cellulose with different morphologies, J. Appl. Polym. Sci. 132 (2015), https://doi.org/10.1002/app.41237.
[45] M.T. Luo, C. Zhao, C. Huang, X.F. Chen, Q.L. Huang, G.X. Qi, et al., Efficient using durian shell hydrolysate as low-cost substrate for bacterial cellulose production by *Gluconacetobacter xylinus*, Indian J. Microbiol. (2017), https://doi.org/10.1007/s12088-017-0681-1.
[46] L. Chen, F. Hong, X. Yang, S. Han, Biotransformation of wheat straw to bacterial cellulose and its mechanism, Bioresour. Technol. 135 (2013) 464–468, https://doi.org/10.1016/j.biortech.2012.10.029.
[47] M.U. Rani, K.A.A. Appaiah, Production of bacterial cellulose by *Gluconacetobacter hansenii* UAC09 using coffee cherry husk, J. Food Sci. Technol. 50 (2013) 755–762,-https://doi.org/10.1007/s13197-011-0401-5.

[48] B. Adebayo-Tayo, M. Akintunde, J. Sanusi, Effect of different fruit juice media on bacterial cellulose production by Acinetobacter sp. BAN1 and *Acetobacter pasteurianus* PW1, J. Adv. Biol. Biotechnol. 14 (2017) 1–9, https://doi.org/10.9734/JABB/2017/34171.
[49] G. Pacheco, C.R. Nogueira, A.B. Meneguin, E. Trovatti, M.C.C. Silva, R.T.A. Machado, et al., Development and characterization of bacterial cellulose produced by cashew tree residues as alternative carbon source, Ind. Crop Prod. 107 (2017) 13–19, https://doi.org/10.1016/j.indcrop.2017.05.026.
[50] S. Lotfiman, D.R. Awang Biak, T.B. Ti, S. Kamarudin, S. Nikbin, Influence of date syrup as a carbon source on bacterial cellulose production by *Acetobacter xylinum* 0416, Adv. Polym. Technol. 37 (2018) 1085–1091, https://doi.org/10.1002/adv.21759.
[51] C.-H. Kuo, C.-Y. Huang, C.-J. Shieh, H.-M.D. Wang, C.-Y. Tseng, Hydrolysis of orange peel with cellulase and pectinase to produce bacterial cellulose using *Gluconacetobacter xylinus*, Waste Biomass Valoriz. 10 (2019) 85–93, https://doi.org/10.1007/s12649-017-0034-7.
[52] F. Hong, K. Qiu, An alternative carbon source from konjac powder for enhancing production of bacterial cellulose in static cultures by a model strain *Acetobacter aceti* subsp. xylinus ATCC 23770, Carbohydr. Polym. 72 (2008) 545–549, https://doi.org/10.1016/j.carbpol.2007.09.015.
[53] W.W.Y. Voon, B.J. Muhialdin, N.L. Yusof, Y. Rukayadi, A.S. Meor Hussin, Bio-cellulose production by *Beijerinckia fluminensis* WAUPM53 and *Gluconacetobacter xylinus* 0416 in sago by-product medium, Appl. Biochem. Biotechnol. 187 (2019) 211–220, https://doi.org/10.1007/s12010-018-2807-2.
[54] P. Carreira, J.A.S. Mendes, E. Trovatti, L.S. Serafim, C.S.R. Freire, A.J.D. Silvestre, et al., Utilization of residues from agro-forest industries in the production of high value bacterial cellulose, Bioresour. Technol. 102 (2011) 7354–7360, https://doi.org/10.1016/j.biortech.2011.04.081.
[55] C. Castro, R. Zuluaga, J.L. Putaux, G. Caro, I. Mondragon, P. Gañán, Structural characterization of bacterial cellulose produced by *Gluconacetobacter swingsii* sp. from Colombian agroindustrial wastes, Carbohydr. Polym. 84 (2011) 96–102, https://doi.org/10.1016/j.carbpol.2010.10.072.
[56] A. Casarica, G. Campeanu, M. Moscovici, A. Ghiorghita, V. Manea, Improvement of bacterial cellulose production by aceobacter xyilinum dsmz-2004 on poor quality horticultural substrates using the taguchi method for media optimization. Part i, Cellul. Chem. Technol. 47 (2013) 61–68.
[57] J.V. Kumbhar, J.M. Rajwade, K.M. Paknikar, Fruit peels support higher yield and superior quality bacterial cellulose production, Appl. Microbiol. Biotechnol. 99 (2015) 6677–6691, https://doi.org/10.1007/s00253-015-6644-8.
[58] Z. Cheng, R. Yang, X. Liu, X. Liu, H. Chen, Green synthesis of bacterial cellulose via acetic acid pre-hydrolysis liquor of agricultural corn stalk used as carbon source, Bioresour. Technol. 234 (2017) 8–14, https://doi.org/10.1016/j.biortech.2017.02.131.
[59] F.D.E. Goelzer, P.C.S. Faria-Tischer, J.C. Vitorino, M.-R. Sierakowski, C.A. Tischer, Production and characterization of nanospheres of bacterial cellulose from *Acetobacter xylinum* from processed rice bark, Mater. Sci. Eng. C 29 (2009) 546–551, https://doi.org/10.1016/j.msec.2008.10.013.
[60] F. Hong, Y.X. Zhu, G. Yang, X.X. Yang, Wheat straw acid hydrolysate as a potential cost-effective feedstock for production of bacterial cellulose, J. Chem. Technol. Biotechnol. 86 (2011) 675–680, https://doi.org/10.1002/jctb.2567.

[61] W. Al-Abdallah, Y. Dahman, Production of green biocellulose nanofibers by *Gluconacetobacter xylinus* through utilizing the renewable resources of agriculture residues, Bioprocess Biosyst. Eng. 36 (2013) 1735–1743, https://doi.org/10.1007/s00449-013-0948-9.

[62] E.A. Skiba, O.V. Baibakova, V.V. Budaeva, I.N. Pavlov, M.S. Vasilishin, E.I. Makarova, et al., Pilot technology of ethanol production from oat hulls for subsequent conversion to ethylene, Chem. Eng. J. 329 (2017) 178–186, https://doi.org/10.1016/j.cej.2017.05.182.

[63] E.A. Skiba, V.V. Budaeva, E.V. Ovchinnikova, E.K. Gladysheva, E.I. Kashcheyeva, I.N. Pavlov, et al., A technology for pilot production of bacterial cellulose from oat hulls, Chem. Eng. J. 383 (2020) 123128, https://doi.org/10.1016/j.cej.2019.123128.

[64] C. Huang, X.-Y. Yang, L. Xiong, H.-J. Guo, J. Luo, B. Wang, et al., Evaluating the possibility of using acetone-butanol-ethanol (ABE) fermentation wastewater for bacterial cellulose production by Gluconacetobacter xylinus, Lett. Appl. Microbiol. 60 (2015) 491–496, https://doi.org/10.1111/lam.12396.

[65] J.-M. Wu, R.-H. Liu, Thin stillage supplementation greatly enhances bacterial cellulose production by *Gluconacetobacter xylinus*, Carbohydr. Polym. 90 (2012) 116–121,-https://doi.org/10.1016/j.carbpol.2012.05.003.

[66] V. Revin, E. Liyaskina, M. Nazarkina, A. Bogatyreva, M. Shchankin, Cost-effective production of bacterial cellulose using acidic food industry by-products, Braz. J. Microbiol. 49 (2018) 151–159, https://doi.org/10.1016/j.bjm.2017.12.012.

[67] J.Y. Hyun, B. Mahanty, C.G. Kim, Utilization of Makgeolli sludge filtrate (MSF) as low-cost substrate for bacterial cellulose production by *Gluconacetobacter xylinus*, Appl. Biochem. Biotechnol. 172 (2014) 3748–3760, https://doi.org/10.1007/s12010-014-0810-9.

[68] M. Liu, Optimization of extraction parameters for protein from beer waste brewing yeast treated by pulsed electric fields (PEF), Afr. J. Microbiol. Res. 6 (2012), https://doi.org/10.5897/AJMR12.117.

[69] A. Vazquez, M.L. Foresti, P. Cerrutti, M. Galvagno, Bacterial cellulose from simple and low cost production media by *Gluconacetobacter xylinus*, J. Polym. Environ. 21 (2013) 545–554, https://doi.org/10.1007/s10924-012-0541-3.

[70] C. Huang, H.-J. Guo, L. Xiong, B. Wang, S.-L. Shi, X.-F. Chen, et al., Using wastewater after lipid fermentation as substrate for bacterial cellulose production by *Gluconacetobacter xylinus*, Carbohydr. Polym. 136 (2016) 198–202, https://doi.org/10.1016/j.carbpol.2015.09.043.

[71] R.P. Kona, N. Qureshi, J.S. Pai, Production of glucose oxidase using *Aspergillus niger* and corn steep liquor, Bioresour. Technol. 78 (2) (2001) 123–126, https://doi.org/10.1016/S0960-8524(01)00014-1.

[72] H.-I. Jung, O.-M. Lee, J.-H. Jeong, Y.-D. Jeon, K.-H. Park, H.-S. Kim, et al., Production and characterization of cellulose by Acetobacter sp. V6 using a cost-effective molasses–corn steep liquor medium, Appl. Biochem. Biotechnol. 162 (2010) 486–497,-https://doi.org/10.1007/s12010-009-8759-9.

[73] A.F.S. Costa, F.C.G. Almeida, G.M. Vinhas, L.A. Sarubbo, Production of bacterial cellulose by *Gluconacetobacter hansenii* using corn steep liquor as nutrient sources, Front. Microbiol. 8 (2017), https://doi.org/10.3389/fmicb.2017.02027.

[74] E. Tsouko, C. Kourmentza, D. Ladakis, N. Kopsahelis, I. Mandala, S. Papanikolaou, et al., Bacterial cellulose production from industrial waste and by-product streams, Int. J. Mol. Sci. 16 (2015) 14832–14849, https://doi.org/10.3390/ijms160714832.

[75] Z. Li, L. Wang, J. Hua, S. Jia, J. Zhang, H. Liu, Production of nano bacterial cellulose from waste water of candied jujube-processing industry using *Acetobacter xylinum*, Carbohydr. Polym. 120 (2015) 115–119, https://doi.org/10.1016/j.carbpol.2014.11.061.

[76] C.-H. Kuo, P.-J. Lin, C.-K. Lee, Enzymatic saccharification of dissolution pretreated waste cellulosic fabrics for bacterial cellulose production by *Gluconacetobacter xylinus*, J. Chem. Technol. Biotechnol. 85 (2010) 1346–1352, https://doi.org/10.1002/jctb.2439.
[77] N.F.A. Sanadi, Y. Van Fan, C.W. Leow, J.H. Wong, Y.S. Koay, C.T. Lee, et al., Growth of bacillus coagulans using molasses as a nutrient source, Chem. Eng. Trans. (2017), https://doi.org/10.3303/CET1756086.
[78] A.F.S. de Costa, V.R. Do Nascimento, J.D.P. de Amorim, E.A.S. de Gomes, L.M. de Araújo, L.A. Sarubbo, Residue from the production of sugar cane: an alternative nutrient used in biocellulose production by gluconacetobacter hansenii, Chem. Eng. Trans. (2018), https://doi.org/10.3303/CET1864002.
[79] N. Tyagi, S. Suresh, Production of cellulose from sugarcane molasses using *Gluconacetobacter intermedius* SNT-1: optimization & characterization, J. Clean. Prod. 112 (2016) 71–80, https://doi.org/10.1016/j.jclepro.2015.07.054.
[80] W.A. Khattak, T. Khan, M. Ul-Islam, M.W. Ullah, S. Khan, F. Wahid, et al., Production, characterization and biological features of bacterial cellulose from scum obtained during preparation of sugarcane jaggery (gur), J. Food Sci. Technol. 52 (2015) 8343–8349,-https://doi.org/10.1007/s13197-015-1936-7.
[81] A. Kurosumi, C. Sasaki, Y. Yamashita, Y. Nakamura, Utilization of various fruit juices as carbon source for production of bacterial cellulose by *Acetobacter xylinum* NBRC 13693, Carbohydr. Polym. 76 (2009) 333–335, https://doi.org/10.1016/j.carbpol.2008.11.009.
[82] S.S. Kim, S.Y. Lee, K.J. Park, S.M. Park, H.J. An, J.M. Hyun, et al., *Gluconacetobacter* sp. gel_SEA623-2, bacterial cellulose producing bacterium isolated from citrus fruit juice, Saudi J. Biol. Sci. 24 (2017) 314–319, https://doi.org/10.1016/j.sjbs.2015.09.031.
[83] H.L.S. Lima, E.S. Nascimento, F.K. Andrade, A.I.S. Brígida, M.F. Borges, A.R. Cassales, et al., Bacterial cellulose production by *Komagataeibacter hansenii* ATCC 23769 using sisal juice—an agroindustry waste, Braz. J. Chem. Eng. 34 (2017) 671–680,-https://doi.org/10.1590/0104-6632.20170343s20150514.
[84] M.R. Kosseva, M. Li, J. Zhang, Y. He, N.A.S. Tjutju, Study on the Bacterial Cellulose Production From Fruit Juices, 2017, pp. 36–42, https://doi.org/10.17501/biotech.2017.2104.
[85] B. Hungund, Production of bacterial cellulose from *Gluconacetobacter persimmonis* GH-2 using dual and cheaper carbon sources, J. Microb. Biochem. Technol. 05 (2013), https://doi.org/10.4172/1948-5948.1000095.
[86] P. Lestari, N. Elfrida, A. Suryani, Y. Suryadi, Study on the production of bacterial cellulose from *Acetobacter xylinum* using agro—waste, Jordan J. Biol. Sci. 7 (2014) 75–80,-https://doi.org/10.12816/0008218.
[87] A.W. Indrianingsih, V.T. Rosyida, T.H. Jatmiko, D.J. Prasetyo, C.D. Poeloengasih, W. Apriyana, et al., Preliminary study on biosynthesis and characterization of bacteria cellulose films from coconut water, in: IOP Conference Series: Earth and Environmental Science, 2017, https://doi.org/10.1088/1755-1315/101/1/012010.
[88] M. Usha Rani, K.A. Anu Appaiah, Gluconacetobacter hansenii UAC09-mediated transformation of polyphenols and pectin of coffee cherry husk extract, Food Chem. 130 (2012) 243–247, https://doi.org/10.1016/j.foodchem.2011.07.021.
[89] S. Emanuele, M. Lauricella, G. Calvaruso, A. D'Anneo, M. Giuliano, Litchi chinensis as a functional food and a source of antitumor compounds: an overview and a description of biochemical pathways, Nutrients 9 (2017) 992, https://doi.org/10.3390/nu9090992.
[90] X.-Y. Yang, C. Huang, H.-J. Guo, L. Xiong, J. Luo, B. Wang, et al., Bacterial cellulose production from the litchi extract by *Gluconacetobacter xylinus*, Prep. Biochem. Biotechnol. 46 (2016) 39–43, https://doi.org/10.1080/10826068.2014.958163.

[91] S.M. Yim, J.E. Song, H.R. Kim, Production and characterization of bacterial cellulose fabrics by nitrogen sources of tea and carbon sources of sugar, Process Biochem. 59 (2017) 26–36, https://doi.org/10.1016/j.procbio.2016.07.001.
[92] W.N. Goh, A. Rosma, B. Kaur, A. Fazilah, A.A. Karim, R. Bhat, Fermentation of black tea broth (kombucha): I. effects of sucrose concentration and fermentation time on the yield of microbial cellulose, Int. Food Res. J. 19 (2012) 109–117.
[93] T. Heinze, T. Liebert, Unconventional methods in cellulose functionalization, Prog. Polym. Sci. (2001), https://doi.org/10.1016/S0079-6700(01)00022-3.
[94] A. Carambassis, M.W. Rutland, Interactions of cellulose surfaces: effect of electrolyte, Langmuir (1999), https://doi.org/10.1021/la9815852.
[95] S. Khan, M. Ul-Islam, M.W. Ullah, M. Israr, J.H. Jang, J.K. Park, Nano-gold assisted highly conducting and biocompatible bacterial cellulose-PEDOT: PSS films for biology-device interface applications, Int. J. Biol. Macromol. 107 (2018) 865–873.
[96] R.A.N. Pertile, F.K. Andrade, C. Alves, M. Gama, Surface modification of bacterial cellulose by nitrogen-containing plasma for improved interaction with cells, Carbohydr. Polym. 82 (2010) 692–698, https://doi.org/10.1016/j.carbpol.2010.05.037.
[97] S. Khan, M. Ul-Islam, W.A. Khattak, M.W. Ullah, J.K. Park, Bacterial cellulose-titanium dioxide nanocomposites: nanostructural characteristics, antibacterial mechanism, and biocompatibility, Cellul. (2015), https://doi.org/10.1007/s10570-014-0528-4.
[98] M.W. Ullah, M. Ul-Islam, S. Khan, Y. Kim, J.H. Jang, J.K. Park, In situ synthesis of a bio-cellulose/titanium dioxide nanocomposite by using a cell-free system, RSC Adv. 6 (2016) 22424–22435.
[99] M. Zaborowska, A. Bodin, H. Bäckdahl, J. Popp, A. Goldstein, P. Gatenholm, Microporous bacterial cellulose as a potential scaffold for bone regeneration, Acta Biomater. (2010), https://doi.org/10.1016/j.actbio.2010.01.004.
[100] U. Bora, K. Kannan, P. Nahar, A simple method for functionalization of cellulose membrane for covalent immobilization of biomolecules, J. Membr. Sci. (2005), https://doi.org/10.1016/j.memsci.2004.10.028.
[101] W.K. Czaja, D.J. Young, M. Kawecki, R.M. Brown, The future prospects of microbial cellulose in biomedical applications, Biomacromolecules 8 (2007) 1–12, https://doi.org/10.1021/bm060620d.
[102] C.F. De Souza, N. Lucyszyn, M.A. Woehl, I.C. Riegel-Vidotti, R. Borsali, M.R. Sierakowski, Property evaluations of dry-cast reconstituted bacterial cellulose/tamarind xyloglucan biocomposites, Carbohydr. Polym. (2013), https://doi.org/10.1016/j.carbpol.2012.04.062.
[103] N. Chiaoprakobkij, N. Sanchavanakit, K. Subbalekha, P. Pavasant, M. Phisalaphong, Characterization and biocompatibility of bacterial cellulose/alginate composite sponges with human keratinocytes and gingival fibroblasts, Carbohydr. Polym. 85 (2011) 548–553, https://doi.org/10.1016/j.carbpol.2011.03.011.
[104] A.L. Da Róz, F.L. Leite, L.V. Pereiro, P.A.P. Nascente, V. Zucolotto, O.N. Oliveira, et al., Adsorption of chitosan on spin-coated cellulose films, Carbohydr. Polym. (2010), https://doi.org/10.1016/j.carbpol.2009.10.062.
[105] S. Khan, M. Ul-Islam, M. Ikram, M.W. Ullah, M. Israr, F. Subhan, et al., Three-dimensionally microporous and highly biocompatible bacterial cellulose-gelatin composite scaffolds for tissue engineering applications, RSC Adv. (2016), https://doi.org/10.1039/C6RA18847H.
[106] G. Juncu, A. Stoica-Guzun, M. Stroescu, G. Isopencu, S.I. Jinga, Drug release kinetics from carboxymethylcellulose-bacterial cellulose composite films, Int. J. Pharm. (2016), https://doi.org/10.1016/j.ijpharm.2015.11.053.

[107] W. Czaja, A. Krystynowicz, S. Bielecki, R.M. Brown, Microbial cellulose—the natural power to heal wounds, Biomaterials 27 (2006) 145–151, https://doi.org/10.1016/j.biomaterials.2005.07.035.
[108] T. Maneerung, S. Tokura, R. Rujiravanit, Impregnation of silver nanoparticles into bacterial cellulose for antimicrobial wound dressing, Carbohydr. Polym. (2008), https://doi.org/10.1016/j.carbpol.2007.07.025.
[109] W.G.P. Eardley, S.A. Watts, J.C. Clasper, Extremity trauma, dressings, and wound infection: should every acute limb wound have a silver lining? Int. J. Low. Extrem. Wounds (2012), https://doi.org/10.1177/1534734612457028.
[110] Y. Li, H. Jiang, W. Zheng, N. Gong, L. Chen, X. Jiang, et al., Bacterial cellulose–hyaluronan nanocomposite biomaterials as wound dressings for severe skin injury repair, J. Mater. Chem. B 3 (2015) 3498–3507, https://doi.org/10.1039/C4TB01819B.
[111] J.K. Schluesener, H.J. Schluesener, Nanosilver: application and novel aspects of toxicology, Arch. Toxicol. (2013), https://doi.org/10.1007/s00204-012-1007-z.
[112] S.A. Hutchens, R.S. Benson, B.R. Evans, H.M. O'Neill, C.J. Rawn, Biomimetic synthesis of calcium-deficient hydroxyapatite in a natural hydrogel, Biomaterials 27 (2006) 4661–4670, https://doi.org/10.1016/j.biomaterials.2006.04.032.
[113] A. Gupta, S.M. Briffa, S. Swingler, H. Gibson, V. Kannappan, G. Adamus, et al., Synthesis of silver nanoparticles using curcumin-cyclodextrins loaded into bacterial cellulose-based hydrogels for wound dressing applications, Biomacromolecules (2020), https://doi.org/10.1021/acs.biomac.9b01724.
[114] M. Yang, J. Ward, K.L. Choy, Nature-inspired bacterial cellulose/methylglyoxal (BC/MGO) nanocomposite for broad-spectrum antimicrobial wound dressing, Macromol. Biosci. (2020), https://doi.org/10.1002/mabi.202000070.
[115] Y. Qiu, L. Qiu, J. Cui, Q. Wei, Bacterial cellulose and bacterial cellulose-vaccarin membranes for wound healing, Mater. Sci. Eng. C 59 (2016) 303–309, https://doi.org/10.1016/j.msec.2015.10.016.
[116] O.A. Saibuatong, M. Phisalaphong, Novo aloe vera-bacterial cellulose composite film from biosynthesis, Carbohydr. Polym. (2010), https://doi.org/10.1016/j.carbpol.2009.08.039.
[117] M. Ul-Islam, F. Ahmad, A. Fatima, N. Shah, S. Yasir, M.W. Ahmad, et al., Ex situ synthesis and characterization of high strength multipurpose bacterial cellulose-*Aloe vera* hydrogels, Front. Bioeng. Biotechnol. 9 (2021) 1–12, https://doi.org/10.3389/fbioe.2021.601988.
[118] A. Fatima, S. Yasir, M.S. Khan, S. Manan, M.W. Ullah, M. Ul-Islam, Plant extract-loaded bacterial cellulose composite membrane for potential biomedical applications, J. Bioresour. Bioprod. 6 (2021) 26–32.
[119] T. Oshima, S. Taguchi, K. Ohe, Y. Baba, Phosphorylated bacterial cellulose for adsorption of proteins, Carbohydr. Polym. (2011), https://doi.org/10.1016/j.carbpol.2010.09.005.
[120] G.F. Picheth, M.R. Sierakowski, M.A. Woehl, C.L. Pirich, W.H. Schreiner, R. Pontarolo, et al., Characterisation of ultra-thin films of oxidised bacterial cellulose for enhanced anchoring and build-up of polyelectrolyte multilayers, Colloid Polym. Sci. (2014), https://doi.org/10.1007/s00396-013-3048-0.
[121] S.J. Kiprono, M.W. Ullah, G. Yang, Surface engineering of microbial cells: strategies and applications, Eng. Sci. 1 (2018) 33–45.
[122] K.C. Krogman, J.L. Lowery, N.S. Zacharia, G.C. Rutledge, P.T. Hammond, Spraying asymmetry into functional membranes layer-by-layer, Nat. Mater. (2009), https://doi.org/10.1038/nmat2430.

[123] N.H.C.S. Silva, A.F. Rodrigues, I.F. Almeida, P.C. Costa, C. Rosado, C.P. Neto, et al., Bacterial cellulose membranes as transdermal delivery systems for diclofenac: in vitro dissolution and permeation studies, Carbohydr. Polym. (2014), https://doi.org/10.1016/j.carbpol.2014.02.014.
[124] M.L. Cacicedo, G. Pacheco, G.A. Islan, V.A. Alvarez, H.S. Barud, G.R. Castro, Chitosan-bacterial cellulose patch of ciprofloxacin for wound dressing: preparation and characterization studies, Int. J. Biol. Macromol. (2020), https://doi.org/10.1016/j.ijbiomac.2019.10.082.
[125] M.L. Cacicedo, E.I. León, S.J. Gonzalez, M.L. Porto, A.V. Alvarez, G.R. Castro, Modified bacterial cellulose scaffolds for localized doxorubicin release in human colorectal HT-29 cells, Colloids Surf. B Biointerfaces 140 (2016) 421–429, https://doi.org/10.1016/j.colsurfb.2016.01.007.
[126] W. Shao, H. Liu, S. Wang, J. Wu, M. Huang, H. Min, et al., Controlled release and antibacterial activity of tetracycline hydrochloride-loaded bacterial cellulose composite membranes, Carbohydr. Polym. (2016), https://doi.org/10.1016/j.carbpol.2016.02.065.
[127] S. Moritz, C. Wiegand, F. Wesarg, N. Hessler, F.A. Müller, D. Kralisch, et al., Active wound dressings based on bacterial nanocellulose as drug delivery system for octenidine, Int. J. Pharm. (2014), https://doi.org/10.1016/j.ijpharm.2014.04.062.
[128] M.L. Cacicedo, G.A. Islan, M.F. Drachemberg, V.A. Alvarez, L.C. Bartel, A.D. Bolzán, et al., Hybrid bacterial cellulose-pectin films for delivery of bioactive molecules, New J. Chem. (2018), https://doi.org/10.1039/c7nj03973e.
[129] A. Stoica-Guzun, M. Stroescu, F. Tache, T. Zaharescu, E. Grosu, Effect of electron beam irradiation on bacterial cellulose membranes used as transdermal drug delivery systems, Nucl. Instruments Methods Phys. Res. Sect. B Beam Interact. Mater. Atoms. (2007), https://doi.org/10.1016/j.nimb.2007.09.036.
[130] J. Wang, C. Gao, Y. Zhang, Y. Wan, Preparation and in vitro characterization of BC/PVA hydrogel composite for its potential use as artificial cornea biomaterial, Mater. Sci. Eng. C 30 (2010) 214–218, https://doi.org/10.1016/j.msec.2009.10.006.
[131] S. Gonçalves, J. Padrão, I.P. Rodrigues, J.P. Silva, V. Sencadas, S. Lanceros-Mendez, et al., Bacterial cellulose as a support for the growth of retinal pigment epithelium, Biomacromolecules 16 (2015) 1341–1351, https://doi.org/10.1021/acs.biomac.5b00129.
[132] I. Anton-Sales, J.C. D'Antin, J. Fernández-Engroba, V. Charoenrook, A. Laromaine, A. Roig, et al., Bacterial nanocellulose as a corneal bandage material: a comparison with amniotic membrane, Biomater. Sci. (2020), https://doi.org/10.1039/d0bm00083c.
[133] P.J. Ferrone, J.J. Dutton, Rate of vascularization of coralline hydroxyapatite ocular implants, Ophthalmology (1992), https://doi.org/10.1016/S0161-6420(92)31975-X.
[134] G.F. Picheth, C.L. Pirich, M.R. Sierakowski, M.A. Woehl, C.N. Sakakibara, C.F. de Souza, et al., Bacterial cellulose in biomedical applications: a review, Int. J. Biol. Macromol. (2017), https://doi.org/10.1016/j.ijbiomac.2017.05.171.
[135] M. Cavicchioli, C.T. Corso, F. Coelho, L. Mendes, S. Saska, C.P. Soares, et al., Characterization and cytotoxic, genotoxic and mutagenic evaluations of bacterial cellulose membranes incorporated with ciprofloxacin: a potential material for use as therapeutic contact lens, World J. Pharm. Pharm. Sci. 4 (2015) 1626–1647.
[136] I. Ullah, A. Gloria, W. Zhang, M.W. Ullah, B. Wu, W. Li, et al., Synthesis and characterization of sintered Sr/Fe-modified hydroxyapatite bioceramics for bone tissue engineering applications, ACS Biomater Sci. Eng. 6 (2020) 375–388, https://doi.org/10.1021/acsbiomaterials.9b01666.
[137] I. Ullah, W. Zhang, L. Yang, M.W. Ullah, O.M. Atta, S. Khan, et al., Impact of structural features of Sr/Fe co-doped HAp on the osteoblast proliferation and osteogenic

differentiation for its application as a bone substitute, Mater. Sci. Eng. C 110 (2020) 110633, https://doi.org/10.1016/j.msec.2020.110633.
[138] D. Aki, S. Ulag, S. Unal, M. Sengor, N. Ekren, C.C. Lin, et al., 3D printing of PVA/hexagonal boron nitride/bacterial cellulose composite scaffolds for bone tissue engineering, Mater. Des. (2020), https://doi.org/10.1016/j.matdes.2020.109094.
[139] J.C. Fricain, P.L. Granja, M.A. Barbosa, B. De Jéso, N. Barthe, C. Baquey, Cellulose phosphates as biomaterials. In vivo biocompatibility studies, Biomaterials (2002), https://doi.org/10.1016/S0142-9612(01)00152-1.
[140] T. Oshima, K. Kondo, K. Ohto, K. Inoue, Y. Baba, Preparation of phosphorylated bacterial cellulose as an adsorbent for metal ions, React. Funct. Polym. (2008), https://doi.org/10.1016/j.reactfunctpolym.2007.07.046.
[141] P. Favi, R. Benson, N. Neilsen, C. Ehinger, M. Dhar, Novel biodegradable microporous bacterial cellulose scaffolds engineered for bone and cartilage regeneration, FASEB J. (2013).
[142] Y. Hu, Y. Zhu, X. Zhou, C. Ruan, H. Pan, J.M. Catchmark, Bioabsorbable cellulose composites prepared by an improved mineral-binding process for bone defect repair, J. Mater. Chem. B (2016), https://doi.org/10.1039/c5tb02091c.
[143] C. Xu, X. Ma, S. Chen, M. Tao, L. Yuan, Y. Jing, Bacterial cellulose membranes used as artificial substitutes for dural defection in rabbits, Int. J. Mol. Sci. (2014), https://doi.org/10.3390/ijms150610855.
[144] E. Goldschmidt, M. Cacicedo, S. Kornfeld, M. Valinoti, M. Ielpi, P.M. Ajler, et al., Construction and in vitro testing of a cellulose dura mater graft, Neurol. Res. (2016), https://doi.org/10.1080/01616412.2015.1122263.
[145] P. Cazón, M. Vázquez, Bacterial cellulose as a biodegradable food packaging material: a review, Food Hydrocoll. 113 (2021) 106530, https://doi.org/10.1016/j.foodhyd.2020.106530.
[146] Z. Shi, Y. Zhang, G.O. Phillips, G. Yang, Utilization of bacterial cellulose in food, Food Hydrocoll. 35 (2014) 539–545, https://doi.org/10.1016/j.foodhyd.2013.07.012.
[147] S.B. Lin, L.C. Chen, H.H. Chen, Physical characteristics of surimi and bacterial cellulose composite gel, J. Food Process Eng. (2011), https://doi.org/10.1111/j.1745-4530.2009.00533.x.
[148] C.F. Chau, P. Yang, C.M. Yu, G.C. Yen, Investigation on the lipid- and cholesterol-lowering abilities of biocellulose, J. Agric. Food Chem. (2008), https://doi.org/10.1021/jf7035802.
[149] A. Okiyama, M. Motoki, S. Yamanaka, Bacterial cellulose IV. Application to processed foods, Top. Catal. (1993), https://doi.org/10.1016/S0268-005X(09)80074-X.
[150] J. Padrao, S. Gonçalves, J.P. Silva, V. Sencadas, S. Lanceros-Méndez, A.C. Pinheiro, et al., Bacterial cellulose-lactoferrin as an antimicrobial edible packaging, Food Hydrocoll. 58 (2016) 126–140.
[151] I.M. Jipa, A. Stoica-Guzun, M. Stroescu, Controlled release of sorbic acid from bacterial cellulose based mono and multilayer antimicrobial films, LWT–Food Sci. Technol. (2012), https://doi.org/10.1016/j.lwt.2012.01.039.
[152] H. Haghighi, M. Gullo, S. La China, F. Pfeifer, H.W. Siesler, F. Licciardello, et al., Characterization of bio-nanocomposite films based on gelatin/polyvinyl alcohol blend reinforced with bacterial cellulose nanowhiskers for food packaging applications, Food Hydrocoll. (2021), https://doi.org/10.1016/j.foodhyd.2020.106454.
[153] B.J. Kuswandi, R. Oktaviana, A. Abdullah, L.Y. Heng, A novel on-package sticker sensor based on methyl red for real-time monitoring of broiler chicken cut freshness, Packag. Technol. Sci. (2014), https://doi.org/10.1002/pts.2016.

[154] S. Pourjavaher, H. Almasi, S. Meshkini, S. Pirsa, E. Parandi, Development of a colorimetric pH indicator based on bacterial cellulose nanofibers and red cabbage (*Brassica oleraceae*) extract, Carbohydr. Polym. (2017), https://doi.org/10.1016/j.carbpol.2016.09.027.
[155] W. Hu, S. Chen, B. Zhou, L. Liu, B. Ding, H. Wang, Highly stable and sensitive humidity sensors based on quartz crystal microbalance coated with bacterial cellulose membrane, Sens. Actuators B (2011), https://doi.org/10.1016/j.snb.2011.07.014.
[156] W. Hu, S. Chen, L. Liu, B. Ding, H. Wang, Formaldehyde sensors based on nanofibrous polyethyleneimine/bacterial cellulose membranes coated quartz crystal microbalance, Sens. Actuators B 157 (2011) 554–559, https://doi.org/10.1016/j.snb.2011.05.021.
[157] M.W. Ahmad, B. Dey, G. Sarkhel, D.S. Bag, A. Choudhury, Exfoliated graphene reinforced polybenzimidazole nanocomposite with improved electrical, mechanical and thermal properties, Mater. Chem. Phys. 223 (2019) 426–433, https://doi.org/10.1016/j.matchemphys.2018.11.026.
[158] B. Dey, M.W. Ahmad, A. Almezeni, G. Sarkhel, D.S. Bag, A. Choudhury, Enhancing electrical, mechanical, and thermal properties of polybenzimidazole by 3D carbon nanotube@graphene oxide hybrid, Compos. Commun. 17 (2020) 87–96, https://doi.org/10.1016/j.coco.2019.11.012.
[159] W. Wang, T.J. Zhang, D.W. Zhang, H.Y. Li, Y.R. Ma, L.M. Qi, et al., Amperometric hydrogen peroxide biosensor based on the immobilization of heme proteins on gold nanoparticles-bacteria cellulose nanofibers nanocomposite, Talanta (2011), https://doi.org/10.1016/j.talanta.2010.12.015.
[160] M.W. Ahmad, S. Verma, D.-J. Yang, M.U. Islam, A. Choudhury, Synthesis of silver nanoparticles-decorated poly(m-aminophenol) nanofibers and their application in a non-enzymatic glucose biosensor, J. Macromol. Sci. Part A Pure Appl. Chem. (2021), https://doi.org/10.1080/10601325.2021.1886585.
[161] W. Wang, H.Y. Li, D.W. Zhang, J. Jiang, Y.R. Cui, S. Qiu, et al., Fabrication of bienzymatic glucose biosensor based on novel gold nanoparticles-bacteria cellulose nanofibers nanocomposite, Electroanalysis (2010), https://doi.org/10.1002/elan.201000235.
[162] P. Sukhavattanakul, H. Manuspiya, Fabrication of hybrid thin film based on bacterial cellulose nanocrystals and metal nanoparticles with hydrogen sulfide gas sensor ability, Carbohydr. Polym. 230 (2020) 115566.
[163] J. He, H. Zhao, X. Li, D. Su, F. Zhang, H. Ji, et al., Superelastic and superhydrophobic bacterial cellulose/silica aerogels with hierarchical cellular structure for oil absorption and recovery, J. Hazard. Mater. (2018), https://doi.org/10.1016/j.jhazmat.2017.12.045.
[164] V. Dubey, L.K. Pandey, C. Saxena, Pervaporative separation of ethanol/water azeotrope using a novel chitosan-impregnated bacterial cellulose membrane and chitosan-poly(vinyl alcohol) blends, J. Membr. Sci. (2005), https://doi.org/10.1016/j.memsci.2004.11.009.
[165] Y.J. Choi, Y. Ahn, M.S. Kang, H.K. Jun, I.S. Kim, S.H. Moon, Preparation and characterization of acrylic acid-treated bacterial cellulose cation-exchange membrane, J. Chem. Technol. Biotechnol. (2004), https://doi.org/10.1002/jctb.942.
[166] F.P. Wang, X.J. Zhao, F. Wahid, X.Q. Zhao, X.T. Qin, H. Bai, et al., Sustainable, superhydrophobic membranes based on bacterial cellulose for gravity-driven oil/water separation, Carbohydr. Polym. (2021), https://doi.org/10.1016/j.carbpol.2020.117220.
[167] K. Qiu, A.N. Netravali, A review of fabrication and applications of bacterial cellulose based nanocomposites, Polym. Rev. 54 (2014) 598–626, https://doi.org/10.1080/15583724.2014.896018.

[168] B.M. Novak, Hybrid nanocomposite materials—between inorganic glasses and organic polymers, Adv. Mater. (1993), https://doi.org/10.1002/adma.19930050603.
[169] H. Yano, J. Sugiyama, A.N. Nakagaito, M. Nogi, T. Matsuura, M. Hikita, et al., Optically transparent composites reinforced with networks of bacterial nanofibers, Adv. Mater. 17 (2005) 153–155.
[170] M. Nogi, S. Ifuku, K. Abe, K. Handa, A.N. Nakagaito, H. Yano, Fiber-content dependency of the optical transparency and thermal expansion of bacterial nanofiber reinforced composites, Appl. Phys. Lett. (2006), https://doi.org/10.1063/1.2191667.
[171] S. Ifuku, M. Nogi, K. Abe, K. Handa, F. Nakatsubo, H. Yano, Surface modification of bacterial cellulose nanofibers for property enhancement of optically transparent composites: dependence on acetyl-group DS, Biomacromolecules 8 (2007) 1973–1978, https://doi.org/10.1021/bm070113b.
[172] S.C.M. Fernandes, L. Oliveira, C.S.R. Freire, A.J.D. Silvestre, C.P. Neto, A. Gandini, et al., Novel transparent nanocomposite films based on chitosan and bacterial cellulose, Green Chem. 11 (2009) 2023–2029.
[173] B.R. Evans, H.M. O'Neill, V.P. Malyvanh, I. Lee, J. Woodward, Palladium-bacterial cellulose membranes for fuel cells, Biosens. Bioelectron. 18 (2003) 917–923.
[174] J. Yang, D. Sun, J. Li, X. Yang, J. Yu, Q. Hao, et al., In situ deposition of platinum nanoparticles on bacterial cellulose membranes and evaluation of PEM fuel cell performance, Electrochim. Acta 54 (2009) 6300–6305.
[175] B. Wang, X. Li, B. Luo, J. Yang, X. Wang, Q. Song, et al., Pyrolyzed bacterial cellulose: a versatile support for lithium ion battery anode materials, Small (2013), https://doi.org/10.1002/smll.201300692.
[176] J. Huang, D. Li, M. Zhao, P. Lv, L. Lucia, Q. Wei, Highly stretchable and bio-based sensors for sensitive strain detection of angular displacements, Cellul. (2019), https://doi.org/10.1007/s10570-019-02313-3.
[177] L.F. Chen, Z.H. Huang, H.W. Liang, Q.F. Guan, S.H. Yu, Bacterial-cellulose-derived carbon nanofiber@MnO2 and nitrogen-doped carbon nanofiber electrode materials: an asymmetric supercapacitor with high energy and power density, Adv. Mater. 25 (2013) 4746–4752, https://doi.org/10.1002/adma.201204949.
[178] Y. Shu, Q. Bai, G. Fu, Q. Xiong, C. Li, H. Ding, et al., Hierarchical porous carbons from polysaccharides carboxymethyl cellulose, bacterial cellulose, and citric acid for supercapacitor, Carbohydr. Polym. 227 (2020) 115346.

# Polymer-metal oxide composites from renewable resources for agricultural and environmental applications

**11**

*Manuel Palencia[a,b], Andrés Otálora[a,b], and Arturo Espinosa-Duque[a,b]*
[a]Research Group in Science with Technological Applications (GI-CAT), Department of Chemistry, Universidad del Valle, Cali, Colombia, [b]Mindtech Research Group (Mindtech-RG), Mindtech s.a.s., Cali, Colombia

## 1 Introduction

Currently, most chemical substances used in the synthetic industry are obtained from a nonrenewable resource, i.e., fossil fuels. Many chemical compounds are involved in the making of polymers. Thus, by monomers, solvents, and auxiliary reagents derived from fossil fuels, it is possible to obtain different polymeric materials with wide use in almost every aspect of modern society, e.g., plastics such as poly(ethylene), poly(propylene), poly(ethylene terephthalate), poly(styrene), poly(vinyl chloride); synthetic fibers composed of poly(amide) like Nylon; coatings and paints made from poly(urethane) or poly(urethane-urea); and thermosetting materials made from phenolic or epoxy resins. The high demand for these polymers has been estimated to be 360 million metric tons by 2018, with a continuous annual increase of 9% since the beginning of the 21st century [1,2].

Despite the wide use of these materials, their production and employment turn out to be unsustainable in the long term. First, these materials are mainly produced from fossil fuels, which have limited availability in the near future due to their excessive use for energy generation (more specifically, 80% of world energy) as well as for obtaining raw materials. It is estimated that its consumption rate is higher than its regeneration rate by a factor of almost $10^6$; consequently, this resource will only be available for a few more decades. In addition, there are negative socio-environmental implications associated with its extraction and use, including soil degradation, damage to ecosystems, death of different species, and release of greenhouse gases [2–4]. On the contrary, the structural design of polymeric materials has focused mainly on their applications and not on their residuality and postuse impacts. The aforementioned is related to the accumulation of plastic materials in different ecosystems due to their low rate of degradation, difficulty of reuse, and bad practices in waste management [2,5,6].

**Renewable Polymers and Polymer-Metal Oxide Composites. https://doi.org/10.1016/B978-0-323-85155-8.00007-8**

The problem of plastic materials has reached a strong echo today; the dissemination of shameful images of ecosystems strongly affected by plastic pollution, the irresponsible use of single-use plastics, islands of plastic garbage, among others, have impacted the sensitivity of society and promoted a powerful level of consciousness globally [5,6]. In this sense, plastic pollution has directed the attention to consider three important goals in the design of polymeric materials: (i) the need to identify and implement other renewable raw resources for the making of polymers, including commons plastics, as well as new polymeric and polymer-based materials, (ii) advancing the design of new materials with an approach of minimal residuality and/or high recyclability, and (iii) advancing the implementation of new synthetic processes and methodologies under green chemistry principles [7–9]. Polymeric materials that seek to satisfy these aspects are the focus of this chapter and they have been called "sustainable polymers."

The concept of sustainability is really very broad, such that, it may even seem fuzzy in scope. Sustainability can be considered a chimera in chemistry since energy consumption, waste generation, and environmental disturbance are unavoidable. The impact of a material often lies in its excessive use, which is directly connected with its production, and likewise, with the economy behind each production process. An illustration of the aforementioned is found in the so-called unused useful waste paradox. In this paradox, proposed by the Research Group in Sciences with Technological Applications (GI-CAT) of the Universidad del Valle, in Colombia, the question is posed: how much does pollution by petroleum-derived substances decrease if all polymers, obtained from this type of derivatives, are completely replaced with biodegradable materials obtained from biomass? Although it might be thought that the answer is 100%, in reality, it is 0%, or what is the same, the problem does not decrease, but changes. Well, it happens to have the same amount of waste without an immediate function or use, only in the form of different chemical species. This paradox reflects the importance of the context when addressing this type of problem, and therefore, the specificity of the solution itself is not a solution, but rather a strategy to try to reach a solution through a change of focus [10].

## 2 Sustainable polymers

The concept of sustainability can be approached in different ways depending on the source; however, its meaning and objective are the same in all cases. This refers to the ability of a process to meet current needs without compromising the ability to meet future needs. In other words, every sustainable process seeks to use efficiently the resources at its disposal, ensuring at the same time that, within its development, resources and capacities are not compromised to satisfy any need in the future [11–13]. In this way, the current population makes and implements responsible decisions that do not affect new generations. We could ask ourselves what are the criteria to establish what is responsible and what is not. Since the future is inaccessible to us with total certainty, there is always a margin of error that must be corrected. Clearly, the best tool for decision-making that we currently have is the knowledge that emerges from science, and the stronger this knowledge, the better our predictions and decisions will be. In the context of the issue that concerns us, replacing poly(propylene) bags with cellulose bags could seem like a good alternative since a nonbiodegradable waste is replaced by one that is biodegradable. However, we currently know that the pulp industry is highly polluting and affecting the air, water, and logging ecosystems [14,15]. Therefore, it can be concluded from the previous example that limited knowledge could lead to wrong decisions, or not totally satisfactory. Consequently, it is not only important to make decisions that we believe are correct, but that belief must be based on objective scientific aspects, not influenced by economic, social, or political biases.

Sustainability includes the use of clean energy within the process, e.g., solar and wind energy, the inclusion of important properties such as biodegradability and recycling, as well as the reuse of natural resources, such as water and soil, and minimizing of negative impact on ecosystems, communities, and health [11,12]. Also, the appropriate use of natural resources must be carried out; therefore, it is important to keep in mind that, a sustainable process must be a productive cycle of environmental systems, and consequently, the extraction rate of the resource always should be lower than its corresponding regeneration rate [11,16,17].

In this way, the term "sustainable polymers" has been used to refer to polymeric materials that, through their obtaining method, functionality, or intrinsic properties, seek to generate a sustainable solution to different problems associated with the use of this type of materials; in consequence, they are called "eco-friendly polymers" [18,19]. In general, an ideal sustainable polymer should satisfy a cyclical life cycle at an appropriate time within its productive framework. However, it is difficult to satisfy all conditions at the same time, since, depending on the final application of the material, generally one or more of their properties must be sacrificed for the improvement of another, e.g., susceptibility to microbial colonization, mechanical properties, hydrophobicity, and availability of functional groups. For this reason, the concept of "sustainable polymer" is still subjected to individual interpretations [7,11,20,21].

Usually, sustainable polymers are called biological biopolymers when they are obtained from the biomass of living organisms, such as plants, insects, yeast, bacteria, and mammalians. They can be obtained directly from residual biomass (e.g., cellulose, lignin, hemicellulose, chitin, pullulan, collagen, etc.) or directly produced from living organisms without its obtaining implies its death (e.g., silk, wool, starches, milk proteins, etc.); also, biological biopolymers can be obtained by biotechnological methods and sometimes they are named to be biotechnological biopolymers (e.g., poly(hydroxyalkanoates) and bacterial cellulose) [12,22–24]. On the contrary, sustainable polymers can also be obtained by synthetic procedures. In this case, raw materials for their synthesis are obtained from natural sources, e.g., chitosan is obtained from deacetylation of chitin, cellulose acetate is obtained by acetylation of cellulose, poly(sorbitol citrate) is obtained from crosslinking of sorbitol and citric acid, and poly (starch-citrate) is obtained from starch and citric acid. Consequently, these polymeric materials are often referred as biobased polymers [25–30]. Even synthetic polymers obtained from petroleum derivatives can be sustainable if recyclability and biodegradability properties are introduced into them, or if they fulfill a specific environmental function with great impact (i.e., sustainable polymers according to their functionality) [31,32], and they can be biobased by different synthetic strategies [33,34]. For example, it is possible to obtain substances such as ethylene, ethylene glycol, lactic acid, succinic acid, fatty acids, terpenes, epoxides, among others, which are used to produce widely known polymers such as poly(ethylene), poly(ethylene terephthalate), polycarbonates, polyamides, epoxy resins, elastomers, among others [3,7,9,12,35] (see Fig. 1A). These materials derived directly or indirectly from biomass are categorized as "bio," e.g., "bio-poly(ethylene)" or simply "bio-PE" [13,33,34]. However, this name should not be confused with the intrinsic properties of materials such as biodegradability or biocompatibility, which strictly depend on the chemical structure of the material. As will be discussed later, this approach has greater advantages and better satisfies the requirements associated with the design of materials from a sustainable point of view.

The use of carbon dioxide as a starting material for the development of polymeric materials has also been the subject of extensive research (see Fig. 1B). From this gas, it has been possible to obtain polycarbonates and polyols with optimal mechanical properties for their application in automotive, coatings, and adhesives [36–38]. Likewise, the chemical modification of $CO_2$-based polymers or the generation of copolymers and blends has made it possible to design materials with different mechanical properties, expanding the application possibilities [39–41]. Despite the fact that the use of $CO_2$ is a sustainable alternative to obtaining polymers, and though, as a starting material, allows us to face the problems associated with its contamination, its use has several limitations. For example, the great kinetic and thermodynamic stabilities of

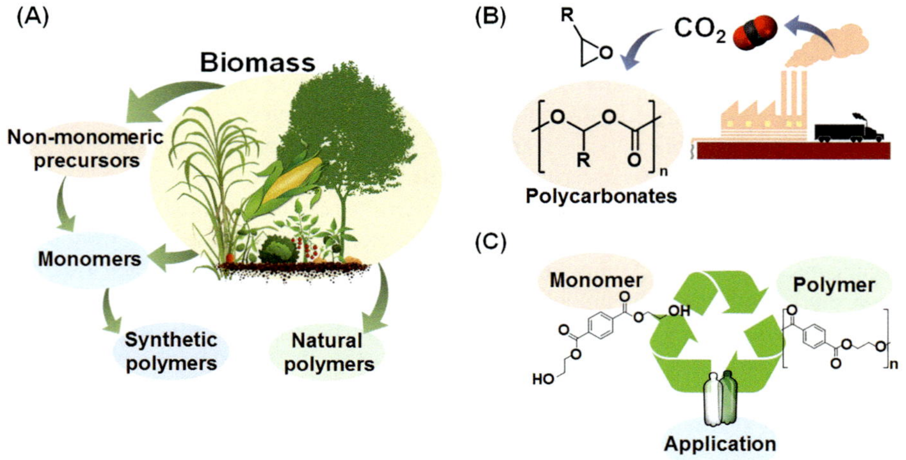


**Fig. 1** Different approaches in the design and production of sustainable polymers.
(A) Approach to obtaining monomers and polymers from biomass, (B) approach to using carbon dioxide to obtain polycarbonates, and (C) approach to improving the recyclability of polymers (exemplified with PET).

$CO_2$ have the requirement of high energy supply in their reactions due to the fact that they are highly endothermic processes, and in many cases, they use catalysts (and their design) to obtain good yields [36,39,42]. Currently, these are limitations that chemists try to overcome, for example, by improving the electrophilic character of $CO_2$ and the design of better and cheaper catalysts for these processes [40,43].

On the contrary, the need to solve other environmental problems, such as accumulation of solid wastes in different ecosystems, the limitations associated with an effective recycling, and, in some cases, their nonexistence, have promoted advances related to the improvement of their properties such as degradability or recyclability (see Fig. 1C). In this way, degradable polymers, either through hydrolytic processes or by microorganism action, such as poly(glycolic acid), poly(trimethyl carbonate), poly(lactic acid), and various poly(hydroxyalkanoates) (PHAs) have been investigated for their application in the design of single-use materials, including those in the biomedical field [20,44–46]. For instance, currently there are on the market poly(lactic acid)-based single-use products, such as glasses, cutlery, and food storage containers, which can be degraded in a relatively short time. However, there are limitations associated with the environmental conditions to which this type of materials is subjected, which in many cases are not optimal to produce a rapid degradation of the material [47–49]. Also, materials with better recycling properties have been designed, which are structurally made by bonds that can be formed or reverted under controlled conditions (e.g., disulfide bonds, Diels-Alder adducts, imine bonds, esters, boronic esters, among others). These bonds are called "dynamic covalent bonds" and they allow the design of polymeric materials, even those thermostable, that can satisfy a closed life cycle, or also called "circular economy." In other words, these polymeric materials once used can be returned to their starting materials or reformed into new useful materials [50–52]. Some limitations associated with this approach are the use of fossil fuels as raw material, high energy consumption associated with recycling processes, use of catalysts, and environmental pollution due to the generation of by-products [7].

## 2.1 Economic aspects

The economy associated with polymeric materials is one of the largest in the world due to the wide use of these materials, mainly plastics, in almost all sectors of modern life. As mentioned earlier, in 2018, the production of polymeric materials reached 360 million metric tons, with approximately 385 million tons consumed in the same year. Of this amount, 44% corresponds to polymeric materials applied in packaging of basic market materials, while the rest corresponds to materials used for manufacturing purposes [53,54]. According to this, in 2020, the global market for polymeric materials reached a total value of over USD 650 trillion, with an estimated projected annual growth of 3.2% from 2020 to 2027 [2,55].

The world polymer production is dominated by materials obtained from fossil fuels, while about 1% of the total is produced from renewable raw materials [3,9]. In addition, currently, the market of polymeric materials is estimated to be priced at approximately USD 2.2 trillion per year due to its negative environmental and social implications [56]. As can be intuitively concluded, the reserves of fossil fuels are finite and their future availability is strongly conditioned to factors such as their consumption and price. From an energy point of view, this form of energy is useful, economical, and efficient in terms of serving to fulfill the objectives, but from an environmental point of view, it is dirty energy, with a high percentage of residuality.

On the contrary, biomass is used as a renewable raw material, which is produced annually in quantities of approximately 12 billion tons, of which 61% is generated by agriculture and 39% by forestry. The main uses of biomass are the generation of energy, food, and fertilizer; besides, around 8% is destined to the generation of biomaterials, while 8% are losses, which is translated into approximately 0.96 billion tons that are not utilized for any useful purpose [57]. At this point, it is important to emphasize the economic implication of biobased polymers, as a recent economic and technological sector, within a market close to USD 15 billion [58]. It is clear that the promise of replacing plastic materials with others that can be labeled as eco-friendly is strongly stimulating investment in these sectors, both at scientific and technological levels. Perhaps the biggest problem in the generation of biobased polymeric materials is that the focus is often on the product and not on the process. For example, the mean composition of cassava roots that are used to obtain starch, a biopolymer with multiple applications, is 24% of starch (moisture, fiber, protein, and other substances including minerals are 70%, 2%, 1%, and 3%, respectively) [59]. In consequence, since fiber and proteins are not useful raw materials because are in low proportion, and not including the moisture, 6% of roots are wastes whereas only 24% is useful (note that, here, water content was not considered a waste since its discharge strictly not affect the environment when it is referred to "pure" or "clean" water). However, these percentages of useful raw materials decrease strongly when the aerial part of the plant is included, i.e., stem and leaves. It is possible to affirm that stem, leaves, fiber, protein, and minerals are reintroduced to the environment due to their biological nature, and therefore, no negative impact is produced in the environment. But really, reintroduction of exogenous biological materials produces a perturbation when it is not carried out correctly. It is well known that, for instance, excess nutrients in aquatic ecosystems can produce environmental disequilibria resulting in a strong affectation of aquatic life organisms (eutrophication) [60,61].

Another important aspect that sometimes is not taken into consideration from a scientific perspective is the increase in economic conditions and the generation of jobs. The use of biomass is an intensive work, both for obtaining energy and for the manufacture of new materials, while the use of fossil fuels is a capital-intensive work. Therefore, using biomass a greater number of jobs can be generated, resulting in a better economic investment and representing a sustainable alternative to the generation of polymeric materials [62–64].

## 2.2 Environmental aspects

Although comments and examples in the environmental field have been discussed previously, it will go a little deeper in the environmental context since this is the main engine behind the obtaining and development of biomaterials. The high production of polymeric materials and the linear economy style adopted for these materials, i.e., production-use-disposal, have triggered serious environmental problems. Thus, the accumulation of plastic materials and their microparticles in different ecosystems, death of animals due to direct consumption or by being trapped between structures made of these materials, and serious health problems for the human being are typical examples [2,65]. Although the design of sustainable polymeric materials does not radically solve all these problems, it does seek to address them in the best possible way, either through different government programs, change in the consumption habits of the population, design of eco-friendly production routes in all its stages, substitution of sources of raw materials, inclusion of new properties such as biodegradability and recyclability in products, secondary raw materials, or promotion of implementation of production systems based on circular economy, among others [12,66]. In the case of circular economy, the impact associated with the disposal of wastes during the production of biomaterials is reduced, returning them to the same stages in the processes, or stages of secondary processes [46,50].

One of the main challenges that the production of biopolymers must face is precisely related to their primary-obtaining sources. Aspects such as the type of source, its production process, and the amount of material required for a significant investment to be considered economically viable are decisive factors, in some cases, over the advantages of the material. For instance, if the polymeric material is extracted directly from a vegetable source, to meet the demand, large areas of land initially associated with food production will change their use. Therefore, a competition between the production of materials and the production of food for land use could be triggered, or worse still, the expansion of the agricultural frontier and the deterioration of natural ecosystems can be forced in order to obtain better conditions for obtaining biomaterials [67–69]. This apparent antagonism is not absolute, since the last century there have been large areas of land used exclusively for the production of biopolymers, the most emblematic case being the production of cotton fiber, which corresponds to cellulose fibers. Furthermore, the soil can be used for the usage of food or substances for application in the food and cosmetic sector to the extent that biopolymeric fibers are simultaneously produced [68,70]. In this case, the example again is cotton, since from its seeds it is possible to obtain oils for use in food and cosmetics [71,72]. However, for every gram of cotton fiber produced, a vastly greater amount of plant residue is produced. If this is not used, the waste is burned, which triggers the release of $CO_2$ into the atmosphere and the progressive deterioration of the soils when the burning is in situ. If the burning is carried out in boilers, the economic viability of the crop decreases due to logistical costs without benefit; therefore, in almost all cases, in situ burning in a controlled way of all or part of the biomass is a common practice. The use of part of these residues has been directed toward obtaining chemicals, however, only at the laboratory level because it has not reached its industrial materialization due to economic criteria, as well as in its use as a substrate for other crops [73,74].

## 2.3 Social aspects

The implementation of a more sustainable economy in the design and use of polymeric materials has several positive social impacts. In the first place, the use of renewable resources makes it possible to strengthen an entire productive chain associated with cultivating and harvesting of raw materials; this is translated into an increase in jobs associated with it, a diversification of the productive sectors, and a change in the economic dynamics of many places. In addition, similar approaches can be implemented for improving polymer-recycling processes or composting of biodegradable polymers [75,76]. Likewise, the increase in the use of land for the exploitation of biomass as a renewable resource is a possible solution to problems such as land abandonment, increasing unemployment, and the exodus from rural areas. However, there are also some effects that must be carefully analyzed in each specific case, such as the opposition of local inhabitants to the use of land for nontraditional purposes, less logistics in aspects such as transportation, less technical training in the field of new materials, and less availability of the infrastructure required for these purposes. Additionally, it is necessary to establish a production rate that does not interfere with the preparation of food crops [9,62].

# 3 Biomass: A renewable resource for the making of polymers

In technical words, biomass is understood to be all material of biological origin, i.e., it is the product resulting from the growth of living organisms, one of their components, or one of their products [77]. This means that biomass is all organic material obtained from plants, animals, or the industrial sector, which has the potential to be renewed. Due to its origin, biomass is derived from plants, composed of cellulose, lignin, hemicellulose, starch, and other polysaccharides; in addition, it currently represents one of the major approaches for obtaining biomaterials (see Table 1) [93,94]. Likewise, due to the large amount of waste material from the animal feed industry, the use of this biomass, mainly composed of proteins, has begun to be considered as raw material for obtaining different chemicals and materials, including polymers [8,75].

The potential and important use of biomass as a raw material for the production of polymeric materials and, in general, of useful chemical substances underlies in the diversity of its composition and the availability to be transformed into new high-value substances, for example, thermal treatment in the presence or absence of oxygen [95], biological treatments based on using enzymes, or microorganisms, and extraction of biocomposites using chemical reactions like transesterification [75,96]. This type of process requires several and interconnected steps, allowing an optimal transformation of the raw material into products of interest. For example, in the production of ethylene glycol from starch, it is required to carry out a starch degradation, glucose fermentation, ethanol dehydration, ethane oxidation, and finally, the hydrolysis of the product [9].

**Table 1** Production of monomers and polymers from biomass as an approach to a more sustainable economy of polymeric materials.

| Raw material | Monomer or polymer | Uses | References |
|---|---|---|---|
| Pine trees | Polyterpenes (e.g., poly(isoprene)) | Obtaining elastomers applied in automotive, nanostructures, composites, etc. | [78] |
| | Terpenes and derivatives (e.g., α-pinene, β-pinene, limonene, myrcene, among others) | Production of thermoplastic elastomers. | [79,80] |
| Soybeans, sunflowers, castor oil plant, palm tree | Triglycerides, glycerol, and fatty acids ($C_{12}$-$C_{22}$) | Obtaining epoxy resins, polyesters, polyols, and polyamides. | [81–83] |
| Maize, sugarcane, wheat, agricultural waste | Polysaccharides (e.g., starch, cellulose, hemicellulose, glycogen) | Production of fibers, paper, composites, gels, among others. | [84–86] |
| | Carboxylic acids (e.g., succinic acid, lactic acid, furan derivatives), lactones, and sugars | Obtaining thermoplastic biodegradable polyesters, which can be used as packaging or fibers. | [87–89] |
| Bacterial metabolism | Poly(hydroxyalkanoates) | Design of thermoplastic materials based on bio-derived polyesters. | [44,45,49] |
| Animal tissue waste | Amino acids and proteins (e.g., collagen, keratin) | Design of biomaterials, drug carriers, gels, polymeric matrices, among others. | [90–92] |

It is important to highlight the need to consider the abundance of renewable sources used to obtain sustainable polymers, as well as the ease of their recovery and processing. For example, carbohydrates are the most abundant and easiest to process raw material which is obtained from plant biomass; therefore, employing the methodologies for the use and transformation of carbohydrates, including an exploitation of their functionality, to obtain high-value monomers and polymers is its main goal [9]. Production of monomers and polymers from biomass is summarized in Table 1.

On the contrary, lignin is the second most abundant polymer in plant biomass representing an opportunity to obtain chemical substances and polymeric materials. This polymer serves as a structural network in plant cell walls and is composed of coniferyl, sinapyl, or cinnamyl alcohol units, complexly crosslinked with each other [97,98]. Different thermal, reducing, oxidative, and acid- or base-catalyzed methodologies have been designed to carry out the depolymerization of its structural network in order to obtain high-value monomers (e.g., dienes, alcohols, and aromatic compounds) [99,100]. Another important approach associated with the use of biomass for the production of sustainable polymers is the obtaining of monomers (e.g., ethylene, propylene, butadiene, and isoprene) of common polyolefins, generally through processes that involve cracking, fermentation, metathesis, and dehydrogenation [12,94]. In this way, research on biomass as renewable raw materials currently is increasing; therefore, it is expected that optimal methodologies will be established in future years for the use of this resource as a precursor of sustainable polymer materials at an industrial level. Examples of routes for obtaining biobased polymers are shown in Fig. 2.

# 4 Polymer for agricultural and environmental applications

Prior to advances toward applications of polymer-metal-oxide composite in agriculture and environment, a short description of biopolymer applications in these fields is shown. In the first case, agriculture is a very important activity for food production. Typical problems, or goals, of agricultural production are the adequate pest control, the supply and efficient use of water, the conservation of natural resources, the decrease of use of fertilizer and pesticides, and the increase of production of crops [41,101,102]. Clearly, between the goals of agriculture, the environmental aspects play an important role; however, the environmental sphere goes beyond agricultural production and extends to all other productive sectors. Environmental applications are focused on the study, diagnosis, conservation, and recovery of ecosystems regardless of the cause. Consequently, the monitoring of wastes or disturbances of the natural systems, removal of pollutants from water, soil, or air, and restoration of ecosystemic function of affected zones are typical environmental problems where polymers are used. Characteristic examples are contamination by heavy metals, pesticides, fertilizers, drugs, volatile organic compounds, eutrophication, loss of biotic components by overexploitation, bad conservation practices, gaseous emissions from combustion

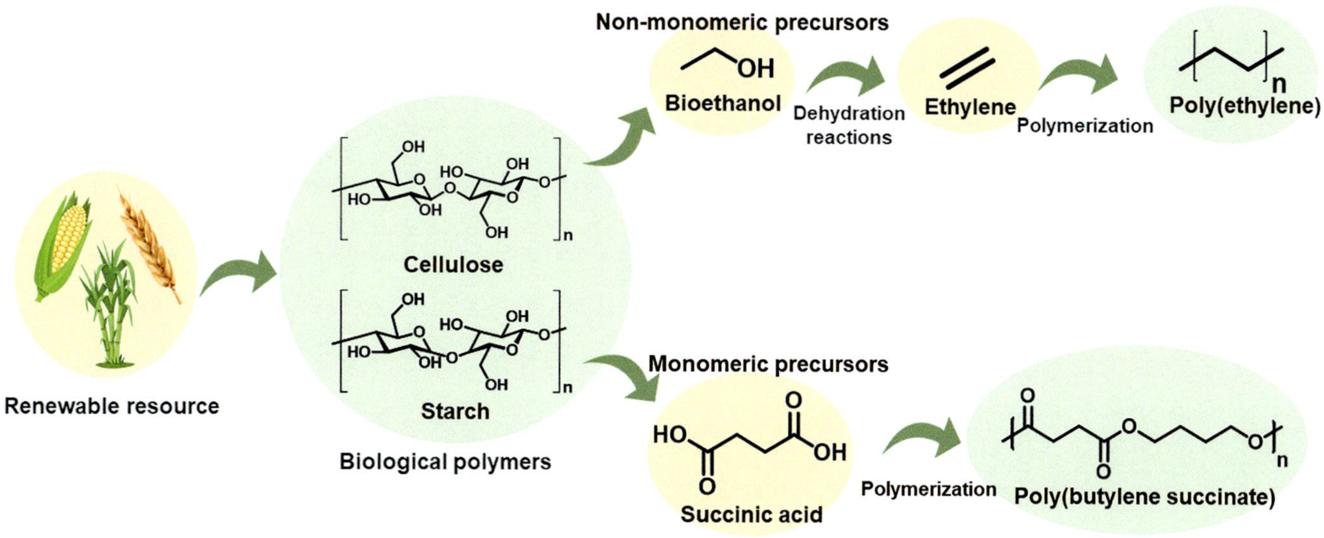


**Fig. 2** Example of routes for obtaining biobased polymers.
© No permission required. Own authorship image.

or industrial processes, global warming, and climate change, among others. A wide number of technologies have been developed to try to resolve these problems [103–105].

Polymers from renewable sources, or sustainable biopolymers, are characterized to contain, in their structure, functional groups such as carboxylic acids, hydroxyls, ethers, amines, amides, esters, sulfonates, phosphates, among others. In general, these groups contributed are related to a greater hydrophilicity, higher adsorption capacity on charged surfaces, more significant number of interactions with ions and other species in aqueous dissolution, and lower mechanical properties compared with structural polymers like poly(ethylene), poly(propylenes), poly(styrene), etc. Typical examples of sustainable biopolymers are chitosan, starch, cellulose, among others [8,9,75]. These properties allow the use of biopolymers in multiple applications in both agriculture and environmental sciences. Some examples are summarized in Table 2.

# 5 Polymer-metal-oxide composites for agricultural applications

Several types of nanomaterials are used in agriculture, however, contrary to other applications, agricultural systems are characterized to be complex, dynamic, with "hard" conditions like high ionic strength, temperature, and elevated concentration of substances. In addition, sometimes some components can be adsorbents and promote the immobilization of substances added to soil as well as systems can be strongly heterogeneous. In consequence, substance added for any agricultural function should be stable in a wide spectrum of conditions; for instance, the rhizosphere is a very different environment compared to the aqueous phase of soil, the surface of clay particles forming the solid phase, or supramolecular aggregates resulting of humic substances [128,129]. From the aforementioned, it is important to remember that metal-oxide nanoparticles are usually obtained to be the disperse phase of colloidal systems where the continuous phase is commonly water [130,131]. These systems can be destabilized by flocculation, precipitation, and chemical reactions changing their nature; adsorbed on surfaces; or retained by components existing in the application environment [132,133]. Therefore, their use combined with biopolymers forming nanocomposites allows the design and making of multifunctional materials where polymeric phase and dispersed phase act synergistically to achieve one or more objectives, but at the same time, the polymer phase helps to stabilize the dispersed phase.

Important nutrients for plants are the micronutrients, also called minor nutrients. These are required by the plants in small amounts, e.g., Cu, Fe, Mn, Zn, Mo, and B. Except for the latter, metal-oxide nanoparticles for all micronutrients have been synthesized (see Table 3). In Table 3 can be observed that direct application of nanomaterials must be carefully analyzed. However, it is described that metal-oxide nanoparticles are used as pesticides, herbicides, fertilizers, additives for soil remediation, and growth regulators [137,141].

In order to improve the micronutrient uptake, nanostructuration of corresponding metal oxide has been used as an alternative to achieve this goal. However, in order to

**Table 2** Example of biopolymers used for agricultural and environmental applications.

| Sector | Biopolymer | Applications | References |
|---|---|---|---|
| Agriculture | Carboxymethylcellulose and hydroxyethylcellulose, citric acid and sorbitol, whey protein concentrate, functionalized starches, among others | Hydrogels for the supply of water | [10,26,27,106] |
| | Starch, cellulose, chitin, chitosan, dextran, alginate, casein, albumin, gelatin, rubbers and waxes, polylatate, polyglycolate, poly(hydroxybutirate), tannins, among others | Control of plagues is by the development of micro- and nanocapsules containing herbicides, insecticides, or fungicide | [107,108] |
| | Cellulose acetate in conjunction with CuO nanoparticles | Agrochemical encapsulation such as fertilizers, pesticides, antimicrobials | [41] |
| | Functional polyurethanes based on *N*-methyl-D-glucamine | Smart release of boron | [109,110] |
| | Poly(sorbitol citrate) | Bacterial inoculum of *Azotobacter chroococcum* | [111,112] |
| | Carboxymethylcellulose and zeolite | Smart materials for macro- and micronutrients delivery | [113] |
| | Bacterial polyhydroxyalkanoates | Release of pesticides, phytohormones, and plant growth promoting rhizobacteria | [114] |
| | Polyesters from citric acid and sorbitol, poly(acrylic acid) and sorbitol, among other | Geomimetic soil conditioner for the growth of plants under no adequate conditions | [112,115,116] |
| | Starch, gelatin, poly(vinyl alcohol), cellulose diacetate, chitosan, cyclodextrins, chitosan cinnamic acid, | Seed coating and encapsulation | [117–121] |

*Continued*

**Table 2** Continued

| Sector | Biopolymer | Applications | References |
|---|---|---|---|
| | sodium alginate + calcium chloride, starch-alginate-clay-rice husk, alginate + lignin, lignin | | |
| Environmental Sci. | Polyesters from citric acid and sorbitol, poly(acrylic acid) and sorbitol, among other | Geo-mimicry systems for remediation of degraded soils | [112,115,116] |
| | Hydrophobic polymers analogous to polystyrene or polyethylene | Passive samplers for water, soils and airs with different polymer as active phase | [122–124] |
| | Poly(vinyl-benzyl-*N*-methyl-D-glucamine) | Boron removal from water | [109,110] |
| | Cellulose acetate | Active layer of ultrafiltration membrane for removal of pollutants | [32,109,110,125,126] |
| | Sodium salt of oleoyl carboxymethyl chitosan, chitosan microspheres modified with n-butyl acrylate, carboxymethyl chitosan modified with monochloroacetic acid, blends of poly(vinyl alcohol) nanoparticles with chitosan or starch | Remediation of aquatic ecosystems | [127] |

ensure the nanoparticle stability for the fulfillment of its function, the coating by the use of biopolymers has been proposed [128,129]. It is important to remember that micronutrients are absorbed in small amounts for the plants, consequently, though they are not nanostructured, an excess of any of the micronutrients produces a toxic effect in plants.

For example, if the values in the first row of Table 3 are carefully analyzed, it is possible to realize that the relevant information is the type of crops (maize), the size of nanoparticles (20–40 nm), the applied dose (100 mg/L), dissolution kinetic of nanoparticles, and concentration range of micronutrient in term of essentiality and toxicity (the typical dose for maize is 110 g Cu/Ha, whereas for foliar fertilization is commonly used 125 g Cu/Ha) [134,142,143]. Therefore, the dose of nanoparticles should be defined with respect to the requirements of crops. A fast uptake could clearly produce toxicity only as a result of the application of excess Cu in the form of nanoparticles.

**Table 3** Example of metal-oxide nanoparticles.

| Element | Nanoparticle | Observations | References |
|---|---|---|---|
| Cu | CuO | Crops: maize. Nanoparticle size: 20–40 nm. Conclusion: inhibition of root elongation | [134,135] |
| | | Crops: rice. Nanoparticle size: <50 nm. Conclusion: percentage of germination was decreased | [135] |
| | | Crops: cucumber. Nanoparticle size: not specified. Conclusion: inhibition of root elongation and germination of seeds | [135] |
| | | Particle size: $57.0 \pm 18$ nm. Negative impact on microbial activities in soils | [136] |
| | | High copper uptake by foliar tissue was observed after 15 days. Accumulation of nanoparticles was observed in necrotic Cu-rich areas. Use of nanoparticles should not be on edible plants or fruits due to potential health risk | [137] |
| Fe | $Fe_2O_3$ | Crops: spinach. Nanoparticle size: not specified. Conclusion: decrease of dry and wet biomass (dose: 200 mg/L) in hydroponic crops | [138] |
| | | Crops: ryegrass pumpkin. Nanoparticle size: 25 nm. Conclusion: induction of oxidative stress (dose: 30–100 mg/L) in hydroponic crops | [135] |
| Zn | ZnO | Crops: green pea. Particle size: 10 nm. Decrease of chlorophyll content by up to 77% | [139] |
| | | Crops: maize. Particle size: 10 nm. Significant increase of chlorophyll *a* (in presence of alginate and 400 mg/L) | [140] |

An interesting example of polymer-metal-oxide composites applied to agriculture is the use of polymer coating based on humic acids. Humified organic matter can be extracted, separated, and classified by differences in solubility. These substances are named as humic acids, fulvic acids, and humin. Humic acids and fulvic acids can be described to be organic molecules with acid functional groups such as carboxylic acid and hydroxyl on aromatic rings as well as other heteroatoms can be identified in them from elemental analysis and spectroscopic techniques [144,145]. On the contrary, humin is an insoluble organic fraction because it is found to be anchored to the clay surface [144]. Humic and fulvic acids can be used like coats of nanoparticles [146,147]. A clear advantage of this strategy is that exogenous organic substances are not added to the soil since, though humic and fulvic acids are different depending on localization, history, and type of processes occurring in it along much time, they are substances exclusively existing in soils. A second advantage is that coating based on humic and fulvic acids facilitates ion transport, microbiological colonization, and promotes the dissolution of nanoparticles and uptake of nutrient by plants [130,148,149]. Other polymers with great potential for the improvement of fertilization with micronutrients are biodegradable hydrogels, and these can be loaded with nanoparticles of respective oxides in order to protect them from environmental conditions promoting their instability (see Fig. 3) [150–152]. Some potential examples of hydrogel-based coatings are poly(sorbitol citrate) which can be obtained by esterification of sorbitol and citrate in absence of solvent [26], and starch-citrate hydrogels which are synthesized by cross-linking of starch using citrate as cross-linker [27].

On the contrary, clays are a geomaterial with very small particle size, which are formed by meteorization process of rocks and can be easily dispersed in water, producing colloid systems. They are inorganic polymers with different shapes, including particles, chains, layers, and networks. The types of clays common in soils are those formed by aluminosilicates, which are mixed oxides: aluminum oxides ($Al_2O_3$) and silicon ($SiO_2$) [153,154]. When clays are exfoliated, sheets of clays are separated, evidencing a nanometric scale in its thickness. Clays have been modified with synthetic polymers and biopolymers for the production of novel nanocomposites with

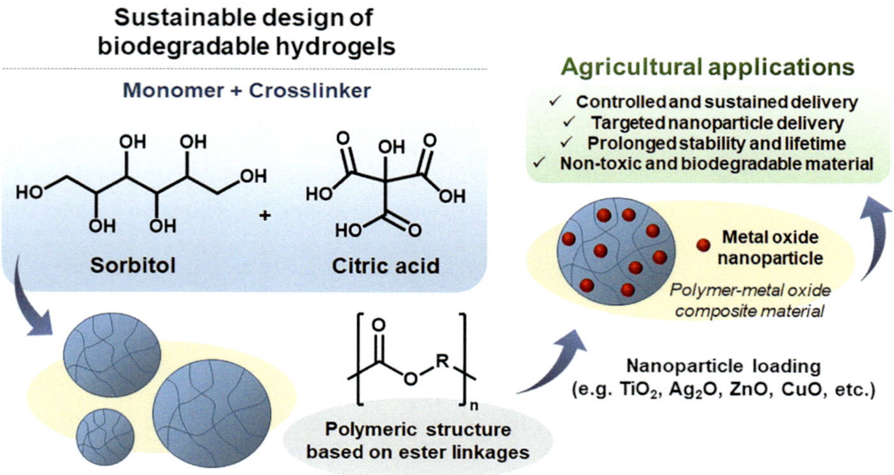


**Fig. 3** An example of the sustainable design of polymer-metal-oxide composite materials using by-products from renewable raw matter, such as sorbitol and citric acid, with potential agricultural applications (adapted from [27]). In this case, hydrogels can serve as a reservoir of nanoparticles and promote their controlled, sustained, and targeted delivery on soils, for example, as well as their protection against environmental degradation.
Adapted from M. Palencia, T.A. Lerma, E.M. Combatt, Hydrogels based in cassava starch with antibacterial activity for controlled release of cysteamin-silver nanostructures agents. Curr. Chem. Biol. 11 (2017) 28–35. https://doi.org/10.2174/2212796810666161108152319. © No permission required. Own authorship image.

a geomimetic approach. These composites have been evaluated and proposed to promote the soil remediation and act as soil conditioner, with multiple functions like hydrogels for the water storage, release of nutrients, improvement of soil structure, airing, and support for the growth of bacteria with importance in agriculture [112,115,116].

# 6 Polymer-metal-oxide composites for environmental applications

Metal-oxide nanoparticles have a wide use in environmental applications since the spectrum of potential use includes monitoring of pollutants, treatment of water, and remediation activities. These materials can be used as adsorbents, photocatalysts, sensors, etc. For instance, metal-oxide nanoparticles based on $TiO_2$, ZnO, CuO, $SnO_2$, $ZrO_2$, and $Fe_3O_4$ have been used in photocatalytic processes for the advanced oxidation of organic pollutants, such as pharmaceutical wastes, dyes, and dissolved organic matter, and even they can be used in microbiological disinfection activities [155,156]. An interesting environmental application is the use of chitosan-$Al_2O_3$ composites, composites $Fe_3O_4$ based on honeycomb briquette cinders, and calcium alginate and activated carbon, for the removal of arsenic ions from aqueous solutions (see Fig. 4) [157].

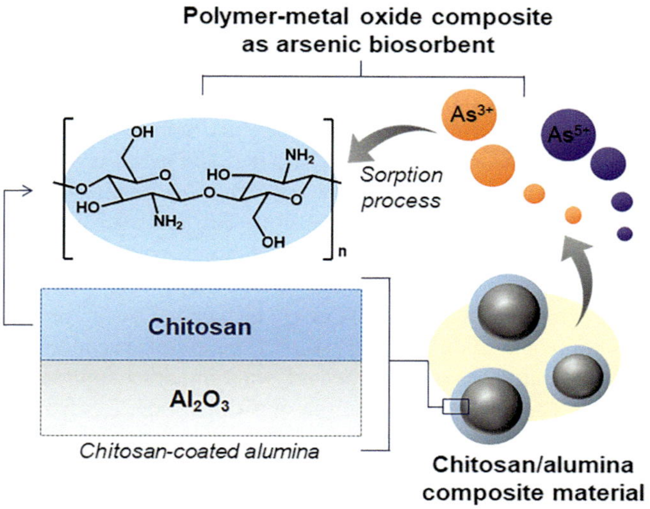


**Fig. 4** Chitosan-$Al_2O_3$ composite material for the sorption of arsenic in water environment. Adapted from Rahim M., Hakim M.R. Application of biopolymer composites in arsenic removal from aqueous medium: a review. J. Radiat. Res. Appl. Sci. 8 (2015) 255–263. doi:10.1016/j.jrras.2015.03.001Rahim M., Hakim M.R. Application of biopolymer composites in arsenic removal from aqueous medium: A review. Journal of Radiation Research and Applied Sciences. 8 (2015) 255-263. doi:10.1016/j.jrras.2015.03.001 ). © No permission required. Own authorship image.

Some of the most used polymer-metal-oxide composites for environmental applications are those based on $TiO_2$. Several forms of $TiO_2$ have been identified in nature as rutile and anatase as well as in combination with Fe (ilmenite). The main characteristic of nanoscale $TiO_2$ is the photocatalytic activity by action of ultraviolet light, which produces the formation of radicals that are able to oxidize organic species in water [158]. Among synthetic biopolymers used for the making of polymer-metal-oxide composites based on nanoscaled $TiO_2$ are poly(hydroxybutyrate) and chitosan [159,160]. Also, its use is more frequent with synthetic polymers like poly(aniline), poly(dimethylsiloxane), poly(3-hexylthiophene), poly(tetrafluoroethylene), poly(fluorene-*co*-thiophene), poly(1-naphthylamine), polyethylene, Naflon, and polypropylene [161–164]. Between target pollutants for photocatalytic degradation are dyes like methylene blue, methyl orange, and Victoria blue R, as well as organic compounds like phenol, trichlorobenzene, and acetaldehyde [165,166]. In addition, preparation methods and efficiency described in specialized literature are very wide, including in the first case, casting, direct polymerization in the presence of metal-oxide nanoparticles, sol–gel, electrophoretic deposition, and hot pressing [167]. Other examples are the $Fe_2O_3$/cellulose and $Fe_2O_3$/alginate for removal of As(III) and As(V) [167–169].

# 7 Conclusions and remarks

Polymer-metal-oxide composites have shown a great potential for agriculture and environmental applications. However, studies focused on the use of renewable resources are limited, being more common the use of nanocomposites based on biopolymers in agricultural applications, while, in the environmental context, nanocomposites based on synthetic polymers are the most used.

Particularly in the case of metallic oxides, where the metal is a micronutrient, the effects described in many of the experiments evidence a generalized toxicity in plants. However, these results must be looked at carefully because micronutrients have well-defined ranges of essentiality and, outside of this range, both deficiency and toxicity are usually observed phenomena. In addition, these ranges are characteristic of the type of plant, growth stage, and will be strongly influenced by factors associated with transport, dissolution, and interaction with biotic and abiotic components of the study systems. Based on the analysis of published information, other applications in the agricultural field seem more promising than the nutritional one, for example, the use of these materials as herbicides, insecticides, and a reservoir of bioelements. On the contrary, for environmental applications, photocatalytic applications for the degradation of pollutants are the most studied. However, the fact that studies using biopolymers and other environmental applications with a different approach to photocatalysis are scarce is a reflection of how much progress still needs to be made in this type of application.

One of the most crucial aspects to take into account is the release of nanostructured waste, since, the very important fact that biopolymers are a phase susceptible to biodegradation, the direct consequence of this is related to the release into the environment of synthetic material nanostructured that can have great mobility and interaction with the different entities of ecosystems.

## References


[1] Y. Chen, A.K. Awasthi, F. Wei, Q. Tan, J. Li, Single-use plastics: production, usage, disposal, and adverse impacts, Sci. Total Environ. 752 (2021) 141772, https://doi.org/10.1016/j.scitotenv.2020.141772.

[2] R. Geyer, J.R. Jambeck, K.L. Law, Production, use, and fate of all plastics ever made, Sci. Adv. 3 (2017), https://doi.org/10.1126/sciadv.1700782, e1700782.

[3] A. Pellis, M. Malinconico, A. Guarneri, L. Gardossi, Renewable polymers and plastics: performance beyond the green, N. Biotechnol. 60 (2021) 146–158, https://doi.org/10.1016/j.nbt.2020.10.003.

[4] L. Reijnders, Fuels for the future, J. Integr. Environ. Sci. 6 (2009) 279–294, https://doi.org/10.1080/19438150903068596.

[5] A. Chamas, H. Moon, J. Zheng, Y. Qiu, T. Tabassum, J.H. Jang, M. Abu-Omar, S.L. Scott, S. Suh, Degradation rates of plastics in the environment, ACS Sustain. Chem. Eng. 8 (2020) 3494–3511, https://doi.org/10.1021/acssuschemeng.9b06635.

[6] L. Peng, D. Fu, H. Qi, C.Q. Lan, H. Yu, C. Ge, Micro- and nano-plastics in marine environment: source, distribution and threats—a review, Sci. Total Environ. 698 (2020) 134254, https://doi.org/10.1016/j.scitotenv.2019.134254.

[7] D.K. Schneiderman, M.A. Hillmyer, 50th anniversary perspective: there is a great future in sustainable polymers, Macromolecules 50 (2017) 3733–3749, https://doi.org/10.1021/acs.macromol.7b00293.

[8] J. Wang, W. Qian, Y. He, Y. Xiong, P. Song, R.M. Wang, Reutilization of discarded biomass for preparing functional polymer materials, Waste Manag. 65 (2017) 11–21, https://doi.org/10.1016/j.wasman.2017.04.025.

[9] Y. Zhu, C. Romain, C.K. Williams, Sustainable polymers from renewable resources, Nature 540 (2016) 354–362, https://doi.org/10.1038/nature21001.

[10] M. Palencia, A. Espinosa-Duque, A. Otálora, A. García-Quintero, Cellulose-based stimuli-responsive hydrogels, in: T.K. Giri, B. Ghosh (Eds.), Plant and Algal Hydrogels for Drug Delivery and Regenerative Medicine, Woodhead Publishing, 2021, pp. 423–470, https://doi.org/10.1016/B978-0-12-821649-1.00002-7.

[11] D.E. Fagnani, J.L. Tami, G. Copley, M.N. Clemons, Y.D.Y.L. Getzler, A.J. McNeil, 100th anniversary of macromolecular science viewpoint: redefining sustainable polymers, ACS Macro Lett. 10 (2021) 41–53, https://doi.org/10.1021/acsmacrolett.0c00789.

[12] Z. Wang, M.S. Ganewatta, C. Tang, Sustainable polymers from biomass: bridging chemistry with materials and processing, Prog. Polym. Sci. 101 (2020) 101197, https://doi.org/10.1016/j.progpolymsci.2019.101197.

[13] M.A. Wolf, M. Baitz, J. Kreissig, Assessing the sustainability of polymer products, in: P. Eyerer, M. Weller, C. Hübner (Eds.), Polymers—Opportunities and Risks II. The Handbook of Environmental Chemistry, vol. 12, Springer, Berlin, Heidelberg, 2009, https://doi.org/10.1007/698_2009_10.

[14] J. Antonkiewicz, R. Pełka, M. Bik-Małodzińska, G. Żukowska, K. Gleń-Karolczyk, The effect of cellulose production waste and municipal sewage sludge on biomass and heavy metal uptake by a plant mixture, Environ. Sci. Pollut. Res. 25 (2018) 31101–31112,- https://doi.org/10.1007/s11356-018-3109-5.

[15] M.A. Hubbe, J.R. Metts, D. Hermosilla, M.A. Blanco, L. Yerushalmi, et al., Wastewater treatment and reclamation: a review of pulp and paper industry practices and opportunities, Bioresources 11 (2016) 7953–8091.
[16] P. Glavič, R. Lukman, Review of sustainability terms and their definitions, J. Clean. Prod. 15 (2007) 1875–1885, https://doi.org/10.1016/j.jclepro.2006.12.006.
[17] J.P. Holdren, Science and technology for sustainable well-being, Science 319 (2008) 424–434, https://doi.org/10.1126/science.1153386.
[18] W. Fan, Y. Jin, L. Shi, W. Du, R. Zhou, Transparent, eco-friendly, super-tough "living" supramolecular polymers with fast room-temperature self-healability and reprocessability under visible light, Polymer 190 (2020) 122199, https://doi.org/10.1016/j.polymer.2020.122199.
[19] V. Lebedev, T. Tykhomyrova, I. Litvinenko, S. Avina, Z. Saimbetova, Design and research of eco-friendly polymer composites, Mater. Sci. Forum 1006 (2020) 259–266,-https://doi.org/10.4028/www.scientific.net/MSF.1006.259.
[20] M. Hong, E.Y.X. Chen, Future directions for sustainable polymers, Trends Chem. 1 (2019) 148–151, https://doi.org/10.1016/j.trechm.2019.03.004.
[21] S.A. Miller, Sustainable polymers: opportunities for the next decade, ACS Macro Lett. 2 (2013) 550–554, https://doi.org/10.1021/mz400207g.
[22] H. Nosrati, S. Pourmotabed, E. Sharifi, A review on some natural biopolymers and their applications in angiogenesis and tissue engineering, J. Appl. Biotechnol. Rep. 5 (2018) 81–91, https://doi.org/10.29252/JABR.05.03.01.
[23] A. Sionkowska, Current research on the blends of natural and synthetic polymers as new biomaterials: review, Prog. Polym. Sci. 36 (2011) 1254–1276, https://doi.org/10.1016/j.progpolymsci.2011.05.003.
[24] K.M. Zia, S. Tabasum, M. Nasif, N. Sultan, N. Aslam, A. Noreen, M. Zuber, A review on synthesis, properties and applications of natural polymer based carrageenan blends and composites, Int. J. Biol. Macromol. 96 (2017) 282–301, https://doi.org/10.1016/j.ijbiomac.2016.11.095.
[25] C. Casadidio, D. Vargas, M.R. Gigliobianco, S. Deng, R. Censi, P. Di Martino, Chitin and chitosans: characteristics, eco-friendly processes, and applications in cosmetic science, Mar. Drugs 17 (2019) 369, https://doi.org/10.3390/md17060369.
[26] M. Palencia, T.A. Lerma, E.M. Combatt, Hydrogels based in cassava starch with antibacterial activity for controlled release of cysteamin-silver nanostructures agents, Curr. Chem. Biol. 11 (2017) 28–35, https://doi.org/10.2174/2212796810666161108152319.
[27] M. Palencia, M. Mora, S.L. Palencia, Biodegradable polymer hydrogel based in sorbitol and citric acid for controlled release of bioactive substances from plants (polyphenols), Curr. Chem. Biol. 11 (2017) 36–43, https://doi.org/10.2174/2212796810666161028114432.
[28] T. Saito, R.H. Brown, M.A. Hunt, D.L. Pickel, J.M. Pickel, J.M. Messman, F.S. Baker, M. Keller, A.K. Naskar, Turning renewable resources into value-added polymer: development of lignin-based thermoplastic, Green Chem. 14 (2012) 3295–3303, https://doi.org/10.1039/C2GC35933B.
[29] D.J. Saxon, A.M. Luke, H. Sajjad, W.B. Tolman, T.M. Reineke, Next-generation polymers: isosorbide as a renewable alternative, Prog. Polym. Sci. 101 (2020) 101196, https://doi.org/10.1016/j.progpolymsci.2019.101196.
[30] Y. Srithep, L.S. Turng, J. Morris, D. Pholharn, O. Veangi, Sustainable polymers: from recycling of non-biodegradable to renewable resources composites and foams, Mahasarakham Int. J. Eng. Technol. 1 (2015) 24–28.
[31] B.L. Rivas, E.D. Pereira, M. Palencia, J. Sánchez, Water-soluble functional polymers in conjunction with membranes to remove pollutant ions aqueous solutions, Prog. Polym. Sci. 36 (2011) 294–322, https://doi.org/10.1016/j.progpolymsci.2010.11.001.

[32] B.L. Rivas, J. Sanchez, M. Palencia, Organic membranes and polymers for removal of polutants, in: P.M. Vikash, O. Nazarenko (Eds.), Nanostructured Polymer Membranes: Volume 1, Scrivener Publiching LLC, 2016, pp. 203–236.
[33] V. Siracusa, I. Blanco, Bio-polyethylene (bio-PE), bio-polypropylene (bio-PP) and bio-poly(ethylene terephthalate) (bio-PET): recent developments in bio-based polymers analogous to petroleum-derived ones for packaging and engineering applications, Polymers 12 (2020) 1641, https://doi.org/10.3390/polym12081641.
[34] S. Walker, R. Rothman, Life cycle assessment of bio-based and fossil-based plastic: a review, J. Clean. Prod. 261 (2020) 121158, https://doi.org/10.1016/j.jclepro.2020.121158.
[35] M.R. Thomsett, T.E. Storr, O.R. Monaghan, R.A. Stockman, S.M. Howdle, Progress in the Synthesis of Sustainable Polymers From Terpenes and Terpenoids, vol. 4, 2016, pp. 115–134, https://doi.org/10.1680/jgrma.16.00009.
[36] H. Cao, X. Wang, Carbon dioxide copolymers: emerging sustainable materials for versatile applications, Sustain. Mat. 1 (2021) 88–104, https://doi.org/10.1002/sus2.2.
[37] A.M. Chapman, C. Keyworth, M.R. Kember, A.J.J. Lennox, C.K. Williams, Adding value to power station captured CO2: tolerant Zn and Mg homogeneous catalysts for polycarbonate polyol production, ACS Catal. 5 (2015) 1581–1588, https://doi.org/10.1021/cs501798s.
[38] C.K. Williams, M.A. Hillmyer, Polymers from renewable resources: a perspective for a special issue of polymer reviews, Polym. Rev. 48 (2008) 1–10, https://doi.org/10.1080/15583720701834133.
[39] D.H. Lee, J.H. Ha, I. Kim, J.H. Baik, S.C. Hong, Carbon dioxide based poly(ether carbonate) polyol in bi-polyol mixtures for rigid polyurethane foams, J. Polym. Environ. 28 (2020) 1160–1168, https://doi.org/10.1007/s10924-020-01668-0.
[40] R. Muthuraj, T. Mekonnen, Recent progress in carbon dioxide (CO2) as feedstock for sustainable materials development: co-polymers and polymer blends, Polymer 145 (2018) 348–373, https://doi.org/10.1016/j.polymer.2018.04.078.
[41] T. Xu, C. Ma, Z. Aytac, X. Hu, K. Woei, J.C. White, P. Demokritou, Enhancing agrichemical delivery and seedling development with biodegradable, tunable, biopolymer-based nanofiber seed coatings, ACS Sustain. Chem. Eng. 8 (2020) 9537–9548, https://doi.org/10.1021/acssuschemeng.0c02696.
[42] C. Song, Global challenges and strategies for control, conversion and utilization of CO2 for sustainable development involving energy, catalysis, adsorption and chemical processing, Catal. Today 115 (2006) 2–32, https://doi.org/10.1016/j.cattod.2006.02.029.
[43] Y.C. Su, B.T. Ko, Alternating copolymerization of carbon dioxide with epoxides using highly active dinuclear nickel complexes: catalysis and kinetics, Inorg. Chem. 60 (2021) 852–865, https://doi.org/10.1021/acs.inorgchem.0c02902.
[44] V. Garcés, M. Palencia, Biosynthesis and biotransformations of biopolymers: starch and poly(hydroxyalkanoates), J. Sci. Technol. Appl. 9 (2020) 4–9, https://doi.org/10.34294/j.jsta.20.9.60.
[45] J. Lu, R.C. Tappel, C.T. Nomura, Mini-review: biosynthesis of poly(hydroxyalkanoates), Polym. Rev. 49 (2009) 226–248, https://doi.org/10.1080/15583720903048243.
[46] I. Vroman, L. Tighzert, Biodegradable polymers, Materials (Basel) 2 (2009) 307–344, https://doi.org/10.3390/ma2020307.
[47] A.C. Albertsson, M. Hakkarainen, Designed to degrade, Science 358 (2017) 872–873, https://doi.org/10.1126/science.aap8115.
[48] K.J. Jem, B. Tan, The development and challenges of poly (lactic acid) and poly (glycolic acid), Adv. Ind. Eng. Polymer Res. 3 (2020) 60–70, https://doi.org/10.1016/j.aiepr.2020.01.002.

[49] Z. Li, J. Yang, X.J. Loh, Polyhydroxyalkanoates: opening doors for a sustainable future, NPG Asia Mater. 8 (2016), https://doi.org/10.1038/am.2016.48, e265.
[50] D. Ayre, Technology advancing polymers and polymer composites towards sustainability: a review, Curr. Opinion Green Sustain. Chem. 13 (2018) 108–112, https://doi.org/10.1016/j.cogsc.2018.06.018.
[51] M. Hong, E.Y.X. Chen, Chemically recyclable polymers: a circular economy approach to sustainability, Green Chem. 19 (2017) 3692–3706, https://doi.org/10.1039/C7GC01496A.
[52] A. Otálora, M. Palencia, Polymeric hydrogels based on dynamic covalent bonds for potential biomedical applications, J. Sci. Technol. Appl. 8 (2020) 55–72, https://doi.org/10.34294/j.jsta.20.8.55.
[53] P. Rai, S. Mehrotra, S. Priya, E. Gnansounou, S.K. Sharma, Recent advances in the sustainable design and applications of biodegradable polymers, Bioresour. Technol. 124739 (2021), https://doi.org/10.1016/j.biortech.2021.124739.
[54] Z. Su, S. Huang, Y. Wang, H. Ling, X. Yang, Y. Jin, X. Wang, W. Zhang, Robust, high-barrier, and fully recyclable cellulose-based plastic replacement enabled by a dynamic imine polymer, J. Mater. Chem. A 8 (2020) 14082–14090, https://doi.org/10.1039/D0TA02138E.
[55] A.C. Niloy, Jute: solution to global challenges and opportunities of Bangladesh, Seisense Bus. Rev. 1 (2021) 59–75, https://doi.org/10.33215/sbr.v1i2.633.
[56] A. Forrest, L. Giacovazzi, S. Dunlop, J. Reisser, D. Tickler, A. Jamieson, J.J. Meeuwig, Eliminating plastic pollution: how a voluntary contribution from industry will drive the circular plastics economy, Front. Mar. Sci. 6 (2019) 627, https://doi.org/10.3389/fmars.2019.00627.
[57] J. Popp, S. Kovács, J. Oláh, Z. Divéki, E. Balázs, Bioeconomy: biomass and biomass-based energy supply and demand, New Technol. 60 (2021) 76–84, https://doi.org/10.1016/j.nbt.2020.10.004.
[58] F.C. De Paula, C.B.C. De Paula, J. Contiero, Prospective biodegradable plastics from biomass conversion processes, in: K. Biernat (Ed.), Biofuels-State of Development, IntechOpen, 2018, https://doi.org/10.5772/intechopen.75111.
[59] N.J. Tonukari, Cassava and the future of starch, Electron. J. Biotechnol. 7 (2004).
[60] B. Bhagowati, K.U. Ahamas, A review on lake eutrophication dynamics and recent developments in lake modeling, Ecohydrol. Hydrobiol. 19 (2019) 155–166, https://doi.org/10.1016/j.ecohyd.2018.03.002.
[61] B. Vincon-Leite, C. Casenave, Modelling eutrophication in lake ecosystems: a review, Sci. Total Environ. 651 (2019) 2985–3001, https://doi.org/10.1016/j.scitotenv.2018.09.320.
[62] P. Carneiro, P. Ferreira, The economic, environmental and strategic value of biomass, Renew. Energy 44 (2012) 17–22, https://doi.org/10.1016/j.renene.2011.12.020.
[63] K. Refsgaard, M. Kull, E. Slätmo, M.W. Meijer, Bioeconomy—a driver for regional development in the Nordic countries, N. Biotechnol. 60 (2021) 130–137, https://doi.org/10.1016/j.nbt.2020.10.001.
[64] M. Rutkowska, A. Sulich, Green jobs on the background of industry 4.0, Proc. Comput. Sci. 176 (2020) 1231–1240, https://doi.org/10.1016/j.procs.2020.09.132.
[65] A.E. Schwarz, N. Ligthart, E. Boukris, T. Van Harmelen, Sources, transport, and accumulation of different types of plastic litter in aquatic environments: a review study, Mar. Pollut. Bull. 143 (2019) 92–100, https://doi.org/10.1016/j.marpolbul.2019.04.029.
[66] T.P. Haider, C. Völker, J. Kramm, K. Landfester, F.R. Wurm, Plastics of the future? The impact of biodegradable polymers on the environment and on society, Angew. Chem. Int. Ed. 58 (2019) 50–62, https://doi.org/10.1002/anie.201805766.

[67] J.K.A. Benhin, Agriculture and deforestation in the tropics: a critical theoretical and empirical review, Ambio 35 (2006) 9–16. *https://www.jstor.org/stable/4315676.*
[68] D. Byerlee, J. Stevenson, N. Villoria, Does intensification slow crop land expansion or encourage deforestation? Global Food Secur. 3 (2014) 92–98, https://doi.org/10.1016/j.gfs.2014.04.001.
[69] E.M. Ordway, G.P. Asner, E.F. Lambin, Deforestation risk due to commodity crop expansion in sub-Saharan Africa, Environ. Res. Lett. 12 (2017), https://doi.org/10.1088/1748-9326/aa6509, 044015.
[70] A. Oberlintner, M. Bajić, G. Kalčíková, B. Likozar, U. Novak, Biodegradability study of active chitosan biopolymer films enriched with *Quercus* polyphenol extract in different soil types, Environ. Technol. Innov. 21 (2021) 101318, https://doi.org/10.1016/j.eti.2020.101318.
[71] M. Kumar, J. Potkule, S. Patil, V. Mageshwaran, V. Radha Satankar, M.K. Berwal, A. Mahapatra, S. Saxena, N. Ashtaputre, C. D Souza, Evaluation of detoxified cottonseed protein isolate for application as food supplement, Toxin Rev. (2021), https://doi.org/10.1080/15569543.2021.1889605.
[72] M. Kumar, M. Tomar, S. Punia, S. Grasso, F. Arrutia, J. Choudhary, S. Singh, P. Verma, A. Mahapatra, S. Patil, S. Radha Dhumal, J. Potkule, S. Saxena, R. Amarowicz, Cottonseed: a sustainable contributor to global protein requirements, Trends Food Sci. Technol. 111 (2021) 100–113, https://doi.org/10.1016/j.tifs.2021.02.058.
[73] S. Petropoulos, A. Fernandes, S. Plexida, C. Pereira, M.I. Dias, et al., The sustainable use of cotton, hazelnut and ground peanut waste in vegetable crop production, Sustainability 12 (2020) 8511, https://doi.org/10.3390/su12208511.
[74] S. Radhakrishnan, Sustainable cotton production, in: S. Senthikannan Muthu (Ed.), Sustainable Fibres and Textiles, Elsevier, 2017, pp. 21–67, https://doi.org/10.1016/B978-0-08-102041-8.00002-0.
[75] S.M. Ioannidou, C. Pateraki, D. Ladakis, H. Papapostolou, M. Tsakona, A. Vlysidis, I.K. Kookos, A. Koutinas, Sustainable production of bio-based chemicals and polymers via integrated biomass refining and bioprocessing in a circular bioeconomy context, Bioresour. Technol. 307 (2020) 123093, https://doi.org/10.1016/j.biortech.2020.123093.
[76] S. Spierling, E. Knüpffer, H. Behnsen, M. Mudersbach, H. Krieg, S. Springer, S. Albrecht, C. Herrman, H.-J. Endres, Bio-based plastics—a review of environmental, social and economic impact assessments, J. Clean. Prod. 185 (2018) 476–491,-https://doi.org/10.1016/j.jclepro.2018.03.014.
[77] N. Afanasjeva, L.C. Castillo, J.C. Sinisterra, Lignocellulosic biomass. Part I: Biomass transformation, J. Sci. Technol. Appl 3 (2017) 27–43, https://doi.org/10.34294/j.jsta.17.3.22.
[78] L. Bokobza, Natural rubber nanocomposites: a review, Nanomaterials (Basel). 9 (2012) 12, https://doi.org/10.3390/nano9010012.
[79] J.M. Bolton, M.A. Hillmyer, T.R. Hoye, Sustainable thermoplastic elastomers from terpene-derived monomers, ACS Macro Lett. 3 (2014) 717–720, https://doi.org/10.1021/mz500339h.
[80] J.L. González-Zapata, F.J. Enríquez-Medrano, H.R.L. Gonzáles, J. Revilla-Vázquez, R. M. Carrizales, D. Georgouvelas, L. Valencia, R.E.D. de León Gómez, Introducing random bio-terpene segments to high cis-polybutadiene: making elastomeric materials more sustainable, RSC Adv. 10 (2020) 44096–44102, https://doi.org/10.1039/D0RA09280K.
[81] A. Behr, J.P. Gomes, The refinement of renewable resources: new important derivatives of fatty acids and glycerol, Eur. J. Lipid Sci. Technol. 112 (2010) 31–50, https://doi.org/10.1002/ejlt.200900091.

[82] L. Maisonneuve, T. Lebarbé, E. Grau, H. Cramail, Structure–properties relationship of fatty acid-based thermoplastics as synthetic polymer mimics, Polym. Chem. 4 (2013) 5472–5517, https://doi.org/10.1039/C3PY00791J.
[83] F. Stempfle, P. Ortmann, S. Mecking, Long-chain aliphatic polymers to bridge the gap between semicrystalline polyolefins and traditional polycondensates, Chem. Rev. 116 (2016) 4597–4641, https://doi.org/10.1021/acs.chemrev.5b00705.
[84] V. Gopinath, S. Saravanan, A.R. Al-Maleki, M. Ramesh, J. Vadivelu, A review of natural polysaccharides for drug delivery applications: special focus on cellulose, starch and glycogen, Biomed. Pharmacother. 107 (2018) 96–108, https://doi.org/10.1016/j.biopha.2018.07.136.
[85] N. Lin, J. Huang, A. Dufresne, Preparation, properties and applications of polysaccharidenanocrystals in advanced functional nanomaterials: a review, Nanoscale 4 (2012) 3274–3294, https://doi.org/10.1039/C2NR30260H.
[86] V. Rana, S. Malik, G. Joshi, N.K. Rajput, P.K. Gupta, Preparation of alpha cellulose from sugarcane bagasse and its cationization: synthesis, characterization, validation and application as wet-end additive, Int. J. Biol. Macromol. 170 (2021) 793–809, https://doi.org/10.1016/j.ijbiomac.2020.12.165.
[87] R. Auras, B. Harte, S. Selke, An overview of polylactides as packaging materials, Macromol. Biosci. 4 (2004) 835–864, https://doi.org/10.1002/mabi.200400043.
[88] S.K. Burgess, J.E. Leisen, B.E. Kraftschik, C.R. Mubarak, R.M. Kriegel, W.J. Koros, Chain mobility, thermal, and mechanical properties of poly(ethylene furanoate) compared to poly(ethylene terephthalate), Macromolecules 47 (2014) 1383–1391, https://doi.org/10.1021/ma5000199.
[89] I. Delidovich, P.J.C. Hausol, L. Deng, R. Pfützenreuter, M. Rose, R. Palkovits, Alternative monomers based on lignocellulose and their use for polymer production, Chem. Rev. 116 (2016) 1540–1599, https://doi.org/10.1021/acs.chemrev.5b00354.
[90] R.M. Broyer, G.N. Grover, H.D. Maynard, Emerging synthetic techniques for protein-polymer conjugations, Chem. Commun. 47 (2011) 2212–2226, https://doi.org/10.1039/C0CC04062B.
[91] R. Khan, M.H. Khan, Use of collagen as a biomaterial: an update, J. Indian Soc. Periodontol. 17 (2013) 539–542, https://doi.org/10.4103/0972-124X.118333.
[92] M.W.T. Werten, G. Eggink, M.A.C. Stuart, F.A. De Wolf, Production of protein-based polymers in Pichia pastoris, Biotechnol. Adv. 37 (2019) 642–666, https://doi.org/10.1016/j.biotechadv.2019.03.012.
[93] A.C. Opia, M.K.B.A. Hamid, S. Syahrullail, A.B.A. Rahim, C.A.N. Johnson, Biomass as a potential source of sustainable fuel, chemical and tribological materials—overview, Mater. Today Proc. (2020), https://doi.org/10.1016/j.matpr.2020.04.045.
[94] P. Zong, Y. Jiang, Y. Tian, J. Li, M. Yuan, Y. Ji, M. Chen, D. Li, Y. Qiao, Pyrolysis behavior and product distributions of biomass six group components: starch, cellulose, hemicellulose, lignin, protein and oil, Energ. Conver. Manage. 216 (2020) 112777, https://doi.org/10.1016/j.enconman.2020.112777.
[95] M. Antar, D. Lyu, M. Nazari, A. Shah, X. Zhou, D.L. Smith, Biomass for a sustainable bioeconomy: an overview of world biomass production and utilization, Renew. Sustain. Energy Rev. 139 (2021) 110691, https://doi.org/10.1016/j.rser.2020.110691.
[96] P. Choudhary, P.P. Assemany, F. Naaz, A. Bhattacharya, J.S. Castro, E.A.C. Couto, M.L. Calijuri, K.K. Pant, A. Malik, A review of biochemical and thermochemical energy conversion routes of wastewater grown algal biomass, Sci. Total Environ. 726 (2020) 137961, https://doi.org/10.1016/j.scitotenv.2020.137961.

[97] M. Gou, X. Yang, Y. Zhao, X. Ran, Y. Song, C.J. Liu, Cytochrome $b_5$ is an obligate electron shuttle protein for syringyl lignin biosynthesis in Arabidopsis, Plant Cell 31 (2019) 1344–1366, https://doi.org/10.1105/tpc.18.00778.
[98] Y. Zhang, S. Lin, Y. Zhou, J. Wen, X. Kang, X. Han, C. Liu, W. Yin, X. Xia, PdNF-YB21 positively regulated root lignin structure in poplar, Ind. Crop Prod. 168 (2021) 113609, https://doi.org/10.1016/j.indcrop.2021.113609.
[99] M. Parit, Z. Jiang, Towards lignin derived thermoplastic polymers, Int. J. Biol. Macromol. 165 (2020) 3180–3197, https://doi.org/10.1016/j.ijbiomac.2020.09.173.
[100] S. Raza, J. Zhang, I. Ali, X. Li, C. Liu, Recent trends in the development of biomass-based polymers from renewable resources and their environmental applications, J. Taiwan Inst. Chem. Eng. 115 (2020) 293–303, https://doi.org/10.1016/j.jtice.2020.10.013.
[101] A. Bhargava, S. Srivastava, Human civilization and agriculture, in: Participatory Plant Breeding: Concept and Applications, Springer, Singapore, 2019, pp. 1–27, https://doi.org/10.1007/978-981-13-7119-6_1.
[102] J. Reganold, J. Wachter, Organic agriculture in the twenty-first century, Nat. Plants 2 (2016) 15221, https://doi.org/10.1038/nplants.2015.221.
[103] N. Kumar, T. Fatima, I. Mishra, M. Verma, J. Mishra, V. Mishra, Environmental sustainability: challenges and viable solutions, Environ. Sustain. 1 (2018) 309–340, https://doi.org/10.1007/s42398-018-00038-w.
[104] M. Palencia, Liquid-phase polymer-based retention: theory, modeling, and application for the removal of pollutant inorganic ions, J. Chem. 965624 (2015), https://doi.org/10.1155/2015/965624.
[105] D.A. Vallero, T.M. Letcher, Regulation of wastes, in: D.A. Vallero, T.M. Letcher (Eds.), Wastes: A Handbook for Management, ScienceDirect, 2011, pp. 23–59, https://doi.org/10.1016/B978-0-12-381475-3.10003-8.
[106] S. Durpekova, K. Filatova, J. Cisar, A. Ronzova, E. Kutalkova, V. Sedlarik, A novel hydrogel based on renewable materials for agricultural application, Int. J. Polym. Sci. 8363418 (2020), https://doi.org/10.1155/2020/8363418.
[107] D. Sopeña, C. Maqueda, E. Morillo, Controlled release formulations of herbicides based on micro-encapsulation, Cienc. Investig. Agrar 35 (2009) 27–42.
[108] M. Nuruzzaman, M.M. Rahman, Y. Liu, R. Naidu, Nanoencapsulation, nano-guard for pesticides: a new window for safe application, J. Agric. Food Chem. 64 (2016) 1447–1483, https://doi.org/10.1021/acs.jafc.5b05214.
[109] M. Palencia, A. Córdoba, M. Vera, Membrane technology and chemistry, in: P.M. Visakh, O. Nazarenko (Eds.), Nanostructured Polymer Membranes, vol. 1, Wiley, 2016, pp. 27–54.
[110] M. Palencia, D.F. Restrepo, E. Combatt, Functional polymer from high molecular weight linear polyols and polyurethane-based crosslinking units: synthesis, characterization, and boron retention properties, J. Appl. Polym. Sci. 133 (2016) 43895–43905, https://doi.org/10.1002/app.43895.
[111] V. Garcés, M. Palencia, E. Combatt, Development of bacterial inoculums based on biodegradable hydrogels for agricultural applications, J. Sci. Technol. Appl. 3 (2017) 5–14,-https://doi.org/10.34294/j.jsta.17.2.11.
[112] T. Lerma, V. Garces, M. Palencia, Novel multi-and bio-functional hybrid polymer hydrogels based on bentonite-poly (acrylic acid) composites and sorbitol polyesters: structural and functional characterization, Eur. Polym. J. 128 (2020) 109627, https://doi.org/10.1016/j.eurpolymj.2020.109627.

[113] T. Pereira, V.D. Dias, R. Faez, Multilayer films of carboxymethylcellulose/zeolite as smart materials for macro and micronutrients delivery, Microporous Mesoporous Mater. 302 (2020) 110195, https://doi.org/10.1016/j.micromeso.2020.110195.
[114] A. Dash, S. Maity, S. Pati, S. Mohapatra, D.P. Samantaray, Biopolymer (polyhydroxyalkanoates)production by bacteria and its application in seed coating for sustainable agriculture, in: Conference: International Conference On Organic Farming for Sustainable Agriculture, 2017, https://doi.org/10.13140/RG.2.2.36671.82085.
[115] T. Lerma, E. Combatt, M. Palencia, Soil-mimicking hybrid composites based on clay, polymers and nitrogen-fixing bacteria for the development of remediation systems of degraded soil, J. Sci. Technol. Appl. 4 (2018) 17–27, https://doi.org/10.34294/j.jsta.18.4.27.
[116] T. Lerma, M. Palencia, E.M. Combatt, Soil polymer conditioner based on montmorillonite-poly(acrylic acid) composites, J. Appl. Polym. Sci. 135 (2018), https://doi.org/10.1002/app.46211.
[117] J. Hu, Y. Guan, The application of high polymer materials in the aspects of seed technology, in: V. Thakur, M.V.S. Thakur (Eds.), Polymeric Gels. Gels Horizons From Sci. To smart mater, Springer, Singapore, 2018, pp. 55–69.
[118] B.V. Farias, T. Pirzada, R. Mathew, T.L. Sit, C. Opperman, S.A. Khan, Electrospun polymer nanofibers as seed coatings for crop protection, ACS Sustain. Chem. Eng. 7 (2019) 19848–19856, https://doi.org/10.1021/acssuschemeng.9b05200. 19848–56.
[119] V. Pathak, R.P.K. Ambrose, Starch-based biodegradable hydrogel as seed coating for corn to improve early growth under water shortage, J. Appl. Polym. Sci. 48523 (2019) 1–12, https://doi.org/10.1002/app.48523.
[120] A.E.S. Vercelheze, B.M. Marim, A.L.M. Oliveira, S. Mali, Development of biodegradable coatings for maize seeds and their application for *Azospirillum brasilense* immobilization, Appl. Microbiol. Biotechnol. 103 (2019) 2193–2203, https://doi.org/10.1007/s00253-019-09646-w.
[121] T. Pirzada, B.V. de Francis, R. Mathew, R.H. Guenther, M.V. Byrd, et al., Recent advances in biodegradable matrices for active ingredient release in crop protection: towards attaining sustainability in agriculture, Curr. Opin. Colloid Interface Sci. 48 (2020) 121–136, https://doi.org/10.1016/j.cocis.2020.05.002.
[122] J. Huang, S.N. Lyman, J. Stamenkovic, M. Sexauer, A review of passive sampling systems for ambient air mercury measurements, Environ. Sci.: Processes Impacts 16 (2013) 374–392, https://doi.org/10.1039/c3em00501a.
[123] J.M. Martínez, M.I. Paez, Evaluation of the QuEChERS method with GC-MS detection for the determination of pesticides into white grain corn, J. Sci. Technol. Appl. 1 (2016) 15–29, https://doi.org/10.34294/j.jsta.16.1.2.
[124] V. Garcés, T.A. Lerma, M. Palencia, E. Combatt, Building a probe-type passive sampler to study the contents of heavy metals in the aqueous phase in the soil, J. Sci. Technol. Appl. 4 (2018) 4–16, https://doi.org/10.34294/j.jsta.18.4.26.
[125] M. Palencia, T. Lerma, M. Córdoba, Polyurethanes with boron retention properties for the development of agricultural fertilization smart systems, J. Sci. Technol. Appl. 1 (2016) 39–52, https://doi.org/10.34294/j.jsta.16.1.4.
[126] Y. Huang, X. Feng, Polymer-enhanced ultrafiltration: fundamentals, applications and recent developments, J. Membr. Sci. 586 (2019) 53–83, https://doi.org/10.1016/j.memsci.2019.05.037.
[127] B. Doshi, M. Sillanpaa, S. Kalliola, A review of bio-based materials for oil spill treatment, Water Res. 135 (2018) 262–277, https://doi.org/10.1016/j.watres.2018.02.034.
[128] S.S. Dhaliwal, R.K. Naresh, A. Mandal, R. Singh, M.K. Shaliwal, Dynamics and transformations of micronutrients in agricultural soils as influenced by organic matter build-

up: a review, Environ. Sustain. Indic. 1 (2019) 100007, https://doi.org/10.1016/j.indic.2019.100007.

[129] R. Jacoby, M. Peukert, A. Succurro, A. Koprivova, S. Kopriva, The role of soil microorganisms in plant mineral nutrition—current knowledge and future directions, Front. Plant Sci. 8 (2017) 1617, https://doi.org/10.3389/fpls.2017.01617.

[130] A. Cartwright, K. Jackson, C. Morgan, A. Anderson, D.W. Britt, A review of metal and metal-oxide nanoparticle coating technologies to inhibit agglomeration and increase bioactivity for agricultural applications, Agronomy 10 (2020) 1018, https://doi.org/10.3390/agronomy10071018.

[131] A. Rastogi, M. Zivcak, O. Sytar, H. Kalaji, X. He, S. Mbarki, M. Brestic, Impact of metal and metal oxide nanoparticles on plant: a critical review, Front. Chem. 5 (2017) 78, https://doi.org/10.3389/fchem.2017.00078.

[132] R.P. Sigh, K. Sharma, K. Mausam, Dispersion and stability of metal oxide nanoparticles in aqueous suspension: a review, Mater. Today Proc. 26 (2020) 2021–2025, https://doi.org/10.1016/j.matpr.2020.02.439.

[133] C.P. Tso, C.M. Zhung, Y.H. Shih, Y.M. Tseng, S.C. Wu, R.A. Doong, Stability of metal oxide nanoparticles in aqueous solutions, Water Sci. Technol. 61 (2010) 127–133,-https://doi.org/10.2166/wst.2010.787.

[134] Z. Wang, X. Xie, J. Zhao, X. Liu, W. Feng, J.C. White, B. Xing, Xylem- and phloem-based transport of CuO nanoparticles in maize (*Zea mays* L.), Environ. Sci. Technol. 46 (2012) 4434–4441, https://doi.org/10.1021/es204212z.

[135] W. Du, W. Tan, J. Peralta-Videa, J. Galdea-Torresdey, R. Ji, et al., Interaction of metal oxide nanoparticles with higher terrestrial plants: physiological and biochemical aspects, Plant Physiol. Biochem. 110 (2016), https://doi.org/10.1016/j.plaphy.2016.04.024.

[136] M. Simonin, A. Cantarel, A. Crouzet, J. Garvaiz, J. Martins, A. Richaume, Negative effects of copper oxide nanoparticles on carbon and nitrogen cycle microbial activities in contrasting agricultural soils and in presence of plants, 13 (2018), https://doi.org/10.3389/fmicb.2018.03102.

[137] T.T. Xiong, C. Dumat, V. Dappe, H. Vezin, E. Schreck, M. Shahid, A. Piearart, S. Sobanska, Copper oxide nanoparticle foliar uptake, phytotoxicity, and consequences for sustainable urban agriculture, Environ. Sci. Technol. 51 (2017) 5242–5251, https://doi.org/10.1021/acs.est.6b05546.

[138] K. Jeyasubramanian, U. Thoppey, G.S. Hikku, N. Selvakumar, A. Subramania, K. Krishnamoorthy, Enhancement in growth rate and productivity of spinach grown in hydroponics with iron oxide nanoparticles, RSC Adv. 6 (2016) 15451–15459,-https://doi.org/10.1039/C5RA23425E.

[139] A. Mukherjee, J.R. Peralta-Videa, S. Bandyopadhyay, C.M. Rico, L. Zhao, J.L. Gardea-Torresdey, Physiological effects of nanoparticulate ZnO in green peas (*Pisum sativum* L.) cultivated in soil, Matallomics 6 (2014) 132–138, https://doi.org/10.1039/c3mt00064h.

[140] L.J. Zhao, J.A. Hernandez-Viezcas, J.R. Peralta-Videa, S. Bandyopadhyay, B. Peng, B. Muñoz, A.A. Keller, J.L. Gardea-Torresdey, ZnO nanoparticles fate in soil and zinc bioaccumulation in corn plants (*Zea mays*) influenced by alginate, Environ. Sci.: Processes Impacts 15 (2013) 260–266, https://doi.org/10.1039/C2EM30610G.

[141] A. García-Quintero, M. Palencia, Interactions among biobased iron nanomaterials and soil abiotic components—a review, J. Sci. Technol. Appl. 10 (2021) 4–26, https://doi.org/10.34294/j.jsta.21.10.65.

[142] R. Hidalgo, L. Almeri, F. Miyazaki, M. Pilecco, S. Oliveira, D. Bigaton, Foliar copper uptake by maize plants: effects on growth and yield, Cienc. Rural 43 (2013), https://doi.org/10.1590/S0103-84782013000900005.

[143] M.A. Neves-Dias, S.M. Cicero, A.D.L. Coelho, Uptake of seed-applied copper by maize and the effects on seed vigor, Bragantia 74 (2015), https://doi.org/10.1590/1678-4499.0044.
[144] A. Garcia, E. Combatt, M. Palencia, Structural study of humin and its interaction with humic acids by Fourier-transform mid-infrared spectroscopy, J. Sci. Technol. Appl. 4 (2018) 28–39, https://doi.org/10.34294/j.jsta.18.4.28.
[145] D. Restrepo, M. Palencia, V. Palencia, Study by attenuated total reflectance spectroscopy of structural changes of humified organic matter by chemical perturbations via alkaline dissolution, J. Sci. Technol. Appl. 4 (2018) 49–59, https://doi.org/10.34294/j.jsta.18.4.30.
[146] P. Pham, M. Rashid, Y. Cai, M. Yoshinaga, D.D. Dionysiou, K. O'Shea, Removal of As(III) from water using the adsorptive and photocatalytic properties of humic acid-coated magnetite nanoparticles, Nanomaterials 10 (2020) 1604, https://doi.org/10.3390/nano10081604.
[147] Y. Zhang, J.B. Fein, Y. Li, Q. Yu, B. Zu, C. Zheng, U(VI) adsorption to Fe3O4 nanoparticles coated with lignite humic acid: experimental measurements and surface complexation modeling, Colloids Surf. A Physicochem. Eng. Asp. 614 (2021) 126150, https://doi.org/10.1016/j.colsurfa.2021.126150.
[148] M. Berrio, S. Palencia, T. Lerma, M. Mora, Bacterial colonization modelling on soil particles: effect of humic acids on the formation of nitrogen-fixing bacteria biofilms, J. Sci. Technol. Appl. 5 (2018) 33–44, https://doi.org/10.34294/j.jsta.18.5.33.
[149] M. Mora, M. Palencia, E. Combatt, Effect of dissolved organic matter and humic substances on transport of ions and low molecular weight molecules by liquid membranes, J. Sci. Technol. Appl. 4 (2018) 40–48, https://doi.org/10.34294/j.jsta.18.4.29.
[150] G.U. Badranova, P.M. Gotovtsev, Y.V. Zubavichus, I.A. Staroselsky, A.L. Vasiliev, I.N. Trunkin, M.V. Fedorov, Biopolymer-based hydrogels for encapsulation of photocatalytic $TiO_2$ nanoparticles prepared by the freezing/thawing method, J. Mol. Liq. 223 (2016) 16–20, https://doi.org/10.1016/j.molliq.2016.07.135.
[151] W. Gao, Y. Zhang, Q. Zhang, L. Zhang, Nanoparticle-hydrogel: a hybrid biomaterial system for localized drug delivery, Ann. Biomed. Eng. 44 (2016) 2049–2061, https://doi.org/10.1007/s10439-016-1583-9.
[152] F. Wahid, C. Zhong, H.S. Wang, X.H. Hu, L.Q. Chu, Recent advances in antimicrobial hydrogels containing metal ions and metals/metal oxide nanoparticles, Polymers 9 (2017) 636, https://doi.org/10.3390/polym9120636.
[153] A. Abdulmalik, N.A.M. Sani, A. Mohamed, A.R.M. Sam, J. Usman, N.H.A. Khalid, Characterization of marine clay under microstructure examination as a potential Pozzolana, J. Comput. Theor. Nanosci. 17 (2020) 1026–1031, https://doi.org/10.1166/jctn.2020.8761.
[154] S. Iftikhar, K. Rashid, E.U. Haq, I. Zafar, F.K. Alqahtani, M.I. Khan, Synthesis and characterization of sustainable geopolymer green clay bricks: an alternative to burnt clay brick, Construct. Build Mater. 259 (2020) 119659, https://doi.org/10.1016/j.conbuildmat.2020.119659.
[155] T.A. Dontsova, S. Nahirniak, I.M. Astrelin, Metaloxide nanomaterials and nanocomposites of ecological purpose, J. Nanomater. 5942194 (2019), https://doi.org/10.1155/2019/5942194.
[156] N.A. Yusoff, S.A. Ong, L.N. Ho, Y.S. Wong, W.F. Khalik, Degradation of phenol through solar-photocatalytic treatment by zinc oxide in aqueous solution, Desalin. Water Treat. 54 (2014) 1–8, https://doi.org/10.1080/19443994.2014.908414.

[157] M. Rahim, M.R. Hakim, Application of biopolymer composites in arsenic removal from aqueous medium: a review, J. Radiat. Res. Appl. Sci. 8 (2015) 255–263, https://doi.org/10.1016/j.jrras.2015.03.001.
[158] Y. Nam, J.H. Lim, K.C. Ko, J.Y. Lee, Photocatalytic activity of TiO2 nanoparticles: a theoretical aspect, J. Mater. Chem. A 7 (2019) 13833–13859, https://doi.org/10.1039/C9TA03385H.
[159] L.M. Anaya-Esparza, J.M. Ruvalcaba-Gómez, C.I. Maytorena-Verdugo, N. Gonzáles-Silva, R. Romero-Toledo, S. Aguilera-Aguirre, A. Pérez-Larios, E. Montalvo-Gonzáles, Chitosan-TiO2: a versatile hybrid composite, Materials 13 (2020) 811, https://doi.org/10.3390/ma13040811.
[160] N.F. Braga, A.P. Da Silva, T.M. Arantes, A.P. Lemes, F.H. Cristovan, Physical–chemical properties of nanocomposites based on poly (3-hydroxybutyrate-co-3-hydroxyvalerate) and titanium dioxide nanoparticles, Mater. Res. Express. 5 (2018), https://doi.org/10.1088/2053-1591/aa9f7a, 015303.
[161] M. Awang, W.R.W. Mohd, N. Sarifuddin, Study the effects of an addition of titanium dioxide (TiO2) on the mechanical and thermal properties of polypropylene-rice husk green composites, Mater. Res. Express. 6 (2019), https://doi.org/10.1088/2053-1591/ab1173, 075311.
[162] S. Jadoun, U. Riaz, J. Yáñez, N.P.S. Chauhan, Synthesis, characterization and potential applications of poly(o-phenylenediamine) based copolymers and nanocomposites: a comprehensive review, Eur. Polym. J. 110600 (2021), https://doi.org/10.1016/j.eurpolymj.2021.110600.
[163] N.K. Sethy, Z. Arif, P.K. Mishra, P. Kumar, Nanocomposite film with green synthesized TiO2 nanoparticles and hydrophobic polydimethylsiloxane polymer: synthesis, characterization, and antibacterial test, J. Polym. Eng. 40 (2020) 211–220, https://doi.org/10.1515/polyeng-2019-0257.
[164] N. Wang, J. Li, W. Lv, J. Feng, W. Yan, Synthesis of polyaniline/$TiO_2$ composite with excellent adsorption performance on acid red G, RSC Adv. 5 (2015) 21132–21141,-https://doi.org/10.1039/C4RA16910G.
[165] M. Sangareswari, M.M. Sundaram, Development of efficiency improved polymer-modified $TiO_2$ for the photocatalytic degradation of an organic dye from wastewater environment, Appl Water Sci 7 (2017) 1781–1790, https://doi.org/10.1007/s13201-015-0351-6.
[166] S. Sarkar, N.T. Ponce, A. Banerjee, R. Bandopadhyay, S. Rajendran, E. Lichtfouse, Green polymeric nanomaterials for the photocatalytic degradation of dyes: a review, Environ. Chem. Lett. 18 (2020) 1569–1580, https://doi.org/10.1007/s10311-020-01021-w.
[167] X. Zhao, L. Lv, B. Pan, W. Zhang, S. Zhang, Q. Zhang, Polymer-supported nanocomposites for environmental application: a review, Chem. Eng. J. 170 (2011) 381–394, https://doi.org/10.1016/j.cej.2011.02.071.
[168] X.J. Guo, F.H. Chen, Removal of arsenic by bead cellulose loaded with iron oxyhydroxide from groundwater, Environ. Sci. Technol. 39 (2005) 6808–6818,-https://doi.org/10.1021/es048080k.
[169] A.I. Zouboulis, I.A. Katsoyiannis, Arsenic removal using iron oxide loaded alginate beads, Ind. Eng. Chem. Res. 41 (2002) 6149–6155, https://doi.org/10.1021/ie0203835.

# Further reading

X. Chen, A. Pizzi, E. Fredon, C. Gerardin, J. Li, X. Zhou, G. Du, Preparation and properties of a novel type of tannin-based wood adhesive, J. Adhes. (2020) 1–18, https://doi.org/10.1080/00218464.2020.1863215.

S.J. Eichhorn, A. Gandini, Materials from renewable resources, MRS Bull. 35 (2010) 187–193,-https://doi.org/10.1557/mrs2010.650.

A. Jha, A. Kumar, Biobased technologies for the efficient extraction of biopolymers from waste biomass, Bioprocess Biosyst. Eng. 42 (2019) 1893–1901, https://doi.org/10.1007/s00449-019-02199-2.

M. Palencia, Unused useful waste paradox, J. Sci. Technol. Appl. (2020) 1–4, https://doi.org/10.34294/j.jsta.20.9.0 (in press).

M. Podzora, Y. Tertyshnaya, A. Popov, Eco-friendly polymer materials for agricultural purposes, MATEC Web Conf. 298 (2019), https://doi.org/10.1051/matecconf/201929800130, 00130.

D.Z. Ye, L. Jiang, C. Ma, M.H. Zhang, X. Zhang, The graft polymers from different species of lignin and acrylic acid: synthesis and mechanism study, Int. J. Biol. Macromol. 63 (2014) 43–48, https://doi.org/10.1016/j.ijbiomac.2013.09.024.

# Polysaccharides-metal oxide composite: A green functional material

12

*Nasrullah Shah[a,b], Wajid Ali Khan[b], Touseef Rehan[c], Dong Lin[a], Halil Tetik[a], and Sajjad Haider[d]*
[a]Department of Industrial and Manufacturing Systems Engineering, Kansas State University, Manhattan, KS, United States, [b]Department of Chemistry, Abdul Wali Khan University Mardan, Mardan, KP, Pakistan, [c]Department of Biochemistry, Shaheed Benazir Bhutto Women University, Peshawar, KP, Pakistan, [d]Department of Chemical Engineering, College of Engineering, King Saud University, Riyadh, Saudi Arabia

## 1 Introduction

A composite is composed of two or more than two constituents with significantly different natures. The properties of a composite material are different from their constituents. The main aim of composite synthesis is to obtain materials with improved properties and performance compared to the individual components [1]. Wood industries have been synthesizing composites for a century and every year, new composite materials are produced. The composite components range from meters to nanosized [2]. One of the most important composite materials is polysaccharides based, which has drawn the attention of researchers due to their interesting properties. Polysaccharides replaced synthetic polymers because of easy availability, low cost, environment friendly nature, renewability, biodegradability, and easy chemical modification [3,4]. Moreover, the polysaccharide component is mostly obtained from natural sources like plants, animals, algae, and microorganisms [5]. The demand of metal oxides also increased in different research areas like adsorption, optoelectronics, catalysis, sensors formation, drug delivery system, and magnetic composites [6–12]. Polysaccharides-based magnetic composites are one of the important classes of composites. The interaction between metal oxide and polysaccharides is not common in nature [13]. But polysaccharides-metal oxide composites have potential applications in medical, environmental, and agricultural fields due to properties such as biocompatibility, biodegradability, and eco-friendly nature [14]. In this chapter, we will discuss different types of polysaccharides-metal oxide composites and their application in different fields.

**Renewable Polymers and Polymer-Metal Oxide Composites. https://doi.org/10.1016/B978-0-323-85155-8.00013-3**

# 2 Cellulose-metal oxide composites

Cellulose was discovered by French chemist Anselme Payen in 1838. At that time, it was used only as an energy source. The natural cellulose is widely obtained from cotton, flax, jute, and hemp [15]. Green polymers replaced the synthetic polymers due to the biocompatibility and eco-friendly nature. These limit the nonbiodegradable part to a large extent and thus reduce the pollution [16]. Cellulose can also be obtained from the cell wall of plants and produced by microorganisms [1,17].

## *2.1 Bacterial cellulose-metal oxide composites*

Various types of cellulose are used in metal oxide composites. One of them is bacterial cellulose (BC), which is produced by microorganisms like algae, fungi, and bacteria. BC is produced by microorganisms in the form of a nanosized network by the fermentation process [1,18]. Best BC is produced by the fermentation process in the temperature range of 25–30°C and at pH of3–7. BC shows biodegradability, solubility in water, and chemical modification abilities due to the presence of a greater number of -OH groups. Moreover, BC has unique properties due to microbial synthesis processes like nanosized structure, high crystallinity, more purity, high water-absorbing capacity, high polymerization capability, and high mechanical strength. These advantages make BC a preferred component in composite synthesis [19,20]. The use of metal oxide as a constituent in the BC composite plays a key role and enhances the efficiency of the composite material. For example, $TiO_2$ can be used for the treatment of water because of antibacterial activity against Gram-negative and Gram-positive bacteria [21]. Khalid et al. synthesized BC-$ZnO_2$ nanocomposites to determine their antimicrobial capacity. Zinc oxide nanoparticles (NPs) were used in the nanocomposite due to the lack of antimicrobial activity of BC. The prepared nanocomposites were used against burn pathogens. The synthesized nanocomposites show outstanding performance against various bacteria and are used as novel dressing for burns (Fig. 1) [22]. BC-metal oxide composites are reported for the applications in energy [23], water treatment [24], burn wounds dressing [22], dyes degradation [25], antibacterial activity [26], bioadsorbents for heavy metals [27], and cancer chemotherapy [28].

## *2.2 Cellulose nanocrystals and cellulose nanofibers metal oxide composites*

Cellulose nanocrystals (CNCs) are the nanosized cellulose that are obtained from plants. The only difference between this nanosized cellulose and BC is the preparation mode. CNCs and cellulose nanofibrils (CNFs) are produced through the chemical or mechanical procedure using the top-down method, while BC is produced by microorganisms via a bottom-up approach [17,29]. The primary cell wall produced longer and thinner CNFs than the secondary cell wall. The properties of CNFs such as level of crystallinity, fibrillation intensity, and morphology depend on the method of preparation. The most widely extraction technique applied for CNFs is TEMPO oxidation [30].

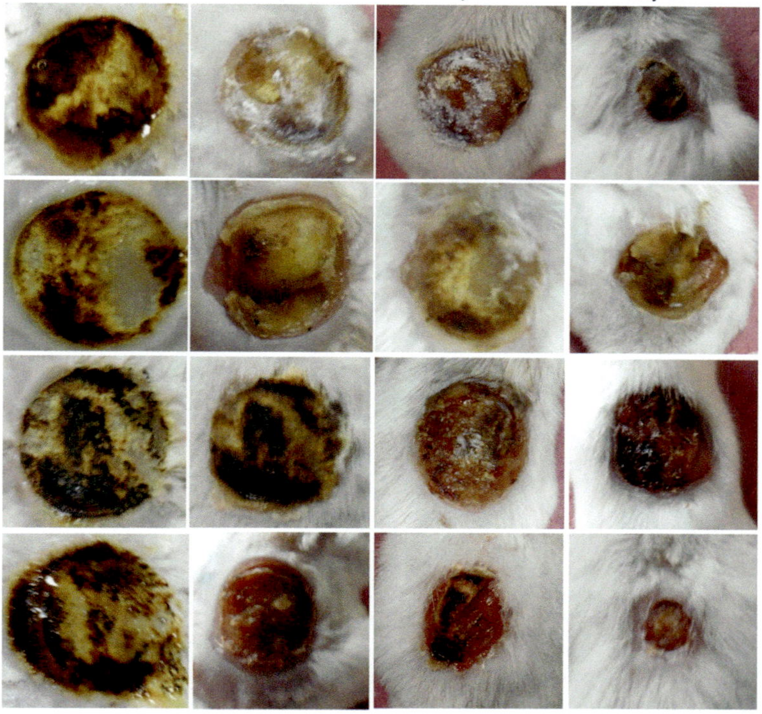


**Fig. 1** Representative wound photographs of BC-ZnO nanocomposites treated group in comparison with negative control, BC, and positive control group on different treatment days. Reprinted with permission from reference A. Khalid, et al., Bacterial cellulose-zinc oxide nanocomposites as a novel dressing system for burn wounds, Carbohydr. Polym. 164 (2017) 214–221.

Various natural sources are reported for the preparation and extraction of CNFs. CNCs are obtained from cellulose with the use of the biosynthesis process. CNCs are much more rigid than CNFs due to high crystallinity (58%–88%), which was obtained during acid hydrolysis. The thermal stability and density values are approximately 260°C and 1.5–1.6 $g\,cm^{-3}$, respectively. CNCs are also reported with various names such as nanorods, nanowhiskers, and rod crystals of cellulose [31]. Since metal oxide has photocatalytic, antimicrobial as well as UV barrier properties, hybrid materials of these nanocellulose and metal oxide attract the attention of researchers from different fields for the applications of UV absorption [32], adsorption [33,34], sensors formation [35], antimicrobial activities [36,37], food packaging [38], skin care items [39], and photocatalysis [37,40].

There are other several cellulose-based metal oxide composites reported in the literature. In one example, Khatri et al. synthesize cellulose fiber sheets modified with saccharide capped-ZnO NPs. The modified sheets were used for the immobilization of

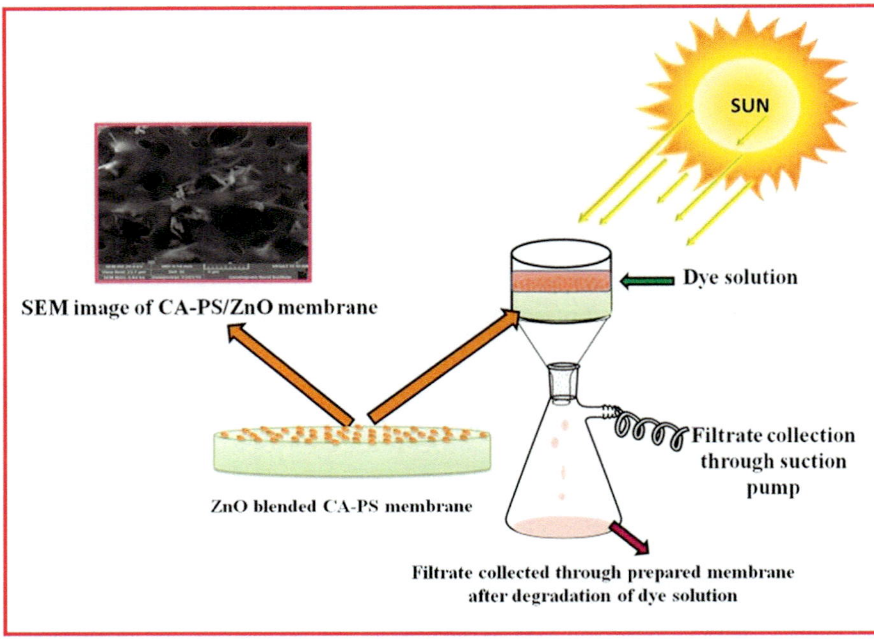


**Fig. 2** Photodegradation process of dyes.
Reprinted with permission from reference A. Rajeswari, E.J.S. Christy, A. Pius, New insight of hybrid membrane to degrade Congo red and Reactive yellow under sunlight, J. Photochem. Photobiol. B Biol. 179 (2018) 7–17.

antibodies [41]. The composites of cellulose fibers with $TiO_2$ were also reported as the dye-sensitized material for solar cells [42], UV filters [43], adsorbent materials [44], and for evaluating the effect of humidity on conductivity [45]. Cellulose can be processed as a film with metal oxide NPs. Rajeswari et al. synthesized the membrane of cellulose acetate-polystyrene with and without ZnO. The membranes were used for the degradation of pollutants (Congo red and Reactive yellow 105) in the presence of sunlight. The efficiency of the hybrid membrane was higher than 90% for the degradation of the target dyes. Similarly, the efficiency of the modified membrane was better in terms of permeability and antifouling. The schematic photodegradation process of the developed method is shown in Fig. 2 [46]. These composites materials have applications in medicine [47], fabrics modification [48], antibacterial activities [49], sensor formation [50,51], and photocatalytic activities [52].

# 3 Chitin/chitosan-metal oxide composites

Chitin is the linear and second most abundant biopolymer of polysaccharides on the earth after cellulose. Chitin is found in the arthropods skeletal system, yeasts, fungi, green algae, bacteria, and mushroom envelops [53–56]. Chitin shows excellent adsorption capability toward heavy metals and dyes due to the presence of oxygen and nitrogen in the hydroxyl and amine groups, respectively. These atoms act as reaction or chelating sites [57,58]. Chitin and its derivatives have application in the field of the food industry due to excellent properties such as easy degradation, solubility in weak acids, sensitivity to pH, film producing property, biocompatibility, nontoxicity, and nonantigenic properties [59–61]. Chitosan is derived from chitin by the deacetylation of chitin. The presence of multi-functional groups and the antibacterial, antifungal, antiviral, antioxidant, antihypertensive, anticoagulant, antiallergic, antiinflammatory, anticancer, antihemostatic, and mucoadhesive properties makes chitosan a potential candidate for application in biomedical field [62]. The transition metal oxide NPs shows excellent surface properties, microstructural properties, and large surface area, which attract much more attention for environmental researchers. The active sites and large surface area increased the adsorption capability. Thus, metal oxide NPs show much more efficiency than bulk materials for the adsorption of different pollutants [63]. The application of metal oxide NPs is not limited to the adsorption of pollutants. Literature reported numerous applications for metal oxide and its chitin/chitosan composites. In one example, Vaseeharan et al. synthesized chitosan-ZnO composites. The chitosan was extracted from mud crab Scylla serrata shells. The physiochemical characteristics of the synthesized composite were studied by different instrumental techniques. The developed composites show excellent efficiency against aquatic bacteria [64]. Chitin-metal oxide composites applications were reported for the adsorption of pollutants [58,63,65–68], food preservation [62], antibacterial activity [64], and sensor development [69].

# 4 Alginate-metal oxide composites

Alginate is a class of linear polysaccharides composed of β-D-mannuronic acid (M-units) and α-L-guluronic acid (G-units). Alginate is present in both forms. It may be formed from homo-polyuronate blocks (MM or GG) or alternating blocks (MG). Natural sources of alginate are brown algae and some species of bacteria [70,71]. Alginates have unique features such as biocompatibility, biological activity, low cost, biodegradability, water-solubility, excellent binding capability, easy production, and renewability [72]. There are a lot of applications of water-soluble alginate that is obtained from macroalgae. The reason behind it, is the rheological properties such as gelling, viscosifying, and stabilization of dispersions [73]. Alginates have a wide range of applications in medicine, food, textile, agriculture, and waste-water treatment [74]. The unique features of alginate have drawn much more attention, and hence, alginate-metal oxide composites are reported in the literature for different applications. In one example, Satriaji et al. synthesize bio-nanocomposite films of alginate and ZnO NPs. In this study, calcium sulfate was used as a crosslinker. The advantages like increase in thickness and tensile strength are obtained with the use of a crosslinker. The constituent ZnO NPs boost the mechanical and UV barrier properties of the synthesized bio-nanocomposite films. The films were also found effective against bacteria [75]. Mahmoud et al. prepared alginate-based nanocomposites for the

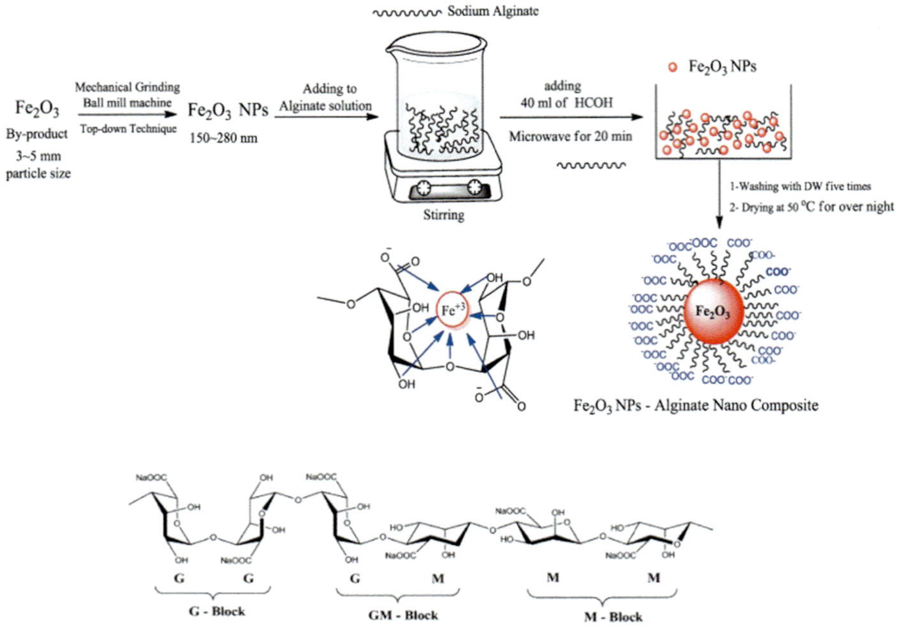


**Fig. 3** Preparation process of $Fe_2O_3$NPs-Alginate nanocomposites.
Reprinted with permission from reference M.E. Mahmoud, et al., A sustainable nanocomposite for removal of heavy metals from water based on crosslinked sodium alginate with iron oxide waste material from steel industry, J. Environ. Chem. Eng. 8 (4) (2020) 104015.

adsorption of heavy metals from water. In their study, they synthesized iron oxide NPs and cross-linked them with sodium alginate (Fig. 3). The characteristic properties of the synthesized nanocomposite were studied via FTIR, SEM, TEM, TGA, and XRD. The prepared nanocomposite was effectively used for the adsorption of heavy metals from waste-water [76]. Alginate-metal oxide composites were reported in the literature for a variety of applications. These include catalytic reduction of organic dyes [77], adsorption of pollutants [76,78–80], tissue engineering [81], photocatalytic degradation [82], antibacterial activities [75,83], food packaging [84], and coating [85].

# 5 Lignocellulosic-metal oxide composites

Lignocellulose is composed of cellulose and lignin. Lignocellulose is a plant-based material, obtained from different parts of plants. Lignocellulose replaced the synthetic fiber due to its wide range of advantages such as easy availability, environment friendly, biodegradability, low cost, and high stiffness. As we have already mentioned, these are plant-based materials; therefore, these can also be obtained from wood crops and plants [86]. The concentration of lignocellulose varies from species to species. The role of lignocellulose in plants is similar to bone. Lignocellulose has pores that help in the nutrient's mobilization. The hierarchical structure of lignocellulose is from nano to macro scale. Lignocellulose is present in plants like a porous network, which supports the exchange and mobilization of nutrients [87]. Based on plant parts, lignocellulose can be obtained from the bast, or bark, leaves, or bamboo or different straws. Different methods are used for their extraction which may be manual, machine-based, or microbial [88]. The use of lignocellulose in composite formation can open a new window for a wide range of applications. However, the hydrophilic nature, low mechanical strength, and sensitivity to thermal degradation of lignocellulose pose problems [86]. These limitations can be overcome with composite formation using metal oxide. Therefore, lignocellulose-metal oxide composites are prepared for various purposes. Djellabi et al. fabricated $TiO_2$/lignocellulosic biomass composites for the photo reduction of Cr (VI). The synthesized composite material shows outstanding efficiency for the photo-reduction of Cr(VI). The study shows that the Ti—O—C bonding bridge between $TiO_2$ clusters and olive pits lignocellulosic surface shows an important role in the visible light response. The developed method is shown in Fig. 4 [89]. The same research group synthesized a water-floating lignocellulosic biomass-$TiO_2$-aerogel for the photo reduction of Cr (VI). The main idea behind this experiment was to overcome the limitation of suspended $TiO_2$ photocatalysis. In this study, two floating forms were designed. The results showed better response and recycling ability [90].

# 6 Starch-metal oxide composites

Starch is one of the most favorable among natural polymers due to its low-cost, high availability, edibility, biodegradability, and biocompatibility. However, due to the hydration property, it swells easily and leads to rupturing, loss of viscosity, and paste production [91]. Similarly, poor mechanical barrier, and processing properties make it

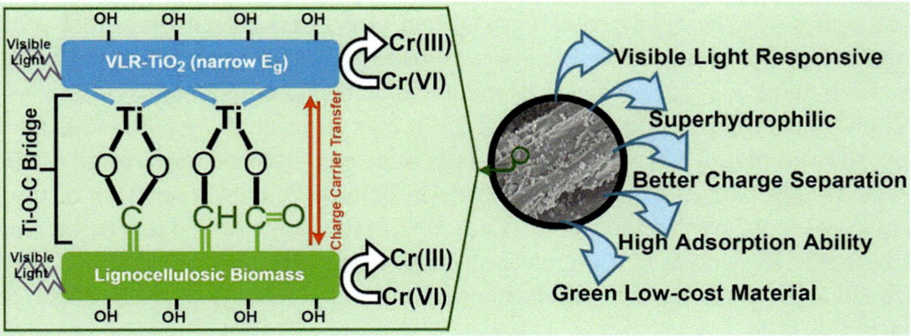


**Fig. 4** Photoreduction of Cr(VI) on $TiO_2$-lignocellulose composites.
Reprinted with permission from reference R. Djellabi, et al., Unravelling the mechanistic role of TiOC bonding bridge at titania/lignocellulosic biomass interface for Cr (VI) photoreduction under visible light, J. Colloid Interface Sci. 553 (2019) 409–417.

not suitable for the packaging industry [92]. Starch NPs, however, have shown some attraction to researchers due to their nano size, larger surface area, better biocompatibility, biological activity, unique biodegradability, and adsorptive properties [93]. Recently, packaging industry paid much more attention to the synthesis of starch-metal oxide nanocomposites as an antimicrobial agent. However, the applications of starch-metal oxide nanocomposites are not limited to the packaging industry. ZnO-chitosan NPs were synthesized and incorporated in a starch-based composite for the anti-microbial study. The results showed that the introduction of ZnO-chitosan NPs reduced water permeability from 51% to 43.7%. Moreover, an increase in tensile strength from 4.11 to 12.79 MPa was recorded. The anti-microbial activity showed that the synthesized films had strongly suppressed Gram-positive *Staphylococcus aureus* compared to Gram-negative *Escherichia coli* [94]. Other applications of starch-metal oxide were reported for photocatalytic activities [95,96], light-soaking stability [97], heavy metals adsorption [98], and food packaging [99]. Our group has recently reported a study of coating natural magnetic particles coated with starch for methyl orange adsorption from aqueous media. The coated magnetic nanoparticles showed excellent results and were easily collected from the adsorption medium with the help of an external magnet [100].

# 7 Agar-metal oxide composites

Agar is a natural polysaccharide biopolymer obtained from the red algae. Agar is environmentally friendly and that's why replaced with petroleum-based plastic packaging materials. The applications of agar are also reported in pharmaceuticals [101,102] and agriculture [103]. Agar biopolymer is used in packaging materials due to its advantages such as renewability, biodegradability, better film-forming ability, easy availability, and biocompatibility. Moreover, agar is also used as an emulsifier, conservative, stabilizer, gelling and film-forming, and coagulant agent in various industries [104]. That's why, the use of agar production increased from 6800 to 9600 tons in between 2002 and 2009 [105,106]. However, the hydrophilic and medium mechanical properties of agar made the final product costly and also interfere with the final product. These limitations of agar-based films/composites can be overcome with the use of nano-sized materials [107]. The use of metal oxide in agar composites synthesis is also common. Hou et al. studied the effect of $TiO_2$ NPs addition in agar-based films. The results showed that the addition of $TiO_2$ NPs improved hydrophobicity and increased the water vapor barrier ability up to 35%. An increase in tensile strength of the film was also observed by the addition of $TiO_2$ NPs in the range of 0–1 g. The increase in NPs beyond 1 g, decreased the tensile strength due to agglomeration of $TiO_2$ NPss, which led to the discontinuity in polymer molecules and reduction in the weak Vander Waal forces [108].

Magesh et al. synthesize pure zinc oxide and Agar/ZnO nanocomposites. The structure, morphology, elemental composition, and optical properties of the synthesized materials were studied. The antibacterial study of agar and pure ZnO (AZ0, AZ1, AZ2, and AZ3) nanocomposites was carried out against Gram-positive (*Bacillus*

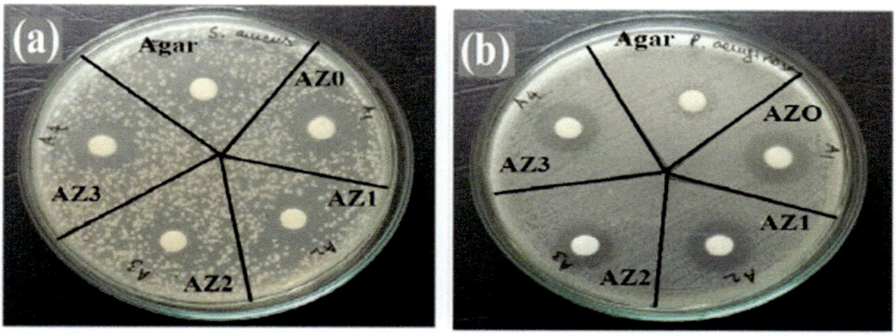


**Fig. 5** (A and B) Antibacterial activity against Gram-positive and Gram-negative bacteria of (A). Agar (B) pure ZnO (AZ0), (C) AZ1, (D) AZ2, and (E) AZ3.
Reprinted with permission from reference G. Magesh, et al., Structural, morphological, optical and biological properties of pure ZnO and agar/zinc oxide nanocomposites, Int. J. Biol. Macromol. 117 (2018) 959–966.

*subtilis*) and Gram-negative (*Pseudomonas aeruginosa*) bacteria (Fig. 5). The Agar-ZnO nanocomposite was also used against normal and breast cancer cell lines. The result demonstrates that Agar-ZnO nanocomposites showed a dose-dependent toxicity in both cell lines [109]. The agar-metal oxide composites are also reported for various applications in a variety of fields [110,111].

# 8 Pectin-metal oxide composites

The polymer used in the synthesis of the composite should be biocompatible and nonhazardous if it is intended for use in a biosystem. Furthermore, the polymer constituent of the composite material should be biodegradable. All these criteria for the composite material were fulfilled by pectin polymers [112]. Pectin polymers are obtained from the cell wall of citrus plants especially lemon, orange, and grapes. Other sources include potatoes, sugar beets, tomatoes, and carrots. Pectin polymer composed of a poly-D-galacturonic acid is bonded by alpha-glycosidic linkages. The presence of the carboxylic acid group in the structure made it more suitable for drug delivery, proteins, and other macromolecules [113]. The hydrophilic and all the other properties of pectin mentioned above make it a preferred candidate for food and other products and a widely used texture modifier, thickener, and coating and gelling agent [114]. However, pectin swells and eroded due to its hydrophilic nature. These limitations limit its application, especially in drug delivery [115]. These drawbacks can be overcome with the use of metal oxide NPs. For example, ZnO and other metal oxide NPs have been reported for drug delivery and tumor imaging applications [116–118]. Hira et al. synthesized pectin-guar gum-ZnO nanocomposite (Figs. 6 and 7) and used it as an immunomodulator for human peripheral blood lymphocytes (PBL) to enhance its cancer cells killing efficiency. Moreover, the use of nanocomposite pretreated human PBL

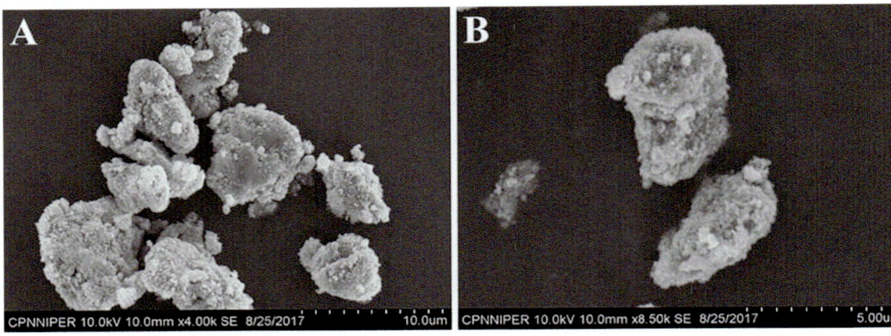


**Fig. 7** SEM images of PEC-GG-ZnO.
Reprinted with permission from reference I. Hira, et al., Pectin-guar gum-zinc oxide nanocomposite enhances human lymphocytes cytotoxicity towards lung and breast carcinomas, Mater. Sci. Eng. C 90 (2018) 494–503.

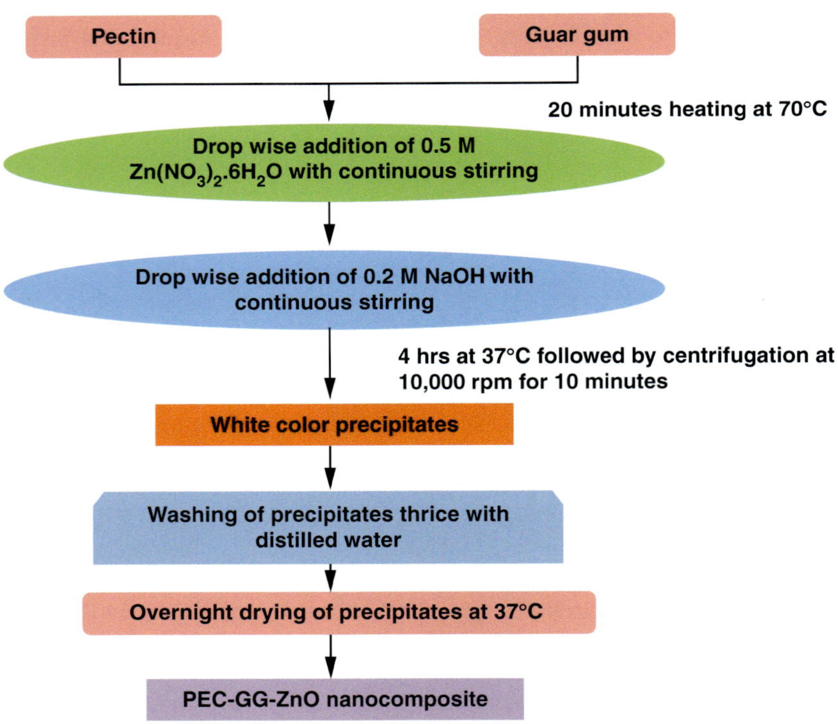


**Fig. 6** Synthesis process of PEC-GG-ZnO nanocomposites.
Reprinted with permission from reference I. Hira, et al., Pectin-guar gum-zinc oxide nanocomposite enhances human lymphocytes cytotoxicity towards lung and breast carcinomas, Mater. Sci. Eng. C 90 (2018) 494–503.

shows a boosting effect of cytotoxicity toward lung and breast carcinoma cells compared to untreated PBL. The results also showed an increase in cancer cell death from 2.5:1 to 20:1[119]. Pectin-metal oxide composite materials are found in literature for the electrochemical senor fabrication [120,121], drug delivery [113], wound healing [122], bioelectrocatalysis of microbial fuel cells [123], and electrocatalytic sensing [112].

# 9 Methods for composite synthesis

Various methods are used for composite synthesis. Electrospinning, film casting, dip coating, physical mixing, layer-by-layer assembly, ionotropic gelation, colloidal assembly, co-precipitation, in situ preparation, and covalent coupling are the most commonly used techniques. The synthesized composites through these techniques have hydrogen bonding and Vander Waals forces [124].

## *9.1 Electrospinning*

enerally, an electrospinning setup consists of a voltage source, a spinneret, and a collection plate (Fig. 8) [125]. A spinneret ejects polymer solution, which is collected on a collector plate as a fiber, when voltage is applied across the spinneret and collector. A voltage of 15–30kV is needed to generate fibers. However this varies from polymer to polymer [126]. Typically, a volatile solvent is used for the preparation of polymer solution. This is important because the solvent will evaporates rapidly during the fiber formation in the wet-dry electrospinning process. Nonvolatile solvents like room temperature ionic liquids are preferred in the wet-wet spinning process [127]. Several factors affect the formation of fiber in the electorspinning process. These include applied potential, the temperature and humidity, the concentration of the polymer solution, the conductivity, the flow rate, and so on. Generally, the process is carried out at room temperature. The material size is in the range of nano to microns [128]. Co-axial electrospinning is a new setup, which simultaneously produces fibers from two different components.

The wet-dry method is normally used for water-soluble polysaccharides such as starch. However, crystalline polysaccharides such as cellulose and chitin do not dissolve easily in common solvent and therefore cannot be used readily . This issue can be overcome with the use of room temperature ionic liquids (RTILs) [127]. RTILs have tremendous properties such as high thermal stability, a wide electrochemical window, and more stability. Furthermore, these solvents can be recycled and are eco-friendly, that's why they are regarded as green solvents. With the use of wet-wet spinning, cellulose fibers can be fabricated in the ethanol or water coagulation bath. The properties of the obtained cellulose fibers can be improved with inorganic metal oxide NPs or organic molecules [124].

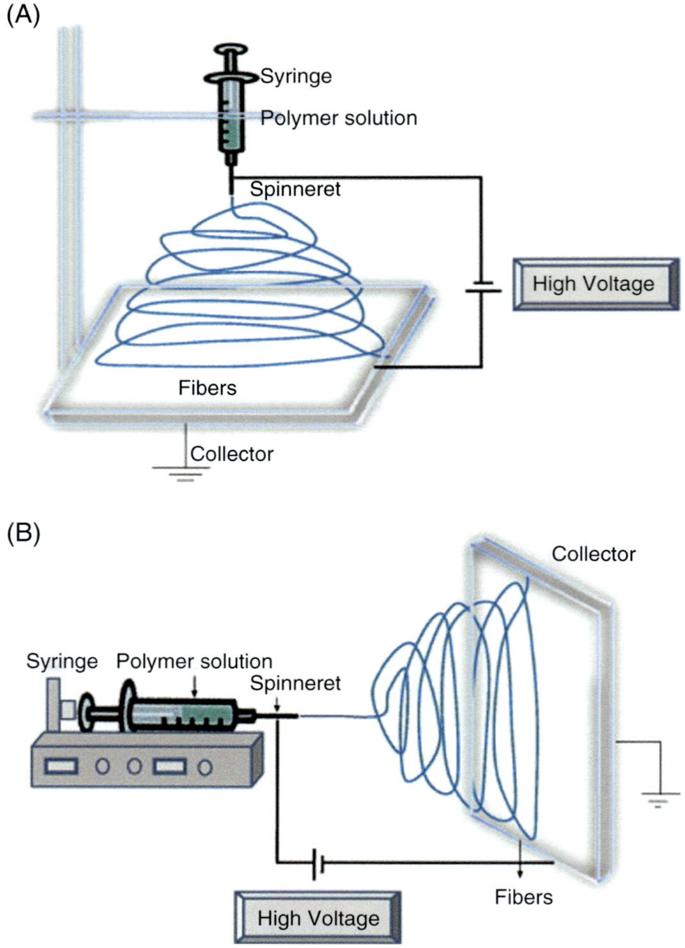


**Fig. 8** Set-up of electrospinning apparatus.
Reprinted with permission from reference N. Bhardwaj, S.C. Kundu, Electrospinning: a fascinating fiber fabrication technique, Biotechnol. Adv. 28 (3) (2010) 325–347.

## *9.2 Film casting*

Film casting is the most widely used technique for the synthesis of polysaccharides-based nanocomposites. In this method, thin layers or films are synthesized. The solution is passed through a moving belt having smooth-surfaces which expand. As a result, the solvent completely evaporates and dry. Consequently, the film is formed which is removed from the belt (Fig. 9). With the use of this method, an elongated sheet (a few hundred cm) can be produced with the thickness of micro-size. The physical

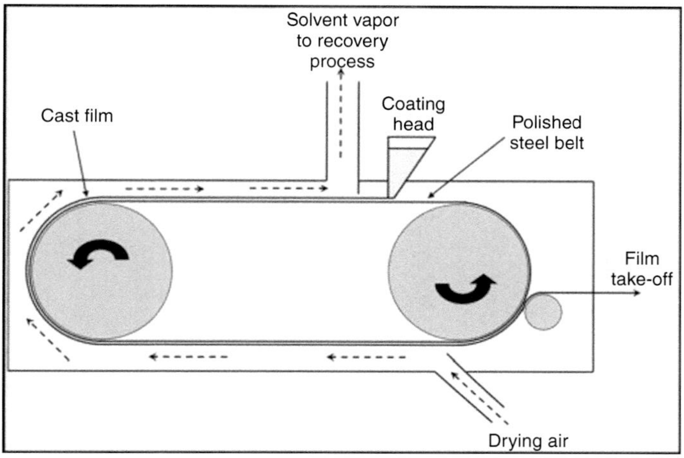


**Fig. 9** Schematic illustration of film casting process.
Reprinted with permission from reference S. Ebnesajjad, Polyvinyl Fluoride: Technology and Applications of PVF, William Andrew, 2012.

characteristics of the obtained film depend on the rate of film casting operation. Generally, this method of preparation of the film is used to enhance the antimicrobial properties, transparency, and thermal conductivity of composite material [129].

## 9.3 Dip coating

Dip-coating is a simple method which is commercially used for thin films since 1939 [130]. This method is commonly used for the synthesis of polysaccharide-based composites [124]. In this technique, a solution of interest is prepared and then a substrate is immersed in it to form a layer on it. The layer that forms on the substrate is formed by chemical interaction of the solution layer with the substrate surface. This interaction can be hydrophobic, electrostatic or ionic. Generally, the process consists of five steps [131]. In the first step, the substrate is immersed in the solution of interest. In the second step, the substrate is left in the solution for some time which is called startup. In the third step, a layer is formed on the substrate, while in the fourth step, the substrate is separated from the solution. In the final step, the substrate is dried by evaporating the solvent from the resulting material (Fig. 10). In the past, the process is carried out manually. However, to increase reproducibility, automation of dip-coating has been developed and used commercially. The advantages of this method include low cost, ease of handling in layer thickness, and convenience. The limitation of this technique is the blocking of sites in the substrate, which can affect the efficiency of the product material [132].

Several factors affect the efficiency of the resulting material, such as the rate of condensation and evaporation, the rate of ablation, density, surface tension, and

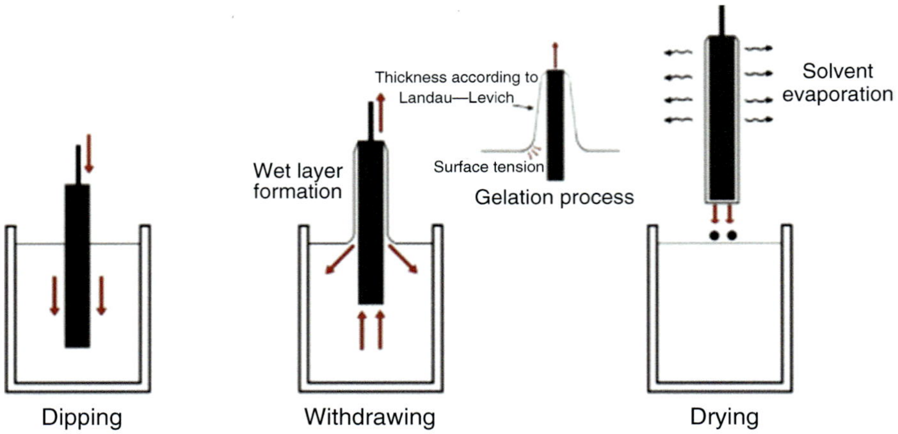


**Fig. 10** Schematic illustration of dip coating process.
Reprinted with permission from reference A. Mohammadzadeh, et al., Mechanical engineering of solid oxide fuel cell systems: geometric design, mechanical configuration, and thermal analysis, in: Design and Operation of Solid Oxide Fuel Cells, Elsevier, 2020, pp. 85–130.

capillary pressure [133–135]. The composites synthesized by dip-coating have been used for antimicrobial applications [136], biomedical applications [124,137], etc.

## 9.4 Layer by layer assembly

Layer-by-layer is a technique used for coating the substrate with charged materials such as colloids, biomolecules, polymers, NPs, and even cells [138,139]. This technique was introduced by Decher and his colleagues [140]. The advantages of this method are its simplicity, excellent control, and versatility, which make it superior to thin-film deposition approaches [141]. Moreover , this technique does not need highly purified components. The films are synthesized using the electrostatic force of attraction. It has also been investigated that interactions such as hydrogen bonding, covalent bonding, biological recognition, charge transfer, and hydrophobic forces also lead to film formation. These different types of interactions increase the stability of the films even when expose to harsh conditions [142].

## 9.5 Thermo-physico-mechanical casting and drying

Shah et al. reported the fabrication of agar-based composites containing magnetic nanoparticles and some other additives by physical mixing and heating followed by casting and drying. The resulting composites exhibited improved physical, mechanical, and thermal properties and were successfully used for biological application [143]. Other techniques for the synthesis of polysaccharide composites have also been reported in the literature. Some of these include colloidal assembly, ionotropic gelation, covalent coupling, in situ preparation, co-precipitation, etc.

# 10 Applications

## 10.1 Food packaging

The increase in population boomed the use of fresh and healthy food across the world [144]. Consequently, the demand for plastic-based packaging material to preserve and increase the shelf-life of food products increased [145]. The literature review shows that petroleum-based packaging material increases by 8% annually and about approximately 5% is recycled [146]. As a result, pollution was increasing day by day [147]. This problem has increased the demand for biodegradable material for packaging purposes. It has been studied that the most widely used biopolymers for packaging films are polysaccharides, proteins, and lipids. Among these biopolymers, polysaccharides have advantages over proteins and lipids in easy availability, low cost, network-formation property, and other tremendous properties [148,149]. The extensively used polysaccharides for packaging films are starch, alginate, chitosan, pectin, and carrageenan [145].

Sarvanakumar et al. combined different ratios of Na-alginate, cellulose nano-whisker (CNW), and embedded them with CuO nanoparticles. Analysis of the resulting composite showed outstanding performance in terms of barrier properties and moisture penetration [84]. Another agar-$TiO_2$ composite film has been reported for food packaging [108].

## 10.2 As an adsorbent material for removal of contaminants

Usually, water and foods are contaminated by heavy metals or other pollutants. In food, these pollutants are present either in chemical or biological form. The measurement and removal of these pollutants is a tedious process as they are present in low concentration [150]. In the last decade, various adsorption techniques have been developed for the determination and removal of pollutants. However, these techniques are labor-intensive, costly, and do not provide sufficient clean-up [151]. Moreover, the adsorbent is expensive and nonbiodegradable. Polysaccharide-based composites are the alternative adsorbents for the extraction of heavy metals and dyes. These composites overcome the limitations of traditional sorbents and have tremendous advantages. These advantages are nontoxicity, ease of preparation, biodegradability, cheapness, and renewability. Therefore they are used as alternatives for the removal of heavy metals and dyes [105]. Mahmoud et al. prepared a biodegradable composite from waste iron oxide nanoparticles and alginate polymer (IOWNPs-Alg). The synthesized polymer composite was used for the adsorption of cadmium, lead, and copper from wastewater samples. The results show that the prepared nanocomposite has an effective adsorption capacity for the target analytes [76]. Other polysaccharide-metal oxide composite materials were used for the removal of lead, copper [152], arsenic, cadmium [79], fluoride [80], lead [27], nickel, cadmium, lead [58], copper, lead [68], and nickel [67].

Le et al. prepared the magnetic graphene oxide-chitosan composite for the removal of nickel and blue 19 dye from water samples. In this experiment, various influencing factors were optimized. The removal of heavy metals and dyes was carried out using the synthesized composite under the optimum conditions. The results showed that the prepared material has an effective adsorption capacity for both analytes.

### 10.3 Biomedical applications

Nowadays, the fields of tissue engineering and regenerative medicine are mainly focused finding out material-based therapeutic solution. For this purpose, the polymers are widely used due to their ease of synthesis, structural control, economic feasibility, and easy availability. Cellulose and chitin are the most prominent polysaccharides used as bio-polymers in these areas of study [153]. Similarly, the BC nanofibers also have the same microscopic shape as collagen fibers, which makes them a promising component for tissue engineering [154]. BC is also used as an essential component for scaffold in bone engineering [155]. Cellulose acetate/gel nanofibers were prepared to have the same structural and morphological properties as human skin. The prepared nanofibers were effectively used for wound healing [156].

The biomedical application of polysaccharide-based composite is not limited to tissue engineering or wound healing, but it also plays a key role in the biosensor formation. Kaushik et al. used chitosan-$Fe_3O_4$ films for the determination of glucose. The developed biosensor was able to quantify the glucose level in a range of 0.5–22 mM [157]. Similarly, Au-cellulose acetate [158] and cellulose-$SnO_2$ [159] nanocomposites are used for the development of glucose biosensors. A chitosan-$SnO_2$ nanocomposite-based biosensor was reported for the quantification of cholesterol [160].

## 11 Conclusion

Polysaccharide-based composites have attracted the attention of researchers in various fields due to their interesting properties. The polysaccharide component is usually derived from natural sources such as plants, animals, algae, and microorganisms and is biocompatible and readily biodegradable. Polysaccharide composites have replaced synthetic polymers because they are readily available, inexpensive, environmentally friendly, renewable, biodegradable, and easily chemically modified. Various methods are used for the preparation of polysaccharide and metal oxide-based composites. Although the interaction between metal oxides and polysaccharides is not common in nature, the polysaccharides and metal oxide-based composites are increasingly being considered for various important applications such as medicine, environmental protection, and agriculture due to their human friendly nature and efficient material.

## References

[1] N. Shah, et al., Overview of bacterial cellulose composites: a multipurpose advanced material, Carbohydr. Polym. 98 (2) (2013) 1585–1598.
[2] N.M. Stark, Z. Cai, C. Carll, Wood-based composite materials: panel products, glued-laminated timber, structural composite lumber, and wood-nonwood composite materials, in: Wood Handbook: Wood as an Engineering Material: Chapter 11. Centennial ed.

General Technical Report FPL; GTR-190, US Dept. of Agriculture, Forest Service, Forest Products Laboratory, Madison, WI, 2010, pp. 11.1–11.28. 190 (2010) 11.1–11.28.
[3] I. Šimkovic, What could be greener than composites made from polysaccharides? Carbohydr. Polym. 74 (4) (2008) 759–762.
[4] A. Mohanty, M.A. Misra, G. Hinrichsen, Biofibres, biodegradable polymers and biocomposites: an overview, Macromol. Mater. Eng. 276 (1) (2000) 1–24.
[5] S. Dumitriu, Polysaccharides: Structural Diversity and Functional Versatility, CRC Press, 2004.
[6] C. Vilela, et al., Polysaccharides-based hybrids with metal oxide nanoparticles, in: Polysaccharide Based Hybrid Materials, Springer, 2018, pp. 31–68.
[7] S. Gul, et al., Fabrication of magnetic core shell particles coated with phenylalanine imprinted polymer, Polym. Test. 75 (2019) 262–269.
[8] N. Shah, et al., Effective role of magnetic core-shell nanocomposites in removing organic and inorganic wastes from water, Recent Pat. Nanotechnol. 10 (3) (2016) 202–212.
[9] N. Shah, S. Gul, M. Ul-Islam, Core-shell molecularly imprinted polymer nanocomposites for biomedical and environmental applications, Curr. Pharm. Des. 25 (34) (2019) 3633–3644.
[10] M. Ul-Islam, et al., Current advancements of magnetic nanoparticles in adsorption and degradation of organic pollutants, Environ. Sci. Pollut. Res. 24 (14) (2017) 12713–12722.
[11] N. Shah, et al., Magnetic aerogel: an advanced material of high importance, RSC Adv. 11 (13) (2021) 7187–7204.
[12] M. Ul-Islam, et al., Recent advancement in cellulose based nanocomposite for addressing environmental challenges, Recent Pat. Nanotechnol. 10 (3) (2016) 169–180.
[13] B. Boury, S. Plumejeau, Metal oxides and polysaccharides: an efficient hybrid association for materials chemistry, Green Chem. 17 (1) (2015) 72–88.
[14] D. Visinescu, et al., Polysaccharides route: a new green strategy for metal oxides synthesis, in: Environmental Chemistry for a Sustainable World, Springer, 2012, pp. 119–169.
[15] D. Klemm, et al., Cellulose: fascinating biopolymer and sustainable raw material, Angew. Chem. Int. Ed. 44 (22) (2005) 3358–3393.
[16] R. Song, et al., Current development of biodegradable polymeric materials for biomedical applications, Drug Des. Devel. Ther. 12 (2018) 3117.
[17] M. Oprea, S.I. Voicu, Recent advances in composites based on cellulose derivatives for biomedical applications, Carbohydr. Polym. (2020) 116683.
[18] M.T. Islam, et al., Preparation of nanocellulose: a review, AATCC J. Res. 1 (5) (2014) 17–23.
[19] F. Esa, S.M. Tasirin, N. Abd Rahman, Overview of bacterial cellulose production and application, Agric. Agric. Sci. Procedia 2 (2014) 113–119.
[20] M. Moniri, et al., Production and status of bacterial cellulose in biomedical engineering, Nanomaterials 7 (9) (2017) 257.
[21] Y.H. Tsuang, et al., Studies of photokilling of bacteria using titanium dioxide nanoparticles, Artif. Organs 32 (2) (2008) 167–174.
[22] A. Khalid, et al., Bacterial cellulose-zinc oxide nanocomposites as a novel dressing system for burn wounds, Carbohydr. Polym. 164 (2017) 214–221.
[23] N. Dal'Acqua, et al., Characterization and application of nanostructured films containing Au and TiO2 nanoparticles supported in bacterial cellulose, J. Phys. Chem. C 119 (1) (2015) 340–349.
[24] N. Janpetch, C. Vanichvattanadecha, R. Rujiravanit, Photocatalytic disinfection of water by bacterial cellulose/N–F co-doped TiO2 under fluorescent light, Cellulose 22 (5) (2015) 3321–3335.

[25] G. Li, et al., Laccase-immobilized bacterial cellulose/TiO2 functionalized composite membranes: evaluation for photo-and bio-catalytic dye degradation, J. Membr. Sci. 525 (2017) 89–98.
[26] P. Wang, et al., Flexible and monolithic zinc oxide bionanocomposite foams by a bacterial cellulose mediated approach for antibacterial applications, Dalton Trans. 43 (18) (2014) 6762–6768.
[27] A. Shoukat, et al., Titanium oxide-bacterial cellulose bioadsorbent for the removal of lead ions from aqueous solution, Int. J. Biol. Macromol. 129 (2019) 965–971.
[28] L. Chaabane, et al., Functionalization of developed bacterial cellulose with magnetite nanoparticles for nanobiotechnology and nanomedicine applications, Carbohydr. Polym. (2020) 116707.
[29] D.M. Panaitescu, et al., Structural and morphological characterization of bacterial cellulose nano-reinforcements prepared by mechanical route, Mater. Des. 110 (2016) 790–801.
[30] J. Desmaisons, et al., A new quality index for benchmarking of different cellulose nanofibrils, Carbohydr. Polym. 174 (2017) 318–329.
[31] A. Farooq, et al., Cellulose from sources to nanocellulose and an overview of synthesis and properties of nanocellulose/zinc oxide nanocomposite materials, Int. J. Biol. Macromol. 154 (2020) 1050–1073.
[32] F. Grüneberger, et al., Nanofibrillated cellulose in wood coatings: dispersion and stabilization of ZnO as UV absorber, Prog. Org. Coat. 87 (2015) 112–121.
[33] D.-C. Wang, et al., Superfast adsorption–disinfection cryogels decorated with cellulose nanocrystal/zinc oxide nanorod clusters for water-purifying microdevices, ACS Sustain. Chem. Eng. 5 (8) (2017) 6776–6785.
[34] B. Nath, et al., Synthesis and characterization of ZnO: CeO2: nanocellulose: PANI bionanocomposite. A bimodal agent for arsenic adsorption and antibacterial action, Carbohydr. Polym. 148 (2016) 397–405.
[35] Z. Pang, et al., A room temperature ammonia gas sensor based on cellulose/TiO2/PANI composite nanofibers, Colloids Surf. A Physicochem. Eng. Asp. 494 (2016) 248–255.
[36] N.C. Martins, et al., Antibacterial paper based on composite coatings of nanofibrillated cellulose and ZnO, Colloids Surf. A Physicochem. Eng. Asp. 417 (2013) 111–119.
[37] H.-Y. Yu, et al., A facile one-pot route for preparing cellulose nanocrystal/zinc oxide nanohybrids with high antibacterial and photocatalytic activity, Cellulose 22 (1) (2015) 261–273.
[38] N.A. El-Wakil, et al., Development of wheat gluten/nanocellulose/titanium dioxide nanocomposites for active food packaging, Carbohydr. Polym. 124 (2015) 337–346.
[39] N. Shandilya, I. Capron, Safer-by-design hybrid nanostructures: an alternative to conventional titanium dioxide UV filters in skin care products, RSC Adv. 7 (33) (2017) 20430–20439.
[40] R. Liu, et al., Preparation and photocatalytic property of mesoporous ZnO/SnO2 composite nanofibers, J. Alloys Compd. 503 (1) (2010) 103–110.
[41] V. Khatri, et al., ZnO-modified cellulose fiber sheets for antibody immobilization, Carbohydr. Polym. 109 (2014) 139–147.
[42] L. Csóka, et al., Photo-induced changes and contact relaxation of the surface AC-conductivity of the paper prepared from poly (ethyleneimine)–TiO2–anthocyanin modified cellulose fibers, Cellulose 22 (1) (2015) 779–788.
[43] K. Fujiwara, et al., Fabrication of photocatalytic paper using TiO2 nanoparticles confined in hollow silica capsules, Langmuir 33 (1) (2017) 288–295.
[44] Y. Li, et al., In situ growing directional spindle TiO2 nanocrystals on cellulose fibers for enhanced Pb2+ adsorption from water, J. Hazard. Mater. 289 (2015) 140–148.

[45] B. Škipina, et al., Generation of photo charge in poly (ethyleneimine)-TiO2-anthocyanin modified papers conditioned at different humidities, Dyes Pigments 149 (2018) 51–58.
[46] A. Rajeswari, E.J.S. Christy, A. Pius, New insight of hybrid membrane to degrade Congo red and Reactive yellow under sunlight, J. Photochem. Photobiol. B Biol. 179 (2018) 7–17.
[47] C. Dumitriu, et al., Production and characterization of cellulose acetate–titanium dioxide nanotubes membrane fraxiparinized through polydopamine for clinical applications, Carbohydr. Polym. 181 (2018) 215–223.
[48] N.A. Ibrahim, et al., Multifunctional cellulose-containing fabrics using modified finishing formulations, RSC Adv. 7 (53) (2017) 33219–33230.
[49] F. Fu, et al., Construction of cellulose based ZnO nanocomposite films with antibacterial properties through one-step coagulation, ACS Appl. Mater. Interfaces 7 (4) (2015) 2597–2606.
[50] A.J. Gimenez, et al., based ZnO oxygen sensor, IEEE Sensors J. 15 (2) (2014) 1246–1251.
[51] S. Mun, et al., Preparation and characterisation of cellulose ZnO hybrid film by blending method and its glucose biosensor application, Mater. Technol. 30 (sup7) (2015) B150–B154.
[52] M. Xu, et al., Study of synergistic effect of cellulose on the enhancement of photocatalytic activity of ZnO, J. Mater. Sci. 52 (14) (2017) 8472–8484.
[53] O. Kalinkevich, et al., Chitosan-based composite materials comprising metal or metal oxide nanoparticles: synthesis, characterization and antimicrobial activity, in: 2018 IEEE 8th International Conference Nanomaterials: Application & Properties (NAP), IEEE, 2018.
[54] N. Doiphode, et al., Biotechnological applications of dimorphic yeasts, in: Yeast Biotechnology: Diversity and Applications, Springer, 2009, pp. 635–650.
[55] R. Amorim, et al., Chitosan from Syncephalastrum racemosum using sugar cane substrates as inexpensive carbon sources, Food Biotechnol. 20 (1) (2006) 43–53.
[56] J. Dutta, Non-Medical Applications of Chitosan Nanocomposite Coatings, 2019. https://encyclopedia.pub/219.
[57] H. Ge, Z. Ma, Microwave preparation of triethylenetetramine modified graphene oxide/chitosan composite for adsorption of Cr (VI), Carbohydr. Polym. 131 (2015) 280–287.
[58] M. Keshvardoostchokami, et al., Synthesized Chitosan/Iron Oxide Nanocomposite and Shrimp Shell in Removal of Nickel, Cadmium and Lead From Aqueous Solution, Global J. Environ. Sci. Manage. 3 (2017) (2017) 267–278.
[59] J. Bonilla, et al., Physical, structural and antimicrobial properties of poly vinyl alcohol–chitosan biodegradable films, Food Hydrocoll. 35 (2014) 463–470.
[60] F. Croisier, C. Jérôme, Chitosan-based biomaterials for tissue engineering, Eur. Polym. J. 49 (4) (2013) 780–792.
[61] S. Kumari, P.K. Rath, Extraction and characterization of chitin and chitosan from (Labeo rohit) fish scales, Procedia Mater. Sci. 6 (2014) 482–489.
[62] Y. Shahbazi, N. Shavisi, Current advancements in applications of chitosan based nanometal oxides as food preservative materials, Nanomed. Res. J. 4 (3) (2019) 122–129.
[63] R. Gusain, et al., Adsorptive removal and photocatalytic degradation of organic pollutants using metal oxides and their composites: a comprehensive review, Adv. Colloid Interf. Sci. 272 (2019) 102009.
[64] B. Vaseeharan, J. Sivakamavalli, R. Thaya, Synthesis and characterization of chitosan-ZnO composite and its antibiofilm activity against aquatic bacteria, J. Compos. Mater. 49 (2) (2015) 177–184.
[65] L. Zemskova, et al., New chitosan/iron oxide composites: fabrication and application for removal of Sr2+ radionuclide from aqueous solutions, Biomimetics 3 (4) (2018) 39.

[66] B. Zhang, et al., Fabrication of chitosan/magnetite-graphene oxide composites as a novel bioadsorbent for adsorption and detoxification of Cr (VI) from aqueous solution, Sci. Rep. 8 (1) (2018) 1–12.
[67] T.T.N. Le, et al., Preparation of magnetic graphene oxide/chitosan composite beads for effective removal of heavy metals and dyes from aqueous solutions, Chem. Eng. Commun. 206 (10) (2019) 1337–1352.
[68] H.F. Heiba, et al., Preparation and characterization of novel mesoporous chitin blended MoO3-montmorillonite nanocomposite for Cu (II) and Pb (II) immobilization, Int. J. Biol. Macromol. 152 (2020) 554–566.
[69] S. Bano, et al., Preparation and study of ternary polypyrrole-tin oxide-chitin nanocomposites and their potential applications in visible light photocatalysis and sensors, J. Environ. Chem. Eng. 7 (2) (2019) 103012.
[70] H.-L. Ma, et al., Preparation and characterization of superparamagnetic iron oxide nanoparticles stabilized by alginate, Int. J. Pharm. 333 (1–2) (2007) 177–186.
[71] Y. Nishio, et al., Preparation and magnetometric characterization of iron oxide-containing alginate/poly (vinyl alcohol) networks, Polymer 45 (21) (2004) 7129–7136.
[72] G.A. Kloster, et al., Magnetic composite films based on alginate and nano-iron oxide particles obtained by synthesis "in situ", Eur. Polym. J. 94 (2017) 43–55.
[73] K. Varaprasad, et al., Alginate-based composite materials for wound dressing application: a mini review, Carbohydr. Polym. (2020) 116025.
[74] K.Y. Lee, D.J. Mooney, Alginate: properties and biomedical applications, Prog. Polym. Sci. 37 (1) (2012) 106–126.
[75] K.P. Satriaji, et al., Antibacterial bionanocomposite films based on CaSO4-crosslinked alginate and zinc oxide nanoparticles, Food Packag. Shelf Life 24 (2020) 100510.
[76] M.E. Mahmoud, et al., A sustainable nanocomposite for removal of heavy metals from water based on crosslinked sodium alginate with iron oxide waste material from steel industry, J. Environ. Chem. Eng. 8 (4) (2020) 104015.
[77] M. Hachemaoui, et al., Composites beads based on Fe3O4@ MCM-41 and calcium alginate for enhanced catalytic reduction of organic dyes, Int. J. Biol. Macromol. 164 (2020) 468–479.
[78] E. Alver, A.Ü. Metin, F. Brouers, Methylene blue adsorption on magnetic alginate/rice husk bio-composite, Int. J. Biol. Macromol. 154 (2020) 104–113.
[79] J. Shim, et al., Sustainable removal of pernicious arsenic and cadmium by a novel composite of MnO2 impregnated alginate beads: a cost-effective approach for wastewater treatment, J. Environ. Manag. 234 (2019) 8–20.
[80] T. Wu, L. Mao, H. Wang, Adsorption of fluoride from aqueous solution by using hybrid adsorbent fabricated with Mg/Fe composite oxide and alginate via a facile method, J. Fluor. Chem. 200 (2017) 8–17.
[81] B.K. Shanmugam, et al., Biomimetic TiO2-chitosan/sodium alginate blended nanocomposite scaffolds for tissue engineering applications, Mater. Sci. Eng. C 110 (2020) 110710.
[82] S. Mohamed, et al., Coupled adsorption-photocatalytic degradation of crystal violet under sunlight using chemically synthesized grafted sodium alginate/ZnO/graphene oxide composite, Int. J. Biol. Macromol. 108 (2018) 1185–1198.
[83] A. Salama, et al., Crosslinked alginate/silica/zinc oxide nanocomposite: a sustainable material with antibacterial properties, Compos. Commun. 7 (2018) 7–11.
[84] K. Saravanakumar, et al., Physical and bioactivities of biopolymeric films incorporated with cellulose, sodium alginate and copper oxide nanoparticles for food packaging application, Int. J. Biol. Macromol. 153 (2020) 207–214.

[85] W. Wu, et al., Rheological and antibacterial performance of sodium alginate/zinc oxide composite coating for cellulosic paper, Colloids Surf. B: Biointerfaces 167 (2018) 538–543.

[86] S. Kalia, Lignocellulosic Composite Materials, Springer, 2017.

[87] E.M. Fernandes, et al., Bionanocomposites from lignocellulosic resources: properties, applications and future trends for their use in the biomedical field, Prog. Polym. Sci. 38 (10–11) (2013) 1415–1441.

[88] S.G. Kestur, L.P. Ramos, F. Wypych, Comparative study of Brazilian natural fibers and their composites with other, in: Natural Fibre Reinforced Polymer Composites: From Macro to Nanoscale, Old City Publishing, Inc., Philadelphia, USA, 2009.

[89] R. Djellabi, et al., Unravelling the mechanistic role of TiOC bonding bridge at titania/lignocellulosic biomass interface for Cr (VI) photoreduction under visible light, J. Colloid Interface Sci. 553 (2019) 409–417.

[90] R. Djellabi, et al., Sustainable self-floating lignocellulosic biomass-TiO2@ Aerogel for outdoor solar photocatalytic Cr (VI) reduction, Sep. Purif. Technol. 229 (2019) 115830.

[91] A. Kamenan, A. Rolland-Sabaté, P. Colonna, Stability of yam starch gels during processing, Afr. J. Biotechnol. 4 (1) (2005) 94–101.

[92] P. Kanmani, J.-W. Rhim, Properties and characterization of bionanocomposite films prepared with various biopolymers and ZnO nanoparticles, Carbohydr. Polym. 106 (2014) 190–199.

[93] V. Nain, et al., Development, characterization, and biocompatibility of zinc oxide coupled starch nanocomposites from different botanical sources, Int. J. Biol. Macromol. 162 (2020) 24–30.

[94] X. Hu, et al., Improving the properties of starch-based antimicrobial composite films using ZnO-chitosan nanoparticles, Carbohydr. Polym. 210 (2019) 204–209.

[95] S.-J. Bao, et al., Environmentally-friendly biomimicking synthesis of TiO 2 nanomaterials using saccharides to tailor morphology, crystal phase and photocatalytic activity, CrystEngComm 15 (23) (2013) 4694–4699.

[96] B. Khodadadi, Facile sol–gel synthesis of Nd, Ce-codoped TiO2 nanoparticle using starch as a green modifier: structural, optical and photocatalytic behaviors, J. Sol-Gel Sci. Technol. 80 (3) (2016) 793–801.

[97] X. Wang, et al., Sodium carboxymethyl starch-based highly conductive gel electrolyte for quasi-solid-state quantum dot-sensitized solar cells, Res. Chem. Intermed. 44 (2) (2018) 1161–1172.

[98] A. Baysal, C. Kuznek, M. Ozcan, Starch coated titanium dioxide nanoparticles as a challenging sorbent to separate and preconcentrate some heavy metals using graphite furnace atomic absorption spectrometry, Int. J. Environ. Anal. Chem. 98 (1) (2018) 45–55.

[99] V. Goudarzi, I. Shahabi-Ghahfarrokhi, A. Babaei-Ghazvini, Preparation of ecofriendly UV-protective food packaging material by starch/TiO2 bio-nanocomposite: characterization, Int. J. Biol. Macromol. 95 (2017) 306–313.

[100] N. Shah, et al., Synthesis and Characterization of Starch Coated Natural Magnetic Iron Oxide Nanoparticles for the Removal of Methyl Orange Dye from Water, Lett. Appl. NanoBioSci. 10 (2021) (2021) 2750–2759.

[101] D. Diaz-Bleis, et al., Thermal characterization of magnetically aligned carbonyl iron/agar composites, Carbohydr. Polym. 99 (2014) 84–90.

[102] G. Kavoosi, et al., Microencapsulation of zataria essential oil in agar, alginate and carrageenan, Innovative Food Sci. Emerg. Technol. 45 (2018) 418–425.

[103] V. Hasija, et al., Green synthesis of agar/gum Arabic based superabsorbent as an alternative for irrigation in agriculture, Vacuum 157 (2018) 458–464.

[104] Q.-Q. Ouyang, et al., Thermal degradation of agar: mechanism and toxicity of products, Food Chem. 264 (2018) 277–283.
[105] O. Duman, et al., Agar/κ-carrageenan composite hydrogel adsorbent for the removal of methylene blue from water, Int. J. Biol. Macromol. 160 (2020) 823–835.
[106] W.-K. Lee, et al., Biosynthesis of agar in red seaweeds: a review, Carbohydr. Polym. 164 (2017) 23–30.
[107] N. Radovanovic, et al., Influence of different concentrations of Zn-carbonate phase on physical-chemical properties of antimicrobial agar composite films, Mater. Lett. 255 (2019) 126572.
[108] X. Hou, et al., Characterization and property investigation of novel eco-friendly agar/carrageenan/TiO2 nanocomposite films, J. Appl. Polym. Sci. 136 (10) (2019) 47113.
[109] G. Magesh, et al., Structural, morphological, optical and biological properties of pure ZnO and agar/zinc oxide nanocomposites, Int. J. Biol. Macromol. 117 (2018) 959–966.
[110] E. Abdollahzadeh, H. Mahmoodzadeh Hosseini, A.A. Imani Fooladi, Antibacterial activity of agar-based films containing nisin, cinnamon EO, and ZnO nanoparticles, J. Food Saf. 38 (3) (2018), e12440.
[111] A. Vejdan, et al., Effect of TiO2 nanoparticles on the physico-mechanical and ultraviolet light barrier properties of fish gelatin/agar bilayer film, LWT- Food Sci. Technol. 71 (2016) 88–95.
[112] K. Yazhini, S. Suja, Synthesis and characterization of hetero-metal oxide nano-hybrid composite on pectin scaffold, Appl. Surf. Sci. 491 (2019) 195–205.
[113] A.K. Kodoth, et al., Application of pectin-zinc oxide hybrid nanocomposite in the delivery of a hydrophilic drug and a study of its isotherm, kinetics and release mechanism, Int. J. Biol. Macromol. 115 (2018) 418–430.
[114] A. Noreen, et al., Pectins functionalized biomaterials; a new viable approach for biomedical applications: a review, Int. J. Biol. Macromol. 101 (2017) 254–272.
[115] E. Ruoslahti, S.N. Bhatia, M.J. Sailor, Targeting of drugs and nanoparticles to tumors, J. Cell Biol. 188 (6) (2010) 759–768.
[116] M. Pandurangan, G. Enkhtaivan, D.H. Kim, Anticancer studies of synthesized ZnO nanoparticles against human cervical carcinoma cells, J. Photochem. Photobiol. B Biol. 158 (2016) 206–211.
[117] G. Plascencia-Villa, et al., Imaging interactions of metal oxide nanoparticles with macrophage cells by ultra-high resolution scanning electron microscopy techniques, Integr. Biol. 4 (11) (2012) 1358–1366.
[118] J.W. Rasmussen, et al., Zinc oxide nanoparticles for selective destruction of tumor cells and potential for drug delivery applications, Expert Opin. Drug Deliv. 7 (9) (2010) 1063–1077.
[119] I. Hira, et al., Pectin-guar gum-zinc oxide nanocomposite enhances human lymphocytes cytotoxicity towards lung and breast carcinomas, Mater. Sci. Eng. C 90 (2018) 494–503.
[120] F.J. Arévalo, et al., Development of an electrochemical sensor for the determination of glycerol based on glassy carbon electrodes modified with a copper oxide nanoparticles/multiwalled carbon nanotubes/pectin composite, Sensors Actuators B Chem. 244 (2017) 949–957.
[121] A. Di Tocco, et al., Development of an electrochemical biosensor for the determination of triglycerides in serum samples based on a lipase/magnetite-chitosan/copper oxide nanoparticles/multiwalled carbon nanotubes/pectin composite, Talanta 190 (2018) 30–37.
[122] A. Soubhagya, A. Moorthi, M. Prabaharan, Preparation and characterization of chitosan/pectin/ZnO porous films for wound healing, Int. J. Biol. Macromol. 157 (2020) 135–145.

[123] X. Wu, et al., Pectin assisted one-pot synthesis of three dimensional porous NiO/ graphene composite for enhanced bioelectrocatalysis in microbial fuel cells, J. Power Sources 378 (2018) 119–124.
[124] Y. Zheng, J. Monty, R.J. Linhardt, Polysaccharide-based nanocomposites and their applications, Carbohydr. Res. 405 (2015) 23–32.
[125] X. Lu, C. Wang, Y. Wei, One-dimensional composite nanomaterials: synthesis by electrospinning and their applications, Small 5 (21) (2009) 2349–2370.
[126] N. Bhardwaj, S.C. Kundu, Electrospinning: a fascinating fiber fabrication technique, Biotechnol. Adv. 28 (3) (2010) 325–347.
[127] L. Meli, et al., Electrospinning from room temperature ionic liquids for biopolymer fiber formation, Green Chem. 12 (11) (2010) 1883–1892.
[128] J. Yu, et al., Production of aligned helical polymer nanofibers by electrospinning, Eur. Polym. J. 44 (9) (2008) 2838–2844.
[129] E.J. Bealer, et al., Protein and polysaccharide-based magnetic composite materials for medical applications, Int. J. Mol. Sci. 21 (1) (2020) 186.
[130] P. Yimsiri, M.R. Mackley, Spin and dip coating of light-emitting polymer solutions: matching experiment with modelling, Chem. Eng. Sci. 61 (11) (2006) 3496–3505.
[131] K. Kakaei, M.D. Esrafili, A. Ehsani, Graphene and anticorrosive properties, in: Interface Science and Technology, Elsevier, 2019, pp. 303–337.
[132] S.K. Sahoo, B. Manoharan, N. Sivakumar, Introduction: why perovskite and perovskite solar cells? in: Perovskite Photovoltaics, Elsevier, 2018, pp. 1–24.
[133] C. Brinker, et al., Fundamentals of sol-gel dip coating, Thin Solid Films 201 (1) (1991) 97–108.
[134] C.J. Brinker, Dip coating, in: Chemical Solution Deposition of Functional Oxide Thin Films, Springer, 2013, pp. 233–261.
[135] I. Strawbridge, P. James, The factors affecting the thickness of sol-gel derived silica coatings prepared by dipping, J. Non-Cryst. Solids 86 (3) (1986) 381–393.
[136] B.-L. Wang, et al., Fast and long-acting antibacterial properties of chitosan-Ag/ polyvinylpyrrolidone nanocomposite films, Carbohydr. Polym. 90 (1) (2012) 8–15.
[137] S.K. Samal, et al., Biomimetic magnetic silk scaffolds, ACS Appl. Mater. Interfaces 7 (11) (2015) 6282–6292.
[138] F. Chen, et al., Chitosan-based layer-by-layer assembly: towards application on quality maintenance of lemon fruits, Adv. Polym. Technol. 2020 (2020).
[139] J.J. Richardson, et al., Innovation in layer-by-layer assembly, Chem. Rev. 116 (23) (2016) 14828–14867.
[140] G. Decher, J.D. Hong, Buildup of ultrathin multilayer films by a self-assembly process, 1 consecutive adsorption of anionic and cationic bipolar amphiphiles on charged surfaces, in: Makromolekulare Chemie. Macromolecular Symposia, Wiley Online Library, 1991.
[141] S. Fujita, S. Shiratori, Waterproof anti reflection films fabricated by layer-by-layer adsorption process, Jpn. J. Appl. Phys. 43 (4S) (2004) 2346.
[142] Z. Tang, et al., Biomedical applications of layer-by-layer assembly: from biomimetics to tissue engineering, Adv. Mater. 18 (24) (2006) 3203–3224.
[143] N. Shah, et al., Preparation and characterization of agar based magnetic nanocomposite for potential biomedical applications, Curr. Pharm. Des. 25 (34) (2019) 3672–3680.
[144] P.J.P. Espitia, et al., Edible films from pectin: physical-mechanical and antimicrobial properties—a review, Food Hydrocoll. 35 (2014) 287–296.
[145] F.S. Mostafavi, D. Zaeim, Agar-based edible films for food packaging applications—a review, Int. J. Biol. Macromol. 159 (2020) 1165–1176.

[146] S. Muizniece-Brasava, L. Dukalska, I. Kantike, Consumer's knowledge and attitude to traditional and environmentally friendly food packaging materials in market of Latvia, in: The 6th Baltic Conference on Food Science and Technology "FoodBalt-2011", 2011.
[147] E. Tavassoli-Kafrani, H. Shekarchizadeh, M. Masoudpour-Behabadi, Development of edible films and coatings from alginates and carrageenans, Carbohydr. Polym. 137 (2016) 360–374.
[148] A. Kurt, T. Kahyaoglu, Characterization of a new biodegradable edible film made from salep glucomannan, Carbohydr. Polym. 104 (2014) 50–58.
[149] M.E. Embuscado, K.C. Huber, Edible Films and Coatings for Food Applications, vol. 222, Springer, 2009.
[150] W.A. Khan, M.B. Arain, M. Soylak, Nanomaterials-based solid phase extraction and solid phase microextraction for heavy metals food toxicity, Food Chem. Toxicol. 145 (2020) 111704.
[151] W.A. Khan, et al., A new microfluidic-chip device for selective and simultaneous extraction of drugs with various properties, New J. Chem. 43 (24) (2019) 9689–9695.
[152] L. Pan, et al., Efficient removal of lead and copper ions from water by enhanced strength-toughness alginate composite fibers, Int. J. Biol. Macromol. 134 (2019) 223–229.
[153] P.I. Soares, et al., Hybrid polysaccharide-based systems for biomedical applications, in: Hybrid Polymer Composite Materials, Elsevier, 2017, pp. 107–149.
[154] W. Jing, et al., Laser patterning of bacterial cellulose hydrogel and its modification with gelatin and hydroxyapatite for bone tissue engineering, Soft Mater. 11 (2) (2013) 173–180.
[155] B. Fang, et al., Proliferation and osteoblastic differentiation of human bone marrow stromal cells on hydroxyapatite/bacterial cellulose nanocomposite scaffolds, Tissue Eng. A 15 (5) (2009) 1091–1098.
[156] E. Vatankhah, et al., Development of nanofibrous cellulose acetate/gelatin skin substitutes for variety wound treatment applications, J. Biomater. Appl. 28 (6) (2014) 909–921.
[157] A. Kaushik, et al., Iron oxide nanoparticles–chitosan composite based glucose biosensor, Biosens. Bioelectron. 24 (4) (2008) 676–683.
[158] X. Ren, et al., Amperometric glucose biosensor based on a gold nanorods/cellulose acetate composite film as immobilization matrix, Colloids Surf. B: Biointerfaces 72 (2) (2009) 188–192.
[159] S.K. Mahadeva, J. Kim, Conductometric glucose biosensor made with cellulose and tin oxide hybrid nanocomposite, Sensors Actuators B Chem. 157 (1) (2011) 177–182.
[160] A.A. Ansari, et al., Electrochemical cholesterol sensor based on tin oxide-chitosan nanobiocomposite film, Electroanalysis 21 (8) (2009) 965–972.

# Recent advances in renewable polymer/metal oxide systems used for tissue engineering

13

*Rawaiz Khan[a,b], Sajjad Haider[c], Saiful Izwan Abd Razak[b,d], Adnan Haider[e], Muhammad Umar Aslam Khan[b,d], Mat Uzir Wahit[a,d], Nausheen Bukhari[f], and Ashfaq Ahmad[g]*

[a]Department of Polymer Engineering, Faculty of Engineering, School of Chemical and Energy, Universiti Teknologi Malaysia, Johor Bahru, Johor, Malaysia, [b]BioInspired Device and Tissue Engineering Research Group, School of Biomedical Engineering and Health Sciences, Faculty of Engineering, Universiti Teknologi Malaysia, Skudai, Johor, Malaysia, [c]Department of Chemical Engineering, College of Engineering, King Saud University, Riyadh, Saudi Arabia, [d]Centre for Advanced Composite Materials, Universiti Teknologi Malaysia, Skudai, Johor, Malaysia, [e]Department of Biological Sciences, National University of Medical Sciences, Rawalpindi, Punjab, Pakistan, [f]Mohammad College of Medicine, Peshawar, Pakistan, [g]Department of Chemistry, College of Science, King Saud University Riyadh, Riyadh, Saudi Arabia

## 1 Introduction

In today's consumption-driven society, plastic is required for the production of millions of products. Packaging materials largely contribute to this demand for plastics. The high demand for packaging materials can be easily understood by today's fast-paced lifestyle, where handiness and prepackaged food in individual portions are important. The cost-effective packaging material that can effectively protect a product can only be perceived using plastic [1]. Plastic is usually synthesized from crude oil (a non-renewable resource) products. Since the precursors of plastic are non-renewable, plastic is generally recycled, after its first use, to make the material cost effective. In recent years, the intensified research on developing plastic from sources other than oil has resulted in the discoveries of some natural and renewable precursors for plastics. Plastic is successfully produced from these precursors. Natural plastic has the advantage of not posing a waste problem as is the case with synthetic plastic, as natural plastic is biodegradable under temperature and humidity [2]. Biodegradable plastic is an important material innovation as it has not only reduced dependence on crude oil (to avoid fossil fuels depletion), but also provided a potential solution to the huge waste disposal problem without compromising the properties and benefits offered by synthetic/traditional plastic [3].

Natural polymers are usually produced by microorganisms or extracted from plants and animals [4–6]. There are three categories of natural polymers: Polysaccharides,

**Renewable Polymers and Polymer-Metal Oxide Composites. https://doi.org/10.1016/B978-0-323-85155-8.00010-8**

polypeptides, and polyesters. The interest in the production of natural polymers was mainly directed toward their use in biomedical applications, cosmetics, pharmaceutical materials, and regenerative medicine. These polymers have special properties such as high stability, manageable solubility, excellent structural properties, 3D morphology, remarkable biocompatibility, cytocompatibility, and antigenicity. Moreover, they can precisely target tissue/cell [7,8]. It has been widely reported in the literature that natural polymers not only mimic the ECM better, but also show good interaction with tissues than semisynthetic or synthetic polymers. Both these properties are attributed to their resemblance with the tissue environment [9]. However, after mentioning the benefits of natural polymers and their widely reported application, these polymers also do have many limitations. These include production costs, which are high in the case of collagen and hyaluronic acid, batch to batch discrepancy due to the structural, compositional macromolecular architecture and morphological complexity, hydrophilicity, limited resources, and potential risk of microbial infection at the implant site [10]. Besides, natural polymers also have the limitation of being difficult to process (cellulose and chitosan), have lower mechanical properties (polypeptides), and poor stability (in some cases). The poor stability leads to a higher rate of degradation and catabolization compared to the rate of host tissue regeneration. Considering these limitations, work on the molecular architecture and functionalities of natural polymers could lead to controlled degradation and sufficient mechanical, structural, and compositional properties, which may open newavenues for the development of biomaterials suitable for successful tissue engineering application [11]. Various strategies have been proposed in the last decade. These include chemical modification [12], fabrication of blends using synthetic or semisynthetic and natural polymers [13], cross-linking of the blends to give more stability to the resulting material [14], and physical modification [15].

Natural polymers include cellulose, lignin, starch, etc. Cellulose is present in all plants in sufficient quantity. Similarly, lignin is obtained from wood, and starch is obtained from plants like potatoes, wheat, and corn. The difference between synthetic and natural polymers is in their functionalities. Natural polymers contain oxygen and nitrogen functionalities, which allow the polymer to biodegrade. Fig. 1 shows biodegradable polymers and their sources [16].

Recently, polymers derived from natural resources (PFNRs) have received considerable attention mainly for two main reasons: (i) environmental safety and (ii) the sensitivity to depleting petroleum resources. However, recently these have also been considered for a third reason, namely to add value to agricultural products [17]. These products are economically vital for many countries. The main advantage of polymers derived from PFNR is their biodegradability. Therefore, any product produced from these polymers will have a positive impact on the environment (environmentally friendly). Moreover, various properties of the PFNRs can be enhanced by employing blending and composite formulation techniques, which will expend their use in the manufacturing more valuable products. Synthetic polymers, especially polyolefins, are not biodegraded by microorganisms. This biodegradation of polyolefins is prevented by their stable chemical structure and high hydrophobicity. Since these materials cause a major waste problem in many countries, researchers in industrially

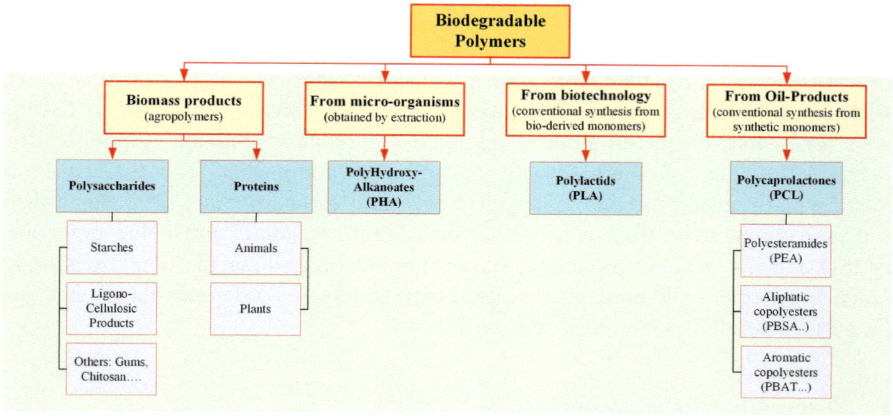


**Fig. 1** Biodegradable polymer classification based on the source.

advanced countries are engaged in developing novel biodegradable polymeric materials. They are also trying to tune the biodegradability of synthetic polymers by adding natural fillers to the polymer matrix and changing structural bonding, thus enabling the development of polymer materials with improved biodegradability. Such materials can partially replace the nonbiodegradable polymers obtained from petrochemical resources. Also, in composite materials, the synthetic polymer will offer optimal mechanical strength and recyclability. The composite material has shown biodegradability that proves the importance of filler, choice of filler type, and the continued study [17].

## 2 History

Henry Ford first found an interest in creating plastic from renewable resources around 1910. His interest was in creating plastic from agricultural waste. He was successful in producing a "plastic car" in 1941 using soybean waste, which was blended with other components to increase its strength [18]. Although this was a new method, interest in the production of biodegradable polymers did not accelerate until the 1960s. During this era, researcher found that too much plastic was being produced and its waste was having a negative impact on the environment [19]. This gave a reason to an environmental movement that spurred research into the production of biodegradable plastic. Despite these efforts, the adoption of biodegradable plastic in real use has taken a long time and only recently have biodegradable plastics derived from renewable resources been used for packaging. As discussed earlier, the polymer derived from renewable resources does not always retain the desired properties of the plastic. Therefore, it is common for polymers derived from renewable resources to be mixed with either oil-based polymers or other filler materials to obtain desired properties. For instance, the polymers that are derived from starch are generally not very strong. Therefore, to increase their strength, they were mixed with oil-based polymers such as polyvinyl alcohol (PVA) and polyethylene (PE) [20]. These blends were marketed in the 1980s as biodegradable plastics, which were quickly accepted by consumers. However, after a short period of time, they were subjected to strict analysis due to their incomplete biodegradation (starch was degraded, but PE and PVA were not). Polymers derived from renewable resources showed complete degradation and are not misleading in their claims, as was the case with PE, and PVA composites. The degradation of plastic yields carbon dioxide, nitrogen, water, and other minerals [21]. Since there is disagreement over what constitutes degradation, standards are being developed by ISO and the ASTM. Guidelines have been established for the conditions under which a polymer will biodegrades, the composition of biodegradable plastic, and the number of additives used in the plastic.

# 3 Renewable materials

The definition of the bioplastic that is recognized and practically used today signifies bioplastics as biodegradable plastics and plastics derived from renewable resources. However, this definition in the industry does not necessarily show that all bioplastics are biodegradable. Thus, this definition also includes bioplastics that are not biodegradable, even though these are obtained from renewable resources [21]. Fig. 2 will give us a better idea of the biodegradable polymers and the types of bioplastics.

## *3.1 Natural polymers*

In the 1970s, the oil crisis motivated polymer scientists and engineers to look for a more sustainable source of polymers. The only sustainable source was nature. They felt that the production/extraction of polymers from this source was not only cost-effective but also did not create any environmental issues as they were easily degradable compared to oil-based polymers. The similar increasing attention toward regenerative therapeutic and controlled targeted pharmacological strategies using biodegradable polymers has also stimulated research on polymers that are derived from sustainable sources. Natural polymers derived from sustainable sources offer advantages over synthetic polymers for biomedical applications. These advantages of natural polymers include essential biological signaling, surface chemistry for stimulating cell adhesion, and cell-responsive degradation. The surface chemistry of the building blocks of the natural polymer (proteins and polysaccharides) provides opportunities for chemical derivatization/bioactive functionalization [23]. The interesting challenges faced by natural polymers in biomedical applications are their shaping into a specific geometry and the development of controlled porous structures under physiological and aqueous conditions. To overcome these challenges, much work has been reported in literature on modifying the physical and surface chemistry of natural polymers. The work aims to engineer natural polymers into structures with suitable physical properties. One attractive feature of the natural polymers found in these studies was the biocompatible hydrogels. Hydrogels are the hydrated polymers that upon freeze-drying gives 3D porous structures. Thus, natural polymers will give a highly

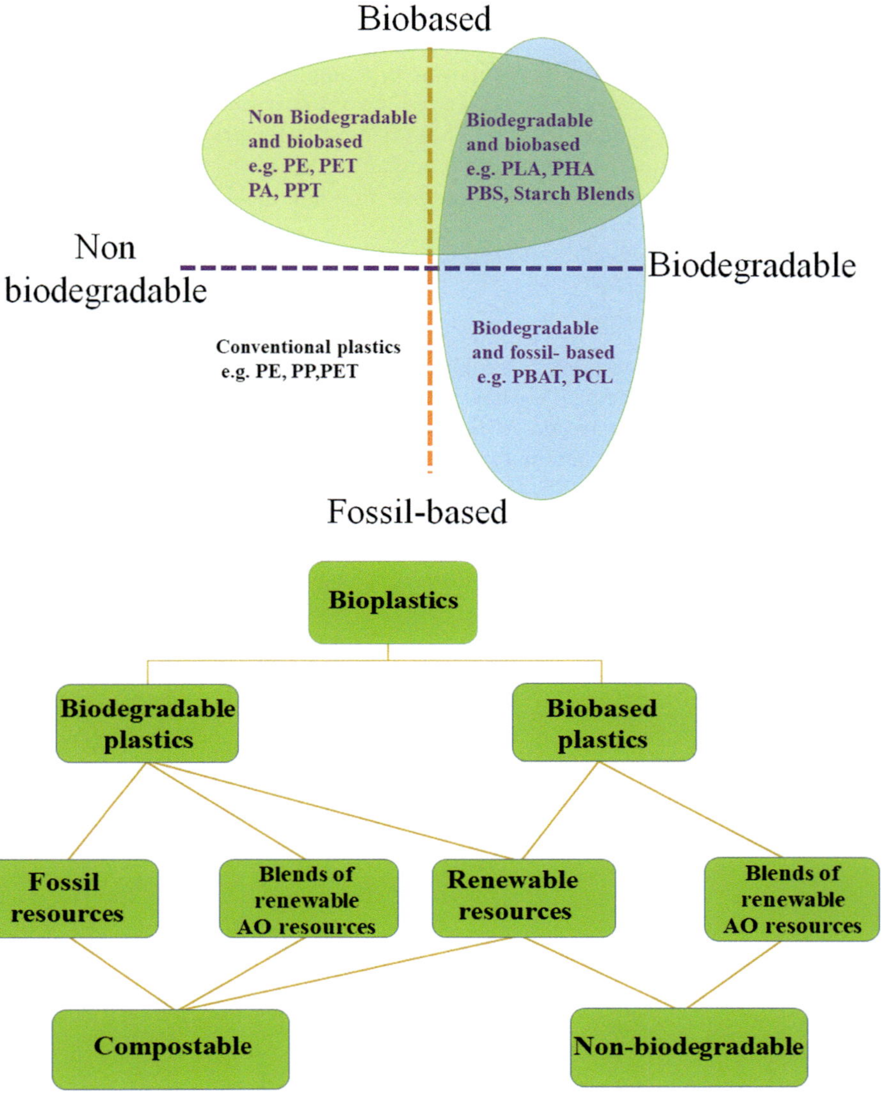


**Fig. 2** Bioplastic coordinated and types of bioplastics [22].

hydrated atmosphere, more like soft natural tissue, to the adhering cells. Hydrogels that are fabricated from collagen and gelatin (protein), alginate, and chitosan (polysaccharides) are not only frequently employed in tissue engineering, but also used in various biomedical applications, especially in drug delivery. As discussed earlier that one of the limitations of natural polymers is their poor processibility into scaffolds. However, this limitation has now been overcome thanks to cutting-edge research reported on tailoring the physical-chemical properties of hydrogels. The research explored the potential of natural polymers for fabricating structures that were more reproducible and had superior biocompatibility and bioactivity. The most studied natural polymers in this area are (i) polysaccharides, (ii) proteins, and (iii) microbial polymers [24,25]. These natural polymers exhibited better biodegradation properties, enhanced biocompatibility, lower toxicity, and good absorption of bioactive molecules. Therefore, they are ideal for tissue engineering applications.

### 3.1.1 Polysaccharides

Polysaccharides are macromolecules that consist of monosaccharides connected by glycosidic linkages (Scheme 1). They have diverse physiological functions, which make them important materials in tissue engineering and regenerative medicine [9,26]. Polysaccharides are among the first choice polymers that are renewable, environmentally friendly, and qualify for green chemistry in the fabrication of biomedical materials. Different sources such as vegetables, animals, and microbes can be used to extract/produce polysaccharides with characteristics chemical nature, physical-chemical properties, and lower production costs compared to other natural polymers in particular proteins. Below are several polysaccharides that have been studied for tissue engineering.

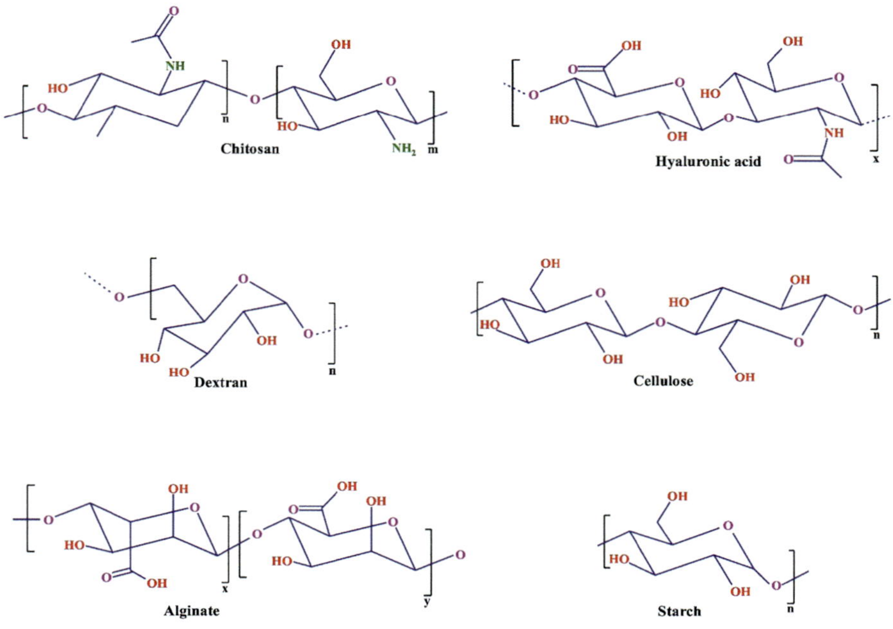


**Scheme 1** The polysaccharides family used in the fabrication of scaffold.

## Chitosan/chitin

Chitosan is a natural bioactive polymer of polysaccharides that has been extensively studied for its excellent properties, including its high biocompatibility, biodegradability, and antibacterial properties. Among different biodegradable polymers, it is widely used for biosensors, drug delivery, and packaging purposes. However, the applications of chitosan have been seriously hampered by its poor mechanical properties. The incorporation of fillers into the chitosan matrix has considerably enhanced its physicochemical properties. For example, chitosan-carbonaceous nanoparticle composites were prepared by dissolving chitosan in an aqueous solution of acetic acid and adding graphene to the solution under sonication. The incorporation of carbonaceous NPs did not alter the crystallinity of chitosan and also considerably enhanced the Young's modulus and tensile strength of the resulting material [27]. The use of chitosan in wound healing is very common. Chitosan has shown potential for pulp capping. Many in vitro studies illustrated that chitosan offered a 3D structure for postnatal human dental pulp stem cells (DPSCs). The 3D structure improved cell adhesion, proliferation, and differentiation. However, as discussed in the natural polymer section, since chitosan lacks mechanical strength, it must be blended with other polymers or filled with NPs (to form a composite) not only to enhance its biological activity but also to give mechanical strength. In one of the studies, chitosan/carboxymethyl cellulose (CMC)-blended composite was fabricated through freeze-drying, and the resulting material was used for pulp tissue engineering. Adding CMC showed a significant effect on the porosity development and size reduction. As a result, the modified scaffold exhibited improved biocompatibility compared to the control scaffold (pure chitosan). The new scaffold illustrated an improved proliferation and upregulated expression of osteonectin (ON) and dentin sialophosphoprotein (DSPP). The chitosan-collagen composite was also studied for the formation of a dentin-pulp complex. Promoting vascular growth, in vivo, was also observed for human periodontal ligament cells (HPLC) when chitosan was filled with tricalcium phosphate (ceramic). In addition, the preparation of the 3D multilayer coculture systems, in which type I collagen and chitosan were mixed and seeded with DPSCs and HAT-7 dental epithelial cells, showed the deposition of calcium ions after 24 days of coculture. The layered macroscale structure with tunable mechanical properties allowed the movement of cells in all directions. Polysaccharides are normally either acidic or neutral. However, chitin and chitosan are highly basic polysaccharides and because of their higher content of nitrogen (5%–8%), these are useful chelating agents. Chitin is highly hydrophobic and is immiscible with water and many organic solvents [28,29]. Chitin can be converted to chitosan by deacetylation. The percentage of deacetylation depends on the nature of the origin. Deacetylation makes chitosan more soluble in aqueous acidic media. Chitosan demonstrates a pH-responsive nature that mainly depends on the number of amino group polycation. The polycation helps ease the processing of chitosan. Chitosan is highly soluble when the pH is below six due to the protonation of amine group. That is why chitosan easily dissolves in the aqueous solutions of acetic acid and hydrochloric acid. At the basic pH, the neutralization of amino groups causes the hydrated polymer chains to form a gel. The characteristics of chitosan in a solution depend on the degree of deacetylation, molecular weight, and arrangement of acetyl groups in the backbone [30,31]. The properties of chitosan are often altered by the chemical cross-linking using the most common cross-linking agents, such as reactive dialdehydes and carbodiimide compounds. These properties include biodegradation, water sorption, the permeability of oxygen, antibacterial activity, blood coagulation, and cytokine induction [32,33]. Chitosan, from chitin, is obtained via deacetylation and chitin is derived from the exoskeleton of crustaceans, insects, and some fungi. Chitosan exhibits pseudoplastic behavior [34,35]. Chitosan is highly biocompatible, shows low immunogenicity at the interface with living organisms, and accepts host response when implanted [36]. Chitosan is also susceptible to biodegradation by an enzyme in in vivo [37]. The biodegradation rate and biocompatibility of chitosan are affected by the degree of deacetylation, the free amino groups promote cell interaction [38,39]. Apart from tissue engineering, chitosan is widely studied in the controlled drug delivery (growth factors, antibiotics, and antiinflammatory drugs) [35,40]. A huge amount of literature reported chitosan-based scaffolds for human tissues engineering. These include bone, cartilage, skin, liver, and nerves [41]. In addition to cellulose, chitosan was also blended with synthetic polymers (PVA [42] and PLLA [43]) or natural (alginate [44] and collagen [45] (as discussed in the previous text)) and ceramics (Ca-P [46] and natural coralline) to improve its mechanical properties and maintain a predefined shape once processed [47].

## Starch

Starch-based materials have received significant attention due to their low economic cost, biodegradability, wide availability, and renewability. Therefore, these materials have great potential to replace conventional polymers. These types of materials have been exploited in several sectors including medicine, agriculture, packaging industries, etc. However, its poor barrier properties, poor water resistance at high moisture content, and lower mechanical strength restrain its frequent use for medical applications. To overcome these problems, researchers have combined starch-based biopolymers with various types of NPs that have resulted in improved mechanical, thermal, and antibacterial properties. Zheng et al. have produced starch-grafted graphene nanosheets (GN-starch). The resulting nanocomposite has demonstrated good solubility and stability in water. Besides, the nanocomposites were also prepared by utilizing GN-starch as filler in the plasticized-starch (PS) matrix. Due to the strong interactions between the components involved, the tensile strength of the resulting material has increased significantly [48]. Starch is among the most abundantly available and cost-effective polysaccharides [49]. Its granules are hydrophilic and induce strong intermolecular attraction due to the hydrogen bonds formed by the presence of hydroxyl groups (-OH) on the surface of the granules. The hydroxyl groups that existed on the backbone of starch exhibit specific reactivity with alcohols, resulting in the creation of hydrogen bonds, ethers, and esters by oxidation or reduction. Starch can enhance the degradability of polymers with hydrophobic nature that is why starch-based materials have been widely used as biomaterials in biomedical applications and tissue engineering scaffolds [50,51].

### Hyaluronic acid

Hyaluronic acid (HAc) is a sustainable linear polysaccharide. It is composed of D-glucuronic acid and D-*N*-acetylglucosamine units that are alternatively connected by −1.4 and −1,3 glycosidic bonds. HAc is not inert structurally and has a high molecular weight that can be broken down to small molecular species through the action of glycosidase controlled by environmental cues. These cues include pH and reactive oxygen. The biological activity, half-life, and rheology of HAc are affected by molecular weight. HAc is the main constituent of the skin (in the intracellular space (about 2.5 g/L)), cartilage, and vitreous humor's ECM. HAc and its derivatives are extensively reported in the literature for their potential use in the biomedical field, particularly in tissue engineering. Derivatives are structurally and chemically altered form of HAc. HAc is reported to be the most vital glycosaminoglycans in the extracellular matrix (ECM). It has retained the morphologic organization by preserving extracellular space. As discussed in the previous section, the dental pulp (a connective tissue) comprises glycosaminoglycans, proteoglycans, and collagens in significant quantities. A study carried out by Felszeghy and coworkers has shown that the expression of HAc in the dental pulp decreases during the growth of tooth germs. This made it clear that HAc is only vital in the initial growth of the dental pulp and dentin matrix. Both growth factors and HAc sponges are needed to develop restorative materials that induce the formation of dentin. Besides, HAc sponges illustrated an appropriate structure (physicochemical), good cytocompatibility, and biodegradation when implanted to regenerate dental pulp. Having discussed the benefits of HAc, it is also important to mention that HAc has also major limitations. During the biodegradation of HAc, molecules with low molecular weight are liberated, which tempers the inflammatory process. Particularly, it is described that these low molecular weight molecules prevent leukocyte migration and neutrophil adhesion. Furthermore, It is reported that ~50% of the total HA affects cell proliferation and differentiation that helps in the regeneration of tissue [52,53]. As HAc is highly soluble in an aqueous medium, fabricating HAc-based scaffolds for tissue engineering needs stable structure. This issue can be overcome by cross-linking and esterification, which avert water entry and reduces HAc solubility. Various techniques are available for the cross-linking of HAc. These include indirect cross-linking, where HAc is cross-linked by the attachment of thiols, methacrylates, or tyramines, and direct cross-linking, where formaldehyde or divinyl sulfone is used as cross-linkers. As discussed, HAc is an essential functional constituent of nearly all vertebrate tissues. Consequently, diverse animal tissues (e.g., shark skin, rooster combs, and bovine eyeballs) are used to isolate HAc with a high molar weight [54]. HAc is extensively used as an injectable and biomedical polymer implant mainly due to its inherent immunogenic properties, ease of chain size manipulation, capability to engage with cell-surface receptors, and common availability [55]. Besides, HAc helps to regulate the water content, provide sufficient lubrication, absorb shocks, and function as a free radicals scavenger [56]. HAc blending with polysaccharides, proteins, and ceramics is studied and reported in the literature as an artificial ECM to promote cell interaction as HAc is polyanionic and does not support cell adhesion [52]. HAc is prone to agglomerate in an aqueous medium due to the hydrophobic interactions and acetamido and carboxylate groups' hydrogen bonding [57]. However, agglomeration can be dissociated under physiological conditions.

### Alginate

Alginate is another polysaccharide that exists in nature and is extensively used in various biomedical fields. These include cell encapsulation, drug delivery, scaffold tissue engineering, and antiadhesion material [58]. The in vivo studies on alginate reported in literature found immunogenicity and inflammatory responses. These are ascribed to the impurities present in alginate. These impurities include heavy metals, endotoxins, proteins, and polyphenolic compounds [59,60]. Hence, appropriate decontamination/purification is recommended before its use in the biomedical field. The impurities might alter biocompatibility, grip over physiochemical properties, stability in aqueous mediums, and rate of biodegradability [61]. The obvious hydrophilic nature of alginate, which is mostly due to its carboxylic groups, does not facilitate the proteins adsorption required for mediated biological processes, such as immunogenicity and cell adhesion [62]. Alginate is prone to form hydrogels in aqueous mediums. The gelation is caused by the interactions between carboxylic acid ions and chelating cations [63]. The biocompatible nature and the effortless gelation made alginate an attractive material for a wide range of biomedical applications. The most known of these biomedical applications are wound healing, drug delivery, and tissue engineering. As mentioned, alginate occurs naturally as an anionic polymer derived from brown seaweed after treating it with an alkali aqueous solution. Many reports are available in the literature where alginate has been used in the biomedical application mainly because of its biocompatible nature, low toxicity, and fairly low cost [64]. Further, the application of alginate can be increased by modifying its chemical structure. The modified structure will have different structures, properties, functions, and applications [65,66]. As discussed earlier, the immunogenic responses of alginate as an injectable or at the implantation site are ascribed to impurities that remained in the polymer after it is extracted from the brown seaweed. It is, therefore, important to adopt a multistep extraction procedure to obtain a highly purified form of alginate [58,67,68].

## 3.1.2 Proteins

Proteins are biological molecules comprised of one or more long-chain amino acids arranged in a 3D-folded structure. Proteins show typical motility, stabilization, elasticity, protection of cells, tissues, and organisms. Due to these properties, amino acids are considerably employed in drug delivery and fabrication of scaffolds for tissue engineering [69,70]. Proteins are the constituent of the ECMs of many human tissues [71]. Proteins derived from the animals have low mechanical strength, poor thermal stability, good immunogenicity, high enzymatic degradation, and nonspecific cell adhesion; therefore, they are notsuitable for the fabrication of scaffold for tissue engineering [72].

## Collagen

Collagen is a vital biomaterial being considered for tissue engineering and delivery of bioactive molecules. These applications largely depend on its biocompatibility, low antigenicity, biodegradability, high mechanical strength. The mechanical strength, biodegradability, and water uptake can be controlled by the use of cross-linking [73]. Cross-linking can be used to produce collagen networks with 3D architectures suitable for use as scaffolds. Collagen scaffolds are susceptible to enzymatic degradation in the body over time, and as discussed, the rate of its degradation can be tuned by controlling the cross-linking [74]. Collagen can be converted into gelatin, which is a soluble protein compound, by partial hydrolysis. The properties of gelatin are largely affected by source, animal age, and type of collagen. Gelatin is used as a thermoreversible gelling agent in pharmaceuticals due to its inherited biodegradation and biocompatibility under physiological conditions. The conversion of collagen to gelatin depends on pretreatment and extraction in hot water is affected by temperature, pH, and duration of extraction [75]. Processing at elevated temperature causes the release of hydrogen and breakage of covalent bonds, which causes the destabilization of the triple helix of collagen (helix-to-coil transition takes place) and transformation to gelatin. Gelatin scaffolds are fabricated by chemically cross-linking gelatin carbodiimide or glutaraldehyde as cross-linkers [76,77]. Collagen has attracted the interest of researchers in the last decade for a variety of biomedical applications including tissue engineering, not only due to its biodegradability, low antigenicity, good cell binding, and unusual mechanical properties [73], but also because it is a structural protein of the native ECM of various tissues. These tissues include skin, bone, cartilage, teeth, tendons, and blood vessels [78]. There are more than 22 types of collagen in mammals; however, types I–IV are most commonly used in biomedical applications. Collagen is usually extracted from bovine or porcine skin, as well as from Achilles tendon of bovine or equine origin using neutral/slightly alkaline salt, acidic, alkaline, or acidic/proteolytic enzyme solutions followed by precipitation [79]. Bacterial collagen was investigated as a reliable, inexpensive, and nonimmunogenic material [80]. Collagen has been studied in various forms for the tissue engineering of bone and cartilage. These include sponges [81], gels [82,83], or electrospun ultrafine fibers [84]. There also studies on the use of collagen for skin repair and blood vessel engineering [85–87]. The remarkable progress made in the use of collagen has lead to the development of the many devices that are globally accessible. These include dermal membranes, localized delivery of antibiotics sponges [88], skin substitutes [89,90], and hemostatic sealants. The first tissue engineering clinical trials done with the implantation of autologous bladder [91] and vaginal organs [92] used scaffolds that were made of collagen and collagen/PGA composite.

## Gelatin

As discussed in the previous section, gelatin is extracted from collagen (Scheme 2). It is comprised of 19 various varieties of amino acids. These amino acids are connected by peptide bonds. Due to its polyelectrolyte nature/structure, it is widely used in many industries such as pharmaceutical, food, and photographic. However, since gelatin is

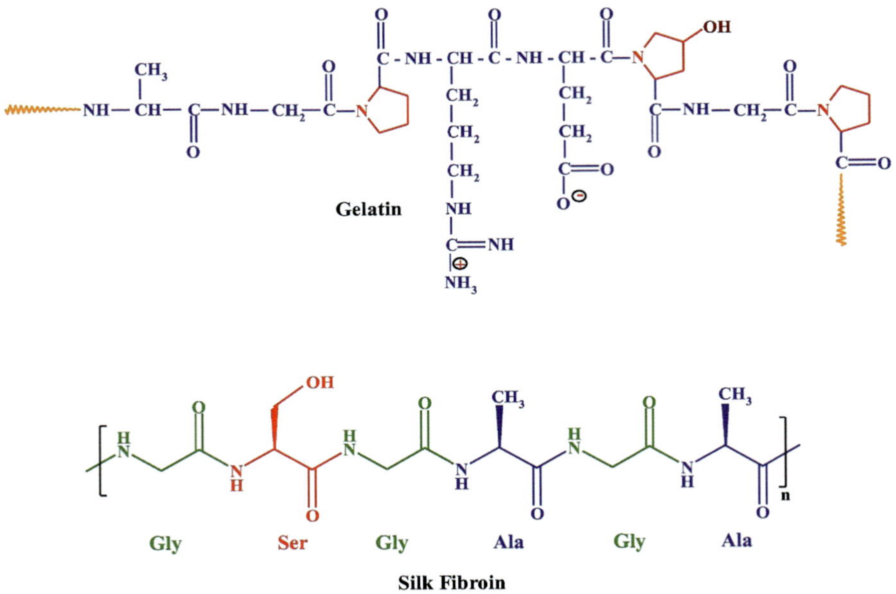


**Scheme 2** The chemical structures of gelatin and silk fibroin as scaffolding polypeptides.

hydrophilic, its large-scale industrial application is affected by hygroscopicity and low wet strength. To solve these issues and expand its industrial applications, gelatin is usually reinforced with filler materials. Many such studies are previously reported. Wang et al. reported the fabrication of graphene-gelatin composite films using solution casting technique. In their study, they used genipin as a cross-linker. They observed that adding graphene to gelatin enhanced the tensile strength of the composite films. The incorporation of graphene and cross-linking also decreased the water swelling ability of the films [93]. Besides, gelatin has also been cross-linked into hydrogel using physiological conditions. It was effective in its wide applications in biomedical applications [94,95]. By varying the cross-linking density, in vitro degradation of gelatin can be controlled. The in vivo degradation of gelatin is carried out by matrix metalloproteinases (e.g., collagenase) [96]. Furthermore, gelatin has also been reported as discs, 3D hydrogels, carriers of growth factors [97–99], encapsulating chondrocytes [100,101], osteoblasts [102], mesenchymal stem cells (MSCs) [103], and preadipocytes [104].

## Synthetic polypeptides

Synthetic polypeptides are polymers with structures that mimic natural proteins. They are biocompatible and biodegradable [105]. The polypeptide structure can be engineered to enhance its thermal stability, immunogenicity, and enzymatic degradability [106]. Natural elastin is the key protein of ECM. It is an insoluble protein, and therefore, its processability is complicated. The insolubility of elastin is attributed to its extensive cross-linking. Therefore, similar synthetic polypeptides such as elastins like proteins (ELPs) are soluble below their phase transition temperature in an aqueous solution [107]. Currently, many synthetic polypeptides such as ubiquitin and antimicrobial proteins are being produced [108,109]. Scaffolds fabricated from peptides contain not only a metalloproteinase-2-sensitive enzyme-cleavable site, but also cell-adhesion motif tripeptide arginine-glycine-aspartic acid (RGD). Moreover, growth factors can also be inserted into them by a mechanism found in the natural ECM. Heparin, a negatively charged glycosaminoglycan, binds growth factors, protect them from proteolytic degradation, and makes them available to cells during rebuilding the ECM. Heparin, exhibited a slow-release profile for transforming growth factor-beta 1 (TGF-β1), basic fibroblast growth factor 2 (FGF-2), and vascular endothelial growth factor (VEGF). Some studies revealed that released and bound growth factors in hydrogel scaffolds supported the differentiation of DPSCs and angiogenesis.

#### Silk

Silk is a natural protein usually obtained from the cocoon (protective casing spun from silk) of silkworm (*Bombyx mori*) (Scheme 2). It is made up of glycine, alanine, and serine [110,111]. The polymeric chains of a silk fiber are mostly arranged parallel, which results in a strong and rather inextensible fiber. Silk proteins can be easily fabricated into films, nanofibers, scaffolds, gels, powders, and membranes that are extremely suitable for barrier membrane and drug delivery applications [112–116]. As silk proteins are produced in fiber by silkworms and spiders, the amino acid sequence differs in silks produced by different species (also within a species). This difference results in different functional properties. The properties are also affected by extraction, purification, processing, and environmental conditions. Silk fibroin is extensively studied for biomedical applications mainly due to its easy processability at room temperature in an aqueous medium, biodegradability (to amino acids), biocompatibility, cross-linking that gives high tensile strength, and elasticity [115,117]. The silk proteins used in biomedical applications in the form of hydrogels, fibers, sponges, particles, and membranes are extensively documented in the literature. These applications include sutures, wound dressing, hemostatic devices, and scaffolds for tissue engineering (e.g., cartilage, vascular, neural, ligament, and bone tissue) [118]. Besides electrospun, nanofibers and nonwoven meshes are also used in the in vitro growth of different human cells. These include endothelial cells, epithelial cells, keratinocytes, and bone marrow-derived MSCs [119–123].

### 3.1.3 *Microbial biopolymers*

Microorganisms are known for producing several biopolymers. These include polysaccharides, polyesters, and polyamides. Genetic manipulation of the microorganisms lets the researchers produce biopolymers having tailored properties suitable for various medical applications, in particular for tissue engineering and drug delivery [114]. Various microbial polymers obtained via fermentation are commercially available [124].

### Poly(lactic acid)

Poly(lactic acid) (PLA) is an attractive biopolymer because its monomers are derived from nontoxic recycled feedstock [125]. The key substrate in the synthesis of PLA is lactic acid, which is obtained in large quantities by bacterial fermentation of a carbon source (about 90%), whereas lactic acid is obtained synthetically by hydrolysis of lactonitrile accounts for only 10%. PLA's mechanical properties and crystallization activity are determined by its molecular weight and the stereochemical composition of its backbone [126]. PLA is a semicrystalline hydrophobic polymer with a glass transition temperature of 40–70°C and a melting temperature of 130–180°C. It has a tensile strength of 44–59 MPa and a decay time of 18–24 months. PLA's melting temperature and degree of crystallinity are proportional to its molar mass and purity level. PLA degrades via easy hydrolysis of the ester bonds, which are not catalyzed by enzymes [127]. PLA and its copolymers are commonly used throughout biomedical applications such as implants, sutures, stent matrix, drug delivery, and tissue engineering due to their excellent biocompatibility and mechanical properties [128,129].

### Poly-(-glutamic acid)

Poly-(-glutamic acid) (PGA) is a polypeptide (anionic) consisting of D- and/or L-glutamic acid units [130]. PGA comprises repeating units of carboxylic acid groups. The presence of this group in the polymer is susceptible to functionalization [131]. Several articles are documented in the literature investigating the capability of its ester derivatives for fabricating films and fibers using standard polymer processing methods. The PGA scaffold for the biomedical application can be fabricated by either irradiation cross-linking or by cross-linker, e.g., hexamethylene diisocyanate and poly(ethylene imine). Even though PGA is not widely studied for its potential in biomedical application yet, several PGA composites with various materials are being studied in biomedical applications considering the useful properties of PGA. These properties include nontoxicity, biodegradability, hydrophilicity, and adsorption capacity (cationic material).

### Polyhydroxyalkanoates

Polyhydroxyalkanoates (PHAs) are thermoplastic, biocompatible, and biodegradable materials. These properties make PHAs an interesting biomaterial for the development of both conventional medicine and tissue engineering devices [132]. The crystallinity of PHAs is in the range of 30%–70%, their melting temperature is in the range of 50–180°C, and the molecular weight of PHAs is in the range of $5 \times 10^4$ to $3 \times 10^6$ Da. The molecular weight is, however, source and growth condition dependent [133]. Furthermore, their mechanical properties are analogous to those of polypropylene (PP) or polyethylene (PE). PHAs can be processed via extrusion, molded, made into films, and spun into fibers. They can be used to synthesize heteropolymers using other synthetic polymers [134]. A homopolymer obtained from monomers with short side chains, such as PHB, has low crystallinity, good flexibility, and is easy to process. These properties make it a suitable carrier for cells in tissue engineering. Despite their several advantages, PHAs have not yet substituted conventional plastics on a large scale.

This is attributed to their high cost and lack of degradability in the body. Research is still ongoing and we believe PHA-derived polymers may be used in the body in the future.

## Bacterial cellulose

Bacterial cellulose (BC) was reported for the first time by Brown in 1988. During his study, he observed the growth of an unbranched pellicle. A study of the chemical structure of this unbranched pellicle revealed that its chemical structure was similar to plant cellulose. BC is produced through the oxidative fermentation of sugars (mainly used as carbon sources) in both synthetic and nonsynthetic media. Various acetic acid bacteria are used for this purpose. These include *Acetobacter*, *Rhizobium*, *Agrobacterium*, *Sarcina*, *Acanthamoeba*, *Achromobacter*, and *Zoogloea* [135]. Since the structure of BC comprises glucose monomers, therefore, BC has many beneficial properties. These include a characteristic nanostructure and excellent water-holding ability, high degree of polymerization, good mechanical properties, and crystalline nature. The well-arranged 3D nanofiber network results in the development of highly porous hydrogel [136]. BC can be processed into a gel (as discussed) by freeze-drying and can also be hot-pressed into dry films. BC can also be used to fabricate composite with nanomaterials or with other polymers of both natural and synthetic origins. The ease of processability, biocompatibility, good mechanical strength, purity, high water holding capacity, and chemical and morphological control make BC a choice for diverse medical applications. However, having mentioned the advantageous properties of BC, one should also keep in mind that it showed poor in vitro degradation [137,138]. Furthermore, its commercial production and use have been hampered by its high costs and low yield. Plant cellulose, which is almost similar to BC, has been extensively studied for various biomedical applications to utilize not only its surface chemistry and biocompatibility but also its biodegradation properties. It has been studied in the repair of the skin due to accelerated epithelialization [139]. Plant cellulose is the most plentiful renewable polymer on the planet. It is obtained from lignocellulosic material in the forests. The primary sources for plant cellulose are algae, agricultural waste, and chemical synthesis [140]. Like other polysaccharides, it is not thermoplastic and degrades with thermal treatment. Additionally, the processing of cellulose is a challenge as the high intramolecular and intermolecular hydrogen bonding renders it insoluble in water and many organic solvents. However, there are many studies where cellulose is dissolved in ionic liquid solutions and then regenerated in a nonsolvent [141]. This enables cellulose to be processed in solution-casting, molding, wet-spinning, and electrospinning techniques.

## Dextran

Dextran is a branched glucan, a polysaccharide, produced by the condensation of glucose. However, its mass production on a commercial scale is only possible by bacteria, e.g., by culturing *Leuconostoc mesenteroides* in a sucrose-rich media, in the presence of organic nitrogen sources like growth agents, peptone, trace minerals, and phosphate. *L. mesenteroides* is an anaerobic/microaerophilic bacterium that promotes anaerobic fermentation. Furthermore, dextransucrase is a glucan sucrose, which is a glycoside hydrolase superfamily member, and is considered the main enzyme in the synthesis of dextran [142]. The molecular weight distribution of dextran varies with the source. Both branching degree and molecular weight distribution affect the physical and chemical properties of dextran. Since dextrans contain a large number of –OH groups, drugs and protein can be easily attached to it. In addition, it is a cheap biomaterial having low immunogenicity and clinical history. All of these properties make it an appealing material for biomedical and drug delivery applications. More studies are needed to use dextrin in the fabrication of tissue engineering scaffolds.

# 4 Techniques for the preparation scaffolds

Tissue engineering is by far the most widely studied biomedical field. Indeed, tissue engineering banks on the implantation of biodegradable scaffolds. Scaffolds, as mentioned earlier, are biocompatible, biodegradable materials with interconnected porous structures that in some cases contain growth factors that promote cell adhesion. Generally, scaffolds act as a temporary template that not only provides mechanical support but also serves as a substrate for cell adhesion during tissue regeneration process [143]. It has been reported in the literature that scaffold chemistry, porosity, topography, and mechanical stiffness greatly affect cell characteristics including cell differentiation, shape, function, cytoskeletal organization, protein expression, and differentiation [144–146]. Numerous methods for fabricating scaffolds have been described in the literature some are listed below.

## 4.1 Solvent casting and particulate leaching

This is a very convenient technique for the fabrication of a porous scaffold. This technique involve the casting of a polymer and particulate mixed solution and then leaching of the particulate porogen (salt-, sugar-, particulate-, or powder-based material) in the solution. After casting, the solution solidifies due to solvent evaporation, cross-linking, or other reactions (Fig. 3). After drying, the scaffold is immersed in a solvent in which the porogens dissolve leaving pores in the scaffold. The most common porogens include sodium chloride, sugar crystals, carbohydrates, gelatin, sand, and ice particles. All these materials are used to produce interconnecting pores in the scaffold upon leaching during scaffold development. Sodium bicarbonate and ammonium bicarbonate salts at high temperature releases carbon dioxide and/or ammonia in an acidic aqueous solution [147,148]. The usage of these porogens and the process have certain advantages such as flexibility, simplicity, regularity, and control over pore size and shape. The shape and size of the pores can also be controlled by selecting porogens with appropriate shape and size [149]. Even though this technique has been intensely studied and reported in the literature, it has some limitations. These include the following: (i) Complete removal of porogens is not possible, as it generates membranes with a dense surface layer. (ii) Leaching of porogen particles needs long soaking periods in the solvent. Long-time soaking solvent can hurt both pore structure

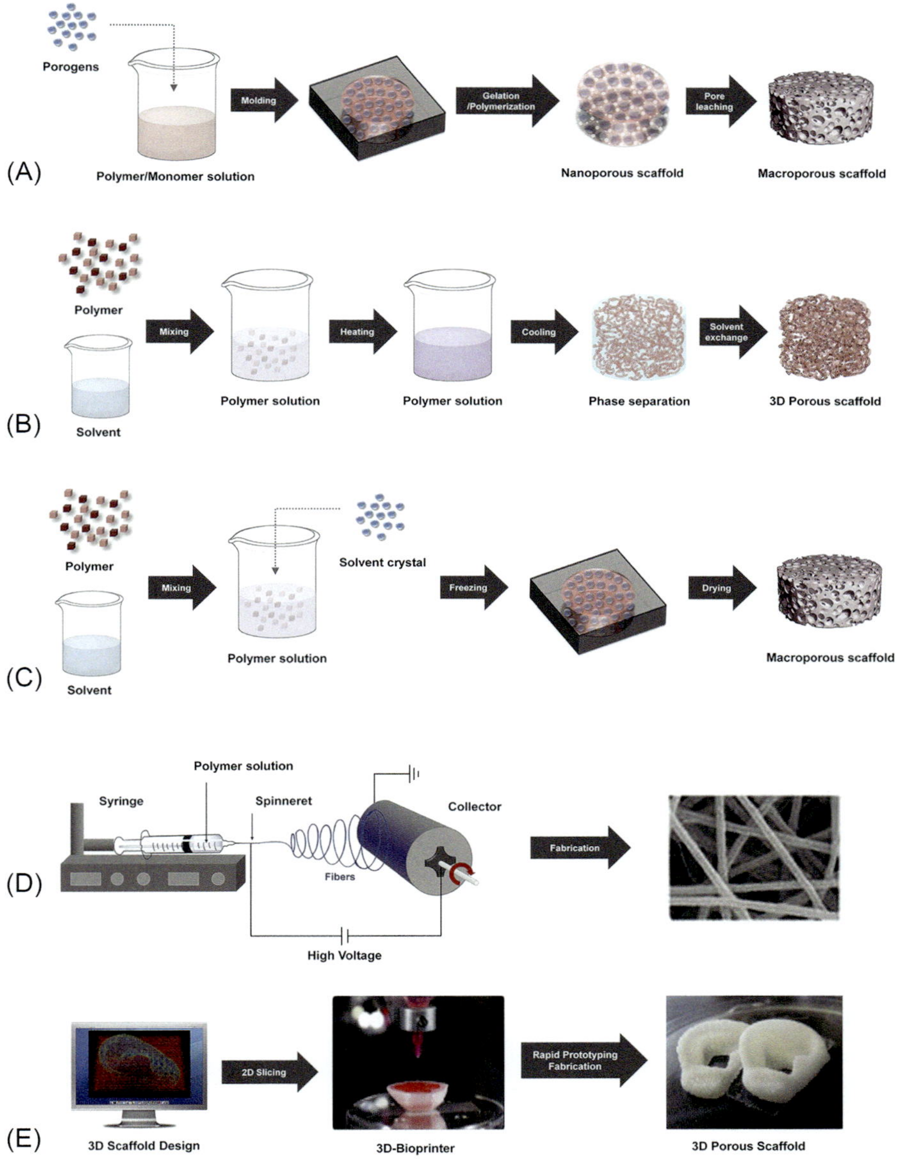


**Fig. 3** Producing porous scaffolds using various techniques: (A) solvent casting and particulate leaching, (B) thermally induced phase separation, (C) emulsion freeze-drying, (D) electrospinning [153], and (E) 3D bioprinting [154].
(D) Copyright 2008. Reproduced with permission from Wiley-VCH Verlag GmbH & Co. KGaA, Weinheim. Reproduced with permission from Nature Publishing Group S.-B. Park, E. Lih, K.-S. Park, Y.K. Joung, D.K. Han, Biopolymer-based functional composites for medical applications, Prog. Polym. Sci. 68 (2017) 77–105, Copyright 2014.

and mechanical properties. (iii) At times, the residual solvent and porogens in the scaffolds may damage tissues and proteins. To tackle these issues, many researchers used gas instead of toxic organic solvents and porogens. The use of gas to form porous foam can eliminate the disadvantages of the use of porogens as it will cause no impurities and will generate scaffolds with high porosity [150,151]. The gas foaming method may be employed to obtain a thin scaffold with an open-cell architecture and up to 93% porosity with typical pore diameters up to 500 μm. To fabricate a thick 3D scaffold, thin porous membranes are bonded into multilayer structures [152].

## 4.2 Thermally induced phase separation

This technique is also described in the literature to produce highly porous scaffolds for biomedical applications. It can produce 3D polymer scaffolds with approximately 97% porosity. This technique provides more control over the microstructures of the scaffold (Fig. 3). As mentioned earlier, the scaffolds produced by this technique showed high porosity, pore interconnectivity, pore anisotropy, and tubular pore morphology [155]. Thermally induced phase separation (TIPS) technique involves the thermodynamic segregation of the polymer/solvent solution into a polymer-rich phase and a polymer-poor phase during the cooling step. During mixing, the polymer-poor phase is removed from the polymer-rich solidified phase. The removal of the polymer-poor phase results in a very porous network of polymer. Usually, the removal of solvent is carried out by freeze-drying or extraction [156,157]. The process parameters play a vital role in regulating the macro- and microstructure porosity, pore morphology, mechanical strength, degradation rate, and bioactivity of the obtained scaffolds. These include polymer concentration, secondary-phase volume fraction, quenching temperature, and quenching rate [158]. Using this technique, nanoscale fibrous structures can be constructed mimicking the architecture of natural ECM. This structure provides an enhanced environment and support for cell attachment and its normal functioning. Many research studies using natural biodegradable polymers and composites are reported in the literature to produce scaffolds for biomedical applications using this technique. These include PGA, gelatin, chitosan, and alginate composites. The key benefit of TIPS is the creation of a 3D pore arrangement. Having said this, it is a complicated procedure lacking control over fiber configuration that limits its widespread use in the production of scaffold for tissue engineering [159–161].

## 4.3 Emulsion freeze-drying

The emulsion freeze-drying technique is based on immiscibility and is carried out at low temperatures. The use of low temperature is beneficial to incorporate bioactive molecules in the scaffold that are heat-sensitive (Fig. 3). As mentioned, this method is based on immiscibility and the creation of ice crystals. These ice crystals through ice sublimation and desorption induce porosity in the scaffolds. An emulsion solution comprises water dispersed and an organic continuous phase. The biodegradable polymer is dissolved in the continuous phase. In the freeze-drying stage, both the solvents are removed to produce porous scaffolds with 90% porosity, pores of various sizes (up to 200 μm), and interconnectivity [162]. In the freezing stage, the solvent forms ice crystals, that force the polymer molecules to cluster in interstitial spaces. The solvent is evaporated at a pressure lower than the equilibrium vapor pressure of the frozen solvent. This leads sublimation (transition of a substance directly from the solid to the gas state) of the solvent, which results in the construction of a dry polymer scaffold with an interconnected porous network. By controlling polymer concentration, freezing time, and freezing temperature, one can control pore size, porosity, and interconnectivity [163,164]. However, the key drawback of this procedure is to modify the structural stability and mechanical characteristics of the porous structures after hydration.

## 4.4 Electrospinning

Electrospinning is a well-known simple, most economical, scalable, and flexible technique used for the fabrication of scaffolds. Since the last decade, this technique has attracted the attention of various research groups working in biomedical fields around the world. The technique has allowed the construction of biopolymer and synthetic polymer-based nanofibrous architectures that mimic natural ECM (Fig. 3). Nanofibrous architectures significantly supported cell binding and spreading compared to microfibrous architectures and that is one of the reasons that this technique has gained attraction in tissue engineering research [153,165]. The electrospinning process is regulated by a high-intensity electric field generated by applying a voltage between the syringe needle and the collector drum. The polymer solution is held in the syringe, which is placed on the pump. Applying the voltage (5–50 kV) draws the polymer solution from a nozzle under gravity and an electric field to form a Taylor cone. However, as the electric charge surpasses the surface tension of the polymer solution droplet, a very fine jet is produced from the needle tip. This jet fly between the needle tip and drum that collects the nanofibers. Subsequently, the solvent evaporates during the flight of the jet resulting in the formation of solid nanofibers [166]. The diameters of the nanofibers and the scaffold architectures are best controlled by altering electrospinning parameters. There are three types of parameters (i) solution- and polymer-related parameters (solution viscosity and surface tension, polymer's molecular weight, and conductivity), (ii) process-related parameters (flow rate, electrospinning voltage, needle tip to collector distance), and (iii) environmental conditions (relative humidity and temperature) [167]. A huge amount of literature is available on the use of electrospinning in the construction of scaffolds for tissue engineering. However, besides its usefulness, this technique also has its limitations. These include the use of toxic solvents, two-dimensional (2D) arrangement of microstructures, and poor infiltration of the cells to the core of scaffolds. A 2D culture condition provides limited nutrient exchange on only one side for the growth of the cell. However, in a 3D nanofibrous structure, the supply of nutrients and use of the receptors takes place throughout the surface. Hence, the construction of micro- and/or macroporous and nanofibrous 3D scaffolds is desirable for tissue regeneration.

## 4.5 3D bioprinting

The recently developed technique of solid free-form fabrication, also known as rapid prototyping (RP) for the fabrication of classy tissue engineering scaffolds (Fig. 3), is attracting the focus of the researcher around the world. RP employs a computer-aided design model to construct a layer-by-layer 3D architecture. The technique has shown accurate control of morphology, chemical composition, and mechanical properties. This is accomplished by sequentially depositing the layers of a material in various forms. 3D bioprinting permits the construction of exceedingly reproducible scaffolds. The size and shape are tailored exactly as per requirements [168,169]. Several additive processes are available for 3D printing. These include selective laser sintering, stereolithography, and fused deposition modeling [154,170]. The literature on the use of 3D printing in biomedical applications is gradually accumulating. Currently, only a few biopolymers can be used in 3Dpriniting to construct scaffolds for use in regenerative medicine [154]. More research is needed on the 3D printing technique for the construction of scaffold before it causes to produce scaffold commercially.

## 4.6 Melt mixing

Melt processing is another technique that has been used for the fabrication of composite material. During melt mixing, NPs are added to the polymer and then mixing of the mixture is carried under shear in molten state [171,172]. This method has advantages over other methods such as in situ intercalative polymerization and polymer solution intercalation techniques, particularly in the preparation of clay-containing bio-nanocomposites. Furthermore, this method is environmentally safe as no organic solvent is used. Also, melt processing is compatible with extrusion and injection molding, which are commonly used in industries. This method permits the utilization of biopolymers that were not appropriate for in situ polymerization. The properties of the bio-nanocomposites fabricated through melt processing depend on the thermodynamic of the interaction between the polymer and the NPs. For instance, the interaction is dependent on the transport/diffusion of polymer chains into the silicate interlayers from the bulk melt [171,173]. To obtain bio-nanocomposites with improved properties, polymer needs to be adequately compatible with the surface of the NPs. Improvement can only be expected if they are homogenously dispersed into the polymer matrix. Having said this, it is also important to note that other intermediate structures, such as intercalated and flocculated, are also formed. Two key factors determine the dispersion of NPs into the polymer matrix during melt processing: (i) processing conditions and (ii) enthalpic interaction between the polymer and the NPs. Favorable enthalpic interactions will lead to proper melt compounding. The need to optimize melt processing conditions is as important as favorable enthalpic interactions. This technique has not been fully explored in the fabrication of scaffold due to the lack of porosity in the final device. More studies are required to understand and fully explore the potential of melt mixing for the fabrication of the scaffold that could be useful in biomedical applications.

# 5 Modification of renewable polymers

## 5.1 Chemical modification

Chemical modification is an effective tool used to enhance the mechanical strength, and biological characteristics of natural polymers [174,175]. Since natural polymers contain several functionalities (e.g., amine and hydroxyl groups) that can be altered, numerous approaches have been used to alter these functionalities. These include polymer grafting [176], small molecule attachment [177], and chemical reactions (etherification, esterification, acetylation, silylation, quaternization, alkylation, and oxidation) [178]. Among these approaches, the grafting of polymer to the obtained tailored surface with desired properties is highly recommended [179]. Grafting is largely applied in polymers such as cellulose [180], chitosan [181,182], gelatin [183], and collagen [184]. This is usually done to obtain more suitable biomaterials for scaffolding. Three graftings approaches are in use to graft natural polymers and synthesize copolymers. These are "grafting from," "grafting to," and "grafting through." Out of these, the "grafting from" is a more regularly used approach to synthesize natural polymer-based copolymers [185]. In the "grafting from" approach, the use of reversible-deactivation radical polymerization (RDRP) has attracted more attention. The RDRP has three types; (a) nitroxide-mediated radical polymerization (NMRP) [186–188], (b) reversible addition of fragmentation chain transfer (RAFT) polymerization [189,190], and (c) atom transfer radical polymerization (ATRP) [191–193]. These techniques are employed to synthesize copolymers with regulated molecular weight, narrow molecular weight distribution, and complicated macromolecular structures. RAFT and ATRP are commonly reported in the literature. NMRP, however, did not get much popularity for the synthesis of natural polymer-based copolymers [185].

## 5.2 Blending natural polymers

Blend refers to a polymeric matrix composed of two or more polymers. The polymers in the mixture are well mixed processed either by solution casting technique or melting processing. Blending is usually carried out to enhance the physical and chemical properties of the pristine polymer [194]. Each polymer in the blend retains its specific chemical and biological properties. The blend produced by mixing polymers usually has higher strength and stiffness, low density, and weight compared to pristine polymers [12]. Natural polymers generally suffer due to low mechanical strength and are sensitive to humidity and temperature [12]. To tackle these issues, polymers are usually blended as mentioned either by solution mixing or melting mixing. Depending on the interactions of the polymers, blends are divided into two major classes: (i) miscible blends and (ii) immiscible blends. In miscible blends, polymers show likeness to each other. They usually show one glass transition temperature ($T_g$). These are analogous to random copolymers and homopolymers. Whereas in immiscible blends, polymers segregate from each other. They have multiple $T_g$ [195]. To make polymers miscible, compatibilizers are considered. Compatibilizers effectively reduce the interfacial tension and consequently increase the interaction between the polymers [196]. Strategies such as melt blending, freeze-drying, and electrospinning are used to prepare natural polymer-based blends [197]. Having said this, it must be kept in mind that melt blending is not an appropriate choice for protein-based natural polymer, since high temperature leads to denaturation and degradation of protein biopolymers [70]. Electrospinning of blends of natural polymers is an interesting technique to obtain nanofibrous scaffolds because they have unique characteristics: (a) similar morphology to ECM; (b) porosity and high surface area; (c) highly thin fibers (5–500 nm); (d) easy fabrication, scalable, and cheap; and (f) use of both organic and inorganic materials [198].

All of these properties meet the criteria for a scaffold that can be successfully used in tissue engineering. Several studies have been reported in literature using polymer blends to fabricate scaffolds that can be used for tissue engineering. Lin et al. studied the blending of keratin/chitosan to fabricate a scaffold for tissue engineering that retained the bioactivity of keratin and improved physicochemical properties of chitosan [199]. Another study showed improved thermal stability and physical properties for the spongy blend of silk fibroin/chitosan polymers compared to the sponge pure silk fibroin or pure CS [200]. A blend of gelatin and collagen accelerated the formation of an apatite layer on the blend films in bone tissue engineering, demonstrating its importance as a trigger for apatite nucleation [201]. Furthermore, hydroxyapatite (HAp) and chitosan interact poorly in an HAp/chitosan blend scaffold, resulting in weak physicochemical properties. The combination of CMC and chitosan could be a good option for scaffold fabrication because of the ionic crosslinking between two polyions of CS and CMC. HAp is incorporated into the CS-CMC matrix without aggregating. [202]. Fibronectin is a polypeptide that promotes cell adhesion, and chitosan is known for promoting stem cell differentiation into multiple lineages. The combination of the natural polymers as described already results in a more efficient scaffolding biomaterial [203,204]. To summarize, mixing two or more naturally occurring polymers is a cost-effective and flexible strategy for producing biomaterials with synergistic physicochemical and biological properties that qualify them for tissue engineering.

## 5.3 Blending natural and synthetic polymers

Synthetic polymers outperform natural polymers in terms of mechanical strength and thermal stability. The biological effects of synthetic polymers, on the contrary, are the most serious concerns (e.g., biocompatibility and biodegradability) [205]. Based on the basic properties of both forms of polymers, a new class of biomaterials will emerge by combining renewable and synthetic polymers. These mixtures are referred to as bioartificial or biosynthetic polymeric materials [205]. Biocompatible synthetic polymers including poly (vinyl alcohol) (PVA) and thermoplastic polyurethane (TPU) can therefore boost the mechanical strength of the resulting blends, making them candidates for tissue engineering [206]. Poly(-caprolactone) (PCL), an aliphatic and synthetic biodegradable polyester, is a commonly used polymer in tissue engineering along with other natural polymers, e.g., starch, gelatin, collagen, and chitosan, due to its superior mechanical strength and tunable degradation kinetics [207–209]. However, tests have shown that it has low cell affinity, shows a negative foreign body reaction in vivo, and lacks surface cell recognition sites [210]. Gelatin, on the contrary, is a naturally occurring polymer that is widely used in tissue engineering. Hydrophilic gelatin is distinguished by biological recognition, low immunogenicity, and antigenicity. Low mechanical strength and accelerated degradability, on the other hand, are its main drawbacks. Combining gelatin with PCL is an effective way to resolve the limitations of each polymer in tissue engineering [211]. Scaffolds for vascular grafts have been prepared from collagen and elastin (the main structural constituents of the ECM in vascular tissues) [212,213]. Although the scaffolds produced increase cell adhesion, proliferation, and showed efficient cell migration, they however lack the requisite mechanical properties. In contrast to neat collagen, data suggest that blending collagen with PCL or poly(L-lactide-*co*-caprolactone) (PLCL) improves the functional properties of the resulting scaffold [214]. Furthermore, combining chitosan with PCL combines chitosan's biological affinity (e.g., facilitation of cell adhesion and proliferation, provision of hydrophilicity and cell recognition sites, and the formation of a porous structure) with PCL's physicochemical properties (e.g., improving the mechanical strength) [215]. PVA is another significant synthetic polymer that is used to make a polymer-bioglass colloid device. Some studies have identified its biocompatibility and widespread use in the successful development of controlled delivery systems and tissue engineering [216–218]. Various studies, however, have suggested that PVA has a low capacity to penetrate living tissue [219]. Silk fibroin (SF) combined with PVA and a number of other synthetic biopolymers offers a good combination of properties for scaffold applications. These include enhanced biocompatibility, good mechanical strength, biodegradation, and improved biological properties [220–222]. PLGA is a commonly used scaffolding material due to its good mechanical strength, favorable biocompatibility, and biodegradable nature [223,224]. Synthetic polyesters are hydrophobic, and the lack functional groups limits their interaction with cell surfaces, which limits their use as scaffolds for tissue engineering [225,226]. To resolve the problem, electrospinning may be paired with natural and synthetic polymers blends. This will result in biomaterials with improved surface roughness, hydrophilicity, and cell adhesion [225].

## 5.4 Cross-linking approach

When the long chains of polymer are mutually linked together through either covalent bonding or intramolecular interaction, the process is called cross-linking. Cross-linking approach is adopted in the case of some natural polymers, particularly for polypeptides. It is usually considered a best option for the modification of polymers properties [227,228]. The ease, efficiency, and cost advantages of this approach have sparked intense interest in its implementation. Cross-linking increases mechanical properties and aqueous stability to some degree [14,229]. However, cross-linking reduces the degree and rate of degradation of polymers as well as accessibility to their functional groups. Other drawbacks of this approach include modifications in functionality, rheology and increased cytotoxicity [229]. Cross-linking methods are broadly divided into three types: chemical, physical, and enzymatic [230,231]. In chemical cross-linking, the long chains of polymers are linked via a covalent bond. Since some of the cross-linking agents are volatile, they cannot be dissolved in solvent [232]. Both small molecules (such as glutaraldehyde) and macromolecules (such as poly(carboxylic acids)) could be used for chemical cross-linking. Glutaraldehyde is by far the most commonly utilized cross-linking agent for renewable polymers, owing to its intrinsic properties such as reaction with different functional groups (e.g., amine and hydroxyl) and the ability to produce mechanically stable products [233,234]. However, in some instances, the glutaraldehyde-cross-linked compounds were cytotoxic [235]. Accordingly , green chemicals and more effective cross-linking agents can be used to develop biomaterials with physicochemical and biological properties appropriate for biomedical applications. Physical cross-linking should be used to prevent any of the harmful effects of chemical cross-linking. Physical cross-linking of natural polymers can be achieved by a range of processes, including ionic and hydrogen bonding interactions. The cross-linking of alginate with divalent cations such as calcium ($Ca^{2+}$) is a good example of physical cross-linking. Furthermore, the combination of CMC/starch [236], agar/gelatin [237], and methylcellulose/hyaluronic acid [238] results in physical cross-linking and forms a gel-like framework. Enzyme-catalyzed cross-linking is a comparatively recent and effective technique that has gained popularity owing to its improved characteristics such as high cross-linking performance, fast reaction, mild reaction parameters, and improved biocompatibility. This method is appropriate for in situ gelation systems [239]. More lately, transglutaminases (TGase; protein glutamine gamma-glutamyltransferase) [240] and horseradish peroxidase (HRP)/hydrogen peroxide ($H_2O_2$) [241] are also used as enzyme-based agents for the construction of diverse kinds of scaffolds.

# 6 Metal oxides for tissue engineering

## 6.1 *Silver/gold*

Gold (Au) NPs, known as colloidal gold with diameters ranging from 3 to 200nm may be easily produced with different shapes (nanorods, nanocubes, and nanostars). In general, the shape of Au NP is quasispherical (surface energy promotes the production of spherical particles). On the contrary, the shape of silver (Ag) NPs (diameter ranging from 1 to 100nm) varies from spherical to cubic. In the past two decades, Au and Ag NPs have been widely studied for various applications. The interest in the study of these NPs developed due to their unique and high surface-to-volume ratio, broad optical characteristics, ease of synthesis, and simple surface chemistry for functionalization. These properties have opened gates for their use in various biomedical applications. These applications include but are not limited to diagnostic assays, thermal ablation, radiotherapy enhancement, tissue engineering, and drug delivery. The detailed in vitro and in vivo studies of the Au and Ag NPs on different types of cancers revealed their promise as anticancer agents primarily due to their success against drug-resistant tumor cells via different routes. Besides, Au and Ag NPs were also studied against microbes. Ag NPs showed a wide-ranging antimicrobial activity against bacteria strains. These also include the antibiotic-resistant strain. The antimicrobial mechanism follows the inhibition of bacterial enzymatic activities, disruption of bacterial cell membranes, and disrupting DNA replication. Au NPs, on the contrary, did gain attention for antimicrobial activity, however, their antimicrobial mechanism is not clear. More studies are needed to fully understand their antimicrobial mechanism. Knowing the effectiveness of these NPs in the antimicrobial activity, recently, Au and Ag NPs were also considered as potential materials for applications in tissue regeneration. The advantage of using these NPs in tissue engineering lies in their tiny size, which enables them to pass through the cell membranes. Moreover, the size and surface properties can be modified as per requirements. In some studies, Au NPs were found to be potentially osteogenic for the regeneration of bone tissue. Au NPs encourage the differentiation of osteoblasts and mesenchymal stem cells via the instigation of extracellular signal-regulated kinases (ERK)/mitogen-activated protein kinases (MAPK), and p38 MAPK pathways, respectively. Cultural test studies of Au NPs/polymer composites (chitosan/Au NPs and gellan gum-coated Au nanorods) on mesenchymal stem cells (human adipose-derived) and osteoblast-like cells revealed a significant effect on stem cells. In another study, a thermoresponsive hydrogel composite was synthesized using chitosan and chitosan-stabilized Au NPs. The results of the study showed that the addition of Au NPs has improved the differentiation of the MSCs into cardiac lineages, which was attributed to the electrical cues in the matrix due to Au NP's electroconductive property. Thus, it was concluded that NP-hydrogel composites due to their conductivity boost connexin 43 expression, which could open the gate to new therapeutic opportunities in cardiac tissue engineering. Similarly, studies on Ag NPs also suggested that Ag NPs promoted human mesenchymal stem cell (MSC) proliferation. They concluded that this effect was associated with the hypoxia-inducible factor 1-alpha (HIF-1$\alpha$)-mediated upregulation of interleukin-8 (IL-8) expression. MSCs can differentiate into adipocytes, chondrocytes, and osteocytes. Another study showed Ag NP-induced osteogenic differentiation of urine-derived stem cells. Having discussed the positive aspect of the Ag NPs in the cell differentiation, it is also imperative to study the release of the Ag from the scaffold around the implant. Alarcon et al. studied Ag NP-collagen hydrogel scaffold and revealed that the Ag NPs were released after scaffold implantation. The released NPs accumulated in the tissues around the implant, and within 24 h of the implant, the NPs were found in the kidney, liver, and spleen. The study suggests extensive safety analysis before scaffold implant. Other than Ag and Au, the applications of other metals in tissue engineering were also studied. These will be discussed in the later sections. However, these studies are fewer in number as compared to Au and Ag NPs. The synthesis strategy potential of Au and Ag-polymer composite hydrogels and their application in skin, bone, and cardiac regeneration have been particularly highlighted and summarized in Fig. 4.

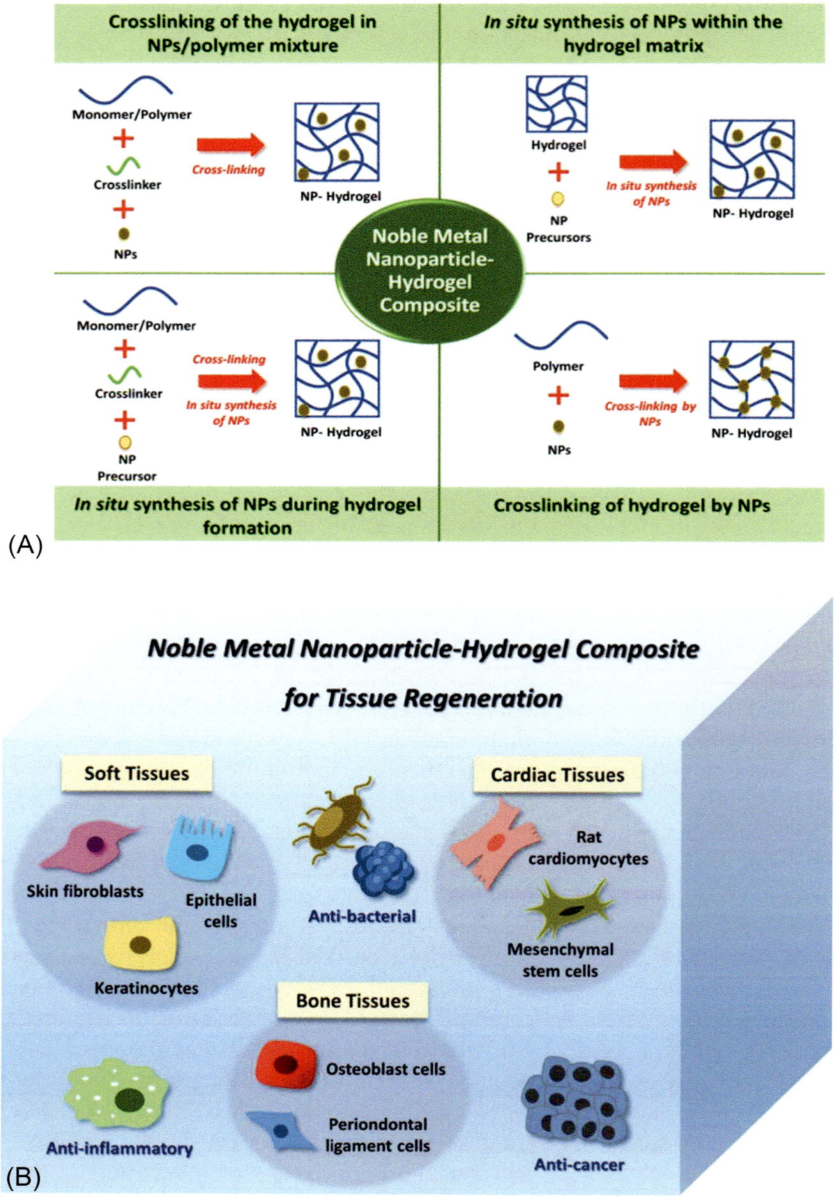


**Fig. 4** (A) Various methods can be used to integrate noble metal nanoparticles (NPs) into the cross-linked hydrogel matrix, and (B) noble metal nanoparticle (NP)-hydrogel composites for tissue regeneration. The composites have been shown to have the ability to regenerate tissues such as soft tissues, bone tissues, and cardiac tissues. Around the same time, the composites have some interesting bioactivities, such as antibacterial, antiinflammatory, and anticancer effects [242].

## 6.2 Zinc oxide

Zinc oxide (ZnO) is another metal oxide that has attracted the focus of researchers for its use in biomedical applications. This interest in the ZnO NPs is attributed to their anticancer, antibacterial, antidiabetic, antiinflammatory properties, as well as to their application in drug delivery, and bioimaging. Due to the characteristic toxicity of ZnO NPs, these NPs robustly inhibit cancerous cells and bacteria. The mechanisms are based on the generation of intracellular reactive oxygen species (ROS) and triggering apoptotic-signaling pathway. Apoptosis is a characteristic form of cell death demonstrating precise morphological and biochemical features. This includes cell membrane blebbing (sadden expansion of a part of the plasma membrane), chromatin condensation, genomic DNA destruction, and exposure of phagocytosis signaling molecules on the cell surface [243]. Besides, ZnO NPs are also known to stimulate the bioavailability of therapeutic drugs, when used as drug carriers, to attain improved therapy efficiency. ZnO NPs have also been found to increase insulin levels and decrease blood glucose. This further increases their importance in the treatment of diabetes. But to be used in the biomedical fields, the material should be approved by the food and drug authority (FDA). It is already reported in the literature that ZnO NPs are on the list of materials approved by the FDA. Having discussed the beneficial effects of the ZnO NPs, some critical issues still need to be explored. These include the nonexistence of studies on the comparative analysis of the biological advantages of ZnO NPs with other metal NPs. New research has revealed that ZnO NP is biologically toxic, there is no random research evidence on the therapeutic roles of ZnO NPs as anticancer, antibacterial, antiinflammatory, and antidiabetic activities, and finally, there is lack of complete understanding of in vivo anticancer, antibacterial, antiinflammatory, and antidiabetic activity. In order to confidently use ZnO in biomedicine, thorough research in these areas is required [244]. The development of new bone tissue requires the use of biomaterials with the ability to not only stimulate the rapid adhesion and proliferation of bone cells but also show a fast bioactive response. To achieve these objectives, various materials including bioglass matrices (biocompatible properties and fast bioactive response) are being investigated extensively. With regards to the safe use of ceramic material in tissue engineering, ZnO nanostructures are also being studied to achieve the aforementioned objectives. In one of the studies, Augustine et al. incorporated ZnO NPs into polycaprolactone (PCL) and showed fast degradation of the PCL. The degradation was investigated in vitro, by soaking the samples in simulated body fluid (SBF) for 30 days. They observed a change in crystalline structure and surface wettability due to degradation and formation of functional OH and COOH groups. Both these changes indicated that biodegradation has occurred. Furthermore, the mechanical properties of PCL decreased, when ZnO NPs were incorporated (especially at high ZnO loading). The faster degradation of the PCL/ZnO composite was attributed first to the decrease in the crystallinity of the composite and secondly to the ability of ZnO NPs to generate ROS. These species participated in the hydrolysis of PCL and accelerated the degradation of the composite in the aqueous environment. Besides, they also studied the effect of the ZnO NPs on the bioactive properties of composite foams. They did not observe HAp precipitation by soaking PCL/ZnO foams in SBF solution for 15 days. However, as the time was increased to 30 days, they observed the formation of a mineral phase. The precipitates of HAp were mostly observed in the ZnO-doped samples assuming the form of small dots as compared to the continuous layer. Also increased loading of ZnO led to lower precipitation of HAp. The creation of HAp was attributed to the existence of the characteristics OH group of ZnO NPs and the availability of more functional groups as a result of hydroxylation of PCL in SBF solution [245]. Guo et al. investigated the bioactive response of the glass/ZnO whiskers/gelatine composite scaffolds. Regardless of apatite layer creation, the release of Zn ions slowed the development of carbonated hydroxyapatite (CHAp). The formation of hydroxo-carbonate apatite (HCA) as a function of soaking time for the undoped composite scaffold (panels A–D) and doped scaffold (panels E–H) is shown in Fig. 5. The incorporation of ZnO into the composite resulted in the HCA layer appearing as a worm-like amorphous layer instead of the sponge-like crystalline. This change in morphology and crystallinity could be due to the fact that ZnO slows the release of Si ions from the bioglass system and accelerates the formation of a $CaO\text{-}P_2O_5$ barrier layer, preventing the precipitation of HCA, accelerating the precipitation of phosphates, and poorer crystallinity and crystal size could result from the substitution of Ca in HAp by Zn [246].

The incorporation of ZnO NPs and multiwall carbon nanotubes (MWCNTs) into polyurethane (PU) scaffold significantly accelerates osteogenic differentiation of MC3T3-E1 preosteoblast. The higher proliferation of cells was observed onto PU-ZnO NPs-MWCNTs composite compared to PU control after 2, 5, and 7 days of culture. Fig. 6 clearly shows preosteoblast cells, continuous cell proliferation, differentiation, and migration on the scaffold. PU-ZnO NPs-MWCNTs also improved cell viability as compared to PU control [247].

## 6.3 Titanium dioxide

Titanium dioxide ($TiO_2$) NPs are white solids. They are technologically considered important due to their various unique properties. These include abundant availability, easy synthesis, low elastic modulus, good tensile strength, good biocompatibility, antibacterial activity, photocatalytic activity, and corrosion resistance. $TiO_2$ NPs are widely used as catalyst support in biomedicine, water and air purification, pigments, cosmetics, solar cells, and tissue engineering. $TO_2$ NPs occurs in three polymorphic forms. These polymorphic forms include rutile, anatase, and brookite. The polymorphs of the $TiO_2$ attributed to the $Ti^{4+}$ cation coordinated by six oxygen atoms. This coordination results in a distorted octahedron. The shapes of these polymorphs are dictated by the combination of these octahedra in various patterns. The unit cell of anastas polymorph has four titanium dioxide molecules and a tetragonal structure. The rutile polymorph has a tetragonal structure and its unit cell has two titanium dioxide molecules. Brookite polymorph has an orthorhombic structure and its unit cell contains eight titanium dioxide molecules. Brookite is the least dense form of $TiO_2$ as compared to anatase and rutile. All these polymorphs have different properties. For example, the most stable polymorph is rutile, whereas anatase and brookite are metastable polymorphs. At high temperatures, these can irreversibly transform into rutile.

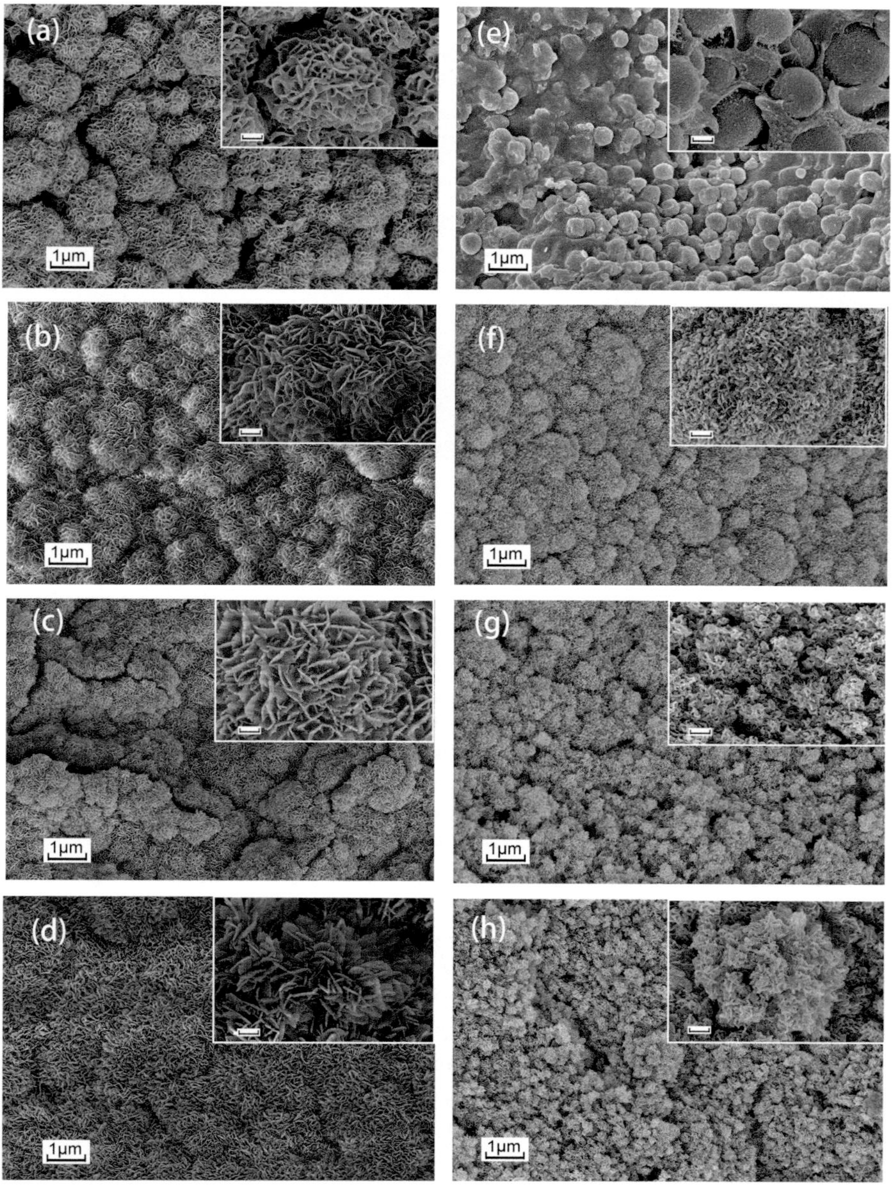


**Fig. 5** SEM images of the scaffolds after they had been saturated in SBF, undoped scaffold: (A) 1 day, (B) 3 days, (C) 5 days, and (D) 7 days; (E) 1 day, (F) 3 days, (G) 5 days, and (H) 7 days; and ZnO-doped scaffold: (E) 1 day, (F) 3 days, (G) 5 days, and (H) 7 days. Higher magnification is used in the insets (scale bar indicates 200 nm).
Reproduced with permission from W. Guo, F. Zhao, Y. Wang, J. Tang, X. Chen, Characterization of the mechanical behaviors and bioactivity of tetrapod ZnO whiskers reinforced bioactive glass/gelatin composite scaffolds, J. Mech. Behav. Biomed. Mater. 68 (2017) 8–15.

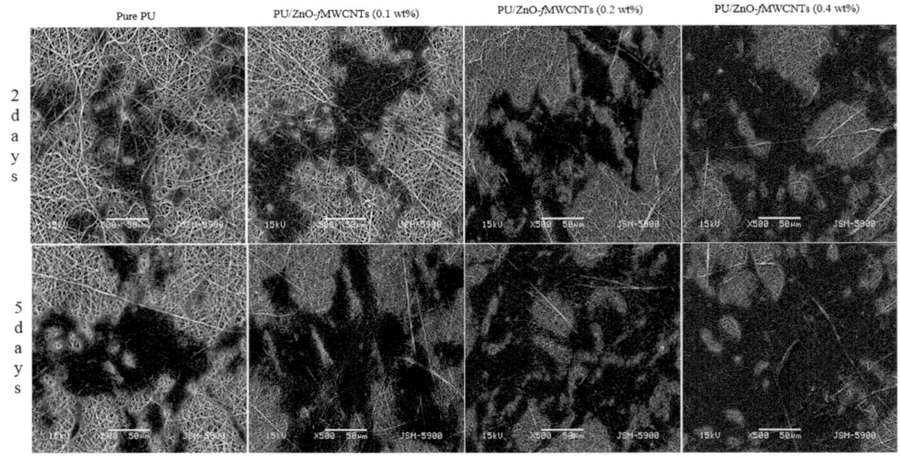


**Fig. 6** SEM micrographs of MC3T3-E1 preosteoblast adhesion and proliferation on various nanofibrous scaffolds for 2 and 5 days. The scale bar is 50 m high.
Reproduced with permission from B.K. Shrestha, S. Shrestha, A.P. Tiwari, J.-I. Kim, S.W. Ko, H.-J. Kim, C.H. Park, C.S. Kim, Bio-inspired hybrid scaffold of zinc oxide-functionalized multi-wall carbon nanotubes reinforced polyurethane nanofibers for bone tissue engineering, Mater. Des. 133 (2017) 69–81. Copyright (2017) Elsevier.

The phase transition of these polymorphs, and hence their stability, is affected by impurities, defects, grain size, and synthesis conditions. Anatase is reported to be stable at crystallite sizes smaller than 11–45 nm. This stability is attributed to the size associated with a high surface free energy. Recently, nanostructured $TiO_2$ has attracted the attention of researcher due to its aforementioned properties and its potential to be employed in various fields of science, engineering, and technology. Various methods are reported in the literature for the synthesis of $TiO_2$ nanostructure. These include anodization, sol-gel, hydrothermal, and electrospinning. However, among these methods, anodization, and sol-gel are more focused on biomedical applications primarily due to their low cost, simplicity, and ability to produce highly organized nanostructures with a high surface area to volume ratio. $TiO_2$ is considered a promising and nonresorbable material in bone implants. The recent reports in the literature showed that altering the surface topography of $TiO_2$ at the nanoscale has positively influenced cell behavior, improved osseointegration of the implant, and aided in the initial biointegration with surrounding tissues. The topography of the nanoscale $TiO_2$ mimics the extracellular matrix (ECM) of bone tissue, offering cellular support, and sufficient porosity for cell adhesion, proliferation, and differentiation during the formation of bone formation [248]. Other studies developed methods for the fabrication of scaffolds for bone tissue by combining $TiO_2$ NPs with biocompatible and biodegradable polymers [249]. In these studies, $TiO_2$ NPs were added to polymeric materials. These polymers include collagen and chitosan. The aim was to improve their mechanical strength, antibacterial (to avoid infection at the implant site) properties, and bioactivity [250]. These polymers were selected due to their similarity with

ECM and glycosaminoglycan. $TiO_2$ modified with vitamin $B_6$ (pyridoxal 5′-phosphate) has also significantly improved the hemophilic property at the implant-blood interface and inactivated platelet for cytokine growth and migration of osteoblasts [251].

## 6.4 Iron oxides

Iron oxide is a widely used material in different applications. Literature reported iron oxide in 17 forms as oxides, hydroxides, or oxide-hydroxides. So far six nonhydrated crystalline iron oxide phases are identified. These are classified into their crystal structures based on iron valence state. Among these crystal structures, magnetite ($Fe_3O_4$) and maghemite ($\gamma$-$Fe_2O_3$) have the potential for biomedical application primarily due to their functionalities and promising magnetic properties. The crystal structure and crystallographic data of hematite, magnetite, and maghemite are shown in Fig. 7. Magnetite ($Fe_3O_4$) is a ferromagnetic material with an inverse spinel structure, black. Magnetite exists in the $Fe^{2+}$ and $Fe^{3+}$ states. Similar to magnetite, maghemite ($\gamma$-$Fe_2O_3$) is also ferromagnetic but reddish brown. Magnetite and maghemite are isostructural. However, maghemite is cation-deficient. The important property of magnetite and maghemite is hidden in the crystal size. Both exhibit superparamagnetism, when their NPs are smaller than 20 nm. At low-dosage iron oxide NPs are reported non-toxic, biodegradable, and biocompatible.

The control of the topography of the scaffolds for tissue engineering offers biochemical cues. This has been widely reported in the literature. Magnetic scaffolds have recently been studied for the regeneration of injured tissues. The loading of magnetic nanoparticles (MNPs) into scaffolds significantly improves the growth and differentiation of bone cell. This potential of MNPs is attributed to the ability of the tissue to sense the mechanoelectric conversion that leads to improved cellular

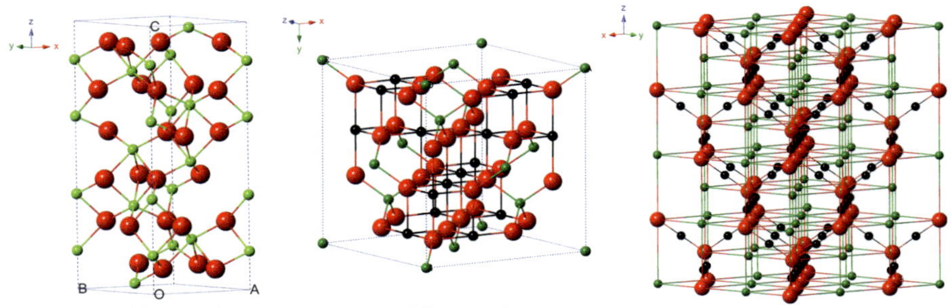


**Fig. 7** (A) Hematite, (B) magnetite, and (C) maghemite crystal structures and crystallographic details (the *black ball* is $Fe^{2+}$, the *green ball* is $Fe^{3+}$, and the *red ball* is $O_2$).
Reproduced with permission from W. Wu, Z. Wu, T. Yu, C. Jiang, W.-S. Kim, Recent progress on magnetic iron oxide nanoparticles: synthesis, surface functional strategies and biomedical applications, Sci. Technol. Adv. Mater. 16 (2) (2015) 023501. Copyright 2015, IOP Publishing.

spreading and differentiation. Over the period, several methods have been developed for the fabrication of magnetic scaffolds. However, in this section, we will limit ourselves to the methods using biopolymer and MNPs. The most common method for the fabrication of magnetic scaffolds is electrospinning. Several mechanisms of action are proposed for the MNP-based scaffolds, but in all cases the magnetic properties of the MNPs are utilized. A group of researchers fabricated magnetic scaffolds by dipping the composite of HAps/collagen in a MNP dispersion. Their in vitro study revealed that the scaffolds supported adhesion and proliferation of human bone marrow-derived stem cells. In another study, superparamagnetic MNPs/PLGA composite nanofibers with an average diameter in the range of 400–600 nm were prepared via electrospinning. Rosc17/2.8 (osteosarcoma cell lines) and MC3T3-E1 (osteoblast cell lines) were used in this study. The results revealed that MNPs/PLGA composite nanofibers promoted rapid and improved cell adhesion and proliferation compared to control (PLGA nanofibers). A research carried out with a modified emulsion technique and electrospinning using HAps/MNPs/PLA indicated that under an external magnetic field, the fabricated scaffolds prompted earlier and improved amounts of osteocalcin positive cells in situ, resulting in earlier and faster bone formation. Besides improved bioactivity, the magnetic field caused faster degradation of the scaffold [252].

## 6.5 Aluminum oxides

Alumina is an extremely stable oxide. The stability of alumina is attributed to the strong ionic and covalent bonds between aluminum and oxygen. These bonds make alumina corrosion resistant. This further makes the surface smooth and that is why alumina has exceptional corrosion resistance, low wear, and friction coefficient. Considering these properties, it is quite clear that this material will remain inert. This property of alumina was proved in 1975. The bonds also give alumina great hardness and abrasion resistance. These bonds make alumina insoluble at room temperature in most chemical reagent. Alumina has a hexagonal structure, where aluminum ions occupy the interstitial sites. The above properties have promoted the use of alumina in hard tissue engineering. The first usage of alumina in artificial implants was reported in 1970. Since then, studies have suggested no toxic effect of alumina implants in the bone marrow. Further, the implants based on alumina survive for a longer time [253]. As discussed bioceramics are widely employed in research mainly due to their better tissue responses compared to other materials (e.g., polymers and metals). Bioceramics such as HAp and alumina are envisioned as permanent devices, since they do not erode and do not release any of their components into the human body to cause any toxicity. Alumina is a highly interesting material. Alumina has also been used with polymers. In one of the studies, Peroglio et al. coated alumina scaffolds with PCL. The coating was carried out with two different methods, PCL solution and PCL nanodispersion. In both cases, PCL infiltrated the scaffolds. The presence of PCL on the surface and cracks in the alumina microstructure were confirmed. They suggested that elastic behavior is controlled by the ceramic scaffold and fracture energy by polymer phase [254]. In another case, thin PCL films were fabricated by adding different carbonaceous and ceramic particles to PCL (alumina, graphene, carbonated

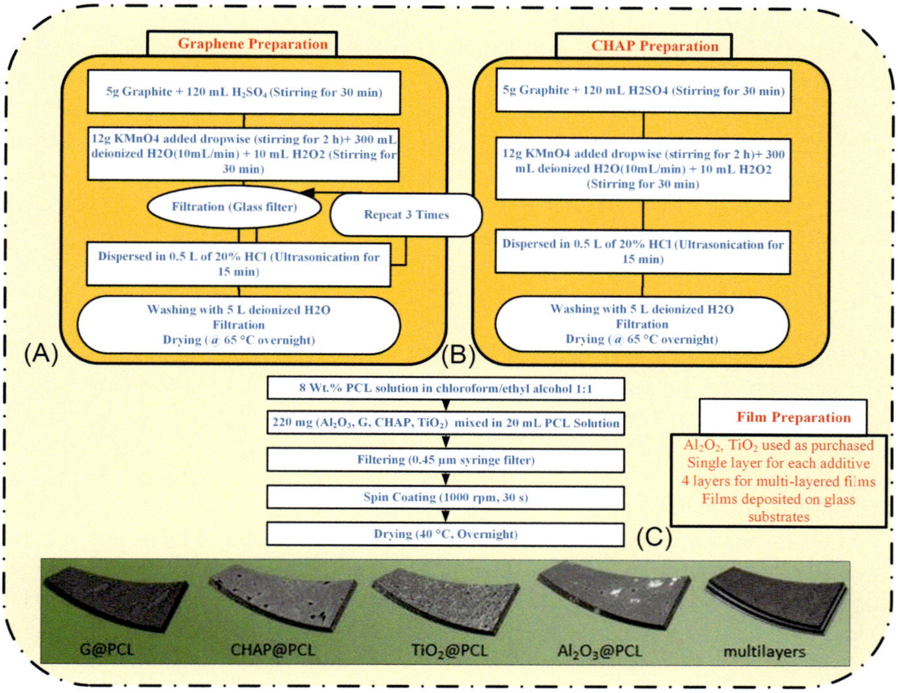


**Fig. 8** Flowchart for the synthesis and processing of $Al_2O_3$@PCL, G@PCL, CHAP@PCL, $TiO_2$@PCL, and ML-PCL thin films [255].

HAp, and $TiO_2$) using the spin coating technique. The samples were studied for composition, grain size, morphology, surface roughness, porosity, cell response, mechanical properties, and electrochemical performance. A highly rough surface was found for $TiO_2$@PCL while a low roughness was observed for gaphene@PCL (roughness averaging at 55 and 26 nm). As far the grain size, it was largest for alumina in $Al_2O_3$@PCL (70–145 μm) and smallest for $TiO_2$ in $TiO_2$@PCL (under a micrometer) (Fig. 8). The microhardness of $Al_2O_3$@PCL was the highest. Similarly, the lowest porosity was recorded for $Al_2O_3$@PCL compared to the others. Thanks to the highest porosity of $TiO_2$@PCL, more cells adhered, spread, and grew on the surface of $TiO_2$@PCL. The resistance to degradation based on electrochemical testing was higher for $Al_2O_3$@PCL, and this was also true for the properties of the alumina [255].

## 6.6 *Zirconium oxides*

Martin Heinrich Klaproth first introduced zirconia in 1789. Since then, it has been used as a pigment for ceramics. It occurs in three crystalline forms at different temperatures, e.g., at normal temperature it is monoclinic, which mostly occurs naturally, at elevated temperature it occurs in cubic and tetragonal crystalline forms. The toughness of zirconia increases during phase transformation from tetragonal to monoclinic (a stable phase) with a decrease in temperature. This property of zirconia can be used to tune its mechanical properties. Since only partially stable zirconia (with normal mechanical strength) can be employed for implant fabrication, it had to be stabilized partially. Yttria stabilized zirconia is a good example of partially stabilized zirconia. Yttria avoids phase transformation and stabilizes the tetragonal and cubic phases of zirconia. The stabilized zirconia-based implants usually have improved static and fatigue strengths. The presence of a single water molecule prompts phase transformation from tetragonal to monoclinic, which in turn leads to surface roughness and microcracking. An example of this is the slow degradation of femoral heads in the body after their implantation. Nonmetallic compounds (MgO, CaO, and $Y_2O_3$) can be used to improve the stability of zirconia. Due to good mechanical strength and wear performance zirconia is superior to alumina. Zirconia induces early bone growth and development because it is biocompatible [256]. Josset et al. studied the behavior of human osteoblasts cultured with zirconia discs. They observed improved osteoblasts adhesion, proliferation, and osteogenic differentiation [257]. Similarly, Scarano et al. investigated the response to zirconia implants introduced into the tibia of rabbits. They observed new bone formation, osteoblasts on zirconia and no inflammation. This good biocompatibility of zirconia is credited to its hydrophilicity and better protein adsorption [258]. Zirconia has also been used along with polymers in recent studies. One of the studies reported the preparation of porous PLGA/zirconia scaffolds, stabilized by amorphous calcium phosphate (Zr-ACP), scaffolds. The study observed an increase in mechanical strength, amorphous, and pore structures. The pore size for these scaffolds ranged from 3.6 to 154.4 μm (25% (w/v) Zr-ACP and 10% (w/v) PLGA/dioxane mixture). The compressive modulus increased to $51.9 \pm 5.6$ MPa (50:50 PLGA/Zr-ACP) from $14.7 \pm 6.8$ MPa (pristine PLGA). The scaffolds were well suited for biomedical applications [259]. Another study reported the development of a composite system where the growth factor BMP2 entrapped into PLGA microspheres, were integrated into a $ZrO_2$ scaffold. The study reported that the fully released BMP2 not only retained its biological activity but also induced osteoblastic marker expression hMSCs. Moreover, the study observed that coculture of hMSCs with endothelial cells overcame the limitations of nutrient and oxygen diffusion into the scaffold. Further studies are needed to determine the potential of these scaffolds for regeneration of bone tissue [260]. PLGA was also used for the preparation of a composite scaffold by loading a nanodiamond phospholipid compound (NDPC) by physical mixing. The results obtained in the study were very interesting. The loading of 10 wt% of NDPC into the PLGA resulted in a 100% increase in Young's modulus, an approximate 550% increase in hardness and a decrease in the water contact angle from 80 to 55 degrees. These results showed that this material could be a replacement implant for the human cortical bone. The study further observed that composite implanted into mice (for 8 weeks) induced satisfactory immune response and reduced the biodegradation of PLGA. The results of the in vitro and in vivo studies suggested the composite scaffold could have the potential bone tissue engineering [261].

# 7 Metal oxide cytotoxicity

Although the techniques to analyze the cytotoxic activity of NPs are not well understood, it is believed that the reactive oxygen species (ROS) are by far the most important factors. The cell defense system interacts with limited quantities of ROS; nevertheless, when the number of ROS increases the antioxidative potential of cells, the biomolecules are degraded during oxidation reactions that kill the cell. The effect of ROS on cells has three levels. In the first level, antioxidant enzyme activity increases to protect the cell. In the second level, there is an increase in proinflammatory cytokines that causes inflammation, and in the third level, there is mitochondrial dysfunctions, which cause cell death [262]. The key reason for the poisonous metallic NPs could be the release of their ions into the body environment, which migrate to other sensitive parts of the body. Therefore, it can be inferred that the antimicrobial activity of the metallic NPs is proportional to the amount of biologically active ions produced and their accessibility for contact with the bacterial cell wall. Metal ions will diffuse through cells, including healthy ones, after entering the bloodstream. Metal ion release from NPs is possible due to several factors, namely particle size, form, the nature of external surface coatings, and the conditions they are placed. Free ions have an effect on cell viability, trigger oxidative stress, and trigger cytokines to be released [263]. The manner in which NPs reach the cell is also essential. NPs that enter the cell via endocytosis can cause much more damage than NPs that enter via a distinct pathway. It is synonymous with the acidic environment in lysosomesfavoring the release of ions from NPs. Those that react with the cytoplasm lead to the formation of oxygen radicals (superoxide radicals). Through the protective action of the antioxidant enzyme superoxide dismutase (SOD) , these radicals are transformed into hydrogen peroxide ($H_2O_2$), which is then converted to water in the presence of another enzyme, catalase (CAT). However, as mentioned earlier, If the quantity of superoxide radicals is too high in comparison with CAT capability, a portion of the $H_2O_2$ does not decompose, reaches the nucleus, and undergoes Fenton-type reactions wherein ROS are formed, leading to the destruction of biomolecules and consequently to cell death [264]. Another concern that affects the stabilization of NPs and restricts their use is aggregation. Brownian motion induces particle collisions, which result in aggregation and precipitation. To tackle this issue, methods for producing stable NPs at neutral or acidic pH are developed. Shakiba et al. prepared Au NPs with high acidic stability by forming a self-assembled monolayer to enhance the particles interfacial properties. The creation of an aqueous membrane across the particles leads to high particle dispersion in the biological medium, resulting in increased stability [265]. Particles smaller than 30 nm are being used to achieve clarity in tanning UV safety materials and creams, where both ZnO and $TiO_2$ are commonly used for cosmetic and marketing purposes. Given the very tiny particle size, the risk of NPs penetrating the skin surface layers and entering viable cells should be addressed. Thankfully, in this case, in vitro, and in vivo studies showed that these NPs did not penetrate the layers of the skin [266]. Once within the body, NPs can easily disperse across the body and individual organs as well as invade cells via phagocytic and endocytic mechanisms. Wang et al. discovered that ZnO NPs given to mice orally gradually enter the liver, heart, spleen, pancreas, and bone. As a result, careful consideration is needed to analyze exposure to mild dosage of small-scale ZnO NPs [267]. The lack of applicable scientific and theoretical evidence is an undeniable limitation in the use of metal NPs and metal oxides [268]. However within a comparatively short period of time, NPs have achieved considerable acceptance in the biomedical field. Understanding the biological compatibility of NPs requires further studies, particularly with regard to the effects of metallic NPs after prolonged use.

# 8 Conclusions and prospects

Biodegradable polymers have great potential for environmental safety and biomedical applications. Currently, only a few biopolymers are of market value. Industry-sponsored research has intensified at universities to find good substitutes for synthetic polymers. Standards organizations such as ASTM and ISO have published methods for materials testing. NPs have also been considered as potential materials for applications in tissue regeneration. The advantage of the use of these NPs in tissue engineering is their tiny size that aids in their passage across the cell membranes. ZnO is a metal oxide that has attracted the focus of researchers for its use in biomedical applications. ZnO NPs have anticancer, antibacterial, antidiabetic, antiinflammatory properties and are used for drug delivery and bioimaging. ZnO NPs have been approved by the Food and Drug Authority (FDA) for use in biomedical application. Alumina is an extremely stable oxide with exceptional corrosion resistance, low wear, and friction coefficient, and is a good material for biomedical application. $TiO_2$ are white solids. $TiO_2$ NPs are commonly used in biomedicine as catalyst support for purification. MNPs have also been used in tissue engineering in combination with polymers. MNP-based scaffolds were prepared by dipping the composite of HAps/collagen in MNP dispersion. The fabricated scaffolds supported the adhesion and spreading of human bone marrow stem cells. Alumina is an extremely stable oxide. The stability of alumina is attributed to the strong ionic and covalent bonds between aluminum. Alumina has also been used with polymer in tissue engineering. Zirconia induces early bone growth and development as it is biocompatible and has good mechanical properties. However, it should be noted that the toxicity of NPs is gaining attention. The key reason why metallic NPs can be toxic is that their ions are released into the environment and dispersed across the body. Metal ions will diffuse through cells, including healthy ones, after entering the bloodstream. Metal ions release from NPs is possible due to several factors, namely particle size, form, the nature of external surface coatings, and the environment in which they are located. NPs that enter the cell via endocytosis can do more damage than NPs that enter via a different pathway. Studies have found that while using NPs in biomedical fields careful consideration should be given to small doses of small scale.

## References

[1] R.C. Thompson, C.J. Moore, F.S. vom Saal, S.H. Swan, Plastics, the environment and human health: current consensus and future trends, Philos. Trans. R. Soc. Lond. B Biol. Sci. 364 (1526) (2009) 2153–2166.
[2] J. Hopewell, R. Dvorak, E. Kosior, Plastics recycling: challenges and opportunities, Philos. Trans. R. Soc. Lond. B Biol. Sci. 364 (1526) (2009) 2115–2126.
[3] I. Vroman, L. Tighzert, Biodegradable polymers, Materials 2 (2) (2009) 307–344.
[4] I. Titorencu, M. Georgiana Albu, M. Nemecz, V. V Jjinga, Natural polymer-cell bioconstructs for bone tissue engineering, Curr. Stem Cell Res. Ther. 12 (2) (2017) 165–174.
[5] J.M. Dang, K.W. Leong, Natural polymers for gene delivery and tissue engineering, Adv. Drug Deliv. Rev. 58 (4) (2006) 487–499.
[6] K. Khoshnevisan, H. Maleki, H. Samadian, S. Shahsavari, M.H. Sarrafzadeh, B. Larijani, F.A. Dorkoosh, V. Haghpanah, M.R. Khorramizadeh, Cellulose acetate electrospun nanofibers for drug delivery systems: applications and recent advances, Carbohydr. Polym. 198 (2018) 131–141.
[7] P.B. Malafaya, G.A. Silva, R.L. Reis, Natural–origin polymers as carriers and scaffolds for biomolecules and cell delivery in tissue engineering applications, Adv. Drug Deliv. Rev. 59 (4-5) (2007) 207–233.
[8] X. Zhao, B. Guo, H. Wu, Y. Liang, P.X. Ma, Injectable antibacterial conductive nanocomposite cryogels with rapid shape recovery for noncompressible hemorrhage and wound healing, Nat. Commun. 9 (1) (2018) 1–17.
[9] J. Mano, G. Silva, H.S. Azevedo, P. Malafaya, R. Sousa, S.S. Silva, L. Boesel, J.M. Oliveira, T. Santos, A. Marques, Natural origin biodegradable systems in tissue engineering and regenerative medicine: present status and some moving trends, J. R. Soc. Interface 4 (17) (2007) 999–1030.
[10] E. Lavik, R. Langer, Tissue engineering: current state and perspectives, Appl. Microbiol. Biotechnol. 65 (1) (2004) 1–8.
[11] G. Kaur, R. Adhikari, P. Cass, M. Bown, P. Gunatillake, Electrically conductive polymers and composites for biomedical applications, RSC Adv. 5 (47) (2015) 37553–37567.
[12] K.M. Zia, S. Tabasum, M. Nasif, N. Sultan, N. Aslam, A. Noreen, M. Zuber, A review on synthesis, properties and applications of natural polymer based carrageenan blends and composites, Int. J. Biol. Macromol. 96 (2017) 282–301.
[13] D. Ozdil, H.M. Aydin, Polymers for medical and tissue engineering applications, J. Chem. Technol. Biotechnol. 89 (12) (2014) 1793–1810.
[14] K.-H. Jeong, D. Park, Y.-C. Lee, Polymer-based hydrogel scaffolds for skin tissue engineering applications: a mini-review, J. Polym. Res. 24 (7) (2017) 1–10.
[15] L. Moroni, M.K. Gunnewiek, E.M. Benetti, Polymer brush coatings regulating cell behavior: passive interfaces turn into active, Acta Biomater. 10 (6) (2014) 2367–2378.
[16] B. Gervet, The Use of Crude Oil in Plastic Making Contributes to Global Warming, Lulea University of Technology, Lulea, 2007.
[17] B.G.H. Almasi, B. Ghanbarzadeh, Biodegradable polymers, in: R. Chamy (Ed.), Biodegradation: Life of Science, InTech, Rijeka, Croatia, 2013.
[18] L. Yu, S. Petinakis, K. Dean, A. Bilyk, D. Wu, Green polymeric blends and composites from renewable resources, in: Macromolecular Symposia, Wiley Online Library, 2007, pp. 535–539.
[19] J. Farrin, Biodegradable Plastics from Natural Resources, Institute of Technology, Rochester, 2005.

[20] G.E. Luckachan, C. Pillai, Biodegradable polymers—a review on recent trends and emerging perspectives, J. Polym. Environ. 19 (3) (2011) 637–676.
[21] T.F. Garrison, A. Murawski, R.L. Quirino, Bio-based polymers with potential for biodegradability, Polymers 8 (7) (2016) 262.
[22] A. Folino, A. Karageorgiou, P.S. Calabrò, D. Komilis, Biodegradation of wasted bioplastics in natural and industrial environments: a review, Sustainability 12 (15) (2020) 6030.
[23] A. Morelli, D. Puppi, F. Chiellini, Polymers from renewable resources, J. Renew. Mater. 1 (2) (2013) 83–112.
[24] S. Kulkarni Vishakha, D. Butte Kishor, S. Rathod Sudha, Natural polymers—a comprehensive review, Int. J. Res. Pharmaceut. Biomed. Sci. 3 (4) (2012) 1597–1613.
[25] G.K. Jani, D.P. Shah, V.D. Prajapati, V.C. Jain, Gums and mucilages: versatile excipients for pharmaceutical formulations, Asian J. Pharm. Sci. 4 (5) (2009) 309–323.
[26] K.Y. Lee, L. Jeong, Y.O. Kang, S.J. Lee, W.H. Park, Electrospinning of polysaccharides for regenerative medicine, Adv. Drug Deliv. Rev. 61 (12) (2009) 1020–1032.
[27] X. Wang, H. Bai, Z. Yao, A. Liu, G. Shi, Electrically conductive and mechanically strong biomimetic chitosan/reduced graphene oxide composite films, J. Mater. Chem. 20 (41) (2010) 9032–9036.
[28] M.N.R. Kumar, A review of chitin and chitosan applications, React. Funct. Polym. 46 (1) (2000) 1–27.
[29] M. Rinaudo, Chitin and chitosan: properties and applications, Prog. Polym. Sci. 31 (7) (2006) 603–632.
[30] J.-K.F. Suh, H.W. Matthew, Application of chitosan-based polysaccharide biomaterials in cartilage tissue engineering: a review, Biomaterials 21 (24) (2000) 2589–2598.
[31] N. Kubota, Y. Eguchi, Facile preparation of water-soluble N-acetylated chitosan and molecular weight dependence of its water-solubility, Polym. J. 29 (2) (1997) 123–127.
[32] K. Tomihata, Y. Ikada, In vitro and in vivo degradation of films of chitin and its deacetylated derivatives, Biomaterials 18 (7) (1997) 567–575.
[33] S.-H. Chen, C.-T. Tsao, C.-H. Chang, Y.-T. Lai, M.-F. Wu, C.-N. Chuang, H.-C. Chou, C.-K. Wang, K.-H. Hsieh, Assessment of reinforced poly (ethylene glycol) chitosan hydrogels as dressings in a mouse skin wound defect model, Mater. Sci. Eng. C 33 (5) (2013) 2584–2594.
[34] M.G. Peter, Chitin and chitosan from animal sources, in: A. Steinbüchel, S.K. Rhee (Eds.), Polysaccharides and Polyamides in the Food Industry: Properties, Production, and Patents, WILEY-VCH Verlag GMBH & Co. KGaA, 2005, pp. 115–208.
[35] M. Dash, F. Chiellini, R.M. Ottenbrite, E. Chiellini, Chitosan—a versatile semi-synthetic polymer in biomedical applications, Prog. Polym. Sci. 36 (8) (2011) 981–1014.
[36] K. Kurita, Chitin and chitosan: functional biopolymers from marine crustaceans, Marine Biotechnol. 8 (3) (2006) 203–226.
[37] T. Kean, M. Thanou, Biodegradation, biodistribution and toxicity of chitosan, Adv. Drug Deliv. Rev. 62 (1) (2010) 3–11.
[38] C. Chatelet, O. Damour, A. Domard, Influence of the degree of acetylation on some biological properties of chitosan films, Biomaterials 22 (3) (2001) 261–268.
[39] D. Ren, H. Yi, W. Wang, X. Ma, The enzymatic degradation and swelling properties of chitosan matrices with different degrees of N-acetylation, Carbohydr. Res. 340 (15) (2005) 2403–2410.
[40] K. Divya, M. Jisha, Chitosan nanoparticles preparation and applications, Environ. Chem. Lett. 16 (1) (2018) 101–112.

[41] F. Croisier, C. Jérôme, Chitosan-based biomaterials for tissue engineering, Eur. Polym. J. 49 (4) (2013) 780–792.
[42] B. Duan, X. Yuan, Y. Zhu, Y. Zhang, X. Li, Y. Zhang, K. Yao, A nanofibrous composite membrane of PLGA–chitosan/PVA prepared by electrospinning, Eur. Polym. J. 42 (9) (2006) 2013–2022.
[43] X. Cai, H. Tong, X. Shen, W. Chen, J. Yan, J. Hu, Preparation and characterization of homogeneous chitosan–polylactic acid/hydroxyapatite nanocomposite for bone tissue engineering and evaluation of its mechanical properties, Acta Biomater. 5 (7) (2009) 2693–2703.
[44] Z. Li, H.R. Ramay, K.D. Hauch, D. Xiao, M. Zhang, Chitosan–alginate hybrid scaffolds for bone tissue engineering, Biomaterials 26 (18) (2005) 3919–3928.
[45] J.L. Cuy, B.L. Beckstead, C.D. Brown, A.S. Hoffman, C.M. Giachelli, Adhesive protein interactions with chitosan: consequences for valve endothelial cell growth on tissue-engineering materials, J. Biomed. Mater. Res. A 67 (2) (2003) 538–547.
[46] Y. Zhang, M. Zhang, Microstructural and mechanical characterization of chitosan scaffolds reinforced by calcium phosphates, J. Non Cryst. Solids 282 (2-3) (2001) 159–164.
[47] M. Gravel, R. Vago, M. Tabrizian, Use of natural coralline biomaterials as reinforcing and gas-forming agent for developing novel hybrid biomatrices: microarchitectural and mechanical studies, Tissue Eng. 12 (3) (2006) 589–600.
[48] P. Zheng, T. Ma, X. Ma, Fabrication and properties of starch-grafted graphene nanosheet/plasticized-starch composites, Ind. Eng. Chem. Res. 52 (39) (2013) 14201–14207.
[49] A. Buleon, P. Colonna, V. Planchot, S. Ball, Starch granules: structure and biosynthesis, Int. J. Biol. Macromol. 23 (2) (1998) 85–112.
[50] D. Lu, C. Xiao, S. Xu, Starch-based completely biodegradable polymer materials, Express Polym. Lett. 3 (6) (2009) 366–375.
[51] S. Ali Akbari Ghavimi, M.H. Ebrahimzadeh, M. Solati-Hashjin, N.A. Abu Osman, Polycaprolactone/starch composite: fabrication, structure, properties, and applications, J. Biomed. Mater. Res. A 103 (7) (2015) 2482–2498.
[52] M.N. Collins, C. Birkinshaw, Hyaluronic acid based scaffolds for tissue engineering—a review, Carbohydr. Polym. 92 (2) (2013) 1262–1279.
[53] B.F. Chong, L.M. Blank, R. Mclaughlin, L.K. Nielsen, Microbial hyaluronic acid production, Appl. Microbiol. Biotechnol. 66 (4) (2005) 341–351.
[54] G. Kogan, L. Šoltés, R. Stern, P. Gemeiner, Hyaluronic acid: a natural biopolymer with a broad range of biomedical and industrial applications, Biotechnol. Lett. 29 (1) (2007) 17–25.
[55] D.D. Allison, K.J. Grande-Allen, Hyaluronan: a powerful tissue engineering tool, Tissue Eng. 12 (8) (2006) 2131–2140.
[56] L. Liu, Y. Liu, J. Li, G. Du, J. Chen, Microbial production of hyaluronic acid: current state, challenges, and perspectives, Microb. Cell Fact. 10 (1) (2011) 1–9.
[57] J.E. Scott, F. Heatley, Biological properties of hyaluronan in aqueous solution are controlled and sequestered by reversible tertiary structures, defined by NMR spectroscopy, Biomacromolecules 3 (3) (2002) 547–553.
[58] K.Y. Lee, D.J. Mooney, Alginate: properties and biomedical applications, Prog. Polym. Sci. 37 (1) (2012) 106–126.
[59] M. Otterlei, K. Ostgaard, G. Skjåk-Bræk, O. Smidsrød, P. Soon-Shiong, T. Espevik, Induction of cytokine production from human monocytes stimulated with alginate, J. Immunother. 10 (4) (1991) 286–291.
[60] U. Zimmermann, G. Klöck, K. Federlin, K. Hannig, M. Kowalski, R.G. Bretzel, A. Horcher, H. Entenmann, U. Sieber, T. Zekorn, Production of mitogen-contamination

free alginates with variable ratios of mannuronic acid to guluronic acid by free flow electrophoresis, Electrophoresis 13 (1) (1992) 269–274.
[61] H.K. Holme, L. Davidsen, A. Kristiansen, O. Smidsrød, Kinetics and mechanisms of depolymerization of alginate and chitosan in aqueous solution, Carbohydr. Polym. 73 (4) (2008) 656–664.
[62] M. Machluf, Protein therapeutic delivery using encapsulated cell platform, in: Applications of Cell Immobilisation Biotechnology, Springer, 2005, pp. 197–209.
[63] G.T. Grant, E.R. Morris, D.A. Rees, P.J. Smith, D. Thom, Biological interactions between polysaccharides and divalent cations: the egg-box model, FEBS Lett. 32 (1) (1973) 195–198.
[64] J. Sun, H. Tan, Alginate-based biomaterials for regenerative medicine applications, Materials 6 (4) (2013) 1285–1309.
[65] K.L. Douglas, M. Tabrizian, Effect of experimental parameters on the formation of alginate–chitosan nanoparticles and evaluation of their potential application as DNA carrier, J. Biomater. Sci. Polym. Ed. 16 (1) (2005) 43–56.
[66] A. Tampieri, M. Sandri, E. Landi, G. Celotti, N. Roveri, M. Mattioli-Belmonte, L. Virgili, F. Gabbanelli, G. Biagini, HA/alginate hybrid composites prepared through bio-inspired nucleation, Acta Biomater. 1 (3) (2005) 343–351.
[67] J. Lee, K.Y. Lee, Local and sustained vascular endothelial growth factor delivery for angiogenesis using an injectable system, Pharm. Res. 26 (7) (2009) 1739–1744.
[68] V. Sridhar, I. Lee, H. Chun, H. Park, Graphene reinforced biodegradable poly (3-hydroxybutyrate-co-4-hydroxybutyrate) nano-composites, Express Polym. Lett. 7 (4) (2013) 320–328.
[69] N. Ninan, M. Muthiah, I.-K. Park, T.W. Wong, S. Thomas, Y. Grohens, Natural polymer/inorganic material based hybrid scaffolds for skin wound healing, Polym. Rev. 55 (3) (2015) 453–490.
[70] P. Gupta, K.K. Nayak, Characteristics of protein-based biopolymer and its application, Polym. Eng. Sci. 55 (3) (2015) 485–498.
[71] C. He, X. Zhuang, Z. Tang, H. Tian, X. Chen, Stimuli-sensitive synthetic polypeptide-based materials for drug and gene delivery, Adv. Healthc. Mater. 1 (1) (2012) 48–78.
[72] S.S. Amruthwar, A.V. Janorkar, In vitro evaluation of elastin-like polypeptide–collagen composite scaffold for bone tissue engineering, Dent. Mater. 29 (2) (2013) 211–220.
[73] C.H. Lee, A. Singla, Y. Lee, Biomedical applications of collagen, Int. J. Pharm. 221 (1-2) (2001) 1–22.
[74] H. Chajra, C. Rousseau, D. Cortial, M. Ronziere, D. Herbage, F. Mallein-Gerin, A. Freyria, Collagen-based biomaterials and cartilage engineering. Application to osteochondral defects, Biomed. Mater. Eng. 18 (s1) (2008) 33–45.
[75] J.W. Nichol, S.T. Koshy, H. Bae, C.M. Hwang, S. Yamanlar, A. Khademhosseini, Cell-laden microengineered gelatin methacrylate hydrogels, Biomaterials 31 (21) (2010) 5536–5544.
[76] Y.-H. Lee, G. Bhattarai, S. Aryal, N.-H. Lee, M.-H. Lee, T.-G. Kim, E.-C. Jhee, H.-Y. Kim, H.-K. Yi, Modified titanium surface with gelatin nano gold composite increases osteoblast cell biocompatibility, Appl. Surf. Sci. 256 (20) (2010) 5882–5887.
[77] A.J. Kuijpers, G.H. Engbers, J. Krijgsveld, S.A. Zaat, J. Dankert, J. Feijen, Cross-linking and characterisation of gelatin matrices for biomedical applications, J. Biomater. Sci. Polym. Ed. 11 (3) (2000) 225–243.
[78] H. Li, C. Tan, L. Li, Review of 3D printable hydrogels and constructs, Mater. Des. 159 (2018) 20–38.

[79] L.M. Delgado, N. Shologu, K. Fuller, D.I. Zeugolis, Acetic acid and pepsin result in high yield, high purity and low macrophage response collagen for biomedical applications, Biomed. Mater. 12 (6) (2017), 065009.
[80] B. An, D.L. Kaplan, B. Brodsky, Engineered recombinant bacterial collagen as an alternative collagen-based biomaterial for tissue engineering, Front. Chem. 2 (2014) 40.
[81] M. Fujioka-Kobayashi, B. Schaller, N. Saulacic, B.E. Pippenger, Y. Zhang, R.J. Miron, Absorbable collagen sponges loaded with recombinant bone morphogenetic protein 9 induces greater osteoblast differentiation when compared to bone morphogenetic protein 2, Clin. Exp. Dent. Res. 3 (1) (2017) 32–40.
[82] H. Pulkkinen, V. Tiitu, P. Valonen, J. Jurvelin, M. Lammi, I. Kiviranta, Engineering of cartilage in recombinant human type II collagen gel in nude mouse model in vivo, Osteoarthr. Cartil. 18 (8) (2010) 1077–1087.
[83] F. Chamieh, A.-M. Collignon, B.R. Coyac, J. Lesieur, S. Ribes, J. Sadoine, A. Llorens, A. Nicoletti, D. Letourneur, M.-L. Colombier, Accelerated craniofacial bone regeneration through dense collagen gel scaffolds seeded with dental pulp stem cells, Sci. Rep. 6 (1) (2016) 1–11.
[84] C. Dhand, S.T. Ong, N. Dwivedi, S.M. Diaz, J.R. Venugopal, B. Navaneethan, M.H. Fazil, S. Liu, V. Seitz, E. Wintermantel, Bio-inspired in situ crosslinking and mineralization of electrospun collagen scaffolds for bone tissue engineering, Biomaterials 104 (2016) 323–338.
[85] E.D. Boland, J.A. Matthews, K.J. Pawlowski, D.G. Simpson, G.E. Wnek, G.L. Bowlin, Electrospinning collagen and elastin: preliminary vascular tissue engineering, Front. Biosci. 9 (1422) (2004) e32.
[86] J. Venugopal, L. Ma, T. Yong, S. Ramakrishna, In vitro study of smooth muscle cells on polycaprolactone and collagen nanofibrous matrices, Cell Biol. Int. 29 (10) (2005) 861–867.
[87] J. Stitzel, J. Liu, S.J. Lee, M. Komura, J. Berry, S. Soker, G. Lim, M. Van Dyke, R. Czerw, J.J. Yoo, Controlled fabrication of a biological vascular substitute, Biomaterials 27 (7) (2006) 1088–1094.
[88] N.M. Blanco, J. Edwards, W.A. Zamboni, Dermal substitute (Integra) for open nasal wounds, Plast. Reconstr. Surg. 113 (7) (2004) 2224–2225.
[89] S.G. Priya, H. Jungvid, A. Kumar, Skin tissue engineering for tissue repair and regeneration, Tissue Eng. Part B Rev. 14 (1) (2008) 105–118.
[90] Y. Bello, A. Falabella, The role of graftskin (Apligraf) in difficult-to-heal venous leg ulcers, J. Wound Care 11 (5) (2002) 182–183.
[91] A. Atala, S.B. Bauer, S. Soker, J.J. Yoo, A.B. Retik, Tissue-engineered autologous bladders for patients needing cystoplasty, Lancet 367 (9518) (2006) 1241–1246.
[92] A.M. Raya-Rivera, D. Esquiliano, R. Fierro-Pastrana, E. López-Bayghen, P. Valencia, R. Ordorica-Flores, S. Soker, J.J. Yoo, A. Atala, Tissue-engineered autologous vaginal organs in patients: a pilot cohort study, Lancet 384 (9940) (2014) 329–336.
[93] Y. Hu, X. Sun, Chemically functionalized graphene and their applications in electrochemical energy conversion and storage, in: Advances in Graphene Science, IntechOpen, 2013, pp. 161–189.
[94] Y. Li, J. Rodrigues, H. Tomás, Injectable and biodegradable hydrogels: gelation, biodegradation and biomedical applications, Chem. Soc. Rev. 41 (6) (2012) 2193–2221.
[95] Y. Tabata, Y. Ikada, Protein release from gelatin matrices, Adv. Drug Deliv. Rev. 31 (3) (1998) 287–301.
[96] D. Hellio, M. Djabourov, Physically and chemically crosslinked gelatin gels, in: Macromolecular Symposia, Wiley Online Library, 2006, pp. 23–27.

[97] K. Yamada, Y. Tabata, K. Yamamoto, S. Miyamoto, I. Nagata, H. Kikuchi, Y. Ikada, Potential efficacy of basic fibroblast growth factor incorporated in biodegradable hydrogels for skull bone regeneration, J. Neurosurg. 86 (5) (1997) 871–875.
[98] M. Yamamoto, Y. Takahashi, Y. Tabata, Controlled release by biodegradable hydrogels enhances the ectopic bone formation of bone morphogenetic protein, Biomaterials 24 (24) (2003) 4375–4383.
[99] Z.S. Patel, M. Yamamoto, H. Ueda, Y. Tabata, A.G. Mikos, Biodegradable gelatin microparticles as delivery systems for the controlled release of bone morphogenetic protein-2, Acta Biomater. 4 (5) (2008) 1126–1138.
[100] H. Park, J.S. Temenoff, T.A. Holland, Y. Tabata, A.G. Mikos, Delivery of TGF-β1 and chondrocytes via injectable, biodegradable hydrogels for cartilage tissue engineering applications, Biomaterials 26 (34) (2005) 7095–7103.
[101] J. Malda, E. Kreijveld, J. Temenoff, C. Van Blitterswijk, J. Riesle, Expansion of human nasal chondrocytes on macroporous microcarriers enhances redifferentiation, Biomaterials 24 (28) (2003) 5153–5161.
[102] R.G. Payne, M.J. Yaszemski, A.W. Yasko, A.G. Mikos, Development of an injectable, in situ crosslinkable, degradable polymeric carrier for osteogenic cell populations. Part 1. Encapsulation of marrow stromal osteoblasts in surface crosslinked gelatin microparticles, Biomaterials 23 (22) (2002) 4359–4371.
[103] Y. Liu, X.Z. Shu, G.D. Prestwich, Osteochondral defect repair with autologous bone marrow–derived mesenchymal stem cells in an injectable, in situ, cross-linked synthetic extracellular matrix, Tissue Eng. 12 (12) (2006) 3405–3416.
[104] Y. Kimura, M. Ozeki, T. Inamoto, Y. Tabata, Adipose tissue engineering based on human preadipocytes combined with gelatin microspheres containing basic fibroblast growth factor, Biomaterials 24 (14) (2003) 2513–2521.
[105] S.J. Shirbin, F. Karimi, N.J.-A. Chan, D.E. Heath, G.G. Qiao, Macroporous hydrogels composed entirely of synthetic polypeptides: biocompatible and enzyme biodegradable 3D cellular scaffolds, Biomacromolecules 17 (9) (2016) 2981–2991.
[106] P. Zorlutuna, N.E. Vrana, A. Khademhosseini, The expanding world of tissue engineering: the building blocks and new applications of tissue engineered constructs, IEEE Rev. Biomed. Eng. 6 (2012) 47–62.
[107] D. Chow, M.L. Nunalee, D.W. Lim, A.J. Simnick, A. Chilkoti, Peptide-based biopolymers in biomedicine and biotechnology, Mater. Sci. Eng. R. Rep. 62 (4) (2008) 125–155.
[108] A.E. Kieffer, Y. Goumon, O. Ruh, S. Chasserot-Golaz, G. Nullans, C. Gasnier, D. Aunis, M.H. Metz-Boutigue, The N-and C-terminal fragments of ubiquitin are important for the antimicrobial activities, FASEB J. 17 (6) (2003) 776–778.
[109] J. Wu, Advances in the production and functional properties of corn protein peptides, in: IOP Conference Series: Earth and Environmental Science, IOP Publishing, 2020, p. 012089.
[110] M. Mondal, K. Trivedy, K.S. Nirmal, The silk proteins, sericin and fibroin in silkworm, Bombyx mori Linn.,-a review, Casp. J. Environ. Sci. 5 (2007) 63–76.
[111] M.B. Hinman, J.A. Jones, R.V. Lewis, Synthetic spider silk: a modular fiber, Trends Biotechnol. 18 (9) (2000) 374–379.
[112] L. Uebersax, T. Apfel, K.M. Nuss, R. Vogt, H.Y. Kim, L. Meinel, D.L. Kaplan, J.A. Auer, H.P. Merkle, B. von Rechenberg, Biocompatibility and osteoconduction of macroporous silk fibroin implants in cortical defects in sheep, Eur. J. Pharm. Biopharm. 85 (1) (2013) 107–118.
[113] T. Chlapanidas, S. Faragò, F. Mingotto, F. Crovato, M.C. Tosca, B. Antonioli, M. Bucco, G. Lucconi, A. Scalise, D. Vigo, Regenerated silk fibroin scaffold and infrapatellar

adipose stromal vascular fraction as feeder-layer: a new product for cartilage advanced therapy, Tissue Eng. Part A 17 (13-14) (2011) 1725–1733.
[114] J. Plank, Applications of biopolymers and other biotechnological products in building materials, Appl. Microbiol. Biotechnol. 66 (1) (2004) 1–9.
[115] C. Vepari, D.L. Kaplan, Silk as a biomaterial, Prog. Polym. Sci. 32 (8-9) (2007) 991–1007.
[116] D.N. Rockwood, R.C. Preda, T. Yücel, X. Wang, M.L. Lovett, D.L. Kaplan, Materials fabrication from Bombyx mori silk fibroin, Nat. Protoc. 6 (10) (2011) 1612.
[117] F. Vollrath, D.P. Knight, Liquid crystalline spinning of spider silk, Nature 410 (6828) (2001) 541–548.
[118] B. Kundu, R. Rajkhowa, S.C. Kundu, X. Wang, Silk fibroin biomaterials for tissue regenerations, Adv. Drug Deliv. Rev. 65 (4) (2013) 457–470.
[119] R.E. Unger, M. Wolf, K. Peters, A. Motta, C. Migliaresi, C.J. Kirkpatrick, Growth of human cells on a non-woven silk fibroin net: a potential for use in tissue engineering, Biomaterials 25 (6) (2004) 1069–1075.
[120] F. Zhang, Z. Zhang, X. Zhu, E.-T. Kang, K.-G. Neoh, Silk-functionalized titanium surfaces for enhancing osteoblast functions and reducing bacterial adhesion, Biomaterials 29 (36) (2008) 4751–4759.
[121] C. Pignatelli, G. Perotto, M. Nardini, R. Cancedda, M. Mastrogiacomo, A. Athanassiou, Electrospun silk fibroin fibers for storage and controlled release of human platelet lysate, Acta Biomater. 73 (2018) 365–376.
[122] L. Meinel, S. Hofmann, O. Betz, R. Fajardo, H.P. Merkle, R. Langer, C.H. Evans, G. - Vunjak-Novakovic, D.L. Kaplan, Osteogenesis by human mesenchymal stem cells cultured on silk biomaterials: comparison of adenovirus mediated gene transfer and protein delivery of BMP-2, Biomaterials 27 (28) (2006) 4993–5002.
[123] X. Zhang, C. Cao, X. Ma, Y. Li, Optimization of macroporous 3-D silk fibroin scaffolds by salt-leaching procedure in organic solvent-free conditions, J. Mater. Sci. Mater. Med. 23 (2) (2012) 315–324.
[124] B.H. Rehm, Bacterial polymers: biosynthesis, modifications and applications, Nat. Rev. Microbiol. 8 (8) (2010) 578–592.
[125] M. Jamshidian, E.A. Tehrany, M. Imran, M. Jacquot, S. Desobry, Poly-lactic acid: production, applications, nanocomposites, and release studies, Compr. Rev. Food Sci. Food Saf. 9 (5) (2010) 552–571.
[126] D. Garlotta, A literature review of poly (lactic acid), J. Polym. Environ. 9 (2) (2001) 63–84.
[127] D.L. Kaplan, Introduction to biopolymers from renewable resources, in: Biopolymers from Renewable Resources, Springer, 1998, pp. 1–29.
[128] L. Avérous, Polylactic acid: synthesis, properties and applications, in: Monomers, Polymers and Composites from Renewable Resources, Elsevier, 2008, pp. 433–450.
[129] A.J. Lasprilla, G.A. Martinez, B.H. Lunelli, A.L. Jardini, R. Maciel Filho, Poly-lactic acid synthesis for application in biomedical devices—a review, Biotechnol. Adv. 30 (1) (2012) 321–328.
[130] E.C. King, A.J. Blacker, T.D. Bugg, Enzymatic breakdown of poly-γ-D-glutamic acid in Bacillus licheniformis: identification of a polyglutamyl γ-hydrolase enzyme, Biomacromolecules 1 (1) (2000) 75–83.
[131] A. Ogunleye, A. Bhat, V.U. Irorere, D. Hill, C. Williams, I. Radecka, Poly-γ-glutamic acid: production, properties and applications, Microbiology 161 (1) (2015) 1–17.
[132] K. Grage, A.C. Jahns, N. Parlane, R. Palanisamy, I.A. Rasiah, J.A. Atwood, B.H. Rehm, Bacterial polyhydroxyalkanoate granules: biogenesis, structure, and potential use as

nano-/micro-beads in biotechnological and biomedical applications, Biomacromolecules 10 (4) (2009) 660–669.
[133] K. Sudesh, H. Abe, Y. Doi, Synthesis, structure and properties of polyhydroxyalkanoates: biological polyesters, Prog. Polym. Sci. 25 (10) (2000) 1503–1555.
[134] S. Khanna, A.K. Srivastava, Recent advances in microbial polyhydroxyalkanoates, Process Biochem. 40 (2) (2005) 607–619.
[135] N.A. Hoenich, Cellulose for medical applications: past, present, and future, BioResources 1 (2) (2006) 270–280.
[136] M. Ul-Islam, T. Khan, J.K. Park, Water holding and release properties of bacterial cellulose obtained by in situ and ex situ modification, Carbohydr. Polym. 88 (2) (2012) 596–603.
[137] J. Wu, Y. Zheng, W. Song, J. Luan, X. Wen, Z. Wu, X. Chen, Q. Wang, S. Guo, In situ synthesis of silver-nanoparticles/bacterial cellulose composites for slow-released antimicrobial wound dressing, Carbohydr. Polym. 102 (2014) 762–771.
[138] N. Tazi, Z. Zhang, Y. Messaddeq, L. Almeida-Lopes, L.M. Zanardi, D. Levinson, M. Rouabhia, Hydroxyapatite bioactivated bacterial cellulose promotes osteoblast growth and the formation of bone nodules, AMB Express 2 (1) (2012) 1–10.
[139] G.F. Picheth, C.L. Pirich, M.R. Sierakowski, M.A. Woehl, C.N. Sakakibara, C.F. de Souza, A.A. Martin, R. da Silva, R.A. de Freitas, Bacterial cellulose in biomedical applications: a review, Int. J. Biol. Macromol. 104 (2017) 97–106.
[140] D. Klemm, B. Heublein, H.P. Fink, A. Bohn, Cellulose: fascinating biopolymer and sustainable raw material, Angew. Chem. Int. Ed. 44 (22) (2005) 3358–3393.
[141] K. Jedvert, T. Heinze, Cellulose modification and shaping—a review, J. Polym. Eng. 37 (9) (2017) 845–860.
[142] M. Naessens, A. Cerdobbel, W. Soetaert, E.J. Vandamme, Leuconostoc dextransucrase and dextran: production, properties and applications, J. Chem. Technol. Biotechnol. 80 (8) (2005) 845–860.
[143] D. Puppi, F. Chiellini, A.M. Piras, E. Chiellini, Polymeric materials for bone and cartilage repair, Prog. Polym. Sci. 35 (4) (2010) 403–440.
[144] M.M. Stevens, J.H. George, Exploring and engineering the cell surface interface, Science 310 (5751) (2005) 1135–1138.
[145] A.J. Engler, S. Sen, H.L. Sweeney, D.E. Discher, Matrix elasticity directs stem cell lineage specification, Cell 126 (4) (2006) 677–689.
[146] H.N. Kim, A. Jiao, N.S. Hwang, M.S. Kim, D.-H. Kim, K.-Y. Suh, Nanotopography-guided tissue engineering and regenerative medicine, Adv. Drug Deliv. Rev. 65 (4) (2013) 536–558.
[147] C.J. Liao, C.F. Chen, J.H. Chen, S.F. Chiang, Y.J. Lin, K.Y. Chang, Fabrication of porous biodegradable polymer scaffolds using a solvent merging/particulate leaching method, J. Biomed. Mater. Res. 59 (4) (2002) 676–681.
[148] L. Draghi, S. Resta, M.G. Pirozzolo, M.C. Tanzi, Microspheres leaching for scaffold porosity control, J. Mater. Sci. Mater. Med. 16 (12) (2005) 1093–1097.
[149] M.J. Moore, E. Jabbari, E.L. Ritman, L. Lu, B.L. Currier, A.J. Windebank, M.J. Yaszemski, Quantitative analysis of interconnectivity of porous biodegradable scaffolds with micro-computed tomography, J. Biomed. Mater. Res. A 71 (2) (2004) 258–267.
[150] J. Zhang, L. Wu, D. Jing, J. Ding, A comparative study of porous scaffolds with cubic and spherical macropores, Polymer 46 (13) (2005) 4979–4985.
[151] M. Okamoto, Biodegradable polymer/layered silicate nanocomposites: a review, J. Ind. Eng. Chem. 10 (7) (2004) 1156–1181.

[152] A.G. Mikos, G. Sarakinos, S.M. Leite, J.P. Vacant, R. Langer, Laminated three-dimensional biodegradable foams for use in tissue engineering, Biomaterials 14 (5) (1993) 323–330.
[153] J. Xie, X. Li, Y. Xia, Putting electrospun nanofibers to work for biomedical research, Macromol. Rapid Commun. 29 (22) (2008) 1775–1792.
[154] S.V. Murphy, A. Atala, 3D bioprinting of tissues and organs, Nat. Biotechnol. 32 (8) (2014) 773–785.
[155] A.R. Boccaccini, V. Maquet, Bioresorbable and bioactive polymer/Bioglass® composites with tailored pore structure for tissue engineering applications, Compos. Sci. Technol. 63 (16) (2003) 2417–2429.
[156] J. Shao, C. Chen, Y. Wang, X. Chen, C. Du, Early stage structural evolution of PLLA porous scaffolds in thermally induced phase separation process and the corresponding biodegradability and biological property, Polym. Degrad. Stab. 97 (6) (2012) 955–963.
[157] K.-W.D. Lee, P.K. Chan, X. Feng, Morphology development and characterization of the phase-separated structure resulting from the thermal-induced phase separation phenomenon in polymer solutions under a temperature gradient, Chem. Eng. Sci. 59 (7) (2004) 1491–1504.
[158] P.X. Ma, R. Zhang, Synthetic nano-scale fibrous extracellular matrix, J. Biomed. Mater. Res. 46 (1) (1999) 60–72.
[159] Y.S. Nam, T.G. Park, Porous biodegradable polymeric scaffolds prepared by thermally induced phase separation, J. Biomed. Mater. Res. 47 (1) (1999) 8–17.
[160] S.-B. Park, J. Sakamoto, M.-H. Sung, H. Uyama, pH-controlled degradation and thermal stability of a porous poly (γ-glutamic acid) monolith crosslinked with an oxazoline-functionalized polymer, Polym. Degrad. Stab. 99 (2014) 99–104.
[161] F. Zhao, Y. Yin, W.W. Lu, J.C. Leong, W. Zhang, J. Zhang, M. Zhang, K. Yao, Preparation and histological evaluation of biomimetic three-dimensional hydroxyapatite/chitosan-gelatin network composite scaffolds, Biomaterials 23 (15) (2002) 3227–3234.
[162] A. Hottot, S. Vessot, J. Andrieu, A direct characterization method of the ice morphology. Relationship between mean crystals size and primary drying times of freeze-drying processes, Drying Technol. 22 (8) (2004) 2009–2021.
[163] F.J. O'Brien, B.A. Harley, I.V. Yannas, L. Gibson, Influence of freezing rate on pore structure in freeze-dried collagen-GAG scaffolds, Biomaterials 25 (6) (2004) 1077–1086.
[164] S.V. Madihally, H.W. Matthew, Porous chitosan scaffolds for tissue engineering, Biomaterials 20 (12) (1999) 1133–1142.
[165] A. Haider, S. Haider, M.R. Kummara, T. Kamal, A.-A.A. Alghyamah, F.J. Iftikhar, B. Bano, N. Khan, M.A. Afridi, S.S. Han, Advances in the scaffolds fabrication techniques using biocompatible polymers and their biomedical application: a technical and statistical review, J. Saudi Chem. Soc. 24 (2) (2020) 186–215.
[166] Z. Ma, M. Kotaki, R. Inai, S. Ramakrishna, Potential of nanofiber matrix as tissue-engineering scaffolds, Tissue Eng. 11 (1-2) (2005) 101–109.
[167] X. Xu, M. Zhou, Antimicrobial gelatin nanofibers containing silver nanoparticles, Fibers Polym. 9 (6) (2008) 685–690.
[168] S.M. Peltola, F.P. Melchels, D.W. Grijpma, M. Kellomäki, A review of rapid prototyping techniques for tissue engineering purposes, Ann. Med. 40 (4) (2008) 268–280.
[169] W.-Y. Yeong, C.-K. Chua, K.-F. Leong, M. Chandrasekaran, Rapid prototyping in tissue engineering: challenges and potential, Trends Biotechnol. 22 (12) (2004) 643–652.
[170] X. Li, R. Cui, L. Sun, K.E. Aifantis, Y. Fan, Q. Feng, F. Cui, F. Watari, 3D-printed biopolymers for tissue engineering application, Int. J. Polym. Sci. 2014 (2014), 829145.

[171] R.A. Vaia, K.D. Jandt, E.J. Kramer, E.P. Giannelis, Microstructural evolution of melt intercalated polymer – organically modified layered silicates nanocomposites, Chem. Mater. 8 (11) (1996) 2628–2635.

[172] R.A. Vaia, E.P. Giannelis, Lattice model of polymer melt intercalation in organically-modified layered silicates, Macromolecules 30 (25) (1997) 7990–7999.

[173] R.A. Vaia, E.P. Giannelis, Polymer melt intercalation in organically-modified layered silicates: model predictions and experiment, Macromolecules 30 (25) (1997) 8000–8009.

[174] Q. Chen, H. Yu, L. Wang, Z. ul Abdin, Y. Chen, J. Wang, W. Zhou, X. Yang, R.U. Khan, H. Zhang, Recent progress in chemical modification of starch and its applications, RSC Adv. 5 (83) (2015) 67459–67474.

[175] J. Qu, X. Zhao, Y. Liang, T. Zhang, P.X. Ma, B. Guo, Antibacterial adhesive injectable hydrogels with rapid self-healing, extensibility and compressibility as wound dressing for joints skin wound healing, Biomaterials 183 (2018) 185–199.

[176] V.K. Thakur, M.K. Thakur, R.K. Gupta, Graft copolymers from natural polymers using free radical polymerization, Int. J. Polym. Anal. Charact. 18 (7) (2013) 495–503.

[177] Q.-S. Zhao, L.-L. Hu, Z.-D. Wang, Z.-P. Li, A.-W. Wang, J. Liu, Resveratrol-loaded folic acid-grafted dextran stearate submicron particles exhibits enhanced antitumor efficacy in non-small cell lung cancers, Mater. Sci. Eng. C 72 (2017) 185–191.

[178] N.M. Alves, J.F. Mano, Chitosan derivatives obtained by chemical modifications for biomedical and environmental applications, Int. J. Biol. Macromol. 43 (5) (2008) 401–414.

[179] O.Y. Mansour, A. Nagaty, Grafting of synthetic polymers to natural polymers by chemical processes, Prog. Polym. Sci. 11 (1-2) (1985) 91–165.

[180] A. Esmaeili, M. Haseli, Optimization, synthesis, and characterization of coaxial electrospun sodium carboxymethyl cellulose-graft-methyl acrylate/poly (ethylene oxide) nanofibers for potential drug-delivery applications, Carbohydr. Polym. 173 (2017) 645–653.

[181] B.H. Oh, A. Bismarck, M.B. Chan-Park, High internal phase emulsion templating with self-emulsifying and thermoresponsive chitosan-graft-PNIPAM-graft-oligoproline, Biomacromolecules 15 (5) (2014) 1777–1787.

[182] Y. Liang, X. Zhao, P.X. Ma, B. Guo, Y. Du, X. Han, pH-responsive injectable hydrogels with mucosal adhesiveness based on chitosan-grafted-dihydrocaffeic acid and oxidized pullulan for localized drug delivery, J. Colloid Interface Sci. 536 (2019) 224–234.

[183] S. Gautam, C.-F. Chou, A.K. Dinda, P.D. Potdar, N.C. Mishra, Surface modification of nanofibrous polycaprolactone/gelatin composite scaffold by collagen type I grafting for skin tissue engineering, Mater. Sci. Eng. C 34 (2014) 402–409.

[184] A. Ospina-Orejarena, R. Vera-Graziano, M.M. Castillo-Ortega, J.P. Hinestroza, M. -Rodriguez-Gonzalez, L. Palomares-Aguilera, M. Morales-Moctezuma, A. Maciel-Cerda, Grafting collagen on poly (lactic acid) by a simple route to produce electrospun scaffolds, and their cell adhesion evaluation, Tissue Eng. Regen. Med. 13 (4) (2016) 375–387.

[185] S.G. Karaj-Abad, M. Abbasian, M. Jaymand, Grafting of poly[(methyl methacrylate)-block-styrene] onto cellulose via nitroxide-mediated polymerization, and its polymer/clay nanocomposite, Carbohydr. Polym. 152 (2016) 297–305.

[186] Z. Yang, H. Peng, W. Wang, T. Liu, Crystallization behavior of poly(ε-caprolactone)/layered double hydroxide nanocomposites, J. Appl. Polym. Sci. 116 (5) (2010) 2658–2667.

[187] M. Hatamzadeh, M. Jaymand, B. Massoumi, Graft copolymerization of thiophene onto polystyrene synthesized via nitroxide-mediated polymerization and its polymer – clay nanocomposite, Polym. Int. 63 (3) (2014) 402–412.

[188] M. Jaymand, Synthesis and characterization of an exfoliated modified syndiotactic polystyrene/Mg–Al-layered double-hydroxide nanocomposite, Polymer J. 43 (2) (2011) 186–193.
[189] A. Ghamkhari, B. Massoumi, M. Jaymand, Novel 'schizophrenic' diblock copolymer synthesized via RAFT polymerization: poly (2-succinyloxyethyl methacrylate)-b-poly [(N-4-vinylbenzyl), N, N-diethylamine], Des. Monomers Polym. 20 (1) (2017) 190–200.
[190] S. Davaran, A. Ghamkhari, E. Alizadeh, B. Massoumi, M. Jaymand, Novel dual stimuli-responsive ABC triblock copolymer: RAFT synthesis, "schizophrenic" micellization, and its performance as an anticancer drug delivery nanosystem, J. Colloid Interface Sci. 488 (2017) 282–293.
[191] B. Massoumi, M. Shafagh-kalvanagh, M. Jaymand, Soluble and electrically conductive polyaniline-modified polymers: incorporation of biocompatible polymeric chains through ATRP technique, J. Appl. Polym. Sci. 134 (16) (2017), 44720.
[192] B. Massoumi, M. Jaymand, Chemical and electrochemical grafting of polythiophene onto poly(methyl methacrylate), and its electrospun nanofibers with gelatin, J. Mater. Sci. Mater. Electron. 27 (12) (2016) 12803–12812.
[193] M. Jaymand, Synthesis and characterization of well-defined poly (4-chloromethyl styrene-g-4-vinylpyridine)/TiO2 nanocomposite via ATRP technique, J. Polym. Res. 18 (6) (2011) 1617–1624.
[194] V. Dikshit, S.K. Bhudolia, S.C. Joshi, Multiscale polymer composites: a review of the interlaminar fracture toughness improvement, Fibers 5 (4) (2017) 38.
[195] N. Panapitiya, S. Wijenayake, D. Nguyen, C. Karunaweera, Y. Huang, K. Balkus, I. Musselman, J. Ferraris, Compatibilized immiscible polymer blends for gas separations, Materials 9 (8) (2016) 643.
[196] A. Taguet, P. Cassagnau, J.-M. Lopez-Cuesta, Structuration, selective dispersion and compatibilizing effect of (nano) fillers in polymer blends, Prog. Polym. Sci. 39 (8) (2014) 1526–1563.
[197] L.A. Utracki, Compatibilization of polymer blends, Can. J. Chem. Eng. 80 (6) (2002) 1008–1016.
[198] G. Wang, D. Yu, A.D. Kelkar, L. Zhang, Electrospun nanofiber: emerging reinforcing filler in polymer matrix composite materials, Prog. Polym. Sci. 75 (2017) 73–107.
[199] Y.-H. Lin, K.-W. Huang, S.-Y. Chen, N.-C. Cheng, J. Yu, Keratin/chitosan UV-crosslinked composites promote the osteogenic differentiation of human adipose derived stem cells, J. Mater. Chem. B 5 (24) (2017) 4614–4622.
[200] A. Sionkowska, A. Płanecka, Preparation and characterization of silk fibroin/chitosan composite sponges for tissue engineering, J. Mol. Liq. 178 (2013) 5–14.
[201] A. Haroun, H. Beherei, M.A. El-Ghaffar, Preparation, characterization, and in vitro application of composite films based on gelatin and collagen from natural resources, J. Appl. Polym. Sci. 116 (4) (2010) 2083–2094.
[202] J. Liuyun, L. Yubao, X. Chengdong, Preparation and biological properties of a novel composite scaffold of nano-hydroxyapatite/chitosan/carboxymethyl cellulose for bone tissue engineering, J. Biomed. Sci. 16 (1) (2009) 1–10.
[203] M. Chen, D.Q. Le, A. Baatrup, J.V. Nygaard, S. Hein, L. Bjerre, M. Kassem, X. Zou, C. Bünger, Self-assembled composite matrix in a hierarchical 3-D scaffold for bone tissue engineering, Acta Biomater. 7 (5) (2011) 2244–2255.
[204] T.-W. Chung, T. Limpanichpakdee, M.-H. Yang, Y.-C. Tyan, An electrode of quartz crystal microbalance decorated with CNT/chitosan/fibronectin for investigating early adhesion and deforming morphology of rat mesenchymal stem cells, Carbohydr. Polym. 85 (4) (2011) 726–732.

[205] A. Asti, L. Gioglio, Natural and synthetic biodegradable polymers: different scaffolds for cell expansion and tissue formation, Int. J. Artif. Organs 37 (3) (2014) 187–205.
[206] F. Zou, R. Li, J. Jiang, X. Mo, G. Gu, Z. Guo, Z. Chen, Mechanical enhancement and in vitro biocompatibility of nanofibrous collagen-chitosan scaffolds for tissue engineering, J. Biomater. Sci. Polym. Ed. 28 (18) (2017) 2255–2270.
[207] R. Xiong, N. Hameed, Q. Guo, Cellulose/polycaprolactone blends regenerated from ionic liquid 1-butyl-3-methylimidazolium chloride, Carbohydr. Polym. 90 (1) (2012) 575–582.
[208] A.G.S. Ali, M. Ebrahimzadeh, M. Solati-Hashjin, O.N. Abu, Polycaprolactone/starch composite: fabrication, structure, properties, and applications, J. Biomed. Mater. Res. A 103 (7) (2015) 2482.
[209] T. Prasad, E. Shabeena, D. Vinod, T. Kumary, P.A. Kumar, Characterization and in vitro evaluation of electrospun chitosan/polycaprolactone blend fibrous mat for skin tissue engineering, J. Mater. Sci. Mater. Med. 26 (1) (2015) 28.
[210] O. Suwantong, Biomedical applications of electrospun polycaprolactone fiber mats, Polym. Adv. Technol. 27 (10) (2016) 1264–1273.
[211] Q. Zhou, H. Zhang, Y. Zhou, Z. Yu, H. Yuan, B. Feng, P. van Rijn, Y. Zhang, Alkali-mediated miscibility of gelatin/polycaprolactone for electrospinning homogeneous composite nanofibers for tissue scaffolding, Macromol. Biosci. 17 (12) (2017) 1700268.
[212] C. Dong, Y. Lv, Application of collagen scaffold in tissue engineering: recent advances and new perspectives, Polymers 8 (2) (2016) 42.
[213] S.G. Wise, S.M. Mithieux, A.S. Weiss, Engineered tropoelastin and elastin-based biomaterials, Adv. Protein Chem. Struct. Biol. 78 (2009) 1–24.
[214] W. Fu, Z. Liu, B. Feng, R. Hu, X. He, H. Wang, M. Yin, H. Huang, H. Zhang, W. Wang, Electrospun gelatin/PCL and collagen/PLCL scaffolds for vascular tissue engineering, Int. J. Nanomedicine 9 (2014) 2335.
[215] D. Jhala, H. Rather, R. Vasita, Polycaprolactone–chitosan nanofibers influence cell morphology to induce early osteogenic differentiation, Biomater. Sci. 4 (11) (2016) 1584–1595.
[216] X.-H. Qin, X. Wang, M. Rottmar, B.J. Nelson, K. Maniura-Weber, Near-infrared light-sensitive polyvinyl alcohol hydrogel photoresist for spatiotemporal control of cell-instructive 3D microenvironments, Adv. Mater. 30 (10) (2018) 1705564.
[217] S.-H. Xia, S.-H. Teng, P. Wang, Synthesis of bioactive polyvinyl alcohol/silica hybrid fibers for bone regeneration, Mater. Lett. 213 (2018) 181–184.
[218] J. Jalvandi, M. White, Y. Gao, Y.B. Truong, R. Padhye, I.L. Kyratzis, Polyvinyl alcohol composite nanofibres containing conjugated levofloxacin-chitosan for controlled drug release, Mater. Sci. Eng. C 73 (2017) 440–446.
[219] S.N. Alhosseini, F. Moztarzadeh, M. Mozafari, S. Asgari, M. Dodel, A. Samadikuchaksaraei, S. Kargozar, N. Jalali, Synthesis and characterization of electrospun polyvinyl alcohol nanofibrous scaffolds modified by blending with chitosan for neural tissue engineering, Int. J. Nanomedicine 7 (2012) 25.
[220] B. Singh, K. Pramanik, Development of novel silk fibroin/polyvinyl alcohol/sol–gel bioactive glass composite matrix by modified layer by layer electrospinning method for bone tissue construct generation, Biofabrication 9 (1) (2017), 015028.
[221] J. Melke, S. Midha, S. Ghosh, K. Ito, S. Hofmann, Silk fibroin as biomaterial for bone tissue engineering, Acta Biomater. 31 (2016) 1–16.
[222] D. Wang, H. Liu, Y. Fan, Silk fibroin for vascular regeneration, Microsc. Res. Tech. 80 (3) (2017) 280–290.

[223] C. Martins, F. Sousa, F. Araujo, B. Sarmento, Functionalizing PLGA and PLGA derivatives for drug delivery and tissue regeneration applications, Adv. Healthc. Mater. 7 (1) (2018) 1701035.
[224] P. Gentile, V. Chiono, I. Carmagnola, P.V. Hatton, An overview of poly(lactic-co-glycolic) acid (PLGA)-based biomaterials for bone tissue engineering, Int. J. Mol. Sci. 15 (3) (2014) 3640–3659.
[225] A.R. Sadeghi-Avalshahr, M. Khorsand-Ghayeni, S. Nokhasteh, A.M. Molavi, H. Naderi-Meshkin, Synthesis and characterization of PLGA/collagen composite scaffolds as skin substitute produced by electrospinning through two different approaches, J. Mater. Sci. Mater. Med. 28 (1) (2017) 1–10.
[226] Y. Boukari, O. Qutachi, D.J. Scurr, A.P. Morris, S.W. Doughty, N. Billa, A dual-application poly (dl-lactic-co-glycolic) acid (PLGA)-chitosan composite scaffold for potential use in bone tissue engineering, J. Biomater. Sci. Polym. Ed. 28 (16) (2017) 1966–1983.
[227] F. Naeem, S. Khan, A. Jalil, N.M. Ranjha, A. Riaz, M.S. Haider, S. Sarwar, F. Saher, S. Afzal, pH responsive cross-linked polymeric matrices based on natural polymers: effect of process variables on swelling characterization and drug delivery properties, Bioimpacts 7 (3) (2017) 177.
[228] A.M. Elbarbary, H.A. Abd El-Rehim, N.M. El-Sawy, E.-S.A. Hegazy, E.-S.A. Soliman, Radiation induced crosslinking of polyacrylamide incorporated low molecular weights natural polymers for possible use in the agricultural applications, Carbohydr. Polym. 176 (2017) 19–28.
[229] B. Guo, L. Glavas, A.-C. Albertsson, Biodegradable and electrically conducting polymers for biomedical applications, Prog. Polym. Sci. 38 (9) (2013) 1263–1286.
[230] T. Rudolph, F.H. Schacher, Selective crosslinking or addressing of individual domains within block copolymer nanostructures, Eur. Polym. J. 80 (2016) 317–331.
[231] A. Oryan, A. Kamali, A. Moshiri, H. Baharvand, H. Daemi, Chemical crosslinking of biopolymeric scaffolds: current knowledge and future directions of crosslinked engineered bone scaffolds, Int. J. Biol. Macromol. 107 (2018) 678–688.
[232] S. Billiet, X.K. Hillewaere, R.F. Teixeira, F.E. Du Prez, Chemistry of crosslinking processes for self-healing polymers, Macromol. Rapid Commun. 34 (4) (2013) 290–309.
[233] J.M. Frick, A. Ambrosi, L.D. Pollo, I.C. Tessaro, Influence of glutaraldehyde crosslinking and alkaline post-treatment on the properties of chitosan-based films, J. Polym. Environ. 26 (7) (2018) 2748–2757.
[234] F.-H. Lin, C.-H. Yao, J.-S. Sun, H.-C. Liu, C.-W. Huang, Biological effects and cytotoxicity of the composite composed by tricalcium phosphate and glutaraldehyde cross-linked gelatin, Biomaterials 19 (10) (1998) 905–917.
[235] M. Kumorek, O. Janoušková, A. Höcherl, M. Houska, E. Mázl-Chánová, N. Kasoju, L. Cuchalová, R. Matějka, D. Kubies, Effect of crosslinking chemistry of albumin/heparin multilayers on FGF-2 adsorption and endothelial cell behavior, Appl. Surf. Sci. 411 (2017) 240–250.
[236] O.S. Kittipongpatana, S. Burapadaja, N. Kittipongpatana, Carboxymethyl mungbean starch as a new pharmaceutical gelling agent for topical preparation, Drug Dev. Ind. Pharm. 35 (1) (2009) 34–42.
[237] Y. Wang, M. Dong, M. Guo, X. Wang, J. Zhou, J. Lei, C. Guo, C. Qin, Agar/gelatin bilayer gel matrix fabricated by simple thermo-responsive sol-gel transition method, Mater. Sci. Eng. C 77 (2017) 293–299.

[238] D. Gupta, C.H. Tator, M.S. Shoichet, Fast-gelling injectable blend of hyaluronan and methylcellulose for intrathecal, localized delivery to the injured spinal cord, Biomaterials 27 (11) (2006) 2370–2379.
[239] X. Zhang, S. Malhotra, M. Molina, R. Haag, Micro-and nanogels with labile crosslinks–from synthesis to biomedical applications, Chem. Soc. Rev. 44 (7) (2015) 1948–1973.
[240] R. Jin, L.S. Moreira Teixeira, P.J. Dijkstra, Z. Zhong, C.A. van Blitterswijk, M. Karperien, J. Feijen, Enzymatically crosslinked dextran-tyramine hydrogels as injectable scaffolds for cartilage tissue engineering, Tissue Eng. Part A 16 (8) (2010) 2429–2440.
[241] F. Chen, S. Yu, B. Liu, Y. Ni, C. Yu, Y. Su, X. Zhu, X. Yu, Y. Zhou, D. Yan, An injectable enzymatically crosslinked carboxymethylated pullulan/chondroitin sulfate hydrogel for cartilage tissue engineering, Sci. Rep. 6 (1) (2016) 1–12.
[242] H.-L. Tan, S.-Y. Teow, J. Pushpamalar, Application of metal nanoparticle–hydrogel composites in tissue regeneration, Bioengineering 6 (1) (2019) 17.
[243] J.F. Kerr, A.H. Wyllie, A.R. Currie, Apoptosis: a basic biological phenomenon with wideranging implications in tissue kinetics, Br. J. Cancer 26 (4) (1972) 239–257.
[244] J. Jiang, J. Pi, J. Cai, The advancing of zinc oxide nanoparticles for biomedical applications, Bioinorg. Chem. Appl. 2018 (2018), 1062562.
[245] R. Augustine, N. Kalarikkal, S. Thomas, Effect of zinc oxide nanoparticles on the in vitro degradation of electrospun polycaprolactone membranes in simulated body fluid, Int. J. Polym. Mater. Polym. Biomater. 65 (1) (2016) 28–37.
[246] W. Guo, F. Zhao, Y. Wang, J. Tang, X. Chen, Characterization of the mechanical behaviors and bioactivity of tetrapod ZnO whiskers reinforced bioactive glass/gelatin composite scaffolds, J. Mech. Behav. Biomed. Mater. 68 (2017) 8–15.
[247] B.K. Shrestha, S. Shrestha, A.P. Tiwari, J.-I. Kim, S.W. Ko, H.-J. Kim, C.H. Park, C.S. Kim, Bio-inspired hybrid scaffold of zinc oxide-functionalized multi-wall carbon nanotubes reinforced polyurethane nanofibers for bone tissue engineering, Mater. Des. 133 (2017) 69–81.
[248] I.A.B. Neta, M.F. Mota, H.L. Lira, G.A. Neves, R.R. Menezes, Nanostructured titanium dioxide for use in bone implants: a short review, Cerâmica 66 (2020) 440–450.
[249] L.C. Gerhardt, G.M.R. Jell, A.R. Boccaccini, Titanium dioxide (TiO2) nanoparticles filled poly(d,l lactid acid) (PDLLA) matrix composites for bone tissue engineering, J. Mater. Sci. Mater. Med. 18 (7) (2007) 1287–1298.
[250] J. Lewandowska-Łańcucka, S. Fiejdasz, Ł. Rodzik, A. Łatkiewicz, M. Nowakowska, Novel hybrid materials for preparation of bone tissue engineering scaffolds, J. Mater. Sci. Mater. Med. 26 (9) (2015) 1–15.
[251] J.S. Lee, K. Kim, J.P. Park, S.-W. Cho, H. Lee, Role of Pyridoxal 5′-phosphate at the titanium implant interface in vivo: increased hemophilicity, inactive platelet adhesion, and osteointegration, Adv. Healthc. Mater. 6 (5) (2017) 1600962.
[252] W. Wu, Z. Wu, T. Yu, C. Jiang, W.-S. Kim, Recent progress on magnetic iron oxide nanoparticles: synthesis, surface functional strategies and biomedical applications, Sci. Technol. Adv. Mater. 16 (2) (2015), 023501.
[253] J. Miller, J. Talton, S. Bhatia, Total hip replacement: metal-on-metal systems, in: L.L. Hench, J. Wilson (Eds.), Clinical Performance of Skeletal Prostheses, Chapman and Hall, London, 1996, pp. 41–56.
[254] M. Peroglio, L. Gremillard, J. Chevalier, L. Chazeau, C. Gauthier, T. Hamaide, Toughening of bio-ceramics scaffolds by polymer coating, J. Eur. Ceram. Soc. 27 (7) (2007) 2679–2685.

[255] M. Afifi, M.K. Ahmed, A.M. Fathi, V. Uskoković, Physical, electrochemical and biological evaluations of spin-coated ε-polycaprolactone thin films containing alumina/graphene/carbonated hydroxyapatite/titania for tissue engineering applications, Int. J. Pharm. 585 (2020) 119502.
[256] P. Kumar, B.S. Dehiya, A. Sindhu, Bioceramics for hard tissue engineering applications: a review, Int. J. Appl. Eng. Res. 13 (5) (2018) 8.
[257] Y. Josset, Z. Oum'Hamed, A. Zarrinpour, M. Lorenzato, J.-J. Adnet, D. Laurent-Maquin, In vitro reactions of human osteoblasts in culture with zirconia and alumina ceramics, J. Biomed. Mater. Res. 47 (4) (1999) 481–493.
[258] A. Scarano, F. Di Carlo, M. Quaranta, A. Piattelli, Bone response to zirconia ceramic implants: an experimental study in rabbits, J. Oral Implantol. 29 (1) (2003) 8–12.
[259] B.M. Whited, A.S. Goldstein, D. Skrtic, B.J. Love, Fabrication and characterization of poly (DL-lactic-co-glycolic acid)/zirconia-hybridized amorphous calcium phosphate composites, J. Biomater. Sci. Polym. Ed. 17 (4) (2006) 403–418.
[260] Y. Lupu-Haber, O. Pinkas, S. Boehm, T. Scheper, C. Kasper, M. Machluf, Functionalized PLGA-doped zirconium oxide ceramics for bone tissue regeneration, Biomed. Microdevices 15 (6) (2013) 1055–1066.
[261] F. Zhang, Q. Song, X. Huang, F. Li, K. Wang, Y. Tang, C. Hou, H. Shen, A novel high mechanical property PLGA composite matrix loaded with nanodiamond–phospholipid compound for bone tissue engineering, ACS Appl. Mater. Interfaces 8 (2) (2016) 1087–1097.
[262] J.W. Rasmussen, E. Martinez, P. Louka, D.G. Wingett, Zinc oxide nanoparticles for selective destruction of tumor cells and potential for drug delivery applications, Expert Opin. Drug Deliv. 7 (9) (2010) 1063–1077.
[263] L. Wei, J. Lu, H. Xu, A. Patel, Z.-S. Chen, G. Chen, Silver nanoparticles: synthesis, properties, and therapeutic applications, Drug Discov. Today 20 (5) (2015) 595–601.
[264] C. Sudakar, A. Dixit, R. Regmi, R. Naik, G. Lawes, V.M. Naik, P.P. Vaishnava, U. Toti, J. Panyam, $Fe_3O_4$ incorporated AOT-alginate nanoparticles for drug delivery, IEEE Trans. Magn. 44 (11) (2008) 2800–2803.
[265] A. Shakiba, O. Zenasni, M.D. Marquez, T.R. Lee, Advanced drug delivery via self-assembled monolayer-coated nanoparticles, AIMS Bioeng. 4 (2017) 275–299.
[266] A.V. Zvyagin, X. Zhao, A. Gierden, W. Sanchez, J. Ross, M.S. Roberts, Imaging of zinc oxide nanoparticle penetration in human skin in vitro and in vivo, J. Biomed. Opt. 13 (6) (2008), 064031.
[267] B. Wang, W. Feng, M. Wang, T. Wang, Y. Gu, M. Zhu, H. Ouyang, J. Shi, F. Zhang, Y. Zhao, Acute toxicological impact of nano-and submicro-scaled zinc oxide powder on healthy adult mice, J. Nanopart. Res. 10 (2) (2008) 263–276.
[268] O. Długosz, K. Szostak, A. Staroń, J. Pulit-Prociak, M. Banach, Methods for reducing the toxicity of metal and metal oxide NPs as biomedicine, Materials 13 (2) (2020) 279.

# Lignin-metal oxide composite for photocatalysis and photovoltaics

14

*Farzana Yeasmin[a], Rifat Ara Masud[a], Adib H. Chisty[b], Md. Arif Hossain[c], Abul K. Mallik[b], and Mohammed Mizanur Rahman[b]*
[a]Department of Applied Chemistry and Chemical Engineering, Faculty of Engineering and Technology, Bangabandhu Sheikh Mujibur Rahman Science & Technology University, Gopalganj, Bangladesh, [b]Department of Applied Chemistry and Chemical Engineering, Faculty of Engineering and Technology, University of Dhaka, Dhaka, Bangladesh, [c]Department of Chemistry, Faculty of Science, Dhaka University of Engineering & Technology, Gazipur, Bangladesh

## 1 Introduction

Fossil fuel sources are being consumed day by day with cumulative demand, disturbing earth's climate in return. So why, the maximum utilization of renewable energy sources is the new center of attention. Scientists are trying to look for replacements with eco-friendly substances to face the imminent exhaustion of fossil energy reserve on the earth.

The second highest abundant organic material in complex form is lignin, which is originated in plant cell. Lignin composition fluctuates immensely from one plant species to another [1]. For instance, in softwood conifers, hardwood lignin has a lignin content 25%–40% and 18%–25% of dry weight, respectively. Conventionally, it is recognized as low-value waste product. Despite having complex chemical structure and unusual reactivity, lignin is utilized as a raw material for many valuable organic chemicals like multifunctional hydrocarbons, phenolic compounds, numerous oxidized products, and carbon fiber [2,3].

The term "lignin" has come up from Latin "lignum" that indicates wood. Its empirical formula is $C_{31}H_{34}O_{11}$. Plants generate 150 billion tons of lignin every year, contributing to one-third of the whole biomass content throughout the world. Lignin holds around 95 billion tons of carbon, demonstrating the unexploited high carbon energy storage in the earth [4]. Lignin can be of two types: One is lignin that contains sulfur and another one is lignin without sulfur. Paper and pulp industry is the chief source where sulfur-containing lignin can be extracted [3]. Soda lignin and organosolv lignin are two types of sulfur-free lignin, which can be derived from pulping with sodium hydroxide and organosolv pulping [5]. Biomass conversion industries and paper industries can be run more economically with the help of lignin-derived coproducts.

Renewable Polymers and Polymer-Metal Oxide Composites. https://doi.org/10.1016/B978-0-323-85155-8.00012-1

Amid all the divisions of biomass, lignin is moderately less exploited due to its structure, exceptional reactivity making its consumptions remarkably critical. Lignin encompasses 30% of organic carbon content in the earth. Furthermore, many industries generate plentiful volumes of lignin as byproduct for utilization in energy sector. Lignin is a possible ancestor of several valuable chemicals as well as carbon-rich materials. Lignin is found to have widespread applications together with the production of valuable chemicals, various types of polymers, adsorbing materials, electrode surface materials, etc. [6,7]. This chapter will review the current advancement in lignin-metal oxide composites to understand their function in photovoltaics and photocatalysis.

Generally, photoreactions occurred through a photocatalytic degradation mechanism. Photocatalytic mechanism is a successful method to tackle the existing global ecological problem and power catastrophe around the earth, which reduces the poisonous contaminants entirely and generates hydrogen ($H_2$) from the breakdown of water by means of solar energy [8]. Carbon-based materials exhibit better-quality electron conductivity owing to copious $sp^2$-hybridized carbon atoms. Consequently, these materials are frequently utilized to pair semiconductor photocatalyst. Additionally, photoelectronic association occurs among rays of light and carbon-based materials in the interface, which can expand the range of optical absorption [9]. Wang et al. studied a new hybrid composite with lignin-based carbon/ZnO, which was found to exhibit exceptional photocatalytic performance. This prepared photocatalyst was found to have enormous functions, particularly in organic dye degradation [10]. Gómez-Avilés et al. prepared lignin-based C-modified $TiO_2$ photocatalyst materials, which eliminate emerging pollutants from water by solar light. The increasing amount of pollutants thus can be reduced by photodegradation with the help of this material [11]. Lignin has been combined with $CeO_2$, CuO separately, and their photocatalytic performance has also been evaluated in the literature [12,13].

In order to solve the problems of energy crisis as well as environmental pollution, attention has been given to the production of organic photovoltaic cells. Photovoltaic devices produce electricity straight from sunlight through an electronic process that arises naturally in certain types of material, called semiconductors. Considerable endeavors have been performed to evidently comprehend the working principle of photovoltaic cells along with reformed structural patterns as well as apparatus arrangement providing enrichment of effectiveness of power alteration via silicon. But their production and installation costs have been found to be very high [14]. A familiar conducting polymer is PEDOT:PSS, which is implemented in anode interfaces. Nevertheless, this polymeric material has numerous negative aspects, for example, its acidity that stimulates deterioration and inconsistent conductivities because of its heterogeneity in structural and electrical properties [15].

# 2 Lignin

## 2.1 Sources of lignin

Lignin is a natural aromatic biopolymer abundantly found in nature along with cellulose, hemicellulose, and chitin. Lignocellulosic biomass, mainly collected from the agricultural wastes, woody, and nonwoody biomass resources, is the chief natural source of lignin. Lignin, in combination with both cellulose and hemicellulose, encompasses the major component of a plant's secondary cell wall, which provides rigidity and facilitates the erect growth of plants by reinforcing the strength of middle lamella and crystalline cellulose. Different plant species have different percentages of lignin content and typically are 20%–35% in cell walls of terrestrial plants. Lignin content is commonly reported to be highest in softwood and decline from hardwood to grasses. Around $5 \times 10^7$ tons of processed lignin has been produced per annum by chemical treatment processing, collectively from paper industries and bioethanol refineries [16,17].

## 2.2 Physical properties associated with the chemical structure of lignin

On the whole, lignins are highly branched, almost colorless, three-dimensional, complex, and amorphous macromolecules with molecular masses between 1000 and 20,000 $g\,mol^{-1}$, varying with the process of extraction employed. In lignin macromolecule, the three basic monomeric units of monolignols, viz., *p*-hydroxyphenyl, syringyl, and guaiacyl, are linked by mainly β-O-4 ether linkages (>50% of the linkage structures in lignin), which are typically targeted during the degradation mechanisms. Smaller percentages of other bonds like α-O-4′ ether, β-5 phenylcoumaran, 4-O-5′ diphenyl ether, β-β′ resinol, β-1′ diphenyl methane, and 5–5 biphenyl are also found. The isolation and extraction methods of lignin largely interrupt the hydroxyl, carboxyl, phenolic, and sulfonate functional group contents. That's why, the complicated molecular structure of lignin creates bewildering issues in different lignin valorization processes [18,19].

Lignin possesses ample of potentially valuable physicochemical properties, which made it a promising component in different lignin-based products and in renewable energy field. The major advantageous properties of lignin are [18]:

- **I.** Antioxidant property, antimicrobial, and antifungal activities,
- **II.** Availability as byproducts of industrial waste in profusion,
- **III.** Biodegradability and higher resistance to chemical and biological attacks than those of cellulose/hemicellulose,
- **IV.** Capability of UV radiation absorption and fire retardance,
- **V.** Either hydrophilicity or hydrophobicity, varying with the source of lignin,
- **VI.** Better rheological properties, compatibility to several engineering materials, comparative viscoelastic characteristics, capacity of film formation, and
- **VII.** Thermoplasticity with glass transition temperature at the range of 50–150°C varying with lignin extraction process.

## 2.3 Classification of lignin

There are three main categories of lignin, according to its origin, viz. [14,18]:

- **I. Softwood lignin or guaiacyl lignin**: It contains coniferyl alcohols with fewer sinapyl alcohol-derived units and can provide the highest amount of lignin during extraction. Coniferous tree like pine is softwood.

II. **Hardwood lignin or guaiacyl-syringyl lignin**: Such lignin contains substantial volume of coniferyl alcohol and sinapyl alcohol. Deciduous plants, like maple and shrubs, are source of this lignin.

III. **Grass lignin or guaiacyl-syringyl *p*-hydroxybenzaldehyde lignin**: It comprises leading masses of p-coumaryl alcohol-derived structural elements with around 30% of *p*-hydroxybenzaldehyde. Monocotyledons like bamboo, banana contain grass lignin.

## 2.4 Extraction process of lignin

Because of the complicated chemical and physical linkages among lignin and other components of plants, extraction of lignin from plants is challenging. Depending on the variation of sources of lignocellulosic feedstock, different lignin isolation methodologies are practiced, where polymeric lignins degraded to lower molecular weights (with differences in their physicochemical properties) by chemically mediated cleavages are obtained.

### 2.4.1 Pretreatment methods for lignin isolation

Pretreatment methods are employed to the biomass in order to disrupt the bonding linkage between lignin and carbohydrates, which reduces the biomass size with disassembled physical structure suitable for posttreatments. Any one of the physical pretreatment methods like ball milling or mechanical grinding, physicochemical, or chemical pretreatment methods, viz., auto hydrolysis, alkali treatments, acid treatments, steam digestion, hydrogen peroxide treatment etc., or biological enzyme-based hydrolysis method is accomplished to isolate lignin from carbohydrates.

When a mechanical pretreatment method is followed, the chemical structure of lignin is changed due to depolymerization via cleavages of the β-O-4 linkages and augmentation of carbonyl and phenolic —OH groups. In case of other pretreatment methods, depolymerization of lignin is ensued by higher degree of cleavage of lignin carbohydrate bonds, accompanied by breaking of lignin-lignin ether bonds to a certain extent. During acid pretreatment, depolymerization of lignin yields a condensed phase in the lignin matrix via breakdown of β-O-4 bonds and side chain carbonium ion electrophilic attack to the aromatic rings, generating different types of carbon-carbon bonds [20].

### 2.4.2 Analytical-scale process

#### Milled wood lignin (MWL) process

MWL or milled wood lignin process isolates lignin, by ball milling with neutral solvents (commonly dioxane to water with 9:1 volume ratio) and following Bjorkman process with a low extraction yield of 25%–50%. Due to the mild extraction practice along with the use of neutral solvent, the structure of MWL remains almost similar to that of the native lignin and is considered the superlative model for the interpretation of unaltered native lignin structure. Extensive ball milling during pretreatment can newly produce free phenolic hydroxyl groups by breaking of the β-aryl ether-type linkages to increase α-carbonyl groups due to oxidations occurring at the side chain [21].

#### Cellulolytic enzyme lignin (CEL) process

In CEL process of lignin isolation, cellulolytic enzymes hydrolyze the carbohydrate fraction of the hemicellulose feedstock at mild condition, to produce residue of solid cellulolytic enzyme lignin (CEL) with higher molecular weight than that for MWL, reduced phenolic —OH group content as well as increased β-O-4 bonds. Lengthy processing time (of several hours to few days) required for higher production of the lignin mass and the presence of impurities of protein and carbohydrate are the main concerns of this method [21].

#### Enzymatic mild acidolysis lignin (EMAL) process

To overcome the disadvantages of CEL, both mild acidolysis and enzymatic hydrolysis methods can be amalgamated to extract lignin from milled wood. The final lignin content from this process reveals higher yield percentage with enhanced purity than both MWL and CEL [21].

### *2.4.3 Industrial process*

#### Kraft process

Kraft process is the utmost practice scheme of recovering lignocellulose of almost 85% of the produced lignin. This process produces sulfur-containing lignin, the chief portion (98%) of which is used for energy production purposes through combustion, whereas the rest amount (2%) is utilized for material or chemical synthesis. Conventionally, lignin is dissolved in NaOH and $Na_2S$ at high temperature (around 170°C) and alkaline condition (pH 13–14) by breaking the ether bonds to increase the number of phenolic —OH groups. Lignin is then isolated by acid-facilitated precipitation (mostly by $H_2SO_4$ acid) from the remaining alkaline solution (by lowering pH to 5–7.5) [19].

#### Sulfite process

This process is commonly used in pulp and paper industry, which involves the chemical interaction of lignin, metal sulfite, sulfur dioxide with counter ions of Ca, Mg, or Na having a wide pH range of 2–12 at 120–180°C for 1 to 5 h. Breakdown of β-ether (β-O-4′) as well as α-ether (α-O-4′) bonds in lignin structure produces lignosulfonates with a sulfur content of 4%–8% with molecular weight higher than that of lignin extracted through kraft process [20].

#### Soda process

A primitive lignin isolation process, soda pulping, is used (for 5% of total lignin production) usually for treating nonwoody biomass, like grass, sugarcane bagasse, and straw. Solubilization mechanism of lignin involves treating the biomass in any aqueous solution of high pH at 160°C for cleavages of both α-O-4′ and β-O-4 linkages.

The finally found soda lignin contains a mean MW of 2400 Da with a range of 1000–3000 Da, varying with its carbon content and no sulfur [14,20].

#### Organosolv process

This process is the best-fitting route of lignin extraction in recent days at an industrial scale, employing treatment of the biomass by common solvents, typically formic acid, acetic acid, acetone, methanol, and ethanol as the delignifying agents. This technique involves heating the biomass at a high temperature ranging from 170°C to 190°C with a mixture of polar organic solvents in the presence of acidic or basic catalyst, followed by precipitation under suitable conditions for higher yield of lignin. Both the structure and properties of finally sequestered lignin are affected by the nature of solvent used [5,14].

## 2.5 Characterization techniques of lignin

Characterization of lignin is necessary so that one may determine lignin's properties, namely, chemical structure, types of reaction it will undergo, and types of products that can be obtained. Comprehension in structure would be helpful in deciding the origin and extraction method of lignin, corresponding links among physical and thermal characteristics, also in choosing suitable conversion techniques, and in conducting effectual conversion. Some methods are discussed below for providing information about chemical structure and thermal properties.

### 2.5.1 Molecular weight determination by HPLC

HPLC can be employed to investigate the molecular weight distribution. HPLC column is packed with adsorbent through which the lignin sample trickles down and constituents travel with distinct velocities counting on its chemical characteristics. Retention time of the component is counted. Lignin from particular source has a narrow range of molecular weight distribution with a small number of oligomers. Determination of the number-average molecular weight ($M_n$) and weight-average molecular weight ($M_w$) gives an idea about polydispersity index ($M_w/M_n$) of lignin. Polydispersity index of kraft lignin was found to be 1.144 and 1.176 for two peaks, which are attributed to the mid-range polydispersity of lignin [16].

### 2.5.2 UV spectroscopy

UV spectroscopy is generally carried out using 200–400 nm of UV region. Baeza et al. identified the suitable solvents for lignin dissolution, including water, dimethyl sulfoxide (DMSO), hexafluoro-propanol, and so on [17]. Hexafluoro-propanol has been acknowledged as appropriate solvent for UV and IR spectroscopic analyses of lignin. Generally, aromatic polymer lignin shows strong absorption bands that appeared at 200 nm and 280 nm of UV region [18].

### 2.5.3 FTIR spectroscopy

IR spectroscopy (infrared spectroscopy) deals with the infrared region (4000–400 $cm^{-1}$) of the electromagnetic spectrum. The emitted radiation can be recorded, and frequencies absorbed should be documented. Boeriu et al. reported IR spectra of five types of lignin. A broad band at around 3410–3460 $cm^{-1}$ was found for all types of lignin due to the presence of the hydroxyl groups, and that at around 2938 and 2842 $cm^{-1}$ was found for the stretching of C—H bond. These bands were less intense. For unconjugated carbonyl/carboxyl bonds, stretching bands were found at 1705–1720 $cm^{-1}$. Aromatic ring vibrations appeared at 1600, 1515, and 1426 $cm^{-1}$. A peak appeared at 1462 $cm^{-1}$ for all types of lignin with nonidentical intensity, due to C—H deformation associated with aromatic skeleton vibration. For guaiacyl unit, bands at 1269 $cm^{-1}$, 1140 $cm^{-1}$, 854 $cm^{-1}$, and 817 $cm^{-1}$ were observed. At 1370–1375 $cm^{-1}$, a weak peak was originated because of the presence of phenolic O—H as well as aliphatic C—H in lignin. For C—C plus C—O plus C=O stretching, a peak was aroused strongly at 1215–1220 $cm^{-1}$ [19].

### 2.5.4 Raman spectroscopy

In Raman spectroscopy, IR radiation beam is used to make a contact with the sample. Agarwal et al. studied black spruce along with milled wood lignins and reported Raman spectra, where significant bands were identified at 3068, 2939, 1661, 1595, and 1331 $cm^{-1}$ for different types of vibrations, such as aromatic C—H stretching, asymmetric C—H stretching, conjugated ring C=C stretching, C=O stretching, symmetric aryl ring stretching, and aliphatic O—H bending vibrations, respectively [20].

### 2.5.5 NMR spectroscopy

NMR spectroscopy is used to find out the properties of organic molecules by analyzing the interaction of magnetic field with the particular sample and obtaining information such as basic structure and molecular conformation. Different parts of lignin provide distinct chemical shift in proton NMR spectroscopy; for instance, carboxylic acid, aldehyde, phenolic hydroxyl, β-5 phenolic hydroxyl, syringyl $C_5$ phenolic hydroxyl, aromatic protons, and aliphatic protons show a chemical shit of around 12.6–13.5, 9.4–10.0, 8.0–9.4, 8.99, 8.0–8.5, and 3–7.7 ppm, respectively. Carbon (13)-NMR, phosphorous-NMR, and fluorine-NMR are also used for lignin characterization.

### 2.5.6 Thermal property analysis by differential scanning calorimetry (DSC)

DSC is an analytical method of rising temperature of sample at a constant rate and measuring the heat flow. Phase transition of lignin can be determined by this technique. Mostly changes in heat capacity and glass transition temperature ($T_g$) of the sample are derived from DSC. Lignin from alcell, pine, bagasse, eucalyptus, aspen, mixed oak, tulip poplar, switchgrass, black locust, corn stover, and newsprint has $T_g$ of 108°C, 107°C, 116°C, 136°C, 162°C, 174°C, 106°C, 119°C, 121°C, 105°C, 131°C, and 140°C, respectively, and found to have $DC_p$ of 0.403, 0.417, 0.386, 0.182, 0.102, 0.141, 0.315, 0.239, 0.248, 0.060, and 0.105 J/g°C, respectively.

## 2.6 Application of lignin

Studies on lignin and lignin-derived polymers have been going on since decades. It is possible to convert lignin to high-value products as well as it can be functioned as a precursor of carbon-based materials. Here, we are going to discuss these two categories briefly.

### 2.6.1 Valorization of lignin into high-value chemicals

Implementation of lignin and its derivatives is flourishing day by day. Nonecofriendly and nonrenewable chemicals or polymeric materials are needed to be removed, or the use of those chemicals should be limited.

Lignin combustion with coal can produce more heating value than separate combustion of lignin or coal, which boosts up the boiler efficiency as fuel. Lignin-coal firing improves the boiler efficiency of 38%, along with lessening of carbon emission by 60%.

Pyrolysis of lignin yields solid biochar, bio-oil, or gases based on temperature, heating rate, and residence time. Different process temperatures and times are required for the manufacturing of desired products from same biomass. Bio-oils are liquid products from biomass that can be obtained at elevated heating rates along with small residence time; on the other hand, char is produced when temperature and heating rates are minimum. Most gaseous products can be obtained at extended temperature, high residence time, and low heating conditions. Bio-oils show some drawbacks but their performance can be improved by catalytic upgrading.

A significant approach for lignin valorization can be achieved by gasification, yielding syngas and carbon dioxide. For the production of DME (dimethyl ether) and green diesel, syngas is used [21]. Syngas from lignin can also be applied for the purpose of heating and producing electricity. The successive reactions of lignin are required for complete gasification of lignin using alkali and alkali salt as catalysts [22].

The use of lignin-derived binders in pigment printing compositions, ceramics, briquetting, silicon anode, plywood, or particle boards has been discussed by many researchers [23].

Lignin is being evaluated as an antecedent aromatic polymer to convert it into chemicals with significantly important applications. Lignin can be depolymerized or chemically active groups can be introduced into lignin, or hydroxyl group of lignin can be functionalized by various techniques. In addition, the application of lignin is found in biomedical sectors, bio-nanocomposites, removal of contaminants, energy-storing devices, etc.

Several other applications of lignin include the controlled or modified release of fertilizers and herbicides in cultivation [24], as adsorbents or dispersants in bioplastics and in biofuel [25]. Vanillin, DMSO, aromatic and aliphatic acids, and cyclohexanol are the oxidized products derived from lignin. It is possible to transform lignin by means of hydroxyl-alkylation into phenolic compounds. Nowadays, phenol formaldehyde resins are being replaced by lignin phenol formaldehyde.

Lignin can be converted into valuable hydrocarbons, like benzene-BTX, by fast pyrolysis. Depolymerization and hydrodeoxygenation are the major steps for the conversion of lignin into BTX. Lignin-derived materials are turning into interesting products like hydrogels, drug delivery materials for biomedical applications. Lignin-based synthetic polymers and biopolymers are inflammable and can discharge smoke. As nanoparticles of lignin are found to carry excellent antioxidant properties along with UV protection ability, so these are anticipated in the manufacturing of ointment, emollient, cosmetics, etc. [26].

### *2.6.2 Lignin as precursor for carbon material*

Varieties of expensive materials, namely, carbon fiber, polymer alloy, nanomaterial, etc., are being developed from recent studies of lignin.

Carbon fiber is a kind of fiber obtained from carbon atoms and used as a reinforcing material. It is evaluated that by 2022, the overall need for carbon fiber is expected to be 120.5 thousand metric tons [27].

Carbon fiber is used to be derived from PAN. The cause for high price of carbon fiber is because of the inflated cost of PAN. Otani, S., et al. patented for the first time on lignin-derived carbonized material manufacturing and two spinning techniques that were used to prepare alkali-modified lignin solution [19]. In the 1990s, Sudo et al. prepared a novel, melted, viscous material as a precursor for carbon fibers using hydrogenolysis and phenolysis of lignin, and both are followed by heat treatment [28]. Mainka et al. showed lignin as a pure typical precursor rich in carbon content in their studies, which can be a supreme inexpensive resource to manufacture carbon fiber [29]. Uraki et al. used melt spinning and prepared lignin from organosolv lignin devoid of chemical change [30]. Carbonization of lignin produces high carbon-containing materials without any discharge of poisonous products, which makes lignin a superior candidate for cheap carbon fiber production. Frank et al. also compared numerous sources of carbon fiber precursor, concluding that lignin has been the most suitable carbon fiber precursor [31].

Sports items and aerospace industry is the major consumer of carbon fiber. Kadla et al. used kraft lignin to produce carbon fibers, not performing any conversion chemically with the application of thermal spinning and carbonization. Carbon fibers can be derived from kraft lignin via a single step, combining stabilization along with carbonization by varying time and temperature. From this study, it was attributed that lignin has possibility to use for the commercial production of carbon fiber precursor economically [32]. Lignin is becoming popular to the researchers day by day for carbon fiber production and other products, due to its ampleness and less ecological influence.

On the other hand, lignin faces obstacles in synthesizing carbon-loaded materials such as heterogeneous structure, high ash percentage, and thermoplastic foaming manners.

Dallmeyer and some other scientists studied seven different technical lignins on fiber formation by electrospinning. Unexpectedly, no continuous fibers were observed, rather than beads of fibers obtained after electrospinning of various lignins, which were improved further by the inclusion of poly (ethylene glycol) [33].

Lignin micro- and nanoparticles (LMNPs) are manufactured as byproducts from pulp and paper industry. LMNP's feasibility was assessed based on various potential applications, for example, stabilizers, ultraviolet-protective materials, nanofillers, and so on [34].

## 2.7 Drawbacks of the application of lignin

Though the production of lignin is vast, its complete utilization has not been fulfilled.

On the contrary, products derived from lignin, for instance, adhesive, sorbent, dispersants, are not always better than the feature of existing products. Undoubtedly, these drawbacks are required to be curbed and lignin-based materials production should be increased in wide range without compromising the quality. The prevalence of these complications is generated by structural properties of technical lignin and their extraction from sources along with uncommon reactivity. Drawbacks of applications of lignin have been classified into four categories as discussed below:

### 2.7.1 Recovery from products stream

The biorefinery plants and pulp mills generate technical lignin as by-products. Lignin-containing sources like kraft lignin and soda lignin carry dry solid dissolved lignin approximately 20%–40%, while undissolved pulp and extracted portions are present in 60%–80%.

Separation of black liquor from kraft lignin is carried out by a simple method using precipitation and filtration, and then, it is deodorized and decolorized by the incorporation of bleaching agents like hydrogen peroxide, ozone, and chlorine dioxide.

Lignosulfonates are conventionally detached from spent liquor by ultrafiltration. Lignin separation can only be obtained in an industrial scale for kraft lignin and lignosulfonates. 100% pure lignin cannot be obtained by any of these above-mentioned separation methods, which can be promptly exposed to chemical conversion.

### 2.7.2 Purification of lignin

Sugars, sulfur, silicates, proteins, ash, and some additional compounds are various impurities found in technical lignin emerging from any of raw material or lignin separation method [35]. For the transformation of lignin to high-value material, lignin must be decontaminated from foreign matters.

If lignin with impurities is used in any further process, this would lead to produce low-quality product and the yield would be smaller than expected along with some unwanted byproducts.

Sugar can be separated from lignin by the enzymatic hydrolysis process, followed by acid hydrolysis. This method was found to be effective enough to remove 100% sugars from lignin. Fermentation, ultrafiltration, and precipitation are used to separate sugars from sulfonated lignin [36].

Ash along with some other contaminations can be cleared from lignin, merely by rinsing with water. With enzymatic treatment, nitrogen-containing substances, namely, protein and amino acids, are separated. 70% sulfur chemically bound to lignin can poison industrial catalyst, which is tough to remove, while 30% sulfur can be removed by washing. The bulk portion of sulfur can be isolated using Raney nickel reduction method, making it a pricey method and troublesome to execute in industry. Thiobacillus bacteria has been feasibly utilized for microbiological elimination of sulfur content. The above-mentioned methods cannot certainly remove sulfur completely [37].

Carbonation and desilication can be performed to eliminate silica portions from lignin [38]. But silica components of lignin are neutral, and so why, they are not always necessary to be removed for most uses.

Though many techniques are available for the removal of sulfur, nitrogen, ash, silica, and sugar, these purification stages are not always economical, thus increasing the process cost.

### 2.7.3 Nonuniform structure

Lignin has heterogeneous structure because of the disparity in size, composition, type of functional groups, and amount of cross-linking. Hydroxyl, carbonyl, methoxyl, and carboxyl groups are the chief functional groups present in technical lignin. Controlled degradation method can be performed using the chemical or enzymatic pathway in order to attain a specific range of molecular weight ($M_w$) of lignin. Controlled degradation of lignin can also be achieved by pyrolysis and chemical oxidation along with microwave-range radiation, producing a narrow distribution of molecular weight of lignin.

A number of applications need high molecular weight lightweight lignin. Areskogh et al. reported that oxidative polymerization by laccases can amplify the molecular weight of lignin polymers [39]. Homogenization of lignin is necessary for further application, but any of the aforementioned methods have not been 100% successful. So, to attain the best utilization of lignin, a more effectual method should be developed for the homogenization of lignin.

### 2.7.4 Exceptional reactivity

Because of the heterogeneous structure, lignin does not give a preferred reaction to produce the desired product. Lignin with uniform structure is used in the synthesis of resins. The reactivity of lignin is restricted since it contains few portions of active site with limited accessibility [40]. Reactivity can be boosted up by lessening molecular weight, enhancing number of functional groups by phenolation, and altering the structure.

Depolymerization and introduction to the reactive site can upgrade lignin's reactivity. Depolymerizations transform lignin into smaller molecules. Laccase enzyme can support oxidation reactions of lignin. These reactions guide to the evolvement of fresh reactive positions [41]. Depolymerization with a phenolation reaction produces hydroxyl groups, which assures lignin's desirable reactivity. Using sodium dithionite, more phenolic hydroxyl groups can be incorporated into the lignin skeleton. Genetic modification of lignin had been carried out by some researchers, though it is not utilized in a large scale [42,43].

Four most important categories of problems have been discussed. Though various methods were found to solve these problems, monetary factors bound their extensive uses. Prevailing techniques can resolve these complications only moderately. However, propitious techniques are developing, such as genetic alteration of lignin to allow precise usages of lignin.

# 3 Lignin-based metal oxide composite preparation methods

Utilizing the porosity of lignin, lignin-metal oxide composites can be fabricated, where photoactive components are introduced to be immobilized in an assembled or aligned manner onto the porous lignin-based support. Different techniques mostly followed to prepare lignin-supported metal oxide composites for photocatalyst and photovoltaics are discussed as follows.

## *3.1 Incipient wetness impregnation method*

Incipient wetness impregnation, alternatively called the dry/capillary impregnation method, is one of the popularly used, easier and economical, limited waste generation approaches to fabricate heterogeneous catalysts. Typically in this method, the metal salts to be acted as the catalyst precursor are dissolved by any organic impregnating solution. The produced metal solution is then added to a dry biomass material support drop by drop in a controlled volume, which corresponds to the volume of pores in the support. Due to the capillary action of the inner pores, a slurry is resulted through the absorption of the solution into the pores, and later on, the metal ions are bonded to mostly the oxygen-containing functional groups of the biomass support. The desired pure catalyst is then further obtained after the postcalcination or reduction process, which removes the volatile compounds inside the solution [44,45].

The particle size of active metals (oxide), their dissemination, and accessibility into the support affect the efficiency of the finally dried catalyst. So the determination of the optimum conditions of synthesis parameters is vital [46], as a low dispersion is usually observed in case of metal-kraft lignin composites (due to hydrophobicity of the kraft lignin), fabricated by the wet impregnation method [47].

### 3.2 Template method

Template synthesis is a cutting-edge technology, which is widely used as an effective micro- or nanomaterial synthesis method in recent years, which employs the preparation of the chosen materials (by principle, any solid matter) inside the pores or channels of a nanoporous template, effectively controlling the morphology, structure, and particle size of nanomaterials by adjusting the crystal nucleation and growth during the fabrication process. The template is usually a porous stable structure with respect to the reaction conditions, within which a network of the micro-/nanomaterials is arranged and the removal of the template ultimately produces the filled cavity of the desired micro-/nanomaterials bearing physical/stereochemical properties relevant to the template. The initial step of the template synthesis is to select the template in between the hard template (like polymeric microspheres, ion-exchange resin, plastic foam, carbon fiber, porous membrane) and the soft template (surfactant, nanomineral, biological molecules, cells, and biopolymer), considering the requirement of the template structure. Once the template is prepared, the nanostructure synthesis of the target material is carried out under the function of template commonly by any or combination of several synthetic approaches such as sol-gel method, hydrothermal method, coprecipitation method, etc. Finally, the template is removed by different methods, for instance, etching, dissolution, and sintering without affecting the physicochemical characteristics of the final products [48,49]. Soft template-based synthesis of lignin-metal oxide composites with controlled surface area and pore volume is usually preferred for dodging the necessity of removing hard template [50].

### 3.3 Coprecipitation method

Coprecipitation method of nanomaterial or catalyst synthesis is a facile and convenient, fast, cost-effective approach to easily convert into a larger industrial scale and to produce metallic nanoparticles, especially, metal oxides and metal chalcogenides with uniform component distribution and high purity through an ecofriendly route, without the use of harmful organic solvents, or involvement of high pressure or temperature treatments. The synthesis of nanomaterial takes place from the supersaturated solutions of the metal-organic precursors with templates in suitable solvent, and metal chalcogenides by the reactions of molecular precursors, followed by simultaneous nucleation, growth and coarsening, along with agglomeration by Ostwald ripening, which affect the morphology, particle size, shape, and other application properties. However, products achieved by following this approach sometimes may exhibit poor magnetic properties and low crystallinity as the process shows high sensitivity to reaction conditions and is often not kinetically controlled [51,52].

### 3.4 pH-assisted precipitation method

pH-assisted precipitation method is a newer concept for the fabrication of lignin metal oxide composites developed after considering the influence of pH in precipitate formation (from lignin, extractives, and metals) on surfaces of kraft pulp. Acidification performs the key role in precipitation of lignin as well as extractives on the surfaces of pulp at low pH [53]. Lignin nanoparticles can be obtained from alkaline lignin (pH 11, adjusted by NaOH solution) through the acid precipitation method, by mixing HCl upon stirring and adjusting the pH range at 2.5. Nanoparticles of lignin can also be synthesized from high pH lignin solutions through the rapid addition of $HNO_3$ solutions following this method, as reported by Frangville et al. [54].

## 3.5 Solvent evaporation method

Solvent evaporation is one of the extensively used, popular schemes of polymeric nanoparticle synthesis, which comprises aqueous-phase emulsification of polymer by using any one of the suitable volatile organic solvents from ethyl acetate, dichloromethane, chloroform, and so on, to disperse the polymer by using speedy homogenization/ultrasonication technique, followed by evaporation of solvent (by vacuum or high-temperature treatment) and constant magnetic stirring maintaining room temperature. Then, undesired compounds are removed from the synthesized nanoparticles by ultracentrifugation and washing away with water. Change of evaporation temperature, alteration of evaporation rate, or stirring rate etc. are the significant aspects in altering the particle size of fabricated nanomaterial [55,56]. Lignin-$TiO_2$ composites have been reported to be prepared from different lignins like kraft lignin, sodium lignin, alkali lignin, and organosolv lignin successfully by following this solvent evaporation method [57].

## 3.6 Cocalcination method

Calcination is a technique comprising heating the solids at a higher temperature to remove volatile substances or oxidizing a specific amount of mass. Nanoparticles of metals or metal oxides for photocatalytic applications are recently being prepared by a combination of conventional nanosynthesis methods like sol-gel method, coprecipitation method etc. and cocalcination method, where the calcination temperature is quite impactful in tailoring the characteristics of the fabricated materials. Lignin-supported porous carbon-$CeO_2$ composites can be prepared using lignin template via the co-calcination method at a temperature of 400–600°C [58].

## 3.7 Solid-phase synthesis method

Solid-phase synthesis is another easier approach of photocatalytic composite fabrication, in which the reacting molecules are bonded covalently on a solid support material. At first, the substrate is deposited on any suitable polymeric support through grinding for well mixing. Once the reaction is completed, the attained precipitates are washed with suitable solvents repeatedly to remove the excess of reagents and filtrated to get the final product. CuO nanoparticles based on aminated lignin to be used as a photocatalyst can be synthesized through the reaction between NaOH and $Cu(NO_3)_2$ by the solid-phase synthesis method. CuO-ZnO nanocomposites can also be prepared using zinc carbonate as the precursor and sodium lignosulfonate as the support for the semiconductor component by this method [57].

### 3.8 One-pot in situ method

A cheap and environmentally friendly approach of fabricating industrial alkali lignin-based composite photocatalyst is one-pot in situ method. In this technique, photoactive core particles are produced initially through homogenization and sonication; the template precursor or the coating agent is then added for precipitation at controlled pH. The final composite is swept away with water and desiccated carefully to avoid the risk of trapping of any impurity in between the photoactive core surface and the template. In recent studies, carbon-ZnO composites based on lignin from alkali lignin of pulp liquor have been reported to be synthesized by utilizing zinc nitrate precursor as support and for ZnO nanoparticles [59].

Besides the above methods, dry or wet ball milling or a combination of both the techniques in a ball mill is another easier approach of lignin-based metal oxide composite preparation for photocatalytic application, involving direct milling with or without the use of common solvents like hexane, acetone, water, etc., followed by filtration and drying.

## 4 Lignin-based metal oxide composites in photocatalysis and photovoltaics

### 4.1 General idea of photocatalysis and photovoltaics

**Photocatalysis** is a unique type of accelerated photoreaction process involving the presence of a catalyst that typically has the capacity of absorption of light to generate electron-hole pairs (e −/h+), which allows the chemical alteration of the reactants and restores its pristine chemical configuration at the end of each cycle of interactions. In semiconductor, there remains an occupied valence band, VB, along with an unoccupied conduction band, CB, having a band gap in between the bands. When irradiation of a semiconductor by photons of energy (hv) equals or higher than the energy ($E_g$) of band gap of the initially not radiated semiconductor occurs, an excited electron ($e^-$) imports the movement to the CB from the VB, parting a hole (h+) behind. This photogenerated pair of CB electron and VB hole (e −/h+) is capable of recombining to be deactivated with the release of heat/light or being trapped on metastable surface or oxidizing and/or reducing the surrounding materials adsorbed on the surface of photocatalyst [58].

Photocatalytic reactions are two kinds, viz., homogeneous photocatalysis and heterogeneous photocatalysis. Profuse solar light-crafted semiconductor-based heterogeneous photocatalysis as an effective and promising technology with significant features, like suitable morphology with desired band gap, enhanced surface area, improved stability, and reusability, is receiving much attention in recent days. Metal oxides having such properties, like $TiO_2$, $Cu_2O$, ZnO, $CeO_2$ etc., can follow primary photocatalytic processes, viz., light absorption by visible light or ultraviolet light or a unification of both, to bring out a charge separation process in order to produce positive holes (powerful oxidizing agent) and photoexcited electrons (good reducing agent), forming an electron/hole pair. Band energy positions of the semiconducting material and redox potential for the species adsorbed mainly determine the transmission of photo-induced electrons to its surface adsorbate. The more positive potential value of the acceptor species (sited below) than that of the CB in semiconductor and the more negative potential value of the donor (sited above) than that of the VB of the semiconductor are a must to give away an electron to the unoccupied hole [58,60].

**Photovoltaics** (**PV**) is a method to generate electricity through photovoltaic effect involving the conversion of light into electricity by using solar cell. When exposed to light, the semiconducting materials of the solar cell can generate electric current and voltage. Photovoltaic technology was utilized to power orbiting satellites and spacecraft at the very beginning. But photovoltaic technology has been receiving much attention to advance exceedingly over the past decade. Today, the greater part of photovoltaic modules are chosen for power generation by grid-connected systems in combination with an inverter to alter the DC to AC.

In any solar cell, the photovoltaic energy conversion yields an electron-hole pair by the absorption of light. The device structure is designed to separate the electron and the hole, holding electrons at the negative terminal and holes at the positive terminal in order to generate electrical power [61]. Inside a photovoltaic cell, an ITO (indium-tin-oxide) electrode along with another electrode mostly composed of metals like Al, Au, Ca-Mg etc. is designed to hold the light-absorbing layers arranged in a packed array in between them. When exposed to light/photon, absorption of a certain wavelength of radiation shifts electrons from the highest occupied molecular orbital, shortly named as HOMO, toward the lowest unoccupied molecular orbital, shortly named as LUMO. Free electron-hole pairs ($e-/h+$) produced over the exciton dissociation are then transferred to Al and ITO in turn, and the movement of electrons thus generated through the external circuit yields a flow of electric current. The asymmetry of the two electrodes ensures the electrons flow in between the low work function zone and the high work function zone [62].

Easily available and cheap semiconductor materials as well as their facile manufacture routes for photovoltaic (PV) junctions have been an enduring target of photovoltaic materials research. Semiconductors derived from organic polymer-based metal oxide with low cost, renewability, and conjugated structure are being explored as a potential alternative to the inorganic semiconductive materials in the field of photovoltaics.

## 4.2 Approaches to increase the performances of metal oxide composites in photocatalysis and photovoltaics

$TiO_2$, ZnO, $SnO_2$, $CeO_2$, etc. are the broadly used photocatalysts, particularly in heterogeneous photocatalysis over the decades, due to their good biocompatibility, abundant availability, exceptional stability under different conditions, and capacity to yield charge carriers upon stimulation by the requisite quantity of light energy. This promising blending of electronic configuration and structure along with properties related to light absorption as well as charge transport of metal oxides is quite vital for its wide applications in photocatalysis and photovoltaics [60]. Biomass-derived carbon-based materials like lignin, cellulose, hemicellulose, wood, and biochar are being extensively used as doping agents and templates for semiconductors in the field of photocatalytic applications due to enhanced visible light-responsive performance. Due to the porosity with augmented surface area and tendency to increase the rate of interface charge transfer as well as to lessen the rate of electron-hole recombination, carbon materials can impart high adsorption rate to adjust the mechanism of photochemical reaction in a synergistic way [63].

In photovoltaic technologies, oxides have been functionally used as conductors, semiconductors, and insulators in solar cells, due to their characteristics and viability to be synthesized in facile, cheap, and easily mountable methods (to industrial scale). It also deliberates most of the metal oxides an exclusive opportunity in the next-generation photovoltaics (NGPVs). A noteworthy point in NGPVs industrialization and commercialization is that both the undoped and doped semiconductor oxides have imparted enhanced stability of service life to advanced hi-tech PVs, especially halide perovskite solar cells (HPSCs) along with organic photovoltaics (OPV) [64].

So, the fabrication of metal oxide composites based on lignin as potential candidates in photocatalysis and photovoltaics still needs be investigated consistently as their physicochemical properties like size, shape, composition, and morphology, which are imperative for efficient photocatalytic performance, can be controlled well by different synthetic procedures. The development of thin films or powder forms of catalyst having the compulsory photoactive characteristics is thus facilitated.

## 4.3 Lignin-based metal oxide composites in photocatalysis and photovoltaics

### 4.3.1 Lignin-$TiO_2$ composite

Xiaoyun et al. studied the fabrication of mesoporous $TiO_2$ using lignin as template and $TiCl_4$ as reactant via the hydrolysis precipitation method, to investigate the photocatalytic sensitivity of $TiO_2$ by phenolic degradation upon UV light irradiation. In this study, the $TiO_2$ showed around 11.1 nm of average crystallite size having 165.8 $m^2/g$ of specific area and this mesoporous structure resulted in 0.312 $cm^3/g$ of pore volume, when calcined at 500°C. The $TiO_2$ thus fabricated on lignin-based template resulted in enhanced photocatalytic activity, which might be due to uniformly distributed $TiO_2$ precursor to produce $TiO_2$ mesoporous particles with the support of the electronegativity and the network frame of lignin [65].

The pH-assisted precipitation technique to prepare $TiO_2$-lignin composite by solubilizing alkali lignin in suitable solvent and homogeneously adding $TiO_2$ coating via sonication was studied by Morsella et al. The final $TiO_2$-lignin clusters were precipitated out at low pH controlled by the addition of acids to the solution. Lignin-$TiO_2$ composites were also prepared by dry milling followed by wet milling inside the same ball mill, where lignin and $TiO_2$ were milled at dry condition at 120 rpm for 6 h with a mass ratio of 1:1, and subsequently wet milling in the presence of varied solvents, viz., hexane, water, or acetone maintaining solvent to feedstock mass ratio of 1:2. At last, filtration and drying at 40°C of the final composites were carried out sequentially [57].

$TiO_2$/lignin composite photocatalyst was also reported to be fabricated by sol-gel microwave technique by Nattida et al., who found that kraft lignin-based carbon improved the UVA-irradiated photocatalytic action of $TiO_2$-lignin composite when the ratio of $TiO_2$ to lignin is 1:0.5 [66].

### 4.3.2 Lignin-ZnO composite

Zinc oxide (ZnO) nanoparticles of particle size 15–44 nm were synthesized using lignin-amine (LA) template having high sunlight photocatalytic activity comparable with $TiO_2$, as reported by Xiaohong et al. by a solid-phase method [67]. Another recent report investigated the simplified fabrication technique of ZnO nanocomposites with lignin-based carbon by one-pot in situ method using zinc nitrate precursor for support and ZnO nanoparticles, respectively, and alkali lignin of pulping liquor waste. LC/ZnO with 6.38% of lignin-based carbon revealed admirable proficiency of photodegradation up to 98.9% for methyl orange, greater than the efficiency of pure ZnO due to the uniform structure along with higher absorption capacity for organic dyes [59]. Binpeng et al. reported the fabrication of uniformly anchored ZnO composites with lignin-derived flower-like carbon by cocalcination method, which showed excellent photocatalytic performance with superior recyclability [68].

### 4.3.3 Lignin-CuO composites

Copper oxide (CuO) with a narrow band gap is highly competitive to $TiO_2$ and ZnO as an important p-type semiconductor due to its distinctive physicochemical properties. Wang et al. described a unique but easy solid-phase reaction scheme for the fabrication of mesoporous nano-CuO photocatalyst based on aminated lignin. The solid-state reaction of NaOH with $Cu(NO_3)_2$ on the aminated lignin template yields CuO nanoparticles upon calcination with better electrochemical properties in KOH electrolyte and improved photocatalytic performance than pure CuO [56].

### 4.3.4 Lignin-CuO/ZnO composites

Hybrid nanocomposites of CuO/ZnO were synthesized through a simplified and economic method of solid-state grinding by Fangsheng et al. using $ZnCO_3$ as the precursor for the semiconductor component and Na-lignosulfonate as the base support. The $ZnCO_3$-SLS precipitates and other chemicals were initially mixed followed by centrifugation, removing excess reagents through water rinsing, then vacuum drying at a temperature of 60°C, and finally annealing at 400°C. The supercapacitive performances and superior photocatalytic efficiency of lignin-based nanocomposites of CuO/ZnO upon visible light irradiation might be due to the efficient separation of photogenerated charge carriers in addition to the compatible CuO p-type band edge and ZnO n-type band edge [69].

### 4.3.5 Lignin-based porous carbon-$CeO_2$ composites

The porous composites of carbon-$CeO_2$ have been fabricated using sodium lignin sulfonate as template following the cocalcination method, as reported by Wang et al. Calcination at 600°C for 4h yielded porous carbon-$CeO_2$ composites with enhanced photocatalytic activity compared with $CeO_2$. Decomposition of lignin resulted in the porous and uniform $CeO_2$ nanorods, which were grown in the carbon pores, ultimately generating the composites of porous carbon-based $CeO_2$. The porous composites thus fabricated had been applied to remove $SO_2$ by photocatalysis under light irradiation at room temperature, and the carbon-$CeO_2$ composite revealed the highest photocatalytic transformation of $SO_2$ up to 51.89%, which was much greater than that of only $CeO_2$ [12].

### 4.3.6 Sodium lignosulfonate-functionalized MWNTs/$SnO_2$ hybrids

Sodium lignosulfonate (SLS) was also reported to be involved as a dispersing agent, to functionalize the surface of multiwalled carbon nanotubes (MWNTs) by adsorption via Π-Π noncovalent stacking interactions. Lie et al. reported the SLS-functionalized MWNTs/$SnO_2$ hybrids prepared by grinding in situ formation technique. The interaction of positively charged nanoparticle ions ($Sn^{4+}$) with groups of sodium lignosulfonates having negative charge preceded homogeneous distribution of $SnO_2$ nanoparticles throughout the template of MWNTs functionalized by SLS. The SLS-functionalized MWNTs provided a remarkable support in the synthesis of stable quantum dot hybrids [57].

## 4.4 Mechanism of interaction between lignin and semiconductor

Photons irradiate semiconductor with energy equals to or higher than its band gap energy, ($E_g$) and then photochemical process is commenced. Fig. 1 illustrates a photocatalysis process, where photon strikes in the first place, then an electron goes to conduction band from valence band after being excited while dropping a hole in the valence band.

Second, electrons perform as a reductant and combine with A, thus producing $A^-$ regardless of whether holes can act at that moment as oxidant to react with D as electron donors to $D^+$. The produced $A^-$ and $D^+$ are accumulated either in the surface of semiconductor or within the double layer. $A^-$ and $D^+$ take part in the mineralization of organic compounds.

In mesoporous $TiO_2$ nanoparticles fabricated using lignin template, the surface of lignin has electronegative hydroxyl groups, which build up an intense affinity for electropositive metal ions (Fig. 2). The cation, $Ti(OH)_n^{(4-n)+}$, has attraction for the nucleophilic group; as a consequence, these cations are adsorbed in the exterior of lignin by means of electrostatic attraction forces. The accumulated ions later decomposed by the addition of water and transferred themselves to Ti-(O-lignin)$_4$ over the surface. After whole breakdown of Ti-(O-lignin)$_4$ with the addition of water, desired $TiO_2$ nanoparticles are produced.

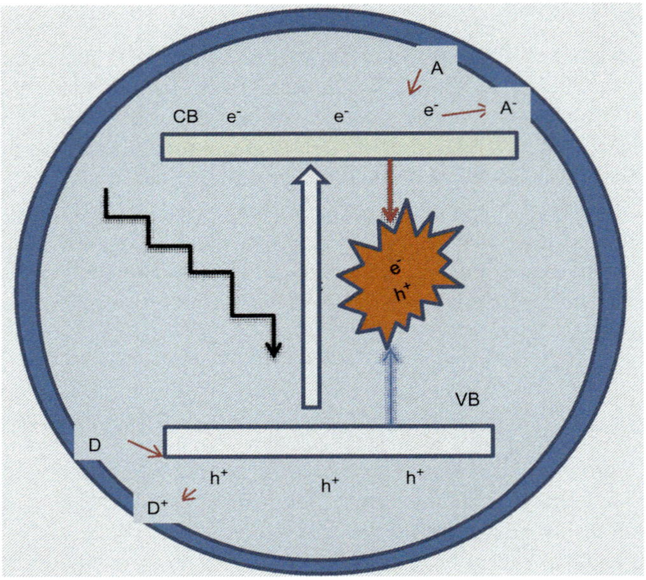


**Fig. 1** Illustration of the progress of photocatalysis [70].

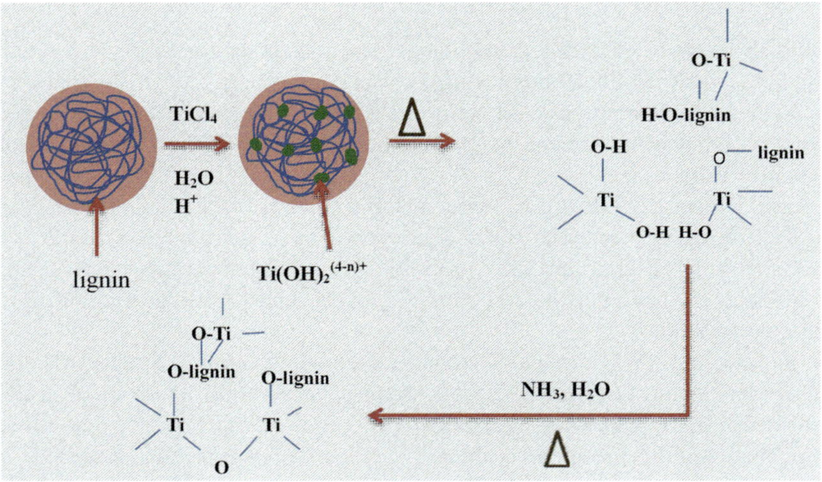


**Fig. 2** Mechanism of template formation of mesoporous $TiO_2$ with lignin [71].

# 5 Application of lignin-based metal oxide composites

## 5.1 Application of lignin-metal oxide composite materials in photocatalysis

The importance of utilizing lignin in material engineering is growing tremendously along with fabricating various substances of composite character in the application of lignin as photocatalysts [65]. The successful utilization of lignin as a support of photocatalysts permits proper monitoring of the particles size to attain an identical particle distribution of the photocatalyst [13]. Furthermore, lignin is usually obtained from various renewable biomass resources as a side product in pulp and paper industries, which subsequently increases the use of lignin in materials science through the reduction of price in order to aid the fabrication of the composites that are environmental friendly. Henceforth, it is always well-thought-out to be a prominent support during the fabrication of composite photocatalysts. In recent times, several efforts have been taken to improve photosensitive composites containing lignin, which would further develop the efficacy of photocatalytic processes [10]. To overcome the restrictions associated with the photocatalytic reactions, photochemical and photophysical mechanisms of the processes are needed to be understand. Generally, the interaction of photoactive materials with light plays significant features in evaluating the efficacy of the reactions involving photocatalysts. However, both the oxidation and reduction reactions can also be hindered by the extended differences in conduction band and valence band present in semiconductors [72]. Hence, UV radiation with higher energy is considered as one of the prime requirements to carry out the photochemical reactions [73].

Photocatalysts comprising lignin have pronounced prospective in the treatment of polluted water and consumed much more considerations in current years [59]. The proper implementation of photocatalysts containing lignin support would approve characteristic properties, resulting in prominent physical and chemical activities in the specific uses attributing to the collaboration between constituents. According to numerous studies in recent times, the combination between metal oxides (such as zinc oxide, copper oxide, titanium oxide) and lignin has been found to upsurge the degradation of pollutant in comparison with pristine metal oxide alone [74].

In a study, hybrid comprising mesoporous zinc oxide and lignin-amine (LA) showed higher sunlight photocatalytic activity as a photocatalyst. Successful amination leading to the addition of amine groups to the lignin has been found to improve the surface activity together with the flocculation and decolorization proficiency during wastewater treatment [75]. Calcination temperature, during the preparation of catalysts, holds a major part to control the parameters, including dimension, microstructure, morphology, and the activity of nanosized ZnO photocatalyst as well. Composite comprising ZnO and lignin-amine, calcined at 400 °C, was reported to exhibit extended photocatalytic efficacy compared to those calcined at higher temperature [57]. With the successive increase in temperature, the size of the catalysts was found to increase while decreasing the specific surface area of the photocatalyst. Moreover, the incorporation of lignin amine in zinc oxide originator pays lesser dimension together with higher specific exterior of nanosized ZnO particles through avoiding its agglomeration. Composite of zinc oxide and lignin amine showed photocatalytic degradation in case of methyl orange ($20\,mg\,L^{-1}$) with the efficacy of nearly 96.4% and 99.2% in the presence of solar radiation (6h) and ultraviolet light radiation, respectively [57]. Photocatalytic performance of zinc oxide and lignin amine composite was found nearly identical as of $TiO_2$ in the presence of radiation generated by solar system [67]. In case of a photochemical reaction, the water molecules get attached onto the surface and undergoes oxidation to hydroxide free radicals, while oxygen molecules get attached onto the exterior of zinc oxide after getting reduced by the electrons. Radicals thus formed further result in a proper modification of methyl orange through mineralization.

Furthermore, quaternized hybrid comprising alkali lignin and zinc oxide has also been found to be implemented for the successful deterioration of dyes called methyl orange (MO) and rhodamine B (Rh B). The efficacy of such hybrid was too much superior compared to unadulterated zinc oxide. Methyl orange ($15\,mg\,L^{-1}$) was found to be totally deteriorated by the composite with the exposure of light for 30min although unadulterated ZnO exhibited a degradation efficacy of about 75.3%, regardless of the time. Hybrid of alkali lignin and zinc oxide showed inferior deterioration efficacy intended for rhodamine B ($15\,mg\,L^{-1}$) than that of methyl orange but was greater compared to zinc oxide. Methyl orange was found to be degraded mostly by holes and can be attributed to the presence of negative charge of the dye pressed in the direction of ZnO. However, deterioration of the dye Rhodamine B was accelerated by hydroxide free radicals and superoxides owing to their robust oxidation capabilities [10]. Such useful hybrids were moderately even, and not any evident of reduction in photodegradation efficacy was detected in three consecutive reprocessing assessments [10]. Hence, incorporation with carbon materials was found to improve the proficiency of ZnO through the successful prevention of zinc oxide corrosion and to diminish the restrictions related to the unadulterated zinc oxide as photocatalyst [76].

## 5.2 Application of lignin-metal oxide composite materials in photovoltaic

Suitable production of photovoltaics is the prime requirement designed for the optimal light collecting ability and developed photo-sensitive transmission related to charge. Production of an anode interface is of extreme importance to develop the cell efficacy [57]. Conversion system related to power, with the efficacy of around 10%, has been found to attain following the cathode modification technique. Nonetheless, in the anode compartment, the water-soluble conductive polymers, known as poly (3,4-ethylene dioxythiophene) and poly(styrene sulfonic acid), have been reported to be frequently used. Proficiency of such polymers is mainly focused by both the highest occupied molecular orbital (HOMO) energy level and the conduction ability of polymers. HOMO energy level was found to be controlled by doping perfluorinated ionomer (PFI) into poly(3,4-ethylene dioxythiophene), which further revealed improved effectiveness compared to poly(3,4-ethylene dioxythiophene) and poly(styrene sulfonic acid) [57]. Correspondingly, the usage of poly(styrene sulfonic acid) as dopant enhanced the conduction ability of poly(3,4-ethylene dioxythiophene) as well. However, both the structural and electrical inhomogeneities due to poly(styrene sulfonic acid) were found to take place owing to its nonconjugated nature [57].

In a study, Wu et al. described the implementation of hybrid made of grafted sulfonated acetone-formaldehyde lignin (GSL), as a p-type dopant for hole-extracting film [77]. Stated hybrid is a semiconductor of polymer obtained by the successful addition of sulfonated acetone-formaldehyde (SAF) to lignin of alkaline nature (AL). Extended aliphatic chain and higher number of sulfonic groups present on GSL convert it to a well dispersant, which can be added as a dopant for poly (3,4-ethylene dioxythiophene). The arrangement of GSL makes it a noble aspirant for the mobility of electron-hole like various polymers utilized in organic electronics.

## 5.3 Environmental and economic aspects of lignin-metal oxide composites

Lignins are chemical species containing several phenolic hydroxyl functionalities together with a bulky side chain of aliphatic and negatively charged functional groups. Thus, it appears beneficial to associate lignin with inorganic functionalities that have the deficiency of these organic groups. Inorganic constituents can effectively adsorb several contaminants in aqueous solutions. For instance, metal ions and organic pollutants are categorized by a stiff structure, high thermal stability, and specific surface area. In addition, inorganic constituents provide prominent resistance to microbial attack [78,79]. Several approaches such as coprecipitation, adsorption with modified or unmodified lignin, and sol-gel method have been found to be applied in preparing lignin-inorganic composites [80,81]. As the chemical nature of the exterior describes the kind of attraction toward an adsorbate, various sources and characteristics of lignin affect the whole adsorbing mechanism [82,83]. Table 1 summarizes the capabilities of composites comprising lignin and inorganics toward metal ions and organic molecules during the adsorption procedure. Though, a straight assessment of capability during adsorption is impossible for all the time owing to several pH values and temperatures.

Fundamentally, composites consisting of lignin and inorganic constituents have been magnificently implemented for the proper removal of dyes with cationic nature [93]. Modified lignin-silica composites were found to eliminate both the methylene blue and crystal violet dyes with greater capacity (10%–15%) in comparison with the initial pure lignins [86].

According to several studies, the elimination of cations by hybrid comprising lignin and inorganic constituents was found much effective than that of lignins in various types. For example, lignin/MgO-$SiO_2$ and lignin/MgO-$TiO_2$ showed prominent adsorption capabilities of 84 and 70 $mg\,g^{-1}$ for $Cu^{2+}$ and $Cd_2^{2+}$, whereas lower values were reported in case of lignin ($Cu^{2+}$: 1.7–26 $mg\,g^{-1}$, $Cd^{2+}$: 6.7–48 $mg\,g^{-1}$) and activated carbon ($Cu^{2+}$: 9–38 $mg\,g^{-1}$, $Cd^{2+}$: 3.7–146 $mg\,g^{-1}$) [94,95]. Nevertheless,

**Table 1** Lignin-metal oxide composites together with their activity as adsorbent [84].

| Composite | Adsorbate | Max. adsorption capacity $[mg\,g^{-1} \mid mg\,m^{-2}]$ | Ref. |
|---|---|---|---|
| Modified lignin/$SiO_2$ | 2,4-Dichloro-phenoxyacetic acid | 4.9 \| 0.03 | [85] |
| Modified lignin/$SiO_2$ | Crystal violet | 110.1 \| 1.3 | [86] |
| Modified lignin/$SiO_2$ | Methylene blue | 41.6 \| 0.6 | [87] |
| Fractionated kraft lignin/$SiO_2$ | Methylene blue | 60.0 \| 0.7 | [87] |
| Lignosulfonate/ $Fe_3O_4$microspheres | Methylene blue | 283.6 \| 6.3 | [88] |
| Alkali lignin-dopamine/$Fe_3O_4$ NPs | $Cr^{3+}$ | 44.6 | [89] |
| Carboxymethylated lignin-$Fe_3O_4$/$SiO_2$ | $Cu^{2+}$<br>$Pb^{2+}$ | 70.7<br>150.3 | [90] |
| Lignin/MgO-$SiO_2$ | $Cu^{2+}$ | 83.9 \| 0.39 | [91] |
| Lignin/$TiO_2$ | $Cu^{2+}$<br>$Cd^{2+}$ | 20.1 \| 0.5<br>22.4 \| 0.57 | [92] |

significantly progressive adsorption capabilities for $Cu^{2+}$ ($87\,mg\,g^{-1}$) and $Cd^{2+}$ ($137\,mg\,g^{-1}$) have been described for lignin obtained from black liquor produced by pulping of Eucalyptus wood [96]. Alterations in both the sources and cooking variables have been reported to modify the basic adsorption activity of lignins obtained in kraft process [97].

## 5.4 Future prospects of photocatalyst and photovoltaics using lignin-based metal oxides

In the present study, existing developments of lignin-based metal oxide hybrids as photocatalysts and photovoltaics have been emphasized. As a biopolymer, lignin support exhibited a prominent efficacy in heterogeneous photocatalysis, especially in the aspect of degrading pollutants present in the atmosphere. Numerous assessments have already proved the implementation of lignin as a praiseworthy mold for the successful synthesis of photocatalysts owing to its extended exterior. The presence of such extended exterior further modifies physical adsorption together with greater transfer of photoelectron [12]. Furthermore, composites comprising lignin and metal oxides have showed a noticeable potentiality to substitute economical materials identical to graphene as photocatalysts. In numerous studies, hybrid comprising graphene oxide showed its efficacy in degrading methyl orange, methylene blue, and Rhodamine B by 87.2%, 85.1%, and 73.9%, respectively [98].

According to the previous studies, lignin-based materials have also found pronounced implementations in various areas that range from biomedicine to photocatalysts to electrochemical energy devices as well. Involving lignin into multifunctional constituents, and its implementation as a template for solid photocatalysts would outstretch a pronounced accomplishment. However, a profound understanding of the fundamental mechanism regarding the attractions between lignin and the photosensitive materials is in need.

Though the significant improvement has been attained, abundant open questions are still available, which must be specified to realize the upsurge present in the photoconversion proficiency to obtain an economically compatible application. Mostly, both the mobility and the separation of charge carrier present inside the materials have to be developed. Only a few assessments explain the potentiality of lignin as a promising adjuvant of anode in case of photovoltaic devices.

# 6 Conclusion

As a functional polymer of comparatively lower molecular weight, lignin and lignin-based metal oxide composites offer large opportunities for application in polymer materials. This review has methodically explained various prospects of application of lignin-based metal oxide composites in photocatalysis and photovoltaics all through the chapter. The commercialization of lignin and lignin-based metal oxide composites embraces not only the operational and sustainable property improvements but also the cost minimization. Prominent advantages regarding the economic aspects from lignin valorization are assigned to its lower cost and abundant accessibility as a byproduct of both the bioethanol and pulping industries. Therefore, the prospects of employing the unutilized sources of lignin paves way the goal of not only improving effective separation techniques but also of synthesizing new hybrids of lignin with great financial values. Nowadays, lignin metal oxide composites are showing excellent performances in photocatalytic degradation of water contaminants and organic dyes and in photovoltaics of sustainable electric devices with improved anode interface. But still, there is room for research and invention of a variety of lignin metal oxide composites with stable photoactivity, tunable anodic dopant properties as well as distinct understanding of structure-property correlation of lignin with several photovoltaic organic materials. Several state-of-the-art technologies should be explored to fabricate newer lignin metal oxide composites for photocatalytic and photovoltaic applications with clear knowledge of photoactive mechanism through rigorous research.

# References

[1] V.K. Thakur, M.K. Thakur, Recent advances in green hydrogels from lignin: a review, Int. J. Biol. Macromol. 72 (2015) 834–847.

[2] A.M. Puziy, O.I. Poddubnaya, O. Sevastyanova, Carbon materials from technical lignins: recent advances, in: Lignin Chemistry, Springer, 2020, pp. 95–128.

[3] A. Agrawal, N. Kaushik, S. Biswas, Derivatives and applications of lignin—an insight, SciTech J. 1 (7) (2014) 30–36.

[4] J.H. Lora, W.G. Glasser, Recent industrial applications of lignin: a sustainable alternative to nonrenewable materials, J. Polym. Environ. 10 (1–2) (2002) 39–48.

[5] J. Hu, Q. Zhang, D.-J. Lee, Kraft lignin biorefinery: a perspective, Bioresour. Technol. 247 (2018) 1181–1183.
[6] H. Lu, X. Zhao, Biomass-derived carbon electrode materials for supercapacitors, Sustain. Energy Fuels 1 (6) (2017) 1265–1281.
[7] A.P. Richter, et al., Synthesis and characterization of biodegradable lignin nanoparticles with tunable surface properties, Langmuir 32 (25) (2016) 6468–6477.
[8] J. Ran, et al., Earth-abundant cocatalysts for semiconductor-based photocatalytic water splitting, Chem. Soc. Rev. 43 (22) (2014) 7787–7812.
[9] H. Zhang, et al., P25-graphene composite as a high performance photocatalyst, ACS Nano 4 (1) (2010) 380–386.
[10] H. Wang, et al., Facile preparation of well-combined lignin-based carbon/ZnO hybrid composite with excellent photocatalytic activity, Appl. Surf. Sci. 426 (2017) 206–216.
[11] A. Gómez-Avilés, et al., C-modified $TiO_2$ using lignin as carbon precursor for the solar photocatalytic degradation of acetaminophen, Chem. Eng. J. 358 (2019) 1574–1582.
[12] N. Wang, H. Fan, S. Ai, Lignin templated synthesis of porous carbon–$CeO_2$ composites and their application for the photocatalytic desulphuration, Chem. Eng. J. 260 (2015) 785–790.
[13] X. Wang, et al., Lignin-assisted solid-phase synthesis of nano-CuO for a photocatalyst with excellent catalytic activity and high performance supercapacitor electrodes, RSC Adv. 6 (70) (2016) 65644–65653.
[14] W. Wang, M.O. Tadé, Z. Shao, Research progress of perovskite materials in photocatalysis-and photovoltaics-related energy conversion and environmental treatment, Chem. Soc. Rev. 44 (15) (2015) 5371–5408.
[15] N. Hong, et al., Unexpected fluorescent emission of graft sulfonated-acetone–formaldehyde lignin and its application as a dopant of PEDOT for high performance photovoltaic and light-emitting devices, J. Mater. Chem. C 4 (23) (2016) 5297–5306.
[16] D. Kun, B. Pukánszky, Polymer/lignin blends: interactions, properties, applications, Eur. Polym. J. 93 (2017) 618–641.
[17] N.S., Lignin and lignin based materials for the removal of heavy metals from waste water-an overview, Z. Phys. Chem. 233 (3) (2019) 315–345.
[18] A. Naseem, S. Tabasum, K.M. Zia, M. Zuber, M. Ali, A. Noreen, Lignin-derivatives based polymers, blends and composites: a review, Int. J. Biol. Macromol. 93 (2016) 296–313.
[19] P. Figueiredo, K. Lintinen, J.T. Hirvonen, M.A. Kostiainen, H.A. Santos, Properties and chemical modifications of lignin: towards lignin-based nanomaterials for biomedical applications, Prog. Mater. Sci. 93 (2018) 233–269.
[20] S. Chatterjee, T. Saito, Lignin-derived advanced carbon materials, ChemSusChem 8 (23) (2015) 3941–3958.
[21] C. Li, X. Zhao, A. Wang, G.W. Huber, T. Zhang, Catalytic transformation of lignin for the production of chemicals and fuels, Chem. Rev. 115 (21) (2015) 11559–11624.
[22] N.M. Stark, D.J. Yelle, U.P. Agarwal, Techniques for characterizing lignin, in: Lignin in Polymer Composites, Forest Service, US Department of Agriculture, 2016, pp. 49–66.
[23] J. Baeza, Chemical characterization of wood and its components, in: Wood and Cellulosic Chemistry, Marcel Dekker, New York, 2001, pp. 275–384.
[24] A. Sakakibara, A structural model of softwood lignin, Wood Sci. Technol. 14 (2) (1980) 89–100.
[25] U.P. Agarwal, S.A. Ralph, FT-Raman spectroscopy of wood: identifying contributions of lignin and carbohydrate polymers in the spectrum of black spruce (*Picea mariana*), Appl. Spectrosc. 51 (11) (1997) 1648–1655.

[26] X. Zhou, et al., Glass transition of oxygen plasma treated enzymatic hydrolysis lignin, Bioresources 7 (4) (2012) 4776–4785.
[27] J.E. Holladay, et al., Top Value-Added Chemicals From Biomass-Volume II—Results of Screening for Potential Candidates From Biorefinery Lignin, Pacific Northwest National Lab. (PNNL), Richland, WA (United States), 2007.
[28] F.G. Calvo-Flores, J.A. Dobado, Lignin as renewable raw material, ChemSusChem 3 (11) (2010) 1227–1235.
[29] Z.-W. He, Q.-F. Lu, J.-Y. Zhang, Facile preparation of hierarchical polyaniline-lignin composite with a reactive silver-ion adsorbability, ACS Appl. Mater. Interfaces 4 (1) (2012) 369–374.
[30] C. Crestini, et al., Oxidative strategies in lignin chemistry: a new environmental friendly approach for the functionalisation of lignin and lignocellulosic fibers, Catal. Today 156 (1–2) (2010) 8–22.
[31] S.R. Yearla, K. Padmasree, Preparation and characterisation of lignin nanoparticles: evaluation of their potential as antioxidants and UV protectants, J. Exp. Nanosci. 11 (4) (2016) 289–302.
[32] N. Smolarski, High-Value Opportunities for Lignin: Unlocking Its Potential, Frost & Sullivan, 2012, p. 1.
[33] S. Otani, et al., Method for Producing Carbonized Lignin Fiber, 1969. Google Patents.
[34] K. Sudo, et al., A new modification method of exploded lignin for the preparation of a carbon fiber precursor, J. Appl. Polym. Sci. 48 (8) (1993) 1485–1491.
[35] H. Mainka, et al., Lignin—an alternative precursor for sustainable and cost-effective automotive carbon fiber, J. Mater. Res. Technol. 4 (3) (2015) 283–296.
[36] Y. Uraki, et al., Preparation of carbon fibers from organosolv lignin obtained by aqueous acetic acid pulping, Holzforschung 49 (4) (1995) 343–350.
[37] E. Frank, et al., Carbon fibers: precursor systems, processing, structure, and properties, Angew. Chem. Int. Ed. 53 (21) (2014) 5262–5298.
[38] J. Kadla, et al., Lignin-based carbon fibers for composite fiber applications, Carbon 40 (15) (2002) 2913–2920.
[39] M. Lallave, et al., Filled and hollow carbon nanofibers by coaxial electrospinning of alcell lignin without binder polymers, Adv. Mater. 19 (23) (2007) 4292–4296.
[40] I. Dallmeyer, F. Ko, J.F. Kadla, Electrospinning of technical lignins for the production of fibrous networks, J. Wood Chem. Technol. 30 (4) (2010) 315–329.
[41] C. Abbati de Assis, et al., Techno-economic assessment, scalability, and applications of aerosol lignin micro-and nanoparticles, ACS Sustain. Chem. Eng. 6 (9) (2018) 11853–11868.
[42] N.-E. El Mansouri, J. Salvadó, Structural characterization of technical lignins for the production of adhesives: application to lignosulfonate, kraft, soda-anthraquinone, organosolv and ethanol process lignins, Ind. Crop. Prod. 24 (1) (2006) 8–16.
[43] D. Argyropoulos, Y. Sun, E. Palus, Isolation of residual kraft lignin in high yield and purity, J. Pulp Pap. Sci. 28 (2) (2002) 50–54.
[44] J.L. Figueiredo, Functionalization of porous carbons for catalytic applications, J. Mater. Chem. A 1 (33) (2013) 9351–9364.
[45] Y. Wang, B. Yu, K. Liu, X. Yang, M. Liu, T.S. Chan, X. Qiu, J. Li, W. Li, Co single-atoms on ultrathin N-doped porous carbon via a biomass complexation strategy for high performance metal–air batteries, J. Mater. Chem. A 8 (4) (2020) 2131–2139.
[46] J.R. Sietsma, A.J. van Dillen, P.E. de Jongh, K.P. de Jong, Application of ordered mesoporous materials as model supports to study catalyst preparation by impregnation and drying, in: Studies in Surface Science and Catalysis, vol. 162, Elsevier, 2006, pp. 95–102.

[47] Q. Yan, J. Li, X. Zhang, J. Zhang, Z. Cai, Synthetic bio-graphene based nanomaterials through different iron catalysts, Nanomaterials 8 (10) (2018) 840.
[48] Y. Xie, D. Kocaefe, C. Chen, Y. Kocaefe, Review of research on template methods in preparation of nanomaterials, J. Nanomater. 2016 (2016).
[49] A. Huczko, Template-based synthesis of nanomaterials, Appl. Phys. A 70 (4) (2000) 365–376.
[50] S. Kausar, A.A. Altaf, M. Hamayun, M. Danish, M. Zubair, S. Naz, S. Muhammad, M. Zaheer, S. Ullah, A. Badshah, Soft template-based bismuth doped zinc oxide nanocomposites for photocatalytic depolymerization of lignin, Inorg. Chim. Acta 502 (2020) 119390.
[51] Q. Yan, Z. Cai, Issues in preparation of metal-lignin nanocomposites by coprecipitation method, J. Inorg. Organomet. Polym. Mater. (2020) 1–19.
[52] T. Athar, Smart precursors for smart nanoparticles, in: Emerging Nanotechnologies for Manufacturing, William Andrew Publishing, 2015, pp. 444–538.
[53] K. Koljonen, M. Österberg, M. Kleen, A. Fuhrmann, P. Stenius, Precipitation of lignin and extractives on kraft pulp: effect on surface chemistry, surface morphology and paper strength, Cellulose 11 (2) (2004) 209–224.
[54] Q. Tang, Y. Qian, D. Yang, X. Qiu, Y. Qin, M. Zhou, Lignin-based nanoparticles: a review on their preparations and applications, Polymers 12 (11) (2020) 2471.
[55] R.V. Gundloori, A. Singam, N. Killi, Nanobased intravenous and transdermal drug delivery systems, in: Applications of Targeted Nano Drugs and Delivery Systems, Elsevier, 2019, pp. 551–594.
[56] K. Vinothini, M. Rajan, Mechanism for the nano-based drug delivery system, in: Characterization and Biology of Nanomaterials for Drug Delivery, Elsevier, 2019, pp. 219–263.
[57] A. Khan, J.C. Colmenares, R. Gläser, Lignin based composite materials for photocatalysis and photovoltaics, Lignin Chem. (2020) 1–31.
[58] A. Gołąbiewska, M. Kobylański, A. Zaleska-Medynska, Fundamentals of Metal Oxide-Based Photocatalysis, Elsevier, 2018.
[59] H. Wang, X. Qiu, R. Zhong, F. Fu, Y. Qian, D. Yang, One-pot in-situ preparation of a lignin-based carbon/ZnO nanocomposite with excellent photocatalytic performance, Mater. Chem. Phys. 199 (2017) 193–202.
[60] M.M. Khan, S.F. Adil, A. Al-Mayouf, Metal Oxides as Photocatalysts, Elsevier, 2015.
[61] T. Markvart, L. Castañer, Principles of solar cell operation, in: McEvoy's Handbook of Photovoltaics, Academic Press, 2018, pp. 3–28.
[62] H. Spanggaard, F.C. Krebs, A brief history of the development of organic and polymeric photovoltaics, Sol. Energy Mater. Sol. Cells 83 (2–3) (2004) 125–146.
[63] G. Khan, Y.K. Kim, S.K. Choi, D.S. Han, A. Abdel-Wahab, H. Park, Evaluating the catalytic effects of carbon materials on the photocatalytic reduction and oxidation reactions of TiO 2, Bull. Kor. Chem. Soc. 34 (4) (2013) 1137–1144.
[64] A. Pérez-Tomás, A. Mingorance, D. Tanenbaum, M. Lira-Cantú, Metal oxides in photovoltaics: all-oxide, ferroic, and perovskite solar cells, in: The Future of Semiconductor Oxides in Next-Generation Solar Cells, Elsevier, 2018, pp. 267–356.
[65] X. Chen, D.H. Kuo, D. Lu, Y. Hou, Y.R. Kuo, Synthesis and photocatalytic activity of mesoporous $TiO_2$ nanoparticle using biological renewable resource of un-modified lignin as a template, Microporous Mesoporous Mater. 223 (2016) 145–151.
[66] N. Srisasiwimon, S. Chuangchote, N. Laosiripojana, T. Sagawa, $TiO_2$/lignin-based carbon composited photocatalysts for enhanced photocatalytic conversion of lignin to high value chemicals, ACS Sustain. Chem. Eng. 6 (11) (2018) 13968–13976.

[67] X. Wang, Y. Zhang, C. Hao, F. Feng, H. Yin, N. Si, Solid-phase synthesis of mesoporous ZnO using lignin-amine template and its photocatalytic properties, Ind. Eng. Chem. Res. 53 (16) (2014) 6585–6592.
[68] B. Zhang, D. Yang, X. Qiu, Y. Qian, H. Wang, C. Yi, D. Zhang, Fabricating ZnO/lignin-derived flower-like carbon composite with excellent photocatalytic activity and recyclability, Carbon 162 (2020) 256–266.
[69] F. Wu, X. Wang, S. Hu, C. Hao, H. Gao, S. Zhou, Solid-state preparation of CuO/ZnO nanocomposites for functional supercapacitor electrodes and photocatalysts with enhanced photocatalytic properties, Int. J. Hydrog. Energy 42 (51) (2017) 30098–30108.
[70] S.-H. Li, S. Liu, J.C. Colmenares, Y.-J. Xu, A sustainable approach for lignin valorization by heterogeneous photocatalysis, Green Chem. 18 (3) (2016) 594–607, https://doi.org/10.1039/C5GC02109J.
[71] X. Chen, D.-H. Kuo, D. Lu, Y. Hou, Y.R. Kuo, Synthesis and photocatalytic activity of mesoporous $TiO_2$ nanoparticle using biological renewable resource of un-modified lignin as a template, Microporous Mesoporous Mater. 223 (2016) 145–151, https://doi.org/10.1016/j.micromeso.2015.11.005.
[72] S. Chaturvedi, P.N. Dave (Eds.), Environmental application of photocatalysis, in: Materials Science Forum, Trans Tech Publ., 2013.
[73] L. Jiang, Y. Wang, C. Feng, Application of photocatalytic technology in environmental safety, Procedia Eng. 45 (2012) 993–997.
[74] T.-T. Miao, Y.-R. Guo, Q.-J. Pan, The SL-assisted synthesis of hierarchical ZnO nanostructures and their enhanced photocatalytic activity, J. Nanopart. Res. 15 (6) (2013) 1–12.
[75] R. Fang, X. Cheng, X. Xu, Synthesis of lignin-base cationic flocculant and its application in removing anionic azo-dyes from simulated wastewater, Bioresour. Technol. 101 (19) (2010) 7323–7329.
[76] S. Akir, A. Hamdi, A. Addad, Y. Coffinier, R. Boukherroub, A.D. Omrani, Facile synthesis of carbon-ZnO nanocomposite with enhanced visible light photocatalytic performance, Appl. Surf. Sci. 400 (2017) 461–470.
[77] Y. Wu, J. Wang, X. Qiu, R. Yang, H. Lou, X. Bao, et al., Highly efficient inverted perovskite solar cells with sulfonated lignin doped PEDOT as hole extract layer, ACS Appl. Mater. Interfaces 8 (19) (2016) 12377–12383.
[78] P. Ray, H. Shipley, F. Fu, Inorganic nano-adsorbents for the removal of heavy metals and arsenic: a review, RSC Adv. 5 (2015) 29885–29907.
[79] S.S. Gupta, K.G. Bhattacharyya, Kinetics of adsorption of metal ions on inorganic materials: a review, Adv. Colloid Interf. Sci. 162 (1–2) (2011) 39–58.
[80] Ł. Klapiszewski, P. Bartczak, M. Wysokowski, M. Jankowska, K. Kabat, T. Jesionowski, Silica conjugated with Kraft lignin and its use as a novel 'green' sorbent for hazardous metal ions removal, Chem. Eng. J. 260 (2015) 684–693.
[81] O. Fesenko, L. Yatsenko, Nanochemistry, Biotechnology, Nanomaterials, and Their Applications, Springer, 2018.
[82] Y. Fu, X. Liu, G. Chen, Adsorption of heavy metal sewage on nano-materials such as titanate/$TiO_2$ added lignin, Results Phys. 12 (2019) 405–411.
[83] J.G. Calvert, Glossary of atmospheric chemistry terms (recommendations 1990), Pure Appl. Chem. 62 (11) (1990) 2167–2219.
[84] T.M. Budnyak, A. Slabon, M.H. Sipponen, Lignin–inorganic interfaces: chemistry and applications from adsorbents to catalysts and energy storage materials, ChemSusChem 13 (17) (2020) 4344–4355.

[85] G. Telysheva, T. Dizhbite, D. Evtuguin, N. Mironova-Ulmane, G. Lebedeva, A. Andersone, et al., Design of siliceous lignins–novel organic/inorganic hybrid sorbent materials, Scr. Mater. 60 (8) (2009) 687–690.
[86] T.M. Budnyak, I.V. Pylypchuk, M.E. Lindstrom, O. Sevastyanova, Electrostatic deposition of the oxidized kraft lignin onto the surface of aminosilicas: thermal and structural characteristics of hybrid materials, ACS Omega 4 (27) (2019) 22530–22539.
[87] T.M. Budnyak, S. Aminzadeh, I.V. Pylypchuk, D. Sternik, V.A. Tertykh, M.E. Lindström, et al., Methylene blue dye sorption by hybrid materials from technical lignins, J. Environ. Chem. Eng. 6 (4) (2018) 4997–5007.
[88] G. Wang, Q. Liu, M. Chang, J. Jang, W. Sui, C. Si, et al., Novel $Fe_3O_4$@ lignosulfonate/phenolic core-shell microspheres for highly efficient removal of cationic dyes from aqueous solution, Ind. Crop. Prod. 127 (2019) 110–118.
[89] L. Dai, Y. Li, R. Liu, C. Si, Y. Ni, Green mussel-inspired lignin magnetic nanoparticles with high adsorptive capacity and environmental friendliness for chromium (III) removal, Int. J. Biol. Macromol. 132 (2019) 478–486.
[90] Y. Zhang, S. Ni, X. Wang, W. Zhang, L. Lagerquist, M. Qin, et al., Ultrafast adsorption of heavy metal ions onto functionalized lignin-based hybrid magnetic nanoparticles, Chem. Eng. J. 372 (2019) 82–91.
[91] F. Ciesielczyk, P. Bartczak, Ł. Klapiszewski, T. Jesionowski, Treatment of model and galvanic waste solutions of copper (II) ions using a lignin/inorganic oxide hybrid as an effective sorbent, J. Hazard. Mater. 328 (2017) 150–159.
[92] Ł. Klapiszewski, K. Siwińska-Stefańska, D. Kołodyńska, Development of lignin based multifunctional hybrid materials for Cu (II) and Cd (II) removal from the aqueous system, Chem. Eng. J. 330 (2017) 518–530.
[93] M.H. Sipponen, V. Pihlajaniemi, K. Littunen, O. Pastinen, S. Laakso, Determination of surface-accessible acidic hydroxyls and surface area of lignin by cationic dye adsorption, Bioresour. Technol. 169 (2014) 80–87.
[94] F. Li, X. Wang, T. Yuan, R. Sun, A lignosulfonate-modified graphene hydrogel with ultrahigh adsorption capacity for Pb (II) removal, J. Mater. Chem. A 4 (30) (2016) 11888–11896.
[95] S. Babel, T.A. Kurniawan, Low-cost adsorbents for heavy metals uptake from contaminated water: a review, J. Hazard. Mater. 97 (1–3) (2003) 219–243.
[96] D. Mohan, C.U. Pittman Jr., P.H. Steele, Single, binary and multi-component adsorption of copper and cadmium from aqueous solutions on Kraft lignin—a biosorbent, J. Colloid Interface Sci. 297 (2) (2006) 489–504.
[97] J. Werner, A.J. Ragauskas, J.E. Jiang, Intrinsic metal binding capacity of kraft lignins, J. Wood Chem. Technol. 20 (2) (2000) 133–145.
[98] N. Hong, X. Qiu, W. Deng, Z. He, Y. Li, Effect of aggregation behavior and phenolic hydroxyl group content on the performance of lignosulfonate doped PEDOT as a hole extraction layer in polymer solar cells, RSC Adv. 5 (110) (2015) 90913–90921.

# Index

Note: Page numbers followed by *f* indicate figures, *t* indicate tables, and *s* indicate schemes.

## C

## D



## E

## M

## N



## O

## T